CAMBRIDGE STUDIES IN ADVANCED MATHEMATICS 187

THE CHARACTER THEORY OF FINITE GROUPS OF LIE TYPE

Through the fundamental work of Deligne and Lusztig in the 1970s, further developed mainly by Lusztig, the character theory of reductive groups over finite fields has grown into a rich and vast area of mathematics. It incorporates tools and methods from algebraic geometry, topology, combinatorics and computer algebra, and has since evolved substantially.

With this book, the authors meet the need for a contemporary treatment, complementing in core areas the well-established books of Carter and Digne–Michel. Focusing on applications in finite group theory, the authors gather previously scattered results and allow the reader to get to grips with the large body of literature available on the subject, covering topics such as regular embeddings, the Jordan decomposition of characters, d-Harish-Chandra theory and Lusztig induction for unipotent characters.

Requiring only a modest background in algebraic geometry, this useful reference is suitable for beginning graduate students as well as researchers.

Meinolf Geck is Professor of Algebra at the University of Stuttgart. He works in the areas of algebraic groups and representation theory of finite groups and has (co-)authored books including *An Introduction to Algebraic Geometry and Algebraic Groups* (2003), *Representations of Hecke Algebras at Roots of Unity* (2011) and *Representations of Reductive Groups* (1998).

Gunter Malle is Professor of Mathematics at the University of Kaiserslautern. He works in the area of representation theory of finite groups and he co-authored *Linear Algebraic Groups and Finite Groups of Lie Type* (2011) and *Inverse Galois Theory* (1999). Professor Malle received an ERC Advanced Grant on 'Counting conjectures' in 2012.

CAMBRIDGE STUDIES IN ADVANCED MATHEMATICS

All the titles listed below can be obtained from good booksellers or from Cambridge University Press. For a complete series listing, visit www.cambridge.org/mathematics.

Already Published

148 S. R. Garcia, J. Mashreghi & W. T. Ross *Introduction to Model Spaces and Their Operators*
149 C. Godsil & K. Meagher *Erdős–Ko–Rado Theorems: Algebraic Approaches*
150 P. Mattila *Fourier Analysis and Hausdorff Dimension*
151 M. Viana & K. Oliveira *Foundations of Ergodic Theory*
152 V. I. Paulsen & M. Raghupathi *An Introduction to the Theory of Reproducing Kernel Hilbert Spaces*
153 R. Beals & R. Wong *Special Functions and Orthogonal Polynomials*
154 V. Jurdjevic *Optimal Control and Geometry: Integrable Systems*
155 G. Pisier *Martingales in Banach Spaces*
156 C. T. C. Wall *Differential Topology*
157 J. C. Robinson, J. L. Rodrigo & W. Sadowski *The Three-Dimensional Navier–Stokes Equations*
158 D. Huybrechts *Lectures on K3 Surfaces*
159 H. Matsumoto & S. Taniguchi *Stochastic Analysis*
160 A. Borodin & G. Olshanski *Representations of the Infinite Symmetric Group*
161 P. Webb *Finite Group Representations for the Pure Mathematician*
162 C. J. Bishop & Y. Peres *Fractals in Probability and Analysis*
163 A. Bovier *Gaussian Processes on Trees*
164 P. Schneider *Galois Representations and* (φ, Γ)*-Modules*
165 P. Gille & T. Szamuely *Central Simple Algebras and Galois Cohomology* (*2nd Edition*)
166 D. Li & H. Queffelec *Introduction to Banach Spaces, I*
167 D. Li & H. Queffelec *Introduction to Banach Spaces, II*
168 J. Carlson, S. Müller-Stach & C. Peters *Period Mappings and Period Domains* (*2nd Edition*)
169 J. M. Landsberg *Geometry and Complexity Theory*
170 J. S. Milne *Algebraic Groups*
171 J. Gough & J. Kupsch *Quantum Fields and Processes*
172 T. Ceccherini-Silberstein, F. Scarabotti & F. Tolli *Discrete Harmonic Analysis*
173 P. Garrett *Modern Analysis of Automorphic Forms by Example, I*
174 P. Garrett *Modern Analysis of Automorphic Forms by Example, II*
175 G. Navarro *Character Theory and the McKay Conjecture*
176 P. Fleig, H. P. A. Gustafsson, A. Kleinschmidt & D. Persson *Eisenstein Series and Automorphic Representations*
177 E. Peterson *Formal Geometry and Bordism Operators*
178 A. Ogus *Lectures on Logarithmic Algebraic Geometry*
179 N. Nikolski *Hardy Spaces*
180 D.-C. Cisinski *Higher Categories and Homotopical Algebra*
181 A. Agrachev, D. Barilari & U. Boscain *A Comprehensive Introduction to Sub-Riemannian Geometry*
182 N. Nikolski *Toeplitz Matrices and Operators*
183 A. Yekutieli *Derived Categories*
184 C. Demeter *Fourier Restriction, Decoupling and Applications*
185 D. Barnes & C. Roitzheim *Foundations of Stable Homotopy Theory*
186 V. Vasyunin & A. Volberg *The Bellman Function Technique in Harmonic Analysis*
187 M. Geck & G. Malle *The Character Theory of Finite Groups of Lie Type*

The Character Theory of Finite Groups of Lie Type

A Guided Tour

MEINOLF GECK
Universität Stuttgart

GUNTER MALLE
Technische Universität Kaiserslautern, Germany

CAMBRIDGE
UNIVERSITY PRESS

CAMBRIDGE
UNIVERSITY PRESS

University Printing House, Cambridge CB2 8BS, United Kingdom

One Liberty Plaza, 20th Floor, New York, NY 10006, USA

477 Williamstown Road, Port Melbourne, VIC 3207, Australia

314–321, 3rd Floor, Plot 3, Splendor Forum, Jasola District Centre, New Delhi – 110025, India

79 Anson Road, #06–04/06, Singapore 079906

Cambridge University Press is part of the University of Cambridge.

It furthers the University's mission by disseminating knowledge in the pursuit of education, learning, and research at the highest international levels of excellence.

www.cambridge.org
Information on this title: www.cambridge.org/9781108489621
DOI: 10.1017/9781108779081

First published 2020

Printed in the United Kingdom by TJ International Ltd. Padstow Cornwall

A catalogue record for this publication is available from the British Library

ISBN 978-1-108-48962-1 Hardback

Contents

Preface

The subject of this book is the character theory of finite groups of Lie type (or finite reductive groups), following the geometric approach initiated by the fundamental work of Deligne and Lusztig [Lu75], [DeLu76] in the 1970s. Since then, and to its full extent mainly by the monumental work of Lusztig, this has grown into an extremely rich, complex and vast theory, incorporating tools and methods from algebraic geometry, topology, combinatorics and computer algebra. (Lusztig's papers since 1975 on this subject alone comprise already a few thousand densely written pages.) One of the ultimate aims of this theory is to reduce the computation of character tables of whole series of finite groups of Lie type (e.g., the series of groups $E_8(q)$ where q is any prime power) to purely combinatorial tasks that could, for example, be performed automatically on a computer. For the general linear groups $\mathrm{GL}_n(q)$ this was already achieved in principle by Green [Gre55] in 1955, but the analogous problem for the closely related special linear and unitary groups $\mathrm{SL}_n(q)$, $\mathrm{SU}_n(q)$ is still not completely solved.

Within finite group theory, the importance of this subject is highlighted by the classification of finite simple groups: apart from the alternating groups and the 26 sporadic simple groups, all non-abelian finite simple groups are 'of Lie type'. According to Aschbacher [Asch00], [Asch04] when faced with a problem about finite groups, it nowadays seems best to attempt to reduce the problem or a related problem to a question about simple groups or groups closely related to simple groups. The classification then supplies an explicit list of groups which can be studied in detail using the effective description of the groups. In recent years, this programme has led to substantial advances on various long-standing open problems in the representation theory of finite groups: these were shown to reduce to questions on simple groups which could then be solved by applying the deep results on characters of finite reductive groups; see, for example, the book of Navarro [Na18] and the second author's survey [Ma17].

Lusztig's book [Lu84a] is a milestone in the study of representations of finite groups of Lie type, both in terms of conceptual depth and technical complexity. It brings together various deep and rich theories, culminating in the fundamental 'Jordan decomposition of characters'. In particular, this provides a classification of the irreducible characters, and formulae for character degrees, in terms of purely combinatorial data.

The books by Carter [Ca85] and Digne–Michel [DiMi20] provide more background material and have become influential and highly useful references in this area. Further, more recent texts dealing with more specific aspects are the books by Bonnafé [Bo06] and by Cabanes–Enguehard [CE04]. In our book, the primary focus is on explaining Lusztig's classification of the irreducible characters, and surrounding topics like complete root data, regular embeddings, degree and character formulae, Lusztig induction and restriction and Jordan decomposition. Thus, we will complement, enhance and go beyond the above texts in several ways:

- A substantial part of Chapter 1 is concerned with the discussion of explicit constructions involving root data and Steinberg maps; these are at the basis of efficient algorithms and computer implementations for which the CHEVIE system [GHLMP], [MiChv] is our primary reference.
- Chapter 2 provides an introduction (with complete proofs where possible) into the basic formalism of Lusztig's book [Lu84a] leading to the statement of the 'Jordan decomposition of characters', both in the 'connected centre' case and in general. We also discuss the computation of Green functions, a problem which is not yet solved in complete generality.
- In Chapter 3 we present not only the well-established ordinary Harish-Chandra theory but also d-Harish-Chandra theories defined by means of Lusztig induction, which have proved to be of fundamental importance in block theory of finite reductive groups.
- The classification of the all-important unipotent characters of finite reductive groups and various of their properties are described in Chapter 4, including a discussion of Lusztig's 'non-abelian Fourier transforms'. We also discuss the decomposition of the Lusztig functor and its commutation with Jordan decomposition for groups with connected centre. Furthermore, we touch upon some topics in the character theory of disconnected reductive groups.
- An appendix discusses, in a somewhat informal way, various applications, open problems and connections to related theories, with numerous references to further reading.

Throughout, we have tried to design the exposition of the above material with a view towards applications in finite group theory, and to be accessible to a reader with only a modest background in algebraic geometry. In view of the enormous amount of material available in this area, it is clear that we had to make a number of choices concerning the topics that we cover in this book. For example, we have decided not to say anything about the classification of the conjugacy classes of finite groups of Lie type (except for some occasional general statements). Furthermore, as a general rule, we only give proofs for statements for which we could not find any convenient reference in the existing literature. (But sometimes we defer from this rule and give a detailed argument when this appears to be a good illustration for the methods developed so far.)

On the other hand, we have made a serious attempt to provide precise references, thereby giving something like a guided tour through this vast territory. In short, we hope that this text will be a useful addition to the literature on the character theory of finite groups of Lie type, where the choice of topics and the style of exposition have been strongly influenced, of course, by our own experience with the sometimes difficult task of finding appropriate references, or accommodating the existing literature to specific needs in applications.

We are indebted to Marc Cabanes, Bill Casselman, David Craven, Olivier Dudas, Zhicheng Feng, Skip Garibaldi, Jonas Hetz, Jim Humphreys, Radha Kessar, Emil Rotilio, Lucas Ruhstorfer, Jay Taylor for comments on earlier versions. We thank George Lusztig for his interest in this project and for a number of useful conversations about various topics related to it.

Kaiserslautern and Stuttgart, October 2019

1

Reductive Groups and Steinberg Maps

This first chapter is of a preparatory nature; its purpose is to collect some basic results about algebraic groups (with proofs where appropriate) which will be needed for the discussion of characters and applications in later chapters. In particular, one of our aims is to arrive at the point where we can give a precise definition of a 'series of finite groups of Lie type' $\{\mathbf{G}(q)\}$, indexed by a parameter q. We also introduce a number of tools which will be helpful in the discussion of examples.

For a reader familiar with the basic notions about algebraic groups, root data and Frobenius maps, it might just be sufficient to browse through this chapter on a first reading, in order to see some of our notation. There are, however, a few topics and results that are frequently used in the literature on algebraic groups and finite groups of Lie type, but for which we have found the coverage in standard reference texts (like [Bor91], [Ca85], [DiMi20], [Hum91], [Spr98]) not to be sufficient; these will be treated here in a fairly self-contained manner.

Section 1.1 is purely expository: it introduces affine varieties, linear algebraic groups in general, and the first definitions concerning reductive algebraic groups. In Section 1.2, we consider in some detail (abstract) root data, the basic underlying combinatorial structure of the theory of reductive algebraic groups. We present an approach (familiar in the literature concerned with computational aspects, e.g. [CMT04], [BrLu12]) in which root data simply appear as factorisations of the Cartan matrix of a root system. This will be extremely useful for the discussion of examples and the efficient construction of root data from Cartan matrices.

Section 1.3 contains the fundamental existence and isomorphism theorems of Chevalley [Ch55], [Ch05] concerning connected reductive algebraic groups. We also state the more general 'isogeny theorem' and present some of its basic applications. (There is now a quite short proof available, due to Steinberg [St99].) An important class of homomorphisms of algebraic groups to which this more general theorem applies are the Steinberg maps, to be discussed in detail in Section 1.4.

Following [St68], one might just define a Steinberg map of a connected reductive algebraic group $\mathbf{G}$ to be an endomorphism whose fixed point set is finite. But it will be important and convenient to single out a certain subclass of such morphisms to which one can naturally attach a positive real number q (some power of which is a prime power) and such that one can speak of the corresponding finite group $\mathbf{G}(q)$. The known results on Frobenius and Steinberg maps are somewhat scattered in the literature so we treat this in some detail here, with complete proofs.

In Section 1.5, we illustrate the material developed so far by a number of further basic constructions and examples. In Section 1.6, we show how all this leads to the notion of 'generic' reductive groups, in which q will appear as a formal parameter. Finally, Section 1.7 discusses in some detail the first applications to the character theory of finite groups of Lie type: the 'Multiplicity–Freeness' Theorem 1.7.15.

1.1 Affine Varieties and Algebraic Groups

In this section, we introduce some basic notions concerning affine varieties and algebraic groups. We will do this in a somewhat informal way, assuming that the reader is willing to fill in some details from textbooks like [Bor91], [Ca85], [Ge03a], [Hum91], [MaTe11], [Spr98].

1.1.1 Affine varieties. Let k be a field and let $\mathbf{X}$ be a set. Let A be a subalgebra of the k-algebra $\mathscr{A}(\mathbf{X}, k)$ of all functions $f: \mathbf{X} \to k$. Using A we can try to define a topology on $\mathbf{X}$: a subset $\mathbf{X}' \subseteq \mathbf{X}$ is called closed if there is a subset $S \subseteq A$ such that $\mathbf{X}' = \{x \in \mathbf{X} \mid f(x) = 0 \text{ for all } f \in S\}$. This works well, and gives rise to the *Zariski topology* on $\mathbf{X}$, if A is neither too small nor too big. The precise requirements are (see [Car55-56]):

(1) A is a finitely generated k-algebra and contains the identity of $\mathscr{A}(\mathbf{X}, k)$;
(2) A separates points, that is, given $x \neq x'$ in $\mathbf{X}$, there exist some $f \in A$ such that $f(x) \neq f(x')$;
(3) any k-algebra homomorphism $\lambda: A \to k$ is given by evaluation at a point (that is, there exists some $x \in \mathbf{X}$ such that $\lambda(f) = f(x)$ for all $f \in A$).

A pair $(\mathbf{X}, A)$ satisfying the above conditions will be called an *affine variety over* k; the functions in A are called the *regular functions* on $\mathbf{X}$. We define $\dim \mathbf{X}$ to be the supremum of all $r \geqslant 0$ such that there exist r algebraically independent elements in A. Since A is finitely generated, $\dim \mathbf{X} < \infty$. (See [Ge03a, 1.2.18].) If A is an integral domain, then $\mathbf{X}$ is called *irreducible*.

There is now also a natural notion of morphisms. Let $(\mathbf{X}, A)$ and $(\mathbf{Y}, B)$ be affine varieties over k. A map $\varphi: \mathbf{X} \to \mathbf{Y}$ will be called a *morphism* if composition with φ maps B into A (that is, for all $g \in B$, we have $\varphi^*(g) := g \circ \varphi \in A$); in this case, $\varphi^*: B \to A$ is an algebra homomorphism, and every algebra homomorphism $B \to A$ arises in this way. The morphism φ is an *isomorphism* if there is a morphism $\psi: \mathbf{Y} \to \mathbf{X}$ such that $\psi \circ \varphi = \mathrm{id}_{\mathbf{X}}$ and $\varphi \circ \psi = \mathrm{id}_{\mathbf{Y}}$. (Equivalently: the induced algebra homomorphism $\varphi^*: B \to A$ is an isomorphism.)

Starting with these definitions, the basics of (affine) algebraic geometry are developed in [St74], and this is also the approach taken in [Ge03a]. The link with the more traditional approach via closed subsets in affine space (which, when considered as an algebraic set with the Zariski topology, we denote by $\mathbf{k}^n$) is obtained as follows. Let $(\mathbf{X}, A)$ be an affine variety over k. Choose a set $\{a_1, \ldots, a_n\}$ of algebra generators of A and consider the polynomial ring $k[t_1, \ldots, t_n]$ in n independent indeterminates $t_1, \ldots, t_n$. There is a unique algebra homomorphism $\pi: k[t_1, \ldots, t_n] \to A$ such that $\pi(t_i) = a_i$ for $1 \leqslant i \leqslant n$. Then we have a morphism

$$\varphi: \mathbf{X} \to \mathbf{k}^n, \qquad x \mapsto (a_1(x), \ldots, a_n(x)),$$

such that $\varphi^* = \pi$. The image of φ is the 'Zariski closed' set of $\mathbf{k}^n$ consisting of all $(x_1, \ldots, x_n) \in \mathbf{k}^n$ such that $f(x_1, \ldots, x_n) = 0$ for all $f \in \ker(\pi)$.

To develop these matters any further, it is then essential to assume that k is algebraically closed, which we will do from now on. One can go a long way towards those parts of the theory that are relevant for algebraic groups, once the following basic result about morphisms is available (see [St74, §1.13], [Ge03a, §2.2]):

Let $\varphi: \mathbf{X} \to \mathbf{Y}$ be a morphism between irreducible affine varieties such that $\varphi(\mathbf{X})$ is dense in $\mathbf{Y}$. Then there is a non-empty open subset $\mathbf{V} \subseteq \mathbf{Y}$ such that $\mathbf{V} \subseteq \varphi(\mathbf{X})$ and, for all $y \in \mathbf{V}$, we have $\dim \varphi^{-1}(y) = \dim \mathbf{X} - \dim \mathbf{Y}$.

1.1.2 Algebraic groups. In order to define algebraic groups, we need to know that direct products of affine varieties are again affine varieties. So let $(\mathbf{X}, A)$ and $(\mathbf{Y}, B)$ be affine varieties over k. Given $f \in A$ and $g \in B$, we define the function $f \otimes g: \mathbf{X} \times \mathbf{Y} \to k$ by $(x, y) \mapsto f(x)g(y)$. Let $A \otimes B$ be the subspace of $\mathscr{A}(\mathbf{X} \times \mathbf{Y}, k)$ spanned by all $f \otimes g$, where $f \in A$ and $g \in B$. Then $A \otimes B$ is a subalgebra of $\mathscr{A}(\mathbf{X} \times \mathbf{Y}, k)$ (isomorphic to the tensor product of A, B over k) and the pair $(\mathbf{X} \times \mathbf{Y}, A \otimes B)$ is easily seen to be an affine variety over k. Now let $(\mathbf{G}, A)$ be an affine variety and assume that $\mathbf{G}$ is an abstract group where multiplication and inversion are defined by maps $\mu: \mathbf{G} \times \mathbf{G} \to \mathbf{G}$ and $\iota: \mathbf{G} \to \mathbf{G}$. We say that $\mathbf{G}$ is an *affine algebraic group* if μ and ι are morphisms. The first example is the additive group of k which, when considered as an algebraic group, we denote by $\mathbf{k}^+$ (with algebra of regular functions given by the polynomial functions $k \to k$).

Most importantly, the group $\mathrm{GL}_n(k)$ $(n \geqslant 1)$, is an affine algebraic group, with algebra of regular functions given as follows. For $1 \leqslant i, j \leqslant n$ let $f_{ij}: \mathrm{GL}_n(k) \to k$ be the function that sends a matrix $g \in \mathrm{GL}_n(k)$ to its (i, j)-entry; furthermore, let $\delta: \mathrm{GL}_n(k) \to k$, $g \mapsto \det(g)^{-1}$. Then the algebra of regular functions on $\mathrm{GL}_n(k)$

is the subalgebra of $\mathscr{A}(\mathrm{GL}_n(k), k)$ generated by δ and all f_{ij} $(1 \leqslant i, j \leqslant n)$. In particular, $\mathrm{GL}_1(k)$ is an affine algebraic group, which we denote by $\mathbf{k}^\times$.

It is a basic fact that any affine algebraic group $\mathbf{G}$ over k is isomorphic to a closed subgroup of $\mathrm{GL}_n(k)$, for some $n \geqslant 1$; see [Ge03a, 2.4.4]. For this reason, an affine algebraic group is also called a *linear algebraic group*. When we just write 'algebraic group', we always mean a linear algebraic group.

1.1.3 Connected algebraic groups. A topological space is *connected* if it cannot be written as a disjoint union of two non-empty open subsets. A linear algebraic group $\mathbf{G}$ can always be written as the disjoint union of finitely many connected components, where the component containing the identity element is a closed connected normal subgroup of $\mathbf{G}$, denoted by $\mathbf{G}^\circ$; see [Ge03a, 1.3.13]. Thus, $\mathbf{G}$ is connected if and only if $\mathbf{G} = \mathbf{G}^\circ$. (Equivalently: $\mathbf{G}$ is irreducible as an affine variety; see [Ge03a, 1.1.12, 1.3.1].)

What is the significance of this fundamental notion? Every finite group $\mathbf{G}$ can be regarded as a linear algebraic group, with algebra of regular functions given by all of $\mathscr{A}(\mathbf{G}, k)$. Thus, the study of *all* linear algebraic groups is necessarily more complicated than the study of all finite groups. But, as Vogan [Vo07] writes, "a miracle happens" when we consider *connected* algebraic groups: things actually become much less complicated. One reason is that a connected algebraic group is almost completely determined by its Lie algebra (see 1.1.5 and also 1.1.12 below), and the latter can be studied using linear algebra methods.

Combined with our assumption that k is algebraically closed, this gives us some powerful tools. For example, matrices over algebraically closed fields can be put in triangular form. An analogue of this fact for an arbitrary connected algebraic group is the statement that every element is contained in a Borel subgroup (that is, a maximal closed connected solvable subgroup); see [Ge03a, 3.4.9]. A useful criterion for showing the connectedness of a subgroup of $\mathbf{G}$ is as follows.

Let $\{\mathbf{H}_i\}_{i\in I}$ be a family of closed connected subgroups of $\mathbf{G}$. Then the (abstract) subgroup $H = \langle \mathbf{H}_i \mid i \in I\rangle \subseteq \mathbf{G}$ generated by this family is closed and connected; furthermore, $H = \mathbf{H}_{i_1} \cdots \mathbf{H}_{i_n}$ for some n and $i_1, \ldots, i_n \in I$.

The proof uses the result on morphisms mentioned at the end of 1.1.1; see, e.g., [Ge03a, 2.4.6]. Note that, if $\mathbf{U}, \mathbf{V}$ are any closed subgroups of $\mathbf{G}$, then the abstract subgroup $\langle \mathbf{U}, \mathbf{V}\rangle \subseteq \mathbf{G}$ need not even be closed. For example, if $\mathbf{G} = \mathrm{SL}_2(\mathbb{C})$, then it is well known that the subgroup $\mathrm{SL}_2(\mathbb{Z})$ is generated by two elements, of orders 4 and 6, but this subgroup is certainly not closed in $\mathbf{G}$. However, if $\mathbf{V}$ is normalised by $\mathbf{U}$, then $\langle \mathbf{U}, \mathbf{V}\rangle = \mathbf{U}.\mathbf{V}$ is closed; see [Ch05, §3.3, Corollaire].

We will use without further special mention some standard facts (whose proofs also rely on the above-mentioned result on morphisms). For example, if $f\colon \mathbf{G} \to \mathbf{G}'$ is a homomorphism of linear algebraic groups, then the image $f(\mathbf{G})$ is a closed subgroup of $\mathbf{G}'$ (connected if $\mathbf{G}$ is connected), the kernel of f is a closed subgroup of $\mathbf{G}$ and we have $\dim \mathbf{G} = \dim \ker(f) + \dim f(\mathbf{G})$. (See, e.g., [Ge03a, 2.2.14].)

1.1.4 Classical groups. These form an important class of examples of linear algebraic groups. They are closed subgroups of $\mathrm{GL}_n(k)$ defined by certain quadratic polynomials corresponding to a bilinear or quadratic form on the underlying vector space k^n. There is an extensive literature on these groups; see, e.g., [Bou07], [Dieu74], [Gro02], [Tay92]. Since our base field k is algebraically closed, the general theory simplifies considerably and we only need to consider three classes of groups, leading to the Dynkin types B, C, D. First, and quite generally, for any invertible matrix $Q_n \in M_n(k)$, we obtain a linear algebraic group

$$\Gamma(Q_n, k) := \{A \in M_n(k) \mid A^{\mathrm{tr}} Q_n A = Q_n\};$$

note that $\det(A) = \pm 1$ for all $A \in \Gamma(Q_n, k)$. Let us now take Q_n of the form

$$Q_n = \begin{bmatrix} 0 & \cdots & 0 & \pm 1 \\ \vdots & \iddots & \iddots & 0 \\ 0 & \pm 1 & \iddots & \vdots \\ \pm 1 & 0 & \cdots & 0 \end{bmatrix} \in M_n(k) \qquad (n \geqslant 2)$$

where the signs are such that $Q_n^{\mathrm{tr}} = \pm Q_n$. Then Q_n is the matrix of a non-degenerate symmetric or alternating bilinear form on k^n; furthermore, $Q_n^{-1} = Q_n^{\mathrm{tr}}$ and $\Gamma(Q_n, k)$ will be invariant under transposing matrices.

If $Q_n^{\mathrm{tr}} = -Q_n$ and n is even, then $\Gamma(Q_n, k)$ will be denoted $\mathrm{Sp}_n(k)$ and called the *symplectic group*. This group is always connected; see [Ge03a, 1.7.4]. Now assume that $Q_n^{\mathrm{tr}} = Q_n$ and that all signs in Q_n are $+$. Then we also consider the quadratic form on k^n defined by the polynomial

$$\dot{f}_n := \begin{cases} t_1t_{2m+1} + t_2t_{2m} + \ldots + t_mt_{m+2} + t_{m+1}^2 & \text{if } n = 2m+1 \text{ is odd,} \\ t_1t_{2m} + t_2t_{2m-2} + \ldots + t_mt_{m+1} & \text{if } n = 2m \text{ is even,} \end{cases}$$

(where $t_1, \ldots, t_n$ are indeterminates). This defines a function $\dot{f}_n \colon k^n \to k$, where we regard the elements of k^n as column vectors. Using the notation in [MaTe11, §1.2], the *general orthogonal group* is defined as

$$\mathrm{GO}_n(k) := \{A \in M_n(k) \mid \dot{f}_n(Av) = \dot{f}_n(v) \text{ for all } v \in k^n\};$$

furthermore, $\mathrm{SO}_n(k) := \mathrm{GO}_n^\circ(k)$ will be called the *special orthogonal group*. In each case, we have $[\mathrm{GO}_n(k) : \mathrm{SO}_n(k)] \leqslant 2$; see [Ge03a, §1.7], [Gro02] for further details. Note also that, if $\mathrm{char}(k) \neq 2$, then $\mathrm{GO}_n(k) = \Gamma(Q_n, k)$; furthermore, if n is even and $\mathrm{char}(k) = 2$, then $\mathrm{GO}_n(k)$ will be strictly contained in $\Gamma(Q_n, k)$. (See also Example 1.5.5 for the case where n is odd and $\mathrm{char}(k) = 2$.)

The particular choices of Q_n and f_n lead to simple descriptions of a BN-pair in $\mathrm{Sp}_n(k)$ and $\mathrm{SO}_n(k)$; see, e.g., [Ge03a, §1.7] (and also 1.1.14 below). The Dynkin types and dimensions are given as follows.

Group	Type	Dimension
$\mathrm{SO}_{2m+1}(k)$	B_m	$2m^2+m$
$\mathrm{Sp}_{2m}(k)$	C_m	$2m^2+m$
$\mathrm{SO}_{2m}(k)$	D_m	$2m^2-m$

Later on, if there is no danger of confusion, we shall just write GL_n, SL_n, Sp_n, SO_n, GO_n instead of $\mathrm{GL}_n(k)$, $\mathrm{SL}_n(k)$, $\mathrm{Sp}_n(k)$, $\mathrm{SO}_n(k)$, $\mathrm{GO}_n(k)$, respectively.

1.1.5 Tangent spaces and the Lie algebra. Let $(\mathbf{X}, A)$ be an affine variety over k. Then the *tangent space* $T_x(\mathbf{X})$ of $\mathbf{X}$ at a point $x \in \mathbf{X}$ is the set of all k-linear maps $D\colon A \to k$ such that $D(fg) = f(x)D(g) + g(x)D(f)$. (Such linear maps are called *derivations*.) Clearly, $T_x(\mathbf{X})$ is a subspace of the vector space of all linear maps from A to k. Any $D \in T_x(\mathbf{X})$ is uniquely determined by its values on a set of algebra generators of A. Hence, since A is finitely generated, we have $\dim T_x(\mathbf{X}) < \infty$. If $\mathbf{X}' \subseteq \mathbf{X}$ is a closed subvariety, we have a natural inclusion $T_x(\mathbf{X}') \subseteq T_x(\mathbf{X})$ for any $x \in \mathbf{X}'$. For example, we can identify $T_x(\mathbf{k}^n)$ with k^n for all $x \in \mathbf{k}^n$ and so, if $\mathbf{X} \subseteq \mathbf{k}^n$ is a Zariski closed subset, we have $T_x(\mathbf{X}) \subseteq k^n$ for all $x \in \mathbf{X}$ (see [Ge03a, 1.4.10]). More generally, any morphism $\varphi\colon \mathbf{X} \to \mathbf{Y}$ between affine varieties over k naturally induces a linear map $d_x\varphi\colon T_x(\mathbf{X}) \to T_{\varphi(x)}(\mathbf{Y})$ for any $x \in \mathbf{X}$, called the *differential* of φ at x. (See [Ge03a, §1.4].)

Now let $\mathbf{G}$ be a linear algebraic group and denote $L(\mathbf{G}) := T_1(\mathbf{G})$, the tangent space at the identity element of $\mathbf{G}$. Then

$$L(\mathbf{G}) = L(\mathbf{G}^\circ) \qquad \text{and} \qquad \dim \mathbf{G} = \dim L(\mathbf{G});$$

see [Ge03a, 1.5.2]. Furthermore, there is a Lie product $[\ ,\ ]$ on $L(\mathbf{G})$ which can be defined as follows. Consider a realisation of $\mathbf{G}$ as a closed subgroup of $\mathrm{GL}_n(k)$ for some $n \geqslant 1$. We have a natural isomorphism of $L(\mathrm{GL}_n(k))$ onto $M_n(k)$, the vector space of all $n \times n$-matrices over k; see [Ge03a, 1.4.14]. Hence we obtain an embedding $L(\mathbf{G}) \subseteq M_n(k)$ where $M_n(k)$ is endowed with the usual Lie product $[A, B] = AB - BA$ for $A, B \in M_n(k)$. Then one shows that $[L(\mathbf{G}), L(\mathbf{G})] \subseteq L(\mathbf{G})$ and so $[\ ,\ ]$ restricts to a Lie product on $L(\mathbf{G})$; see [Ge03a, 1.5.3]. (Of course, there is also an intrinsic description of $L(\mathbf{G})$ in terms of the algebra of regular functions on $\mathbf{G}$ which shows, in particular, that the product does not depend on the choice of the realisation of $\mathbf{G}$; see [Ge03a, 1.5.4].)

1.1.6 Quotients. Let $\mathbf{G}$ be a linear algebraic group and $\mathbf{H}$ be a closed normal subgroup of $\mathbf{G}$. We have the abstract factor group $\mathbf{G}/\mathbf{H}$ and we would certainly like to know if this can also be viewed as an algebraic group. More generally, let $\mathbf{X}$ be an affine variety and $\mathbf{H}$ be a linear algebraic group such that we have a morphism $\mathbf{H} \times \mathbf{X} \to \mathbf{X}$ which defines an action of $\mathbf{H}$ on $\mathbf{X}$. The question of whether we can view the set of orbits $\mathbf{X}/\mathbf{H}$ as an algebraic variety leads to 'geometric invariant theory'; in general, these are quite delicate matters. Let us begin by noting that there is a natural candidate for the algebra of functions on the orbit set $\mathbf{X}/\mathbf{H}$: If A is the algebra of regular functions on $\mathbf{X}$, then

$$A^{\mathbf{H}} := \{f \in A \mid f(h.x) = f(x) \text{ for all } h \in \mathbf{H} \text{ and all } x \in \mathbf{X}\}$$

can naturally be regarded as an algebra of k-valued functions on $\mathbf{X}/\mathbf{H}$. However, the three properties in 1.1.1 will not be satisfied in general. There are two particular situations in which this is the case, and these will be sufficient for most parts of this book:

- $\mathbf{H}$ is a finite group, and
- $\mathbf{X} = \mathbf{G}$ is an algebraic group and $\mathbf{H}$ is a closed normal subgroup (acting by left multiplication).

(For the proofs, see [Fo69, 5.25] or [Ge03a, 2.5.12] in the first case, and [Fo69, 2.26] or [Spr98, §5.5] in the second case.) Now let us assume that $(\mathbf{X}/\mathbf{H}, A^{\mathbf{H}})$ is an affine variety. Then, first of all, the natural map $\mathbf{X} \to \mathbf{X}/\mathbf{H}$ is a morphism of affine varieties. Furthermore, we have the following universal property:

If $\varphi\colon \mathbf{X} \to \mathbf{Y}$ *is any morphism of affine varieties that is constant on the orbits of* $\mathbf{H}$ *on* $\mathbf{X}$, *then there is a unique morphism* $\bar{\varphi}\colon \mathbf{X}/\mathbf{H} \to \mathbf{Y}$ *such that* φ *is the composition of* $\bar{\varphi}$ *and the natural map* $\mathbf{X} \to \mathbf{X}/\mathbf{H}$.

(Indeed, if B is the algebra of regular functions on $\mathbf{Y}$, then the induced algebra homomorphism $\varphi^*\colon B \to A$ has image in $A^{\mathbf{H}}$, hence it factors through an algebra homomorphism $\bar{\varphi}^*\colon B \to A^{\mathbf{H}}$ for a unique morphism $\bar{\varphi}\colon \mathbf{X}/\mathbf{H} \to \mathbf{Y}$.)

For example, if we are in the second of the above two cases, then the universal property shows that the induced multiplication and inversion maps on $\mathbf{G}/\mathbf{H}$ are morphisms of affine varieties. Thus, $\mathbf{G}/\mathbf{H}$ is an affine algebraic group.

1.1.7 Algebraic groups in positive characteristic. The finite groups that we shall study in this book are obtained as

$$\mathbf{G}^F := \{g \in \mathbf{G} \mid F(g) = g\}$$

where the $F\colon \mathbf{G} \to \mathbf{G}$ are certain bijective endomorphisms with finitely many fixed points, called '*Steinberg maps*'. (This will be discussed in detail in Section 1.4.) Such maps F will only exist if k has prime characteristic, so we will usually assume that p is a prime number and $k = \overline{\mathbb{F}}_p$ is an algebraic closure of the field $\mathbb{F}_p = \mathbb{Z}/p\mathbb{Z}$. Now, algebraic geometry over fields with positive characteristic is, in some respects, more tricky than algebraic geometry over $\mathbb{C}$, say (because of the inseparability of certain field extensions; see also 1.1.8 below). However, some things are actually easier. For example, using an embedding of $\mathbf{G}$ into some $\mathrm{GL}_n(k)$ as in 1.1.2, we see that every element $g \in \mathbf{G}$ has finite order. Thus, we can define g to be *semisimple* if the order of g is prime to p; we define g to be *unipotent* if the order of g is a power of p. Then, clearly, any $g \in \mathbf{G}$ has a unique decomposition

$$g = us = su \qquad \text{where } s \in \mathbf{G} \text{ is semisimple and } u \in \mathbf{G} \text{ is unipotent,}$$

called the *Jordan decomposition of elements*. (In characteristic 0, this certainly requires more work; see [Spr98, §2.4].) Another example: An algebraic group $\mathbf{G}$ is called a *torus* if $\mathbf{G}$ is isomorphic to a direct product of a finite number of copies of $\mathbf{k}^\times$. Then $\mathbf{G}$ is a torus if and only if $\mathbf{G}$ is connected, abelian and consists entirely of elements of order prime to p; see [Ge03a, 3.1.9]. (To formulate this in characteristic 0, one would need the general definition of semisimple elements.)

1.1.8 Some things that go wrong in positive characteristic. Here we collect a few items which show that, when working over $k = \overline{\mathbb{F}}_p$ as above, things may not work as one might hope or expect. First note that a bijective homomorphism of algebraic groups $\varphi\colon \mathbf{G}_1 \to \mathbf{G}_2$ need not be an isomorphism. A standard example is the Frobenius map $\mathbf{k}^+ \to \mathbf{k}^+, x \mapsto x^p$. (Note that, over $\mathbb{C}$, a bijective homomorphism between connected algebraic groups is an isomorphism; see [GoWa98, 11.1.16].) In general, we have (see, e.g., [Ge03a, 2.3.15], [Spr98, 5.3.3]):

(a) *A bijective homomorphism of linear algebraic groups $\varphi\colon \mathbf{G}_1 \to \mathbf{G}_2$ is an isomorphism if and only if the differential $d_1\varphi\colon T_1(\mathbf{G}_1) \to T_1(\mathbf{G}_2)$ between the tangent spaces is an isomorphism.*

The next item concerns the Lie algebra of an algebraic group. Let $\mathbf{G}$ be a linear algebraic group and $\mathbf{U}, \mathbf{H}$ be closed subgroups of $\mathbf{G}$. As already noted in 1.1.5, we have natural inclusions of $L(\mathbf{U})$, $L(\mathbf{H})$ and $L(\mathbf{U} \cap \mathbf{H})$ into $L(\mathbf{G})$. It is always true that $L(\mathbf{U} \cap \mathbf{H}) \subseteq L(\mathbf{U}) \cap L(\mathbf{H})$.

(b) *When considering the intersection of closed subgroups $\mathbf{U}, \mathbf{H}$ of an algebraic group $\mathbf{G}$, it is not always true that $L(\mathbf{U} \cap \mathbf{H}) = L(\mathbf{U}) \cap L(\mathbf{H})$.*

A good example to keep in mind is as follows. Let $\mathbf{G} = \mathrm{GL}_n(\overline{\mathbb{F}}_p)$, $\mathbf{H} = \mathrm{SL}_n(\overline{\mathbb{F}}_p)$ and $\mathbf{Z}$ be the centre of $\mathbf{G}$ (the scalar matrices in $\mathbf{G}$). Then $\mathbf{Z}, \mathbf{H}$ are closed subgroups of $\mathbf{G}$. As in 1.1.5, we can identify $L(\mathbf{G}) = M_n(k)$; then $L(\mathbf{H})$ consists of all matrices of trace 0 and $L(\mathbf{Z})$ consists of all scalar matrices. (For these facts see, for example, [Ge03a, §1.5].) Assume now that p divides n. Then, clearly, $\{0\} \neq L(\mathbf{Z}) \subseteq L(\mathbf{H})$, whereas $\mathbf{Z} \cap \mathbf{H}$ is finite and so $L(\mathbf{Z} \cap \mathbf{H}) = L((\mathbf{Z} \cap \mathbf{H})^\circ) = \{0\}$. (This phenomenon cannot happen in characteristic 0; see [Bor91, 6.12] or [Hum91, 12.5].)

Closely related to the above item is the next item: semidirect products. Let $\mathbf{G}$ be an algebraic group and $\mathbf{U}, \mathbf{N}$ be closed subgroups such that $\mathbf{N}$ is normal, $\mathbf{G} = \mathbf{U}.\mathbf{N}$ and $\mathbf{U} \cap \mathbf{N} = \{1\}$. Following [Bor91, 1.11], we say that $\mathbf{G}$ is the *semidirect product (of algebraic groups)* of $\mathbf{U}, \mathbf{N}$ if the natural map $\mathbf{U} \times \mathbf{N} \to \mathbf{G}$ given by multiplication is an isomorphism of affine varieties. If this holds, we have an inverse isomorphism $\mathbf{G} \to \mathbf{U} \times \mathbf{N}$ and the first projection will induce an isomorphism of algebraic groups $\mathbf{G}/\mathbf{N} \cong \mathbf{U}$. We have the criterion:

(c) *$\mathbf{G}$ is the semidirect product (of algebraic groups) of $\mathbf{U}, \mathbf{N}$ if and only if $L(\mathbf{U}) \cap L(\mathbf{N}) = \{0\}$. For example, this holds if $\mathbf{U}$ or $\mathbf{N}$ is finite.*

(This easily follows from (a) and the description of the differential of the product map $\mathbf{U} \times \mathbf{N} \to \mathbf{G}$; see, e.g., [Ge03a, 1.5.6], [Spr98, 4.4.12].) Take again the above example where $\mathbf{G} = \mathrm{GL}_n(\overline{\mathbb{F}}_p)$, $\mathbf{U} = \mathrm{SL}_n(\overline{\mathbb{F}}_p)$ and $\mathbf{N}$ is the centre of $\mathbf{G}$ (the scalar matrices in $\mathbf{G}$). Assume now that $n = p$. Then $\mathbf{U}, \mathbf{N}$ are closed connected normal subgroups such that $\mathbf{G} = \mathbf{U}.\mathbf{N}$ and $\mathbf{U} \cap \mathbf{N} = \{1\}$. However, $\{0\} \neq L(\mathbf{N}) \subseteq L(\mathbf{U})$ and so this is not a semidirect product of algebraic groups!

For working with fixed points of groups under automorphisms as in 1.1.7, the following completely general result will be useful on several occasions.

Lemma 1.1.9 ([St68, 4.5]) *Let A, B be groups and $f : A \to B$ be a surjective homomorphism with $\ker(f) \subseteq Z(A)$. Let $\sigma : A \to A$ and $\tau : B \to B$ be automorphisms such that $f \circ \sigma = \tau \circ f$. Then $C := \{a^{-1}\sigma(a) \mid a \in \ker(f)\}$ is a subgroup of $\ker(f)$. Furthermore, let*

$$A^\sigma := \{a \in A \mid \sigma(a) = a\} \qquad \text{and} \qquad B^\tau := \{b \in B \mid \tau(b) = b\}.$$

Then $f(A^\sigma)$ is normal in B^τ and there is a canonical injective homomorphism $\delta : B^\tau/f(A^\sigma) \hookrightarrow \ker(f)/C$, with image $(\{a^{-1}\sigma(a) \mid a \in A\} \cap \ker(f))/C$.

Proof Since $\ker(f) \subseteq Z(A)$ and $f \circ \sigma = \tau \circ f$, it is clear that $\sigma(\ker(f)) \subseteq \ker(f)$ and that C is a subgroup of $\ker(f)$. We define $\delta' : B^\tau \to \ker(f)/C$ as follows. Let $b \in B^\tau$ and choose $a \in A$ such that $f(a) = b$. We have $f(\sigma(a)) = \tau(f(a)) = \tau(b) = b$ and so $c := a^{-1}\sigma(a) \in \ker(f)$. Then set $\delta'(b) = cC \in \ker(f)/C$. One verifies that δ' is well defined and a group homomorphism; furthermore, $\ker(\delta') = f(A^\sigma)$ and the image of δ' is as stated above. Thus, we obtain an induced map δ with the required properties. □

1.1.10 The unipotent radical. Let $\mathbf{G}$ be a linear algebraic group over $k = \overline{\mathbb{F}}_p$, where p is a prime. We can now define the *unipotent radical* $R_u(\mathbf{G}) \subseteq \mathbf{G}$, as follows. An abstract subgroup of $\mathbf{G}$ is called *unipotent* if all of its elements are unipotent. Since every element in $\mathbf{G}$ has finite order, one easily sees that the product of two normal unipotent subgroups is again a normal unipotent subgroup of $\mathbf{G}$. If $\mathbf{G}$ is finite, then this immediately shows that there is a unique maximal normal unipotent subgroup in $\mathbf{G}$. (In the theory of finite groups, this is denoted $O_p(\mathbf{G})$.) In the general case, we define

$$R_u(\mathbf{G}) := \text{subgroup of } \mathbf{G} \text{ generated by all } \mathbf{U} \in \mathscr{S}_{\text{unip}}(\mathbf{G}),$$

where $\mathscr{S}_{\text{unip}}(\mathbf{G})$ denotes the set of all closed connected normal unipotent subgroups of $\mathbf{G}$. It is clear that $R_u(\mathbf{G})$ is an abstract normal subgroup of $\mathbf{G}$. By the criterion in 1.1.3, $R_u(\mathbf{G})$ is a closed connected subgroup of $\mathbf{G}$; furthermore, $R_u(\mathbf{G}) = \mathbf{U}_1 \dots \mathbf{U}_n$ for some $n \geqslant 1$ and $\mathbf{U}_1, \dots, \mathbf{U}_n \in \mathscr{S}_{\text{unip}}(\mathbf{G})$. As already remarked before, this product will consist of unipotent elements. Thus, $R_u(\mathbf{G})$ is the unique maximal closed connected normal unipotent subgroup of $\mathbf{G}$. (The analogous definition also works when k is an arbitrary algebraically closed field, using the slightly more complicated characterisation of unipotent elements in that case.)

We say that $\mathbf{G}$ *is reductive if* $R_u(\mathbf{G}) = \{1\}$.

(Thus, connected reductive groups can be regarded as analogues of finite groups G with $O_p(G) = \{1\}$.) These are the groups that we will be primarily concerned with. In an arbitrary algebraic group $\mathbf{G}$, we always have the closed connected normal subgroups $R_u(\mathbf{G}) \subseteq \mathbf{G}^\circ \subseteq \mathbf{G}$, and $\mathbf{G}/R_u(\mathbf{G})$ will be reductive. Note also that, clearly, we have the implication

$$\mathbf{G} \text{ simple} \quad \Rightarrow \quad \mathbf{G} \text{ reductive (and connected).}$$

Here, $\mathbf{G} \neq \{1\}$ is called a *simple algebraic group* if $\mathbf{G}$ is connected, non-abelian and if $\mathbf{G}$ has no closed connected normal subgroups other than $\{1\}$ and $\mathbf{G}$. (For example, $\mathrm{SL}_n(k)$ is a simple algebraic group, although in general it is not simple as an abstract group; $\mathrm{GL}_n(k)$ is reductive, but not simple.)

Even if one is mainly interested in studying a simple group $\mathbf{G}$, one will also have to look at subgroups with a geometric origin, like Levi subgroups or centralisers of semisimple elements. These subgroups tend to be reductive, not just simple. For example, if $\mathbf{G}$ is connected and reductive and $s \in \mathbf{G}$ is a semisimple element, then the centraliser $C_{\mathbf{G}}(s)$ will be a closed reductive (not necessarily connected or simple) subgroup; see [Ca85, 3.5.4].

1.1.11 Characters and co-characters of tori. The simplest examples of connected reductive algebraic groups are tori, and it will be essential to understand some basic constructions with them. First, a general definition. A homomorphism of algebraic groups $\lambda \colon \mathbf{G} \to \mathbf{k}^\times$ will be called a *character* of $\mathbf{G}$. The set $X = X(\mathbf{G})$ of all characters of $\mathbf{G}$ is an abelian group (which we write additively), called the *character group* of $\mathbf{G}$. Similarly, a homomorphism of algebraic groups $\nu \colon \mathbf{k}^\times \to \mathbf{G}$

will be called a *co-character* of $\mathbf{G}$. If $\mathbf{G}$ is abelian, then the set $Y = Y(\mathbf{G})$ of all co-characters of $\mathbf{G}$ also is an abelian group (written additively), called the *co-character group* of $\mathbf{G}$. Now let $\mathbf{G} = \mathbf{T}$ be a torus over k; recall that this means that $\mathbf{T}$ is isomorphic to a direct product of a finite number of copies of $\mathbf{k}^\times$. It is an easy exercise to show that every homomorphism of algebraic groups of $\mathbf{k}^\times$ into itself is given by $\xi \mapsto \xi^n$ for a well-defined $n \in \mathbb{Z}$. Thus, $X(\mathbf{k}^\times) = Y(\mathbf{k}^\times) \cong \mathbb{Z}$ and so

$$X(\mathbf{T}) \cong Y(\mathbf{T}) \cong \mathbb{Z}^r \qquad \text{where} \qquad \mathbf{T} \cong \mathbf{k}^\times \times \ldots \times \mathbf{k}^\times \quad (r \text{ factors}).$$

Hence, $X(\mathbf{T})$ and $Y(\mathbf{T})$ are free abelian groups of the same finite rank. Furthermore, we obtain a natural bilinear pairing

$$\langle\, , \,\rangle \colon X(\mathbf{T}) \times Y(\mathbf{T}) \to \mathbb{Z},$$

defined by the condition that $\lambda(\nu(\xi)) = \xi^{\langle \lambda, \nu \rangle}$ for all $\lambda \in X(\mathbf{T})$, $\nu \in Y(\mathbf{T})$ and $\xi \in \mathbf{k}^\times$. This pairing is a *perfect pairing*, that is, it induces group isomorphisms

$$\begin{aligned} X(\mathbf{T}) &\xrightarrow{\sim} \operatorname{Hom}(Y(\mathbf{T}), \mathbb{Z}), \qquad & \lambda &\mapsto \big(\nu \mapsto \langle \lambda, \nu \rangle\big), \\ Y(\mathbf{T}) &\xrightarrow{\sim} \operatorname{Hom}(X(\mathbf{T}), \mathbb{Z}), \qquad & \nu &\mapsto \big(\lambda \mapsto \langle \lambda, \nu \rangle\big), \end{aligned}$$

(see [MaTe11, 3.6]). Here, Hom just stands for homomorphisms of abstract abelian groups. The pair $(X(\mathbf{T}), Y(\mathbf{T}))$, together with the above pairing, is the simplest example of a 'root datum', which will be considered in more detail in Section 1.2. The assignment $\mathbf{T} \mapsto X(\mathbf{T})$ has the following fundamental property: if $\mathbf{T}'$ is another torus over k, then we have a natural bijection

$$\{\text{homomorphisms of algebraic groups } \mathbf{T} \to \mathbf{T}'\} \;\overset{1-1}{\longleftrightarrow}\; \operatorname{Hom}(X(\mathbf{T}'), X(\mathbf{T})).$$

The correspondence is defined by sending a homomorphism of algebraic groups $f \colon \mathbf{T} \to \mathbf{T}'$ to the map $\varphi \colon X(\mathbf{T}') \to X(\mathbf{T})$, $\lambda' \mapsto \lambda' \circ f$. For future reference, we state the following basic properties of this correspondence:

(a) $f \colon \mathbf{T} \to \mathbf{T}'$ is a closed embedding (that is, an isomorphism onto a closed subgroup of $\mathbf{T}'$) if and only if $\varphi \colon X(\mathbf{T}') \to X(\mathbf{T})$ is surjective; in this case, we have a canonical isomorphism $\ker(\varphi) \cong X(\mathbf{T}'/f(\mathbf{T}))$.

(b) $f \colon \mathbf{T} \to \mathbf{T}'$ is surjective if and only if $\varphi \colon X(\mathbf{T}') \to X(\mathbf{T})$ is injective; in this case, the restriction map $X(\mathbf{T}) \to X(\ker(f))$ is surjective with kernel given by $\varphi(X(\mathbf{T}')) = \{\lambda' \circ f \mid \lambda' \in X(\mathbf{T}')\}$.

See [Bor91, Chap. III, §8], [Ch05, §4.3] and [St74, §2.6] for proofs and further details. Furthermore, by [Ca85, §3.1], $\mathbf{T}$ can be recovered from $X(\mathbf{T})$ through the isomorphism

(c) $\mathbf{T} \xrightarrow{\sim} \operatorname{Hom}(X(\mathbf{T}), k^\times), \qquad t \mapsto (\lambda \mapsto \lambda(t)).$

(Here again Hom just stands for abstract homomorphisms of abelian groups.)

1.1.12 Weight spaces. Characters of tori play a major role in the following context. Let $\mathbf{G}$ be a linear algebraic group and V be a finite-dimensional vector space over k. Note that V is an affine variety with algebra of regular functions given by the subalgebra generated by the dual space $V^* = \operatorname{Hom}(V, k) \subseteq \mathscr{A}(V, k)$. Assume that we have a *representation* of $\mathbf{G}$ on V, that is, we are given a morphism of affine varieties $\mathbf{G} \times V \to V$ which defines a linear action of $\mathbf{G}$ on V. Let $\mathbf{T} \subseteq \mathbf{G}$ be a maximal torus. (Any torus of maximum dimension is maximal.) For each character $\lambda \in X(\mathbf{T})$ we define the subspace

$$V_\lambda := \{v \in V \mid t.v = \lambda(t)v \text{ for all } t \in \mathbf{T}\}.$$

Let $\Psi(\mathbf{T}, V)$ be the set of all $\lambda \in X(\mathbf{T})$ such that $V_\lambda \neq \{0\}$. Since $\mathbf{T}$ consists of pairwise commuting semisimple elements, we have

$$V = \bigoplus_{\lambda \in \Psi(\mathbf{T}, V)} V_\lambda$$

(see [Ge03a, 3.1.5]); in particular, this shows that $\Psi(\mathbf{T}, V)$ is finite. The characters in $\Psi(\mathbf{T}, V)$ are called *weights* and the corresponding subspaces V_λ called *weight spaces* (relative to $\mathbf{T}$). Now, we always have the *adjoint representation* of $\mathbf{G}$ on its Lie algebra $L(\mathbf{G})$, defined as follows. For $g \in \mathbf{G}$, consider the inner automorphism γ_g of $\mathbf{G}$ defined by $\gamma_g(x) = gxg^{-1}$. Taking the differential, we obtain a linear map $d_1\gamma_g \colon L(\mathbf{G}) \to L(\mathbf{G})$, which is a vector space isomorphism. Hence, we obtain a linear action of $\mathbf{G}$ on $L(\mathbf{G})$ such that $g.v = d_1(\gamma_g)(v)$ for all $g \in \mathbf{G}$ and $v \in L(\mathbf{G})$. (The corresponding map $\mathbf{G} \times L(\mathbf{G}) \to L(\mathbf{G})$ is indeed a representation; see, for example, [Hum91, 10.3].) Then the finite set

$$R := \Psi(\mathbf{T}, L(\mathbf{G})) \setminus \{0\} \quad \subseteq X(\mathbf{T})$$

is called the set of *roots* of $\mathbf{G}$ *relative to* $\mathbf{T}$; we have the *root space decomposition*

$$L(\mathbf{G}) = L(\mathbf{G})_0 \oplus \bigoplus_{\alpha \in R} L(\mathbf{G})_\alpha.$$

This works in complete generality, for any algebraic group $\mathbf{G}$. If $\mathbf{G}$ is connected and reductive, then it is possible to obtain much more precise information about the root space decomposition. It turns out that then

$$L(\mathbf{G})_0 = L(\mathbf{T}), \qquad R = -R \quad \text{and} \quad \dim L(\mathbf{G})_\alpha = 1 \quad \text{for all } \alpha \in R.$$

So, in this case, the picture is analogous to that in the theory of complex semisimple Lie algebras and, quite surprisingly, it shows that some crucial aspects of the theory do not depend on the underlying field! This fundamental result, first proved in the *Séminaire Chevalley* [Ch05], will be discussed in more detail in Section 1.3.

1.1.13 General structure of connected reductive algebraic groups. Let $\mathbf{G}$ be a connected linear algebraic group. Denote by $\mathbf{Z} = \mathbf{Z}(\mathbf{G})$ the centre of $\mathbf{G}$. Then we have $\mathbf{Z}^\circ = R_u(\mathbf{Z}) \times \mathbf{S}$ where $\mathbf{S}$ is a torus; see [Ge03a, 3.5.3]. Since $R_u(\mathbf{Z})$ is a characteristic subgroup of $\mathbf{Z}$ and $\mathbf{Z}$ is a characteristic subgroup of $\mathbf{G}$, we see that $R_u(\mathbf{Z})$ is normal in $\mathbf{G}$. Hence, if $\mathbf{G}$ is reductive, then $\mathbf{Z}^\circ$ is a torus. In this case, the above-mentioned results about the root space decomposition lead to the following product decomposition of $\mathbf{G}$ (see [MaTe11, §8.4], [Spr98, §8.1]):

$$\mathbf{G} = \mathbf{Z}^\circ.\mathbf{G}_1 \ldots \mathbf{G}_n \quad \text{where } \mathbf{G}_1, \ldots, \mathbf{G}_n \text{ are closed normal simple subgroups}$$

and $\mathbf{G}_i$, $\mathbf{G}_j$ pairwise commute with each other for $i \neq j$; furthermore, this decomposition of $\mathbf{G}$ has the following properties.

- The subgroups $\{\mathbf{G}_1, \ldots, \mathbf{G}_n\}$ are uniquely determined in the sense that every closed normal simple subgroup of $\mathbf{G}$ is equal to some $\mathbf{G}_i$.
- We have $\mathbf{G}_1 \ldots \mathbf{G}_n = \mathbf{G}_{\text{der}} :=$ commutator (or derived) subgroup of $\mathbf{G}$.

(Recall from 1.1.10 that simple algebraic groups are assumed to be connected and non-abelian; note also that the commutator subgroup of a connected algebraic group is always a closed connected normal subgroup; see [Ge03a, 2.4.7].) A connected reductive algebraic group $\mathbf{G}$ will be called *semisimple* if $\mathbf{Z}^\circ = \{1\}$ (or, equivalently, if the centre of $\mathbf{G}$ is finite). Thus, in the above setting, $\mathbf{G}_{\text{der}}$ is semisimple.

The above product decomposition can be used to prove general statements about connected reductive algebraic groups by a reduction to simple algebraic groups; see, for example, Lemma 1.6.9 and Theorem 1.7.15.

1.1.14 Algebraic BN-pairs (or Tits systems). The concept of BN-pairs has been introduced by Tits [Ti62], and it has turned out to be extremely useful. It applies to connected algebraic groups and to finite groups, and it allows us to give uniform proofs of many results, instead of going through a large number of case-by-case proofs. Recall that two subgroups B, N of an arbitrary (abstract) group G form a *BN-pair* (or a *Tits system*) if the following conditions are satisfied.

(BN1) G is generated by B and N.

(BN2) $H := B \cap N$ is normal in N and the quotient $W := N/H$ is a finite group generated by a set S of elements of order 2.

(BN3) $n_s B n_s \neq B$ if $s \in S$ and n_s is a representative of s in N.

(BN4) $n_s B n \subseteq B n_s n B \cup B n B$ for any $s \in S$ and $n \in N$.

The group W is called the corresponding *Weyl group*. We have a *length function* on W, as follows. We set $l(1) = 0$. If $w \neq 1$, we define $l(w)$ to be the length of a shortest possible expression of w as a product of generators in S. (Note that we don't have to take into account inverses, since $s^2 = 1$ for all $s \in S$.) Thus, any $w \in W$ can be written in the form $w = s_1 \ldots s_p$ where $p = l(w)$ and $s_i \in S$ for all i. Such an expression (which is by no means unique) will be called a *reduced expression* for w.

For any $w \in W$, we set $C(w) := Bn_wB$, where $n_w \in N$ is a representative of w in N. Since any two representatives of w lie in the same coset of $H \subseteq B$, we see that $C(w)$ does not depend on the choice of the representative. The double cosets $C(w)$ are called *Bruhat cells* of G. Then the above axioms imply the fundamental *Bruhat decomposition* (see [Bou68, Chap. IV, n° 2.3]):

$$G = \bigsqcup_{w \in W} Bn_wB.$$

As Lusztig [Lu10] notes, by allowing one to reduce many questions about G to questions about the Weyl group W, the Bruhat decomposition is indispensable for the understanding of both the structure and representations of G. A key role in this context will be played by the Iwahori–Hecke algebra (introduced in [Iw64]); this is a deformation of the group algebra of W whose definition is based on the Bruhat decomposition. (We will come back to this in Section 3.2.)

Now let $\mathbf{G}$ be a linear algebraic group over k and let $\mathbf{B}, \mathbf{N}$ be closed subgroups of $\mathbf{G}$ which form a BN-pair. Following [Ca85, §2.5], we shall say that this is an *algebraic BN-pair* if $\mathbf{H} = \mathbf{B} \cap \mathbf{N}$ is abelian and consists entirely of semisimple elements, and we have an abstract semidirect product decomposition $\mathbf{B} = \mathbf{U}.\mathbf{H}$ where $\mathbf{U}$ is a closed normal unipotent subgroup of $\mathbf{B}$ such that $\mathbf{U} \cap \mathbf{H} = \{1\}$. (If $\mathbf{B}$ is connected, then this is automatically a semidirect product of algebraic groups as in 1.1.8; see [Spr98, 6.3.5].) We do not assume that $\mathbf{G}$ is connected, so the definition can apply in particular to finite algebraic groups. We now have:

Proposition 1.1.15 *Let $\mathbf{B}, \mathbf{N}$ be closed subgroups of $\mathbf{G}$ that form an algebraic BN-pair in $\mathbf{G}$, where $\mathbf{B} = \mathbf{U}.\mathbf{H}$ as above. Assume that $\mathbf{H}$, $\mathbf{U}$ are connected and that $C_{\mathbf{G}}(\mathbf{H}) = \mathbf{H}$. Then the following hold.*

(a) *$\mathbf{G}$ is connected and reductive.*
(b) *$\mathbf{B}$ is a Borel subgroup (that is, a maximal closed connected solvable subgroup of $\mathbf{G}$); we have $\mathbf{B} = N_{\mathbf{G}}(\mathbf{U})$ and $[\mathbf{B}, \mathbf{B}] = \mathbf{U} = R_u(\mathbf{B})$.*
(c) *$\mathbf{H}$ is a maximal torus of $\mathbf{G}$ and we have $\mathbf{N} = N_{\mathbf{G}}(\mathbf{H})$.*

(See [Ca85, §2.5] and [Ge03a, 3.4.6, 3.4.7].) As in [Ge03a, 3.4.5], a BN-pair satisfying the conditions in Proposition 1.1.15 will be called a *reductive BN-pair*. Much more difficult is the converse of the above result, which comes about as the culmination of a long series of arguments. Namely, if $\mathbf{G}$ is a connected reductive algebraic group, then $\mathbf{G}$ has a reductive BN-pair in which $\mathbf{B}$ is a Borel subgroup and $\mathbf{N}$ is the normaliser of a maximal torus contained in $\mathbf{B}$. (We will discuss this in more detail in Section 1.3.) For our purposes here, the realisation of connected reductive algebraic groups in terms of algebraic BN-pairs as above is sufficient for many purposes. For example, if $\mathbf{G}$ is a 'classical group' as in 1.1.4, then algebraic BN-pairs as above are explicitly described in [Ge03a, §1.7]. In these cases, one can always find an algebraic BN-pair in which $\mathbf{B}$ consists of upper triangular matrices and $\mathbf{H}$ consists of diagonal matrices. See also the relevant chapters in [GLS94], [GLS96], [GLS98].

1.2 Root Data

We now introduce abstract root data and prove some basic properties of them. As we shall see in later sections, these form the combinatorial skeleton of connected reductive algebraic groups, that is, they capture those features which do not depend on the underlying field k. (A reader who wishes to see a much more systematic discussion of root data is referred to [DG70/11, Exposé XXI].)

1.2.1 Let X, Y be free abelian groups of the same finite rank; assume that there is a bilinear pairing $\langle\ ,\ \rangle\colon X \times Y \to \mathbb{Z}$ which is perfect, that is, it induces group isomorphisms $Y \cong \mathrm{Hom}(X, \mathbb{Z})$ and $X \cong \mathrm{Hom}(Y, \mathbb{Z})$ (as in 1.1.11). Furthermore, let $R \subseteq X$ and $R^\vee \subseteq Y$ be finite subsets. Then the quadruple $\mathscr{R} = (X, R, Y, R^\vee)$ is called a *root datum* if the following conditions are satisfied.

(R1) There is a bijection $R \to R^\vee$, $\alpha \mapsto \alpha^\vee$, such that $\langle \alpha, \alpha^\vee \rangle = 2$ for all $\alpha \in R$.
(R2) For every $\alpha \in R$, we have $2\alpha \notin R$.
(R3) For $\alpha \in R$, we define endomorphisms $w_\alpha\colon X \to X$ and $w_\alpha^\vee\colon Y \to Y$ by

$$w_\alpha(\lambda) = \lambda - \langle \lambda, \alpha^\vee \rangle \alpha \qquad \text{and} \qquad w_\alpha^\vee(\nu) = \nu - \langle \alpha, \nu \rangle \alpha^\vee$$

for all $\lambda \in X$ and $\nu \in Y$. Then we require that $w_\alpha(R) = R$ and $w_\alpha^\vee(R^\vee) = R^\vee$ for all $\alpha \in R$.

We shall see in 1.2.5 that the concept of root data is, in a very precise sense, an enhancement of the more traditional concept of root systems (related to finite reflection groups; see [Bou68]). First, we need some preparations.

The defining formula immediately shows that $w_\alpha^2 = \mathrm{id}_X$ and $(w_\alpha^\vee)^2 = \mathrm{id}_Y$. Hence, we have $w_\alpha \in \mathrm{Aut}(X)$ and $w_\alpha^\vee \in \mathrm{Aut}(Y)$ for all $\alpha \in R$. We set

$$\mathbf{W} := \langle w_\alpha \mid \alpha \in R \rangle \subseteq \mathrm{Aut}(X) \quad \text{and} \quad \mathbf{W}^\vee := \langle w_\alpha^\vee \mid \alpha \in R \rangle \subseteq \mathrm{Aut}(Y);$$

these groups are called the *Weyl groups* of R and $R^\vee$, respectively.[1] By (R3), we have an action of $\mathbf{W}$ on R and an action of $\mathbf{W}^\vee$ on $R^\vee$.

[1] For the time being, we keep a separate notation for these two Weyl groups; in Remark 1.2.12, we will identify them using the isomorphism in Lemma 1.2.3(a).

1.2.2 Let $\mathscr{R} = (X, R, Y, R^\vee)$ and $\mathscr{R}' = (X', R', Y', R'^\vee)$ be root data. Let $\varphi\colon X' \to X$ be a group homomorphism. The corresponding *transpose map* $\varphi^{\mathrm{tr}}\colon Y \to Y'$ is uniquely defined by the condition that

$$\langle \varphi(\lambda'), \nu\rangle = \langle \lambda', \varphi^{\mathrm{tr}}(\nu)\rangle' \qquad \text{for all } \lambda' \in X' \text{ and } \nu \in Y,$$

where $\langle\ ,\ \rangle$ is the bilinear pairing for $\mathscr{R}$ and $\langle\ ,\ \rangle'$ is the bilinear pairing for $\mathscr{R}'$. We say that φ is a *homomorphism of root data* if φ maps R' bijectively onto R and φ^{tr} maps $R^\vee$ bijectively onto $R'^\vee$. It then automatically follows that $\varphi^{\mathrm{tr}}(\varphi(\beta)^\vee) = \beta^\vee$ for all $\beta \in R'$; see [DG70/11, XXI, 6.1.2]. If φ is a bijective homomorphism of root data, we say that $\mathscr{R}$ and $\mathscr{R}'$ are *isomorphic*.

Lemma 1.2.3 *Let $\mathscr{R} = (X, R, Y, R^\vee)$ be a root datum.*

(a) *There is a unique group isomorphism $\delta\colon \mathbf{W} \xrightarrow{\sim} \mathbf{W}^\vee$ such that $\delta(w_\alpha) = w_\alpha^\vee$ for all $\alpha \in R$; we have*

$$\langle w^{-1}(\lambda), \nu\rangle = \langle \lambda, \delta(w)(\nu)\rangle \quad \textit{for all } w \in \mathbf{W},\ \lambda \in X,\ \nu \in Y.$$

(b) *The quadruple $(Y, R^\vee, X, R)$ is also a root datum, with pairing $\langle\ ,\ \rangle^*\colon Y \times X \to \mathbb{Z}$ defined by $\langle \nu, \lambda\rangle^* := \langle \lambda, \nu\rangle$ for all $\nu \in Y$ and $\lambda \in X$.*

(c) *For any $\lambda \in X$ and $w \in \mathbf{W}$, we have $\lambda - w(\lambda) \in \mathbb{Z}R$.*

The root datum in (b) is called the *dual root datum* of $\mathscr{R}$.

Proof (a) For any group homomorphism $\varphi\colon X \to X$, we consider its transpose $\varphi^{\mathrm{tr}}\colon Y \to Y$, as defined above. Clearly, we have $\mathrm{id}_X^{\mathrm{tr}} = \mathrm{id}_Y$ and $(\varphi \circ \psi)^{\mathrm{tr}} = \psi^{\mathrm{tr}} \circ \varphi^{\mathrm{tr}}$ if $\psi\colon X \to X$ is another group homomorphism. Thus, $\mathbf{W}^{\mathrm{tr}} := \{w^{\mathrm{tr}} \mid w \in \mathbf{W}\}$ is a subgroup of $\mathrm{Aut}(Y)$ and the map $\delta\colon \mathbf{W} \to \mathbf{W}^{\mathrm{tr}}$, $w \mapsto (w^{-1})^{\mathrm{tr}}$, is an isomorphism. Now, using the defining formulae in (R3), one immediately checks that

$$\langle w_\alpha(\lambda), \nu\rangle = \langle \lambda, w_\alpha^\vee(\nu)\rangle \quad \text{for all } \alpha \in R,\ \lambda \in X,\ \nu \in Y.$$

Hence, we have $w_\alpha^{\mathrm{tr}} = w_\alpha^\vee$ for all $\alpha \in R$ and so $\mathbf{W}^{\mathrm{tr}} = \mathbf{W}^\vee$. This yields (a).

(b) This is a straightforward verification.

(c) The defining formula shows that this is true if $w = w_\alpha$ for $\alpha \in R$. But then it also follows in general, since $\mathbf{W}$ is generated by the w_α $(\alpha \in R)$. □

Lemma 1.2.4 *Let $\mathscr{R} = (X, R, Y, R^\vee)$ be a root datum. We set $X_0 := \{\lambda \in X \mid \langle \lambda, \alpha^\vee\rangle = 0$ for all $\alpha \in R\}$. Then*

$$X_0 \cap \mathbb{Z}R = \{0\} \qquad \textit{and} \qquad |X/(X_0 + \mathbb{Z}R)| < \infty.$$

Consequently, $\mathbf{W}$ is a finite group and the action of $\mathbf{W}$ on R is faithful (that is, if $w \in \mathbf{W}$ is such that $w(\alpha) = \alpha$ for all $\alpha \in R$, then $w = 1$).

Proof Let us extend scalars from $\mathbb{Z}$ to $\mathbb{Q}$. We denote $X_\mathbb{Q} = \mathbb{Q}\otimes_\mathbb{Z} X$ and $Y_\mathbb{Q} = \mathbb{Q}\otimes_\mathbb{Z} Y$. Then $\langle\ ,\ \rangle$ extends to a non-degenerate $\mathbb{Q}$-bilinear form on $X_\mathbb{Q}\times Y_\mathbb{Q}$ which we denote by the same symbol. Since X, Y are free $\mathbb{Z}$-modules, we can naturally regard X as a subset of $X_\mathbb{Q}$ and Y as a subset of $Y_\mathbb{Q}$. Similarly, we can regard $\mathbf{W}$ as a subgroup of $\mathrm{GL}(X_\mathbb{Q})$ and $\mathbf{W}^\vee$ as a subgroup of $\mathrm{GL}(Y_\mathbb{Q})$. So, in order to show the statements about X_0 and $\mathbb{Z}R$, it is sufficient to show that

$$X_\mathbb{Q} = X_{0,\mathbb{Q}} \oplus \mathbb{Q}R \quad \text{where} \quad X_{0,\mathbb{Q}} := \{x \in X_\mathbb{Q} \mid \langle x, \alpha^\vee\rangle = 0 \text{ for all } \alpha \in R\}.$$

For this purpose, following [DG70/11, XXI, §1.2], we consider the linear map

$$f\colon X_\mathbb{Q} \to Y_\mathbb{Q}, \qquad x \mapsto \sum_{\alpha\in R} \langle x, \alpha^\vee\rangle\, \alpha^\vee.$$

Let $\beta \in R$. Using (R3), Lemma 1.2.3(a) and the fact that $(w_\beta^\vee)^2 = \mathrm{id}_Y$, we obtain

$$(f\circ w_\beta)(x) = \sum_{\alpha\in R} \langle w_\beta(x), \alpha^\vee\rangle\, \alpha^\vee = \sum_{\alpha\in R} \langle x, w_\beta^\vee(\alpha^\vee)\rangle\, \alpha^\vee = (w_\beta^\vee \circ f)(x)$$

for all $x \in X_\mathbb{Q}$. This identity in turn implies that, for any $\beta \in R$, we have:

$$\begin{aligned} f(\beta) &= -f(w_\beta(\beta)) = -w_\beta^\vee(f(\beta)) = -\sum_{\alpha\in R} \langle \beta, \alpha^\vee\rangle\, w_\beta^\vee(\alpha^\vee) \\ &= -\sum_{\alpha\in R} \langle \beta, \alpha^\vee\rangle \left(\alpha^\vee - \langle \beta, \alpha^\vee\rangle \beta^\vee\right) = -f(\beta) + \Big(\sum_{\alpha\in R} \langle \beta, \alpha^\vee\rangle^2\Big)\beta^\vee. \end{aligned}$$

Noting that $\langle \beta, f(\beta)\rangle = \sum_{\alpha\in R} \langle \beta, \alpha^\vee\rangle^2$, we deduce that

$$2f(\beta) = \langle \beta, f(\beta)\rangle\, \beta^\vee \quad \text{and} \quad \langle \beta, f(\beta)\rangle > 0 \qquad \text{for all } \beta \in R.$$

This shows that $f(\mathbb{Q}R) = \mathbb{Q}R^\vee$, and so $\dim \mathbb{Q}R \geqslant \dim \mathbb{Q}R^\vee$. By the symmetry expressed in Lemma 1.2.3(b), the reverse inequality also holds and so $\dim \mathbb{Q}R = \dim \mathbb{Q}R^\vee$. Thus, f restricts to an isomorphism $f\colon \mathbb{Q}R \xrightarrow{\sim} \mathbb{Q}R^\vee$. Now, we clearly have $X_{0,\mathbb{Q}} \subseteq \ker(f)$, whence $X_{0,\mathbb{Q}} \cap \mathbb{Q}R = \{0\}$. Since $\langle\ ,\ \rangle$ extends to a non-degenerate bilinear form on $X_\mathbb{Q}\times Y_\mathbb{Q}$, we have $\dim X_\mathbb{Q} = \dim X_{0,\mathbb{Q}} + \dim \mathbb{Q}R^\vee$. Since also $\dim \mathbb{Q}R^\vee = \dim \mathbb{Q}R$, we conclude that $X_\mathbb{Q} = X_{0,\mathbb{Q}} \oplus \mathbb{Q}R$, as desired.

Now we show that the action of $\mathbf{W}$ on R is faithful. Let $w \in \mathbf{W}$ be such that $w(\alpha) = \alpha$ for $\alpha \in R$. Then w acts as the identity on the subspace $\mathbb{Q}R \subseteq X_\mathbb{Q}$. The defining equation shows that all w_α, $\alpha \in R$, act as the identity on $X_{0,\mathbb{Q}}$, so $\mathbf{W}$ is trivial on $X_{0,\mathbb{Q}}$. Hence, $w = 1$ since $X_\mathbb{Q} = X_{0,\mathbb{Q}} + \mathbb{Q}R$. Since R is finite, it follows that $\mathbf{W}$ must be finite, too. □

1.2.5 Let $\mathscr{R} = (X, R, Y, R^\vee)$ be a root datum. As in the above proof, we extend scalars from $\mathbb{Z}$ to $\mathbb{Q}$ and set $X_\mathbb{Q} = \mathbb{Q}\otimes_\mathbb{Z} X$. Following [Bou68, Chap. VI, §1, Prop. 3], we define a symmetric bilinear form $(\ ,\ )\colon X_{\mathbb{Q}} \times X_{\mathbb{Q}} \to \mathbb{Q}$ by setting

$$(x, y) := \sum_{\alpha \in R} \langle x, \alpha^\vee \rangle \langle y, \alpha^\vee \rangle \qquad \text{for all } x, y \in X_{\mathbb{Q}}.$$

Using (R3) and Lemma 1.2.3(a), we see that $(\ ,\ )$ is $\mathbf{W}$-invariant, that is, we have $(w(x), w(y)) = (x, y)$ for all $w \in \mathbf{W}$ and all $x, y \in X_{\mathbb{Q}}$. Clearly, we have $(x, x) \geqslant 0$ for all $x \in X_{\mathbb{Q}}$; furthermore, $(\beta, \beta) > 0$ for all $\beta \in R$ (since $\langle \beta, \beta^\vee \rangle = 2 > 0$). By a standard argument (see [Bou68, Chap. VI, §1, Lemme 2]), this yields that

$$2\frac{(\alpha, \beta)}{(\beta, \beta)} = \langle \alpha, \beta^\vee \rangle \in \mathbb{Z} \qquad \text{for all } \alpha, \beta \in R.$$

We claim that the restriction of $(\ ,\ )$ to $\mathbb{Q}R \times \mathbb{Q}R$ is positive-definite. Indeed, assume that $(x, x) = 0$ where $x \in \mathbb{Q}R$. Then $0 = (x, x) = \sum_{\alpha \in R} \langle x, \alpha^\vee \rangle^2$ and so $\langle x, \alpha^\vee \rangle = 0$ for all $\alpha \in R$. Hence, Lemma 1.2.4 shows that $x = 0$.

Thus, we see that R is a crystallographic root system in the subspace $\mathbb{Q}R$ of $X_{\mathbb{Q}}$; see [Bou68, Chap. VI, §1, Déf. 1]. The Weyl group of R is $\mathbf{W}$; see Lemma 1.2.4. Furthermore, R is *reduced*, in the sense that

$$R \cap \mathbb{Q}\alpha = \{\pm\alpha\} \qquad \text{for all } \alpha \in R.$$

(This is an easy consequence of (R2); see e.g. [Bor91, 14.7].) Similarly, $R^\vee$ is a reduced crystallographic root system in $\mathbb{Q}R^\vee$, by the symmetry in Lemma 1.2.3(b).

1.2.6 Keeping the above notation, we now recall some standard results on root systems (see, e.g., the appendices of [St67] or [MaTe11]).

(a) There is a subset $\Pi \subseteq R$ such that Π is linearly independent in $\mathbb{Q}R$ and every $\alpha \in R$ can be written as $\alpha = \sum_{\beta \in \Pi} x_\beta \beta$, where either $x_\beta \in \mathbb{Q}_{\geqslant 0}$ for all $\beta \in \Pi$ or $x_\beta \in \mathbb{Q}_{\leqslant 0}$ for all $\beta \in \Pi$.

We call Π a *base* for R. The corresponding set of *positive roots* $R^+ \subseteq R$ consists of those $\alpha \in R$ which can be written as $\alpha = \sum_{\beta \in \Pi} x_\beta \beta$, where $x_\beta \in \mathbb{Q}_{\geqslant 0}$ for all $\beta \in \Pi$. The roots in $R^- := -R^+$ are called *negative roots*. Furthermore, if (a) holds, then we also have:

(b) For every $\alpha \in R$, there exists some $w \in \mathbf{W}$ such that $w(\alpha) \in \Pi$.
(c) Every $\alpha \in R$ is a $\mathbb{Z}$-linear combination of the roots in the base Π. (That is, the coefficients x_β in (a) are always integers.)
(d) $\mathbf{W}$ is a Coxeter group, with generators $\{w_\beta \mid \beta \in \Pi\}$ and defining relations $(w_\beta w_\gamma)^{m_{\beta\gamma}} = 1$ for all $\beta, \gamma \in \Pi$, where $m_{\beta\gamma} \geqslant 1$ is the order of $w_\beta w_\gamma \in \mathbf{W}$; furthermore, we have $4\cos^2(\pi/m_{\beta\gamma}) = \langle \gamma, \beta^\vee \rangle \langle \beta, \gamma^\vee \rangle$ for all $\beta, \gamma \in \Pi$.

Finally, any two bases of R can be transformed into each other by a unique element of $\mathbf{W}$. In particular, $r := |\Pi|$ is well defined and called the *rank* of R; furthermore, writing $\Pi = \{\beta_1, \ldots, \beta_r\}$, the matrix

$$C := \big(\langle \beta_j, \beta_i^\vee \rangle\big)_{1 \leqslant i,j \leqslant r}$$

is uniquely determined by $\mathscr{R}$ (up to reordering the rows and columns); it is called the *Cartan matrix* of $\mathscr{R}$. We say that two root data $\mathscr{R}$, $\mathscr{R}'$ have the same *Cartan type* if the corresponding Cartan matrices are the same (up to choosing a bijection between the associated bases Π, Π'). Thus, $\mathscr{R}$ and $\mathscr{R}'$ have the same Cartan type if and only if $R \subseteq X_{\mathbb{Q}}$ and $R' \subseteq X'_{\mathbb{Q}}$ are isomorphic root systems (see [Bou68, Chap. VI, n° 1.5]).

We associate with C a *Dynkin diagram*, defined as follows. It has vertices labelled by the elements in $\Pi = \{\beta_1, \ldots, \beta_r\}$. If $i \neq j$ and $|\langle \beta_j, \beta_i^\vee \rangle| \geqslant |\langle \beta_i, \beta_j^\vee \rangle|$, then the corresponding vertices are joined by $|\langle \beta_j, \beta_i^\vee \rangle|$ lines; furthermore, these lines are equipped with an arrow pointing towards the vertex labelled by β_i if $|\langle \beta_j, \beta_i^\vee \rangle| > 1$. (Note that, in this case, we automatically have $|\langle \beta_i, \beta_j^\vee \rangle| = 1$ by (e).)

We say that C is an *indecomposable Cartan matrix* if the associated Dynkin diagram is a connected graph; otherwise, we say that C is *decomposable*. Clearly, any Cartan matrix can be expressed as a block diagonal matrix with diagonal blocks given by indecomposable Cartan matrices. The classification of indecomposable Cartan matrices is well known (see [Bou68, Chap. VI, §4]); the corresponding Dynkin diagrams are listed in Table 1.1. (The Cartan matrices of type A_n, B_n, C_n, G_2, F_4 are printed explicitly in Examples 1.2.19, 1.3.7, 1.5.5 below.) See [Kac85, Chap. 4] for a somewhat different approach to this classification.

We have the following general characterisation of Cartan matrices.

Proposition 1.2.7 (Cf. [Bou68, Chap. VI, §4]) *Let S be a finite set and $C = (c_{st})_{s,t \in S}$ be a matrix with integer entries. Then C is the Cartan matrix of a reduced crystallographic root system if and only if the following conditions hold:*

(C1) *We have $c_{ss} = 2$ and, for $s \neq t$ we have $c_{st} \leqslant 0$; furthermore, $c_{st} \neq 0$ if and only if $c_{ts} \neq 0$.*
(C2) *For all $s, t \in S$, we have $c_{st}c_{ts} = 4\cos^2(\pi/m_{st})$, where $m_{st} \in \mathbb{Z}_{\geqslant 1}$. (Thus, $m_{ss} = 1$ and $m_{st} = m_{ts} \in \{2, 3, 4, 6\}$ for $s \neq t$.) The symmetric matrix $(-\cos(\pi/m_{st}))_{s,t \in S}$ is positive-definite.*

Remark 1.2.8 Let $C = (c_{st})_{s,t \in S}$ be a Cartan matrix. Let Ω be the free abelian group with basis $\{\omega_s \mid s \in S\}$. Let $\mathbb{Z}C \subseteq \Omega$ be the subgroup generated by the columns of C, that is, by all vectors of the form $\sum_{s \in S} c_{st}\omega_s$ for $t \in S$. Then

$$\Lambda(C) := \Omega/\mathbb{Z}C$$

is called the *fundamental group* of C. (This agrees with the definitions in [MaTe11,

Table 1.1 *Dynkin diagrams of indecomposable Cartan matrices*

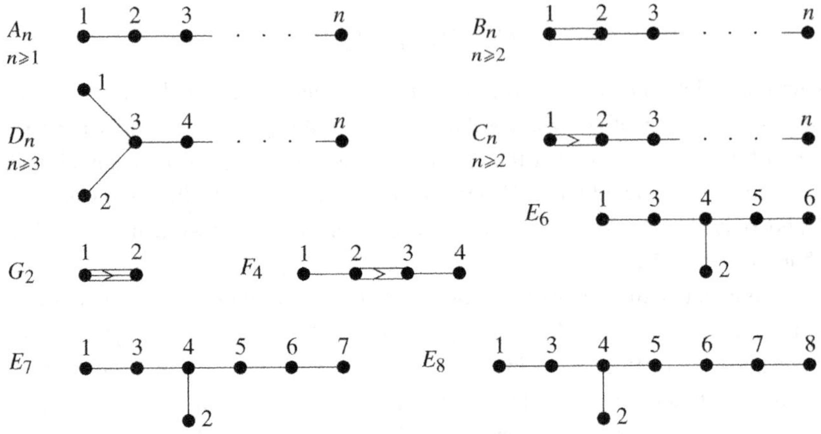

(This labelling will be used throughout this book; it is the same as in CHEVIE [GHLMP], [MiChv]. Note that $B_2 = C_2$ and $D_3 = A_3$, up to relabelling the vertices.)

9.14] or [Spr98, 8.1.11], for example.) If C is indecomposable, then the groups $\Lambda(C)$ are easily computed and listed in Table 1.2.

Table 1.2 *Fundamental groups of indecomposable Cartan matrices*

Type of C	$\Lambda(C)$	
A_{n-1}	$\mathbb{Z}/n\mathbb{Z}$	
B_n, C_n	$\mathbb{Z}/2\mathbb{Z}$	
D_n	$\mathbb{Z}/2\mathbb{Z} \oplus \mathbb{Z}/2\mathbb{Z}$	(n even)
	$\mathbb{Z}/4\mathbb{Z}$	(n odd)
G_2, F_4, E_8	$\{0\}$	
E_6	$\mathbb{Z}/3\mathbb{Z}$	
E_7	$\mathbb{Z}/2\mathbb{Z}$	

In 1.2.2, we defined what it means for two root data to be isomorphic. We shall also need the following, somewhat more general notion.

Definition 1.2.9 Let $\mathscr{R} = (X, R, Y, R^\vee)$ and $\mathscr{R}' = (X', R', Y', R'^\vee)$ be root data. We fix an integer p such that either $p = 1$ or p is a prime number. Then a group homomorphism $\varphi\colon X' \to X$ is called a *p-isogeny of root data* if there exist a bijection $R \to R'$, $\alpha \mapsto \alpha^\dagger$, and positive integers $q_\alpha > 0$, each an integral power of p, such that φ and its transpose $\varphi^{\text{tr}}\colon Y \to Y'$ satisfy the following conditions.

(I1) φ and φ^{tr} are injective.

(I2) We have $\varphi(\alpha^\dagger) = q_\alpha\,\alpha$ and $\varphi^{\mathrm{tr}}(\alpha^\vee) = q_\alpha\,(\alpha^\dagger)^\vee$ for all $\alpha \in R$.

The conditions (I1) and (I2) appear in [Ch05, §18.2]; following Chevalley, we call the numbers $\{q_\alpha\}$ the *root exponents* of φ. Note that $\alpha \mapsto \alpha^\dagger$ and the numbers $\{q_\alpha\}$ are uniquely determined by φ (since R is reduced).

Let $\mathbf{W} \subseteq \mathrm{Aut}(X)$ be the Weyl group of R. Then one easily sees that, for any $\alpha \in R$ and $w \in \mathbf{W}$, we have $q_{w(\alpha)} = q_\alpha$; see [Spr98, 9.6.4]. Hence, by 1.2.6(a), the map $\alpha \mapsto q_\alpha$ is determined by its values on a base of R. We also see that φ is an isomorphism of root data if and only if φ is a bijective isogeny where $q_\alpha = 1$ for all $\alpha \in R$. Finally note that if $p = 1$, then $q_\alpha = 1$ for all $\alpha \in R$.

A simple example of a p-isogeny of a root datum into itself is given by $\varphi\colon X \to X$, $\lambda \mapsto p\lambda$ (scalar multiplication with p); this will be continued in Example 1.3.17.

Remark 1.2.10 Keep the notation in the above definition. Let $\mathbf{W} \subseteq \mathrm{Aut}(X)$ be the Weyl group of $\mathscr{R}$ and $\mathbf{W}' \subseteq \mathrm{Aut}(X')$ be the Weyl group of $\mathscr{R}'$. Then one easily sees that a p-isogeny $\varphi\colon X' \to X$ induces a unique group isomorphism

$$\sigma\colon \mathbf{W} \to \mathbf{W}' \quad \text{such that} \quad \varphi \circ \sigma(w) = w \circ \varphi \quad (w \in \mathbf{W}).$$

We have $\sigma(w_\alpha) = w_{\alpha^\dagger}$ for all $\alpha \in R$ where $w_\alpha \in \mathbf{W}$ is the reflection corresponding to $\alpha \in R$ and $w_{\alpha^\dagger} \in \mathbf{W}'$ is the reflection corresponding to $\alpha^\dagger \in R'$. (See [Ch05, §18.3] for further details; see also 1.2.18 below.) In particular, if $\varphi\colon X \to X$ is a p-isogeny of $\mathscr{R}$ into itself, then $\varphi \circ \mathbf{W} = \mathbf{W} \circ \varphi$, and so φ normalises $\mathbf{W} \subseteq \mathrm{Aut}(X)$.

Remark 1.2.11 Let $\mathscr{R} = (X, R, Y, R^\vee)$ and $\mathscr{R}' = (X', R', Y', R'^\vee)$ be root data. Let us fix a base Π of R and a base Π' of R'; see 1.2.6. Let $\varphi\colon X' \to X$ be a group homomorphism that defines a p-isogeny of root data. Then (I2) shows that $\Pi^\dagger := \{\alpha^\dagger \mid \alpha \in \Pi\}$ also is a base of R', where $\alpha \mapsto \alpha^\dagger$ denotes the bijection $R \to R'$ associated with φ. As already mentioned in 1.2.6, there exists a unique w in the Weyl group $\mathbf{W}'$ of $\mathscr{R}'$ such that $\Pi^\dagger = w(\Pi')$. Now $w \in \mathbf{W}' \subseteq \mathrm{Aut}(X')$ certainly is an isomorphism of $\mathscr{R}'$ into itself. Hence, the composition $\varphi' := \varphi \circ w\colon X' \to X$ is also a p-isogeny of root data and the bijection $R \to R'$ associated with φ' will map Π onto Π'. This shows that, replacing φ by $\varphi \circ w$ for a suitable $w \in \mathbf{W}'$ if necessary, we can always assume that the bijection $R \to R'$ preserves the given bases $\Pi \subseteq R$ and $\Pi' \subseteq R'$.

Remark 1.2.12 Let $\mathscr{R} = (X, R, Y, R^\vee)$ be a root datum and $\mathbf{W} \subseteq \mathrm{Aut}(X)$ be the corresponding Weyl group. Let Π be a base of R. By 1.2.6, $\mathbf{W}$ is a Coxeter group with generating set $S = \{w_\alpha \mid \alpha \in \Pi\}$. We denote by $l\colon \mathbf{W} \to \mathbb{Z}_{\geqslant 0}$ the corresponding length function. To unify the notation, we shall use S as an indexing set for Π, that is, $\Pi = \{\alpha_s \mid s \in S\}$ where α_s is the root of the reflection s.

Using the isomorphism δ in Lemma 1.2.3(a), we can identify $\mathbf{W}^\vee = \mathbf{W} = \langle S \rangle$.

Under this identification, $\mathbf{W}$ will act on both X and Y, and we have

$$\langle w^{-1}.\lambda, \nu\rangle = \langle \lambda, w.\nu\rangle \qquad \text{for all } w \in \mathbf{W},\ \lambda \in X,\ \nu \in Y.$$

In particular, for any $\alpha \in R$, we have a corresponding reflection w_α which acts both on X and on Y: on the one hand, we have $w_\alpha.\lambda = \lambda - \langle \lambda, \alpha^\vee\rangle\alpha$ for all $\lambda \in X$; on the other hand, we have $w_\alpha.\nu = \nu - \langle \alpha, \nu\rangle\alpha^\vee$ for all $\nu \in Y$.

We now turn to the question of actually constructing root data and p-isogenies for a Cartan matrix C. The key idea is contained in the following remark; see [CMT04, §2] and [BrLu12, §2.1].

Remark 1.2.13 Let $\mathscr{R} = (X, R, Y, R^\vee)$ be a root datum and let Π be a base of R; write $\Pi = \{\alpha_s \mid s \in S\}$ as above. Let C be the corresponding Cartan matrix; see 1.2.6. Let $\{\lambda_i \mid i \in I\}$ be a $\mathbb{Z}$-basis of X, where I is some finite indexing set. (We have $|I| \geqslant |S|$.) Let $\{\nu_i \mid i \in I\}$ be the corresponding dual basis of Y (with respect to the pairing $\langle\ ,\ \rangle$). Then we have unique expressions

$$\beta_s = \sum_{i\in I} a_{s,i}\lambda_i \quad \text{and} \quad \beta_s^\vee = \sum_{i\in I} \breve{a}_{s,i}\nu_i \qquad \text{for all } s \in S,$$

where $a_{s,i} \in \mathbb{Z}$ and $\breve{a}_{s,i} \in \mathbb{Z}$ for $i \in I$. Consequently, we have a factorisation

$$C = \breve{A} \cdot A^{\mathrm{tr}} \quad \text{where} \quad A = (a_{s,i})_{s\in S, i\in I} \quad \text{and} \quad \breve{A} = (\breve{a}_{s,i})_{s\in S, i\in I}.$$

Note that all of $R \subseteq X$ and $R^\vee \subseteq Y$ are uniquely determined by C, A and $\breve{A}$. (This follows from 1.2.6(a) and the fact that $\mathbf{W} = \langle S\rangle$; see 1.2.6(d).) Also note that the dual root system $(Y, R^\vee, X, R)$ (see Lemma 1.2.3) has Cartan matrix C^{tr}, with factorisation $C^{\mathrm{tr}} = \breve{B} \cdot B^{\mathrm{tr}}$ where $B = \breve{A}$ and $\breve{B} = A$.

Reversing this argument, one sees that *every* factorisation as above of a Cartan matrix C leads to a root datum. Let us discuss this in some detail.

1.2.14 Let S be a finite set and $C = (c_{st})_{s,t\in S}$ be a Cartan matrix, that is, a matrix satisfying the conditions in Proposition 1.2.7. Let I be another finite index set, with $|I| \geqslant |S|$, and assume that we have a factorisation

$$C = \breve{A} \cdot A^{\mathrm{tr}} \quad \text{where} \quad A = (a_{s,i})_{s\in S, i\in I} \quad \text{and} \quad \breve{A} = (\breve{a}_{s,i})_{s\in S, i\in I};$$

here, A and $\breve{A}$ are matrices with integer coefficients of size $|S| \times |I|$. Let X be a free abelian group with $\mathbb{Z}$-basis indexed by I, say $\{\lambda_i \mid i \in I\}$; also let Y be a free abelian group with $\mathbb{Z}$-basis labelled by I, say $\{\nu_i \mid i \in I\}$. We define a bilinear pairing $\langle\ ,\ \rangle\colon X \times Y \to \mathbb{Z}$ such that $\{\lambda_i \mid i \in I\}$ and $\{\nu_i \mid i \in I\}$ are dual bases to each other. Then $C = \big(\langle \alpha_t, \alpha_s^\vee\rangle\big)_{s,t\in S}$, where we set

$$\alpha_s := \sum_{i\in I} a_{s,i}\lambda_i \qquad \text{and} \qquad \alpha_s^\vee := \sum_{i\in I} \breve{a}_{s,i}\nu_i \qquad \text{for all } s \in S.$$

For $s \in S$, we define endomorphisms $w_s : X \to X$ and $w_s^\vee : Y \to Y$ by

$$w_s(\lambda) = \lambda - \langle \lambda, \alpha_s^\vee \rangle \alpha_s \qquad \text{and} \qquad w_s^\vee(\nu) = \nu - \langle \alpha_s, \nu \rangle \alpha_s^\vee$$

for all $\lambda \in X$ and $\nu \in Y$. Then $w_s^2 = \mathrm{id}_X$ and $(w_s^\vee)^2 = \mathrm{id}_Y$. Hence, we have $w_s \in \mathrm{Aut}(X)$ and $w_s^\vee \in \mathrm{Aut}(Y)$ for all $s \in S$. Finally, let

$$W := \langle w_s \mid s \in S \rangle \subseteq \mathrm{Aut}(X) \quad \text{and} \quad W^\vee := \langle w_s^\vee \mid s \in S \rangle \subseteq \mathrm{Aut}(Y);$$
$$R := \{w(\alpha_s) \mid w \in W, s \in S\} \quad \text{and} \quad R^\vee := \{w^\vee(\alpha_s^\vee) \mid w^\vee \in W^\vee, s \in S\}.$$

Lemma 1.2.15 *The quadruple $\mathscr{R} := (X, R, Y, R^\vee)$ in 1.2.14 is a root datum with Cartan matrix C, where $\{\alpha_s \mid s \in S\}$ is a base of R and $\{\alpha_s^\vee \mid s \in S\}$ is a base of $R^\vee$. Furthermore, we have $W = \mathbf{W}$ and $W^\vee = \mathbf{W}^\vee$ (with the notation of 1.2.1). The bijection $R \to R^\vee$, $\alpha \mapsto \alpha^\vee$, is determined as follows. If $w \in \mathbf{W}$ and $s \in S$ are such that $\alpha = w(\alpha_s)$, then $\alpha^\vee = \delta(w)(\alpha_s^\vee)$ (with δ as in Lemma 1.2.3(a)).*

Proof Let $V \subseteq \mathbb{Q} \otimes_{\mathbb{Z}} X$ be the subspace spanned by $\{\alpha_s \mid s \in S\}$. Then $w_s(\alpha_t) = \alpha_t - c_{st}\alpha_s$ for all $t \in S$. So $w_s(V) \subseteq V$ for all $s \in S$. Let $W_R \subseteq \mathrm{GL}(V)$ be the group generated by the restrictions $w_s : V \to V$; then $R = \{w(\alpha_s) \mid w \in W_R, s \in S\} \subseteq V$. Thus, we are in the setting of [GePf00, §1.1].

The matrix C is *symmetrisable*, that is, there exist positive numbers $\{d_s \mid s \in S\}$ such that $(d_s c_{st})_{s,t \in S}$ is a symmetric matrix. (This easily follows from the fact that there are no closed paths in the Dynkin diagram of C; see also [Kac85, §4.6].) Then we can define a W_R-invariant symmetric bilinear form on V by $(\alpha_s, \alpha_t) = d_s c_{st}/2$ for $s, t \in S$. The W_R-invariance implies that

$$c_{st} = 2\frac{(\alpha_s, \alpha_t)}{(\alpha_s, \alpha_s)} \qquad \text{for all } s, t \in S;$$

see [GePf00, 1.3.2]. By (C2), this form is positive-definite; furthermore, each map $w_s : V \to V$ is an orthogonal reflection with root α_s. So W_R and R are finite; see [Bou68, Chap. V, §8] or [GePf00, 1.3.8]. In fact, R is a root system in V with Weyl group W_R, with $\{\alpha_s \mid s \in S\}$ as base and C as Cartan matrix; see [GePf00, 1.1.10]. Also note that R is reduced, that is, (R2) holds; see [GePf00, 1.3.7]. Let R^+ be the set of positive roots in R defined by $\{\alpha_s \mid s \in S\}$.

Similarly, let $V^\vee \subseteq \mathbb{Q} \otimes_{\mathbb{Z}} Y$ be the subspace spanned by $\{\alpha_s^\vee \mid s \in S\}$. Then $w_s^\vee(\alpha_t^\vee) = \alpha_t^\vee - c_{ts}\alpha_s^\vee$ for all $t \in S$. Let $W_{R^\vee} \subseteq \mathrm{GL}(V^\vee)$ be the group generated by the restrictions $w_s^\vee : V^\vee \to V^\vee$. Again, we are in the setting of [GePf00, §1.1] (with respect to C^{tr}, the transpose of C) and so we can repeat the previous argument. Consequently, $W_{R^\vee}$ is also finite; furthermore, $R^\vee$ is a reduced root system in $V^\vee$ with Weyl group $W_{R^\vee}$, with $\{\alpha_s^\vee \mid s \in S\}$ as base and C^{tr} as Cartan matrix. Let $(R^\vee)^+$ be the set of positive roots in $R^\vee$ defined by the base $\{\alpha_s^\vee \mid s \in S\}$.

Next, we define a linear map $f : V \to V^\vee$ by $f(\alpha_s) = \frac{1}{2}(\alpha_s, \alpha_s)\alpha_s^\vee$ for all $s \in S$.

(This is analogous to the definition in the proof of Lemma 1.2.4.) Clearly, f is bijective. One immediately checks that $w_s^\vee \circ f = f \circ w_s$ for all $s \in S$. So the map $w \mapsto f \circ w \circ f^{-1}$ defines a group isomorphism $\delta\colon W_R \xrightarrow{\sim} W_{R^\vee}$ such that $\delta(w_s) = w_s^\vee$ for all $s \in S$. Consequently, we can define a bijection $R \to R^\vee$, $\alpha \mapsto \alpha^\vee$, as follows. First, let $\alpha \in R^+$. By definition, $\alpha = w(\alpha_s)$ for some $w \in W$, $s \in S$. Then

$$f(\alpha) = f(w(\alpha_s)) = \delta(w)(f(\alpha_s)) = \tfrac{1}{2}(\alpha_s, \alpha_s)\delta(w)(\alpha_s^\vee).$$

Since $\alpha_s^\vee \in R^\vee$, we have $\delta(w)(\alpha_s^\vee) \in R^\vee$ by definition. Thus, $f(\alpha) \in V^\vee$ is a positive scalar multiple of some element of $R^\vee$; since $R^\vee$ is reduced, there is a unique positive root with this property, denoted $\alpha^\vee$, and the above computation shows that $\alpha^\vee = \delta(w)(\alpha_s^\vee)$. The definition of $\alpha^\vee$ for negative α is analogous; we then have $(-\alpha)^\vee = -\alpha^\vee$ for all $\alpha \in R$. Consequently, we obtain a map $R \to R^\vee$, $\alpha \mapsto \alpha^\vee$, which is easily seen to be bijective. Once this is established, the maps $w_\alpha\colon X \to X$ and $w_\alpha^\vee\colon Y \to Y$ are defined for any $\alpha \in R$. The W_R-invariance of $(\ ,\ )$ implies that

$$\langle \beta, \alpha^\vee \rangle = 2\frac{(\alpha, \beta)}{(\alpha, \alpha)} \qquad \text{for all } \alpha, \beta \in R.$$

In particular, $\langle \alpha, \alpha^\vee \rangle = 2$; thus, (R1) holds. Finally, we show (R3). For each $\alpha \in R$, we claim that $w_\alpha(R) = R$. First one notices that $w_\alpha(V) \subseteq V$. But the above formula for $\langle \beta, \alpha^\vee \rangle$ shows that the restriction $w_\alpha\colon V \to V$ is the orthogonal reflection with root α. Hence, since W_R is the Weyl group of R, we have $w_\alpha(R) = R$, as desired. The argument for $w_\alpha^\vee$ is analogous. □

In view of Remark 1.2.13, the above construction yields all root data up to isomorphism. Thus, given a Cartan matrix C, we can think of the various root data of Cartan type C simply as factorisations $C = \breve{A} \cdot A^{\mathrm{tr}}$, where $A, \breve{A}$ are integer matrices of the same size. (This observation, in this explicit form, appears in [CMT04] and [BrLu12]; it is also implicit in [Lu89, §1].)

Example 1.2.16 Let C be a Cartan matrix. We have just seen that any factorisation $C = \breve{A} \cdot A^{\mathrm{tr}}$ as in 1.2.14 gives rise to a root datum $\mathscr{R} = (X, R, Y, R^\vee)$. Obviously, there are two natural choices for such a factorisation, namely,

- either A is the identity matrix and, hence, $\breve{A} = C$;
- or $\breve{A}$ is the identity matrix and, hence, $A = C^{\mathrm{tr}}$.

In the first case, we denote the corresponding root datum by $\mathscr{R} = \mathscr{R}_{\mathrm{ad}}(C)$. We have $X = \mathbb{Z}R$ in this case; any root datum satisfying $X = \mathbb{Z}R$ will be called a *root datum of adjoint type*. In the second case, we denote the corresponding root datum by $\mathscr{R} = \mathscr{R}_{\mathrm{sc}}(C)$. We have $Y = \mathbb{Z}R^\vee$ in this case; any root datum satisfying $Y = \mathbb{Z}R^\vee$ will be called a *root datum of simply connected type*.

Thus, $\mathscr{R}_{\mathrm{ad}}(C)$ and $\mathscr{R}_{\mathrm{sc}}(C)$ may be regarded as the standard models of root data of adjoint and simply connected types, respectively. (See also Example 1.2.21.) The relevance of these notions will become clearer when we consider semisimple algebraic groups in Section 1.5.

Example 1.2.17 There is an obvious notion of a direct product of root data. Indeed, if $\mathscr{R}_i = (X_i, R_i, Y_i, R_i^\vee)$ for $i = 1, \ldots, n$ are root data, then we obtain a new root datum $\mathscr{R} = (X, R, Y, R^\vee)$ as follows. We set

$$X := X_1 \oplus \cdots \oplus X_n, \qquad R := \dot{R}_1 \cup \cdots \cup \dot{R}_n,$$
$$Y := Y_1 \oplus \cdots \oplus Y_n, \qquad R^\vee := \dot{R}_1^\vee \cup \cdots \cup \dot{R}_n^\vee,$$

where, for each i, we let $\dot{R}_i \subseteq X$ denote the image of R_i under the natural embedding $X_i \hookrightarrow X$; similarly, $\dot{R}_i^\vee \subseteq Y$ denotes the image of $R_i^\vee$ under the natural embedding $Y_i \hookrightarrow Y$. Furthermore, the perfect bilinear pairings for the various $\mathscr{R}_i$ define a unique perfect bilinear pairing for $\mathscr{R}$ in a natural way. Also note that, if Π_i is a base of R_i for $i = 1, \ldots, n$, then $\Pi := \dot{\Pi}_1 \cup \cdots \cup \dot{\Pi}_n$ is a base of R.

In terms of the matrix language of 1.2.14, the situation is described as follows. Each $\mathscr{R}_i$ is determined by a factorisation $C_i = \breve{A}_i \cdot A_i^{\text{tr}}$ where C_i is the Cartan matrix of R_i with respect to a base Π_i of R_i. Then $\mathscr{R}$ is determined by the factorisation $C = \breve{A} \cdot A^{\text{tr}}$, where C, A and $\breve{A}$ are block-diagonal matrices with diagonal blocks given by the C_i, A_i and $\breve{A}_i$, respectively. The matrix C is the Cartan matrix of R with respect to the base $\Pi = \dot{\Pi}_1 \cup \cdots \cup \dot{\Pi}_n$.

We now translate the conditions in Definition 1.2.9 into the matrix language of Remark 1.2.13. This will be an extremely efficient tool for constructing isogenies, as it reduces the conditions to be checked to the verification of simple matrix identities.

1.2.18 Let $\mathscr{R} = (X, R, Y, R^\vee)$ and $\mathscr{R}' = (X', R', Y', R'^\vee)$ be root data. Assume that X and X' have the same rank and that R and R' have bases indexed by the same set S. Denote these bases by $\Pi = \{\alpha_s \mid s \in S\}$ and $\Pi' = \{\beta_s \mid s \in S\}$, respectively. Let C and C' be the corresponding Cartan matrices. Let us also fix a $\mathbb{Z}$-basis $\{\lambda_i \mid i \in I\}$ of X and a $\mathbb{Z}$-basis $\{\lambda_j' \mid j \in J\}$ of X'. Then $\mathscr{R}$ and $\mathscr{R}'$ are determined by factorisations as in Remark 1.2.13:

$$C = \breve{A} \cdot A^{\text{tr}} \quad \text{where} \quad A = (a_{s,i})_{s \in S, i \in I} \quad \text{and} \quad \breve{A} = (\breve{a}_{s,i})_{s \in S, i \in I},$$
$$C' = \breve{B} \cdot B^{\text{tr}} \quad \text{where} \quad B = (b_{s,j})_{s \in S, j \in J} \quad \text{and} \quad \breve{B} = (\breve{b}_{s,j})_{s \in S, j \in J}.$$

(Here, $|I| = |J|$, since X, X' have the same rank.) Giving a linear map $\varphi \colon X' \to X$ is the same as giving a matrix $P = (p_{ij})_{i \in I, j \in J}$ with integer coefficients:

$$\varphi(\lambda_j') = \sum_{i \in I} p_{ij} \lambda_i \qquad \text{for all } j \in J.$$

Assume now that $\varphi \colon X' \to X$ is a linear map which is 'base preserving', in the sense that there is a permutation $S \to S$, $s \mapsto s^\dagger$, such that

$$\varphi(\beta_{s^\dagger}) = q_s\alpha_s \qquad \text{where } 0 \neq q_s \in \mathbb{Z} \text{ for all } s \in S.$$

We encode this in a monomial matrix $P^\circ = (p^\circ_{st})_{s,t\in S}$, where $p^\circ_{ss^\dagger} = q_s$ for $s \in S$. Let $p = 1$ or p be a prime number and assume that φ is a p-isogeny. Then the conditions in Definition 1.2.9 immediately imply that the following conditions hold.

(MI1) P° is a monomial matrix whose non-zero entries are all powers of p.
(MI2) P is invertible over $\mathbb{Q}$; furthermore, $P \cdot B^{\text{tr}} = A^{\text{tr}} \cdot P^\circ$, $P^\circ \cdot \breve{B} = \breve{A} \cdot P$.

Conversely, it is straightforward to check that *any* pair of integer matrices (P, P°) satisfying (MI1) and (MI2) defines a p-isogeny of root data. The argument is similar to the proof of Lemma 1.2.15; let us just briefly sketch it. Let $\varphi\colon X' \to X$ be the linear map with matrix P. Condition (I1) holds since P is invertible over $\mathbb{Q}$. Since P° is monomial, there is a permutation $S \to S$, $s \mapsto s^\dagger$, such that

$$q_s := p^\circ_{ss^\dagger} \neq 0 \qquad \text{for all } s \in S.$$

Then (MI2) means that $\varphi(\beta_{s^\dagger}) = q_s\alpha_s$ and $\varphi^{\text{tr}}(\alpha_s^\vee) = q_s\beta_{s^\dagger}^\vee$ for all $s \in S$. Thus, (I2) holds for simple roots and coroots. To see that (I2) holds for all roots and coroots, note that (MI2) implies that $C \cdot P^\circ = P^\circ \cdot C'$. Consequently, using 1.2.6(d) for $\mathbf{W}$ and for $\mathbf{W}'$, there is a unique group isomorphism

$$\sigma\colon \mathbf{W} \to \mathbf{W}' \qquad \text{such that} \qquad w_{\alpha_s} \mapsto w_{\beta_{s^\dagger}} \qquad (s \in S).$$

Using Lemma 1.2.4, one shows that this implies that

$$\varphi \circ \sigma(w) = w \circ \varphi \qquad \text{for all } w \in \mathbf{W}.$$

We can now define a bijection $R \to R'$, $\alpha \mapsto \alpha^\dagger$, with the required properties, as follows. Let $\alpha \in R$ and write $\alpha = w(\alpha_s)$ for some $w \in \mathbf{W}$ and $s \in S$. Then we set $\alpha^\dagger := \sigma(w)(\beta_{s^\dagger}) \in R'$. Now, we have

$$\varphi(\alpha^\dagger) = \varphi(\sigma(w)(\beta_{s^\dagger})) = w(\varphi(\beta_{s^\dagger})) = q_s w(\alpha_s) = q_s\alpha.$$

Since φ is injective, $\alpha^\dagger$ is uniquely determined by α (and does not depend on the choice of w and s); furthermore, the first of the two identities in (I2) holds, where $q_\alpha = q_s$. The argument for the second identity is similar, using the bijection $R \to R^\vee$ (see Lemma 1.2.15). Thus, (I2) is seen to hold for all roots and coroots.

Example 1.2.19 (Cf. [Ch05, §§21.5, 22.4, 23.7]) Let $C = (c_{ij})_{1\leqslant i,j\leqslant r}$ $(r = 2$ or $4)$ be a Cartan matrix of type C_2, G_2 or F_4; see Table 1.1. Explicitly:

$$C_2 : \begin{pmatrix} 2 & -1 \\ -2 & 2 \end{pmatrix}, \quad G_2 : \begin{pmatrix} 2 & -1 \\ -3 & 2 \end{pmatrix}, \quad F_4 : \begin{pmatrix} 2 & -1 & 0 & 0 \\ -1 & 2 & -1 & 0 \\ 0 & -2 & 2 & -1 \\ 0 & 0 & -1 & 2 \end{pmatrix}$$

(Note that $C_2 = B_2$ up to relabelling the two vertices of the Dynkin diagram.) We set $p = 2$ if C is of type C_2 or F_4, and $p = 3$ if C is of type G_2. Let us consider the corresponding root datum $\mathscr{R} = \mathscr{R}_{\text{ad}}(C) = (X, R, Y, R^\vee)$ as in Example 1.2.16; we have $C = \breve{A} \cdot A^{\text{tr}}$, where A is the identity matrix and $C = \breve{A}$. For any $m \geqslant 0$, we define two matrices P_m° and P_m as follows:

$$C_2: \qquad P_m = P_m^\circ := \begin{pmatrix} 0 & 2^m \\ 2^{m+1} & 0 \end{pmatrix},$$

$$G_2: \qquad P_m = P_m^\circ := \begin{pmatrix} 0 & 3^m \\ 3^{m+1} & 0 \end{pmatrix},$$

$$F_4: \qquad P_m = P_m^\circ := \begin{pmatrix} 0 & 0 & 0 & 2^m \\ 0 & 0 & 2^m & 0 \\ 0 & 2^{m+1} & 0 & 0 \\ 2^{m+1} & 0 & 0 & 0 \end{pmatrix}.$$

Now, in the setting of 1.2.18, let $C' = C$, $B = A$, $\breve{B} = \breve{A}$. Then P_m, P_m° satisfy (MI1), (MI2) and, hence, the pair (P_m, P_m°) defines a group homomorphism $\varphi_m \colon X \to X$ which is a p-isogeny of $\mathscr{R}$ into itself, such that $\varphi_m^2 = p^{2m+1} \operatorname{id}_X$.

See [Ca72, 12.3, 12.4] for a more detailed discussion of these 'exceptional' isogenies; they give rise to the finite Suzuki and Ree groups (see Example 1.4.22). Another instance of such an exceptional isogeny will be considered in Example 1.5.5.

Remark 1.2.20 Recall from Definition 1.2.9 that a p-isogeny $\varphi \colon X' \to X$ is an isomorphism of root data if and only if φ is bijective and $q_\alpha = 1$ for all $\alpha \in R$. Hence, in the setting of 1.2.18, φ is an isomorphism if and only if P° is a permutation matrix and P is invertible over $\mathbb{Z}$, where the relations $P \cdot B^{\text{tr}} = A^{\text{tr}} \cdot P^\circ$, $P^\circ \cdot \breve{B} = \breve{A} \cdot P$ hold; note that these imply that $C \cdot P^\circ = P^\circ \cdot C'$.

Example 1.2.21 Assume that C is a Cartan matrix of type G_2, F_4 or E_8. Then C is invertible over $\mathbb{Z}$ and, hence, $\mathscr{R}_{\text{ad}}(C)$ and $\mathscr{R}_{\text{sc}}(C)$ are isomorphic root data. Indeed, $\mathscr{R}_{\text{ad}}(C)$ corresponds to the factorisation $C = \breve{A} \cdot A^{\text{tr}}$ where A is the identity matrix, while $\mathscr{R}_{\text{sc}}(C)$ corresponds to the factorisation $C = \breve{B} \cdot B^{\text{tr}}$ where $\breve{B}$ is the identity matrix. Then the conditions in Remark 1.2.20 hold, where $P = C^{-1}$ and P° is the identity matrix.

1.3 Chevalley's Classification Theorems

Throughout this section, let k be an algebraically closed field and $\mathbf{G}$ be a linear algebraic group over k. We can now explain how one can naturally attach to $\mathbf{G}$ a root datum, when $\mathbf{G}$ is connected reductive.

1.3.1 Let $\mathbf{G}$ be connected reductive. Let $\mathbf{T} \subseteq \mathbf{G}$ be a maximal torus, $X = X(\mathbf{T})$ and $L(\mathbf{G})$ be the Lie algebra. Recall from 1.1.12 that there is a finite subset $R \subseteq X$ and a corresponding root space decomposition of $L(\mathbf{G})$:

$$L(\mathbf{G}) = L(\mathbf{G})_0 \oplus \bigoplus_{\alpha \in R} L(\mathbf{G})_\alpha.$$

As already mentioned in 1.1.12, we have $L(\mathbf{G})_0 = L(\mathbf{T})$, $R = -R$ and $\dim L(\mathbf{G})_\alpha = 1$ for all $\alpha \in R$; in particular,

$$\dim \mathbf{G} = \dim L(\mathbf{G}) = \dim \mathbf{T} + |R|.$$

The roots can be directly characterised in terms of $\mathbf{G}$, as follows. Let $\alpha \in X$. Then $\alpha \in R$ if and only if there exists a homomorphism of algebraic groups $u_\alpha \colon \mathbf{k}^+ \to \mathbf{G}$ such that u_α is an isomorphism onto its image and we have

$$t u_\alpha(\xi) t^{-1} = u_\alpha(\alpha(t)\xi) \qquad \text{for all } t \in \mathbf{T} \text{ and } \xi \in k.$$

Thus, $\mathbf{U}_\alpha := \{u_\alpha(\xi) \mid \xi \in k\} \subseteq \mathbf{G}$ is a one-dimensional closed connected unipotent subgroup normalised by $\mathbf{T}$. It is uniquely determined by α and called the *root subgroup* corresponding to α. Conversely, every one-dimensional closed connected unipotent subgroup normalised by $\mathbf{T}$ is equal to $\mathbf{U}_\alpha$ for some $\alpha \in R$. We have

$$\mathbf{G} = \langle \mathbf{T}, \mathbf{U}_\alpha \mid \alpha \in R \rangle.$$

Now consider also the co-character group $Y = Y(\mathbf{T})$; we wish to define a finite subset $R^\vee \subseteq Y$. Recall from 1.1.11 that X, Y are free abelian groups of the same (finite) rank and that there is a natural pairing $\langle\ ,\ \rangle \colon X \times Y \to \mathbb{Z}$. The Weyl group of $\mathbf{G}$ with respect to $\mathbf{T}$ is defined as $\mathbf{W}(\mathbf{G}, \mathbf{T}) := N_{\mathbf{G}}(\mathbf{T})/\mathbf{T}$. Since $N_{\mathbf{G}}(\mathbf{T})$ acts on $\mathbf{T}$ by conjugation, we have induced actions of $\mathbf{W}(\mathbf{G}, \mathbf{T})$ on X and on Y via

$$(w.\lambda)(t) = \lambda(\dot{w}^{-1} t \dot{w}) \qquad (\lambda \in X,\ t \in \mathbf{T}),$$
$$(w.\nu)(\xi) = \dot{w}\nu(\xi)\dot{w}^{-1} \qquad (\nu \in Y,\ \xi \in k^\times),$$

where, for any $w \in \mathbf{W}(\mathbf{G}, \mathbf{T})$, we denote by $\dot{w}$ a representative in $N_{\mathbf{G}}(\mathbf{T})$. Using these actions, we can identify $\mathbf{W}(\mathbf{G}, \mathbf{T})$ with subgroups of $\mathrm{Aut}(X)$ and of $\mathrm{Aut}(Y)$. Now let $\alpha \in R$. Then $\mathbf{G}_\alpha := C_{\mathbf{G}}(\ker(\alpha)^\circ) = \langle \mathbf{T}, \mathbf{U}_\alpha, \mathbf{U}_{-\alpha} \rangle$ is a closed connected reductive subgroup of $\mathbf{G}$; its Weyl group $\mathbf{W}(\mathbf{G}_\alpha, \mathbf{T}) := N_{\mathbf{G}_\alpha}(\mathbf{T})/\mathbf{T}$ has order 2. Let $w_\alpha \in \mathbf{W}(\mathbf{G}_\alpha, \mathbf{T})$ be the non-trivial element and $\dot{w}_\alpha$ be a representative of w_α in $N_{\mathbf{G}_\alpha}(\mathbf{T}) \subseteq N_{\mathbf{G}}(\mathbf{T})$. Then there exists a unique $\alpha^\vee \in Y$ such that

$$w_\alpha.\lambda = \lambda - \langle \lambda, \alpha^\vee \rangle \alpha \qquad \text{for all } \lambda \in X.$$

Following, e.g., [Con14, 1.2.8], this element $\alpha^\vee$ can also be determined as follows.

The maps $u_{\pm\alpha}\colon \mathbf{k}^+ \to \mathbf{U}_{\pm\alpha}$ can be chosen such that the assignment

$$\begin{pmatrix} 1 & \xi \\ 0 & 1 \end{pmatrix} \mapsto u_\alpha(\xi), \qquad \begin{pmatrix} 1 & 0 \\ \xi & 1 \end{pmatrix} \mapsto u_{-\alpha}(\xi) \qquad (\xi \in k)$$

defines a homomorphism of algebraic groups $\varphi_\alpha\colon \mathrm{SL}_2(k) \to \mathbf{G}$. Then

$$\alpha^\vee(\xi) = \varphi_\alpha \begin{pmatrix} \xi & 0 \\ 0 & \xi^{-1} \end{pmatrix} \in \mathbf{T} \qquad \text{for all } \xi \in k^\times.$$

Thus, we obtain a well-defined subset $R^\vee = \{\alpha^\vee \mid \alpha \in R\} \subseteq Y$; we have

$$\mathbf{W}(\mathbf{G}, \mathbf{T}) = \langle w_\alpha \mid \alpha \in R \rangle.$$

Complete proofs of the above statements can be found in the textbooks [Bor91], [Hum91], [Spr98] and, of course, the original source [Ch05]. A thorough guide through this argument, with indications of the proofs and many worked-out examples, can be found in [MaTe11, §8]. See also [Al09], [Con14], [Jan03, Chap. II], [DG70/11] for further reading.

With this notation, we can now state the following result which shows that we are exactly in the situation described by Proposition 1.1.15.

Theorem 1.3.2 *The quadruple* $\mathscr{R} = (X(\mathbf{T}), R, Y(\mathbf{T}), R^\vee)$ *in 1.3.1 is a root datum as defined in 1.2.1, with Weyl group* $\mathbf{W}(\mathbf{G}, \mathbf{T})$ *(identified with a subgroup of* $\mathrm{Aut}(X(\mathbf{T}))$ *as above). Furthermore, let* $R^+ \subseteq R$ *be the set of positive roots with respect to a base* $\Pi \subseteq R$. *Then*

$$\mathbf{B} := \langle \mathbf{T}, \mathbf{U}_\alpha \mid \alpha \in R^+ \rangle \subseteq \mathbf{G}$$

is a Borel subgroup; the subgroups $\mathbf{B}$ *and* $N_{\mathbf{G}}(\mathbf{T})$ *form a reductive BN-pair in* $\mathbf{G}$, *where* $C_{\mathbf{G}}(\mathbf{T}) = \mathbf{T} = \mathbf{B} \cap N_{\mathbf{G}}(\mathbf{T})$.

Proof In its essence, this is due to Chevalley [Ch05], but the notion of BN-pairs was not yet available at that time. A proof of the fact that $\mathscr{R}$ is a root datum can be found, for example, in [MaTe11, 9.11], [Spr98, 7.4.3]. The BN-pair axioms are shown in [Bor91, 14.15], [MaTe11, 11.16]. For the equality $C_{\mathbf{G}}(\mathbf{T}) = \mathbf{T}$, see [MaTe11, 8.13] or [Spr98, 7.6.4]. □

Theorem 1.3.3 *Assume that* $\mathbf{G}$ *is connected reductive. Then* $\mathbf{G}$ *acts transitively (by simultaneous conjugation) on the set of all pairs* $(\mathbf{T}, \mathbf{B})$ *where* $\mathbf{T} \subseteq \mathbf{G}$ *is a maximal torus and* $\mathbf{B} \subseteq \mathbf{G}$ *is a Borel subgroup such that* $\mathbf{T} \subseteq \mathbf{B}$. *In particular, the root data (as in Theorem 1.3.2) with respect to any two maximal tori of* $\mathbf{G}$ *are isomorphic in the sense of 1.2.2.*

Proof The conjugacy results are due to Borel [Bor91, 10.6,11.1]; see also [Spr98, 6.2.7, 6.3.5]. (See [PaVi10] for a discussion of earlier work of Morozov on Borel subalgebras; we thank George Lusztig for this reference.) A somewhat more elementary proof of the conjugacy of Borel subgroups is given in [St77]. (See also [Ge03a, §3.4].) Once these conjugacy results are shown, the assertion about the isomorphism between root data is clear. □

Remark 1.3.4 Let $\mathbf{G}$, $\mathbf{T}$, $\mathscr{R}$, $\mathbf{B}$ as in Theorem 1.3.2; let $\mathbf{W} := \mathbf{W}(\mathbf{G},\mathbf{T})$. Then the set of *all* Borel subgroups of $\mathbf{G}$ containing $\mathbf{T}$ is described as follows. Let $\mathbf{B}_1$ be any Borel subgroup of $\mathbf{G}$ containing $\mathbf{T}$. By Theorems 1.3.2, 1.3.3 and the BN-pair axioms, there is a unique $w \in \mathbf{W}$ such that $\mathbf{B}_1 = \dot{w}^{-1}\mathbf{B}\dot{w}$. Now, the base $\Pi \subseteq R$ used to define $\mathbf{B}$ is transformed under w to a new base Π_1 of R. Consequently, we have $\mathbf{B}_1 = \langle \mathbf{T}, \mathbf{U}_\alpha \mid \alpha \in R_1^+\rangle$ where $R_1^+ \subseteq R$ is the set of positive roots with respect to Π_1. Further recall from 1.2.6 that any two bases of R can be transformed into each other by a unique element of $\mathbf{W}$. Thus, we obtain bijective correspondences

$$\{\text{Borel subgroups containing } \mathbf{T}\} \quad \stackrel{1-1}{\longleftrightarrow} \quad \mathbf{W} \quad \stackrel{1-1}{\longleftrightarrow} \quad \{\text{bases of } R\}.$$

Remark 1.3.5 Assume that $\mathbf{G}$ is connected reductive. In 1.1.13 we have defined $\mathbf{G}$ to be semisimple if $|\mathbf{Z}| < \infty$ where $\mathbf{Z} = \mathbf{Z}(\mathbf{G})$ denotes the centre of $\mathbf{G}$; alternatively, $\mathbf{G}$ is semisimple if and only if $\mathbf{G} = \mathbf{G}_{\mathrm{der}}$ (see 1.1.13). We also have the following characterisation in terms of the root datum $\mathscr{R} = (X, R, Y, R^\vee)$ (with respect to a maximal torus $\mathbf{T} \subseteq \mathbf{G}$). By [MaTe11, 8.17(h)], [Spr98, 8.1.8], we have

$$\mathbf{Z} = \{t \in \mathbf{T} \mid \alpha(t) = 1 \text{ for all } \alpha \in R\} \tag{a}$$

and the isomorphism $\mathbf{T} \cong \mathrm{Hom}(X(\mathbf{T}), k^\times)$ in 1.1.11 restricts to

$$\mathbf{Z} \cong \mathrm{Hom}(X(\mathbf{T})/\mathbb{Z}R,\ k^\times). \tag{b}$$

Thus, we obtain the equivalences:

$$|\mathbf{Z}| < \infty \quad \Longleftrightarrow \quad |X/\mathbb{Z}R| < \infty \quad \Longleftrightarrow \quad |Y/\mathbb{Z}R^\vee| < \infty. \tag{c}$$

If we consider the factorisation $C = \breve{A} \cdot A^{\mathrm{tr}}$ determined by $\mathscr{R}$ as in Remark 1.2.13, then $\mathbf{G}$ is semisimple if and only if $A, \breve{A}$ are square matrices.

Remark 1.3.6 Assume that $\mathbf{G}$ is connected reductive. As in 1.1.13, we have $\mathbf{G} = \mathbf{Z}^\circ.\mathbf{G}_{\mathrm{der}}$; furthermore, $\mathbf{G}_{\mathrm{der}} = \mathbf{G}_1 \ldots \mathbf{G}_n$ where $\mathbf{G}_1, \ldots, \mathbf{G}_n$ are the closed normal simple subgroups of $\mathbf{G}$; they commute pairwise with each other. These subgroups have the following description in terms of the root datum $\mathscr{R} = (X, R, Y, R^\vee)$ (with respect to a maximal torus $\mathbf{T} \subseteq \mathbf{G}$) and the corresponding root subgroups $\mathbf{U}_\alpha$ ($\alpha \in R$). First note that

$$\mathbf{G}_{\mathrm{der}} = \langle \mathbf{U}_\alpha \mid \alpha \in R\rangle,$$

see [MaTe11, 8.21]. Now let C be the Cartan matrix of the root system R, with respect to a base Π of R. Then C can be expressed as a block diagonal matrix where the diagonal blocks are indecomposable Cartan matrices, $C_1, \dots, C_n$ say. (Thus, $C_1, \dots, C_n$ correspond to the connected components of the Dynkin diagram of C.) Let $\Pi = \Pi_1 \sqcup \cdots \sqcup \Pi_n$ be the corresponding partition of Π. Then we also have $R = R_1 \sqcup \cdots \sqcup R_n$ where R_i consists of all roots in R which can be expressed as linear combinations of simple roots in Π_i. Then we have

$$\mathbf{G}_i = \langle \mathbf{U}_\alpha \mid \alpha \in R_i \rangle \subseteq \mathbf{G} \qquad \text{for } i = 1, \dots, n.$$

A maximal torus of $\mathbf{G}_i$ is given by $\mathbf{T}_i := \mathbf{G}_i \cap \mathbf{T}$ where $\mathbf{T}$ is a fixed maximal torus of $\mathbf{G}$. (See [Bor91, Chap. IV, §11], [MaTe11, §8.4], [Spr98, §8.1] for further details.)

Before continuing with the general theory, we give three concrete examples. We shall see that the point of view in 1.2.14, where root data are described in terms of factorisations of Cartan matrices, provides a particularly efficient and convenient way of encoding the information involved in these examples.

Example 1.3.7 Let $\mathbf{G} = \mathrm{GL}_n(k)$. Let $\mathbf{B} \subseteq \mathbf{G}$ be the subgroup of all upper triangular matrices and $\mathbf{N} \subseteq \mathbf{G}$ the subgroup of all monomial matrices. It is well known that these groups form a BN-pair; see [Bou68, Chap. IV, n° 2.2]. For further details see [Ge03a, 1.6.10, 3.4.5], where it is also shown that this is an algebraic BN-pair satisfying the conditions in Proposition 1.1.15; in particular, $\mathbf{G}$ is connected reductive. Let us describe the root datum of $\mathbf{G}$ with respect to the maximal torus $\mathbf{T} = \mathbf{B} \cap \mathbf{N}$ consisting of all diagonal matrices in $\mathbf{G}$.

It will be convenient to introduce some notation concerning matrices. For $1 \leqslant i \leqslant n-1$, let n_i be the matrix which is obtained by interchanging the i th and the $(i+1)$th row in the identity matrix, which we denote by I_n. More generally, if $w \in \mathfrak{S}_n$ is any permutation, let n_w be the matrix which is obtained by permuting the rows of I_n as specified by w. (Thus, if $\{e_1, \dots, e_n\}$ denotes the standard basis of k^n, then $n_w(e_i) = e_{w(i)}$ for $1 \leqslant i \leqslant n$; we have $n_{ww'} = n_w n_{w'}$ for all $w, w' \in \mathfrak{S}_n$.) Then $\mathbf{N} = \{h n_w \mid h \in \mathbf{T}, w \in \mathfrak{S}_n\}$ and so we have an exact sequence

$$\{1\} \to \mathbf{T} \to \mathbf{N} \to \mathfrak{S}_n \to \{1\},$$

where $\mathbf{N} \to \mathfrak{S}_n$ sends n_w to w. Next, for $1 \leqslant i, j \leqslant n$ let E_{ij} be the 'elementary' matrix with coefficient 1 at the position (i, j) and 0 otherwise. We define

$$\mathbf{U}_{ij} := \{I_n + \xi E_{ij} \mid \xi \in k\} \qquad \text{where } 1 \leqslant i, j \leqslant n, i \neq j.$$

All of these are one-dimensional, closed connected subgroups of $\mathbf{G}$. Finally, if $\xi_1, \dots, \xi_n$ are non-zero elements of k, we denote by $h(\xi_1, \dots, \xi_n) \in \mathbf{T}$ the diagonal matrix with $\xi_1, \dots, \xi_n$ along the diagonal. Then the map

$$(\mathbf{k}^\times)^n \to \mathbf{T}, \qquad (\xi_1, \dots, \xi_n) \mapsto h(\xi_1, \dots, \xi_n),$$

is certainly an isomorphism of algebraic groups. Hence $X = X(\mathbf{T})$ is the free abelian group with basis $\lambda_1, \ldots, \lambda_n$, where $\lambda_i(h(\xi_1, \ldots, \xi_n)) = \xi_i$ for all i.

Each subgroup $\mathbf{U}_{ij}$ is normalised by $\mathbf{T}$. Let $u_{ij} : \mathbf{k}^+ \to \mathbf{G}$ be the homomorphism given by $u_{ij}(\xi) = I_n + \xi E_{ij}$ for $\xi \in \mathbf{k}^+$. Then $\mathbf{U}_{ij}$ is the image of this homomorphism, u_{ij} is an isomorphism onto its image and we have

$$tu_{ij}(\xi)t^{-1} = u_{ij}(\xi_i\xi_j^{-1}\xi) \qquad \text{where } t = h(\xi_1, \ldots, \xi_n) \in \mathbf{T} \text{ and } \xi \in \mathbf{k}^+.$$

Hence, $\alpha_{ij} := \lambda_i - \lambda_j \in X$ is a root and $\mathbf{U}_{ij}$ is the corresponding root subgroup. To see that these are all the roots, we can use the formula $\dim \mathbf{G} = \dim \mathbf{T} + |R|$ in 1.3.1. Thus, since $\dim \mathbf{G} = n^2$ and $\dim \mathbf{T} = n$, we conclude that $R = \{\alpha_{ij} \mid 1 \leqslant i, j \leqslant n, i \neq j\}$ is the root system of $\mathbf{G}$ with respect to $\mathbf{T}$. We also see that

$$\Pi := \{\alpha_{i,i+1} = \lambda_i - \lambda_{i+1} \mid 1 \leqslant i \leqslant n-1\} \subseteq R$$

is a base of R and that $\mathbf{B}$ is the Borel subgroup associated with this base (as in Remark 1.3.4). Now consider coroots. The dual basis of $Y = Y(\mathbf{T})$ is given by the co-characters $\nu_j : \mathbf{k}^\times \to \mathbf{T}$ such that $\nu_j(\xi)$ is the diagonal matrix with coefficient ξ at position j, and coefficient 1 otherwise. For $i \neq j$, we have a unique embedding of algebraic groups $\varphi_{ij} : \mathrm{SL}_2(k) \hookrightarrow \mathbf{G}$ such that

$$\varphi_{ij}\begin{pmatrix} 1 & \xi \\ 0 & 1 \end{pmatrix} = u_{ij}(\xi) \quad \text{and} \quad \varphi_{ij}\begin{pmatrix} 1 & 0 \\ \xi & 1 \end{pmatrix} = u_{ji}(\xi) \quad \text{for all } \xi \in k.$$

Hence, φ_{ij} satisfies the condition in 1.3.1 and so we obtain $\alpha_{ij}^\vee \in Y$ such that

$$\alpha_{ij}^\vee(\xi) = \varphi_{ij}\begin{pmatrix} \xi & 0 \\ 0 & \xi^{-1} \end{pmatrix} \in \mathbf{T}$$

is the diagonal matrix with coefficient ξ at position i and coefficient ξ^{-1} at position j. Thus, we have $R^\vee = \{\alpha_{ij}^\vee = \nu_i - \nu_j \mid 1 \leqslant i, j \leqslant n, i \neq j\}$. We also see that the set

$$\Pi^\vee := \{\alpha_{i,i+1}^\vee = \nu_i - \nu_{i+1} \mid 1 \leqslant i \leqslant n-1\} \subseteq R^\vee$$

is a base of $R^\vee$. The Cartan matrix $C = (c_{ij})_{1 \leqslant i,j \leqslant n-1}$ with respect to this base is given by

$$C = \begin{pmatrix} 2 & -1 & 0 & \ldots & & 0 \\ -1 & 2 & -1 & 0 & \ldots & 0 \\ 0 & -1 & 2 & -1 & \ddots & \vdots \\ \vdots & \ddots & \ddots & \ddots & \ddots & 0 \\ 0 & \ldots & 0 & -1 & 2 & -1 \\ 0 & & \ldots & 0 & -1 & 2 \end{pmatrix}.$$

Thus, C is of type A_{n-1}. The factorisation in Remark 1.2.13 is given by

$$C = \breve{A} \cdot A^{\text{tr}} \quad \text{where} \quad A = \breve{A} = \begin{pmatrix} 1 & -1 & 0 & \dots & & 0 \\ 0 & 1 & -1 & 0 & \dots & 0 \\ \vdots & \ddots & \ddots & \ddots & \ddots & \vdots \\ 0 & \dots & 0 & 1 & -1 & 0 \\ 0 & & \dots & 0 & 1 & -1 \end{pmatrix}.$$

(Here, $A = \breve{A}$ has $n-1$ rows and n columns.)

Example 1.3.8 Let $n \geqslant 2$ and $\mathbf{G}' = \mathrm{SL}_n(k)$, the *special linear group*. We keep the notation $\mathbf{G} = \mathrm{GL}_n(k)$, $\mathbf{U}_{ij}$, $\mathbf{B}$, $\mathbf{N}$, $\mathbf{T}$, $X = X(\mathbf{T})$, $Y = Y(\mathbf{T})$ from the previous example. Then an algebraic BN-pair satisfying the conditions in Proposition 1.1.15 is given by the subgroups $\mathbf{B}' := \mathbf{B} \cap \mathbf{G}'$ and $\mathbf{N}' := \mathbf{N} \cap \mathbf{G}'$; see [Bou68, Chap. IV, §2, Exercise 10], [Ge03a, 1.6.11, 3.4.5]. Let us describe the root datum of $\mathbf{G}'$ with respect to the maximal torus $\mathbf{T}' = \mathbf{T} \cap \mathbf{G}'$. Let

$$X' = X(\mathbf{T}') \qquad \text{and} \qquad Y' = Y(\mathbf{T}').$$

For $1 \leqslant i, j \leqslant n$, $i \neq j$, the subgroup $\mathbf{U}_{ij}$ of $\mathbf{G}$ is already contained in $\mathbf{G}'$. So, if α'_{ij} denotes the restriction of $\alpha_{ij} \in X$ to $\mathbf{T}'$, then $\alpha'_{ij} \in X'$ and α'_{ij} is a root of $\mathbf{G}'$ with corresponding root subgroup $\mathbf{U}_{ij} \subseteq \mathbf{G}'$. Since $\dim \mathbf{G}' = n^2 - 1$ and $\dim \mathbf{T}' = n - 1$, it follows as above that $R' = \{\alpha'_{ij} \mid 1 \leqslant i, j \leqslant n, i \neq j\}$ is the root system of $\mathbf{G}'$ with respect to $\mathbf{T}'$ and that

$$\Pi' = \{\alpha'_{i,i+1} \mid 1 \leqslant i \leqslant n-1\} \text{ is a base for } R'.$$

Furthermore, the image of the embedding $\varphi_{ij}\colon \mathrm{SL}_2(k) \hookrightarrow \mathbf{G}$ is clearly contained in $\mathbf{G}'$. Consequently, any coroot $\alpha^\vee_{ij} \in Y$ also is a coroot in Y'. Thus, we have $R'^\vee = R^\vee = \{\alpha^\vee_{ij} \mid 1 \leqslant i, j \leqslant n, i \neq j\}$ and

$$\Pi'^\vee = \{\alpha^\vee_{i,i+1} \mid 1 \leqslant i \leqslant n-1\} \text{ is a base for } R'^\vee.$$

In particular, we obtain the same Cartan matrix C of type A_{n-1} as in Example 1.3.7. Now note that we have an isomorphism of algebraic groups

$$(\mathbf{k}^\times)^{n-1} \to \mathbf{T}', \qquad (\xi_1, \dots, \xi_{n-1}) \mapsto h(\xi_1, \dots, \xi_{n-1}, (\xi_1 \cdots \xi_{n-1})^{-1})$$

(with inverse sending $h(\xi_1, \dots, \xi_n) \in \mathbf{T}'$ to $(\xi_1, \dots, \xi_{n-1}) \in (\mathbf{k}^\times)^{n-1}$). Hence, if we define co-characters $\nu'_j\colon \mathbf{k}^\times \to \mathbf{T}'$ (for $1 \leqslant j \leqslant n-1$) such that $\nu'_j(\xi)$ is the diagonal matrix with ξ at position j and ξ^{-1} at position n, then $\{\nu'_1, \dots, \nu'_{n-1}\}$ is a $\mathbb{Z}$-basis of Y'. But then $\Pi'^\vee$ also is a $\mathbb{Z}$-basis of Y'. If we consider the corresponding dual basis of X', then the factorisation in Remark 1.2.13 is given by

$$C = \breve{A} \cdot A^{\text{tr}} \qquad \text{where } \breve{A} = I_{n-1} \text{ and } A = C^{\text{tr}}.$$

Thus, $\mathbf{G}'$ is semisimple and the root datum of $\mathrm{SL}_n(k)$ is of simply connected type (see Example 1.2.16).

Example 1.3.9 Let $\mathbf{G} = \mathrm{GL}_n(k)$ and $\mathbf{Z} \subseteq \mathbf{G}$ be the centre of $\mathbf{G}$, consisting of all non-zero scalar matrices. Assume that $n \geqslant 2$ and let $\bar{\mathbf{G}} = \mathrm{PGL}_n(k) := \mathbf{G}/\mathbf{Z}$, the *projective general linear group*. (This is a linear algebraic group by 1.1.6.) Let us denote the canonical map $\mathbf{G} \to \bar{\mathbf{G}}$ by $g \mapsto \bar{g}$. In particular, we obtain subgroups $\bar{\mathbf{B}}$ and $\bar{\mathbf{N}}$ of $\bar{\mathbf{G}}$ which form a BN-pair since $\mathbf{Z} \subseteq \mathbf{B}$; see [Bou68, Chap. IV, §2, Exercise 2]. One easily checks that this is an algebraic BN-pair satisfying the conditions in Proposition 1.1.15. Let us describe the root datum of $\bar{\mathbf{G}}$ with respect to the maximal torus $\bar{\mathbf{T}}$ of $\bar{\mathbf{G}}$. Let

$$\bar{X} = X(\bar{\mathbf{T}}) \qquad \text{and} \qquad \bar{Y} = Y(\bar{\mathbf{T}}).$$

For every root α of $\mathbf{G}$, we clearly have $\mathbf{Z} \subseteq \ker(\alpha)$. So, using the universal property of quotients, there is a well-defined $\bar{\alpha} \in \bar{X}$ such that $\alpha(t) = \bar{\alpha}(\bar{t})$ for all $t \in \mathbf{T}$. Now, for $1 \leqslant i, j \leqslant n$, $i \neq j$, the image $\bar{\mathbf{U}}_{ij}$ of the subgroup $\mathbf{U}_{ij} \subseteq \mathbf{G}$ in $\bar{\mathbf{G}}$ is still closed, connected, isomorphic to $\mathbf{k}^+$ and normalised by $\bar{\mathbf{T}}$. Hence, $\bar{\alpha}_{ij}$ is a root with corresponding root subgroup $\bar{\mathbf{U}}_{ij} \subseteq \bar{\mathbf{G}}$. As above, it follows that $\bar{R} = \{\bar{\alpha}_{ij} \mid 1 \leqslant i, j \leqslant n, i \neq j\}$ is the root system of $\bar{\mathbf{G}}$ with respect to $\bar{\mathbf{T}}$ and that

$$\bar{\Pi} = \{\bar{\alpha}_{i,i+1} \mid 1 \leqslant i \leqslant n-1\} \text{ is a base for } \bar{R}.$$

On the other hand, we obtain morphisms of algebraic groups $\bar{\varphi}_{ij} \colon \mathrm{SL}_2(k) \to \bar{\mathbf{G}}$, simply by composing $\varphi_{ij} \colon \mathrm{SL}_2(k) \hookrightarrow \mathbf{G}$ with the canonical map $\mathbf{G} \to \bar{\mathbf{G}}$. Thus, every coroot $\alpha^\vee$ of $\mathbf{G}$ determines a coroot $\bar{\alpha}^\vee \in \bar{Y}$. Consequently, we have $\bar{R}^\vee = \{\bar{\alpha}_{ij}^\vee \mid 1 \leqslant i, j \leqslant n, i \neq j\}$ and

$$\bar{\Pi}^\vee = \{\bar{\alpha}_{i,i+1}^\vee \mid 1 \leqslant i \leqslant n-1\} \text{ is a base for } \bar{R}^\vee.$$

In particular, we obtain the same Cartan matrix C of type A_{n-1} as in Example 1.3.7. Now consider the homomorphism of algebraic groups

$$\mathbf{T} \to (\mathbf{k}^\times)^{n-1}, \qquad h(\xi_1, \ldots, \xi_n) \mapsto (\xi_1\xi_n^{-1}, \ldots, \xi_{n-1}\xi_n^{-1}).$$

It has $\mathbf{Z}$ in its kernel so there is an induced homomorphism of algebraic groups $\bar{\mathbf{T}} \to (\mathbf{k}^\times)^{n-1}$. The latter homomorphism is an isomorphism: its inverse is given by sending $(\xi_1, \ldots, \xi_{n-1}) \in (\mathbf{k}^\times)^{n-1}$ to the image of the element $h(\xi_1, \ldots, \xi_{n-1}, 1) \in \mathbf{T}$ in $\bar{\mathbf{T}}$. It follows that

$$\{\bar{\alpha}_{i,n} \mid 1 \leqslant i \leqslant n-1\} \quad \text{is a } \mathbb{Z}\text{-basis of } \bar{X}.$$

But then $\bar{\Pi}$ also is a $\mathbb{Z}$-basis of $\bar{X}$. If we consider the corresponding dual basis of $\bar{Y}$, then the factorisation in Remark 1.2.13 is given by

$$C = \breve{A} \cdot A^{\mathrm{tr}} \qquad \text{where } \breve{A} = C \text{ and } A = I_{n-1}.$$

Thus, $\mathrm{PGL}_n(k)$ is semisimple and the root datum of $\mathrm{PGL}_n(k)$ is of adjoint type (see Example 1.2.16). In particular, we see that the root data of $\mathrm{PGL}_n(k)$ and $\mathrm{SL}_n(k)$ are not isomorphic. (In the former, $\mathbb{Z}R = X$; in the latter, $\mathbb{Z}R \neq X$.) So, by Theorem 1.3.3, $\mathrm{PGL}_n(k)$ and $\mathrm{SL}_n(k)$ are not isomorphic as algebraic groups (even if $\mathbf{Z}(\mathrm{SL}_n(k)) = \{1\}$, in which case these two groups are isomorphic as abstract groups).

A key feature of the whole theory is the fact that a connected reductive algebraic group is uniquely determined by its root datum up to isomorphism. This follows from a more general result, the '*isogeny theorem*'. As preparation, we cite the following general results concerning surjective homomorphisms of algebraic groups, which will be useful at several places below.

1.3.10 Let $f\colon \mathbf{G} \to \mathbf{G}'$ be a surjective homomorphism of connected algebraic groups over k. Then we have the following *preservation results*.

(a) f maps a Borel subgroup of $\mathbf{G}$ onto a Borel subgroup of $\mathbf{G}'$, and all Borel subgroups of $\mathbf{G}'$ arise in this way; a similar statement holds for maximal tori. (See [Bor91, 11.14].)

(b) f maps the unipotent radical of $\mathbf{G}$ onto the unipotent radical of $\mathbf{G}'$; in particular, if $\mathbf{G}$ is reductive, then so is $\mathbf{G}'$. (See [Bor91, 14.11].)

(c) If $\mathbf{G}$ is reductive, then f maps the centre of $\mathbf{G}$ onto the centre of $\mathbf{G}'$. (This follows from (a) and the fact that the centre of a reductive group is the intersection of all its maximal tori; see [Bor91, 11.11].)

(d) Assume that $\mathbf{G}$ and $\mathbf{G}'$ are reductive. Let $\mathbf{T}$ be a maximal torus of $\mathbf{G}$; by (a), $\mathbf{T}' := f(\mathbf{T})$ is a maximal torus of $\mathbf{G}'$. Then f induces a surjective homomorphism $\mathbf{W}(\mathbf{G},\mathbf{T}) \to \mathbf{W}(\mathbf{G}',\mathbf{T}')$, and this is an isomorphism if $\ker(f)$ is contained in the centre of $\mathbf{G}$. (See [Bor91, 11.20].)

There is a further property which is certainly well known and which we will need in Chapter 2, but it is not easy to find a reference with a proof:

(e) If $\mathbf{G}$ is reductive and $\ker(f)$ is contained in the centre of $\mathbf{G}$, then we have $f\big(C_{\mathbf{G}}^\circ(g)\big) = C_{\mathbf{G}'}^\circ(f(g))$ for any $g \in \mathbf{G}$.

Here is a proof of (e). Let $g' = f(g)$. It is clear that $f\big(C_{\mathbf{G}}(g)\big) \subseteq C_{\mathbf{G}'}(g')$ and $f\big(C_{\mathbf{G}}^\circ(g)\big) \subseteq C_{\mathbf{G}'}^\circ(g')$. So it is sufficient to show that the index of $f(C_{\mathbf{G}}(g))$ in $C_{\mathbf{G}'}(g')$ is finite. To see this, we apply Lemma 1.1.9 with $A = \mathbf{G}$, $B = \mathbf{G}'$, $\sigma(x) = g^{-1}xg$ $(x \in \mathbf{G})$, $\tau(x') = g'^{-1}x'g'$ $(x' \in \mathbf{G}')$. Then $A^\sigma = C_{\mathbf{G}}(g)$ and $B^\tau = C_{\mathbf{G}'}(g')$; since $\ker(f) \subseteq Z(A)$, we have $C = \{a^{-1}\sigma(a) \mid a \in \ker(f)\} = \{1\}$. So the image of the map δ in Lemma 1.1.9 is given by

$$\{a^{-1}\sigma(a) \mid a \in A\} \cap \ker(f) \subseteq \mathbf{G}_{\mathrm{der}} \cap Z(\mathbf{G}).$$

Hence, this image is finite since $\mathbf{G}$ is reductive. Consequently, $C_{\mathbf{G}'}(g')/f(C_{\mathbf{G}}(g))$ is finite, as required. Thus, (e) is proved.

1.3.11 Let $\mathbf{G}$ and $\mathbf{G}'$ be connected reductive algebraic groups over k. Let $f\colon \mathbf{G} \to \mathbf{G}'$ be an *isogeny*, that is, a surjective homomorphism of algebraic groups such that $\ker(f)$ is finite. Note that then $\ker(f)$ is automatically contained in the centre of $\mathbf{G}$. Further to the properties in 1.3.10, we can describe quite precisely the effect of f on roots and coroots. For the following discussion, we refer to [Ch05, §18.2], [MaTe11, §11], [Spr98, §9.6], [St99, §1] for further details.

Let $\mathbf{T}$ be a maximal torus of $\mathbf{G}$; then $\ker(f) \subseteq \mathbf{T}$ and $\mathbf{T}' = f(\mathbf{T})$ is a maximal torus of $\mathbf{G}'$. Let $(X, R, Y, R^\vee)$ and $(X', R', Y', R'^\vee)$ be the corresponding root data. The map f induces a homomorphism $\varphi\colon X' \to X$ such that $\varphi(\lambda') = \lambda' \circ f|_{\mathbf{T}}$ for all $\lambda' \in X'$. The transpose map is given by $\varphi^{\mathrm{tr}}\colon Y \to Y'$, $\nu \mapsto f \circ \nu$. Then it follows that φ is a p-isogeny of root data as in Definition 1.2.9, where p is the *characteristic exponent* of k. (Recall that the characteristic exponent of k is 1 in case $\mathrm{char}(k) = 0$ and is equal to $\mathrm{char}(k)$ otherwise.) The numbers $\{q_\alpha \mid \alpha \in R\}$ and the bijection $R \to R'$, $\alpha \mapsto \alpha^\dagger$, in (I2) come about as follows.

Let $\alpha \in R$ and consider the corresponding root subgroup $\mathbf{U}_\alpha \subseteq \mathbf{G}$; see 1.3.1. Then $f(\mathbf{U}_\alpha)$ is a one-dimensional closed connected unipotent subgroup of $\mathbf{G}'$ normalised by $\mathbf{T}'$. Hence, there is a well-defined $\alpha^\dagger \in R'$ such that $f(\mathbf{U}_\alpha)$ equals the root subgroup $\mathbf{U}'_{\alpha^\dagger}$ in $\mathbf{G}'$. Let $u_\alpha\colon \mathbf{k}^+ \to \mathbf{U}_\alpha$ and $u'_{\alpha^\dagger}\colon \mathbf{k}^+ \to \mathbf{U}'_{\alpha^\dagger}$ be the corresponding isomorphisms. Then the map $f\colon \mathbf{U}_\alpha \to \mathbf{U}'_{\alpha^\dagger}$ has the following form. There is some $c_\alpha \in k^\times$ such that

$$f(u_\alpha(\xi)) = u'_{\alpha^\dagger}(c_\alpha \xi^{q_\alpha}) \qquad \text{for all } \xi \in \mathbf{k}^+.$$

In this situation, the numbers $\{q_\alpha\}$ will also be called the *root exponents* of f. The above discussion shows that an isogeny of connected reductive groups induces a p-isogeny of root data. Conversely, we have the following fundamental result.

Theorem 1.3.12 (Isogeny Theorem) *Let $\mathbf{G}$ and $\mathbf{G}'$ be connected reductive algebraic groups over k, let $\mathbf{T} \subseteq \mathbf{G}$ and $\mathbf{T}' \subseteq \mathbf{G}'$ be maximal tori, and let $\varphi\colon X(\mathbf{T}') \to X(\mathbf{T})$ be a p-isogeny of their root data (see Definition 1.2.9), where p is the characteristic exponent of k. Then there exists an isogeny $f\colon \mathbf{G} \to \mathbf{G}'$ which maps $\mathbf{T}$ onto $\mathbf{T}'$ and induces φ. If $f'\colon \mathbf{G} \to \mathbf{G}'$ is another isogeny with these properties, then there exists some $t \in \mathbf{T}$ such that $f'(g) = f(tgt^{-1})$ for all $g \in \mathbf{G}$.*

See [St99] for a recent, quite short proof of this fundamental result which, for semisimple groups, is one of the main results of the *Séminaire Chevalley*; see [Ch05, §18.2]. As a first consequence, we have:

Corollary 1.3.13 (Isomorphism Theorem) *In the setting of the Isogeny Theorem,* *assume that φ is an isomorphism of root data. Then the isogeny $f\colon \mathbf{G} \to \mathbf{G}'$ is an isomorphism of algebraic groups.*

Proof We use the notation in 1.3.11. Since φ is an isomorphism, the inverse map $\varphi^{-1}\colon X \to X'$ also defines an isogeny of root data. By Theorem 1.3.12 there exist isogenies $f\colon \mathbf{G} \to \mathbf{G}'$ and $g\colon \mathbf{G}' \to \mathbf{G}$ corresponding to φ and φ^{-1}. Then $g \circ f$ induces the identity isogeny of the root datum of $\mathbf{G}$ and hence equals the inner automorphism ι_t for some $t \in \mathbf{T}$. Thus $g' \circ f = \mathrm{id}_{\mathbf{G}}$ with $g' = \iota_t^{-1} \circ g$, and then $f \circ g' \circ f = f$ and $f \circ g' = \mathrm{id}_{\mathbf{G}'}$ because f is surjective. Hence f is an isomorphism with g' as its inverse. □

The general theory is completed by the following existence result.

Theorem 1.3.14 (Existence Theorem) *Let $\mathscr{R} = (X, R, Y, R^\vee)$ be a root datum. Then there exists a connected reductive algebraic group $\mathbf{G}$ over k and a maximal torus $\mathbf{T} \subseteq \mathbf{G}$ such that $\mathscr{R}$ is isomorphic to the root datum of $\mathbf{G}$ relative to $\mathbf{T}$.*

For semisimple groups, this is originally due to Chevalley; see [Ch55] and the comments in [Ch05, §24]. See [Ca72], [St67, §5, Theorem 6] where this is explained in detail, following and extending Chevalley's original approach; see also [Ge17] where the question of choosing signs in a Chevalley basis for the underlying semisimple Lie algebra is resolved. The general case can be reduced to this one; see [Spr98, §10.1] and [DG70/11, Exposé XXV]. Only recently, Lusztig [Lu09c] found a new approach to the general case based on the theory of 'canonical bases' of quantum groups (completing earlier results of Kostant [Ko66]).

Example 1.3.15 Let us see what the above results mean in the simplest non-trivial case where $\mathscr{R} = (X, R, Y, R^\vee)$ is a root datum of Cartan type A_1. Let $\mathbf{G}$ be a corresponding connected reductive algebraic group over k. Now, since $C = (2)$ is the Cartan matrix in this case, $\mathscr{R}$ is determined by an equation

$$2 = \sum_{1 \leqslant i \leqslant d} \breve{a}_i a_i \qquad \text{where } d = \operatorname{rank} X = \operatorname{rank} Y \text{ and } a_i, \breve{a}_i \in \mathbb{Z} \text{ for all } i;$$

see Remark 1.2.13. Up to isomorphism (where isomorphisms are determined by an invertible matrix P over $\mathbb{Z}$ as in Remark 1.2.20), there are three possible cases:

(1) $(a_1, \ldots, a_d) = (2, 0, \ldots, 0)$ and $(\breve{a}_1, \ldots, \breve{a}_d) = (1, 0, \ldots, 0)$, in which case $\mathbf{G} \cong \mathrm{SL}_2(k) \times (\mathbf{k}^\times)^{d-1}$.
(2) $(a_1, \ldots, a_d) = (1, 0, \ldots, 0)$ and $(\breve{a}_1, \ldots, \breve{a}_d) = (2, 0, \ldots, 0)$, in which case $\mathbf{G} \cong \mathrm{PGL}_2(k) \times (\mathbf{k}^\times)^{d-1}$.
(3) $d \geqslant 2$, $(a_1, \ldots, a_d) = (1, 1, 0, \ldots, 0)$ and $(\breve{a}_1, \ldots, \breve{a}_d) = (1, 1, 0, \ldots, 0)$, in which case $\mathbf{G} \cong \mathrm{GL}_2(k) \times (\mathbf{k}^\times)^{d-2}$.

This is contained in [St99, 2.2]; we leave it as an exercise to the reader. In particular, for $d = 1$ (that is, $\mathbf{G}$ semisimple), we have either $\mathbf{G} \cong \mathrm{SL}_2(k)$ or $\mathbf{G} \cong \mathrm{PGL}_2(k)$.

Besides its fundamental importance for the classification of connected reductive algebraic groups, the Isogeny Theorem is an indispensable tool for showing the existence of homomorphisms with prescribed properties. Here are the first examples.

Example 1.3.16 Let $\mathbf{G}$ be a connected reductive algebraic group over k and $\mathbf{T}$ be a maximal torus of $\mathbf{G}$. Let $(X, R, Y, R^\vee)$ be the corresponding root datum. Then there exists an automorphism of algebraic groups $\tau \colon \mathbf{G} \to \mathbf{G}$ such that

$$\tau(t) = t^{-1} \quad (t \in \mathbf{T}) \qquad \text{and} \qquad \tau(\mathbf{U}_\alpha) = \mathbf{U}_{-\alpha} \quad (\alpha \in R).$$

Indeed, $\varphi\colon X \to X$, $\lambda \mapsto -\lambda$, certainly is a p-isogeny, where $q_\alpha = 1$ for all $\alpha \in R$. Hence, since φ is bijective, Corollary 1.3.13 shows that there exists an automorphism $\tau\colon \mathbf{G} \to \mathbf{G}$ such that $\tau(\mathbf{T}) = \mathbf{T}$ and such that φ is the map induced by τ on X. Now, as discussed in 1.1.11, there is a natural bijection between group homomorphisms of X into itself and algebraic homomorphisms of $\mathbf{T}$ into itself. Under this bijection, φ clearly corresponds to the map $t \mapsto t^{-1}$ on $\mathbf{T}$. Hence, τ has the required properties.

Example 1.3.17 Let p be a prime number and $\mathbf{G}$ be a connected reductive algebraic group over $k = \overline{\mathbb{F}}_p$. Let $\mathbf{T}$ be a maximal torus of $\mathbf{G}$ and $(X, R, Y, R^\vee)$ be the corresponding root datum. Then $\varphi\colon X \to X$, $\lambda \mapsto p\lambda$, certainly is a p-isogeny of root data, where $q_\alpha = p$ for all $\alpha \in R$. Hence, by Theorem 1.3.12, there exists an isogeny $F_p\colon \mathbf{G} \to \mathbf{G}$ such that $F_p(\mathbf{T}) = \mathbf{T}$ and such that F_p induces φ on X. Arguing as in the previous example, it follows that

$$F_p(\mathbf{U}_\alpha) = \mathbf{U}_\alpha \quad (\alpha \in R) \qquad \text{and} \qquad F_p(t) = t^p \quad (t \in \mathbf{T}).$$

For example, $F_p\colon \mathrm{GL}_n(k) \to \mathrm{GL}_n(k)$, $(a_{ij}) \mapsto (a_{ij}^p)$, is an isogeny satisfying the above conditions.

We shall see in Section 1.4 that the fixed point set of $\mathbf{G}$ under F_p is a finite group. More generally, we shall consider isogenies $F\colon \mathbf{G} \to \mathbf{G}$ such that $F^d = F_p^m$ for some $d, m \geqslant 1$. The finite groups arising as fixed point sets of connected reductive groups under such isogenies are the *finite groups of Lie type*; see Definition 1.4.7.

Example 1.3.18 Let $\mathscr{R}_i = (X_i, R_i, Y_i, R_i^\vee)$ (for $i = 1, \ldots, n$) be root data. Let $\mathscr{R} = (X, R, Y, R^\vee)$ be the direct product of these root data; see Example 1.2.17. For $i = 1, \ldots, n$, let $\mathbf{G}_i$ be a connected reductive algebraic group with root datum isomorphic to $\mathscr{R}_i$ (relative to a maximal torus $\mathbf{T}_i \subseteq \mathbf{G}_i$). Then, using Corollary 1.3.13, one easily sees that the direct product $\mathbf{G} := \mathbf{G}_1 \times \cdots \times \mathbf{G}_n$ has root datum isomorphic to $\mathscr{R}$ (relative to the maximal torus $\mathbf{T} := \mathbf{T}_1 \times \cdots \times \mathbf{T}_n$ of $\mathbf{G}$).

Example 1.3.19 Let $\mathbf{G} = \mathrm{GL}_n(k)$, with root datum $\mathscr{R} = (X, R, Y, R^\vee)$ as in

Example 1.3.7. It is given by a factorisation $C = \breve{A} \cdot A^{\mathrm{tr}}$ where $C = (c_{ij})_{1 \leqslant i,j \leqslant n-1}$ is the Cartan matrix of type A_{n-1} and $A = \breve{A}$ is a certain matrix of size $(n-1) \times n$. Then, by the procedure in 1.2.18, we obtain an isogeny $\varphi\colon X \to X$ via the pair of matrices $(P, P^\circ) = (-J_n, J_{n-1})$ where, for any $m \geqslant 1$, we set

$$J_m := \begin{pmatrix} 0 & \cdots & 0 & 1 \\ \vdots & \iddots & \iddots & 0 \\ 0 & 1 & \iddots & \vdots \\ 1 & 0 & \cdots & 0 \end{pmatrix} \in M_m(k).$$

Then φ has order 2. So there is a corresponding automorphism of algebraic groups $\gamma\colon \mathrm{GL}_n(k) \to \mathrm{GL}_n(k)$ which maps the maximal torus $\mathbf{T}$ into itself and induces φ on X. Concretely, the map given by

$$\gamma\colon \mathrm{GL}_n(k) \to \mathrm{GL}_n(k), \qquad g \mapsto J_n (g^{\mathrm{tr}})^{-1} J_n,$$

is an automorphism with this property.

Remark 1.3.20 Let $f\colon \mathbf{G} \to \mathbf{G}'$ be an isogeny of connected reductive algebraic groups over k. In the setting of 1.3.11, let $\{q_\alpha \mid \alpha \in R\}$ be the root exponents of f. Following [Spr98, 9.6.3], we say that f is a *central isogeny* if $q_\alpha = 1$ for all $\alpha \in R$. The terminology is justified as follows. Consider the corresponding homomorphism of Lie algebras $d_1 f\colon L(\mathbf{G}) \to L(\mathbf{G}')$. Then, by [Bor91, 22.4], f is a central isogeny if and only if the kernel of $d_1 f$ is contained in the centre of $L(\mathbf{G})$. For example, the isogeny in Example 1.3.16 is central while that in Example 1.3.17 is not.

There are extensions of the Isogeny Theorem to the case where we consider homomorphisms whose kernel is still central but not finite: We shall only formulate the following version here. (This will be needed, for example, in Section 1.7.)

1.3.21 Let $\mathbf{G}, \mathbf{G}'$ be connected reductive algebraic groups over k, and $f\colon \mathbf{G} \to \mathbf{G}'$ be a homomorphism of algebraic groups.

(a) Following [Bo06, Chap. I, 3.A], we say that f is an *isotypy* if $\ker(f) \subseteq \mathbf{Z}(\mathbf{G})$ and $\mathbf{G}'_{\mathrm{der}} \subseteq f(\mathbf{G})$. If this is the case, then we have $\mathbf{G}' = f(\mathbf{G}).\mathbf{Z}(\mathbf{G}')$, $f(\mathbf{G}_{\mathrm{der}}) = \mathbf{G}'_{\mathrm{der}}$ and f restricts to an isogeny $\mathbf{G}_{\mathrm{der}} \to \mathbf{G}'_{\mathrm{der}}$.

(b) Now let $\mathbf{T} \subseteq \mathbf{G}$ and $\mathbf{T}' \subseteq \mathbf{G}'$ be maximal tori such that $f(\mathbf{T}) \subseteq \mathbf{T}'$. Then f induces a group homomorphism $\varphi\colon X(\mathbf{T}') \to X(\mathbf{T})$, $\lambda \mapsto \lambda \circ f|_{\mathbf{T}}$. In analogy to Remark 1.3.20, we say that f is a *central isotypy* if φ is a homomorphism of root data as in 1.2.2. (Note that, as pointed out in the remarks following [Jan03, II, Prop. 1.14], a central isotypy is automatically an isotypy.)

Theorem 1.3.22 (Extended Isogeny Theorem; cf. [Jan03, II, 1.14, 1.15], [St99, §5]) *Let* $\mathbf{G}$ *and* $\mathbf{G}'$ *be connected reductive algebraic groups over* k*, let* $\mathbf{T} \subseteq \mathbf{G}$

and $\mathbf{T}' \subseteq \mathbf{G}'$ *be maximal tori, and let* $\varphi\colon X(\mathbf{T}') \to X(\mathbf{T})$ *be a homomorphism of their root data (see 1.2.2). Then there exists a central isotypy* $f\colon \mathbf{G} \to \mathbf{G}'$ *such that* $f(\mathbf{T}) \subseteq \mathbf{T}'$ *and* f *induces* φ*. Furthermore, the following hold.*

(a) *If* $f'\colon \mathbf{G} \to \mathbf{G}'$ *is another central isotypy inducing* φ*, then there exists some* $t \in \mathbf{T}$ *such that* $f'(g) = f(tgt^{-1})$ *for all* $g \in \mathbf{G}$.

(b) *If* $f|_{\mathbf{T}}\colon \mathbf{T} \to \mathbf{T}'$ *is an isomorphism, then so is* $f\colon \mathbf{G} \to \mathbf{G}'$.

Proof Let Π be a base of the root system $R \subseteq X(\mathbf{T})$. For $\alpha \in \Pi$, consider the subgroup $\mathbf{G}_\alpha = \langle \mathbf{T}, \mathbf{U}_\alpha, \mathbf{U}_{-\alpha}\rangle \subseteq \mathbf{G}$ defined in 1.3.1. Then $\mathbf{G}_\alpha \cap \mathbf{G}_\beta = \mathbf{T}$ for $\alpha \neq \beta$ in Π. As in [Jan03, II, §1.13], one sees that there exists a map

$$f\colon \bigcup_{\alpha\in\Pi} \mathbf{G}_\alpha \to \mathbf{G}'$$

which is a homomorphism on each $\mathbf{G}_\alpha$ and is such that f maps $\mathbf{T}$ into $\mathbf{T}'$ and induces φ. Now, $\mathbf{U}_\alpha$ and $\mathbf{U}_{-\beta}$ certainly commute with each other for all $\alpha \neq \beta$ in Π (by Chevalley's commutator relations; see [MaTe11, 11.8]). Hence, by [St99, Theorem 5.3], f extends to a homomorphism of algebraic groups from $\mathbf{G}$ to $\mathbf{G}'$. The uniqueness statement in (a) is proved as in the case of Theorem 1.3.12; see [St99, §3]. Finally, (b) holds by [Jan03, II, §1.15]. □

1.4 Frobenius Maps and Steinberg Maps

We assume in this section that $k = \overline{\mathbb{F}}_p$ is an algebraic closure of the finite field with p elements, where p is a prime number. We consider a particular class of isogenies in this context, the '*Steinberg maps*'. This will be treated in some detail, where one aim is to work out explicitly some useful characterisations of Steinberg maps in terms of isogenies of root data. In particular, in Proposition 1.4.18, we recover the set-up in Example 1.3.17. We also establish a precise characterisation of Frobenius maps among all Steinberg maps; see Proposition 1.4.28.

Definition 1.4.1 Let $\mathbf{X}$ be an affine variety over k. Let q be a power of p and $\mathbb{F}_q \subseteq k$ be the finite subfield with q elements. We say that $\mathbf{X}$ has an $\mathbb{F}_q$*-rational structure* (or that $\mathbf{X}$ is *defined over* $\mathbb{F}_q$) if there exists some $n \geqslant 1$ and an isomorphism of affine varieties $\iota\colon \mathbf{X} \to \mathbf{X}'$ where $\mathbf{X}' \subseteq \mathbf{k}^n$ is Zariski closed and stable under the *standard Frobenius map*

$$F_q\colon \mathbf{k}^n \to \mathbf{k}^n, \qquad (\xi_1, \ldots, \xi_n) \mapsto (\xi_1^q, \ldots, \xi_n^q).$$

In this case, there is a unique morphism of affine varieties $F\colon \mathbf{X} \to \mathbf{X}$ such that $\iota \circ F = F_q \circ \iota$; it is called the *Frobenius map* corresponding to the $\mathbb{F}_q$-rational structure of $\mathbf{X}$. Note that F_q is a bijective morphism whose fixed point set is $\mathbb{F}_q^n$. Consequently, F is a bijective morphism such that

$$|\mathbf{X}^F| < \infty \qquad \text{where} \qquad \mathbf{X}^F := \{x \in \mathbf{X} \mid F(x) = x\}.$$

Example 1.4.2 Let $\mathbf{X} \subseteq \mathbf{k}^n$ be Zariski closed. Then $\mathbf{X}$ is called $\mathbb{F}_q$-*closed* if $\mathbf{X}$ is defined by a set of polynomials in $\mathbb{F}_q[T_1, \ldots, T_n]$. If this holds, then $\mathbf{X}$ is stable under F_q and so $\mathbf{X}$ has an $\mathbb{F}_q$-rational structure, as defined above; the fixed point set $\mathbf{X}^{F_q}$ consists precisely of those $x \in \mathbf{X}$ which have all their coordinates in $\mathbb{F}_q$. Conversely, if $F_q(\mathbf{X}) \subseteq \mathbf{X}$, then $\mathbf{X}$ is $\mathbb{F}_q$-closed. (The proof uses the fact that $k \supseteq \mathbb{F}_q$ is a separable field extension; see [Ge03a, 4.1.6], [Bor91, AG.14.4]. In general, the discussion of rational structures is much more complicated.)

Remark 1.4.3 Let $\mathbf{X}$ be an affine variety over k and assume that $\mathbf{X}$ is defined over $\mathbb{F}_q$, with Frobenius map $F\colon \mathbf{X} \to \mathbf{X}$. Here are some basic properties of F. First note that $F^2, F^3, \ldots$ are also Frobenius maps. Furthermore, for any $x \in \mathbf{X}$, we have $F^m(x) = x$ for some $m \geqslant 1$. Hence,

$$\mathbf{X} = \bigcup_{m \geqslant 1} \mathbf{X}^{F^m} \quad \text{where} \quad |\mathbf{X}^{F^m}| < \infty \quad \text{for all } m \geqslant 1.$$

(Note that every element of k lies in a finite subfield of k.) Finally, it is also clear that, if $\mathbf{X}' \subseteq \mathbf{X}$ is a closed subset such that $F(\mathbf{X}') \subseteq \mathbf{X}'$, then $\mathbf{X}'$ is defined over $\mathbb{F}_q$, with Frobenius map given by the restriction of F to $\mathbf{X}'$.

Remark 1.4.4 Let $\mathbf{X}$ be an affine variety over k and let A be the algebra of regular functions on $\mathbf{X}$. There is an intrinsic characterisation of Frobenius maps in terms of A, as follows. A morphism $F\colon \mathbf{X} \to \mathbf{X}$ is the Frobenius map corresponding to an $\mathbb{F}_q$-rational structure of $\mathbf{X}$ if and only if the following two conditions hold for the associated algebra homomorphism $F^*\colon A \to A$:

(a) F^* is injective and $F^*(A) = \{a^q \mid a \in A\}$.
(b) For each $a \in A$, there exists some $e \geqslant 1$ such that $(F^*)^e(a) = a^{q^e}$.

One checks that (a) and (b) are satisfied for the standard Frobenius map $F_q\colon \mathbf{k}^n \to \mathbf{k}^n$. This implies that (a) and (b) hold for any F_q-stable closed subset $\mathbf{X} \subseteq \mathbf{k}^n$ as in Example 1.4.2. The converse requires a bit more work; see [Ge03a, §4.1] or [Sr79, Chap. II] for details. One advantage of this characterisation of Frobenius maps is, for example, that it provides an easy proof of the following statement (see, for example, [Ge03a, Exercise 4.4]):

(c) If F is a Frobenius map (with respect to $\mathbb{F}_q$, as above) and $\gamma\colon \mathbf{X} \to \mathbf{X}$ is an automorphism of affine varieties of finite order which commutes with F, then $\gamma \circ F$ also is a Frobenius map on $\mathbf{X}$ (with respect to $\mathbb{F}_q$, same q).

The above characterisation is equivalent to the definition of an 'abstract affine algebraic $(\mathbb{F}_q, k)$-set' in [Car55-56].

In the sequel, $\mathbf{G}$ will always be a linear algebraic group over $k = \overline{\mathbb{F}}_p$.

1.4.5 Assume that, as an affine variety, $\mathbf{G}$ is defined over $\mathbb{F}_q$ with corresponding Frobenius map F. Then we say that $\mathbf{G}$ (as an algebraic group) is *defined over* $\mathbb{F}_q$ if F is a group homomorphism. In this case, the set of fixed points $\mathbf{G}^F$ is a finite group. There is a more concrete description, similar to Definition 1.4.1. We have the standard Frobenius map

$$F_q\colon \mathrm{GL}_n(k) \to \mathrm{GL}_n(k), \qquad (a_{ij}) \mapsto (a_{ij}^q).$$

Then $\mathbf{G}$ is defined over $\mathbb{F}_q$ if and only if there is a homomorphism

$$\iota\colon \mathbf{G} \to \mathrm{GL}_n(k) \qquad (\text{for some } n \geqslant 1)$$

of algebraic groups such that ι is an isomorphism onto its image and the image is stable under F_q; in this case, the corresponding Frobenius map $F\colon \mathbf{G} \to \mathbf{G}$ is defined by the condition that $\iota \circ F = F_q \circ \iota$. (See [Ge03a, 4.1.11] for further details.) In particular, if $\mathbf{G} \subseteq \mathrm{GL}_n(k)$ is a closed subgroup defined by a collection of polynomials with coefficients in $\mathbb{F}_q$, then F_q restricts to a Frobenius map on $\mathbf{G}$.

Example 1.4.6 Let $\mathbf{T} \subseteq \mathbf{G}$ be an abelian subgroup consisting of semisimple elements (e.g., a torus). We claim that there always exists *some* Frobenius map $F\colon \mathbf{G} \to \mathbf{G}$ (with respect to an $\mathbb{F}_q$-rational structure on $\mathbf{G}$) such that $\mathbf{T}$ is F-stable and $F(t) = t^q$ for all $t \in \mathbf{T}$.

Indeed, we can realise $\mathbf{G}$ as a closed subgroup $\mathbf{G} \subseteq \mathrm{GL}_n(k)$ for some $n \geqslant 1$. Since $\mathbf{T}$ consists of commuting semisimple elements, we can assume that then $\mathbf{T}$ consists of diagonal matrices. Now, the defining ideal of $\mathbf{G}$ (as an algebraic subset of $\mathrm{GL}_n(k)$) is generated by a finite set of polynomials with coefficients in k. So there is some $q = p^m$ ($m \geqslant 1$) such that all these coefficients lie in $\mathbb{F}_q$. Then $\mathbf{G}$ is stable under the standard Frobenius map F_q on $\mathrm{GL}_n(k)$. So F_q restricts to a Frobenius map $F\colon \mathbf{G} \to \mathbf{G}$. Since any $t \in \mathbf{T}$ is a diagonal matrix, we have $F(t) = t^q$.

Definition 1.4.7 Let $F\colon \mathbf{G} \to \mathbf{G}$ be an endomorphism of algebraic groups. Then F is called a *Steinberg map* if some power of F is the Frobenius map with respect to an $\mathbb{F}_q$-rational structure on $\mathbf{G}$, for some power q of p. Note that, in this case, F is a bijective homomorphism of algebraic groups and $\mathbf{G}^F$ is a finite group. If $\mathbf{G}$ is connected and reductive, then $\mathbf{G}^F$ will be called a *finite group of Lie type* or a *finite reductive group*.

Note that, if $F\colon \mathbf{G} \to \mathbf{G}$ is a Steinberg map and $\mathbf{H} \subseteq \mathbf{G}$ is a closed subgroup such that $F(\mathbf{H}) \subseteq \mathbf{H}$, then Remark 1.4.3 implies that $F|_{\mathbf{H}}\colon \mathbf{H} \to \mathbf{H}$ also is a Steinberg map. This will be used frequently without further mention.

The following result is the key tool to pass from properties of $\mathbf{G}$ to properties of the finite group $\mathbf{G}^F$.

Theorem 1.4.8 ([La56], [St68, 10.1]) *Assume that* $\mathbf{G}$ *is connected. Let* $F: \mathbf{G} \to \mathbf{G}$ *be a Steinberg map (or, more generally, any endomorphism such that* $|\mathbf{G}^F| < \infty$*). Then the map* $\mathscr{L}: \mathbf{G} \to \mathbf{G}$, $g \mapsto g^{-1}F(g)$, *is surjective.*

Proof If F is a Steinberg map (and this is the case that we are mainly interested in), then [Mu03] gives a quick proof, as follows. The group $\mathbf{G}$ acts on itself (on the right) where $g \in \mathbf{G}$ sends $x \in \mathbf{G}$ to $g^{-1}xF(g)$. Any action of an algebraic group on an affine variety has a closed orbit; see [Ge03a, 2.5.2]. Let Ω be such a closed orbit and let $x \in \Omega$. Since $\mathbf{G}$ is connected, it will be sufficient to show that $\dim \mathbf{G} = \dim \Omega$ (because then $\mathbf{G} = \Omega$ and so $1 \in \Omega$). We have $\dim \Omega = \dim \mathbf{G} - \dim \mathrm{Stab}_{\mathbf{G}}(x)$ (see [Ge03a, 2.5.3]), so it will be sufficient to show that $\mathrm{Stab}_{\mathbf{G}}(x)$ is finite. Now, an element $g \in \mathbf{G}$ belongs to this stabiliser if and only if $g^{-1}xF(g) = x$, which is equivalent to $f(g) = g$, where $f(g) := xF(g)x^{-1}$. Let $m \geqslant 1$ be such that F^m is a Frobenius map and $F^m(x) = x$ (see Remark 1.4.3). Let $r \geqslant 1$ be the order of the element $xF(x)F^2(x)\ldots F^{m-1}(x) \in \mathbf{G}$. Then $f^{mr}(g) = F^{mr}(g)$ for all $g \in \mathbf{G}$. So $f^{mr}(g) = g$ has only finitely many solutions in $\mathbf{G}$, hence $f(g) = g$ has only finitely many solutions in $\mathbf{G}$. □

For various parts of the subsequent discussion it would be sufficient to work with endomorphisms of $\mathbf{G}$ whose fixed point set is finite. However, we will just formulate everything in terms of Steinberg maps, as defined above. We note that the discussion in [St68, 11.6] in combination with Proposition 1.4.18 below implies that an endomorphism of a *simple* algebraic group with a finite fixed point set is automatically a Steinberg map; see also Example 1.4.20 below.

Here is the prototype of an application of the above theorem.

Proposition 1.4.9 *Assume that* $\mathbf{G}$ *is connected and let* $F: \mathbf{G} \to \mathbf{G}$ *be a Steinberg map. Let* $X \neq \varnothing$ *be a set on which* $\mathbf{G}$ *acts transitively; let* $F': X \to X$ *be a map such that* $F'(g.x) = F(g).F'(x)$ *for* $g \in \mathbf{G}$ *and* $x \in X$.

(a) *There exists some* $x_0 \in X$ *such that* $F'(x_0) = x_0$.
(b) *If* x_0 *is as in (a) and* $\mathrm{Stab}_{\mathbf{G}}(x_0) \subseteq \mathbf{G}$ *is closed and connected, then the set* $\{x \in X \mid F'(x) = x\}$ *is a single* $\mathbf{G}^F$*-orbit.*

Proof (a) Take any $x \in X$. Since $\mathbf{G}$ acts transitively, we have $F'(x) = g^{-1}.x$ for some $g \in \mathbf{G}$. By Theorem 1.4.8, we can write $g = h^{-1}F(h)$. Then one immediately checks that $x_0 := h.x$ is fixed by F'.

(b) Let $\mathbf{H} := \mathrm{Stab}_{\mathbf{G}}(x_0)$. Since $F'(x_0) = x_0$, we have $F(\mathbf{H}) \subseteq \mathbf{H}$ and so F restricts to a Steinberg map on $\mathbf{H}$. Let $x \in X$ be such that $F'(x) = x$ and write $x = g.x_0$ for some $g \in \mathbf{G}$. Then $g.x_0 = x = F'(x) = F'(g.x_0) = F(g).F'(x_0) = F(g).x_0$ and so

$g^{-1}F(g) \in \mathbf{H}$. By Theorem 1.4.8 (applied to $\mathbf{H}$), we can write $g^{-1}F(g) = h^{-1}F(h)$ for some $h \in \mathbf{H}$. Then $gh^{-1} \in \mathbf{G}^F$ and $x = gh^{-1}.x_0$ as desired. □

Example 1.4.10 Assume that $\mathbf{G}$ is connected and let $F\colon \mathbf{G} \to \mathbf{G}$ be a Steinberg map. Let C be a conjugacy class of $\mathbf{G}$ such that $F(C) = C$. Then $\mathbf{G}$ acts transitively on C by conjugation; let F' be the restriction of F to C. Applying Proposition 1.4.9 yields that there exists an element $x \in C$ such that $F(x) = x$. Furthermore, if $C_{\mathbf{G}}(x)$ is connected, then C^F is a single $\mathbf{G}^F$-conjugacy class.

Similarly, if $\mathbf{G}$ is reductive, there exists a pair $(\mathbf{T}, \mathbf{B})$ with $\mathbf{T}$ an F-stable maximal torus of $\mathbf{G}$ and $\mathbf{B}$ an F-stable Borel subgroup with $\mathbf{T} \subseteq \mathbf{B}$. (Just note that, by Theorem 1.3.3, all these pairs are conjugate in $\mathbf{G}$ and, by 1.3.10(a), F preserves the set of all these pairs.)

An F-stable maximal torus of $\mathbf{G}$ which is contained in an F-stable Borel subgroup of $\mathbf{G}$ will be called a *maximally split* torus.

Example 1.4.11 Let $F\colon \mathbf{G} \to \mathbf{G}$ be a Steinberg map. Let $\mathbf{U}, \mathbf{V} \subseteq \mathbf{G}$ be F-stable closed subgroups.

(a) Suppose that $\mathbf{G} = \mathbf{U}.\mathbf{V}$ and $\mathbf{U} \cap \mathbf{V}$ is connected. Then $\mathbf{G}^F = \mathbf{U}^F.\mathbf{V}^F$.

Indeed, let $g \in \mathbf{G}^F$ and write $g = uv$ where $u \in \mathbf{U}$ and $v \in \mathbf{V}$. Then $uv = F(uv) = F(u)F(v)$ and so $x := u^{-1}F(u) = vF(v)^{-1} \in \mathbf{U} \cap \mathbf{V}$. By Theorem 1.4.8 (applied to $\mathbf{U} \cap \mathbf{V}$) we can write $x = y^{-1}F(y)$ for some $y \in \mathbf{U} \cap \mathbf{V}$. It follows that $F(uy^{-1}) = uy^{-1}$ and $F(yv) = yv$. Hence, we have $g = (uy^{-1})(yv) \in \mathbf{U}^F.\mathbf{V}^F$.

(b) Assume that $\mathbf{U}$ is connected and $\mathbf{U} \subseteq \mathbf{V}$. Let $v \in \mathbf{V}$ be such that the coset $\mathbf{U}v$ is F-stable. Then there exists some $u \in \mathbf{U}$ such that $F(uv) = uv$. (Indeed, $\mathbf{U}$ acts transitively on $X := \mathbf{U}v$ by left multiplication; so we can just apply Proposition 1.4.9(a).) Furthermore, if $\mathbf{U}$ is normal in $\mathbf{V}$, then F induces an abstract automorphism of $\mathbf{V}/\mathbf{U}$ (which we denote by the same symbol) and we conclude that $\mathbf{V}^F/\mathbf{U}^F$ is isomorphic to the group of fixed points of F on $\mathbf{V}/\mathbf{U}$.

Proposition 1.4.12 (Cf. [St68, 10.10]) *Let $\mathbf{G}$ be connected reductive and $F\colon \mathbf{G} \to \mathbf{G}$ be a Steinberg map. Then all maximally split tori of $\mathbf{G}$ are $\mathbf{G}^F$-conjugate. More precisely, all pairs $(\mathbf{T}, \mathbf{B})$ consisting of an F-stable Borel subgroup $\mathbf{B}$ and an F-stable maximal torus $\mathbf{T} \subseteq \mathbf{B}$ are conjugate in $\mathbf{G}^F$.*

Proof Let $(\mathbf{T}, \mathbf{B})$ and $(\mathbf{T}_1, \mathbf{B}_1)$ be two pairs as above. By Theorem 1.3.3, there exists some $x \in \mathbf{G}$ such that $x\mathbf{B}x^{-1} = \mathbf{B}_1$ and $x\mathbf{T}x^{-1} = \mathbf{T}_1$. Since $\mathbf{B}, \mathbf{B}_1$ are F-stable, this implies that $x^{-1}F(x) \in N_{\mathbf{G}}(\mathbf{B}) = \mathbf{B}$, where the last equality holds by [Bor91, 11.16] or [Spr98, 6.4.9]. Similarly, since $\mathbf{T}, \mathbf{T}_1$ are F-stable, we have $x^{-1}F(x) \in N_{\mathbf{G}}(\mathbf{T})$. Hence, $x^{-1}F(x) \in \mathbf{B} \cap N_{\mathbf{G}}(\mathbf{T}) = \mathbf{T}$ (see Theorem 1.3.2). Applying Theorem 1.4.8 to the restriction of F to $\mathbf{T}$, we obtain an element $t \in \mathbf{T}$ such that $x^{-1}F(x) = t^{-1}F(t)$. Then $g := xt^{-1} \in \mathbf{G}^F$ and g simultaneously conjugates $\mathbf{B}$ to $\mathbf{B}_1$ and $\mathbf{T}$ to $\mathbf{T}_1$. □

The following result deals with a subtlety concerning Steinberg maps: A surjective homomorphism of algebraic groups will not necessarily induce a surjective map on the level of the fixed point sets under Steinberg maps. But one can say precisely what happens in this situation:

Proposition 1.4.13 (Cf. [St68, 4.5]) *Let $f\colon \mathbf{G} \to \mathbf{G}'$ be a surjective homomorphism of connected algebraic groups such that $\mathbf{K} := \ker(f)$ is contained in the centre of $\mathbf{G}$. Let $F\colon \mathbf{G} \to \mathbf{G}$ and $F'\colon \mathbf{G}' \to \mathbf{G}'$ be Steinberg maps such that $F' \circ f = f \circ F$. We denote $G = \mathbf{G}^F$ and $G' = \mathbf{G}'^{F'}$.*

(a) *Let $\mathscr{L}\colon \mathbf{G} \to \mathbf{G}$, $g \mapsto g^{-1}F(g)$. Then $\mathscr{L}(\mathbf{K})$ is a normal subgroup of $\mathbf{K}$.*
(b) *$f(G) \subseteq G'$ is a normal subgroup and $G'/f(G) \cong \mathbf{K}/\mathscr{L}(\mathbf{K})$. In particular, if $\mathbf{K}$ is connected, then $\mathscr{L}(\mathbf{K}) = \mathbf{K}$ and $f(G) = G'$.*
(c) *If $\mathbf{K}$ is finite (that is, f is an isogeny), then $|G| = |G'|$.*

Proof We apply Lemma 1.1.9 with $A = \mathbf{G}$, $B = \mathbf{G}'$, $\sigma = F$, $\tau = F'$. Then $C = \{g^{-1}F(g) \mid g \in \mathbf{K}\} = \mathscr{L}(\mathbf{K})$; thus, (a) holds. Since $\mathbf{G}$ is connected, we have $\mathscr{L}(\mathbf{G}) = \mathbf{G}$ by Theorem 1.4.8; this yields (b). Furthermore, we have

$$\ker(\mathscr{L}|_{\mathbf{K}}) = \{z \in \mathbf{K} \mid z^{-1}F(z) = 1\} = \mathbf{K}^F = \ker(f|_G).$$

Now assume that $\mathbf{K}$ is finite. Then $|\mathbf{K}/\mathscr{L}(\mathbf{K})| = |\ker(f|_G)|$ and, hence, $|G| = |f(G)||\mathbf{K}/\mathscr{L}(\mathbf{K})|$. But, by (b), $\mathbf{K}/\mathscr{L}(\mathbf{K})$ and $G'/f(G)$ have the same order and so $|G| = |G'|$, that is, (c) holds. □

Lemma 1.4.14 (Cf. [St68, 10.9]) *Let $\mathbf{G}$ be connected and $F\colon \mathbf{G} \to \mathbf{G}$ be a Steinberg map. Let $y \in \mathbf{G}$ and define $F'\colon \mathbf{G} \to \mathbf{G}$ by $F'(g) = yF(g)y^{-1}$ for all $g \in \mathbf{G}$. Then F' is a Steinberg map and we have $\mathbf{G}^{F'} \cong \mathbf{G}^F$.*

Furthermore, if F is a Frobenius map corresponding to an $\mathbb{F}_q$-rational structure, then so is F' (with the same q).

Proof Since $\mathbf{G}$ is connected, Theorem 1.4.8 shows that we can write $y = x^{-1}F(x)$ for some $x \in \mathbf{G}$. Then $F'(g) = x^{-1}F(xgx^{-1})x$ for all $g \in \mathbf{G}$. Thus, we have $F' = \iota_x^{-1} \circ F \circ \iota_x$ where ι_x denotes the inner automorphism of $\mathbf{G}$ defined by x. This formula shows that $F'(g) = g$ if and only if $xgx^{-1} \in \mathbf{G}^F$. Hence, conjugation with x defines a group isomorphism $\mathbf{G}^{F'} \cong \mathbf{G}^F$.

Now we show that F' is a Steinberg map. For $m \geqslant 1$, we have $(F')^m(g) = x^{-1}F^m(xgx^{-1})x$ for all $g \in \mathbf{G}$. By Remark 1.4.3 (and the definition of Steinberg maps), there exists some $m \geqslant 1$ such that $F^m(x) = x$. For this m, we have $(F')^m(g) = F^m(g)$ for all $g \in \mathbf{G}$. Thus, F' is a Steinberg map.

Finally, assume that F is a Frobenius map. We use the characterisation in Remark 1.4.4 to show that F' is a Frobenius map. Since F' is the conjugate of F by an automorphism of $\mathbf{G}$, we have that F'^* is the conjugate of F^* by an algebra automorphism of A. So F'^* is injective and $F'^*(A) = \{a^q \mid a \in A\}$. On the other hand, $(F')^m = F^m$. So, if $a \in A$ and $e \geqslant 1$ are such that $(F^*)^e(a) = a^{q^e}$, then $(F'^*)^{em}(a) = (F^*)^{em}(a) = a^{q^{em}}$, as required. □

Lemma 1.4.15 *Assume that* $\mathbf{G}$ *is connected reductive. Let* $F\colon \mathbf{G} \to \mathbf{G}$ *be a Steinberg map and* $\mathbf{T}$ *be an* F*-stable maximal torus of* $\mathbf{G}$*. Let* $F'\colon \mathbf{G} \to \mathbf{G}$ *be another isogeny of* $\mathbf{G}$ *such that* $F'(\mathbf{T}) = \mathbf{T}$*. If* F *and* F' *induce the same map on* $X(\mathbf{T})$*, then there exists some* $y \in \mathbf{T}$ *such that* $F'(g) = yF(g)y^{-1}$ *for all* $g \in \mathbf{G}$*. In particular, the conclusions of Lemma 1.4.14 apply to* F'*.*

Proof Since F, F' induce the same map on $X(\mathbf{T})$, Theorem 1.3.12 implies that there exists some $y \in \mathbf{T}$ such that $F'(g) = yF(g)y^{-1}$ for all $g \in \mathbf{G}$. □

Example 1.4.16 Assume that $\mathbf{G}$ is connected reductive and let $\mathbf{T} \subseteq \mathbf{G}$ be a maximal torus, with associated root datum $\mathscr{R} = (X, R, Y, R^\vee)$. Let $F_p\colon \mathbf{G} \to \mathbf{G}$ be an isogeny as in Example 1.3.17, such that

$$F_p(\mathbf{U}_\alpha) = \mathbf{U}_\alpha \quad (\alpha \in R) \qquad \text{and} \qquad F_p(t) = t^p \quad (t \in \mathbf{T}).$$

(Note that F_p is only unique up to composition with inner automorphisms defined by elements of $\mathbf{T}$.) Now, there is a stronger version of Theorem 1.3.14 (involving fields of definition), which shows that an isogeny F_p as above is the Frobenius map with respect to an $\mathbb{F}_p$-rational structure on $\mathbf{G}$; see [Spr98, 16.3.3], [DG70/11, Exposé XXV]. (A more direct argument is given by [Klu16, 2.93]; alternatively, one could use Lusztig's approach [Lu09c], as pointed out in [DG70/11, Exposé XXV, footnote 1].) If $\mathbf{G}$ is semisimple, then this is also contained in [St67, Theorem 6 (p. 58)], [Bor70, Part A, §3.3 and §4.3]. Note that, once some F_p as above is known to be a Frobenius map, then Lemma 1.4.15 shows that *any* F_p satisfying the above conditions is a Frobenius map.

We also point out that, in any case, it is easily seen that F_p is a Steinberg map. Indeed, by Example 1.4.6, there exists a Frobenius map $F\colon \mathbf{G} \to \mathbf{G}$ such that $F(t) = t^q$ for all $t \in \mathbf{T}$, where $q = p^m$ for some $m \geqslant 1$. Then F induces multiplication with q on X. Hence, F induces the same map on X as F_p^m. So Lemma 1.4.15 shows that F_p is a Steinberg map.

Lemma 1.4.17 *Let* $\mathbf{T}$ *be a torus over* k *and* $F\colon \mathbf{T} \to \mathbf{T}$ *be the Frobenius map corresponding to an* $\mathbb{F}_q$*-rational structure on* $\mathbf{T}$*, where* q *is a power of* p*. Then the map induced by* F *on* $X = X(\mathbf{T})$ *is given by* $q\psi_0$ *where* $\psi_0\colon X \to X$ *is an invertible endomorphism of finite order.*

Proof (Cf. [DiMi20, Prop. 4.2.3], [Sa71, §I.2.4].) Let A be the algebra of regular functions on $\mathbf{T}$. Let $\lambda \in X$. Composing λ with the inclusion $k^\times \hookrightarrow k$, we can regard λ as a regular function on $\mathbf{T}$, that is, $\lambda \in A$. By Remark 1.4.4, we have

$F^*(A) = \{a^q \mid a \in A\}$. Hence, $\lambda^q = F^*(\lambda^\bullet)$ for some $\lambda^\bullet \in A$. Then

$$\lambda^\bullet(F(t)) = \lambda(t)^q = \lambda(t^q) \qquad \text{for all } t \in \mathbf{T}. \tag{$*$}$$

Since $F\colon \mathbf{T} \to \mathbf{T}$ is a bijective group homomorphism, $\lambda^\bullet$ is uniquely determined by $(*)$; furthermore, $\lambda^\bullet(\mathbf{T}) \subseteq \mathbf{k}^\times$ and $\lambda^\bullet\colon \mathbf{T} \to \mathbf{k}^\times$ is a group homomorphism. Hence, $\lambda^\bullet \in X$. We also see that the map $\psi\colon X \to X$, $\lambda \mapsto \lambda^\bullet$, is linear. Finally, $(*)$ implies that $\big(\psi^m(\lambda)\big)(F^m(t)) = \lambda(t^{q^m})$ for all $m \geqslant 1$. Now, by Example 1.4.6, we can find some $m \geqslant 1$ such that $F^m(t) = t^{q^m}$ for all $t \in \mathbf{T}$. For any such m, we then have $\psi^m(\lambda) = \lambda$ for all $\lambda \in X$. Hence, ψ is an endomorphism of X of order dividing m. Setting $\psi_0 := \psi^{-1}$, the map on X induced by F is given by $q\psi_0$. □

We now obtain the following characterisation of Steinberg maps.

Proposition 1.4.18 *Let $\mathbf{G}$ be connected reductive, $F\colon \mathbf{G} \to \mathbf{G}$ be an isogeny and $\mathbf{T} \subseteq \mathbf{G}$ be an F-stable maximal torus. Then the following are equivalent.*

(i) *F is a Steinberg map.*
(ii) *There exist integers $d, m \geqslant 1$ such that the map induced by F^d on $X = X(\mathbf{T})$ is given by scalar multiplication with p^m.*
(iii) *There is an isogeny F_p as in Example 1.4.16 such that some positive power of F equals some positive power of F_p.*

Proof '(i) ⇒ (ii)' Let $d_1 \geqslant 1$ be such that F^{d_1} is a Frobenius map with respect to some $\mathbb{F}_{q_0}$-rational structure on $\mathbf{G}$ where q_0 is a power of p. Let $\varphi\colon X \to X$ be the map induced by F. By Lemma 1.4.17, we have $\varphi^{d_1} = q_0\psi_0$ where $\psi_0\colon X \to X$ has finite order, $e \geqslant 1$ say. Then $\varphi^{d_1 e} = q_0^e \,\mathrm{id}_X$.

'(ii) ⇒ (iii)' Assume that the map induced by F^d on X is given by scalar multiplication with p^m. Let F_p be as in Example 1.4.16. Then F^d and F_p^m induce the same map on X and so there is some $y \in \mathbf{T}$ such that $F^d(g) = yF_p^m(g)y^{-1}$ for all $g \in \mathbf{G}$; see Lemma 1.4.15. By Theorem 1.4.8, we can write $y = x^{-1}F_p^m(x)$ for some $x \in \mathbf{T}$. As in the proof of Lemma 1.4.14, we have $F^d = \iota_x^{-1} \circ F_p^m \circ \iota_x$. But then we also have $F^d = (\iota_x^{-1} \circ F_p \circ \iota_x)^m$ and it remains to note that $F_p' := \iota_x^{-1} \circ F_p \circ \iota_x$ is a map satisfying the conditions in Example 1.4.16.

'(iii) ⇒ (i)' This is clear by definition, since F_p is known to be a Steinberg map (see Example 1.4.16). □

Proposition 1.4.19 *Assume that $\mathbf{G}$ is connected. Let $F\colon \mathbf{G} \to \mathbf{G}$ be a Steinberg map. Let q be the positive real number defined by $q^d = q_0$, where $d \geqslant 1$ is an integer such that F^d is a Frobenius map with respect to some $\mathbb{F}_{q_0}$-rational structure on $\mathbf{G}$ (where q_0 is a power of p). Then q does not depend on d, q_0. Furthermore, the following hold for every F-stable maximal torus $\mathbf{T}$ of $\mathbf{G}$.*

(a) *We have* $\det(\varphi) = \pm q^{\operatorname{rank} X}$ *where* $\varphi\colon X(\mathbf{T}) \to X(\mathbf{T})$ *is the linear map induced by* F.
(b) *The map induced by* F *on* $X_{\mathbb{R}} := \mathbb{R} \otimes_{\mathbb{Z}} X(\mathbf{T})$ *is of the form* $q\varphi_0$ *where* $\varphi_0 \in \operatorname{GL}(X_{\mathbb{R}})$ *has finite order.*

Proof The independence of q from d, q_0 is clear, once (a) is established. So let $\mathbf{T}$ be any F-stable maximal torus of $\mathbf{G}$ (which exists by Example 1.4.10). Let $X = X(\mathbf{T})$ and $\varphi\colon X \to X$ be the linear map induced by F.

(a) By Remark 1.4.3, the restriction of F^d to $\mathbf{T}$ is a Frobenius map with respect to an $\mathbb{F}_{q_0}$-rational structure on $\mathbf{T}$. So, by Lemma 1.4.17, we have $\varphi^d = q_0\psi_0$ where $\psi_0\colon X \to X$ has finite order. Then $\det(\varphi)^d = q_0^{\operatorname{rank} X}\det(\psi_0)$. Since $\det(\varphi)$ is an integer and $\det(\psi_0)$ is a root of unity, we must have $\det(\varphi)^d = \pm q_0^{\operatorname{rank} X}$ and, hence, $\det(\varphi) = \pm q^{\operatorname{rank} X}$.

(b) Denote by $\varphi_{\mathbb{R}}$ the canonical extension of φ to $X_{\mathbb{R}}$. Then $\varphi_0 := q^{-1}\varphi_{\mathbb{R}}$ is a linear map such that $\varphi_0^d = \psi_0$. Hence, $\varphi_{\mathbb{R}} = q\varphi_0$ where φ_0 has finite order. □

Having defined q, one may also write $\mathbf{G}(q)$ instead of $\mathbf{G}^F$ if there is no danger of confusion. An alternative characterisation of q will be given in Remark 1.6.8(a). The defining formula in Proposition 1.4.19 shows that q is a dth root of a prime power. The examples below will show that all such roots do actually occur.

Example 1.4.20 This example is just meant to give a simple illustration of the difference between Steinberg maps and arbitrary isogenies with a finite fixed point set. Let q, q' be two distinct powers of p. Let $\mathbf{G} = \operatorname{SL}_2(k) \times \operatorname{SL}_2(k)$ and define $F\colon \mathbf{G} \to \mathbf{G}$ by $F(x, y) = (F_q(x), F_{q'}(y))$ where F_q and $F_{q'}$ denote the standard Frobenius maps with respect to q and q', respectively. Then F is a bijective homomorphism of algebraic groups and $\mathbf{G}^F = \operatorname{SL}_2(q) \times \operatorname{SL}_2(q')$ certainly is finite. Let $\mathbf{T} \cong \mathbf{k}^\times$ be the standard maximal torus of $\operatorname{SL}_2(k)$. Then $\mathbf{T}\times\mathbf{T}$ is an F-stable maximal torus of $\mathbf{G}$ and we can identify $X(\mathbf{T}\times\mathbf{T})$ with $\mathbb{Z}^2$. Under this identification, the map induced by F is given by $(n, m) \mapsto (qn, q'm)$ for all $(n, m) \in \mathbb{Z}^2$. Thus, Proposition 1.4.18(ii) shows that F is not a Steinberg map.

Example 1.4.21 Assume that $\mathbf{G}$ is connected reductive. Let $\mathbf{T} \subseteq \mathbf{G}$ be a maximal torus and $\mathcal{R} = (X, R, Y, R^\vee)$ be the root datum relative to $\mathbf{T}$. Assume that we have an automorphism $\varphi_0\colon X \to X$ of finite order such that $\varphi_0(R) = R$ and $\varphi_0^{\mathrm{tr}}(R^\vee) = R^\vee$. (In particular, φ_0 is an isogeny of root data with all root exponents equal to 1.) Let $q = p^m$ for some $m \geqslant 1$. Then $q\varphi_0$ is a p-isogeny and so, by Theorem 1.3.12, there is a corresponding isogeny $F\colon \mathbf{G} \to \mathbf{G}$ such that $F(\mathbf{T}) = \mathbf{T}$. Now F is a Steinberg map by Proposition 1.4.18; the number $q = p^m$ satisfies the conditions in Proposition 1.4.19. If $\mathbf{G}$ is semisimple, then $G = \mathbf{G}^F$ is an untwisted ($\varphi_0 = \mathrm{id}_X$) or twisted Chevalley group; see Steinberg's lecture notes [St67, §11] for further details. We discuss the various possibilities in more detail in Section 1.6.

Let us just give one concrete example. Let $\mathbf{G} = \mathrm{GL}_n(k)$. If $\varphi_0 = \mathrm{id}_X$, then we obtain a 'standard' Frobenius map

$$F\colon \mathrm{GL}_n(k) \to \mathrm{GL}_n(k), \qquad (a_{ij}) \mapsto (a_{ij}^q),$$

such that $\mathrm{GL}_n(k)^F = \mathrm{GL}_n(q)$, the finite general linear group over $\mathbb{F}_q$. On the other hand, the automorphism $\varphi_0\colon X \to X$ of order 2 in Example 1.3.19 also satisfies the above conditions. The corresponding isogeny $F' = F \circ \gamma\colon \mathrm{GL}_n(k) \to \mathrm{GL}_n(k)$ is a Steinberg map such that $\mathrm{GU}_n(q) := \mathrm{GL}_n(k)^{F'}$ is the *finite general unitary group*. Similarly, we have $\mathrm{SL}_n(k)^F = \mathrm{SL}_n(q)$ and $\mathrm{SL}_n(k)^{F'} = \mathrm{SU}_n(q)$.

Example 1.4.22 Assume that $\mathbf{G}$ is connected reductive and that the root datum $\mathscr{R} = (X, R, Y, R^\vee)$ relative to a maximal torus $\mathbf{T} \subseteq \mathbf{G}$ is as in Example 1.2.19, where $p = 2$ or 3. For any $m \geqslant 0$, we have a p-isogeny φ_m on X such that $\varphi_m^2 = p^{2m+1}\,\mathrm{id}_X$. Let $F\colon \mathbf{G} \to \mathbf{G}$ be the corresponding isogeny such that $F(\mathbf{T}) = \mathbf{T}$. Then Proposition 1.4.18 shows that F is a Steinberg map; the number q in Proposition 1.4.19 is given by $q = \sqrt{p}^{2m+1}$. In these cases[2], $G = \mathbf{G}^F$ is the Suzuki group ${}^2B_2(q^2) = {}^2C_2(q^2)$, the Ree group ${}^2G_2(q^2)$ or the Ree group ${}^2F_4(q^2)$, respectively. See [Ca72, Chap. 13] or Steinberg's lecture notes [St67, §11] for further details.

Example 1.4.23 Let $F\colon \mathbf{G} \to \mathbf{G}$ be the Frobenius map corresponding to some $\mathbb{F}_{q_0}$-rational structure on $\mathbf{G}$ where q_0 is a power of p. Consider the direct product $\mathbf{G}' = \mathbf{G} \times \cdots \times \mathbf{G}$ (with r factors) and define a map

$$F'\colon \mathbf{G}' \to \mathbf{G}', \qquad (g_1, g_2, \ldots, g_r) \mapsto (F(g_r), g_1, \ldots, g_{r-1}).$$

Then F' is a homomorphism of algebraic groups and we have

$$(F')^r(g_1, \ldots, g_r) = (F(g_1), F(g_2), \ldots, F(g_r))$$

for all $g_i \in \mathbf{G}$. Clearly, the latter map is a Frobenius map on $\mathbf{G}'$. Thus, F' is a Steinberg map. The number q in Proposition 1.4.19 is given by $q = \sqrt[r]{q_0}$. Note also that we have a group isomorphism

$$\mathbf{G}'^{F'} \xrightarrow{\sim} \mathbf{G}^F, \qquad (g_1, g_2, \ldots, g_r) \mapsto g_1.$$

(This example is mentioned in [DeLu76, §11].)

We have the following extension of the Isogeny Theorem 1.3.12, taking into account the presence of Steinberg maps.

[2] In finite group theory, it is common to write ${}^2B_2(q^2)$ etc., although this is not entirely consistent with the general setting of algebraic groups where the notation should be ${}^2B_2(q)$. See also [DN19] for an interpretation as algebraic groups over $\mathbb{F}_{\sqrt{p}}$.

Lemma 1.4.24 *In the set-up of Theorem 1.3.12 assume, in addition, that there are Steinberg maps $F\colon \mathbf{G} \to \mathbf{G}$ and $F'\colon \mathbf{G}' \to \mathbf{G}'$ such that $F(\mathbf{T}) = \mathbf{T}$, $F'(\mathbf{T}') = \mathbf{T}'$ and $\Phi \circ \varphi = \varphi \circ \Phi'$ where $\Phi\colon X(\mathbf{T}) \to X(\mathbf{T})$ and $\Phi'\colon X(\mathbf{T}') \to X(\mathbf{T}')$ are the maps induced by F and F'. Then there exists an isogeny $f\colon \mathbf{G} \to \mathbf{G}'$ that maps $\mathbf{T}$ onto $\mathbf{T}'$ and induces φ, such that $f \circ F = F' \circ f$.*

Proof By Theorem 1.3.12, there exists an isogeny $f'\colon \mathbf{G} \to \mathbf{G}'$ that maps $\mathbf{T}$ onto $\mathbf{T}'$ and induces φ. Then $f' \circ F$ and $F' \circ f'$ both induce $\Phi \circ \varphi = \varphi \circ \Phi'$. Hence, by Theorem 1.3.12, there exists some $t \in \mathbf{T}$ such that $(F' \circ f')(g) = (f' \circ F)(t^{-1}gt)$ for all $g \in \mathbf{G}$. Then $f'(F(t)) \in \mathbf{T}'$ and so, by Theorem 1.4.8, we can write $f'(F(t)) = x^{-1}F'(x)$ for some $x \in \mathbf{T}'$. We define $f\colon \mathbf{G} \to \mathbf{G}'$ by $f(g) = xf'(g)x^{-1}$ for all $g \in \mathbf{G}$. Then f is an isogeny that maps $\mathbf{T}$ onto $\mathbf{T}'$ and also induces φ. Furthermore,

$$(F' \circ f)(g) = F'(x)(f' \circ F)(t^{-1}gt)F'(x)^{-1} = x(f' \circ F)(g)x^{-1} = (f \circ F)(g)$$

for all $g \in \mathbf{G}$, as required. □

Example 1.4.25 Assume that $\mathbf{G}$ is connected reductive. Let $F\colon \mathbf{G} \to \mathbf{G}$ be a Steinberg map and $\mathbf{T}$ be an F-stable maximal torus of $\mathbf{G}$.

(a) Lemma 1.4.24 immediately shows that an automorphism $\tau\colon \mathbf{G} \to \mathbf{G}$ as in Example 1.3.16 can be chosen such that we also have $\tau \circ F = F \circ \tau$.

(b) Consider an isogeny $F_p\colon \mathbf{G} \to \mathbf{G}$ as in Example 1.4.16 and let φ be the map induced on $X = X(\mathbf{T})$ by F. Since F_p is a Steinberg map, Lemma 1.4.24 shows that there is an isogeny $F'\colon \mathbf{G} \to \mathbf{G}$ which maps $\mathbf{T}$ onto $\mathbf{T}$ and induces φ, and such that $F' \circ F_p = F_p \circ F'$. Since F, F' induce the same map on X, we have that F' is a Steinberg map such that $\mathbf{G}^F \cong \mathbf{G}^{F'}$; see Lemma 1.4.15. (Thus, replacing F by F' if necessary, we can always assume that $F \circ F_p = F_p \circ F$, that is, we are in the setting of [Lu84a, §2.1].)

Lemma 1.4.26 *Assume that $\mathbf{G}$ is connected reductive. Let $\mathbf{K}$ be a closed normal subgroup of $\mathbf{G}$. Then $\mathbf{K}$ is reductive and $\bar{\mathbf{G}} := \mathbf{G}/\mathbf{K}$ is connected and reductive. If, furthermore, $F\colon \mathbf{G} \to \mathbf{G}$ is a Steinberg map such that $F(\mathbf{K}) = \mathbf{K}$, then the map $\bar{F}\colon \bar{\mathbf{G}} \to \bar{\mathbf{G}}$ induced by F is a Steinberg map.*

Proof Since the unipotent radical in a linear algebraic group is invariant under any automorphism of algebraic groups, it is clear that every closed normal subgroup of $\mathbf{G}$ is reductive. Now consider $\bar{\mathbf{G}} = \mathbf{G}/\mathbf{K}$. First recall from 1.1.6 that $\bar{\mathbf{G}}$ is a linear algebraic group; it is also connected since it is the quotient of a connected group. Finally, $\bar{\mathbf{G}}$ is reductive by 1.3.10(b).

Now let $F\colon \mathbf{G} \to \mathbf{G}$ be a Steinberg map and assume that $F(\mathbf{K}) \subseteq \mathbf{K}$. Then we obtain an induced (abstract) group homomorphism $\bar{F}\colon \bar{\mathbf{G}} \to \bar{\mathbf{G}}$, which is bijective. By the universal property of quotients, $\bar{F}$ is a homomorphism of algebraic groups. Let $\mathbf{T} \subseteq \mathbf{G}$ be an F-stable maximal torus. Let $\bar{\mathbf{T}}$ be the image of $\mathbf{T}$ in $\bar{\mathbf{G}}$. By 1.3.10(a),

$\bar{\mathbf{T}}$ is a maximal torus of $\bar{\mathbf{G}}$; we also have $\bar{F}(\bar{\mathbf{T}}) = \bar{\mathbf{T}}$. Let $X = X(\mathbf{T})$, $\bar{X} = X(\bar{\mathbf{T}})$ and $\varphi \colon \bar{X} \to X$ be the map induced by the canonical map $f \colon \mathbf{G} \to \bar{\mathbf{G}}$; note that φ is injective. Let $\psi \colon X \to X$ and $\bar{\psi} \colon \bar{X} \to \bar{X}$ be the maps induced by F and $\bar{F}$. Since $\bar{F} \circ f = f \circ F$, we have $\varphi \circ \bar{\psi} = \psi \circ \varphi$ and so $\varphi \circ \bar{\psi}^m = \psi^m \circ \varphi$ for all $m \geqslant 1$. Now there is some $m \geqslant 1$ such that ψ^m is given by scalar multiplication with a power of p. Since φ is injective, this implies that $\bar{\psi}^m$ is also given by scalar multiplication with a power of p. Hence, $\bar{F}$ is a Steinberg map by Proposition 1.4.18. □

Finally, we address the question of characterising Frobenius maps among all Steinberg maps on $\mathbf{G}$. The results are certainly well known to the experts and are contained in more advanced texts on reductive groups (like [BoTi65], [Sa71]), where they appear as special cases of general considerations of rationality questions. Since in our case the rational structures are given by Frobenius maps, one can give more direct proofs. The key property is contained in the following result.

Lemma 1.4.27 *Let $\mathbf{G}$ be connected reductive and $F \colon \mathbf{G} \to \mathbf{G}$ be a Frobenius map with respect to some $\mathbb{F}_q$-rational structure on $\mathbf{G}$. Let $\mathbf{T}$ be an F-stable maximal torus. Then the root exponents of F (relative to $\mathbf{T}$) are all equal to q.*

Proof (Cf. [BoTi65, 6.2], [Sa71, §II.2.1].) Let R be the set of roots with respect to $\mathbf{T}$. Let $\alpha \in R$ and $u_\alpha \colon \mathbf{k}^+ \to \mathbf{G}$ be the corresponding homomorphism with image $\mathbf{U}_\alpha \subseteq \mathbf{G}$. We have $F(\mathbf{U}_\alpha) = \mathbf{U}_{\alpha^\dagger}$, where $\alpha^\dagger \in R$; see 1.3.11. In order to identify the root exponents, we need to exhibit a homomorphism $u_{\alpha^\dagger} \colon \mathbf{k}^+ \to \mathbf{G}$ whose image is $\mathbf{U}_{\alpha^\dagger}$ and such that $u_{\alpha^\dagger}$ is an isomorphism onto its image. This is done as follows. Let A be the algebra of regular functions on $\mathbf{G}$. The algebra of regular functions on $\mathbf{k}^+$ is given by the polynomial ring $k[c]$ where c is the identity function on $\mathbf{k}^+$. Since u_α is an isomorphism onto its image, the induced algebra homomorphism $u_\alpha^* \colon A \to k[c]$ is surjective (see [Ge03a, 2.2.1]). Now consider the standard Frobenius map $F_1 \colon \mathbf{k}^+ \to \mathbf{k}^+$, $\xi \mapsto \xi^q$. By Remark 1.4.4, we have $F^*(A) = \{a^q \mid a \in A\}$ and $F_1^*(k[c]) = k[c^q]$. Hence, the composition $u_\alpha^* \circ F^*$ sends A onto $k[c^q]$. Since $F_1^* \colon k[c] \to k[c^q]$ is an isomorphism, we obtain an algebra homomorphism $\gamma \colon A \to k[c]$ by setting $\gamma := (F_1^*)^{-1} \circ u_\alpha^* \circ F^*$; note that γ is surjective. Let $u_{\alpha^\dagger} \colon \mathbf{k}^+ \to \mathbf{G}$ be the morphism of affine varieties such that $u_{\alpha^\dagger}^* = \gamma$. Then $F \circ u_\alpha = u_{\alpha^\dagger} \circ F_1$ and so

$$u_{\alpha^\dagger}(\xi^q) = (u_{\alpha^\dagger} \circ F_1)(\xi) = (F \circ u_\alpha)(\xi) = F(u_\alpha(\xi)) \qquad \text{for all } \xi \in k.$$

This shows, first of all, that $u_{\alpha^\dagger}$ is a group homomorphism with image $F(\mathbf{U}_\alpha) = \mathbf{U}_{\alpha^\dagger}$. Furthermore, since γ is surjective, $u_{\alpha^\dagger}$ is an isomorphism onto its image (see again [Ge03a, 2.2.1]). For all $t \in \mathbf{T}$ and $\xi \in k$, we have

$$F(t)u_{\alpha^\dagger}(\xi^q)F(t)^{-1} = F(tu_\alpha(\xi)t^{-1}) = F(u_\alpha(\alpha(t)\xi)) = u_{\alpha^\dagger}(\alpha(t)^q\xi^q),$$

which shows that $\alpha^\dagger(F(t)) = \alpha(t)^q$ for all $t \in \mathbf{T}$, as desired. □

Proposition 1.4.28 *Assume that* $\mathbf{G}$ *is connected reductive. Let* $F\colon \mathbf{G} \to \mathbf{G}$ *be an isogeny and* $\mathbf{T} \subseteq \mathbf{G}$ *be an* F*-stable maximal torus. Let* φ *be the map induced by* F *on* $X = X(\mathbf{T})$*. Then the following conditions are equivalent.*

- (i) F *is a Frobenius map (corresponding to a rational structure on* $\mathbf{G}$ *over a finite subfield of* k*).*
- (ii) *We have* $\varphi = p^m\varphi_0$ *where* $m \in \mathbb{Z}_{\geqslant 1}$ *and* $\varphi_0\colon X \to X$ *is an automorphism of finite order such that* $\varphi_0(R) = R$ *and* $\varphi_0^{\mathrm{tr}}(R^\vee) = R^\vee$*. (In particular,* φ_0 *is an isogeny of root data with all root exponents equal to* 1*.)*
- (iii) *There exists an automorphism of algebraic groups* $\gamma\colon \mathbf{G} \to \mathbf{G}$ *of finite order such that* $\gamma(\mathbf{T}) = \mathbf{T}$ *and some* $m' \geqslant 1$ *such that* $F = \gamma \circ F_p^{m'} = F_p^{m'} \circ \gamma$*, where* F_p *is an isogeny as in Example 1.4.16.*

If these conditions hold, then m *(as in (ii)) equals* m' *(as in (iii)) and* F *is the Frobenius map with respect to an* $\mathbb{F}_q$*-rational structure where* $q = p^m$*. Furthermore, all root exponents of* F *are equal to* q *and* q *is the number defined in Proposition 1.4.19.*

Proof '(i) $\Rightarrow$ (ii)' Let F be a Frobenius map corresponding to an $\mathbb{F}_q$-rational structure on $\mathbf{G}$ where $q = p^m$ for some $m \geqslant 1$. By assumption, $\mathbf{T}$ is F-stable so $\mathbf{T}$ is also defined over $\mathbb{F}_q$; see Remark 1.4.3. Hence, we can apply Lemma 1.4.17 and so $\varphi = q\varphi_0$ where $\varphi_0\colon X \to X$ has finite order. Furthermore, using Lemma 1.4.27, one sees that $\varphi_0(\alpha^\dagger) = \alpha$ and $\varphi_0^{\mathrm{tr}}(\alpha^\vee) = (\alpha^\dagger)^\vee$ for all $\alpha \in R$. Thus, (I1) and (I2) hold for φ_0, where the root exponents of φ_0 are all equal to 1.

'(ii) $\Rightarrow$ (iii)' Let $F_p\colon \mathbf{G} \to \mathbf{G}$ be as in Example 1.4.16. Then F_p^m is a Steinberg map and the map induced by F_p^m on X is scalar multiplication with p^m. So, by Lemma 1.4.24, there exists an isogeny $f\colon \mathbf{G} \to \mathbf{G}$ which maps $\mathbf{T}$ onto itself and induces φ_0, and such that $f \circ F_p^m = F_p^m \circ f$. Now φ_0 has finite order, say d. Then f^d induces the identity on X. Hence, by Theorem 1.3.12, there exists some $t \in \mathbf{T}$ such that $f^d(g) = tgt^{-1}$ for all $g \in \mathbf{G}$. Since t also has finite order, we conclude that some positive power of f^d is the identity. Hence, f itself has finite order. Now, F and $F' := f \circ F_p^m$ are isogenies which induce the same map on X. Hence, by Theorem 1.3.12, there exists some $y \in \mathbf{T}$ such that $F'(g) = yF(g)y^{-1}$ for all $g \in \mathbf{G}$. As in the proof of Lemma 1.4.14, there exists some $x \in \mathbf{T}$ such that $F' = \iota_x^{-1} \circ F \circ \iota_x$, where ι_x denotes the inner automorphism of $\mathbf{G}$ defined by x. Then

$$F = \iota_x \circ F' \circ \iota_x^{-1} = (\iota_x \circ f \circ \iota_x^{-1}) \circ (\iota_x \circ F_p \circ \iota_x^{-1})^m$$

(and the two factors still commute). Now, since $x \in \mathbf{T}$, the isogeny $F_p' := \iota_x \circ F_p \circ \iota_x^{-1}$ also satisfies the conditions in Example 1.4.16. Furthermore, $\gamma := \iota_x \circ f \circ \iota_x^{-1}$ is an automorphism of finite order such that $\gamma(\mathbf{T}) = \mathbf{T}$. Thus, (iii) holds.

'(iii) ⇒ (i)' As discussed in Example 1.4.16, F_p is the Frobenius map corresponding to an $\mathbb{F}_p$-rational structure on $\mathbf{G}$. Then F_p^m is the Frobenius map corresponding to an $\mathbb{F}_q$-rational structure on $\mathbf{G}$ where $q = p^m$. Hence so is $F = \gamma \circ F_p^m$ by Remark 1.4.4(c).

Finally, assume that (i), (ii), (iii) hold. Then the above arguments show that $m = m'$. Furthermore, (ii) shows that $\det(\varphi) = \pm(p^m)^{\text{rank}X}$ and so $q = p^m$ satisfies the conditions in Proposition 1.4.19. □

The following example indicates that Steinberg maps can be much more complicated than Frobenius maps.

Example 1.4.29 (a) In the setting of Example 1.4.23, one easily sees that neither the conclusion of Lemma 1.4.17 nor that of Lemma 1.4.27 hold for F'. Hence, although F is a Frobenius map, the map F' is not.

(b) Let $\mathbf{G} = \mathrm{SL}_2(k) \times \mathrm{PGL}_2(k)$ and $p = 2$. Then $\mathbf{G}$ is semisimple of type $A_1 \times A_1$, with Cartan matrix $C = 2I_2$ where I_2 denotes the identity matrix. The root datum of $\mathbf{G}$ is determined by the factorisation

$$C = \breve{A} \cdot A^{\text{tr}} \quad \text{where} \quad A = \begin{pmatrix} 2 & 0 \\ 0 & 1 \end{pmatrix} \quad \text{and} \quad \breve{A} = \begin{pmatrix} 1 & 0 \\ 0 & 2 \end{pmatrix}.$$

For a fixed $m \geqslant 1$, we define

$$P = \begin{pmatrix} 0 & 2^m \\ 2^m & 0 \end{pmatrix} \quad \text{and} \quad P^\circ = \begin{pmatrix} 0 & 2^{m-1} \\ 2^{m+1} & 0 \end{pmatrix}.$$

Then the pair (P, P°) satisfies the conditions in 1.2.18 and so there is a corresponding isogeny $F\colon \mathbf{G} \to \mathbf{G}$, with root exponents $2^{m+1}, 2^{m-1}$. Since $P^2 = 4^m I_2$, we know that F is a Steinberg map (see Proposition 1.4.18). Furthermore, we have $P = 2^m P_0$ where $P_0 \in M_2(\mathbb{Z})$ has order 2; thus, the conclusion of Lemma 1.4.17 holds for F where $q = 2^m$. Note also that the two projections (on the first and on the second factor) define isomorphisms of finite groups $\mathbf{G}^F \cong \mathrm{SL}_2(q) \cong \mathrm{PGL}_2(q)$.

On the other hand, since not all root exponents are equal, Lemma 1.4.27 shows that F is not a Frobenius map! One easily checks directly that there is no matrix P_0° such that the pair (P_0, P_0°) satisfies the conditions in 1.2.18. Thus, P_0 does not come from an isogeny of root data.

1.5 Working with Isogenies and Root Data; Examples

We now discuss some applications and examples of the theory developed so far. We start with some basic material about semisimple groups. Up until Proposition 1.5.10, k may be any algebraically closed field.

1.5.1 Let us fix a Cartan matrix $C = (c_{st})_{s,t\in S}$. Let $\Lambda(C)$ be the finite abelian group defined in Remark 1.2.8. Then the semisimple algebraic groups with a root datum of Cartan type C are classified in terms of subgroups of $\Lambda(C)$. This works as follows. We have $\Lambda(C) := \Omega/\mathbb{Z}C$ where Ω is the free abelian group with basis $\{\omega_s \mid s \in S\}$ and $\mathbb{Z}C$ is the subgroup generated by $\{\sum_{s\in S} c_{st}\omega_s \mid t \in S\}$. Thus, giving a subgroup of $\Lambda(C)$ is the same as giving a lattice L such that $\mathbb{Z}C \subseteq L \subseteq \Omega$. Such a lattice L is free abelian of the same rank as Ω. We choose a set of free generators $\{x_s \mid s \in S\}$ of L. Since $\mathbb{Z}C \subseteq L$, we have unique expressions

$$\sum_{s\in S} c_{st}\omega_s = \sum_{u\in S} a_{tu}x_u \quad (t \in S), \qquad \text{where} \qquad A = (a_{tu})_{t,u\in S} \qquad (*)$$

and A is a square matrix with integer coefficients. We also write $x_u = \sum_{s\in S} \breve{a}_{su}\omega_s$ where $\breve{A} = (\breve{a}_{su})_{s,u\in S}$ is a square matrix with integer coefficients. Substituting this into $(*)$ and comparing coefficients, we obtain $C = \breve{A} \cdot A^{\text{tr}}$. As in 1.2.14, such a factorisation of C determines a root datum $\mathscr{R}_L = (X, R, Y, R^\vee)$ of Cartan type C, where R has a base given by $\alpha_t := \sum_{s\in S} a_{ts}x_s$ for $t \in S$. We have $|X/\mathbb{Z}R| < \infty$ since $A, \breve{A}$ are square matrices. If we choose a different set of generators of L, say $\{y_t \mid t \in S\}$, then we obtain another factorisation $C = \breve{B} \cdot B^{\text{tr}}$ where $B, \breve{B}$ are square integer matrices. Writing $y_t = \sum_{u\in S} p_{ut}x_u$ where $P = (p_{ut})_{u,t\in S}$ is invertible over $\mathbb{Z}$, we have $P \cdot B^{\text{tr}} = A^{\text{tr}}$ and $\breve{B} = \breve{A} \cdot P$. Hence, the root data defined by $C = \breve{A} \cdot A^{\text{tr}}$ and by $C = \breve{B} \cdot B^{\text{tr}}$ are isomorphic; see Remark 1.2.20. Thus, every lattice L such that $\mathbb{Z}C \subseteq L \subseteq \Omega$ determines a root datum $\mathscr{R}_L$ as above, which is unique up to isomorphism. By Theorem 1.3.14, there exists a corresponding connected reductive algebraic group $\mathbf{G}_L$ over k (unique up to isomorphism by Corollary 1.3.13). The group $\mathbf{G}_L$ is semisimple since $|X/\mathbb{Z}R| < \infty$; see Remark 1.3.5.

Proposition 1.5.2 *Let $\mathbf{G}$ be a semisimple algebraic group over k with root datum $\mathscr{R} = (X, R, Y, R^\vee)$ (relative to some maximal torus of $\mathbf{G}$). Let $C = (c_{st})_{s,t\in S}$ be the Cartan matrix of $\mathscr{R}$. Then there exists a lattice L as in 1.5.1 such that $\mathbf{G} \cong \mathbf{G}_L$. We have $X/\mathbb{Z}R \cong L/\mathbb{Z}C$ and, hence, $\mathbf{Z}(\mathbf{G}) \cong \mathrm{Hom}(L/\mathbb{Z}C, k^\times)$.*

Proof Let Π be a base of R; we have $|\Pi| = \operatorname{rank} X$ since $\mathbf{G}$ is semisimple and, hence, $X/\mathbb{Z}R$ is finite. Also choose a $\mathbb{Z}$-basis of X. By Remark 1.2.13, we have a corresponding factorisation $C = \breve{A} \cdot A^{\text{tr}}$ where A, $\breve{A}$ are *square* integral matrices. In particular, we can use S as an indexing set for both the rows and the columns of these matrices. Then let L be the sublattice of Ω spanned by the elements

$$x_t := \sum_{s\in S} \breve{a}_{st}\omega_s \quad (t \in S), \qquad \text{where} \qquad \breve{A} = (\breve{a}_{st})_{s,t\in S}.$$

We have $\mathbb{Z}C \subseteq L \subseteq \Omega$, since $C = \breve{A} \cdot A^{\text{tr}}$. Applying the construction in 1.5.1 to L, we obtain a group $\mathbf{G}_L$. Then Corollary 1.3.13 shows that $\mathbf{G} \cong \mathbf{G}_L$. Finally, 1.5.1$(*)$ implies that $X/\mathbb{Z}R \cong L/\mathbb{Z}C$ and this yields $\mathbf{Z}(\mathbf{G})$; see Remark 1.3.5. □

Example 1.5.3 Let $C = (c_{st})_{s,t \in S}$ be a Cartan matrix and consider the group $\Lambda(C) = \Omega/\mathbb{Z}C$, as above.

(a) If $L = \mathbb{Z}C$, then we choose the generators $\{x_s \mid s \in S\}$ of L to be the given generators of $\mathbb{Z}C$. So A in 1.5.1(∗) is the identity matrix and $\breve{A} = C$. Thus, $\mathscr{R}_L$ is the root datum $\mathscr{R}_{\mathrm{ad}}(C)$ in Example 1.2.16. The corresponding group $\mathbf{G}_L$ will be denoted by $\mathbf{G}_{\mathrm{ad}}$; we have $\mathbf{Z}(\mathbf{G}_{\mathrm{ad}}) = \{1\}$.

(b) If $L = \Omega$, then we can take $x_s = \omega_s$ for all $s \in S$. So $A = C^{\mathrm{tr}}$ and $\breve{A}$ is the identity matrix. Hence, in this case, $\mathscr{R}_L$ is the root datum $\mathscr{R}_{\mathrm{sc}}(C)$ in Example 1.2.16. The corresponding group $\mathbf{G}_L$ will be denoted by $\mathbf{G}_{\mathrm{sc}}$; we have $\mathbf{Z}(\mathbf{G}_{\mathrm{sc}}) \cong \mathrm{Hom}(\Lambda(C), k^\times)$.

The groups $\mathbf{G}_{\mathrm{sc}}$ and $\mathbf{G}_{\mathrm{ad}}$ have some important universal properties that will be discussed in further detail below. We shall call $\mathbf{G}_{\mathrm{sc}}$ the *semisimple group of simply connected type* C and $\mathbf{G}_{\mathrm{ad}}$ the *semisimple group of adjoint type* C.

Example 1.5.4 Assume that C is an indecomposable Cartan matrix. The isomorphism types of $\Lambda(C)$ are listed in Remark 1.2.8.

(a) If C is of type A_{n-1}, then $\Lambda(C) \cong \mathbb{Z}/n\mathbb{Z}$. Hence, for each divisor d of n, we have a unique lattice $L_d \subseteq \Omega$ such that $|L_d/\mathbb{Z}C| = d$; let $\mathbf{G}_{(d)}$ be the corresponding group. We have $\mathbf{G}_{(1)} = \mathbf{G}_{\mathrm{ad}} \cong \mathrm{PGL}_n(k)$ and $\mathbf{G}_{(n)} = \mathbf{G}_{\mathrm{sc}} \cong \mathrm{SL}_n(k)$; see Examples 1.3.9 and 1.3.8. The remaining groups $\mathbf{G}_{(d)}$ are explicitly constructed in [Ch05, §20.3], as images of $\mathrm{SL}_n(k)$ under certain representations.

(b) If C is of type B_n, C_n, E_6 or E_7, then $\Lambda(C)$ is cyclic of prime order. Hence, either $L = \mathbb{Z}C$ or $L = \Omega$. In this case, the only possible groups are $\mathbf{G}_{\mathrm{ad}}$ and $\mathbf{G}_{\mathrm{sc}}$.
In type B_n, we have $\mathbf{G}_{\mathrm{ad}} \cong \mathrm{SO}_{2n+1}(k)$ and $\mathbf{G}_{\mathrm{sc}} \cong \mathrm{Spin}_{2n+1}(k)$.
In type C_n, we have $\mathbf{G}_{\mathrm{ad}} \cong \mathrm{PCSp}_{2n}(k)$ and $\mathbf{G}_{\mathrm{sc}} \cong \mathrm{Sp}_{2n}(k)$.
(See the references in 1.1.4 for the precise definitions of these groups.)

(c) If C is of type D_n, then $\Lambda(C)$ has order 4 and there are 3 (for n odd) or 5 (for n even) possible lattices L. We have $\mathbf{G}_{\mathrm{ad}} \cong \mathrm{PCO}_{2n}^\circ(k)$ and $\mathbf{G}_{\mathrm{sc}} \cong \mathrm{Spin}_{2n}(k)$. Using the labelling in Table 1.1, the group $\mathrm{SO}_{2n}(k)$ corresponds to the unique L of index 2 in Ω that is invariant under the involution of Ω obtained by exchanging ω_1 and ω_2. In terms of our matrix language in Section 1.2, the root datum of $\mathrm{SO}_{2n}(k)$ is given by the factorisation $C = \breve{A} \cdot A^{\mathrm{tr}}$ where

$$A = \breve{A} = \begin{pmatrix} 1 & 1 & 0 & \dots & & 0 \\ -1 & 1 & 0 & 0 & \dots & 0 \\ 0 & -1 & 1 & 0 & \ddots & \vdots \\ \vdots & \ddots & \ddots & \ddots & \ddots & 0 \\ 0 & \dots & 0 & -1 & 1 & 0 \\ 0 & & \dots & 0 & -1 & 1 \end{pmatrix}.$$

(See [Spr98, Exercise 7.4.7].) If n is even, then there are two further lattices of index 2, which both give rise to the half-spin group $\mathrm{HSpin}_{2n}(k)$.

(d) Finally, if C is of type G_2, F_4 or E_8, then $\Lambda(C) = \{0\}$. So, in this case, all semisimple algebraic groups over a fixed field k with a root datum of Cartan type C are isomorphic to each other; in particular, $\mathbf{G}_{\mathrm{sc}} \cong \mathbf{G}_{\mathrm{ad}}$.

We refer to [Gro02], [Ge03a, §1.7], [Spr98, §7.4], for further details about the various groups of classical type B_n, C_n and D_n.

Example 1.5.5 The Cartan matrices of type B_n and C_n are the $n \times n$-matrices given by

$$B_n : \begin{pmatrix} 2 & -2 & 0 & \dots & & 0 \\ -1 & 2 & -1 & 0 & \dots & 0 \\ 0 & -1 & 2 & -1 & \ddots & \vdots \\ \vdots & \ddots & \ddots & \ddots & \ddots & 0 \\ 0 & \dots & 0 & -1 & 2 & -1 \\ 0 & & \dots & 0 & -1 & 2 \end{pmatrix}, \qquad C_n : \begin{pmatrix} 2 & -1 & 0 & \dots & & 0 \\ -2 & 2 & -1 & 0 & \dots & 0 \\ 0 & -1 & 2 & -1 & \ddots & \vdots \\ \vdots & \ddots & \ddots & \ddots & \ddots & 0 \\ 0 & \dots & 0 & -1 & 2 & -1 \\ 0 & & \dots & 0 & -1 & 2 \end{pmatrix},$$

respectively. Let C denote the second matrix, and C' the first. Let P° be the diagonal matrix of size n with diagonal entries $1, 2, 2, \dots, 2$. Then $CP^\circ = P^\circ C'$. Thus, if we also set $P = P^\circ$, then the two conditions in 1.2.18 are satisfied and so the pair (P, P°) defines a 2-isogeny from $\mathscr{R}_{\mathrm{sc}}(C')$ to $\mathscr{R}_{\mathrm{sc}}(C)$ (see Example 1.2.16). Let k be an algebraically closed field of characteristic 2. Let $\mathbf{G}_{\mathrm{sc}}$ and $\mathbf{G}'_{\mathrm{sc}}$ be the semisimple algebraic groups over k corresponding to $\mathscr{R}_{\mathrm{sc}}(C)$ and $\mathscr{R}_{\mathrm{sc}}(C')$, respectively. We have $\mathbf{G}_{\mathrm{sc}} \cong \mathrm{Sp}_{2n}(k)$ and $\mathbf{G}'_{\mathrm{sc}} \cong \mathrm{Spin}_{2n+1}(k)$. Then Theorem 1.3.12 yields the existence of an isogeny $f \colon \mathbf{G}'_{\mathrm{sc}} \to \mathbf{G}_{\mathrm{sc}}$. This is one of Chevalley's exceptional isogenies considered at the end of [Ch05, §23.7].

Example 1.5.6 Let $\mathbf{G}$ be connected reductive over k. Let $\mathscr{R} = (X, R, Y, R^\vee)$ be the root datum relative to a maximal torus $\mathbf{T}$ of $\mathbf{G}$. Dual to the isomorphism in 1.1.11(c) we have a canonical isomorphism of abelian groups (see [Ca85, §3.1]):

$$k^\times \otimes_{\mathbb{Z}} Y(\mathbf{T}) \xrightarrow{\sim} \mathbf{T}, \qquad \xi \otimes \nu \mapsto \nu(\xi).$$

Hence, if $\{\nu_1, \dots, \nu_n\}$ is a $\mathbb{Z}$-basis of $Y(\mathbf{T})$, then every element $t \in \mathbf{T}$ can be written uniquely in the form $t = \nu_1(\xi_1) \cdots \nu_n(\xi_n)$ where $\xi_1, \dots, \xi_n \in k^\times$.

Now assume that $\mathbf{G}$ is semisimple of simply connected type. Then $Y(\mathbf{T}) = \mathbb{Z}R^\vee$. Let $\Pi = \{\alpha_1, \dots, \alpha_n\}$ be a base for R and $\{\alpha_1^\vee, \dots, \alpha_n^\vee\}$ be the corresponding base for $R^\vee$. Hence, we have

$$\mathbf{T} = \{h(\xi_1, \dots, \xi_n) := \alpha_1^\vee(\xi_1) \cdots \alpha_n^\vee(\xi_n) \mid \xi_1, \dots, \xi_n \in k^\times\}.$$

In this setting, one can now explicitly determine the centre of $\mathbf{G}$ as a subset of $\mathbf{T}$. Indeed, using Remark 1.3.5 and the above description of $\mathbf{T}$, we obtain

$$\mathbf{Z}(\mathbf{G}) = \{h(\xi_1, \ldots, \xi_n) \in \mathbf{T} \mid \xi_1^{\langle \alpha_j, \alpha_1^\vee \rangle} \cdots \xi_n^{\langle \alpha_j, \alpha_n^\vee \rangle} = 1 \text{ for } 1 \leqslant j \leqslant n\}.$$

Now the numbers $c_{ij} = \langle \alpha_j, \alpha_i^\vee \rangle$ $(1 \leqslant i, j \leqslant n)$ are just the entries of the Cartan matrix of $\mathscr{R}$, so this yields an explicit system of n equations which we need to solve for $\xi_1, \ldots, \xi_n$. Let us describe this explicitly in all cases, where we refer to the labelling of the simple roots in Table 1.1; in each case, we also describe a subtorus $\mathbf{S} \subseteq \mathbf{T}$ such that $\mathbf{Z}(\mathbf{G}) \subseteq \mathbf{S}$.

A_n: $\mathbf{G} \cong \mathrm{SL}_{n+1}(k)$ and $\mathbf{Z}(\mathbf{G}) = \{h(\xi, \xi^2, \xi^3, \ldots, \xi^n) \mid \xi^{n+1} = 1\}$; this is contained in the subtorus $\mathbf{S} := \{h(\xi, \xi^2, \xi^3, \ldots, \xi^n) \mid \xi \in k^\times\} \subseteq \mathbf{T}$.

B_n: $\mathbf{G} \cong \mathrm{Spin}_{2n+1}(k)$. If $n \geqslant 2$ is even, then

$$\mathbf{Z}(\mathbf{G}) = \{h(1, \xi, 1, \xi, 1, \xi, 1, \ldots) \mid \xi^2 = 1\},$$

and we may take $\mathbf{S} := \{h(1, \xi, 1, \xi, 1, \xi, 1, \ldots) \mid \xi \in k^\times\}$. For $n \geqslant 3$ odd, we have $\mathbf{Z}(\mathbf{G}) = \{h(\xi, 1, \xi, 1, \xi, 1, \xi, \ldots) \mid \xi^2 = 1\}$, and this is contained in the subtorus $\mathbf{S} := \{h(\xi, 1, \xi, 1, \xi, 1, \xi, \ldots) \mid \xi \in k^\times\}$.

C_n: $\mathbf{G} \cong \mathrm{Sp}_{2n}(k)$ and $\mathbf{Z}(\mathbf{G}) = \{h(\xi, 1, 1, 1, \ldots) \mid \xi^2 = 1\}$; this is contained in the subtorus $\mathbf{S} := \{h(\xi, 1, 1, 1, \ldots) \mid \xi \in k^\times\}$.

D_n: $\mathbf{G} \cong \mathrm{Spin}_{2n}(k)$. If $n \geqslant 4$ is even, then

$$\mathbf{Z}(\mathbf{G}) = \{h(\xi_1, \xi_2, 1, \xi_1\xi_2, 1, \xi_1\xi_2, 1, \xi_1\xi_2, \ldots) \mid \xi_1^2 = \xi_2^2 = 1\};$$

this is contained in $\mathbf{S} := \{h(\xi_1, \xi_2, 1, (\xi_1\xi_2)^{-1}, 1, \xi_1\xi_2, \ldots) \mid \xi_1, \xi_2 \in k^\times\}$ (where we follow [Mas10, Example 4.2]). If $n \geqslant 3$ is odd, then

$$\mathbf{Z}(\mathbf{G}) = \{h(\xi, \xi^{-1}, \xi^2, 1, \xi^2, 1, \xi^2, 1, \xi^2, \ldots) \mid \xi^4 = 1\};$$

this is contained in $\mathbf{S} := \{h(\xi, \xi^{-1}, \xi^2, 1, \xi^2, 1, \xi^2, \ldots) \mid \xi \in k^\times\}$.

G_2: Since $\det(c_{ij}) = 1$, we have $\mathbf{Z}(\mathbf{G}) = \{1\}$; we may take $\mathbf{S} := \{1\}$.

F_4: Since $\det(c_{ij}) = 1$, we have $\mathbf{Z}(\mathbf{G}) = \{1\}$; we may take $\mathbf{S} := \{1\}$.

E_6: We have $\mathbf{Z}(\mathbf{G}) = \{h(\xi, 1, \xi^{-1}, 1, \xi, \xi^{-1}) \mid \xi^3 = 1\}$; this is contained in the subtorus $\mathbf{S} := \{h(\xi, 1, \xi^{-1}, 1, \xi, \xi^{-1}) \mid \xi \in k^\times\}$.

E_7: We have $\mathbf{Z}(\mathbf{G}) = \{h(1, \xi, 1, 1, \xi, 1, \xi) \mid \xi^2 = 1\}$; this is contained in the subtorus $\mathbf{S} := \{h(1, \xi, 1, 1, \xi, 1, \xi) \mid \xi \in k^\times\}$.

E_8: Since $\det(c_{ij}) = 1$, we have $\mathbf{Z}(\mathbf{G}) = \{1\}$; we may take $\mathbf{S} := \{1\}$.

We have $\dim \mathbf{S} \leqslant 1$ except for type D_n with $n \geqslant 4$ even, in which case $\dim \mathbf{S} = 2$. In order to obtain these descriptions, we did not have to rely on explicit realisations of groups of classical type as matrix groups: the abstract setting in terms of root data has actually been more efficient in this context.

For the construction of isogenies between groups of the same Cartan type, the following remarks will be useful.

1.5.7 Let $\mathbf{G}_1$, $\mathbf{G}_2$ be connected reductive algebraic groups over k. For $i = 1, 2$, let $\mathscr{R}_i = (X_i, R_i, Y_i, R_i^\vee)$ be the corresponding root datum, relative to a maximal torus $\mathbf{T}_i \subseteq \mathbf{G}_i$. Furthermore, let us choose Borel subgroups $\mathbf{B}_i \subseteq \mathbf{G}_i$ such that $\mathbf{T}_i \subseteq \mathbf{B}_i$. By Remark 1.3.4, this is equivalent to choosing bases $\Pi_i \subseteq R_i$. We assume that X_1, X_2 have the same rank and that R_1, R_2 have the same Cartan matrix $C = (c_{st})_{s,t\in S}$. If we also choose $\mathbb{Z}$-bases of X_1 and X_2, then $\mathscr{R}_1$ and $\mathscr{R}_2$ are determined by factorisations as in Remark 1.2.13:

$$\mathscr{R}_1: \quad C = \breve{A}_1 \cdot A_1^{\text{tr}} \qquad \text{and} \qquad \mathscr{R}_2: \quad C = \breve{A}_2 \cdot A_2^{\text{tr}},$$

where A_1, A_2, $\breve{A}_1$, $\breve{A}_2$ are integer matrices, all of the same size. Now, this setting gives rise to a correspondence:

$$\left\{ \begin{array}{c} \text{Isogenies } f\colon \mathbf{G}_1 \to \mathbf{G}_2 \text{ with} \\ f(\mathbf{T}_1) = \mathbf{T}_2,\ f(\mathbf{B}_1) = \mathbf{B}_2 \end{array} \right\} \leftrightarrow \left\{ \begin{array}{c} \text{Pairs of integer matrices } (P, P^\circ) \\ \text{satisfying (MI1), (MI2) in 1.2.18} \end{array} \right\}$$

(Note that, here, the relations in (MI2) read $P \cdot A_2^{\text{tr}} = A_1^{\text{tr}} \cdot P^\circ$ and $P^\circ \cdot \breve{A}_2 = \breve{A}_1 \cdot P$.) Indeed, by 1.3.11, each isogeny of groups on the left determines a p-isogeny of root data; since $f(\mathbf{B}_1) = \mathbf{B}_2$, this p-isogeny is 'base preserving' as in 1.2.18 and, hence, it determines a unique pair of matrices on the right. Conversely, a pair of matrices on the right determines a p-isogeny of root data by 1.2.18; by the Isogeny Theorem 1.3.12, there is a corresponding isogeny of groups on the left, which is unique up to inner automorphisms given by elements of $\mathbf{T}_1$.

Proposition 1.5.8 *Let* $\mathbf{G}$ *be semisimple and assume that the root datum of* $\mathbf{G}$ *(relative to some maximal torus* $\mathbf{T}$ *and some Borel subgroup* $\mathbf{B}$ *containing* $\mathbf{T}$*) is of Cartan type C. Let* $\mathbf{G}_{\text{sc}}$ *and* $\mathbf{G}_{\text{ad}}$ *be of Cartan type C, as in Example 1.5.3, relative to* $\tilde{\mathbf{T}} \subseteq \tilde{\mathbf{B}} \subseteq \mathbf{G}_{\text{sc}}$ *and* $\mathbf{T}' \subseteq \mathbf{B}' \subseteq \mathbf{G}_{\text{ad}}$. *Then there exist central isogenies*

$$\tilde{f}\colon \mathbf{G}_{\text{sc}} \longrightarrow \mathbf{G} \qquad \textit{and} \qquad f'\colon \mathbf{G} \longrightarrow \mathbf{G}_{\text{ad}},$$

such that $\tilde{f}(\tilde{\mathbf{T}}) = \mathbf{T}$, $\tilde{f}(\tilde{\mathbf{B}}) = \mathbf{B}$, $f'(\mathbf{T}) = \mathbf{T}'$, $f'(\mathbf{B}) = \mathbf{B}'$.

Proof First consider $\mathbf{G}_{\text{sc}}$. We place ourselves in the setting of 1.5.7, where

$$(\mathbf{G}_1, \mathbf{T}_1, \mathbf{B}_1) = (\mathbf{G}_{\text{sc}}, \tilde{\mathbf{T}}, \tilde{\mathbf{B}}) \qquad \text{and} \qquad (\mathbf{G}_2, \mathbf{T}_2, \mathbf{B}_2) = (\mathbf{G}, \mathbf{T}, \mathbf{B}).$$

The root datum of $\mathbf{G}_1$ is given by a factorisation of C as above where $A_1 = C^{\text{tr}}$ and $\breve{A}_1 = I$ (identity matrix). The only extra information about the root datum of $\mathbf{G}_2$ is that, in the factorisation $C = \breve{A}_2 \cdot A_2^{\text{tr}}$, both $A_2, \breve{A}_2$ are square matrices (since $\mathbf{G}$ is semisimple). In order to find $\tilde{f}\colon \mathbf{G}_1 \to \mathbf{G}_2$, we need to specify a pair of (square) integral matrices $(\tilde{P}, \tilde{P}^\circ)$ where $\tilde{P} \cdot A_2^{\text{tr}} = C \cdot \tilde{P}^\circ$, $\tilde{P}^\circ \cdot \breve{A}_2 = \tilde{P}$ and $\tilde{P}^\circ$ is a monomial matrix whose non-zero entries are powers of p. (Then $\tilde{P}$ is automatically invertible over $\mathbb{Q}$.) There is a natural choice for such a pair, namely, $(\tilde{P}, \tilde{P}^\circ) := (\breve{A}_2, I)$. Thus, the correspondence in 1.5.7 yields the existence of $\tilde{f}$. The root exponents of $\tilde{f}$ (which are the non-zero entries of $\tilde{P}^\circ = I$) are all equal to 1, hence $\tilde{f}$ is a central isogeny.

Now consider $\mathbf{G}_{\mathrm{ad}}$. We argue as above, where now $(\mathbf{G}_1, \mathbf{T}_1, \mathbf{B}_1) = (\mathbf{G}, \mathbf{T}, \mathbf{B})$ and $(\mathbf{G}_2, \mathbf{T}_2, \mathbf{B}_2) = (\mathbf{G}_{\mathrm{ad}}, \mathbf{T}', \mathbf{B}')$. The root datum of $\mathbf{G}_2$ is given by $C = \breve{A}_2 \cdot A_2^{\mathrm{tr}}$ where $A_2 = I$ and $\breve{A}_2 = C$. We need to specify a pair of (square) integral matrices (P', P'°) where $P' = A_1^{\mathrm{tr}} \cdot P'^\circ$, $P'^\circ \cdot C = \breve{A}_1 \cdot P'$ and P'° is a monomial matrix whose non-zero entries are powers of p. Again, there is a natural choice for such a pair, namely, $(P', P'^\circ) := (A_1^{\mathrm{tr}}, I)$. As above, this yields the existence of f'. □

Proposition 1.5.9 (Steinberg [St68, 9.16]) *Let $\mathbf{G}$ be semisimple and consider central isogenies $\tilde{f}\colon \mathbf{G}_{\mathrm{sc}} \to \mathbf{G}$ and $f'\colon \mathbf{G} \to \mathbf{G}_{\mathrm{ad}}$ as in Proposition 1.5.8. Assume, furthermore, that $F\colon \mathbf{G} \to \mathbf{G}$ is an isogeny. Then the following hold.*

(a) *The isogeny F lifts to $\mathbf{G}_{\mathrm{sc}}$; more precisely, there is a unique isogeny $\tilde{F}\colon \mathbf{G}_{\mathrm{sc}} \to \mathbf{G}_{\mathrm{sc}}$ such that $F \circ \tilde{f} = \tilde{f} \circ \tilde{F}$.*
(b) *The isogeny F descends to $\mathbf{G}_{\mathrm{ad}}$; more precisely, there is a unique isogeny $F'\colon \mathbf{G}_{\mathrm{ad}} \to \mathbf{G}_{\mathrm{ad}}$ such that $F' \circ f' = f' \circ F$.*

In both cases, the root exponents of $\tilde{F}$ and of F' are equal to those of F. If, moreover, F is a Steinberg map, then so are F' and $\tilde{F}$.

Proof First note that F', if it exists, is clearly unique. The uniqueness of $\tilde{F}$ (if it exists) is shown as follows. Let $F_1\colon \mathbf{G}_{\mathrm{sc}} \to \mathbf{G}_{\mathrm{sc}}$ be another isogeny such that $F \circ \tilde{f} = \tilde{f} \circ F_1$. Then the map sending $g \in \mathbf{G}_{\mathrm{sc}}$ to $\tilde{F}(g)F_1(g)^{-1}$ is a homomorphism of algebraic groups from $\mathbf{G}_{\mathrm{sc}}$ to the centre of $\mathbf{G}_{\mathrm{sc}}$. Since $\mathbf{G}_{\mathrm{sc}}$ is connected and the centre of $\mathbf{G}_{\mathrm{sc}}$ is finite, that map must be constant and so $\tilde{F}(g)F_1(g)^{-1} = 1$ for all $g \in \mathbf{G}_{\mathrm{sc}}$. We now turn to the problem of showing the existence of $\tilde{F}$ and F'.

Let $\mathbf{T} \subseteq \mathbf{B} \subseteq \mathbf{G}$, $\tilde{\mathbf{T}} \subseteq \tilde{\mathbf{B}} \subseteq \mathbf{G}_{\mathrm{sc}}$, $\mathbf{T}' \subseteq \mathbf{B}' \subseteq \mathbf{G}_{\mathrm{ad}}$ be as in Proposition 1.5.8. We consider the corresponding root data of $\mathbf{G}$, $\mathbf{G}_{\mathrm{sc}}$, $\mathbf{G}_{\mathrm{ad}}$, and write $X = X(\mathbf{T})$, $\tilde{X} = X(\tilde{\mathbf{T}})$, $X' = X(\mathbf{T}')$. Then $\tilde{f}$ induces a p-isogeny $\tilde{\varphi}\colon X \to \tilde{X}$ and f' induces a p-isogeny $\varphi'\colon X' \to X$. Thus, we are in the setting of 1.5.7. Now consider the isogeny $F\colon \mathbf{G} \to \mathbf{G}$. As already pointed out in the proof of [St68, 9.16], one easily sees that (a) and (b) hold for F if and only if (a) and (b) hold for $\iota_g \circ F$, where ι_g is an inner automorphism of $\mathbf{G}$ (for any $g \in \mathbf{G}$). Hence, using Theorem 1.3.3 and replacing F by $\iota_g \circ F$ for a suitable g, we may assume without loss of generality that $F(\mathbf{T}) = \mathbf{T}$ and $F(\mathbf{B}) = \mathbf{B}$. Then F induces a p-isogeny $\Phi\colon X \to X$ and, again, we are in the setting of 1.5.7. Now, if we can show that there exist p-isogenies $\tilde{\Phi}\colon \tilde{X} \to \tilde{X}$

and $\Phi' \colon X' \to X'$ such that

$$\tilde{\varphi} \circ \Phi = \tilde{\Phi} \circ \tilde{\varphi} \qquad \text{and} \qquad \varphi' \circ \Phi' = \Phi \circ \varphi',$$

then the existence of $\tilde{F}$ and F' follows from a general result about isogenies, which can already be found in [Ch05, §18.4] and which is a step in the proof of the Isogeny Theorem 1.3.12 (see also [St99, 3.2, 3.3]). In order to see how $\tilde{\Phi}$ and Φ' can be constructed, we use the correspondence in 1.5.7 to describe everything in terms of pairs of square integral matrices.

(The following part of the proof is somewhat different from the original proof in [St68].) Let $(\tilde{P}, \tilde{P}^\circ)$ and (P', P'°) be the pairs of matrices corresponding to $\tilde{\varphi}$ and φ', respectively. Furthermore, let (Q, Q°) be the pair of matrices corresponding to Φ. Now recall that the root data of $\mathbf{G}_{\text{sc}}$ and $\mathbf{G}_{\text{ad}}$ are given by the factorisations $C = I \cdot (C^{\text{tr}})^{\text{tr}}$ and $C = C \cdot I^{\text{tr}}$, respectively. Let A, $\breve{A}$ be square integral matrices such that the root datum of $\mathbf{G}$ is given by the factorisation $C = \breve{A} \cdot A^{\text{tr}}$. Then the conditions in 1.2.18 imply that

$$\tilde{P} = \tilde{P}^\circ \cdot \breve{A}, \quad P' = A^{\text{tr}} \cdot P'^\circ, \quad Q \cdot A^{\text{tr}} = A^{\text{tr}} \cdot Q^\circ, \quad Q^\circ \cdot \breve{A} = \breve{A} \cdot Q,$$

where $\tilde{P}^\circ$ and P'° are monomial matrices all of whose non-zero entries are equal to 1 (since $\tilde{f}$ and f' are central).

Assume first that $\tilde{\Phi}$ exists and let $(\tilde{Q}, \tilde{Q}^\circ)$ be the corresponding pair of matrices. Since the root datum of $\mathbf{G}_{\text{sc}}$ is given by the factorisation $C = I \cdot (C^{\text{tr}})^{\text{tr}}$, the conditions in 1.2.18 imply that $\tilde{Q} = \tilde{Q}^\circ$. Since $\tilde{\varphi} \circ \Phi = \tilde{\Phi} \circ \tilde{\varphi}$, we must have $\tilde{P} \cdot Q = \tilde{Q} \cdot \tilde{P}$. Using the above relations, we deduce that

$$\tilde{Q} = \tilde{Q}^\circ = \tilde{P}^\circ \cdot Q^\circ \cdot (\tilde{P}^\circ)^{-1}.$$

Thus, if $\tilde{\Phi}$ exists, then $(\tilde{Q}, \tilde{Q}^\circ)$ is determined by Q° and $\tilde{P}^\circ$; in particular, the root exponents of $\tilde{\Phi}$ are equal to those of Φ. Conversely, it is straightforward to check that the map $\tilde{\Phi} \colon \tilde{X} \to \tilde{X}$ defined by the matrix $\tilde{Q}$ given by the above formula has the required properties. Thus, (a) is proved.

Similarly, assume first that Φ' exists and let (Q', Q'°) be the corresponding pair of matrices. Then one deduces that

$$Q' = Q'^\circ = (P'^\circ)^{-1} \cdot Q^\circ \cdot P'^\circ.$$

Conversely, one checks that the map $\Phi' \colon X' \to X'$ defined by the matrix Q' given by the above formula has the required properties. Thus, (b) is proved.

Finally, if F is a Steinberg map, then (using the characterisation in Proposition 1.4.18) one easily sees that F' and $\tilde{F}$ are also Steinberg maps. □

Proposition 1.5.10 (Cf. [St67, p. 46]) *Let C be a Cartan matrix and $\mathbf{G}_{\text{sc}}$, $\mathbf{G}_{\text{ad}}$ be* *as in Example 1.5.3. Assume that C is a block diagonal matrix, with diagonal blocks $C_1, \dots, C_n$. Then we have*

$$\mathbf{G}_{\mathrm{sc}} = \tilde{\mathbf{G}}_1 \times \cdots \times \tilde{\mathbf{G}}_n \qquad \text{and} \qquad \mathbf{G}_{\mathrm{ad}} = \mathbf{G}'_1 \times \cdots \times \mathbf{G}'_n,$$

where $\tilde{\mathbf{G}}_i \subseteq \mathbf{G}_{\mathrm{sc}}$ and $\mathbf{G}'_i \subseteq \mathbf{G}_{\mathrm{ad}}$ are the normal subgroups corresponding to the various diagonal blocks C_i, as in Remark 1.3.6. For each i, the group $\tilde{\mathbf{G}}_i$ is simple of simply connected type C_i and $\mathbf{G}'_i$ is simple of adjoint type C_i.

Proof The definition of $\mathscr{R}_{\mathrm{sc}}(C)$ shows that this root datum is the direct sum of $\mathscr{R}_{\mathrm{sc}}(C_1), \dots, \mathscr{R}_{\mathrm{sc}}(C_n)$. Hence, the assertion concerning $\mathbf{G}_{\mathrm{sc}}$ immediately follows from Remark 1.3.6. The argument for $\mathbf{G}_{\mathrm{ad}}$ is analogous. □

1.5.11 We shall assume from now on that $\mathbf{G}$ is connected reductive over $k = \overline{\mathbb{F}}_p$ (where p is a prime number) and $F\colon \mathbf{G} \to \mathbf{G}$ is a Steinberg map. Let $\mathbf{Z}$ be the centre of $\mathbf{G}$ and $\mathbf{G}_{\mathrm{der}}$ be the derived subgroup of $\mathbf{G}$. Clearly, we have $F(\mathbf{Z}) = \mathbf{Z}$ and $F(\mathbf{G}_{\mathrm{der}}) = \mathbf{G}_{\mathrm{der}}$. Since $\mathbf{G} = \mathbf{Z}^\circ.\mathbf{G}_{\mathrm{der}}$ and $\mathbf{Z}^\circ \cap \mathbf{G}_{\mathrm{der}}$ is finite, we obtain isogenies

$$\begin{array}{ccc} \mathbf{Z}^\circ \times \mathbf{G}_{\mathrm{der}} & \longrightarrow & \mathbf{G} \\ (z, g) & \mapsto & zg \end{array} \qquad \text{and} \qquad \begin{array}{ccc} \mathbf{G} & \longrightarrow & \mathbf{G}/\mathbf{G}_{\mathrm{der}} \times \mathbf{G}/\mathbf{Z}^\circ \\ g & \mapsto & (g\mathbf{G}_{\mathrm{der}},\ g\mathbf{Z}^\circ). \end{array}$$

(Note that these are maps between groups of the same dimension; the first map is clearly surjective and, hence, has a finite kernel; the second map has a finite kernel and, hence, is surjective.) Recall from 1.1.13 that $\mathbf{G}_{\mathrm{der}}$ is semisimple. The group $\mathbf{G}/\mathbf{G}_{\mathrm{der}}$ is a torus. (For, it is connected, abelian and consists of elements of order prime to p; see 1.1.7.) Furthermore, $\mathbf{G}_{\mathrm{ss}} := \mathbf{G}/\mathbf{Z}^\circ$ is reductive (see Lemma 1.4.26) with a finite centre and, hence, is semisimple. Using also the isogenies in Proposition 1.5.8, we obtain isogenies

$$\mathbf{Z}^\circ \times (\mathbf{G}_{\mathrm{der}})_{\mathrm{sc}} \longrightarrow \mathbf{G} \longrightarrow \mathbf{G}/\mathbf{G}_{\mathrm{der}} \times (\mathbf{G}_{\mathrm{ss}})_{\mathrm{ad}}.$$

Now, by Lemma 1.4.26, we have induced Steinberg maps on $\mathbf{G}_{\mathrm{ss}}$ and on $\mathbf{G}/\mathbf{G}_{\mathrm{der}}$. By Proposition 1.5.9, there are also induced Steinberg maps on $(\mathbf{G}_{\mathrm{der}})_{\mathrm{sc}}$ and on $(\mathbf{G}_{\mathrm{ss}})_{\mathrm{ad}}$. Since all these maps are induced and uniquely determined by F, we will now simplify our notation and denote all these induced maps by F as well. Using Proposition 1.4.13(c), we conclude that $|\mathbf{G}_{\mathrm{der}}^F| = |(\mathbf{G}_{\mathrm{der}})_{\mathrm{sc}}^F|$, $|\mathbf{G}_{\mathrm{ss}}^F| = |(\mathbf{G}_{\mathrm{ss}})_{\mathrm{ad}}^F|$ and

$$|\mathbf{G}^F| = |(\mathbf{Z}^\circ)^F|\,|\mathbf{G}_{\mathrm{der}}^F| = |(\mathbf{G}/\mathbf{G}_{\mathrm{der}})^F|\,|\mathbf{G}_{\mathrm{ss}}^F|.$$

Also note that the natural map $\mathbf{G} \to \mathbf{G}_{\mathrm{ss}}$ induces a surjective map $\mathbf{G}^F \to \mathbf{G}_{\mathrm{ss}}^F$ (since the kernel of $\mathbf{G} \to \mathbf{G}_{\mathrm{ss}}$ is connected).

Remark 1.5.12 By a slight abuse of notation, we shall denote $(\mathbf{G}_{\mathrm{ss}})_{\mathrm{ad}}$ simply by $\mathbf{G}_{\mathrm{ad}}$. Thus, as above, we obtain a central isogeny $\mathbf{G}_{\mathrm{ss}} \to \mathbf{G}_{\mathrm{ad}}$ which commutes with the action of F on both sides. Composing this isogeny with the natural map $\mathbf{G} \to \mathbf{G}_{ss} = \mathbf{G}/\mathbf{Z}^\circ$, we obtain a surjective, central isotypy

$$\pi_{ad}\colon \mathbf{G} \to \mathbf{G}_{ad} \qquad \text{with} \qquad \ker(\pi_{ad}) = \mathbf{Z},$$

which commutes with the action of F on both sides and which we call an *adjoint quotient* of $\mathbf{G}$. We certainly have an inclusion $\pi_{ad}(\mathbf{G}^F) \subseteq \mathbf{G}_{ad}^F$ but, in general, equality will not hold. By Proposition 1.4.13, $\pi_{ad}(\mathbf{G}^F)$ is a normal subgroup of $\mathbf{G}_{ad}^F$ and we have an isomorphism

$$\mathbf{G}_{ad}^F/\pi_{ad}(\mathbf{G}^F) \cong \mathbf{Z}/\mathscr{L}(\mathbf{Z}),$$

where $\mathscr{L}\colon \mathbf{G} \to \mathbf{G}$, $g \mapsto g^{-1}F(g)$. One easily sees that $\mathbf{Z}/\mathscr{L}(\mathbf{Z}) = (\mathbf{Z}/\mathbf{Z}^\circ)_F$, where the subscript F denotes 'F-coinvariants', that is, the largest quotient on which F acts trivially. The above isomorphism is explicitly obtained by sending $g \in \mathbf{G}_{ad}^F$ to $\dot{g}^{-1}F(\dot{g}) \in \mathbf{Z}$ where $\dot{g} \in \mathbf{G}$ is any element satisfying $\pi_{ad}(\dot{g}) = g$.

Also note that each $g \in \mathbf{G}_{ad}^F$ defines an automorphism $\alpha_g\colon \mathbf{G}^F \to \mathbf{G}^F$, $g_1 \mapsto \dot{g}g_1\dot{g}^{-1}$ (where, as above, $\dot{g} \in \mathbf{G}$ is such that $\pi_{ad}(\dot{g}) = g$; the map α_g obviously does not depend on the choice of $\dot{g}$). In this way, we obtain a group homomorphism

$$\mathbf{G}_{ad}^F \to \operatorname{Aut}(\mathbf{G}^F), \qquad g \mapsto \alpha_g.$$

The automorphisms α_g are called *diagonal automorphisms* of $\mathbf{G}^F$.

Remark 1.5.13 Again, by a slight abuse of notation, we shall denote $(\mathbf{G}_{der})_{sc}$ simply by $\mathbf{G}_{sc}$. Thus, as above, we obtain a central isotypy

$$\pi_{sc}\colon \mathbf{G}_{sc} \to \mathbf{G} \qquad \text{with} \qquad \pi_{sc}(\mathbf{G}_{sc}) = \mathbf{G}_{der},$$

which commutes with the action of F on both sides and which we call a *simply connected covering* of the derived subgroup of $\mathbf{G}$. A special feature of groups of simply connected type is that

(a) $\mathbf{G}_{sc}^F = \langle u \in \mathbf{G}_{sc}^F \mid u \text{ unipotent}\rangle$; see [St68, 12.4].

We have $\pi_{sc}(\mathbf{G}_{sc}^F) \subseteq \mathbf{G}_{der}^F$ but, again, equality will not hold in general. In fact, applying Proposition 1.4.13 to $\pi_{sc}\colon \mathbf{G}_{sc} \to \mathbf{G}_{der}$, one easily sees that

(b) $\pi_{sc}(\mathbf{G}_{sc}^F) = \langle u \in \mathbf{G}^F \mid u \text{ unipotent}\rangle \subseteq \mathbf{G}_{der}^F$; see [St68, 12.6].

Finally, if $\mathbf{T}_1 \subseteq \mathbf{G}$ is any F-stable maximal torus and $\tilde{\mathbf{T}}_1 = \pi_{sc}^{-1}(\mathbf{T}_1) \subseteq \mathbf{G}_{sc}$, then the inclusion $\mathbf{T}_1 \subseteq \mathbf{G}$ induces an isomorphism

(c) $\mathbf{T}_1^F/\pi_{sc}(\tilde{\mathbf{T}}_1^F) \cong \mathbf{G}^F/\pi_{sc}(\mathbf{G}_{sc}^F)$ (see [DeLu76, 1.23]).

Note that $\pi_{sc}(\mathbf{G}_{sc}^F)$ is a characteristic subgroup of $\mathbf{G}^F$; furthermore, we have the inclusions $[\mathbf{G}^F, \mathbf{G}^F] \subseteq \pi_{sc}(\mathbf{G}_{sc}^F) \subseteq \mathbf{G}_{der}^F$ but these may be strict, as can be seen already in the example where $\mathbf{G} = \mathbf{G}_{der} = \mathrm{PGL}_2(k)$.

Following [St67, p. 45], [St68, 9.1], we call $\ker(\pi_{sc})$ the *fundamental group* of $\mathbf{G}$. For example, if $\mathbf{G}$ is simple of simply connected type, then the fundamental group of $\mathbf{G}$ is trivial (but the converse is not necessarily true). If $\mathbf{G}$ is simple of adjoint type, with corresponding Cartan matrix C, then $\ker(\pi_{sc}) \cong \mathrm{Hom}(\Lambda(C), k^\times)$ where $\Lambda(C)$ is the fundamental group of C; see Remark 1.2.8 and Example 1.5.3.

1.5.14 We keep the above notation. As already stated in 1.1.13, we have $\mathbf{G}_{\mathrm{der}} = \mathbf{G}_1 \ldots \mathbf{G}_n$ where $\mathbf{G}_1, \ldots, \mathbf{G}_n$ are the closed normal simple subgroups of $\mathbf{G}$. (They elementwise commute with each other.) Now, one complication of the theory arises from the fact that, in general, this product decomposition is not stable under the Steinberg map $F\colon \mathbf{G}_{\mathrm{der}} \to \mathbf{G}_{\mathrm{der}}$. What happens is the following. Consider the set-up in Remark 1.3.6, with partitions $\Pi = \Pi_1 \sqcup \ldots \sqcup \Pi_n$ and $R = R_1 \sqcup \ldots \sqcup R_n$; then $\mathbf{G}_i = \langle \mathbf{U}_\alpha \mid \alpha \in R_i \rangle$ for $i = 1, \ldots, n$. Now, F will permute the simple subgroups $\mathbf{G}_i$ and, correspondingly, the permutation $\alpha \mapsto \alpha^\dagger$ of R (induced by F) will permute the subsets R_i. Hence, there is an induced permutation ρ of $\{1, \ldots, n\}$ such that, for all $i = 1, \ldots, n$, we have $R_{\rho(i)} = \{\alpha^\dagger \mid \alpha \in R_i\}$ and

$$F(\mathbf{G}_i) = \langle F(\mathbf{U}_\alpha) \mid \alpha \in R_i \rangle = \langle \mathbf{U}_{\alpha^\dagger} \mid \alpha \in R_i \rangle = \mathbf{G}_{\rho(i)}. \qquad \text{(a)}$$

Following [St68, p. 78], we say that $\mathbf{G}_{\mathrm{der}}$ is *F-simple* if ρ is a cyclic permutation (it has a single orbit). Thus, by grouping together the factors in the various ρ-orbits on $\{1, \ldots, n\}$, we can write $\mathbf{G}_{\mathrm{der}}$ as a product of various F-stable and F-simple semisimple groups. An analogous statement holds for $\mathbf{G}_{\mathrm{ss}}$. Indeed, for each i, let $\bar{\mathbf{G}}_i$ be the image of $\mathbf{G}_i$ under the canonical map $\mathbf{G} \to \mathbf{G}_{\mathrm{ss}}$. Then we have $\mathbf{G}_{\mathrm{ss}} = \bar{\mathbf{G}}_1 \cdots \bar{\mathbf{G}}_n$ and the induced Steinberg map $F\colon \mathbf{G}_{\mathrm{ss}} \to \mathbf{G}_{\mathrm{ss}}$ permutes the factors $\bar{\mathbf{G}}_i$ according to the permutation ρ.

Now consider the isogenies in 1.5.11 and the groups $\mathbf{G}_{\mathrm{sc}}$, $\mathbf{G}_{\mathrm{ad}}$. By Proposition 1.5.10, we have direct product decompositions

$$\mathbf{G}_{\mathrm{sc}} = \tilde{\mathbf{G}}_1 \times \cdots \times \tilde{\mathbf{G}}_n \qquad \text{and} \qquad \mathbf{G}_{\mathrm{ad}} = \mathbf{G}_1' \times \cdots \times \mathbf{G}_n' \qquad \text{(b)}$$

such that, under the isogeny $\mathbf{G}_{\mathrm{sc}} \to \mathbf{G}_{\mathrm{der}}$, the factor $\tilde{\mathbf{G}}_i$ is mapped to $\mathbf{G}_i$ and, under the isogeny $\mathbf{G}_{\mathrm{ss}} \to \mathbf{G}_{\mathrm{ad}}$, the factor $\bar{\mathbf{G}}_i$ is mapped to $\mathbf{G}_i'$. By the compatibility of all of the above isogenies with the various Steinberg maps involved, it follows that

$$F(\tilde{\mathbf{G}}_i) = \tilde{\mathbf{G}}_{\rho(i)} \qquad \text{and} \qquad F(\mathbf{G}_i') = \mathbf{G}_{\rho(i)}' \qquad \text{for } i = 1, \ldots, n. \qquad \text{(c)}$$

The following result deals with F-simple semisimple groups. (If F is a Frobenius map with respect to some $\mathbb{F}_q$-rational structure, then the construction below is related to the operation '*restriction of scalars*'; see [Spr98, §11.4].)

Lemma 1.5.15 *Assume that $\mathbf{G}$ is semisimple and F-simple, as defined in 1.5.14. Then $\mathbf{G} = \mathbf{G}_{\mathrm{der}} = \mathbf{G}_1 \cdots \mathbf{G}_n$ as above. Assume that this product is an abstract direct* *product. Then* $F^n(\mathbf{G}_1) = \mathbf{G}_1$ *and*

$$\iota\colon \mathbf{G}_1 \to \mathbf{G}, \qquad g \mapsto gF(g)\cdots F^{n-1}(g),$$

is an injective homomorphism of algebraic groups which restricts to an isomorphism $\mathbf{G}_1^{F^n} \cong \mathbf{G}^F$. *Furthermore, if* q *is the positive real number attached to* $(\mathbf{G}, F)$ *(see Proposition 1.4.19), then* q^n *is the positive real number attached to* $(\mathbf{G}_1, F^n)$.

Proof By assumption, F cyclically permutes the factors $\mathbf{G}_i$. So we can choose the labelling such that $F^i(\mathbf{G}_1) = \mathbf{G}_{i+1}$ for $i = 1, \ldots, n-1$ and $F^n(\mathbf{G}_1) = \mathbf{G}_1$. The map ι clearly is a morphism of affine varieties, which is injective since the product is direct. This map is a group homomorphism because the groups $\mathbf{G}_i$ elementwise commute with each other. For the same reason, we have $\iota(\mathbf{G}_1^{F^n}) \subseteq \mathbf{G}^F$. Since we have a direct product, every $g \in \mathbf{G}$ can be written uniquely as $g = g_1 \cdots g_n$ with $g_i \in \mathbf{G}_i$. Then $F(g) = g$ if and only if $F^i(g_1) = g_{i+1}$ for $i = 1, \ldots, n-1$ and $F^n(g_1) = g_1$. Thus, $F(g) = g$ if and only if $g = \iota(g_1)$ and $F^n(g_1) = g_1$. The statement concerning q, q^n immediately follows from the definition of these numbers in Proposition 1.4.19. □

Corollary 1.5.16 *Assume that* $\mathbf{G}$ *is connected reductive; let* $\mathbf{G} = \mathbf{Z}^\circ.\mathbf{G}_{\mathrm{der}}$ *and* $\mathbf{G}_{\mathrm{der}} = \mathbf{G}_1 \cdots \mathbf{G}_n$, *as above. Let* $F\colon \mathbf{G} \to \mathbf{G}$ *be a Steinberg map and* $I \subseteq \{1, \ldots, n\}$ *be a set of representatives of the* ρ*-orbits on* $\{1, \ldots, n\}$, *with* ρ *as in 1.5.14(a). For each* $i \in I$, *let* n_i *be the length of the corresponding* ρ*-orbit. Then we have isomorphisms (of abstract finite groups)*

$$\mathbf{G}_{\mathrm{sc}}^F \cong \prod_{i\in I} \tilde{\mathbf{G}}_i^{F^{n_i}} \qquad \textit{and} \qquad \mathbf{G}_{\mathrm{ad}}^F \cong \prod_{i\in I} \mathbf{G}_i'^{F^{n_i}},$$

where $\mathbf{G}_{\mathrm{sc}} = \tilde{\mathbf{G}}_1 \times \cdots \times \tilde{\mathbf{G}}_n$ *and* $\mathbf{G}_{\mathrm{ad}} = \mathbf{G}_1' \times \cdots \times \mathbf{G}_n'$ *are as in 1.5.14(b).*

Proof This is immediate from 1.5.14 and Lemma 1.5.15. Also recall our identification in 1.5.11 of the various Steinberg maps involved. □

See Example 1.4.29 for a good illustration of the above result; the group $\mathbf{G} = \mathrm{SL}_2(k) \times \mathrm{PGL}_2(k)$ (with $\mathrm{char}(k) = 2$) considered there is F-simple! In particular, the simple factors of an F-simple semisimple group need not all be isomorphic to each other as algebraic groups. (They are isomorphic as abstract groups.)

To complete this section, we discuss another fundamental construction involving the formalism of root data: 'dual groups'.

Definition 1.5.17 ([DeLu76, 5.21]; see also [Ca85, §4.3], [Lu84a, 8.4]) Consider two pairs $(\mathbf{G}, F)$ and $(\mathbf{G}^*, F^*)$ where $\mathbf{G}, \mathbf{G}^*$ are connected reductive and $F\colon \mathbf{G} \to \mathbf{G}$, $F^*\colon \mathbf{G}^* \to \mathbf{G}^*$ are Steinberg maps. We say that $(\mathbf{G}, F)$ and $(\mathbf{G}^*, F^*)$ are in *duality* if there is a *maximally split* torus $\mathbf{T}_0 \subseteq \mathbf{G}$ and a *maximally split* torus $\mathbf{T}_0^* \subseteq \mathbf{G}^*$ such that the following conditions hold, where $\mathscr{R} = (X, R, Y, R^\vee)$ is the root datum of $\mathbf{G}$ (with respect to $\mathbf{T}_0$) and $\mathscr{R}^* = (X^*, R^*, Y^*, R^{*\vee})$ is the root datum of $\mathbf{G}^*$ (with respect to $\mathbf{T}_0^*$):

(a) There is an isomorphism $\delta\colon X \to Y^*$ such that $\delta(R) = R^{*\vee}$ and

$$\langle \lambda, \alpha^\vee \rangle = \langle \alpha^*, \delta(\lambda) \rangle \qquad \text{for all } \lambda \in X(\mathbf{T}_0) \text{ and } \alpha \in R,$$

where $\alpha^* \in R^*$ is defined by $\delta(\alpha) = \alpha^{*\vee}$.

(b) We have $\delta(\lambda \circ F|_{\mathbf{T}_0}) = F^*|_{\mathbf{T}_0^*} \circ \delta(\lambda)$ for all $\lambda \in X$. Thus, if $\lambda\colon \mathbf{T}_0 \to \mathbf{k}^\times$ (an element of $X = X(\mathbf{T}_0)$) and $\nu\colon \mathbf{k}^\times \to \mathbf{T}_0^*$ (an element of $Y^* = Y(\mathbf{T}_0^*)$) correspond to each other under δ, then $\lambda \circ F\colon \mathbf{T}_0 \to \mathbf{k}^\times$ and $F^* \circ \nu\colon \mathbf{k}^\times \to \mathbf{T}_0^*$ also correspond to each other under δ.

The relation of being *in duality* is symmetric: the above conditions on $\delta\colon X \to Y^*$ are equivalent to analogous conditions concerning a map $\varepsilon\colon Y \to X^*$ obtained by transposing δ; see [Ca85, 4.2.2, 4.3.1]. In particular, δ defines an isomorphism of root data between $\mathscr{R}$ and the dual of $\mathscr{R}^*$ (where the dual root datum is given as in Lemma 1.2.3(b)). Thus, connected reductive groups in duality have dual root data. Note also that the dual of $(\mathbf{G}^*, F^*)$ can be naturally identified with $(\mathbf{G}, F)$.

1.5.18 It may be worthwhile to reformulate the above definition in terms of the matrix language of Section 1.2; the following discussion will also show that, for any given pair $(\mathbf{G}, F)$, there exists a corresponding dual pair.

So let $\mathbf{G}$ be connected reductive and $F\colon \mathbf{G} \to \mathbf{G}$ be a Steinberg map. Let $\mathbf{T}_0 \subseteq \mathbf{G}$ be a maximally split torus and $\mathbf{B}_0 \subseteq \mathbf{G}$ be an F-stable Borel subgroup such that $\mathbf{T}_0 \subseteq \mathbf{B}_0$. Let $\mathscr{R} = (X, R, Y, R^\vee)$ be the root datum of $\mathbf{G}$ with respect to $\mathbf{T}_0$; let Π be the base for R determined by $\mathbf{B}_0$ (see Remark 1.3.4). As in Remark 1.2.13, we choose a $\mathbb{Z}$-basis of X. Then $\mathscr{R}$ determines a factorisation

$$C = \breve{A} \cdot A^{\mathrm{tr}} \qquad \text{where} \qquad C = \text{ Cartan matrix of } \mathscr{R}.$$

Furthermore, as discussed in 1.5.7, the isogeny $F\colon \mathbf{G} \to \mathbf{G}$ determines a pair of integer matrices (P, P°) satisfying the conditions (MI1), (MI2) in 1.2.18. Now we notice that C^{tr} is also a Cartan matrix and the corresponding factorisation

$$C^{\mathrm{tr}} = \breve{B} \cdot B^{\mathrm{tr}} \qquad \text{where} \qquad \breve{B} := A \quad \text{and} \quad B := \breve{A}$$

gives rise to the root datum $\mathscr{R}^* := (Y, R^\vee, X, R)$ which is dual to $\mathscr{R}$; see Remark 1.2.13. Here, $\Pi^\vee$ is a base for $R^\vee$. Furthermore, setting $Q := P^{\mathrm{tr}}$ and $Q^\circ := (P^\circ)^{\mathrm{tr}}$, the pair (Q, Q°) satisfies the conditions (MI1), (MI2) with respect to $C^{\mathrm{tr}} = \breve{B} \cdot B^{\mathrm{tr}}$. Hence, performing the above argument backwards, there exists a connected reductive group $\mathbf{G}^*$ such that $\mathscr{R}^*$ is isomorphic to the root datum of $\mathbf{G}^*$ with respect to a maximal torus $\mathbf{T}_0^* \subseteq \mathbf{G}^*$; the base $\Pi^\vee$ for $R^\vee$ determines a Borel subgroup $\mathbf{B}_0^* \subseteq \mathbf{G}^*$ containing $\mathbf{T}_0^*$. The pair of matrices (Q, Q°) determines an isogeny $F^*\colon \mathbf{G}^* \to \mathbf{G}^*$ such that $\mathbf{T}_0^*$ and $\mathbf{B}_0^*$ are F-stable. Since F is a Steinberg map, the characterisation in Proposition 1.4.18 immediately shows that F^* also is a Steinberg map. Thus, we conclude that the pairs $(\mathbf{G}, F)$ and $(\mathbf{G}^*, F^*)$ are in duality; the number q in Proposition 1.4.19 is the same for F and for F^*.

Remark 1.5.19 In the setting of 1.5.18, let $\mathbf{W} = N_{\mathbf{G}}(\mathbf{T}_0)/\mathbf{T}_0$ be the Weyl group of $\mathbf{G}$ and $\mathbf{W}^* = N_{\mathbf{G}^*}(\mathbf{T}_0^*)/\mathbf{T}_0^*$ be the Weyl group of $\mathbf{G}^*$. The choice of a base Π for R determines a set S of Coxeter generators for $\mathbf{W}$; thus, using the notational convention in Remark 1.2.12, we have $\mathbf{W} = \langle S \rangle$ and $\Pi = \{\alpha_s \mid s \in S\}$. We have seen above that Π determines a base Π^* for R^*; in fact, we have $\Pi^* = \{\alpha_s^* \mid s \in S\}$, where we use the bijection $R \leftrightarrow R^*$, $\alpha \leftrightarrow \alpha^*$, in Definition 1.5.17(a). (Recall that $\delta(\alpha) = \alpha^{*\vee}$ for all $\alpha \in R$.) Now, by [Ca85, 4.2.3], there is a unique group isomorphism $\mathbf{W} \xrightarrow{\sim} \mathbf{W}^*$, $w \mapsto w^*$, such that:

(a) We have $\delta(w.\lambda) = w^*.\delta(\lambda)$ for all $\lambda \in X(\mathbf{T}_0)$.
(b) If $\alpha \in R$, then $w_\alpha^* \in \mathbf{W}^*$ is the reflection in the root $\alpha^{*\vee} \in R^{*\vee}$.

In particular, $S^* = \{s^* \mid s \in S\}$ is the set of Coxeter generators for $\mathbf{W}^*$ determined by Π^*. This also implies that $l(w) = l^*(w^*)$ for all $w \in \mathbf{W}$, where $l\colon \mathbf{W} \to \mathbb{Z}_{\geqslant 0}$ is the length function with respect to S and $l^*\colon \mathbf{W}^* \to \mathbb{Z}_{\geqslant 0}$ is the length function with respect to S^*.

Example 1.5.20 (a) Assume that $\mathbf{G}$ is semisimple of adjoint type. Thus, in the setting of 1.5.18, we have $C = \breve{A} \cdot A^{\mathrm{tr}}$ where A is the identity matrix and $\breve{A} = C$. Hence, $C^{\mathrm{tr}} = \breve{B} \cdot B^{\mathrm{tr}}$ where $\breve{B}$ is the identity matrix and $B = C$. So $\mathbf{G}^*$ is seen to be semisimple of simply connected type. If C is symmetric, then Proposition 1.5.8 yields a central isogeny $\mathbf{G}^* \to \mathbf{G}$. Similarly, if $\mathbf{G}$ is semisimple of simply connected type, then $\mathbf{G}^*$ is seen to be semisimple of adjoint type.

(b) The examples in (a) seem to indicate that dual groups are related in quite a strong way. However, as pointed out in the introduction of [Lu09d], dual groups in general are related only through a very weak connection (via their root data); in particular there is no direct, elementary construction which produces $\mathbf{G}^*$ from $\mathbf{G}$. Perhaps the most striking example is the case where $\mathbf{G} = \mathrm{SO}_{2n+1}(k)$. Then $\mathbf{G}$ is simple of adjoint type, with Cartan matrix C of type B_n. As in (a), $\mathbf{G}^*$ will be simple of simply connected type. However, since C^{tr} has type C_n, we see that $\mathbf{G}^* \cong \mathrm{Sp}_{2n}(k)$. If $\mathrm{char}(k) \neq 2$, then there is no isogeny from $\mathbf{G}^*$ to $\mathbf{G}$!

Example 1.5.21 Let $\mathbf{G}$ be connected reductive and $F\colon \mathbf{G} \to \mathbf{G}$ be a Steinberg map. It may well happen that $(\mathbf{G}, F)$ is dual to itself. To see this, let $\mathbf{T}_0 \subseteq \mathbf{G}$ be an F-stable maximal torus and $\mathbf{B}_0 \subseteq \mathbf{G}$ be an F-stable Borel subgroup such that $\mathbf{T}_0 \subseteq \mathbf{B}_0$. Let $C = (c_{st})_{s,t \in S}$ be the corresponding Cartan matrix. As in 1.5.18, the root datum $\mathscr{R} = (X, R, Y, R^\vee)$ of $\mathbf{G}$ with respect to $\mathbf{T}_0$ gives rise to a factorisation

$C = \check{A} \cdot A^{\mathrm{tr}}$. The map F gives rise to a pair of integer matrices (P, P°) satisfying the conditions (MI1), (MI2) in 1.2.18. Now, an isomorphism between the dual of $\mathscr{R}$ and $\mathscr{R}$ itself is given by a group isomorphism $\delta\colon X \to Y$ as in Definition 1.5.17. In the setting of 1.2.18, such an isomorphism δ is specified in terms of a permutation matrix $D^\circ = (d^\circ_{st})_{s,t \in S}$ and an integer matrix D (square and invertible over $\mathbb{Z}$, of the appropriate size) such that

$$D^\circ \cdot \check{A} = A \cdot D, \qquad D^\circ \cdot C = C^{\mathrm{tr}} \cdot D^\circ, \qquad D \cdot P = P^{\mathrm{tr}} \cdot D. \qquad (*)$$

(Note that the first two conditions imply that $D \cdot A^{\mathrm{tr}} = \check{A}^{\mathrm{tr}} \cdot D^\circ$; the third condition ensures that the compatibility in Definition 1.5.17(b) holds.) Thus, if $(*)$ holds, then $(\mathbf{G}, F)$ is in duality with itself; in particular, we obtain a group isomorphism

$$\mathbf{W} \xrightarrow{\sim} \mathbf{W}, \qquad w \mapsto w^*,$$

such that $S^* = \{s^* \mid s \in S\} = S$. If $s \in S$, then $s^* \in S$ is uniquely determined by the condition that $d^\circ_{s^*s} = 1$. Let us consider a few concrete examples.

(a) Let $\mathbf{G} = \mathrm{GL}_n(k)$ and F be one of the two Steinberg maps in Example 1.4.21. Then $\mathbf{G}$ has Cartan type A_{n-1} and the Cartan matrix (which is symmetric in this case) factorises as $C = \check{A} \cdot A^{\mathrm{tr}}$ where $A = \check{A}$; see Example 1.3.7. Furthermore, P is q times the identity matrix or q times a permutation matrix of order 2; in both cases, P is symmetric. Hence, $(*)$ holds if we take for D the identity matrix of size $n \times n$ and for D° the identity matrix of size $(n-1) \times (n-1)$. So $(\mathbf{G}, F)$ is dual to itself; the map $w \mapsto w^*$ is the identity.

(b) Let $\mathbf{G}$ be simple of adjoint type G_2 or F_4; let $F\colon \mathbf{G} \to \mathbf{G}$ be such that $F(t) = t^q$ for all $t \in \mathbf{T}_0$. Then both P and P° are q times the identity matrix (of the appropriate size). Furthermore, $C = \check{A} \cdot A^{\mathrm{tr}}$ where A is the identity matrix. Then $(*)$ holds if we take $D = D^\circ \cdot C$ and

$$D^\circ = \begin{pmatrix} 0 & 1 \\ 1 & 0 \end{pmatrix} \quad (\text{type } G_2), \qquad D^\circ = \begin{pmatrix} 0 & 0 & 0 & 1 \\ 0 & 0 & 1 & 0 \\ 0 & 1 & 0 & 0 \\ 1 & 0 & 0 & 0 \end{pmatrix} \quad (\text{type } F_4).$$

Thus, $(\mathbf{G}, F)$ is dual to itself but something non-trivial is going on: in this case, $w \mapsto w^*$ is the non-trivial graph automorphism of $\mathbf{W}$.

We shall have to say more about groups in duality in later sections. It will be useful and important to know how properties of $\mathbf{G}$ translate or connect to properties of $\mathbf{G}^*$. The lemma below contains just one example. (See Lusztig [Lu09d] where a number of such 'bridges' of a much deeper nature are discussed.)

Lemma 1.5.22 *Let $(\mathbf{G}, F)$ and $(\mathbf{G}^*, F^*)$ be two pairs in duality, as in Definition 1.5.17. Then the following conditions are equivalent.*

(i) *The centre of* $\mathbf{G}$ *is connected.*
(ii) *The abelian group* $X/\mathbb{Z}R$ *has no* p'*-torsion.*
(iii) *The fundamental group of* $\mathbf{G}^*$ *(see Remark 1.5.13) is trivial.*

Proof For the equivalence of (i) and (ii), see [Ca85, 4.5.1]. The equivalence of (ii) and (iii) is shown in [Ca85, 4.5.8]. □

1.6 Generic Finite Reductive Groups

Recall from Definition 1.4.7 that a finite group of Lie type is a finite group of the form $G = \mathbf{G}^F$, where $\mathbf{G}$ is a connected reductive algebraic group over $k = \overline{\mathbb{F}}_p$ and $F: \mathbf{G} \to \mathbf{G}$ is a Steinberg map. Then it is common to speak of the (twisted or untwisted) 'type' of $\mathbf{G}^F$: for example, we say that the finite general linear groups are of untwisted type A_{n-1}, the finite unitary groups are of twisted type A_{n-1} (denoted ${}^2A_{n-1}$; see Example 1.4.21), or that the Suzuki groups are of 'very twisted' type B_2 (denoted ${}^2B_2$; see Example 1.4.22). Here, the superscript (as in ${}^2A_{n-1}$) indicates the order of the automorphism of the Weyl group of $\mathbf{G}$ which is induced by F; in particular, $\mathbf{G}^F$ is of 'untwisted' type if F induces the identity map on the Weyl group. Using the machinery developed in the previous sections, we can now give a somewhat more precise definition, as follows.

1.6.1 Assume that $\mathbf{G}$ is connected reductive and let $F: \mathbf{G} \to \mathbf{G}$ be a Steinberg map. Then we can canonically attach to $\mathbf{G}$ and F a pair

$$\mathscr{C}(\mathbf{G}, F) := (C, P^\circ),$$

where $C = (c_{st})_{s,t\in S}$ is a Cartan matrix and $P^\circ = (p_{st})_{s,t\in S}$ is a monomial matrix whose non-zero entries are positive powers of p and such that $CP^\circ = P^\circ C$. Let us recall how this is done. First, we choose a maximally split torus $\mathbf{T}_0 \subseteq \mathbf{G}$. Recall from Example 1.4.10 that this means that $\mathbf{T}_0$ is an F-stable maximal torus of $\mathbf{G}$ which is contained in an F-stable Borel subgroup $\mathbf{B}_0 \subseteq \mathbf{G}$. (By Proposition 1.4.12, the pair $(\mathbf{T}_0, \mathbf{B}_0)$ is unique up to conjugation by elements of $\mathbf{G}^F$.) Let $\mathscr{R} = (X, R, Y, R^\vee)$ be the root datum of $\mathbf{G}$ relative to $\mathbf{T}_0$ and $\varphi: X \to X$ be the p-isogeny induced by F. By Remark 1.3.4, there is a unique base Π of R such that $\mathbf{B}_0 = \langle \mathbf{T}_0, \mathbf{U}_\alpha \mid \alpha \in R^+\rangle$ where R^+ are the positive roots with respect to Π. Let us write $\Pi = \{\alpha_s \mid s \in S\}$ and let $C = (c_{st})_{s,t\in S}$ be the corresponding Cartan matrix.

Since φ is a p-isogeny, there is a permutation $\alpha \mapsto \alpha^\dagger$ of R such that $\varphi(\alpha^\dagger) = q_\alpha\, \alpha$ for all $\alpha \in R$. The fact that $\mathbf{B}_0$ is F-stable implies that this permutation leaves R^+ invariant. Hence, this permutation will also leave the base Π invariant and so there is an induced permutation $S \to S$, $s \mapsto s^\dagger$, such that $\alpha_s^\dagger = \alpha_{s^\dagger}$ for all $s \in S$. Thus, φ is 'base preserving' as in 1.2.18 and we have a corresponding monomial matrix

$P^\circ = (p^\circ_{st})_{s,t\in S}$ whose non-zero entries are given by $p^\circ_{ss^\dagger} = q_s := q_{\alpha_s}$ for all $s \in S$. The condition (MI2) implies that $CP^\circ = P^\circ C$, which means that

$$q_t c_{st} = q_s c_{s^\dagger t^\dagger} \qquad \text{for all } s, t \in S. \tag{a}$$

Since all pairs $(\mathbf{T}_0, \mathbf{B}_0)$ as above are conjugate by elements of $\mathbf{G}^F$, the pair (C, P°) is uniquely determined by $\mathbf{G}, F$ up to relabelling the elements of S.

Now consider the Weyl group $\mathbf{W}$ of $\mathscr{R}$. Recall from Remark 1.2.12 that we identify S with a subset of $\mathbf{W}$ via $s \leftrightarrow w_{\alpha_s}$; thus, we have $\mathbf{W} = \langle S\rangle$. Let $l\colon \mathbf{W} \to \mathbb{Z}_{\geqslant 0}$ be the corresponding length function. By Remark 1.2.10, the p-isogeny φ induces a group automorphism $\sigma\colon \mathbf{W} \to \mathbf{W}$ such that

$$\sigma(s) = s^\dagger \quad (s \in S) \qquad \text{and} \qquad \varphi \circ \sigma(w) = w \circ \varphi \quad (w \in \mathbf{W}). \tag{b}$$

By Theorem 1.3.2, we can naturally identify $\mathbf{W} = N_{\mathbf{G}}(\mathbf{T}_0)/\mathbf{T}_0$. (Under this identification, the reflection $w_\alpha \in \mathbf{W}$ corresponds to the element $\dot{w}_\alpha \in N_{\mathbf{G}}(\mathbf{T}_0)$ in 1.3.1.) Since $\mathbf{T}_0$ and, hence, $N_{\mathbf{G}}(\mathbf{T}_0)$ are F-stable, F naturally induces an automorphism $\sigma_F\colon \mathbf{W} \to \mathbf{W}$, $g\mathbf{T}_0 \mapsto F(g)\mathbf{T}_0$ $(g \in N_{\mathbf{G}}(\mathbf{T}_0))$. It is straightforward to check that all of the above constructions and identifications are compatible, that is, we have $\sigma_F(w) = \sigma(w)$ for all $w \in \mathbf{W}$.

Finally, the numbers $\{q_s\}$ satisfy the following conditions. If $S_1, \ldots, S_r$ are the orbits of the permutation $s \mapsto s^\dagger$ on S, then

$$q^{|S_i|} = \prod_{s\in S_i} q_s \qquad (i = 1, \ldots, r), \tag{c}$$

where $q \in \mathbb{R}_{>0}$ is defined in Proposition 1.4.19. (This easily follows from the equation $\varphi(\alpha_{s^\dagger}) = q_s\alpha_s$ for all $s \in S$; see [St68, 11.17].) Hence, we also have $q^{|S|} = \prod_{s\in S} q_s$, which provides an alternative characterisation of q.

1.6.2 As in [Lu84a, 3.1], we say that $\sigma\colon \mathbf{W} \to \mathbf{W}$ is an *ordinary automorphism* if the following condition is satisfied: whenever $s \neq t$ in S are in the same $\dagger$-orbit on S, then the order of the product st is 2 or 3. With this notion, we have the following distinction of cases. The group $\mathbf{G}^F$ is

- either 'untwisted', that is, σ is the identity (and $q_s = q$ for all $s \in S$);
- or 'twisted', that is, σ is not the identity but ordinary (as defined above);
- or 'very twisted', otherwise.

The typical examples to keep in mind are: the finite general linear groups (untwisted), the finite general unitary groups (twisted) and the finite Suzuki groups (very twisted). Note that these are notions which depend on $\mathbf{G}$ and F used to define $\mathbf{G}^F$, not just on the finite group $\mathbf{G}^F$: In Example 1.4.29, there is a realisation of $\mathrm{SL}_2(q)$ (where q is a power of 2) as a twisted (but not very twisted) group.

Remark 1.6.3 Assume that F is a Frobenius map, with respect to some $\mathbb{F}_q$-rational structure on $\mathbf{G}$. Then, as pointed out by Lusztig [Lu84a, 3.4.1], the induced automorphism $\sigma\colon \mathbf{W} \to \mathbf{W}$ is *ordinary* in the sense defined above.

This is seen as follows. Let $s \neq t$ in S be in the same †-orbit. Assume that $c_{st} \neq 0$. Replacing F by a power of F if necessary, we can assume without loss of generality that $t = s^\dagger$. Now, applying † repeatedly to $\{s, t\}$, we obtain a whole †-orbit of pairs $\{s', t'\}$ where $s' \neq t'$ are in S. Then 1.6.1(a) shows that $c_{s't'} \neq 0$ for all these pairs. So, if this †-orbit of pairs had more than one element, then we would obtain a closed path in the Dynkin diagram of C, which is impossible (see Table 1.1, p. 20). Hence, we must have $t = s^\dagger$ and $s = t^\dagger$. But then 1.6.1(a) implies that $q_t c_{st} = q_s c_{s^\dagger t^\dagger} = q_s c_{ts}$. However, by Lemma 1.4.27, we have $q_s = q_t = q$. Hence, $c_{st} = c_{ts}$ and so $c_{st} = c_{ts} = -1$, which means that st has order 3, as required.

1.6.4 Let $\mathbf{T} \subseteq \mathbf{G}$ be an F-stable maximal torus. We know that $\mathbf{T}$ is conjugate in $\mathbf{G}$ to our reference torus $\mathbf{T}_0$ but it is not immediately clear how the structure of $\mathbf{T}^F$ can be described. For this purpose, one uses the following construction. Let $\mathbf{T} = g\mathbf{T}_0 g^{-1}$ where $g \in \mathbf{G}$. Then $F(g)\mathbf{T}_0 F(g)^{-1} = F(\mathbf{T}) = \mathbf{T} = g\mathbf{T}_0 g^{-1}$ and so $g^{-1}F(g) \in N_{\mathbf{G}}(\mathbf{T}_0)$. Hence, $g^{-1}F(g) = \dot{w}$ for some $w \in \mathbf{W}$. In this situation, we say that $\mathbf{T}$ is a *torus of type* w. Now define

$$F'\colon \mathbf{G} \to \mathbf{G}, \qquad x \mapsto \dot{w}F(x)\dot{w}^{-1}.$$

By Lemma 1.4.14, F' is a Steinberg map and conjugation with g defines an isomorphism $\mathbf{G}^F \cong \mathbf{G}^{F'}$. Note that $F'(\mathbf{T}_0) = \mathbf{T}_0$. For $t \in \mathbf{T}_0$, we have:

$$gtg^{-1} \in \mathbf{T}^F \;\Leftrightarrow\; F(gtg^{-1}) = gtg^{-1} \;\Leftrightarrow\; g\dot{w}F(t)\dot{w}^{-1}g^{-1} = gtg^{-1} \;\Leftrightarrow\; t \in \mathbf{T}_0^{F'}.$$

Thus, we have $\mathbf{T}^F = g\mathbf{T}_0[w]g^{-1}$ where we define

$$\mathbf{T}_0[w] := \mathbf{T}_0^{F'} = \{t \in \mathbf{T}_0 \mid F(t) = \dot{w}^{-1}t\dot{w}\}.$$

Note that $\mathbf{T}_0[w]$ is a finite subgroup of $\mathbf{T}_0$ that only depends on w, but not on the representative $\dot{w}$. (Another common notation for this subgroup is $\mathbf{T}_0^{wF}$.) In particular, $|\mathbf{T}^F| = |\mathbf{T}_0[w]| = |\mathbf{T}_0^{F'}|$. Recall from 1.6.1 that we denote by $\varphi\colon X \to X$ the map induced by F. Let $\varphi'\colon X \to X$, $\lambda \mapsto \lambda \circ F'|_{\mathbf{T}_0}$, be the map induced by F'. Then, for $\lambda \in X$ and $t \in \mathbf{T}_0$, we have

$$\begin{aligned}\varphi'(\lambda)(t) = \lambda\big(\dot{w}F(t)\dot{w}^{-1}\big) &= (w^{-1}.\lambda)\big(F(t)\big)\\ &= \varphi\big(w^{-1}.\lambda\big)(t) = \big((\varphi\circ w^{-1})(\lambda)\big)(t)\end{aligned}$$

and so $\varphi' = \varphi\circ w^{-1}$ (as maps $X \to X$); see also [Ca85, Prop. 3.3.4(i)].

Before we continue with the general discussion, we give an example which illustrates that the above constructions indeed yield an explicit description of the groups $\mathbf{T}^F$. (For more substantial examples, see [Der84], [Miz77].)

Example 1.6.5 Let $\mathbf{G} = \mathrm{GL}_n(k)$, with root datum as in Example 1.3.7, relative to the maximal torus $\mathbf{T}_0 = \{h(\xi_1, \ldots, \xi_n) \mid \xi_i \in k^\times\}$, where we let $h(\xi_1, \ldots, \xi_n)$ denote the diagonal matrix with diagonal entries $\xi_1, \ldots, \xi_n$. Let $F: \mathbf{G} \to \mathbf{G}$ be the 'standard' Frobenius map, such that $\mathbf{G}^F = \mathrm{GL}_n(q)$ (as in Example 1.4.21). We can identify $\mathbf{W}$ with the group of permutation matrices in $\mathbf{G}$. Let $w_c \in \mathbf{W}$ be the permutation matrix corresponding to the n-cycle $(1, 2, \ldots, n) \in \mathfrak{S}_n$. Then $h(\xi_1, \ldots, \xi_n) \in \mathbf{T}_0[w_c]$ if and only if

$$h(\xi_1^q, \ldots, \xi_n^q) = F(h(\xi_1, \ldots, \xi_n)) = w_c^{-1} h(\xi_1, \ldots, \xi_n) w_c = h(\xi_2, \ldots, \xi_n, \xi_1).$$

This yields the following explicit description:

$$\mathbf{T}_0[w_c] = \{h(\xi, \xi^q, \xi^{q^2}, \ldots, \xi^{q^{n-1}}) \mid \xi \in k^\times, \xi^{q^n} = \xi\} \cong \mathbb{F}_{q^n}^\times,$$

which shows that $\mathbf{T}_0[w_c]$ is cyclic of order $q^n - 1$. Thus, if $\mathbf{T} \subseteq \mathbf{G}$ is an F-stable maximal torus of type w_c, then $\mathbf{T}^F \cong \mathbf{T}_0[w_c]$ is a subgroup of $\mathbf{G}^F$ which is usually called a *Singer cycle* in finite group theory (see, e.g., [Hup67, Satz II.7.3]).

Lemma 1.6.6 *In the setting of 1.6.4, we extend scalars to $\mathbb{R}$ and write $\varphi_\mathbb{R} = q\varphi_0$ where $\varphi_0 \in \mathrm{GL}(X_\mathbb{R})$ is a linear map of finite order (see Proposition 1.4.19). Then*

$$|\mathbf{T}^F| = |\mathbf{T}_0[w]| = \pm \det(\varphi \circ w^{-1} - \mathrm{id}_X) = \det(q\, \mathrm{id}_{X_\mathbb{R}} - \varphi_0 \circ w^{-1}),$$

and this equals $\det(q\, \mathrm{id}_{X_\mathbb{R}} - w^{-1} \circ \varphi_0)$. *In particular,* $|\mathbf{T}_0^F| = \det(q\, \mathrm{id}_{X_\mathbb{R}} - \varphi_0)$.

Proof Consider the algebraic homomorphism $f': \mathbf{T}_0 \to \mathbf{T}_0$, $t \mapsto t^{-1}F'(t)$, with $\ker(f') = \mathbf{T}_0^{F'}$. By the Lang–Steinberg theorem, f' is surjective. So, by 1.1.11(b), the restriction map $X \to \mathrm{Hom}(\mathbf{T}_0^{F'}, k^\times)$ is surjective with kernel $\{\lambda \circ f' \mid \lambda \in X\} = (\varphi' - \mathrm{id}_X)(X)$ (where $\varphi': X \to X$ is the map induced by F', as in 1.6.4). Hence, we have

$$|X/(\varphi' - \mathrm{id}_X)(X)| = |\mathrm{Hom}(\mathbf{T}_0^{F'}, k^\times)| = |\mathbf{T}_0^{F'}|$$

where the second equality holds since $\mathbf{T}_0^{F'}$ has order prime to p. By considering the elementary divisors of $\varphi' - \mathrm{id}_X: X \to X$, the above cardinalities are also equal to $\pm \det(\varphi' - \mathrm{id}_X)$. So we conclude that

$$|\mathbf{T}^F| = |\mathbf{T}_0[w]| = |\mathbf{T}_0^{F'}| = |\mathrm{Irr}(\mathbf{T}_0^{F'})| = \pm \det(\varphi' - \mathrm{id}_X).$$

Extending scalars to $\mathbb{R}$, we have $\varphi'_\mathbb{R} = \varphi_\mathbb{R} \circ w^{-1} = q\varphi_0 \circ w^{-1}$, which yields:

$$\det(\varphi' - \mathrm{id}_X) = \det(q\, \varphi_0 \circ w^{-1} - \mathrm{id}_{X_\mathbb{R}}) = \det(\varphi_0 \circ w^{-1}) \det(q\, \mathrm{id}_{X_\mathbb{R}} - w \circ \varphi_0^{-1}).$$

Now the formula in 1.6.1(b) implies an analogous formula with φ replaced by φ_0; thus, φ_0 normalises $\mathbf{W}$. Hence, since φ_0 has finite order, it follows that the map $w \circ \varphi_0^{-1}: X_\mathbb{R} \to X_\mathbb{R}$ has finite order. But then the characteristic polynomials of

$w\circ\varphi_0^{-1}$ and its inverse $\varphi_0\circ w^{-1}$ are equal; furthermore, $\det(\varphi_0\circ w^{-1}) = \pm 1$. Hence, $|\mathbf{T}^F| = \pm\det(q\,\mathrm{id}_{X_\mathbb{R}} - \varphi_0\circ w^{-1})$.

Finally, arguing as in [Ca85, Prop. 3.3.5], one sees that the latter determinant is a strictly positive real number. Hence, the sign must be $+1$. Clearly, $\varphi_0\circ w^{-1}$ and $w^{-1}\circ\varphi_0$ have the same characteristic polynomial. □

We now have all the ingredients to state the order formula for $\mathbf{G}^F$. Since $\mathbf{B}_0$ is F-stable, its unipotent radical $\mathbf{U}_0 = R_u(\mathbf{B}_0)$ is also F-stable. Since $\mathbf{B}_0 = \mathbf{U}_0.\mathbf{T}_0$ where $\mathbf{U}_0 \cap \mathbf{T}_0 = \{1\}$, we obtain

$$|\mathbf{G}^F| = |\mathbf{U}_0^F|\cdot|\mathbf{T}_0^F|\cdot|\mathbf{G}^F/\mathbf{B}_0^F|.$$

The factor $|\mathbf{T}_0^F|$ is given by Lemma 1.6.6. A further evaluation of the remaining two factors in the above expression for $|\mathbf{G}^F|$ leads to the following formula, due to Chevalley [Ch55] (in the case where $\mathbf{G}$ is semisimple and $\dagger$ is the identity) and Steinberg [St68, §11] (in general).

Theorem 1.6.7 (Order formula) *With the above notation, we have*

$$|\mathbf{G}^F| = q^{|R|/2}\det(q\,\mathrm{id}_{X_\mathbb{R}} - \varphi_0)\sum_{w\in\mathbf{W}^\sigma} q^{l(w)},$$

where $\mathbf{W}^\sigma = \{w \in \mathbf{W} \mid \sigma(w) = w\}$ *(which is a finite Coxeter group). When* $\mathbf{G}$ *is simple, formulae for the various possibilities are given in Table 1.3.*

The fact that the list in Table 1.3 exhausts all the possible pairs $(\mathbf{G}, F)$ where $\mathbf{G}$ is simple and $F\colon \mathbf{G}\to\mathbf{G}$ is a Steinberg map is shown in [St68, §11.6]; note that, in this case, we have $|\mathbf{G}^F| = |\mathbf{G}_{\mathrm{ad}}^F| = |\mathbf{G}_{\mathrm{sc}}^F|$ (see 1.5.11).

Remark 1.6.8 (a) The formula shows that $q^{|R|/2}$ is the p-part of the order of $\mathbf{G}^F$; this provides a further characterisation of the number q. We also note that the above expression for $|\mathbf{G}^F|$ can be interpreted as a polynomial in one variable evaluated at q, where the polynomial only depends on the root datum of $\mathbf{G}$ and the maps φ_0, σ derived from F. This will be formalised in Definition 1.6.10 below.

(b) Steinberg [St68, 14.14] shows that the number of F-stable maximal tori of $\mathbf{G}$ is equal to $q^{|R|}$. This equality is in fact equivalent to the following Molien series identity which yields another expression for the order of $\mathbf{G}^F$:

$$|\mathbf{G}^F| = q^{|R|}\Big(\frac{1}{|\mathbf{W}|}\sum_{w\in W}\frac{1}{|\mathbf{T}_0[w]|}\Big)^{-1} = q^{|R|}\Big(\frac{1}{|\mathbf{W}|}\sum_{w\in\mathbf{W}}\frac{1}{\det(q\,\mathrm{id}_{X_\mathbb{R}} - \varphi_0\circ w^{-1})}\Big)^{-1}$$

see [Ca85, §3.4], [MaTe11, Exc. 30.15] for further details. One advantage of this expression is that it does not involve the length function on $\mathbf{W}$ or the induced automorphism σ of $\mathbf{W}$. The inverse of the sum on the right-hand side can be further expressed as a product of various cyclotomic polynomials; in this way, one obtains the familiar formulae for the order of $\mathbf{G}^F$ when $\mathbf{G}$ is simple; see Table 1.3. Also note

Table 1.3 *Order formulae for* $|\mathbf{G}^F|$ *when* $\mathbf{G}$ *is simple*

Type	$\lvert\mathbf{G}^F\rvert$	
A_{n-1}	$q^{n(n-1)/2}(q^2-1)(q^3-1)\dots(q^n-1)$	
B_n	$q^{n^2}(q^2-1)(q^4-1)\dots(q^{2n}-1)$	
C_n	same as B_n	
D_n	$q^{n^2-n}(q^2-1)(q^4-1)\dots(q^{2n-2}-1)(q^n-1)$	
G_2	$q^6(q^2-1)(q^6-1)$	
F_4	$q^{24}(q^2-1)(q^6-1)(q^8-1)(q^{12}-1)$	
E_6	$q^{36}(q^2-1)(q^5-1)(q^6-1)(q^8-1)(q^9-1)(q^{12}-1)$	
E_7	$q^{63}(q^2-1)(q^6-1)(q^8-1)(q^{10}-1)(q^{12}-1)(q^{14}-1)(q^{18}-1)$	
E_8	$q^{120}(q^2-1)(q^8-1)(q^{12}-1)(q^{14}-1)(q^{18}-1)(q^{20}-1)(q^{24}-1)(q^{30}-1)$	
${}^2A_{n-1}$	$q^{n(n-1)/2}(q^2-1)(q^3+1)\dots(q^n-(-1)^n)$	
${}^2D_n$	$q^{n^2-n}(q^2-1)(q^4-1)\dots(q^{2n-2}-1)(q^n+1)$	
${}^3D_4$	$q^{12}(q^2-1)(q^6-1)(q^8+q^4+1)$	
${}^2E_6$	$q^{36}(q^2-1)(q^5+1)(q^6-1)(q^8-1)(q^9+1)(q^{12}-1)$	
${}^2B_2$	$q^4(q^2-1)(q^4+1)$	$(q=\sqrt{2}^{2m+1})$
${}^2G_2$	$q^6(q^2-1)(q^6+1)$	$(q=\sqrt{3}^{2m+1})$
${}^2F_4$	$q^{24}(q^2-1)(q^6+1)(q^8-1)(q^{12}+1)$	$(q=\sqrt{2}^{2m+1})$

(The first 9 are 'untwisted', the next 4 'twisted', the last 3 'very twisted'.)

that, instead of considering the maps $\varphi_0\circ w^{-1}\colon X_{\mathbb{R}} \to X_{\mathbb{R}}$ in the above formula, one can take their transposes $(\varphi_0\circ w^{-1})^{\mathrm{tr}}\colon Y_{\mathbb{R}} \to Y_{\mathbb{R}}$ where $Y_{\mathbb{R}} = \mathbb{R}\otimes_{\mathbb{Z}} Y$: the characteristic polynomials will certainly remain the same.

(c) If F is a Frobenius map, then Theorem 1.6.7 can be proved by a general argument; see [Ge03a, 4.2.5]. The general case (where the root exponents q_s may not all be equal) is treated in [St68, §11] (see also [MaTe11, §24.1]), assuming that $\mathbf{G}$ is semisimple and 'F-simple' (see 1.5.14). But, as already noted in [St68, p. 78], the case of an arbitrary connected reductive $\mathbf{G}$ can be easily recovered from this case. As an illustration of the methods developed in the previous section, let us explicitly work out the reduction argument.

Lemma 1.6.9 *Suppose that the formula in Theorem 1.6.7 is known to hold when* $\mathbf{G}$ *is simple of adjoint type. Then the formula holds in general.*

Proof Since the formula for $|\mathbf{T}_0^F|$ in Lemma 1.6.6 is already known to hold in general, it will be sufficient to consider the cardinality of $\mathbf{G}^F/\mathbf{T}_0^F$. It will be convenient to slightly rephrase this as follows. Let us consider the set of cosets $\mathbf{G}/\mathbf{T}_0 = \{g\mathbf{T}_0 \mid g \in \mathbf{G}\}$ (just as an abstract set, we don't need the notion of a quotient variety here). Since $\mathbf{T}_0$ is F-stable, we have an induced action of F on $\mathbf{G}/\mathbf{T}_0$.

Consequently, we have a natural injective map $\mathbf{G}^F/\mathbf{T}_0^F \to (\mathbf{G}/\mathbf{T}_0)^F$, $g\mathbf{T}_0^F \mapsto g\mathbf{T}_0$. Now the connected group $\mathbf{T}_0$ acts transitively on $g\mathbf{T}_0$ by right multiplication. Hence, if $g\mathbf{T}_0$ is F-stable, then Proposition 1.4.9 shows that $g\mathbf{T}_0$ contains a representative fixed by F. It follows that the above map is surjective and so $|\mathbf{G}^F/\mathbf{T}_0^F| = |(\mathbf{G}/\mathbf{T}_0)^F|$. Thus, it will now be sufficient to consider the identity:

$$|(\mathbf{G}/\mathbf{T}_0)^F| = \mathbb{O}(\mathbf{W}, \sigma, q) \quad \text{where} \quad \mathbb{O}(\mathbf{W}, \sigma, q) := q^{|R|/2} \sum_{w \in \mathbf{W}^\sigma} q^{l(w)}. \qquad (*)$$

Since the formula in Theorem 1.6.7 is assumed to hold when $\mathbf{G}$ is simple of adjoint type, the same is true of the formula $(*)$. We must deduce from this that $(*)$ holds in general. We do this in two steps.

(1) First assume that $\mathbf{G}$ is semisimple of adjoint type. As in 1.5.14(b), we have a direct product decomposition $\mathbf{G} = \mathbf{G}_1 \times \cdots \times \mathbf{G}_n$ where each $\mathbf{G}_i$ is simple of adjoint type. Furthermore, there is a permutation ρ of $\{1, \ldots, n\}$ such that $F(\mathbf{G}_i) = \mathbf{G}_{\rho(i)}$ for $i = 1, \ldots, n$. Now note that $\mathbf{T}_0 = \mathbf{T}_1 \times \cdots \times \mathbf{T}_n$ where each $\mathbf{T}_i$ (for $i = 1, \ldots, n$) is a maximal torus of $\mathbf{G}_i$ such that $F(\mathbf{T}_i) = \mathbf{T}_{\rho(i)}$. Hence, we can also identify $\mathbf{G}/\mathbf{T}_0$ with $\mathbf{G}_1/\mathbf{T}_1 \times \cdots \times \mathbf{G}_n/\mathbf{T}_n$. Furthermore, if $I \subseteq \{1, \ldots, n\}$ and n_i $(i \in I)$ are as in Corollary 1.5.16, then $F^{n_i}(\mathbf{G}_i/\mathbf{T}_i) = \mathbf{G}_i/\mathbf{T}_i$ for all $i \in I$ and

$$|(\mathbf{G}/\mathbf{T}_0)^F| = \prod_{i \in I} |(\mathbf{G}_i/\mathbf{T}_i)^{F^{n_i}}|. \qquad (1a)$$

It remains to show that there is a similar factorisation of the right-hand side of $(*)$. Recall from 1.5.14 that we have a partition $R = R_1 \sqcup \cdots \sqcup R_n$. Consequently, we also have a direct product decomposition

$$\mathbf{W} = \mathbf{W}_1 \times \cdots \times \mathbf{W}_n \qquad \text{where} \qquad \mathbf{W}_i := \langle w_\alpha \mid \alpha \in R_i \rangle;$$

furthermore, $\sigma(\mathbf{W}_i) = \mathbf{W}_{\rho(i)}$ for $i = 1, \ldots, n$. Here, $\mathbf{W}_i$ is the Weyl group of the factor $\mathbf{G}_i$ (relative to $\mathbf{T}_i \subseteq \mathbf{G}_i$). Now, if $w \in \mathbf{W}$ and $w = w_1 \cdots w_n$ with $w_i \in \mathbf{W}_i$ for all i, then $l(w) = l(w_1) + \cdots + l(w_n)$. Using this formula, it is straightforward to verify that the expression for $\mathbb{O}(W, \sigma, q)$ is compatible with the above product decomposition, that is, we have $\sigma^{n_i}(\mathbf{W}_i) = \mathbf{W}_i$ for all $i \in I$ and

$$\mathbb{O}(\mathbf{W}, \sigma, q) = \prod_{i \in I} \mathbb{O}(\mathbf{W}_i, \sigma^{n_i}, q^{n_i}). \qquad (1b)$$

By assumption and Lemma 1.5.15, we have $|(\mathbf{G}_i/\mathbf{T}_i)^{F^{n_i}}| = \mathbb{O}(\mathbf{W}_i, \sigma^{n_i}, q^{n_i})$ for $i \in I$. Comparing (1a) and (1b), we see that $(*)$ holds for $\mathbf{G}$ as well.

(2) Now let $\mathbf{G}$ be arbitrary (connected and reductive). As in Remark 1.5.12, we consider an adjoint quotient $\pi_{\mathrm{ad}} \colon \mathbf{G} \to \mathbf{G}_{\mathrm{ad}}$ with kernel $\mathbf{Z} = \mathbf{Z}(\mathbf{G})$. By Remark 1.3.5(a), we have $\mathbf{Z} \subseteq \mathbf{T}_0$; furthermore, $\mathbf{T}' := \pi_{\mathrm{ad}}(\mathbf{T}_0)$ is an F-stable maximal torus of $\mathbf{G}_{\mathrm{ad}}$ (see 1.3.10(a)). So we get a bijective map

$$\mathbf{G}/\mathbf{T}_0 \to \mathbf{G}_{\mathrm{ad}}/\mathbf{T}', \qquad g\mathbf{T}_0 \mapsto \pi_{\mathrm{ad}}(g)\mathbf{T}',$$

which is compatible with the action of F on $\mathbf{G}/\mathbf{T}_0$ and on $\mathbf{G}_{\mathrm{ad}}/\mathbf{T}'$. In particular, $|(\mathbf{G}/\mathbf{T}_0)^F| = |(\mathbf{G}_{\mathrm{ad}}/\mathbf{T}')^F|$ and so the left-hand side of $(*)$ does not change when we pass from $\mathbf{G}$ to $\mathbf{G}_{\mathrm{ad}}$. On the other hand, by 1.3.10(d), π_{ad} induces an F-equivariant isomorphism from the Weyl group of $\mathbf{G}$ (relative to $\mathbf{T}_0$) onto the Weyl group of $\mathbf{G}_{\mathrm{ad}}$ (relative to $\mathbf{T}'$). Hence, the right-hand side of $(*)$ does not change either when we pass from $\mathbf{G}$ to $\mathbf{G}_{\mathrm{ad}}$. Thus, if $(*)$ holds for $\mathbf{G}_{\mathrm{ad}}$, then $(*)$ also holds for $\mathbf{G}$. $\square$

Following [BrMa92], we now formally introduce 'series of finite groups of Lie type'. This relies on the following definition, which is a slight modification of [BrMa92, §1]. (See Example 1.6.11 below for further comments.)

Definition 1.6.10 Let $\mathscr{R} = (X, R, Y, R^\vee)$ be a root datum, with Weyl group $\mathbf{W} \subseteq \mathrm{Aut}(X)$. We set $X_{\mathbb{R}} := \mathbb{R} \otimes_{\mathbb{Z}} X$. We can canonically regard X as a subset of $X_{\mathbb{R}}$; we also regard $\mathbf{W}$ as a subgroup of $\mathrm{GL}(X_{\mathbb{R}})$. Let $\varphi_0 \in \mathrm{GL}(X_{\mathbb{R}})$ be an invertible linear map of finite order which normalises $\mathbf{W}$, and assume that

$$\mathscr{P} := \left\{ q \in \mathbb{R}_{>0} \;\middle|\; \begin{array}{l} q\varphi_0(X) \subseteq X \text{ and the corresponding map} \\ q\varphi_0 \colon X \to X \text{ is a } p\text{-isogeny for some prime } p \end{array} \right\}$$

is non-empty. We form the coset $\varphi_0\mathbf{W} = \{\varphi_0 \circ w \mid w \in \mathbf{W}\} \subseteq \mathrm{GL}(X_{\mathbb{R}})$. Then

$$\mathbb{G} = \big((X, R, Y, R^\vee), \varphi_0\mathbf{W}\big)$$

is called a *complete root datum* or a *generic finite reductive group*; we also write $\mathscr{P} = \mathscr{P}_{\mathbb{G}}$ in this case. We define a corresponding rational function $|\mathbb{G}| \in \mathbb{R}(\mathbf{q})$ (where $\mathbf{q}$ is an indeterminate) by

$$|\mathbb{G}| = \mathbf{q}^{|R|}\left(\frac{1}{|\mathbf{W}|}\sum_{w \in \mathbf{W}} \frac{1}{\det(\mathbf{q}\,\mathrm{id}_{X_{\mathbb{R}}} - \varphi_0 \circ w^{-1})}\right)^{-1}.$$

We call $|\mathbb{G}|$ the *order polynomial* of $\mathbb{G}$; this will be justified in Remark 1.6.15 below. Note that $\mathscr{P}$ is an infinite set: If $q \in \mathscr{P}$ and $q\varphi_0$ is a p-isogeny (where p is a prime), then $p^m q \in \mathscr{P}$ for all integers $m \geqslant 1$.

Example 1.6.11 Let $\mathbf{G}$ be connected reductive and $F \colon \mathbf{G} \to \mathbf{G}$ be a Steinberg map. Then we obtain a corresponding complete root datum by taking the root datum of $\mathbf{G}$ (relative to a maximally split torus $\mathbf{T}_0 \subseteq \mathbf{G}$ as in 1.6.1) together with the linear map φ_0 defined in Proposition 1.4.19(b) (such that $\varphi_{\mathbb{R}} = q\,\varphi_0$). In particular, this includes all the cases discussed in Examples 1.4.21, 1.4.22, 1.4.23. This shows that the above Definition 1.6.10 is somewhat more general than that in [BrMa92], in which cases like those in Example 1.4.23 are not included.

Let us now fix a complete root datum $\mathbb{G} = \big((X, R, Y, R^\vee), \varphi_0\mathbf{W}\big)$.

Remark 1.6.12 Let $q \in \mathscr{P}$ and set $\varphi := q\varphi_0$. Then φ is a p-isogeny (for some prime p) and we have $\varphi^d = q^d \mathrm{id}_X$, where $d \geqslant 1$ is the order of φ_0. In particular, this implies that $q^d = p^m$ for some $m \geqslant 1$. Let $\mathbf{G}$ be a connected reductive algebraic group over $k = \overline{\mathbb{F}}_p$ whose root datum (relative to a maximal torus $\mathbf{T} \subseteq \mathbf{G}$) is isomorphic to $(X, R, Y, R^\vee)$. Then φ gives rise to an isogeny $F: \mathbf{G} \to \mathbf{G}$ which is a Steinberg map by Proposition 1.4.18. We write $\mathbb{G}(q) := \mathbf{G}^F$. Thus, we obtain a family of finite groups

$$\{\mathbb{G}(q) \mid q \in \mathscr{P}\}$$

which we call the *series of finite groups of Lie type* defined by $\mathbb{G}$. There are some choices involved in the definition of $\mathbb{G}(q)$ but we shall see in the remarks below that different choices lead to isomorphic finite groups.

Remark 1.6.13 Since the map $\varphi_0 \in \mathrm{GL}(X_\mathbb{R})$ normalises $\mathbf{W}$, we obtain a group automorphism $\sigma: \mathbf{W} \to \mathbf{W}$ such that

$$\sigma(w) = \varphi_0^{-1} \circ w \circ \varphi_0 \qquad \text{for all } w \in \mathbf{W}.$$

Note that this is compatible with Remark 1.2.10: For any $q \in \mathscr{P}$, the automorphism of $\mathbf{W}$ induced by the p-isogeny $q\varphi_0$ (where p is a prime) is given by σ. (The notation is also compatible with 1.6.1(b).) Let us now see what happens when we replace φ_0 by another map in $\varphi_0\mathbf{W}$. First note that $\varphi_0 \circ w$ has finite order for any $w \in \mathbf{W}$. Furthermore, if $q \in \mathscr{P}$ and $q\varphi_0$ is a p-isogeny (where p is a prime), then $(q\varphi_0)\circ w$ also is a p-isogeny. Thus, if φ_0 satisfies the defining conditions for a complete root datum, then so does $\varphi_0 \circ w$ for any $w \in \mathbf{W}$. We also note the following identity:

$$(\varphi_0 \circ w)^m = \varphi_0^m \circ \big(\sigma^{m-1}(w) \cdots \sigma^2(w)\sigma(w)w\big) \qquad \text{for all } m \geqslant 1.$$

Now let $q \in \mathscr{P}$ and let $\mathbf{G}, \mathbf{T}, F$ be as in Remark 1.6.12. Then $\varphi := q\varphi_0$ is the linear map induced on $X \cong X(\mathbf{T})$ by F. Let $w \in \mathbf{W}$ and $\dot{w}$ be a representative of w in $N_\mathbf{G}(\mathbf{T})$. We define $F': \mathbf{G} \to \mathbf{G}$ by $F'(g) := \dot{w}^{-1}F(g)\dot{w}$ for $g \in \mathbf{G}$. By Lemma 1.4.14, F' also is a Steinberg map and we have $\mathbf{G}^{F'} \cong \mathbf{G}^F$. Now $\mathbf{T}$ is F'-stable and one easily sees that $\varphi \circ w: X \to X$ is the linear map induced by F'. (See once more 1.6.4.) This shows that, if we replace φ_0 by $\varphi_0 \circ w$ for some $w \in \mathbf{W}$, then F changes to F' but we obtain isomorphic finite groups.

Remark 1.6.14 Let $q \in \mathscr{P}$ and set $\varphi := q\varphi_0$. Then φ is a p-isogeny (where p is a prime) and so there is a corresponding permutation $\alpha \mapsto \alpha^\dagger$ of R; we have $\varphi(\alpha^\dagger) = q_\alpha \alpha$ for all $\alpha \in R$, where $\{q_\alpha\}$ are the root exponents of φ. Let $\sigma: \mathbf{W} \to \mathbf{W}$ be the group automorphism in Remark 1.6.13. By Remark 1.2.10, we have $\sigma(w_\alpha) = w_{\alpha^\dagger}$ for all $\alpha \in R$; also recall that the root exponents are positive. Hence, we conclude that the permutation $\alpha \mapsto \alpha^\dagger$ only depends on φ_0, but not on q.

Remark 1.6.15 Let us fix a base Π of R. Since any two bases can be transformed into each other by a unique element of $\mathbf{W}$, there is a unique $w \in \mathbf{W}$ such that, if we replace φ_0 by $\varphi_0' := \varphi_0 \circ w$, then $\Pi^\dagger = \Pi$ where $\alpha \mapsto \alpha^\dagger$ is the permutation induced by φ_0'; see Remarks 1.2.11 and 1.6.14. Assume now that this is the case. Let $q \in \mathscr{P}$ and $\mathbf{G}, \mathbf{T}, F$ be as in Remark 1.6.12. Then $\mathbf{T}$ lies in the F-stable Borel subgroup $\mathbf{B} = \langle \mathbf{T}, \mathbf{U}_\alpha \mid \alpha \in R^+ \rangle$ (where R^+ are the positive roots with respect to Π) and we are in the setting of 1.6.1, where $\mathbf{T}_0 := \mathbf{T}$. So, by Remark 1.6.8, we have $|\mathbf{G}^F| = |\mathbb{G}|(q)$ and this is also equal to the expression in Theorem 1.6.7. Since this holds for all $q \in \mathscr{P}$, we obtain an identity of rational functions in $\mathbf{q}$:

$$|\mathbb{G}| = \mathbf{q}^{|R|/2} \det(\mathbf{q}\,\mathrm{id}_{X_\mathbb{R}} - \varphi_0) \sum_{w \in \mathbf{W}^\sigma} \mathbf{q}^{l(w)}. \tag{a}$$

Thus, the rational function $|\mathbb{G}|$ actually is a polynomial in $\mathbf{q}$ such that $|\mathbf{G}^F| = |\mathbb{G}|(q)$. This provides the justification for calling $|\mathbb{G}|$ the *order polynomial* of $\mathbb{G}$ (see also [BrMa92, 1.12]). Note that $\det(\mathbf{q}\,\mathrm{id}_{X_\mathbb{R}} - \varphi_0)$ has degree $\dim \mathbf{T}$ and the polynomial $\sum_{w \in \mathbf{W}^\sigma} \mathbf{q}^{l(w)}$ has degree $|R|/2$. Hence, we conclude that

$$|\mathbb{G}| \in \mathbb{R}[\mathbf{q}] \text{ has degree } \dim \mathbf{G} = |R| + \dim \mathbf{T}. \tag{b}$$

Now let $K \subseteq \mathbb{R}$ be a subfield such that the polynomial $\det(\mathbf{q}\,\mathrm{id}_{X_\mathbb{R}} - \varphi_0)$ has coefficients in K. Since φ_0 has finite order, all roots of this polynomial are roots of unity. By [St68, 2.1], an analogous result is also true for the term $\sum_w \mathbf{q}^{l(w)}$ in (a). So there is a factorisation

$$|\mathbb{G}| = \mathbf{q}^{|R/2|} \times (\text{product of cyclotomic polynomials in } K[\mathbf{q}]). \tag{c}$$

If $\mathbf{G}$ is simple, then such factorisations can be seen explicitly in Table 1.3 (p. 73). If $F \colon \mathbf{G} \to \mathbf{G}$ is a Frobenius map, then we can take $K = \mathbb{Q}$.

Remark 1.6.16 Let $\Pi = \{\alpha_s \mid s \in S\}$ be a base of R and $C = (c_{st})_{s,t \in S}$ be the corresponding Cartan matrix; also choose a $\mathbb{Z}$-basis of X. Then $\mathscr{R}$ is determined by a factorisation $C = \breve{A} \cdot A^{\mathrm{tr}}$ as in Remark 1.2.13. Assume that φ_0 is chosen so that the permutation of R induced by φ_0 leaves Π invariant (which is possible by Remark 1.6.15). Let Q be the matrix of $\varphi_0 \colon X_\mathbb{R} \to X_\mathbb{R}$ (with respect to the chosen basis of X). If $q \in \mathscr{P}$, then $q\varphi_0$ is a p-isogeny (for some prime p) and the conditions (MI1), (MI2) in 1.2.18 show that $qQA^{\mathrm{tr}} = A^{\mathrm{tr}}P^\circ$ and $P^\circ \breve{A} = q\breve{A}Q$, where P° is a monomial matrix whose non-zero entries are all powers of p. It follows that

$$QA^{\mathrm{tr}} = A^{\mathrm{tr}}Q^\circ \qquad \text{and} \qquad Q^\circ \breve{A} = \breve{A}Q,$$

where $Q^\circ := q^{-1}P^\circ$ is a monomial matrix; each non-zero entry of Q° is a positive real number such that some positive power of it is an integral power of p. Now note that, although the pair (P, P°) is used in the construction, Q° is uniquely determined by Q and, hence, independent of (P, P°). There are two cases:

(I) All non-zero entries of Q° are equal to 1. Then φ_0 is a 1-isogeny, as in Example 1.4.21. Consequently, the set $\mathscr{P}$ consists of *all* prime powers. This is what is called the '*cas général*' in [BrMa92, §1].

(II) Otherwise, there is a unique prime number p such that each non-zero entry of Q° has the property that some positive integral power of it is an integral power of p. In this case, $\mathscr{P}$ will only consist of positive real numbers q such that $q\varphi_0$ is a p-isogeny for this prime p.

In case (I), we will also say that $\mathbb{G}$ is 'ordinary'. Note that then the induced map $\sigma\colon \mathbf{W} \to \mathbf{W}$ is ordinary in the sense of 1.6.2. Here is an example where we are in case (II). Consider the root datum of Cartan type $A_1 \times A_1$ in Example 1.4.29(b), where φ_0 is determined by a certain matrix of order 2, denoted P_0. Then

$$Q = P_0 = \begin{pmatrix} 0 & 1 \\ 1 & 0 \end{pmatrix} \quad \text{and} \quad Q^\circ = \begin{pmatrix} 0 & 1/2 \\ 2 & 0 \end{pmatrix}.$$

So we are in case (II) where $p = 2$ and $\mathscr{P} = \{2^m \mid m \geqslant 1\}$. Similarly, the complete root data of the Suzuki and Ree groups in Example 1.4.22 are of type (II), where $\mathscr{P} = \{\sqrt{2}^{2m+1} \mid m \geqslant 0\}$ (for ${}^2B_2$, ${}^2F_4$) or $\mathscr{P} = \{\sqrt{3}^{2m+1} \mid m \geqslant 0\}$ (for ${}^2G_2$).

Definition 1.6.17 (See [BrMa92, p. 250], [BMM93, 1.5]) The *Ennola dual* of a complete root datum $\mathbb{G} = \big((X, R, Y, R^\vee), \varphi_0\mathbf{W}\big)$ is defined by

$$\mathbb{G}^- := \big((X, R, Y, R^\vee), -\varphi_0\mathbf{W}\big).$$

(Note that $\mathbb{G}^-$ is a complete root datum since, for any p-isogeny of root data $\varphi\colon X \to X$, the map $-\varphi$ also is a p-isogeny of root data; in particular, $\mathscr{P}_{\mathbb{G}^-} = \mathscr{P}_{\mathbb{G}}$.) In this situation, we write $\mathbb{G}(-q) := \mathbb{G}^-(q)$ for any $q \in \mathscr{P}_{\mathbb{G}}$. We have

$$|\mathbb{G}^-|(\mathbf{q}) = (-1)^{\operatorname{rank} X}|\mathbb{G}|(-\mathbf{q}).$$

For the origin of the name 'Ennola dual', see Example 1.6.18 below.

Example 1.6.18 (a) Assume that $-\mathrm{id}_X \in \mathbf{W}$. Then, clearly, we have $\mathbb{G}^- = \mathbb{G}$ and $|\mathbb{G}^-| = |\mathbb{G}| \in \mathbb{R}[\mathbf{q}]$.

(b) Let $\mathbf{G} = \mathrm{GL}_n(k)$ and $\mathbf{T}_0 \subseteq \mathbf{G}$ be the maximal torus consisting of the diagonal matrices in $\mathbf{G}$. We have described the corresponding root datum in Example 1.3.7. If $\varphi_0 := \mathrm{id}_{X_\mathbb{R}}$ and $\mathbb{G}$ is the corresponding complete root datum, then $\mathbb{G}(q) \cong \mathrm{GL}_n(q)$ for all prime powers q. We claim that

$$\mathbb{G}(-q) \cong \mathrm{GU}_n(q) \qquad \text{for all } q \in \mathscr{P}_{\mathbb{G}}.$$

This is seen as follows. Let $\tau\colon \mathbf{G} \to \mathbf{G}$ be the automorphism which sends an invertible matrix to its transpose inverse. Then $\tau(\mathbf{T}_0) = \mathbf{T}_0$ and the induced map on X is $-\mathrm{id}_X$. (Thus, τ is a concrete realisation of the isogeny in Example 1.3.16.) Let $F_q\colon \mathbf{G} \to \mathbf{G}$ be the standard Frobenius map (raising every matrix entry to its qth power). Then τ commutes with F_q and so $F' = \tau \circ F_q$ is a Frobenius map on $\mathbf{G}$; see Remark 1.4.4(c). We have $F'(\mathbf{T}_0) = \mathbf{T}_0$ and the induced map on X is given by

$-q\,\mathrm{id}_X$. Thus, $(\mathbf{G}, F')$ gives rise to the complete root datum $\mathbb{G}^-$; finally, note that $\mathbf{G}^{F'} \cong \mathrm{GU}_n(q)$. (The difference between this realisation of $\mathrm{GU}_n(q)$ and the one in Example 1.4.21 is that, here, $\mathbf{T}_0$ is not a maximally split torus for F'.) Now the identity $|\mathbb{G}^-|(\mathbf{q}) = (-1)^{\mathrm{rank}X}|\mathbb{G}|(-\mathbf{q})$ gives an a priori explanation for the fact that the order formula for $\mathrm{GU}_n(q)$ in Table 1.3 (p. 73) is obtained from that of $\mathrm{GL}_n(q)$ by simply changing q to $-q$ (and fixing the total sign). Ennola [Enn63] observed that a similar statement should even be true for the irreducible characters of these groups; we will discuss this in further detail at the end of Section 2.8.

Example 1.6.19 We define the *dual complete root datum* of the root datum $\mathbb{G} = \big((X, R, Y, R^\vee), \varphi_0\mathbf{W}\big)$ by

$$\mathbb{G}^* := \big((Y, R^\vee, X, R), \varphi_0^{\mathrm{tr}}\mathbf{W}\big),$$

where $\varphi_0^{\mathrm{tr}}\colon Y_{\mathbb{R}} \to Y_{\mathbb{R}}$ is the transpose map defined, as in 1.2.2, through the canonical extension of the pairing $\langle\ ,\ \rangle\colon X \times Y \to \mathbb{Z}$ to a pairing $X_{\mathbb{R}} \times Y_{\mathbb{R}} \to \mathbb{R}$. Here, we also use the identification of $\mathbf{W}^\vee \subseteq \mathrm{Aut}(Y)$ with $\mathbf{W}$, as in Remark 1.2.12. We have $\mathscr{P}_{\mathbb{G}^*} = \mathscr{P}_{\mathbb{G}}$. Now, for each $q \in \mathscr{P}_{\mathbb{G}}$, we obtain a finite group $\mathbb{G}(q)$ (arising from a pair $(\mathbf{G}, F)$ as in Remark 1.6.12) and a finite group $\mathbb{G}^*(q)$ (arising from an analogous pair $(\mathbf{G}^*, F^*)$). We then see that $(\mathbf{G}, F)$ and $(\mathbf{G}^*, F^*)$ are in duality as in Definition 1.5.17. By Theorem 1.6.7 and the argument in [Ca85, 4.4.4], we have

$$|\mathbb{G}^*|(\mathbf{q}) = |\mathbb{G}|(\mathbf{q}) \qquad \text{and} \qquad |\mathbf{G}(q)| = |\mathbf{G}^*(q)| \qquad \text{for all } q \in \mathscr{P}_{\mathbb{G}}.$$

Definition 1.6.20 (See [BrMa92, 1.1]) Let $\mathbb{G} = \big((X, R, Y, R^\vee), \varphi_0\mathbf{W}\big)$ be a complete root datum. For any $w \in \mathbf{W}$, the complete root datum

$$\mathbb{T}_w := \big((X, \varnothing, Y, \varnothing), \varphi_0 \circ w^{-1}\big)$$

is called a *maximal toric sub-datum* of $\mathbb{G}$. (We take $\varphi_0 \circ w^{-1}$ here in order to have consistency with 1.6.4; in [BrMa92], the Weyl group $\mathbf{W}$ acts on the right on X so that w^{-1} is replaced by w there.) In general, $\mathbb{G}$ is said to be a *toric datum* if $R = \varnothing$; in this case, a corresponding connected reductive algebraic group is a torus.

1.6.21 Let $\mathbb{G} = \big((X, R, Y, R^\vee), \varphi_0\mathbf{W}\big)$ be a complete root datum. We assume that φ_0 is chosen such that the permutation of R induced by φ_0 leaves a base of R invariant (which is possible by Remark 1.6.15). Thus, if $q \in \mathscr{P}$ and $\mathbf{G}, \mathbf{T}, F$ are as in Remark 1.6.12, then we are in the setting in 1.6.1, where $\mathbf{T}_0 := \mathbf{T}$.

Now let $w \in \mathbf{W}$ and consider the maximal toric sub-datum $\mathbb{T}_w$ as in Definition 1.6.20. The corresponding order polynomial is just given by

$$|\mathbb{T}_w| = \det(\mathbf{q}\,\mathrm{id}_{X_{\mathbb{R}}} - \varphi_0 \circ w^{-1}) \in \mathbb{R}[\mathbf{q}].$$

Let $q \in \mathscr{P}$ and $\mathbf{G}, \mathbf{T}_0, F$ be as above. Let $\dot{w}$ be a representative of w in $N_{\mathbf{G}}(\mathbf{T}_0)$. By Theorem 1.4.8 (Lang–Steinberg), we can write $\dot{w} = g^{-1}F(g)$ for some $g \in \mathbf{G}$. Then

$\mathbf{T}' := g\mathbf{T}_0 g^{-1}$ is an F-stable maximal torus of $\mathbf{G}$; in fact, $\mathbf{T}'$ is a *torus of type* w, as in 1.6.4. By Lemma 1.6.6, we have

$$|\mathbf{T}'^F| = |\mathbf{T}_0[w]| = |\mathbb{T}_w|(q) = \det(q\,\mathrm{id}_{X_\mathbb{R}} - \varphi_0 \circ w^{-1}).$$

Thus, $\mathbf{T}'^F \cong \mathbf{T}_0[w]$ is a member of the series of finite groups of Lie type defined by the complete root datum $\mathbb{T}_w$. Since $|\mathbf{T}'^F|$ divides $|\mathbf{G}^F|$ and since this holds for all $q \in \mathscr{P}$, we conclude that

$$|\mathbb{T}_w| \quad \text{divides} \quad |\mathbb{G}| \quad \text{in } \mathbb{R}[\mathbf{q}].$$

In fact, if $N = |R|/2$ denotes the number of positive roots of $\mathbf{G}$, then $\mathbf{q}^N|\mathbb{T}_w|$ divides $|\mathbb{G}|$ in $\mathbb{R}[\mathbf{q}]$ since $\mathbf{q}^N$ divides $|\mathbb{G}|$ and $\mathbf{q}$ does not divide $|\mathbb{T}_w|$.

Remark 1.6.22 Assume that $\mathbb{G}$ is 'ordinary', as in Remark 1.6.16. Then $\varphi_0 \colon X_\mathbb{R} \to X_\mathbb{R}$ is an automorphism of finite order which leaves $X \subseteq X_\mathbb{R}$ invariant. It follows that the order polynomials $|\mathbb{G}|$ and $|\mathbb{T}_w|$ ($w \in \mathbf{W}$) are actually polynomials in $\mathbb{Z}[\mathbf{q}]$ in this case (not just in $\mathbb{R}[\mathbf{q}]$).

In later chapters, we will see that various other classes of subgroups of $\mathbf{G}^F$ fit into the framework of complete root data. The general formalism is further developed in [BMM93], [BMM99], [BMM14]. Note that, even with our slightly more general definition, any complete root datum as above defines a *reflection datum* as in [BMM14, Def. 2.6], over a suitable subfield $K \subseteq \mathbb{R}$.

1.7 Regular Embeddings

Lusztig's work [Lu84a], [Lu88] (to be discussed in more detail in later chapters) shows that the character theory of finite groups of Lie type is considerably easier when the centre of the underlying algebraic group is connected. Thus, when trying to prove a result about a general finite group of Lie type, it often happens that one first tries to establish an analogous result in the case where the centre is connected. The concept of 'regular embedding' provides an efficient technical tool for passing back and forth between groups with a connected centre and a non-connected centre.

Definition 1.7.1 Let $\mathbf{G}, \tilde{\mathbf{G}}$ be connected reductive algebraic groups over $k = \overline{\mathbb{F}}_p$ and $F \colon \mathbf{G} \to \mathbf{G}$, $\tilde{F} \colon \tilde{\mathbf{G}} \to \tilde{\mathbf{G}}$ be Steinberg maps. Let $i \colon \mathbf{G} \to \tilde{\mathbf{G}}$ be a homomorphism of algebraic groups such that $i \circ F = \tilde{F} \circ i$. Following [Lu88, §7], we say that i is a *regular embedding* if $\tilde{\mathbf{G}}$ has a connected centre, i is an isomorphism of $\mathbf{G}$ with a closed subgroup of $\tilde{\mathbf{G}}$ and $i(\mathbf{G})$, $\tilde{\mathbf{G}}$ have the same derived subgroup.

Note that $\tilde{\mathbf{G}}_{\mathrm{der}} \subseteq i(\mathbf{G})$ and so $i(\mathbf{G})$ is normal in $\tilde{\mathbf{G}}$ with $\tilde{\mathbf{G}}/i(\mathbf{G})$ abelian. Then the finite group $i(\mathbf{G}^F) = i(\mathbf{G})^{\tilde{F}}$ contains the derived subgroup of the finite group $\tilde{\mathbf{G}}^{\tilde{F}}$ and so $i(\mathbf{G}^F)$ is normal in $\tilde{\mathbf{G}}^{\tilde{F}}$ with $\tilde{\mathbf{G}}^{\tilde{F}}/i(\mathbf{G}^F)$ abelian. Thus, as far as the representation theory of $\mathbf{G}^F$ and of $\tilde{\mathbf{G}}^{\tilde{F}}$ is concerned, we are in a situation where Clifford theory (with abelian factor group) applies.

(See also [Ta19] for a refinement of the notion of regular embeddings.)

Example 1.7.2 (a) Let $\mathbf{G}$ be a connected reductive algebraic group with a connected centre and $F\colon \mathbf{G} \to \mathbf{G}$ be a Steinberg map. Then $\mathbf{G}_{\mathrm{der}}$ is semisimple and, clearly, $\mathbf{G}_{\mathrm{der}} \subseteq \mathbf{G}$ is a regular embedding. A standard example is given by $\mathbf{G} = \mathrm{GL}_n(k)$ where $\mathbf{G}_{\mathrm{der}} = \mathrm{SL}_n(k)$; note that this works for both the Frobenius maps in Example 1.4.21, where either $\mathbf{G}^F = \mathrm{GL}_n(q)$ or $\mathbf{G}^F = \mathrm{GU}_n(q)$.

(b) Let $\mathbf{G} = \mathrm{SL}_n(k)$ and $F\colon \mathbf{G} \to \mathbf{G}$ be a Frobenius map. We can also *construct* a regular embedding $i\colon \mathbf{G} \to \tilde{\mathbf{G}}$, as follows. Let

$$\tilde{\mathbf{G}} := \{(A, \xi) \in M_n(\mathbf{k}) \times \mathbf{k}^\times \mid \xi \det(A) = 1\};$$

then $\mathbf{Z}(\tilde{\mathbf{G}}) = \{(\xi I_n, \xi^{-n}) \mid \xi \in k^\times\}$ is connected and $\dim \mathbf{Z}(\tilde{\mathbf{G}}) = 1$. For $A \in \mathbf{G}$ we set $i(A) := (A, 1) \in \tilde{\mathbf{G}}$; then i is a closed embedding. A Frobenius map $\tilde{F}\colon \tilde{\mathbf{G}} \to \tilde{\mathbf{G}}$ is defined by $\tilde{F}(A, \xi) = (F(A), \xi^q)$ if $\mathbf{G}^F = \mathrm{SL}_n(q)$, and by $\tilde{F}(A, \xi) = (F(A), \xi^{-q})$ if $\mathbf{G}^F = \mathrm{SU}_n(q)$.

(c) Let $n \geqslant 2$ and $\mathbf{G} \subseteq \mathrm{GL}_n(k)$ be one of the classical groups in 1.1.4. Then one can apply a similar construction as in (b). In each case, $\mathbf{Z}(\mathbf{G})$ consists of the scalar matrices in $\mathbf{G}$. If $\mathbf{G} = \mathrm{SO}_2(k)$, then $\mathbf{G} = \mathbf{Z}(\mathbf{G}) \cong \mathbf{k}^\times$; otherwise, we have $\mathbf{Z}(\mathbf{G}) = \{\pm I_n\}$ where I_n is the identity matrix. So let us now assume that $\mathrm{char}(k) \neq 2$ and $\mathbf{Z}(\mathbf{G}) = \{\pm I_n\}$. Then $\mathbf{G} = \Gamma(Q_n, k)$ where $Q_n^{\mathrm{tr}} = \pm Q_n$. We set

$$\tilde{\mathbf{G}} = C\Gamma(Q_n, k) := \{(A, \xi) \in M_n(\mathbf{k}) \times \mathbf{k}^\times \mid A^{\mathrm{tr}} Q_n A = \xi Q_n\};$$

this is called the *conformal group* corresponding to the classical group $\mathbf{G}$. Then $\tilde{\mathbf{G}}$ is a linear algebraic group such that $\mathbf{Z}(\tilde{\mathbf{G}}) = \{(\xi I_n, \xi^2) \mid \xi \in k^\times\}$ is connected and $\dim \mathbf{Z}(\tilde{\mathbf{G}}) = 1$. Consider the closed subgroup $\mathbf{G}_1 = \{(A, 1) \mid A \in \mathbf{G}\} \subseteq \tilde{\mathbf{G}}$. Then we have an injective homomorphism of algebraic groups $i\colon \mathbf{G} \to \mathbf{G}_1$, $A \mapsto (A, 1)$, with inverse given by $(A, 1) \mapsto A$. Hence, i is a closed embedding.

Let $F\colon \mathrm{GL}_n(k) \to \mathrm{GL}_n(k)$ be the standard Frobenius map (raising each entry of a matrix to its qth power). Then F restricts to a Frobenius map on $\mathbf{G}$. The map $\tilde{F}\colon \tilde{\mathbf{G}} \to \tilde{\mathbf{G}}$, $(A, \xi) \mapsto (F(A), \xi^q)$, is easily seen to be a Frobenius map such that $i \circ F = \tilde{F} \circ i$. Thus, i is a regular embedding.

If $n = 2m$ is even and $\mathbf{G} = \mathrm{SO}_n(k)$, then we also have a 'twisted' Frobenius map $F_1\colon \mathbf{G} \to \mathbf{G}$, $A \mapsto t_n^{-1} F(A) t_n$, where

$$t_n := \left[\begin{array}{c|c|c} I_{m-1} & 0 & 0 \\ \hline 0 & \begin{array}{cc} 0 & 1 \\ 1 & 0 \end{array} & 0 \\ \hline 0 & 0 & I_{m-1} \end{array}\right] \in \mathrm{GO}_{2m}(k).$$

This gives rise to the finite 'non-split' orthogonal group $\mathbf{G}^{F_1} = \mathrm{SO}_n^-(q)$; see, e.g., [Ge03a, 4.1.10(d)]. We have a Frobenius map

$$\tilde{F}_1 : \tilde{\mathbf{G}} \to \tilde{\mathbf{G}}, \qquad (A, \xi) \mapsto (F_1(A), \xi^q),$$

such that $i \circ F_1 = \tilde{F}_1 \circ i$. Thus, i also is a regular embedding with respect to F_1. (These examples already appeared in [Lu77a, §8.1].)

Lemma 1.7.3 (Cf. [DeLu76, 1.21]) *Let $\mathbf{G}$ be connected reductive and $F : \mathbf{G} \to \mathbf{G}$ be a Steinberg map. Let $\mathbf{Z}$ be the centre of $\mathbf{G}$ and $\mathbf{S} \subseteq \mathbf{G}$ be an F-stable torus such that $\mathbf{Z} \subseteq \mathbf{S}$. (For example, one could take any F-stable maximal torus of $\mathbf{G}$.) Let $\tilde{\mathbf{G}}$ be the quotient of $\mathbf{G} \times \mathbf{S}$ by the closed normal subgroup $\{(z, z^{-1}) \mid z \in \mathbf{Z}\}$. Let $\tilde{\mathbf{S}}$ be the image of $\{1\} \times \mathbf{S} \subseteq \mathbf{G} \times \mathbf{S}$ in $\tilde{\mathbf{G}}$. Then the map $\tilde{F} : \tilde{\mathbf{G}} \to \tilde{\mathbf{G}}$ induced by F is a Steinberg map and the map $i : \mathbf{G} \to \tilde{\mathbf{G}}$ induced by $\mathbf{G} \to \mathbf{G} \times \mathbf{S}$, $g \mapsto (g, 1)$, is a regular embedding, where $\tilde{\mathbf{S}}$ is the centre of $\tilde{\mathbf{G}}$.*

Proof By Lemma 1.4.26, $\tilde{\mathbf{G}}$ is reductive and $\tilde{F} : \tilde{\mathbf{G}} \to \tilde{\mathbf{G}}$ is a Steinberg map. Furthermore, one easily sees that i is injective, that $i \circ F = \tilde{F} \circ i$ and that $\tilde{\mathbf{S}} = \mathbf{Z}(\tilde{\mathbf{G}})$. (Thus, the centre of $\tilde{\mathbf{G}}$ indeed is connected.) Let $\tilde{\mathbf{Z}} := \{(z, z^{-1}) \mid z \in \mathbf{Z}\}$ and $\mathbf{H} := i(\mathbf{G}) = (\mathbf{G} \times \mathbf{Z})/\tilde{\mathbf{Z}} \subseteq \tilde{\mathbf{G}}$. We have $\tilde{\mathbf{G}} = \mathbf{H}.\tilde{\mathbf{S}}$. Since $\tilde{\mathbf{S}} = \mathbf{Z}(\tilde{\mathbf{G}})$, it follows that $\tilde{\mathbf{G}}_{\mathrm{der}} = \mathbf{H}_{\mathrm{der}} = i(\mathbf{G}_{\mathrm{der}})$.

We claim that $i_1 : \mathbf{G} \to \mathbf{H}$, $g \mapsto i(g)$, is an isomorphism of algebraic groups. To see this, consider the homomorphism $\pi : \mathbf{G} \times \mathbf{Z} \to \mathbf{G}$, $(g, z) \mapsto gz$. Since $\tilde{\mathbf{Z}} \subseteq \ker(\pi)$, we have an induced homomorphism of algebraic groups $\bar{\pi} : \mathbf{H} \to \mathbf{G}$, which is obviously inverse to $i_1 : \mathbf{G} \to \mathbf{H}$. Thus, the claim is proved, and it follows that i is a regular embedding. □

Example 1.7.4 Let $\mathbf{G}$ be simple of simply connected type and $F : \mathbf{G} \to \mathbf{G}$ be a Steinberg map. Assume that $\mathbf{Z} = \mathbf{Z}(\mathbf{G})$ is non-trivial. Let $\mathbf{T}_0 \subseteq \mathbf{G}$ be a maximally split torus and $\mathbf{S} \subseteq \mathbf{T}_0$ be a subtorus as in Example 1.5.6. Then $\mathbf{Z} \subseteq \mathbf{S}$ and we would like to perform the construction in Lemma 1.7.3 using $\mathbf{S}$. For this purpose, we need to check that $\mathbf{S}$ is F-stable. Let us check when this happens. Let $\mathscr{R} = (X, R, Y, R^\vee)$ be the root datum of $\mathbf{G}$ with respect to $\mathbf{T}_0$. Let $\Pi = \{\alpha_1, \ldots, \alpha_n\}$ be a base for R, with a labelling as in Table 1.1. Then $Y = \mathbb{Z}R^\vee$ and $\{\alpha_1^\vee, \ldots, \alpha_n^\vee\}$ is a $\mathbb{Z}$-basis of Y. Now F induces a linear map $\varphi : X \to X$ which is a p-isogeny of $\mathscr{R}$. Hence, there is a permutation $s \mapsto s'$ of $\{1, \ldots, n\}$ and there are integers $q_s > 0$ (each an integral power of p) such that

$$\varphi^{\mathrm{tr}}(\alpha_{s'}^\vee) = q_{s'}\alpha_s^\vee \qquad \text{for } 1 \leqslant s \leqslant n.$$

(Note: The permutation $\alpha_s \mapsto \alpha_{s'}$ is inverse to the permutation $\alpha_s \mapsto \alpha_{s^\dagger}$ in 1.6.1.) Now consider the isomorphism $k^\times \otimes_{\mathbb{Z}} Y \to \mathbf{T}_0$, $\xi \otimes \nu \mapsto \nu(\xi)$. Then, for each $\xi \in k^\times$ and $\nu \in Y$, the element $\xi \otimes \varphi^{\mathrm{tr}}(\nu)$ corresponds to $F(\nu(\xi))$. (See [Ca85, §3.2].) Thus, in terms of the notation

$$\mathbf{T}_0 = \{h(\xi_1, \ldots, \xi_n) \mid \xi_1, \ldots, \xi_n \in k^\times\} \qquad \text{(see Example 1.5.6)},$$

the action of F on $\mathbf{T}_0$ is given by

$$F\big(h(\xi_1, \xi_2, \ldots, \xi_n)\big) = h(\xi_{1'}^{q_{1'}}, \xi_{2'}^{q_{2'}}, \ldots, \xi_{n'}^{q_{n'}}) \qquad \text{for all } \xi_1, \ldots, \xi_n \in k^\times.$$

We do not need to consider the case where F is a Steinberg map but not a Frobenius map. For, this case only occurs in types B_2, G_2, F_4 and, in all three cases, we have $\mathbf{Z} = \{1\}$ (since $\text{char}(k) = 2$ in type B_2). So let now F be a Frobenius map. Then all q_s are equal, to q say, and $s \mapsto s'$ determines a symmetry of the Dynkin diagram of $\mathscr{R}$. If $s \mapsto s'$ is the identity, then

$$F\big(h(\xi_1, \xi_2, \ldots, \xi_n)\big) = h(\xi_1^q, \xi_2^q, \ldots, \xi_n^q) \qquad \text{for all } \xi_s \in k^\times.$$

The cases where there exists a non-trivial permutation $s \mapsto s'$ are as follows.

A_n $(n \geqslant 2)$: $s' = n + 1 - s$ for $1 \leqslant s \leqslant n$ and so

$$F\big(h(\xi_1, \xi_2, \ldots, \xi_n)\big) = h(\xi_n^q, \xi_{n-1}^q, \ldots, \xi_1^q) \qquad \text{for all } \xi_s \in k^\times.$$

In this case, the subtorus $\mathbf{S} \subseteq \mathbf{T}_0$ is not necessarily F-stable.

D_n $(n \geqslant 3)$: $1' = 2$, $2' = 1$, $s' = s$ for $3 \leqslant s \leqslant n$ and so

$$F\big(h(\xi_1, \xi_2, \ldots, \xi_n)\big) = h(\xi_2^q, \xi_1^q, \xi_3^q, \ldots, \xi_n^q) \qquad \text{for all } \xi_s \in k^\times.$$

In this case, the subtorus $\mathbf{S} \subseteq \mathbf{T}_0$ is F-stable.

D_4: $1' = 2$, $2' = 4$, $3' = 3$, $4' = 1$ and so

$$F\big(h(\xi_1, \xi_2, \xi_3, \xi_4)\big) = h(\xi_2^q, \xi_4^q, \xi_3^q, \xi_1^q) \qquad \text{for all } \xi_s \in k^\times.$$

In this case, the subtorus $\mathbf{S} \subseteq \mathbf{T}_0$ is F-stable.

E_6: $1' = 6$, $2' = 2$, $3' = 5$, $4' = 4$, $5' = 3$, $6' = 1$ and so

$$F\big(h(\xi_1, \xi_2, \xi_3, \xi_4, \xi_5, \xi_6)\big) = h(\xi_6^q, \xi_2^q, \xi_5^q, \xi_4^q, \xi_3^q, \xi_1^q) \qquad \text{for all } \xi_s \in k^\times.$$

In this case, the subtorus $\mathbf{S} \subseteq \mathbf{T}_0$ is F-stable.

Proposition 1.7.5 (Cf. [Lu84a, §14.1], [Lu88, §10], [Lu08a, 5.3]) *Let $\mathbf{G}$ be simple of simply connected type and $F\colon \mathbf{G} \to \mathbf{G}$ be a Steinberg map. Then there exists a regular embedding $i\colon \mathbf{G} \to \tilde{\mathbf{G}}$ with the following properties.*

(a) *If $\mathbf{G}$ is of type D_n with n even,* $\text{char}(k) \neq 2$ *and $\mathbf{G}^F$ is 'untwisted' (i.e., the above permutation $s \mapsto s'$ is the identity), then* $\dim \mathbf{Z}(\tilde{\mathbf{G}}) = 2$ *and there is a surjective map $\tilde{\mathbf{G}}^{\tilde{F}}/i(\mathbf{G}^F) \to \mathbb{Z}/2\mathbb{Z} \times \mathbb{Z}/2\mathbb{Z}$.*

(b) *In all other cases, $\tilde{\mathbf{G}}^{\tilde{F}}/i(\mathbf{G}^F)$ is cyclic.*

Proof We begin by noting that, if $\mathbf{Z}(\mathbf{G}) = \{1\}$, then we can take $\tilde{\mathbf{G}} = \mathbf{G}$ and i the identity; we have $\tilde{\mathbf{G}}^{\tilde{F}} = i(\mathbf{G}^F)$ in this case. So, for the remainder of the proof, we can assume that $\mathbf{Z}(\mathbf{G}) \neq \{1\}$. If $\mathbf{G} = \mathrm{SL}_n(k)$, then the inclusion $i\colon \mathbf{G} \hookrightarrow \tilde{\mathbf{G}} := \mathrm{GL}_n(k)$ is a regular embedding with the required properties. We can now assume that $\mathbf{G}$ is not of type A_n. Let $\mathbf{S} \subseteq \mathbf{G}$ be the torus in Example 1.5.6. We have $\mathbf{Z}(\mathbf{G}) \subseteq \mathbf{S}$ and $\mathbf{S}$ is F-stable by the discussion in Example 1.7.4. So, applying Lemma 1.7.3 with $\mathbf{S}$, we obtain a regular embedding $i\colon \mathbf{G} \to \tilde{\mathbf{G}}$. Since $\mathbf{G} = \mathbf{G}_{\mathrm{der}}$, we have $\tilde{\mathbf{G}}_{\mathrm{der}} = i(\mathbf{G})$. Then $i(\mathbf{G}^F) = i(\mathbf{G})^{\tilde{F}} = \tilde{\mathbf{G}}_{\mathrm{der}}^{\tilde{F}}$ and so Example 1.4.11(b) implies that

$$\tilde{\mathbf{G}}^{\tilde{F}}/i(\mathbf{G}^F) \cong \mathbf{K}^{\tilde{F}} \qquad \text{where} \qquad \mathbf{K} := \tilde{\mathbf{G}}/\tilde{\mathbf{G}}_{\mathrm{der}}.$$

(We have an induced action of $\tilde{F}$ on $\mathbf{K}$ by Lemma 1.4.26.) Since $\tilde{\mathbf{G}} = i(\mathbf{G}).\tilde{\mathbf{S}}$ and $\tilde{\mathbf{S}} = \mathbf{Z}(\tilde{\mathbf{G}})$, the inclusion $\tilde{\mathbf{S}} \hookrightarrow \tilde{\mathbf{G}}$ induces an isogeny $\tilde{\mathbf{S}} \to \mathbf{K}$. Composition with the map $\mathbf{S} \to \tilde{\mathbf{S}}$ from Lemma 1.7.3 yields an isogeny $f\colon \mathbf{S} \to \mathbf{K}$ such that $f \circ F = \tilde{F} \circ f$ and $\ker(f) = \mathbf{Z}(\mathbf{G})$. In particular, $\mathbf{K}$ is a torus and $1 \leqslant \dim \mathbf{K} = \dim \mathbf{S} \leqslant 2$.

If $\dim \mathbf{K} = 1$, then $\mathbf{K} \cong \mathbf{k}^\times$ and so $\mathbf{K}^{\tilde{F}}$ is isomorphic to a finite subgroup of $\mathbf{k}^\times$; hence, $\mathbf{K}^{\tilde{F}}$ is cyclic in this case.

Finally, assume that $\dim \mathbf{K} = 2$. This case only occurs in type D_n with n even, where $\mathbf{G} = \mathrm{Spin}_{2n}(k)$ and $\mathbf{Z}(\mathbf{G}) = \{t \in \mathbf{S} \mid t^2 = 1\}$. Since $\mathbf{Z}(\mathbf{G}) \neq \{1\}$, we have $\mathrm{char}(k) \neq 2$ and $\mathbf{Z}(\mathbf{G}) \cong \mathbb{Z}/2\mathbb{Z} \times \mathbb{Z}/2\mathbb{Z}$. If F is 'untwisted', then Remark 1.7.6(b) below will show that $\tilde{\mathbf{G}}^{\tilde{F}}/i(\mathbf{G}^F)$ has a factor group isomorphic to $\mathbb{Z}/2\mathbb{Z} \times \mathbb{Z}/2\mathbb{Z}$. This completes the proof of (a).

It remains to consider the case where F is 'twisted', with all root exponents equal to q. By the description in Example 1.7.4, there is an isomorphism of algebraic groups $\mathbf{S} \cong \mathbf{k}^\times \times \mathbf{k}^\times$ such that the action of F on $\mathbf{S}$ corresponds to the map $(s_1, s_2) \mapsto (s_2^q, s_1^q)$ on $\mathbf{k}^\times \times \mathbf{k}^\times$. Consequently, $\mathbf{S}^F \cong \mathbb{F}_{q^2}^\times$ is cyclic. We want to show that a similar argument works for $\mathbf{K}$. To see this, let $\{\varepsilon_1, \varepsilon_2\}$ be a $\mathbb{Z}$-basis of $X(\mathbf{S})$, such that F induces the linear map $\varphi\colon X(\mathbf{S}) \to X(\mathbf{S})$ with $\varphi(\varepsilon_1) = q\varepsilon_2$ and $\varphi(\varepsilon_2) = q\varepsilon_1$. Now consider the isogeny $f\colon \mathbf{S} \to \mathbf{K}$ mentioned above. Since it has kernel $\mathbf{Z}(\mathbf{G})$, and since $\mathrm{char}(k) \neq 2$, the correspondences in 1.1.11 show that

$$X(\mathbf{S})/f^*(X(\mathbf{K})) \cong X(\mathbf{Z}(\mathbf{G})) \cong \mathbb{Z}/2\mathbb{Z} \times \mathbb{Z}/2\mathbb{Z}.$$

So we have $f^*(X(\mathbf{K})) = 2X(\mathbf{S})$. For $i = 1, 2$, let $\delta_i \in X(\mathbf{K})$ be such that $f^*(\delta_i) = 2\varepsilon_i$. Then $\{\delta_1, \delta_2\}$ is a $\mathbb{Z}$-basis of $X(\mathbf{K})$. Let $\beta\colon X(\mathbf{K}) \to X(\mathbf{K})$ be the linear map induced by $\tilde{F}$. Since $f \circ F = \tilde{F} \circ f$, we also have $\varphi \circ f^* = f^* \circ \beta$. Hence, $\beta(\delta_1) = q\delta_2$ and $\beta(\delta_2) = q\delta_1$. Then $\mathbf{K} \to \mathbf{k}^\times \times \mathbf{k}^\times$, $t \mapsto (\delta_1(t), \delta_2(t))$, is an isomorphism of algebraic groups such that the action of $\tilde{F}$ on $\mathbf{K}$ corresponds to the isogeny

$$\mathbf{k}^\times \times \mathbf{k}^\times \to \mathbf{k}^\times \times \mathbf{k}^\times, \qquad (t_1, t_2) \mapsto (t_2^q, t_1^q).$$

Consequently, $\mathbf{K}^{\tilde{F}} \cong \mathbb{F}_{q^2}^\times$ is also cyclic. □

The following remarks contain a number of useful, purely group-theoretical properties of a regular embedding. (Most of these are taken from [Leh78, §1].)

Remark 1.7.6 Let $i\colon \mathbf{G} \to \tilde{\mathbf{G}}$ be a regular embedding. To simplify the notation, we identify $\mathbf{G}$ with its image in $\tilde{\mathbf{G}}$. Thus, $\mathbf{G} \subseteq \tilde{\mathbf{G}}$ and $\mathbf{G}_{\text{der}} = \tilde{\mathbf{G}}_{\text{der}}$. Let $\mathbf{Z}$ denote the centre of $\mathbf{G}$ and $\tilde{\mathbf{Z}}$ denote the centre of $\tilde{\mathbf{G}}$. Now, since $\tilde{\mathbf{G}}_{\text{der}} = \mathbf{G}_{\text{der}} \subseteq \mathbf{G}$, we certainly have $\tilde{\mathbf{G}} = \mathbf{G}.\tilde{\mathbf{Z}}$. This implies that $\mathbf{Z} = \tilde{\mathbf{Z}} \cap \mathbf{G}$ and $\mathbf{Z}^F = \mathbf{G}^F \cap \tilde{\mathbf{Z}}^{\tilde{F}}$. Furthermore, the inclusion $\mathbf{G} \subseteq \tilde{\mathbf{G}}$ induces bijective homomorphisms $\mathbf{G}/\mathbf{Z} \cong \tilde{\mathbf{G}}/\tilde{\mathbf{Z}}$ and

$$(\mathbf{G}/\mathbf{Z})^F \cong (\tilde{\mathbf{G}}/\tilde{\mathbf{Z}})^{\tilde{F}} \cong \tilde{\mathbf{G}}^{\tilde{F}}/\tilde{\mathbf{Z}}^{\tilde{F}} \tag{a}$$

(where the last isomorphism holds by Example 1.4.11(b), using that $\tilde{\mathbf{Z}}$ is connected). Now let $\mathbf{G}' := \mathbf{G} \times \tilde{\mathbf{Z}}$ and consider the natural map $f\colon \mathbf{G}' \to \tilde{\mathbf{G}}$ given by multiplication; note that f is surjective and $\ker(f) = \{(z, z^{-1}) \mid z \in \mathbf{Z}\} \subseteq \mathbf{Z}(\mathbf{G}')$. Applying Lemma 1.1.9 to f (and $\sigma = F \times \tilde{F}$, $\tau = \tilde{F}$), we obtain a canonical isomorphism

$$\tilde{\mathbf{G}}^{\tilde{F}}/(\mathbf{G}^F.\tilde{\mathbf{Z}}^{\tilde{F}}) \cong (\mathbf{Z}/\mathbf{Z}^\circ)_F \tag{b}$$

where $(\mathbf{Z}/\mathbf{Z}^\circ)_F$ is defined in Remark 1.5.12; explicitly, the above isomorphism is induced by sending $\tilde{g} \in \tilde{\mathbf{G}}^{\tilde{F}}$ to $g^{-1}F(g) \in \mathbf{G} \cap \tilde{\mathbf{Z}} = \mathbf{Z}$ where $g \in \mathbf{G}$ is such that $g \in \tilde{g}\tilde{\mathbf{Z}}$. In particular, if $\mathbf{Z}$ is connected, then $\tilde{\mathbf{G}}^{\tilde{F}} = \mathbf{G}^F.\tilde{\mathbf{Z}}^{\tilde{F}}$.

Lemma 1.7.7 *In the above setting, let* $\mathbf{T}$ *be an* F*-stable maximal torus of* $\mathbf{G}$. *Then* $\tilde{\mathbf{T}} := \mathbf{T}.\tilde{\mathbf{Z}}$ *is an* $\tilde{F}$*-stable maximal torus of* $\tilde{\mathbf{G}}$, *and every* $\tilde{F}$*-stable maximal torus of* $\tilde{\mathbf{G}}$ *is of this form. In this situation, we have*

$$\mathbf{T} = \mathbf{G} \cap \tilde{\mathbf{T}}, \qquad \tilde{\mathbf{G}}^{\tilde{F}} = \mathbf{G}^F.\tilde{\mathbf{T}}^{\tilde{F}}, \qquad \mathbf{T}^F = \mathbf{G}^F \cap \tilde{\mathbf{T}}^{\tilde{F}}.$$

Furthermore, the inclusion $\mathbf{G} \subseteq \tilde{\mathbf{G}}$ *induces an isomorphism* $N_{\mathbf{G}}(\mathbf{T})/\mathbf{T} \cong N_{\tilde{\mathbf{G}}}(\tilde{\mathbf{T}})/\tilde{\mathbf{T}}$ *which is compatible with the actions of* F, $\tilde{F}$.

Proof As in Remark 1.7.6, we have a natural surjective map $f\colon \mathbf{G}\times\tilde{\mathbf{Z}} \to \tilde{\mathbf{G}}$ given by multiplication. Since $\mathbf{T}\times\tilde{\mathbf{Z}}$ is a maximal torus of $\mathbf{G}\times\tilde{\mathbf{Z}}$, it follows that $\tilde{\mathbf{T}} = \mathbf{T}.\tilde{\mathbf{Z}}$ is a maximal torus of $\tilde{\mathbf{G}}$ (see 1.3.10(a)); clearly, $\tilde{\mathbf{T}}$ is $\tilde{F}$-stable. Now $\mathbf{T} \subseteq \mathbf{G}\cap\tilde{\mathbf{T}}$. But, since $\mathbf{G}$ is reductive, we have $C_{\mathbf{G}}(\mathbf{T}) = \mathbf{T}$ and so $\mathbf{T} = \mathbf{G} \cap \tilde{\mathbf{T}}$. This also implies that $\mathbf{T}^F = \mathbf{G}^F \cap \tilde{\mathbf{T}}^{\tilde{F}}$. Since $\tilde{\mathbf{Z}} \subseteq \tilde{\mathbf{T}}$, we have $\tilde{\mathbf{G}} = \mathbf{G}.\tilde{\mathbf{T}}$. Furthermore, since $\mathbf{T} = \mathbf{G} \cap \tilde{\mathbf{T}}$ is connected, it follows that $\tilde{\mathbf{G}}^{\tilde{F}} = \mathbf{G}^F.\tilde{\mathbf{T}}^{\tilde{F}}$; see Example 1.4.11(a).

Now consider the Weyl groups. We have $N_{\mathbf{G}}(\mathbf{T}) \subseteq N_{\tilde{\mathbf{G}}}(\tilde{\mathbf{T}})$ and so we obtain an injective homomorphism $\mathbf{W}(\mathbf{G},\mathbf{T}) \to \mathbf{W}(\tilde{\mathbf{G}},\tilde{\mathbf{T}})$. On the other hand, by 1.3.10(d), $\mathbf{W}(\tilde{\mathbf{G}},\tilde{\mathbf{T}})$ is isomorphic to $\mathbf{W}(\mathbf{G}\times\tilde{\mathbf{Z}},\mathbf{T}\times\tilde{\mathbf{Z}}) \cong \mathbf{W}(\mathbf{G},\mathbf{T})$. Hence, the injection $\mathbf{W}(\mathbf{G},\mathbf{T}) \to \mathbf{W}(\tilde{\mathbf{G}},\tilde{\mathbf{T}})$ is also surjective.

Finally, let $\tilde{\mathbf{T}}' \subseteq \tilde{\mathbf{G}}$ be an arbitrary $\tilde{F}$-stable maximal torus. Then $\tilde{\mathbf{T}}' = \tilde{g}\tilde{\mathbf{T}}\tilde{g}^{-1}$ for some $\tilde{g} \in \tilde{\mathbf{G}}$. Since f is surjective, we can write $\tilde{g} = g\tilde{z}$ where $g \in \mathbf{G}$ and $\tilde{z} \in \tilde{\mathbf{Z}}$. Hence, setting $\mathbf{T}' := g\mathbf{T}g^{-1}$, we have $\tilde{\mathbf{T}}' = \mathbf{T}'.\tilde{\mathbf{Z}}$. But then we have again $\mathbf{T}' = \mathbf{G} \cap \tilde{\mathbf{T}}'$ and so $\mathbf{T}'$ is F-stable. □

Lemma 1.7.8 *In the setting of Remark 1.7.6, let* $\mathbf{H} \subseteq \mathbf{G}$ *be an* F*-stable closed connected subgroup such that* $\mathbf{Z}^\circ \subseteq \mathbf{H}$*. Then* $\tilde{\mathbf{H}} := \mathbf{H}.\tilde{\mathbf{Z}}$ *is an* $\tilde{F}$*-stable closed connected subgroup of* $\tilde{\mathbf{G}}$ *and*

$$\dim \tilde{\mathbf{H}} = \dim \mathbf{H} + \dim \tilde{\mathbf{Z}} - \dim \mathbf{Z}^\circ \quad \text{and} \quad |\tilde{\mathbf{H}}^{\tilde{F}}| = |\mathbf{H}^F||\tilde{\mathbf{Z}}^{\tilde{F}}|/|\mathbf{Z}^{\circ F}|.$$

Proof We have $\mathbf{Z}^\circ \subseteq \mathbf{H} \cap \tilde{\mathbf{Z}} \subseteq \mathbf{G} \cap \tilde{\mathbf{Z}} = \mathbf{Z}$ and so $(\mathbf{H} \cap \tilde{\mathbf{Z}})/\mathbf{Z}^\circ$ is finite; consequently, the natural map $\mathbf{H}/\mathbf{Z}^\circ \times \tilde{\mathbf{Z}}/\mathbf{Z}^\circ \to \tilde{\mathbf{H}}/\mathbf{Z}^\circ$ given by multiplication is an isogeny. This yields the dimension formula. Furthermore, Proposition 1.4.13(c) shows that $|(\mathbf{H}/\mathbf{Z}^\circ)^F||(\tilde{\mathbf{Z}}/\mathbf{Z}^\circ)^{\tilde{F}}| = |(\tilde{\mathbf{H}}/\mathbf{Z}^\circ)^{\tilde{F}}|$. So the formula for the order of $\tilde{\mathbf{H}}^{\tilde{F}}$ follows from Example 1.4.11(b). □

In order to obtain further properties of regular embeddings, it will be useful to characterise these maps entirely in terms of root data. In particular, this will allow us to show how regular embeddings relate to dual groups.

Lemma 1.7.9 *Let* $\mathbf{G}, \mathbf{G}'$ *be connected reductive groups over* k *and* $f : \mathbf{G} \to \mathbf{G}'$ *be an isotypy (see* 1.3.21*). Let* $\mathbf{T} \subseteq \mathbf{G}$ *and* $\mathbf{T}' \subseteq \mathbf{G}'$ *be maximal tori such that* $f(\mathbf{T}) \subseteq \mathbf{T}'$*. Let* $\varphi: X(\mathbf{T}') \to X(\mathbf{T})$, $\chi' \mapsto \chi' \circ f|_{\mathbf{T}}$, *be the induced homomorphism. Then the following two conditions are equivalent.*

(i) f *is an isomorphism of* $\mathbf{G}$ *onto a closed subgroup of* $\mathbf{G}'$*.*
(ii) f *is a central isotypy and* φ *is surjective.*

Proof First note that the assumptions imply that $\mathbf{G}' = f(\mathbf{G}).\mathbf{Z}(\mathbf{G}')$ and $f(\mathbf{G}_{\text{der}}) = \mathbf{G}'_{\text{der}}$. Let $\mathbf{G}_1 := f(\mathbf{G}) \subseteq \mathbf{G}'$; this is a closed subgroup which is connected and reductive (see Lemma 1.4.26); furthermore, $\mathbf{G}'_{\text{der}} = f(\mathbf{G}_{\text{der}}) = (\mathbf{G}_1)_{\text{der}}$. Let $\mathbf{T}_1 := f(\mathbf{T}) \subseteq \mathbf{T}'$; then $\mathbf{T}_1$ is a maximal torus of $\mathbf{G}_1$ (see 1.3.10(a)) and we have $\mathbf{T}_1 = \mathbf{G}_1 \cap \mathbf{T}'$. (We have $\mathbf{G}_1 \cap \mathbf{T}' \subseteq C_{\mathbf{G}_1}(\mathbf{T}_1) = \mathbf{T}_1$ where the last equality holds since $\mathbf{G}_1$ is connected reductive; the reverse inclusion is clear.) Thus, we have $f = i \circ f_1$ where $f_1 : \mathbf{G} \to \mathbf{G}_1$ is the restricted map and $i : \mathbf{G}_1 \hookrightarrow \mathbf{G}'$ is the inclusion; it is clear that i is a central isotypy. (Note that $(\mathbf{G}_1)_{\text{der}}$ contains all the root subgroups of $\mathbf{G}_1$; see Remark 1.3.6.) Correspondingly, we have a factorisation $\varphi = \varphi_1 \circ \varepsilon$ where $\varphi_1 : X(\mathbf{T}_1) \to X(\mathbf{T})$ is induced by f_1 and $\varepsilon : X(\mathbf{T}') \to X(\mathbf{T}_1)$ is given by restriction. Note that ε is surjective; see 1.1.11.

Now suppose that (i) holds, that is, $f_1 : \mathbf{G} \to \mathbf{G}_1$ is an isomorphism of algebraic groups. Then the composition $f = i \circ f_1$ will be a central isotypy. Furthermore, $\varphi_1 : X(\mathbf{T}_1) \to X(\mathbf{T})$ is an isomorphism of abelian groups. Since ε is surjective, it follows that $\varphi = \varphi_1 \circ \varepsilon$ must be surjective. Thus, (ii) holds.

Conversely, assume that (ii) holds. Then φ_1 is also surjective. So the correspondences in 1.1.11 show that $f_1 : \mathbf{T} \to \mathbf{T}_1$ is a closed embedding. But we also have $\mathbf{T}_1 = f_1(\mathbf{T})$, and so $f_1 : \mathbf{T} \to \mathbf{T}_1$ is an isomorphism. Then Theorem 1.3.22(b) shows that $f_1 : \mathbf{G} \to \mathbf{G}_1$ also is an isomorphism. □

Corollary 1.7.10 *Let $\mathbf{G}$ and $\mathbf{G}'$ be connected reductive. Let $F\colon \mathbf{G} \to \mathbf{G}$ and $F'\colon \mathbf{G}' \to \mathbf{G}'$ be Steinberg maps. Let $i\colon \mathbf{G} \to \mathbf{G}'$ be a homomorphism such that $i \circ F = F' \circ i$ and $i(\mathbf{T}) \subseteq \mathbf{T}'$, where $\mathbf{T}$ is an F-stable maximal torus of $\mathbf{G}$ and $\mathbf{T}'$ is an F'-stable maximal torus of $\mathbf{G}'$. Then i is a regular embedding if and only if the following three conditions hold.*

(1) *i is a central isotypy, i.e., the induced map $\varphi\colon X(\mathbf{T}') \to X(\mathbf{T})$ is a homomorphism of root data;*
(2) *the map $\varphi\colon X(\mathbf{T}') \to X(\mathbf{T})$ is surjective; and*
(3) *$X(\mathbf{T}')/\mathbb{Z}R'$ has no p'-torsion, where R' are the roots relative to $\mathbf{T}'$.*

Proof Suppose that i is a regular embedding. Since $i(\mathbf{G}_{\text{der}}) = \mathbf{G}'_{\text{der}}$, we have $\mathbf{G}' = i(\mathbf{G}).\mathbf{Z}(\mathbf{G}')$. So the general assumptions of Lemma 1.7.9 plus condition (i) are satisfied. Hence, the first two conditions hold; the third one holds because $\mathbf{Z}(\mathbf{G}')$ is connected (see Lemma 1.5.22). Conversely, if the above three conditions are satisfied, then $\mathbf{Z}(\mathbf{G}')$ is connected and Lemma 1.7.9 shows that i is an isomorphism of $\mathbf{G}$ onto a closed subgroup of $\mathbf{G}'$. Since i is central, we have $\mathbf{G}' = i(\mathbf{G}).\mathbf{Z}(\mathbf{G}')$ which implies that $\mathbf{G}'_{\text{der}} = i(\mathbf{G}_{\text{der}})$. Hence, i is a regular embedding. □

1.7.11 Assume that $(\mathbf{G}, F)$, $(\mathbf{G}^*, F^*)$ are in duality (see Definition 1.5.17), with respect to maximally split tori $\mathbf{T}_0 \subseteq \mathbf{G}$ and $\mathbf{T}_0^* \subseteq \mathbf{G}^*$. Furthermore, assume that $(\mathbf{G}', F')$ and $(\mathbf{G}'^*, F'^*)$ are in duality, with respect to maximally split tori $\mathbf{T}_0' \subseteq \mathbf{G}'$ and $\mathbf{T}_0'^* \subseteq \mathbf{G}'^*$. Thus, we are given isomorphisms

$$\delta\colon X(\mathbf{T}_0) \xrightarrow{\sim} Y(\mathbf{T}_0^*) \qquad \text{and} \qquad \delta'\colon X(\mathbf{T}_0') \xrightarrow{\sim} Y(\mathbf{T}_0'^*)$$

satisfying the above conditions. Now let $f\colon \mathbf{G} \to \mathbf{G}'$ be a central isotypy such that $f \circ F = F' \circ f$ and $f(\mathbf{T}_0) \subseteq \mathbf{T}_0'$. Thus, the induced map $\varphi\colon X(\mathbf{T}_0') \to X(\mathbf{T}_0)$ is a homomorphism of root data as in 1.2.2. But then the transpose map $\varphi^{\text{tr}}\colon Y(\mathbf{T}_0) \to Y(\mathbf{T}_0')$ defines a morphism of the dual root data. Using the transposed isomorphisms $\delta^{\text{tr}}\colon X(\mathbf{T}_0^*) \to Y(\mathbf{T}_0)$ and $\delta'^{\text{tr}}\colon X(\mathbf{T}_0'^*) \to Y(\mathbf{T}_0')$, we obtain a map

$$\hat{\varphi} = (\delta'^{\text{tr}})^{-1} \circ \varphi^{\text{tr}} \circ \delta^{\text{tr}}\colon X(\mathbf{T}_0^*) \to X(\mathbf{T}_0'^*)$$

which is a homomorphism between the root data of $\mathbf{G}'^*$ and $\mathbf{G}^*$. Now, by Theorem 1.3.22 (extended isogeny theorem), there exists a central isotypy $f^*\colon \mathbf{G}'^* \to \mathbf{G}^*$ which maps $\mathbf{T}_0'^*$ into $\mathbf{T}_0^*$ and induces $\hat{\varphi}$. Arguing as in Lemma 1.4.24, one shows that f^* can be chosen such that $f^* \circ F'^* = F^* \circ f^*$. We have $\hat{\varphi}^{\text{tr}} \circ \delta' = \delta \circ \varphi$. So, for any $\lambda' \in X(\mathbf{T}_0')$, we obtain the compatibility relation:

$$f^* \circ \big(\delta'(\lambda')\big) = \hat{\varphi}^{\text{tr}}\big(\delta'(\lambda')\big) = \big(\hat{\varphi}^{\text{tr}} \circ \delta'\big)(\lambda') = \big(\delta \circ \varphi\big)(\lambda') = \delta(\lambda' \circ f) \in Y(\mathbf{T}_0^*).$$

In this situation, we say that the two central isotypies

$$f\colon \mathbf{G} \to \mathbf{G}' \textit{ and } f^*\colon \mathbf{G}'^* \to \mathbf{G}^* \textit{ correspond to each other by duality.}$$

(This relation is symmetric.) With this notation, we can now state:

Lemma 1.7.12 *Let $f\colon \mathbf{G} \to \mathbf{G}'$ and $f^*\colon \mathbf{G}'^* \to \mathbf{G}^*$ correspond to each other by duality, as above. Assume that $f\colon \mathbf{G} \to \mathbf{G}'$ is an isomorphism with a closed subgroup of $\mathbf{G}'$. Then $f^*\colon \mathbf{G}'^* \to \mathbf{G}^*$ is surjective and $\ker(f^*)$ is a central torus. Furthermore, the restricted map $f^*\colon (\mathbf{G}'^*)^{F'^*} \to (\mathbf{G}^*)^{F^*}$ is surjective.*

Proof We follow [Bo06, 2.5], using the notation in 1.7.11. By restriction, f yields a closed embedding $f\colon \mathbf{T}_0 \to \mathbf{T}_0'$. By 1.1.11(a), the induced map $\varphi\colon X(\mathbf{T}_0') \to X(\mathbf{T}_0)$ is surjective and $\ker(\varphi) \cong X(\mathbf{T}_0'/f(\mathbf{T}_0))$ is torsion-free. Since

$$\operatorname{Hom}(X(\mathbf{T}_0), \mathbb{Z}) \cong Y(\mathbf{T}_0) \cong X(\mathbf{T}_0^*),$$
$$\operatorname{Hom}(X(\mathbf{T}_0'), \mathbb{Z}) \cong Y(\mathbf{T}_0') \cong X(\mathbf{T}_0'^*),$$

we obtain an exact sequence

$$\{0\} \longrightarrow X(\mathbf{T}_0^*) \xrightarrow{\hat{\varphi}} X(\mathbf{T}_0'^*) \longrightarrow \operatorname{Hom}(\ker(\varphi), \mathbb{Z}) \longrightarrow \{0\}$$

where, as above, $\hat{\varphi}$ is induced by $f^*\colon \mathbf{T}_0'^* \to \mathbf{T}_0^*$ and the second map is given by identifying $\lambda \in X(\mathbf{T}_0'^*)$ with an element in $\operatorname{Hom}(X(\mathbf{T}_0'), \mathbb{Z})$ and then restricting this to $\ker(\varphi) \subseteq X(\mathbf{T}_0')$. By 1.1.11(b), we deduce that $f^*\colon \mathbf{T}_0'^* \to \mathbf{T}_0^*$ is surjective and that $X(\ker(f^*)) \cong X(\mathbf{T}_0'^*)/\hat{\varphi}(X(\mathbf{T}_0^*)) \cong \operatorname{Hom}(\ker(\varphi), \mathbb{Z})$. Hence, $X(\ker(f^*))$ is torsion-free and so $\ker(f^*)$ is a torus. Since $\mathbf{G}^* = f^*(\mathbf{G}'^*).\mathbf{T}_0^*$ and $f^*(\mathbf{T}_0'^*) = \mathbf{T}_0^*$, we also have $f^*(\mathbf{G}'^*) = \mathbf{G}^*$. Finally, since $\ker(f^*)$ is connected, the fact that $f^*\big((\mathbf{G}'^*)^{F'^*}\big) = (\mathbf{G}^*)^{F^*}$ follows from Proposition 1.4.13(b). □

The following result is cited, for example, in [Lu84a, 8.8], [Lu88, 8.1], [Lu92b, 0.1] in relation to certain reduction arguments; it appeared in an unpublished manuscript of Asai [As]. See [Ta19] for a further discussion of Asai's reduction techniques.

Proposition 1.7.13 (Cf. [As, §2.3]) *Let $\mathbf{G}$ be connected reductive and $F\colon \mathbf{G} \to \mathbf{G}$ a Steinberg map. Then there exist a connected reductive group $\mathbf{G}^\bullet$, a Steinberg map $F^\bullet\colon \mathbf{G}^\bullet \to \mathbf{G}^\bullet$ and a homomorphism $f\colon \mathbf{G}^\bullet \to \mathbf{G}$, such that the following conditions hold:*

(a) *$\mathbf{G}^\bullet_{\mathrm{der}}$ is semisimple of simply connected type;*
(b) *f is a surjective homomorphism of algebraic groups and $F \circ f = f \circ F^\bullet$;*
(c) *$\ker(f)$ is a central torus of $\mathbf{G}^\bullet$.*

In particular, f induces a surjective homomorphism of finite groups $\mathbf{G}^{\bullet F^\bullet} \to \mathbf{G}^F$. Furthermore, if $\mathbf{G}$ has a connected centre, then $\mathbf{G}^\bullet$ has a connected centre, too.

Proof Asai [As] shows this by explicitly constructing the appropriate root datum for $\mathbf{G}^\bullet$ and then using Theorem 1.3.22 (extended isogeny theorem). Here is a more direct argument. Let $\pi_{sc}\colon (\mathbf{G}_{der})_{sc} \to \mathbf{G}_{der}$ be a simply connected covering of the derived group of $\mathbf{G}$, as in Remark 1.5.13. Assume first that π_{sc} is bijective. Let $\mathbf{Z} := \mathbf{Z}(\mathbf{G})$ and $\mathbf{K} := \{z \in \mathbf{Z}((\mathbf{G}_{der})_{sc}) \mid \pi_{sc}(z) \in \mathbf{Z}^\circ\}$. We have an isogeny

$$f_1\colon (\mathbf{G}_{der})_{sc} \times \mathbf{Z}^\circ \to \mathbf{G}, \qquad (z, z') \mapsto \pi_{sc}(z)z',$$

with $\ker(f_1) = \{(z, \pi_{sc}(z)^{-1}) \mid z \in \mathbf{K}\}$; note that $\ker(f_1)$ is finite. Let $\mathbf{G}^\bullet := ((\mathbf{G}_{der})_{sc} \times \mathbf{Z}^\circ)/\ker(f_1)$. Then f_1 induces a bijective morphism of algebraic groups $f\colon \mathbf{G}^\bullet \to \mathbf{G}$. Now (b), (c) are clear. To prove (a), note that

$$\mathbf{G}^\bullet_{der} = \mathbf{G}'.\ker(f_1)/\ker(f_1) \qquad \text{where} \qquad \mathbf{G}' := (\mathbf{G}_{der})_{sc} \times \{1\};$$

here, $\mathbf{G}'$ is a closed subgroup of $(\mathbf{G}_{der})_{sc} \times \mathbf{Z}^\circ$. We have a bijective homomorphism of algebraic groups $(\mathbf{G}_{der})_{sc} \to \mathbf{G}^\bullet_{der}$, sending $g \in (\mathbf{G}_{der})_{sc}$ to the image of $(g, 1)$ in $\mathbf{G}^\bullet$. On the other hand, since $\mathbf{G}' \cap \ker(f_1) = \{1\}$ and $\ker(f_1)$ is finite, the product $\mathbf{G}'.\ker(f_1)$ is semidirect and so the natural projection $\mathbf{G}'.\ker(f_1) \to \mathbf{G}' \to (\mathbf{G}_{der})_{sc}$ is a homomorphism of algebraic groups; see 1.1.8(c). Passing to the quotient by $\ker(f_1)$, we obtain a homomorphism of algebraic groups $\mathbf{G}^\bullet_{der} \to (\mathbf{G}_{der})_{sc}$ which is inverse to the above map $(\mathbf{G}_{der})_{sc} \to \mathbf{G}^\bullet_{der}$. Thus, (a) holds. Finally, since f is bijective, the centre of $\mathbf{G}^\bullet$ is connected if and only if $\mathbf{Z}$ is connected.

Now consider the general case, where the map π_{sc} may not be bijective. By Lemma 1.7.3, there exists a regular embedding $i\colon \mathbf{G}^* \to \mathbf{H}$. By duality, we obtain a homomorphism of algebraic groups $i^*\colon \mathbf{H}^* \to \mathbf{G}$; note that, as remarked in Definition 1.5.17, we can identify $(\mathbf{G}^*)^*$ with $\mathbf{G}$. By Lemma 1.7.12, i^* is surjective and $\ker(i^*)$ is a central torus of $\mathbf{H}^*$; furthermore, by Lemma 1.5.22, the simply connected covering $(\mathbf{H}^*_{der})_{sc} \to \mathbf{H}^*_{der}$ is bijective. By the previous argument, there exists a bijective homomorphism of algebraic groups $f_1 : \mathbf{G}^\bullet \to \mathbf{H}^*$ such that (a), (b), (c) hold. Then (a), (b), (c) hold for the composition $f = i^* \circ f_1 : \mathbf{G}^\bullet \to \mathbf{G}$. Finally, assume that $\mathbf{Z}$ is connected. Now the derived subgroup of $\mathbf{H}$ is isomorphic to that of $\mathbf{G}^*$. Hence, Lemma 1.5.22 implies that the centre of $\mathbf{H}^*$ is connected as well. Since f_1 is bijective, it follows that $\mathbf{G}^\bullet$ also has a connected centre. □

Example 1.7.14 Assume that $\mathbf{G}$ is semisimple and let $i\colon \mathbf{G} \to \tilde{\mathbf{G}}$ be a regular embedding. Applying Proposition 1.7.13 to $\tilde{\mathbf{G}}$, we obtain a homomorphism of algebraic groups $f\colon \tilde{\mathbf{G}}^\bullet \to \tilde{\mathbf{G}}$ satisfying the above three conditions. Furthermore, since $\mathbf{Z}(\tilde{\mathbf{G}})$ is connected, we have that $\mathbf{Z}(\tilde{\mathbf{G}}^\bullet)$ is connected, too. Now $f(\tilde{\mathbf{G}}^\bullet_{der}) = \mathbf{G}_{der} = \mathbf{G}$ and so, by restriction, we obtain an isogeny $\hat{f}\colon \tilde{\mathbf{G}}^\bullet_{der} \to \mathbf{G}$ which is a simply connected covering of $\mathbf{G}$. We have a commutative diagram:

$$\begin{array}{ccc} \mathbf{G} & \xrightarrow{i} & \tilde{\mathbf{G}} \\ \uparrow & & \uparrow \\ \tilde{\mathbf{G}}^\bullet_{der} & \hookrightarrow & \tilde{\mathbf{G}}^\bullet \end{array}$$

Thus, a simply connected covering of $\mathbf{G}$ can always be chosen to be compatible with the given regular embedding $i\colon \mathbf{G} \to \tilde{\mathbf{G}}$. (This remark appears in [Lu88, 8.1(d)].)

The following result was first stated (for $\mathbb{K}$ of characteristic 0) by Lusztig [Lu88, Prop. 10] (see also [Lu84b]), with an outline of the strategy of the proof. The details, which are surprisingly complicated, were provided much later in [Lu08a]; in the meantime, Cabanes and Enguehard also gave a proof in [CE04, Chap. 16], along similar lines. First, one employs a reduction argument which reduces the proof to the case where $\mathbf{G}$ is simple of simply connected type. As far as such groups are concerned, one can then use Proposition 1.7.5, which shows that type D_n with n even is the only case which requires a special argument, but all the difficulty lies with this case. We state here an extension of Lusztig's original result, valid for irreducible representations over any algebraically closed field.

Theorem 1.7.15 (Multiplicity-Freeness Theorem) *Let $i\colon \mathbf{G} \to \tilde{\mathbf{G}}$ be a regular embedding and $\mathbb{K}$ be any algebraically closed field. Then the restriction of every simple $\mathbb{K}\tilde{\mathbf{G}}^{\tilde{F}}$-module to $\mathbf{G}^F$ (via i) is multiplicity-free.*

Proof We can only sketch the general strategy here, and highlight where the principal difficulty of the proof lies. First we note that the reduction argument described in the proof of [Lu88, Prop. 10] works for simple modules over any algebraically closed field $\mathbb{K}$, not just for $\mathrm{char}(\mathbb{K}) = 0$. (Some adjustments of a different kind are required, since Lusztig considers Frobenius maps, not Steinberg maps in general; see [Ta19].) Hence, it suffices to prove the theorem in the case where $\mathbf{G}$ is simple of simply connected type. Furthermore, the reduction argument shows that it is sufficient to consider only one particular regular embedding $i\colon \mathbf{G} \to \tilde{\mathbf{G}}$, namely, one satisfying the conditions in Proposition 1.7.5. So let us now assume that these conditions are satisfied.

If $\tilde{\mathbf{G}}^{\tilde{F}}/i(\mathbf{G}^F)$ is cyclic, then a standard result on representations of finite groups shows that the desired assertion holds; see, e.g., [Fei82, Theorem III.2.14]. (This uses that $\mathbb{K}$ is algebraically closed, but works without any assumption on $\mathrm{char}(\mathbb{K})$.)

It remains to consider the case where $\mathbf{G}$ is of type D_n with n even, $\mathrm{char}(k) \neq 2$ and F is 'untwisted'. Let us identify $\mathbf{G}$ with $i(\mathbf{G})$ and use the notational conventions in Remark 1.7.6. Writing $G = \mathbf{G}^F$, $\tilde{G} = \tilde{\mathbf{G}}^{\tilde{F}}$, $H := G.\tilde{\mathbf{Z}}^{\tilde{F}}$, we have

$$H \trianglelefteq \tilde{G} \quad \text{and} \quad \tilde{G}/H \cong (\mathbf{Z}/\mathbf{Z}^\circ)_F = \mathbf{Z} \cong \mathbb{Z}/2\mathbb{Z} \times \mathbb{Z}/2\mathbb{Z}.$$

Let V be a simple $\mathbb{K}\tilde{G}$-module and denote by V_H its restriction to H. Since $H = G.\tilde{\mathbf{Z}}^{\tilde{F}}$ and $\tilde{\mathbf{Z}}^{\tilde{F}}$ is contained in the centre of $\tilde{G}$, it is sufficient to show that V_H is multiplicity-free. (To see this, one only needs to show that non-isomorphic simple H-submodules of V_H remain simple and non-isomorphic upon restriction to G. And this easily follows, for example, by the argument in [Ge93a, p. 265].) Now, if $\mathrm{char}(\mathbb{K}) = 2$, then

V_H is multiplicity-free by some general results on representations of finite groups; see, e.g., [KlTi09, Lemma 3.14]. If $\mathrm{char}(\mathbb{K}) = \mathrm{char}(k) = p$, then V_H is even simple by [BrLu12, Lemma 3.4] (see also [Ca88, §B.11]).

So, finally, assume that $\mathrm{char}(\mathbb{K}) \neq \mathrm{char}(k)$ and $\mathrm{char}(\mathbb{K}) \neq 2$. In particular, $\mathrm{char}(\mathbb{K})$ is either 0 or a prime not dividing the index of H in $\tilde{G}$. By Clifford's Theorem (see [HuBl82, Theorem VII.9.18]), V_H is semisimple and there are two possibilities: either V_H is multiplicity-free (with 1, 2 or 4 irreducible constituents) or the direct sum of 2 copies of a simple $\mathbb{K}H$-module. In the case where $\mathrm{char}(\mathbb{K}) = 0$, it is shown by an elaborate counting argument (first published in [CE04]; see also [Lu08a]) that the second type does not occur. This argument involves:

- knowing the action (by tensor product) of the four 1-dimensional representations of $\tilde{G}/H$ on the simple $\mathbb{K}\tilde{G}$-modules;
- counting conjugacy classes and simple modules for $\mathrm{Spin}_{2n}(q)$. (As noted in [Lu88, §13], this is 'very long and unpleasant'.)

Finally, it is shown in [Ge93a, §3], using the results on basic sets of Brauer characters in [GeHi91], that Lusztig's argument can be adapted to work as well when $\mathrm{char}(\mathbb{K}) > 0$ (but still $\mathrm{char}(\mathbb{K}) \neq \mathrm{char}(k)$ and $\mathrm{char}(\mathbb{K}) \neq 2$). □

It would be highly desirable to find a more conceptual proof of this result which does not rely on a case-by-case analysis and the counting arguments for $\mathrm{Spin}_{2n}(q)$.

Remark 1.7.16 It is an intriguing challenge to try to formulate a general condition on a finite group Γ and a normal subgroup $\Gamma' \trianglelefteq \Gamma$ such that the statement of Theorem 1.7.15 can be obtained as a special case of it. In [Bo06, §11.E], Bonnafé poses the following question.

Suppose that Γ/Γ' is a p'-group (for some prime p) and that $\Gamma = C_\Gamma(g)\Gamma'$ for every p'-element $g \in \Gamma$. Is it true that the restriction of every irreducible character of Γ to Γ' is multiplicity-free?

It is shown by Navarro [Na19] that this question has a negative answer, by finding a counter example using the library of groups in [GAP4].

2

Lusztig's Classification of Irreducible Characters

Let $\mathbf{G}$ be a connected reductive algebraic group over $k = \overline{\mathbb{F}}_p$ and $F: \mathbf{G} \to \mathbf{G}$ be a Steinberg map. We will be interested in describing the complex irreducible characters of the finite group $\mathbf{G}^F$, where the term 'describing' is left deliberately vague: it can range from complete knowledge of all character values to rough information about the possible character degrees and their multiplicities, to bounds on character values on specific elements and so on. The Cambridge ATLAS [CCNPW] (see also [Bre18]) contains many character tables of individual finite groups of Lie type, even of some of the large groups of exceptional type (for example, ${}^2E_6(2)$). For small rank cases, complete character tables for a whole series of groups (for example, $\mathrm{Sp}_4(q)$ where q runs through all prime powers) have been determined, in many cases without using any of the machinery arising from the Deligne–Lusztig theory that we are going to introduce in this chapter. The most complete results are available for the groups $\mathrm{GL}_n(q)$ by the pioneering work of Green [Gre55].

Section 2.1 introduces some notation and basic constructions for characters of finite groups. From Section 2.2 on, we consider finite groups of the form $\mathbf{G}^F$ as above. We begin by recalling the fundamental construction of the virtual characters $R_{\mathbf{T}}^{\mathbf{G}}(\theta)$ using the ℓ-adic cohomology approach of Deligne–Lusztig [DeLu76]; in particular, we define the set $\mathrm{Uch}(\mathbf{G}^F)$ of unipotent characters in Section 2.3. Since all this is already well covered and re-worked in existing textbooks (e.g., [Ca85], [DiMi20]), we will only introduce the required notation, state the main results (e.g., scalar product formulae and degree formulae) and illustrate them by examples, but without proofs. Only on a few occasions do we present a detailed argument when this appears to be a good illustration for the methods developed so far.

Sections 2.4 and 2.5 form the technical core of this chapter. Here, we work out in some detail the basic formalism of Lusztig's book [Lu84a] which yields an approach to the partition of $\mathrm{Irr}(\mathbf{G}^F)$ into rational/geometric series of characters which is somewhat different from that developed in [Ca85], [DiMi20]. It also provides the technical language for the formulation of the main result, that is, 'Main

Theorem 4.23' of [Lu84a]. Once this is established, the *Jordan decomposition of characters* can be stated in relatively smooth terms, both in the original case where the centre of $\mathbf{G}$ is connected, and in the general case; see Section 2.6. The importance and impact of this fundamental result can hardly be overstated. It is a tremendous achievement, both conceptually and in terms of technical complexity, which leads to an efficient classification of $\mathrm{Irr}(\mathbf{G}^F)$ in terms of data in a group $\mathbf{G}^*$ dual to $\mathbf{G}$ (as already introduced in Chapter 1). Much of this has been turned into explicit algorithms and computer programs; see [GHLMP], [MiChv], [Lue07].

In the final two Sections 2.7 and 2.8 we give a first introduction to the problem of computing the values of the irreducible characters of $\mathbf{G}^F$ on all elements. This problem is not yet completely solved. We will mainly focus on uniform functions and the determinantion of the virtual characters $R_{\mathbf{T}}^{\mathbf{G}}(\theta)$.

2.1 Generalities about Character Tables

We assume some basic familiarity with the 'ordinary' representation theory of finite groups over fields of characteristic 0 (in which case all of their finite-dimensional representations are semisimple); see, e.g., [FuHa91, Part I], [Hup67, Kap. V] or [Is76]. Our main interest will be in studying the characters of representations, where we work throughout over a fixed subfield $\mathbb{K} \subseteq \mathbb{C}$, which is algebraic over $\mathbb{Q}$, invariant under complex conjugation and 'large enough', that is, $\mathbb{K}$ contains sufficiently many roots of unity and $\mathbb{K}$ is a splitting field for all finite groups under consideration.

2.1.1 Let Γ be a finite group. We note by $\mathrm{CF}(\Gamma)$ the vector space of all $\mathbb{K}$-valued class functions on Γ, that is, functions $f: \Gamma \to \mathbb{K}$ that are constant on the conjugacy classes of Γ. There is an inner product given by

$$\langle f, f' \rangle := \frac{1}{|\Gamma|} \sum_{g \in \Gamma} f(g)\overline{f'(g)} \qquad \text{where} \qquad f, f' \in \mathrm{CF}(\Gamma).$$

(Here, the bar denotes complex conjugation.) Let $\mathrm{Irr}(\Gamma)$ be the set of irreducible characters of Γ (afforded by irreducible representations of Γ over $\mathbb{K}$). Let $\mathrm{Cl}(\Gamma)$ be the set of conjugacy classes of Γ. It is well known that $|\mathrm{Irr}(\Gamma)| = |\mathrm{Cl}(\Gamma)|$ and that $\mathrm{Irr}(\Gamma)$ is an orthonormal basis of $\mathrm{CF}(\Gamma)$. The character table of Γ is the matrix

$$X(\Gamma) = \big(\chi(g_C)\big)_{\chi \in \mathrm{Irr}(\Gamma),\, C \in \mathrm{Cl}(\Gamma)}$$

where g_C is a fixed representative of $C \in \mathrm{Cl}(\Gamma)$. We call $f \in \mathrm{CF}(\Gamma)$ a *virtual character* if f is an integral linear combination of $\mathrm{Irr}(\Gamma)$. We say that f is an actual character (or just a character) if f can be written as an integral combination of $\mathrm{Irr}(\Gamma)$

with non-negative coefficients. It is then the character of an actual representation of Γ.

Explicit examples certainly play an important and useful part in this theory. Table 2.1 displays the character tables of three finite groups of Lie type; they are printed in the output format of GAP [Scho97], [GAP4], which is modelled on the Cambridge ATLAS [CCNPW]. (A dot '.' in the tables stands for the value 0; for the notation concerning irrationalities, see the GAP online help.)

2.1.2 A highly useful method for constructing characters of Γ is given by the process of induction, due to Frobenius. If $\Gamma' \leqslant \Gamma$ is a subgroup and $f \in \mathrm{CF}(\Gamma')$, then we denote by $\mathrm{Ind}_{\Gamma'}^{\Gamma}(f) \in \mathrm{CF}(\Gamma)$ the induced class function. It is well known that, if f is a character of Γ', then $\mathrm{Ind}_{\Gamma'}^{\Gamma}(f)$ is a character of Γ. For $f \in \mathrm{CF}(\Gamma')$, the values of the induced class function are given by the following character formula:

$$\mathrm{Ind}_{\Gamma'}^{\Gamma}(f)(g) = \frac{1}{|\Gamma'|} \sum_{x \in \Gamma \,:\, xgx^{-1} \in \Gamma'} f(xgx^{-1}) \qquad (g \in \Gamma).$$

Thus, in order to work out these values, one needs to know the *class fusion* from Γ' to Γ, that is, the map $\eta\colon \mathrm{Cl}(\Gamma') \to \mathrm{Cl}(\Gamma)$ such that $C' \subseteq \eta(C')$ for all $C' \in \mathrm{Cl}(\Gamma')$. For a fixed $g \in \Gamma$, we denote ${}^g\Gamma' := g\Gamma' g^{-1}$ and define ${}^g f \in \mathrm{CF}({}^g\Gamma')$ by ${}^g f(gg'g^{-1}) = f(g')$ for all $g' \in \Gamma'$. Then, clearly, we have

$$\mathrm{Ind}_{{}^g\Gamma'}^{\Gamma}({}^g f) = \mathrm{Ind}_{\Gamma'}^{\Gamma}(f) \qquad \text{for all } f \in \mathrm{CF}(\Gamma').$$

(When we consider characters of finite groups of Lie type, we will see generalisations of the process of induction, defined by cohomological methods.)

2.1.3 A further useful construction is as follows. Let $\Gamma' \trianglelefteq \Gamma$ be a normal subgroup and $f \in \mathrm{CF}(\Gamma/\Gamma')$. Then we obtain a class function $\tilde{f} \in \mathrm{CF}(\Gamma)$ by setting $\tilde{f}(g) = f(g\Gamma')$ for all $g \in \Gamma$. Clearly, if f is a character of Γ/Γ', then $\tilde{f}$ is a character of Γ, called the *inflation* of f. Conversely, if $f \in \mathrm{CF}(\Gamma)$, then we obtain a class function $f' \in \mathrm{CF}(\Gamma/\Gamma')$ by setting

$$f'(g\Gamma') = \frac{1}{|\Gamma'|} \sum_{g' \in \Gamma'} f(gg') \qquad \text{for all } g \in \Gamma.$$

Again, if f is a character of Γ, then f' is a character of Γ/Γ'. Indeed, if V is a $\mathbb{K}\Gamma$-module affording f, then the subspace of fixed points

$$V^{\Gamma'} = \{v \in V \mid g'.v = v \text{ for all } g' \in \Gamma'\}$$

is still a $\mathbb{K}(\Gamma/\Gamma')$-submodule of V, and f' is the character of $V^{\Gamma'}$.

Remark 2.1.4 We shall need a few basic notions from Clifford theory (see, e.g., [Hup67, §V.17]). Let $\Gamma' \trianglelefteq \Gamma$ be a normal subgroup. Then Γ acts by conjugation

Table 2.1 *Tables for* $\mathrm{Sp}_4(2) \cong \mathfrak{S}_6$, $\mathrm{Suz}(8) = {}^2B_2(8)$, $\mathrm{SU}_4(2) \cong \mathrm{PSp}_4(3)$

Sp4(2)

2	4	4	4	4	1	1	1	3	3	.	1
3	2	1	.	1	2	1	2	.	.	.	1
5	1	.	.	.	.	.	.	.	.	1	.
	1a	2a	2b	2c	3a	6a	3b	4a	4b	5a	6b
X.1	1	-1	1	-1	1	-1	1	-1	1	1	-1
X.2	5	-3	1	1	2	.	-1	-1	-1	.	1
X.3	9	-3	1	-3	.	.	.	1	1	-1	.
X.4	5	-1	1	3	-1	-1	2	1	-1	.	.
X.5	10	-2	-2	2	1	1	1	.	.	.	-1
X.6	16	.	.	.	-2	.	-2	.	.	1	.
X.7	5	1	1	-3	-1	1	2	-1	-1	.	.
X.8	10	2	-2	-2	1	-1	1	.	.	.	1
X.9	9	3	1	3	.	.	.	-1	1	-1	.
X.10	5	3	1	-1	2	.	-1	1	-1	.	-1
X.11	1	1	1	1	1	1	1	1	1	1	1

Suz(8)

2	6	6	4	4	.	.	.	.	.	.	.
5	1	.	.	.	1	.	.	.	.	.	.
7	1	.	.	.	.	1	1	1	.	.	.
13	1	.	.	.	.	.	.	.	1	1	1
	1a	2a	4a	4b	5a	7a	7b	7c	13a	13b	13c
X.1	1	1	1	1	1	1	1	1	1	1	1
X.2	14	-2	A	-A	-1	.	.	.	1	1	1
X.3	14	-2	-A	A	-1	.	.	.	1	1	1
X.4	35	3	-1	-1	.	.	.	.	E	G	F
X.5	35	3	-1	-1	.	.	.	.	F	E	G
X.6	35	3	-1	-1	.	.	.	.	G	F	E
X.7	64	.	.	.	-1	1	1	1	-1	-1	-1
X.8	65	1	1	1	.	B	D	C	.	.	.
X.9	65	1	1	1	.	C	B	D	.	.	.
X.10	65	1	1	1	.	D	C	B	.	.	.
X.11	91	-5	-1	-1	1	.	.	.	.	.	.

A = 2*E(4) = 2ER(-1), B = E(7)+E(7)^6
C = E(7)^3+E(7)^4, D = E(7)^2+E(7)^5
E = -E(13)-E(13)^5-E(13)^8-E(13)^12
F = -E(13)^4-E(13)^6-E(13)^7-E(13)^9
G = -E(13)^2-E(13)^3-E(13)^10-E(13)^11

SU4(2)

2	6	6	5	3	3	2	1	4	3	.	3	3	2	2	1	2	.	.	2	2
3	4	2	1	4	4	3	3	1	.	.	2	2	2	2	2	1	2	2	1	1
5	1	.	.	.	.	.	.	.	.	1	.	.	.	.	.	.	.	.	.	.
	1a	2a	2b	3a	3b	3c	3d	4a	4b	5a	6a	6b	6c	6d	6e	6f	9a	9b	12a	12b
X.1	1	1	1	1	1	1	1	1	1	1	1	1	1	1	1	1	1	1	1	1
X.2	5	-3	1	A	/A	-1	2	1	-1	.	F	/F	I	-I	.	1	J	/J	-J	-/J
X.3	5	-3	1	/A	A	-1	2	1	-1	.	/F	F	-I	I	.	1	/J	J	-/J	-J
X.4	6	-2	2	-3	-3	3	.	2	.	1	1	1	1	1	-2	-1	.	.	-1	-1
X.5	10	2	-2	B	/B	1	1	2	.	.	A	/A	-1	-1	-1	1	-/J	-J	J	/J
X.6	10	2	-2	/B	B	1	1	2	.	.	/A	A	-1	-1	-1	1	-J	-/J	/J	J
X.7	15	-1	-1	6	6	3	.	3	-1	.	2	2	-1	-1	2	-1	.	.	.	.
X.8	15	7	3	-3	-3	.	3	-1	1	.	1	1	-2	-2	1	.	.	.	-1	-1
X.9	20	4	4	2	2	5	-1	.	.	.	-2	-2	1	1	1	1	-1	-1	.	.
X.10	24	8	.	6	6	.	3	.	.	-1	2	2	2	2	-1	.	.	.	.	.
X.11	30	10	2	3	3	3	3	-2	.	.	-1	-1	-1	-1	-1	-1	.	.	1	1
X.12	30	6	2	C	/C	-3	.	2	.	.	/F	F	-I	I	.	-1	.	.	/J	J
X.13	30	6	2	/C	C	-3	.	2	.	.	F	/F	I	-I	.	-1	.	.	J	/J
X.14	40	-8	.	D	/D	-2	1	.	.	.	G	/G	G	/G	1	.	-/J	-J	.	.
X.15	40	-8	.	/D	D	-2	1	.	.	.	/G	G	/G	G	1	.	-J	-/J	.	.
X.16	45	-3	-3	E	/E	.	.	1	1	.	H	/H	.	.	.	.	.	.	-J	-/J
X.17	45	-3	-3	/E	E	.	.	1	1	.	/H	H	.	.	.	.	.	.	-/J	-J
X.18	60	-4	4	6	6	-3	-3	.	.	.	2	2	-1	-1	-1	1	.	.	.	.
X.19	64	.	.	-8	-8	4	-2	.	.	-1	.	.	.	.	.	.	1	1	.	.
X.20	81	9	-3	.	.	.	.	-3	-1	1	.	.	.	.	.	.	.	.	.	.

A = E(3)-2*E(3)^2 = (1+3ER(-3))/2, B = 5*E(3)+2*E(3)^2 = (-7+3ER(-3))/2
C = 6*E(3)-3*E(3)^2 = (-3+9ER(-3))/2, D = 2*E(3)+8*E(3)^2 = -5-3ER(-3)
E = -9*E(3) = (9-9ER(-3))/2, F = E(3)+2*E(3)^2 = (-3-ER(-3))/2
G = -2*E(3) = 1-ER(-3) = 1-i3, H = 3*E(3) = (-3+3ER(-3))/2
I = E(3)-E(3)^2 = ER(-3) = i3, J = -E(3) = (1-ER(-3))/2

on Γ' and, hence, on $\mathrm{Irr}(\Gamma')$. Clearly, every $g \in \Gamma'$ acts trivially on $\mathrm{Irr}(\Gamma')$ and so we have an action of Γ/Γ' on $\mathrm{Irr}(\Gamma')$. It is well known that, if $\chi \in \mathrm{Irr}(\Gamma)$, then the irreducible constituents of $\chi|_{\Gamma'}$ form a single orbit under this action; in particular, all these constituents have the same multiplicity in $\chi|_{\Gamma'}$. (The Multiplicity-Freeness Theorem 1.7.15 describes a situation where these multiplicities are equal to 1.)

Now assume, furthermore, that Γ/Γ' is abelian. Let Θ be the group of all *linear characters* $\eta \in \mathrm{Irr}(\Gamma)$ (that is, group homomorphisms $\eta\colon \Gamma \to \mathbb{K}^\times$) such that $\Gamma' \subseteq \ker(\eta)$. This group acts on $\mathrm{Irr}(\Gamma)$ by the usual pointwise product of class functions. Let $\chi_1, \chi_2 \in \mathrm{Irr}(\Gamma)$. Then the restrictions of χ_1, χ_2 to Γ' have an irreducible constituent in common if and only if χ_1, χ_2 are in the same Θ-orbit. More precisely, we have:

$$\langle \chi_1|_{\Gamma'}, \chi_2|_{\Gamma'} \rangle = |\{\eta \in \Theta \mid \eta \cdot \chi_1 = \chi_2\}|.$$

Indeed, since Γ/Γ' is abelian, $\mathrm{Ind}_{\Gamma'}^{\Gamma}(\chi_1|_{\Gamma'}) = \mathrm{Ind}_{\Gamma'}^{\Gamma}(1_{\Gamma'}) \cdot \chi_1 = \sum_{\eta \in \Theta} \eta \cdot \chi_1$, so the above equality immediately follows using Frobenius reciprocity.

Remark 2.1.5 In the following section, we will construct representations of certain finite groups Γ over $\overline{\mathbb{Q}}_\ell$, an algebraic closure of the field of ℓ-adic numbers, where ℓ is a prime number. These constructions will give rise to virtual characters of Γ over $\overline{\mathbb{Q}}_\ell$, that is, $\mathbb{Z}$-linear combinations of characters afforded by finite-dimensional $\overline{\mathbb{Q}}_\ell\Gamma$-modules. Since $\mathbb{K} \subseteq \mathbb{C}$ is algebraic over $\mathbb{Q}$, we can find an embedding $\mathbb{K} \subseteq \overline{\mathbb{Q}}_\ell$. Assume now that a virtual character f over $\overline{\mathbb{Q}}_\ell$ has the property that $f(g) \in \mathbb{K}$ for all $g \in \Gamma$. Then one easily sees that f is also a virtual character over $\mathbb{K}$, that is, a $\mathbb{Z}$-linear combination of characters afforded by finite-dimensional $\mathbb{K}\Gamma$-modules. (Note that $\mathbb{K}$ is assumed to be a splitting field for Γ.)

2.1.6 Assume now that we are also given a group automorphism $\sigma\colon \Gamma \to \Gamma$. It will be convenient to introduce already at this stage some notation related to the presence of σ and its action on the elements and the characters of Γ. We say that two elements $y, y' \in \Gamma$ are *σ-conjugate* if there exists some $x \in \Gamma$ such that $y' = xy\sigma(x)^{-1}$. This defines an equivalence relation on Γ; the equivalence classes are called the σ-conjugacy classes of Γ. For $g \in \Gamma$, the σ-centraliser of g is defined to be the subgroup

$$C_{\Gamma,\sigma}(g) := \{x \in \Gamma \mid xg = g\sigma(x)\}.$$

This is the stabiliser of g for the action of Γ on itself by σ-conjugation and so $|\Gamma| = |C_{\Gamma,\sigma}(g)||C_g|$, where C_g is the σ-conjugacy class of g. As in [Bo06, 1B], [Leh78, §1], we shall denote by $H^1(\sigma, \Gamma)$ the set of σ-conjugacy classes of Γ. (Of course, if $\sigma = \mathrm{id}_\Gamma$ is the identity, then $H^1(\mathrm{id}_\Gamma, \Gamma) = \mathrm{Cl}(\Gamma)$ is just the set of ordinary conjugacy classes.)

2.1.7 In the setting of 2.1.6, a function $f\colon \Gamma \to \mathbb{K}$ is called a *σ-class function* if

f is constant on the σ-conjugacy classes of Γ. Let $\mathrm{CF}_\sigma(\Gamma)$ be the $\mathbb{K}$-vector space of all σ-class functions on Γ. There also is an inner product given by

$$\langle f, f' \rangle_\sigma := \frac{1}{|\Gamma|} \sum_{g \in \Gamma} f(g)\overline{f'(g)} \qquad \text{where} \qquad f, f' \in \mathrm{CF}_\sigma(\Gamma).$$

We obtain functions in $\mathrm{CF}_\sigma(\Gamma)$ by the following construction. Given $\chi \in \mathrm{Irr}(\Gamma)$, we define $\chi^\sigma : \Gamma \to \mathbb{K}$ by $\chi^\sigma(g) := \chi(\sigma(g))$ for all $g \in \Gamma$; then it is clear that we also have $\chi^\sigma \in \mathrm{Irr}(\Gamma)$. We set

$$\mathrm{Irr}(\Gamma)^\sigma := \{\chi \in \mathrm{Irr}(\Gamma) \mid \chi^\sigma = \chi\}.$$

Now let $\chi \in \mathrm{Irr}(\Gamma)^\sigma$. Let $n = \chi(1)$ and $\mathfrak{X} : \Gamma \to \mathrm{GL}_n(\mathbb{K})$ be a matrix representation affording χ. Since $\chi^\sigma = \chi$, there exists an invertible matrix $E \in \mathrm{GL}_n(\mathbb{K})$ such that the following invariance condition holds:

$$\mathfrak{X}(\sigma(g)) = E \cdot \mathfrak{X}(g) \cdot E^{-1} \qquad \text{for all } g \in \Gamma. \qquad (*)$$

Since $\mathfrak{X}$ is irreducible and $\mathbb{K}$ is a splitting field, E is unique up to multiplication by a non-zero scalar (Schur's Lemma). In general, there is no canonical choice for E, but we can and will always assume that E is a matrix of finite order (see the argument in [Fei82, Theorem III.2.14]). Then we obtain a σ-class function $\tilde{\chi} \in \mathrm{CF}_\sigma(\Gamma)$ by

$$\tilde{\chi}(g) := \mathrm{trace}(\mathfrak{X}(g) \cdot E) = \mathrm{trace}(E \cdot \mathfrak{X}(g)) \qquad \text{for all } g \in \Gamma.$$

Note that $\tilde{\chi}$ is well defined up to multiplication by a root of unity; each such $\tilde{\chi}$ will be called a *σ-extension* of χ. (See Remark 2.1.9 below for further explanations.) By $(*)$, we have $\tilde{\chi}(\sigma(g)) = \tilde{\chi}(g)$ for all $g \in \Gamma$. Furthermore, since E has finite order, $\tilde{\chi}(g)$ is a cyclotomic integer for all $g \in \Gamma$.

There are a few situations where there is a natural choice for E as above.

Example 2.1.8 (a) Assume that σ is the identity. Then the invariance condition $(*)$ in 2.1.7 trivially holds with $E = I_n$ (the identity matrix of size n); so we have $\tilde{\chi} = \chi$ with this choice of E. However, for greater flexibility, we do allow here to take $E = \delta I_n$ where δ is any root of unity; we then have $\tilde{\chi} = \delta\chi$ and $\tilde{\chi}(1) = \delta$.

(b) Assume that σ is an inner automorphism, that is, there exists some $g_0 \in \Gamma$ such that $\sigma(g) = g_0 g g_0^{-1}$ for all $g \in \Gamma$. Then, clearly, we have $\chi = \chi^\sigma$ for all $\chi \in \mathrm{Irr}(\Gamma)$. In this case, the above invariance condition simply holds for $E = \mathfrak{X}(g_0)$. Hence, a natural σ-extension $\tilde{\chi}$ of χ is given by

$$\tilde{\chi}(g) := \chi(g g_0) = \chi(g_0 g) \qquad \text{for all } g \in \Gamma.$$

(c) Let $\chi \in \mathrm{Irr}(\Gamma)^\sigma$ be a linear character. Then χ is a group homomorphism $\Gamma \to \mathbb{K}^\times$

and the invariance condition in 2.1.7 certainly holds for $E = 1$. Hence, in this case, a canonical σ-extension $\tilde{\chi}$ of χ is given by

$$\tilde{\chi}(g) := \chi(g) \qquad \text{for all } g \in \Gamma.$$

This situation occurs, of course, when Γ is abelian. Note that we have $|H^1(\sigma, \Gamma)| = |\Gamma^\sigma|$ in this case, where $\Gamma^\sigma := \{g \in \Gamma \mid \sigma(g) = g\}$; see [Bo06, Chap. I, Exp. 1.1].

Remark 2.1.9 As in [DiMi85, §I.6], let $\tilde{\Gamma}$ be the semidirect product of Γ with the cyclic group $\langle\sigma\rangle \subseteq \mathrm{Aut}(\Gamma)$. We identify Γ with a subgroup of $\tilde{\Gamma}$. If σ has order $d \geqslant 1$, then $\tilde{\Gamma}$ is generated by Γ and an additional element $\tilde{\sigma} \in \tilde{\Gamma}$ (of order d) such that, in $\tilde{\Gamma}$, we have the identity

$$\tilde{\sigma}\cdot g\cdot\tilde{\sigma}^{-1} = \sigma(g) \qquad \text{for all } g \in \Gamma.$$

Note that $\tilde{\sigma}\cdot(g\cdot\tilde{\sigma})\cdot\tilde{\sigma}^{-1} = g^{-1}\cdot(g\cdot\tilde{\sigma})\cdot g$ for $g \in \Gamma$. Hence, two elements of $\tilde{\Gamma}$ are conjugate in $\tilde{\Gamma}$ if and only if they are already conjugate by an element of Γ. Consequently, for $g \in \Gamma$, we have $C_{\Gamma,\sigma}(g) = C_\Gamma(g\cdot\tilde{\sigma})$ and the map $g \mapsto g\cdot\tilde{\sigma}$ induces a bijection between the σ-conjugacy classes of Γ and the usual conjugacy classes of $\tilde{\Gamma}$ that are contained in the coset $\Gamma\cdot\tilde{\sigma} \subseteq \tilde{\Gamma}$. Furthermore, $\mathrm{Irr}(\Gamma)^\sigma$ consists precisely of those irreducible characters of Γ that can be extended to $\tilde{\Gamma}$. Now let $\chi \in \mathrm{Irr}(\Gamma)^\sigma$. Then we can find a matrix E satisfying the invariance condition in 2.1.7(∗) and such that $E^d = I_n$ (see [Fei82, Theorem III.2.14]). In this case, $(\tilde{\chi}(g))_{g\in\Gamma}$ are the values of an extension of χ on the elements in the coset $\Gamma\tilde{\sigma}$.

One can also define a process of σ-induction of characters from σ-invariant subgroups of Γ; see [Bo06, 1.C] for further details.

2.1.10 Let us now assume that a particular σ-extension $\tilde{\chi}$ has been chosen for each $\chi \in \mathrm{Irr}(\Gamma)^\sigma$. Then, by [DiMi85, Rem. I.6.3(iii)], we have the following orthogonality relations:

$$\langle\tilde{\chi}, \tilde{\chi}'\rangle_\sigma = \frac{1}{|\Gamma|}\sum_{g\in\Gamma}\tilde{\chi}(g)\overline{\tilde{\chi}'(g)} = \begin{cases} 1 & \text{if } \chi = \chi', \\ 0 & \text{if } \chi \neq \chi'. \end{cases}$$

Moreover, by the argument in [GKP00, 7.3], we have $|\mathrm{Irr}(\Gamma)^\sigma| = |H^1(\sigma, \Gamma)|$. Hence, for any $g, g' \in \Gamma$, we also have

$$\sum_{\chi\in\mathrm{Irr}(\Gamma)^\sigma}\tilde{\chi}(g)\overline{\tilde{\chi}(g')} = \begin{cases} |C_{\Gamma,\sigma}(g)| & \text{if } g, g' \text{ are } \sigma\text{-conjugate}, \\ 0 & \text{otherwise.} \end{cases}$$

In particular, the set $\{\tilde{\chi} \mid \chi \in \mathrm{Irr}(\Gamma)^\sigma\}$ is an orthonormal basis of $\mathrm{CF}_\sigma(\Gamma)$.

Definition 2.1.11 In the setting of 2.1.7, let us choose a representative $g_C \in C$

for each σ-conjugacy class $C \in H^1(\sigma, \Gamma)$. Furthermore, as above, we assume that a particular σ-extension $\tilde{\chi}$ has been chosen for each $\chi \in \mathrm{Irr}(\Gamma)^\sigma$. Then the matrix

$$X_\sigma(\Gamma) = \big(\tilde{\chi}(g_C)\big)_{\chi \in \mathrm{Irr}(\Gamma)^\sigma,\, C \in H^1(\sigma,\Gamma)}$$

is called the σ-*character table* of Γ. This is a square matrix which does not depend on the choice of $g_C \in C$. (But it does depend on the choice of the σ-extension $\tilde{\chi}$ for each $\chi \in \mathrm{Irr}(\Gamma)^\sigma$; if a different choice of $\tilde{\chi}$ is made, then this will result in multiplying the corresponding row of $X_\sigma(\Gamma)$ by a root of unity.) By 2.1.7, the entries of $X_\sigma(\Gamma)$ are cyclotomic integers; the orthogonality relations in 2.1.10 show that $X_\sigma(\Gamma)$ is invertible.

If we choose σ-extensions as in Remark 2.1.9, then the matrix $X_\sigma(\Gamma)$ is part of the ATLAS 'compound character table' for Γ. If, moreover, σ is the identity, then $X_\sigma(\Gamma) = X(\Gamma)$ is just the ordinary character table of Γ.

We now give two examples. First of all, they illustrate that one cannot expect a natural choice of a σ-extension of a character. Secondly, we discuss this in some detail because a completely analogous problem – at a technically more elaborate level – is one of the key issues in completing the character tables of finite groups of Lie type (see Section 2.8, especially Remark 2.8.8).

Example 2.1.12 Let Γ be a dihedral group of order 8. Thus, Γ is generated by two elements $s_1 \neq s_2$ such that s_1, s_2 have order 2 and the product s_1s_2 has order 4. We have $\mathrm{Irr}(W) = \{1_\Gamma, \varepsilon, \varepsilon', \varepsilon'', \phi_1\}$, where 1_Γ is the trivial character, $\varepsilon, \varepsilon', \varepsilon''$ are linear characters, and ϕ_1 has degree 2. Here, the notation is such that $\varepsilon(s_1) = \varepsilon(s_2) = -1$; furthermore,

$$\varepsilon'(s_1) = \varepsilon''(s_2) = -1 \qquad \text{and} \qquad \varepsilon'(s_2) = \varepsilon''(s_1) = 1.$$

(See also Table 4.1, p. 274.) Now, there is an automorphism $\sigma\colon \Gamma \to \Gamma$ such that $\sigma(s_1) = s_2$ and $\sigma(s_2) = s_1$. The σ-conjugacy classes of Γ are given by

$$\{1, s_1s_2, s_2s_1, s_1s_2s_1s_2\}, \ \{s_1, s_2\}, \ \{s_1s_2s_1, s_2s_1s_2\}.$$

In particular, we see that $\mathrm{Irr}(\Gamma)^\sigma = \{1_\Gamma, \varepsilon, \phi_1\}$. We find σ-extensions for 1_Γ and ε by the construction in Example 2.1.8(c). The character ϕ_1 is afforded by the matrix representation

$$\mathfrak{X}\colon \quad s_1 \mapsto \begin{pmatrix} -1 & 0 \\ \sqrt{2} & 1 \end{pmatrix}, \qquad s_2 \mapsto \begin{pmatrix} 1 & \sqrt{2} \\ 0 & -1 \end{pmatrix}.$$

The invariance condition in 2.1.7 is satisfied with $E = \begin{pmatrix} 0 & 1 \\ 1 & 0 \end{pmatrix}$; note also that $E^2 = I_2$. If we use E to define a σ-extension of ϕ_1, then the resulting σ-character table of Γ is the first table printed in Table 2.2. If we replace E by $-E$, then this will change the signs of the values of $\tilde{\phi}_1$.

Table 2.2 *The σ-character tables of dihedral groups of order* 8 *and* 12

Dih(8)	1	s_1	$s_1s_2s_1$
$\tilde{1}_\Gamma$	1	1	1
$\tilde{\varepsilon}$	1	-1	-1
$\tilde{\phi}_1$	.	$\sqrt{2}$	$-\sqrt{2}$

Dih(12)	1	s_1	$s_1s_2s_1$	$s_1s_2s_1s_2s_1$
$\tilde{1}_\Gamma$	1	1	1	1
$\tilde{\varepsilon}$	1	-1	-1	-1
$\tilde{\phi}_1$	.	$\sqrt{3}$	.	$-\sqrt{3}$
$\tilde{\phi}_2$	.	1	-2	1

Example 2.1.13 Let Γ be a dihedral group of order 12. Thus, Γ is generated by two elements $s_1 \neq s_2$ such that s_1, s_2 have order 2 and the product s_1s_2 has order 6. We have $\mathrm{Irr}(W) = \{1_\Gamma, \varepsilon, \varepsilon', \varepsilon'', \phi_1, \phi_2\}$, where 1_Γ is the trivial character, $\varepsilon, \varepsilon', \varepsilon''$ are linear characters, and ϕ_1, ϕ_2 have degree 2. We use the same notational convention for the linear characters as in the previous example; furthermore, the notation for ϕ_1 and ϕ_2 is such that $\phi_1((s_1s_2)^3) = -2$ and $\phi_2((s_1s_2)^3) = 2$. (See also Table 4.2, p. 276.) As in the previous example, there is an automorphism $\sigma\colon \Gamma \to \Gamma$ such that $\sigma(s_1) = s_2$ and $\sigma(s_2) = s_1$. The σ-conjugacy classes of Γ are given by

$$\{1, s_1s_2, s_2s_1, s_1s_2s_1s_2, s_2s_1s_2s_1, (s_1s_2)^3\},\ \{s_1, s_2\},$$
$$\{s_1s_2s_1, s_2s_1s_2\}, \{s_1s_2s_1s_2s_1, s_2s_1s_2s_1s_2\}.$$

In particular, $\mathrm{Irr}(\Gamma)^\sigma = \{1_\Gamma, \varepsilon, \phi_1, \phi_2\}$. We find σ-extensions by similar arguments as in the previous example. The resulting σ-character table of Γ is the second table printed in Table 2.2 (see also [GKP00, Example 7.6]).

Finally, let $\mathbf{G}$ be a connected reductive algebraic group over $k = \overline{\mathbb{F}}_p$ and $F\colon \mathbf{G} \to \mathbf{G}$ be a Steinberg map, as in Section 1.4. Even if we are primarily interested in the ordinary (untwisted) character theory of $\mathbf{G}^F$, we will naturally encounter σ-character tables as well, where σ is typically induced by F in some way. As a first example, recall from 1.6.1 that F induces an automorphism $\sigma\colon \mathbf{W} \to \mathbf{W}$, where $\mathbf{W}$ is the Weyl group of $\mathbf{G}$; note that $\sigma(S) = S$ where S is a suitable set of simple reflections in $\mathbf{W}$. We can apply the above discussion to $\mathbf{W}, \sigma$.

Proposition 2.1.14 (Lusztig [Lu84a, 3.2, 14.2]) *Let $\sigma\colon \mathbf{W} \to \mathbf{W}$ be as in 1.6.1. For each $\phi \in \mathrm{Irr}(\mathbf{W})^\sigma$, there exists a σ-extension $\tilde{\phi}$ such that $\tilde{\phi}(w) \in \mathbb{R}$ for all $w \in \mathbf{W}$. (Note that our σ-extensions are unique up to multiplication by a root of unity, and so there are exactly two real σ-extensions.) If, moreover, σ is ordinary in the sense of 1.6.2, then $\tilde{\phi}$ can be chosen such that $\tilde{\phi}(w) \in \mathbb{Z}$ for all $w \in \mathbf{W}$.*

(This will play a role, for example, in Remark 2.4.17; see also Section 4.1.)

Proof By exactly the same kind of reduction arguments as in the proof of [Lu84a, 3.2], it is sufficient to consider the case where **W** is irreducible. So let us now assume that this is the case. Now, since **W** is a finite Weyl group, it is well known that all irreducible characters of **W** itself are integer valued; see, e.g., [GePf00, Theorem 6.3.8]. So, if σ is the identity, then there is nothing to prove. Let us now assume that σ is not the identity.

If **W** is of type A_n ($n \geqslant 2$), D_{2n+1} ($n \geqslant 2$) or E_6, then σ is given by conjugation by the longest element of **W**. So the assertion holds by Example 2.1.8(b).

If **W** is of type D_{2n} ($n \geqslant 2$) and σ has order 2, then we use the interpretation of σ-extensions in Remark 2.1.9. In the present situation, the semidirect product $\tilde{\mathbf{W}} = \mathbf{W} \rtimes \langle\sigma\rangle$ can be identified with a Weyl group of type B_{2n}. So the assertion holds again by the known result about the character values of the latter group. (See also Example 4.1.4 for further details.)

If **W** is of type D_4 and σ has order 3, then the assertion is proved in [Lu77b, 3.18] (for a slightly different argument see the proof of [Lu84a, 3.2]).

Thus, it remains to consider the cases where σ is not ordinary and **W** is of type B_2, G_2 or F_4. Then σ has order 2 and σ-extensions can be worked out by explicit computations, as in Examples 2.1.12 and 2.1.13. See Example 2.1.15 (below) for type F_4 and [GKP00, §7] for further details. □

Example 2.1.15 Let **W** be the Weyl group of type F_4, with generators $S = \{s_1, s_2, s_3, s_4\}$ labelled as in Table 1.1 (p. 20). The ordinary character table of **W** is printed in [Ca85, p. 413]; we have $|\text{Irr}(\mathbf{W})| = 25$ in this case. An explicit construction of the 25 irreducible characters can be found in [Lu84a, p. 97]. There is an automorphism $\sigma : \mathbf{W} \to \mathbf{W}$ such that $\sigma(s_1) = s_4$, $\sigma(s_2) = s_3$, $\sigma(s_3) = s_2$ and $\sigma(s_4) = s_1$. We have $|\text{Irr}(\mathbf{W})^\sigma| = 11$ and the table $X_\sigma(\mathbf{W})$ has been determined in [GKP00, §7] (even for the corresponding generic Iwahori–Hecke algebra). This table, with some rescaling of the extensions, is printed in Table 2.3. (The reason for the rescaling will be explained in Remark 2.8.19 below.) In the table, we simply write 232 instead of $s_2s_3s_2$, for example. As in [Lu84a, 14.2] we specify, for each w_j, the characteristic polynomial of $w_j\tilde{\sigma} \in \tilde{\mathbf{W}}$ in the natural reflection representation. (See also [Shi75, Table III].)

The aim of the following sections is to explain how the problem of determining the character table of $\mathbf{G}^F$ can be approached. In the context of Section 1.6, we would like to consider $\mathbf{G}^F$ not just as an individual finite group but as a member of an infinite series of finite groups of Lie type, and obtain a uniform description of the irreducible characters of *all* finite groups in such a series. For certain groups of small rank (see Table 2.4), such a uniform description is explicitly available in the

Table 2.3 *The σ-character table of* $\mathbf{W}$ *of type* F_4

	w_1	w_2	w_3	w_4	w_5	w_6	w_7	w_8	w_9	w_{10}	w_{11}
$\tilde{\phi}_{1,0} = \tilde{1}_1$	1	1	1	1	1	1	1	1	1	1	1
$\tilde{\phi}_{1,24} = \tilde{1}_4$	1	−1	−1	1	1	−1	1	1	1	1	1
$\tilde{\phi}_{4,8} = \tilde{4}_1$	2	.	.	−1	−1	.	2	−1	2	2	2
$\tilde{\phi}_{9,2} = \tilde{9}_1$	1	1	−1	.	.	1	−1	.	−3	3	3
$\tilde{\phi}_{9,10} = \tilde{9}_4$	1	−1	1	.	.	−1	−1	.	−3	3	3
$\tilde{\phi}'_{6,6} = \tilde{6}_1$	.	.	.	1	1	.	2	−1	−4	−2	−2
$\tilde{\phi}''_{6,6} = \tilde{6}_2$	2	.	.	−1	−1	.	.	1	−2	−4	−4
$\tilde{\phi}_{12,4} = \tilde{12}$	2	.	.	1	1	.	−2	−1	2	−2	−2
$\tilde{\phi}_{4,1} = \tilde{4}_2$	.	$-\sqrt{2}$	.	$-\sqrt{2}$	$\sqrt{2}$	$\sqrt{2}$	.	.	.	$-2\sqrt{2}$	$2\sqrt{2}$
$\tilde{\phi}_{4,13} = \tilde{4}_5$	.	$-\sqrt{2}$	.	$\sqrt{2}$	$-\sqrt{2}$	$\sqrt{2}$	.	.	.	$2\sqrt{2}$	$-2\sqrt{2}$
$\tilde{\phi}_{16,5} = \tilde{16}$	.	.	.	$-\sqrt{2}$	$\sqrt{2}$	.	.	.	.	$4\sqrt{2}$	$-4\sqrt{2}$

(Note: the scaling is different from that in [GKP00].)

w_j	char. pol. of $w_j\tilde{\sigma}$	No. in [Shi75], [Ma90]
$w_1 = ()$	$(\mathbf{q}^2-1)^2$	1
$w_2 = 232$	$(\mathbf{q}^2-1)(\mathbf{q}^2+\sqrt{2}\mathbf{q}+1)$	4
$w_3 = 1$	$\mathbf{q}^4-1$	2
$w_4 = 1213214321$	$\mathbf{q}^4+\sqrt{2}\mathbf{q}^3+\mathbf{q}^2+\sqrt{2}\mathbf{q}+1$	11
$w_5 = 12$	$\mathbf{q}^4-\sqrt{2}\mathbf{q}^3+\mathbf{q}^2-\sqrt{2}\mathbf{q}+1$	10
$w_6 = 2$	$(\mathbf{q}^2-1)(\mathbf{q}^2-\sqrt{2}\mathbf{q}+1)$	3
$w_7 = 12132132$	$\mathbf{q}^4+1$	5
$w_8 = 1232$	$\mathbf{q}^4-\mathbf{q}^2+1$	9
$w_9 = 121321324321$	$(\mathbf{q}^2+1)^2$	8
$w_{10} = 121321324321324321$	$(\mathbf{q}^2+\sqrt{2}\mathbf{q}+1)^2$	7
$w_{11} = 121321$	$(\mathbf{q}^2-\sqrt{2}\mathbf{q}+1)^2$	6

form of a *generic character table*, and these will serve as valuable examples. (The CHEVIE system [GHLMP] contains these in electronic form.)

Example 2.1.16 Let us consider the series of groups $\mathbf{G}^F = \mathrm{GL}_2(q)$, where q is any prime power. The character table of $\mathrm{GL}_2(q)$ is explicitly described in [St51b]. This has found its way into textbooks on representation theory; see, for example, [FuHa91, §5.2], [Et11, §5.25]. The conjugacy classes of $\mathrm{GL}_2(q)$ are classified in terms of normal forms of matrices. Let σ be a generator of $\mathbb{F}_q^\times$ and τ be a generator of $\mathbb{F}_{q^2}^\times$. There are four types of conjugacy classes with representatives as follows.

$$A_1(a) := \begin{pmatrix} \sigma^a & 0 \\ 0 & \sigma^a \end{pmatrix} \qquad \text{where } 0 \leqslant a \leqslant q-2,$$

$$A_2(a) := \begin{pmatrix} \sigma^a & 0 \\ 1 & \sigma^a \end{pmatrix} \qquad \text{where } 0 \leqslant a \leqslant q-2,$$

Table 2.4 *Known generic character tables of finite groups of Lie type*

Series	Author(s)
$\mathrm{PSL}_2(p)$	Frobenius [Fro96] (p prime)
$\mathrm{SL}_2(q)$	Jordan [Jor07], Schur [Schu07] (see also [Bo11])
$\mathrm{GL}_2(q)$	Jordan [Jor07], Schur [Schu07], Steinberg [St51b]
$\mathrm{GL}_3(q)$, $\mathrm{GL}_4(q)$	Steinberg [St51b]
${}^2B_2(q^2)$	Suzuki [Suz62]
$\mathrm{GU}_3(q)$	Ennola [Enn63]
${}^2G_2(q^2)$	Ward [War66]
$\mathrm{Sp}_4(q)$	Srinivasan [Sr68] (q odd), Enomoto [Eno72] ($q = 2^m$)
$\mathrm{CSp}_4(q)$	Shinoda [Shi82] (q odd)
$\mathrm{SL}_3(q)$, $\mathrm{SU}_3(q)$	Simpson and Frame [SiFr73]
$G_2(q)$	Chang and Ree [ChRe74] ($q = p^m, p \neq 2, 3$), see also Hiss [Hi90b, Anhang B], Enomoto [Eno76] ($q = 3^m$), Enomoto and Yamada [EnYa86] ($q = 2^m$)
$\mathrm{Sp}_6(q)$	Locker [Loc77] ($q = 2^m$), see also Lübeck [Lue93]
$\mathrm{CSp}_6(q)$	Lübeck [Lue93] (q odd)
${}^3D_4(q)$	Spaltenstein [Spa82b], Deriziotis and Michler [DeMi87]
${}^2F_4(q^2)$	Malle [Ma90] (complete table in CHEVIE [GHLMP])
$\mathrm{SO}_8^+(q)$ (partial)	Geck and Pfeiffer [GePf92] (q odd), Geck [Ge95] ($q = 2^m$)
$\mathrm{SO}_8^-(q)$ (partial)	Lübeck [GHLMP]

(Here, 'partial' means: unipotent characters only, as defined later in this chapter.)

$$A_3(a,b) := \begin{pmatrix} \sigma^a & 0 \\ 0 & \sigma^b \end{pmatrix} \qquad \text{where } 0 \leqslant a < b \leqslant q-2,$$

$$B_1(a) := \begin{pmatrix} 0 & -\tau^{a(q+1)} \\ 1 & \tau^a + \tau^{aq} \end{pmatrix} \qquad \text{where } a \in E_q \text{ and } (q+1) \nmid a.$$

Here, $E_q \subseteq \{0, 1, \ldots, q^2-2\}$ is a set of representatives for the equivalence relation on $\mathbb{Z}$ defined by: $a \sim a'$ if $a' \equiv a \bmod (q^2-1)$ or $a' \equiv qa \bmod (q^2-1)$. Note that the matrix $B_1(a)$ is diagonalisable over $\mathbb{F}_{q^2}$ (but not over $\mathbb{F}_q$), with eigenvalues τ^a and τ^{aq}.

Class type	Number of classes	Size of class
$A_1(a)$	$q-1$	1
$A_2(a)$	$q-1$	$(q-1)(q+1)$
$A_3(a,b)$	$\frac{1}{2}(q-1)(q-2)$	$q(q+1)$
$B_1(a)$	$\frac{1}{2}q(q-1)$	$q(q-1)$

The total number of conjugacy classes is $q^2 - 1$. Hence, there are also $q^2 - 1$ irreducible characters, which can be arranged into four families as follows.

$$\begin{array}{ll} \chi_1^{(n)} & \text{of degree 1, where } 0 \leqslant n \leqslant q-2, \\ \chi_q^{(n)} & \text{of degree } q, \text{ where } 0 \leqslant n \leqslant q-2, \\ \chi_{q+1}^{(m,n)} & \text{of degree } q+1, \text{ where } 0 \leqslant n < m \leqslant q-2, \\ \chi_{q-1}^{(n)} & \text{of degree } q-1, \text{ where } n \in E_q \text{ and } (q+1) \nmid n. \end{array}$$

The values are displayed in Table 2.5. Here, $\varepsilon \in \mathbb{K}$ and $\eta \in \mathbb{K}$ denote primitive roots of unity of order $q-1$ and q^2-1, respectively.

Table 2.5 *The generic character table of* $\mathbf{G}^F = GL_2(q)$

	$A_1(a)$	$A_2(a)$	$A_3(a,b)$	$B_1(a)$
$\chi_1^{(n)}$	ε^{2na}	ε^{2na}	$\varepsilon^{n(a+b)}$	ε^{na}
$\chi_q^{(n)}$	$q\varepsilon^{2na}$	.	$\varepsilon^{n(a+b)}$	$-\varepsilon^{na}$
$\chi_{q+1}^{(m,n)}$	$(q+1)\varepsilon^{(m+n)a}$	$\varepsilon^{(m+n)a}$	$\varepsilon^{ma+nb}+\varepsilon^{na+mb}$	.
$\chi_{q-1}^{(n)}$	$(q-1)\eta^{na(q+1)}$	$-\eta^{na(q+1)}$	.	$-(\eta^{na}+\eta^{naq})$

Example 2.1.17 Let us consider the series of groups $\mathbf{G}^F = \mathrm{SL}_2(q)$, where q is any prime power. The generic character table was determined in [Jor07], [Schu07]; see, for example, [FuHa91, §5.2], for a modern textbook reference. Again, the conjugacy classes are classified in terms of normal forms of matrices, but some additional care is required when q is odd. First, we set

$$I := \begin{pmatrix} 1 & 0 \\ 0 & 1 \end{pmatrix}, \qquad N := \begin{pmatrix} 1 & 1 \\ 0 & 1 \end{pmatrix}, \qquad N' := \begin{pmatrix} 1 & \xi \\ 0 & 1 \end{pmatrix},$$

where N' is only defined for q odd: in this case, $\{x^2 \mid x \in \mathbb{F}_q^\times\} \neq \mathbb{F}_q^\times$ and we fix an element $\xi \in \mathbb{F}_q^\times$ which is not a square. Let σ be a generator of $\mathbb{F}_q^\times$ and $\tau_0 := \tau^{q-1}$ where τ is a generator of $\mathbb{F}_{q^2}^\times$. (So $\tau_0 \in \mathbb{F}_{q^2}^\times$ has order $q+1$.) Then we define:

$$S(a) := \begin{pmatrix} \sigma^a & 0 \\ 0 & \sigma^{-a} \end{pmatrix} \quad \text{where} \quad \begin{cases} 1 \leqslant a \leqslant \frac{q}{2} - 1 & \text{if } q \text{ is even,} \\ 1 \leqslant a \leqslant \frac{1}{2}(q-3) & \text{if } q \text{ is odd,} \end{cases}$$

$$T(b) := \begin{pmatrix} 0 & -1 \\ 1 & \tau_0^b + \tau_0^{bq} \end{pmatrix} \quad \text{where} \quad \begin{cases} 1 \leqslant b \leqslant \frac{q}{2} & \text{if } q \text{ is even,} \\ 1 \leqslant b \leqslant \frac{1}{2}(q-1) & \text{if } q \text{ is odd.} \end{cases}$$

Note that $T(b)$ is diagonalisable over $\mathbb{F}_{q^2}$ (but not over $\mathbb{F}_q$), with eigenvalues τ_0^b and

τ_0^{bq}. Representatives of the conjugacy classes of $\mathrm{SL}_2(q)$ are given by

$$\begin{array}{ll} I,\ N,\ \{S(a)\},\ \{T(b)\} & \text{if } q \text{ is even,} \\ I,\ -I,\ N,\ N',\ -N,\ -N',\ \{S(a)\},\ \{T(b)\} & \text{if } q \text{ is odd,} \end{array}$$

where a, b run over the index ranges specified above. The total number of conjugacy classes is $q+1$ if q is even, and $q+4$ if q is odd. We now use the known character table of $\mathrm{GL}_2(q)$ from Example 2.1.16. We define the following restrictions of characters of $\mathrm{GL}_2(q)$ to $\mathrm{SL}_2(q)$.

$$\begin{aligned} \psi_1 &:= \text{restriction of } \chi_1^{(0)}, \\ \psi_q &:= \text{restriction of } \chi_q^{(0)}, \\ \psi_{q+1}^{(i)} &:= \text{restriction of } \chi_{q+1}^{(0,i)} \text{ for } 1 \leqslant 2i \leqslant q-2, \\ \psi_{q-1}^{(j)} &:= \text{restriction of } \chi_{q-1}^{(j)} \text{ for } 1 \leqslant 2j \leqslant q. \end{aligned}$$

If q is even, then the restriction of every irreducible character of $\mathrm{GL}_2(q)$ to $\mathrm{SL}_2(q)$ is irreducible. If q is odd, then we have reducible restrictions as follows. Let $i_0 = \frac{1}{2}(q-1)$ and $j_0 = \frac{1}{2}(q+1)$. Then

$$\psi_{q+1}^{(i_0)} = \psi'_+ + \psi''_+ \qquad \text{and} \qquad \psi_{q-1}^{(j_0)} = \psi'_- + \psi''_-$$

where $\psi'_\pm$ are distinct irreducible characters of $\mathrm{SL}_2(q)$ of degree $\frac{1}{2}(q+1)$ and $\psi''_\pm$ are distinct irreducible characters of degree $\frac{1}{2}(q-1)$. The values of these characters involve the following algebraic numbers:

$$\omega = \tfrac{1}{2}(1+\sqrt{\delta q}) \quad \text{and} \quad \omega^* = \tfrac{1}{2}(1-\sqrt{\delta q}) \quad \text{where} \quad \delta = (-1)^{(q-1)/2}.$$

The generic character tables are displayed in Table 2.6. Here, $\varepsilon \in \mathbb{K}$ denotes again a primitive root of unity of order $q-1$ and $\eta_0 \in \mathbb{K}$ denotes a primitive root of unity of order $q+1$. (We take $\eta_0 = \eta^{q-1}$ where η is the primitive root of unity of order q^2-1 in the table of $\mathrm{GL}_2(q)$.) In Table 2.7, we print the tables for $q = 2, 3, 4$.

2.2 The Virtual Characters of Deligne and Lusztig

In this section we explain the basic construction of Deligne and Lusztig [Lu75], [DeLu76], [Lu77b] in which the theory of ℓ-adic cohomology is used to obtain representations of finite groups acting on algebraic varieties. (See also [Lu14c, §6] for some historical comments about the origins of this construction.)

2.2.1 Let $\mathbf{X}$ be an algebraic variety over $k = \overline{\mathbb{F}}_p$. Let ℓ be a prime different from p and $\mathbb{Q}_\ell$ be the field of ℓ-adic numbers. It is a deep fact that one can attach to $\mathbf{X}$ a

Table 2.6 *The generic character table of* $\mathbf{G}^F = \mathrm{SL}_2(q)$

q even	I	N	$S(a)$	$T(b)$
ψ_1	1	1	1	1
ψ_q	q	.	1	-1
$\psi_{q+1}^{(i)}$	$q{+}1$	1	$\varepsilon^{ai} + \varepsilon^{-ai}$	.
$\psi_{q-1}^{(j)}$	$q{-}1$	-1	.	$-\eta_0^{bj} - \eta_0^{-bj}$

Here, $1 \leqslant i \leqslant \frac{q}{2} - 1$ and $1 \leqslant j \leqslant \frac{q}{2}$.

q odd	I	$-I$	N	N'	$-N$	$-N'$	$S(a)$	$T(b)$
ψ_1	1	1	1	1	1	1	1	1
ψ_q	q	q	.	.	.	.	1	-1
$\psi_{q+1}^{(i)}$	$q{+}1$	$(-1)^i(q{+}1)$	1	1	$(-1)^i$	$(-1)^i$	$\varepsilon^{ai} + \varepsilon^{-ai}$	.
$\psi_{q-1}^{(j)}$	$q{-}1$	$(-1)^j(q{-}1)$	-1	-1	$(-1)^{j+1}$	$(-1)^{j+1}$	.	$-\eta_0^{bj} - \eta_0^{-bj}$
ψ'_+	$\frac{1}{2}(q{+}1)$	$\frac{1}{2}\delta(q{+}1)$	ω	ω^*	$\delta\omega$	$\delta\omega^*$	$(-1)^a$	.
ψ''_+	$\frac{1}{2}(q{+}1)$	$\frac{1}{2}\delta(q{+}1)$	ω^*	ω	$\delta\omega^*$	$\delta\omega$	$(-1)^a$	.
ψ'_-	$\frac{1}{2}(q{-}1)$	$-\frac{1}{2}\delta(q{-}1)$	$-\omega^*$	$-\omega$	$\delta\omega^*$	$\delta\omega$	.	$(-1)^{b+1}$
ψ''_-	$\frac{1}{2}(q{-}1)$	$-\frac{1}{2}\delta(q{-}1)$	$-\omega$	$-\omega^*$	$\delta\omega$	$\delta\omega^*$	.	$(-1)^{b+1}$

Here, $1 \leqslant i \leqslant \frac{1}{2}(q-3)$ and $1 \leqslant j \leqslant \frac{1}{2}(q-1)$.

Table 2.7 *Character tables of* $\mathbf{G}^F = \mathrm{SL}_2(q)$ *for* $q = 2, 3, 4$

$q{=}2$	I	N	$T(1)$
ψ_1	1	1	1
ψ_2	2	.	-1
$\psi_1^{(1)}$	1	-1	1

$q{=}4$	I	N	$S(1)$	$T(1)$	$T(2)$
ψ_1	1	1	1	1	1
ψ_4	4	.	1	-1	-1
$\psi_5^{(1)}$	5	1	-1	.	.
$\psi_3^{(1)}$	3	-1	.	ζ	ζ^*
$\psi_3^{(2)}$	3	-1	.	ζ^*	ζ

$q{=}3$	I	$-I$	N	N'	$-N$	$-N'$	$T(1)$
ψ_1	1	1	1	1	1	1	1
ψ_3	3	3	.	.	.	.	-1
$\psi_2^{(1)}$	2	-2	-1	-1	1	1	.
ψ'_+	2	-2	ω	ω^*	$-\omega$	$-\omega^*$	.
ψ''_+	2	-2	ω^*	ω	$-\omega^*$	$-\omega$	.
ψ'_-	1	1	$-\omega^*$	$-\omega$	$-\omega^*$	$-\omega$	1
ψ''_-	1	1	$-\omega$	$-\omega^*$	$-\omega$	$-\omega^*$	1

Here, $\omega = \frac{1}{2}(1 + \sqrt{-3})$,
$\zeta = \frac{1}{2}(1 - \sqrt{5})$ and $\zeta^* = \frac{1}{2}(1 + \sqrt{5})$.
(Note that $\mathrm{SL}_2(2) \cong \mathfrak{S}_3$ and $\mathrm{SL}_2(4) \cong A_5$.)

family of $\mathbb{Q}_\ell$-vector spaces $\mathrm{H}_c^i(\mathbf{X}, \mathbb{Q}_\ell)$ ($i \in \mathbb{Z}$), called the *ℓ-adic cohomology groups with compact support*; see [Ca85, Appendix], [Sr79, Chap. V] and the references there. These vector spaces are finite dimensional, zero for $i < 0$ and for large i. They are functorial, in the sense that a finite morphism $f\colon \mathbf{X} \to \mathbf{X}'$ naturally induces a linear map $f^*\colon \mathrm{H}_c^i(\mathbf{X}', \mathbb{Q}_\ell) \to \mathrm{H}_c^i(\mathbf{X}, \mathbb{Q}_\ell)$ for each i. This can be used to construct representations of finite groups. In the following, we extend scalars from

$\mathbb{Q}_\ell$ to an algebraic closure $\overline{\mathbb{Q}}_\ell$ and denote the corresponding cohomology spaces by $\mathrm{H}_c^i(\mathbf{X}, \overline{\mathbb{Q}}_\ell)$.

Let Γ be a finite group acting as a group of algebraic automorphisms of $\mathbf{X}$. Then $\mathrm{H}_c^i(\mathbf{X}, \overline{\mathbb{Q}}_\ell)$ is a $\overline{\mathbb{Q}}_\ell\Gamma$-module for each i, where $g \in \Gamma$ acts via $(g^*)^{-1}$. For $g \in \Gamma$, the alternating sum of traces

$$\mathfrak{L}(g, \mathbf{X}) := \sum_i (-1)^i \mathrm{Trace}\big((g^*)^{-1}, \mathrm{H}_c^i(\mathbf{X}, \overline{\mathbb{Q}}_\ell)\big)$$

is called the *Lefschetz number* of g on $\mathbf{X}$. Note that the sum makes sense, since all $\mathrm{H}_c^i(\mathbf{X}, \overline{\mathbb{Q}}_\ell)$ are finite dimensional and zero for almost all i. Thus, the map $g \mapsto \mathfrak{L}(g, \mathbf{X})$ is a virtual character of Γ over $\overline{\mathbb{Q}}_\ell$. It is known that

$$\mathfrak{L}(g, \mathbf{X}) = \mathfrak{L}(g^{-1}, \mathbf{X}) \text{ is a rational integer and does not depend on } \ell;$$

see [DeLu76, 3.3], [Lu77b, 1.2] for a proof. Hence, by Remark 2.1.5, the map $g \mapsto \mathfrak{L}(g, \mathbf{X})$ is also a virtual character of Γ over $\mathbb{K}$.

2.2.2 In the above setting, there is a way of defining Lefschetz numbers without reference to ℓ-adic cohomology at all and this can be used to compute $\mathfrak{L}(g, \mathbf{X})$ in certain examples. For this purpose, let $F \colon \mathbf{X} \to \mathbf{X}$ be the Frobenius map relative to a rational structure on $\mathbf{X}$ over some finite subfield of k and assume that $g \in \Gamma$ commutes with F. Then $F^n \circ g$ also is a Frobenius map for $n = 1, 2, \ldots$ (see Remark 1.4.4(c)); in particular, $F^n \circ g$ has only finitely many fixed points on $\mathbf{X}$. So we can form the formal power series

$$R(t, g) := -\sum_{n=1}^{\infty} |\mathbf{X}^{F^n \circ g}| t^n \in \mathbb{Z}[[t]].$$

Then $R(t, g)$ is a rational function in t, which is independent of F and which only has simple poles and no pole at ∞; furthermore, the value of $R(t, g)$ at ∞ is just $\mathfrak{L}(g, \mathbf{X})$. (The deduction of these statements from properties of the ℓ-adic cohomology groups can be found, for example, in [Lu77b, 1.2], [Ca85, App. (h)], [DiMi20, §8.1].) Let us give two simple illustrations of this principle, which are taken from [Lu77b, §1].

Example 2.2.3 (a) Assume that $\mathbf{X}$ is a finite set; then

$$\mathfrak{L}(g, \mathbf{X}) = |\mathbf{X}^g| \qquad \text{where} \qquad \mathbf{X}^g = \{x \in \mathbf{X} \mid g.x = x\}.$$

Indeed, consider a realisation of $\mathbf{X}$ as a closed subset of k^d for some $d \geqslant 1$. Since $\mathbf{X}$ is finite, there is a power q of p such that $\mathbf{X} \subseteq \mathbb{F}_q^d$ where $\mathbb{F}_q \subseteq k$ is the subfield with q elements. Then the standard Frobenius map $F_q \colon k^d \to k^d$ restricts to a Frobenius map F on $\mathbf{X}$ such that $F(x) = x$ for all $x \in \mathbf{X}$. Thus, we have $|\mathbf{X}^{F^n \circ g}| = |\mathbf{X}^g|$ for all

$n \geqslant 1$. This yields that

$$R(t,g) = -\sum_{n=1}^{\infty} |\mathbf{X}^{F^n \circ g}| t^n = -|\mathbf{X}^g| \sum_{n=1}^{\infty} t^n = |\mathbf{X}^g| \frac{t}{t-1}$$

and so the value at ∞ is $\mathfrak{L}(g, \mathbf{X}) = |\mathbf{X}^g|$, as required.

(b) Assume that $f \colon \mathbf{X} \to \mathbf{X}'$ is a bijective morphism of affine varieties. If $g : \mathbf{X} \to \mathbf{X}$ and $g' \colon \mathbf{X}' \to \mathbf{X}'$ are automorphisms of finite order such that $f \circ g = g' \circ f$, then

$$\mathfrak{L}(g, \mathbf{X}) = \mathfrak{L}(g', \mathbf{X}').$$

Indeed, we can realise $\mathbf{X}$ and $\mathbf{X}'$ as closed subsets of some affine spaces. Then $\mathbf{X}, \mathbf{X}', g, g', f$ are given by finitely many polynomials all of whose coefficients will lie in some finite subfield of k. Thus, we can find Frobenius maps $F \colon \mathbf{X} \to \mathbf{X}$ and $F' \colon \mathbf{X}' \to \mathbf{X}'$ such that g commutes with F, g' commutes with F' and $f \circ F = F' \circ f$. Then $F^n \circ g$ and $F'^n \circ g'$ have the same number of fixed points for all $n \geqslant 1$. Hence, the corresponding power series coincide and so $\mathfrak{L}(g, \mathbf{X}) = \mathfrak{L}(g', \mathbf{X}')$.

2.2.4 Now let $\mathbf{G}$ be a linear algebraic group over $k = \overline{\mathbb{F}}_p$ and $F \colon \mathbf{G} \to \mathbf{G}$ be a Steinberg map. Let $\mathscr{L} \colon \mathbf{G} \to \mathbf{G}$, $g \mapsto g^{-1}F(g)$, be the Lang–Steinberg map. (Recall from Theorem 1.4.8 that $\mathscr{L}$ is surjective if $\mathbf{G}$ is connected.) Let $\mathbf{Y}$ be a closed subset of $\mathbf{G}$. Following [Lu77b, 2.1], we consider the variety

$$\mathscr{L}^{-1}(\mathbf{Y}) = \{x \in \mathbf{G} \mid x^{-1}F(x) \in \mathbf{Y}\}.$$

This is a closed subset of $\mathbf{G}$ stable under left multiplication by elements of $\mathbf{G}^F$. Assume furthermore that H is a subgroup of $\mathbf{G}$ such that

$$|H| < \infty \qquad \text{and} \qquad h^{-1}\mathbf{Y}F(h) \subseteq \mathbf{Y} \quad \text{for all } h \in H. \tag{$*$}$$

(This is slightly more general than the situation described in [Lu77b, 2.1]; this generalisation will be useful in 2.3.18 below.) Then $\mathscr{L}^{-1}(\mathbf{Y})$ is stable under right multiplication by elements of H. So $\mathbf{G}^F \times H$ acts on $\mathscr{L}^{-1}(\mathbf{Y})$ by

$$(g,h).x = gxh^{-1} \qquad (x \in \mathscr{L}^{-1}(\mathbf{Y}),\ g \in \mathbf{G}^F,\ h \in H).$$

Consequently, the vector spaces $\mathrm{H}^i_c(\mathscr{L}^{-1}(\mathbf{Y}), \overline{\mathbb{Q}}_\ell)$ are $\overline{\mathbb{Q}}_\ell(\mathbf{G}^F \times H)$-modules.

Proposition 2.2.5 (Cf. [Lu77b, 2.1]) *Assume that we are in the setting of 2.2.4, where $H \subseteq \mathbf{G}$ satisfies ($*$). Let $\theta \colon H \to \mathbb{K}$ be a class function. Then*

$$R^{\mathbf{G}}_{H,\mathbf{Y}}(\theta) \colon \mathbf{G}^F \to \mathbb{K}, \qquad g \mapsto \frac{1}{|H|} \sum_{h \in H} \mathfrak{L}\big((g,h), \mathscr{L}^{-1}(\mathbf{Y})\big)\, \theta(h)$$

is a class function of $\mathbf{G}^F$. (Here, $\mathfrak{L}\big((g,h), \mathscr{L}^{-1}(\mathbf{Y})\big) \in \mathbb{Z}$ are Lefschetz numbers as in 2.2.1.) If θ is a virtual character of H, then $R^{\mathbf{G}}_{H,\mathbf{Y}}(\theta)$ is a virtual character of $\mathbf{G}^F$ whose values are algebraic integers in $\mathbb{Q}(\theta(h) \mid h \in H) \subseteq \mathbb{K}$.

If H is of the form $H = \mathbf{H}^F$ for a closed F-stable subgroup $\mathbf{H} \subseteq \mathbf{G}$, then we also write $R^{\mathbf{G}}_{\mathbf{H},\mathbf{Y}}(\theta)$ instead of $R^{\mathbf{G}}_{H,\mathbf{Y}}(\theta)$.

Proof For each i, let $\psi_i \colon \mathbf{G}^F \times H \to \overline{\mathbb{Q}}_\ell$ be the character of the $\overline{\mathbb{Q}}_\ell(\mathbf{G}^F \times H)$-module $\mathrm{H}^i_c(\mathscr{L}^{-1}(\mathbf{Y}), \overline{\mathbb{Q}}_\ell)$. Thus, $\mathfrak{L}((g,h), \mathscr{L}^{-1}(\mathbf{Y})) = \sum_i (-1)^i \psi_i(g,h)$ for all $g \in \mathbf{G}^F$ and $h \in H$. Now assume first that $\theta \colon H \to \mathbb{K}$ is not just a class function but a character of H. Let V be a finite-dimensional $\mathbb{K}H$-module affording θ. Choosing an embedding $\mathbb{K} \subseteq \overline{\mathbb{Q}}_\ell$, we can also regard V as a $\overline{\mathbb{Q}}_\ell(\mathbf{G}^F \times H)$-module, with the elements of $\mathbf{G}^F$ acting trivially. We now use the construction in 2.1.3 (where we identify H with a normal subgroup of the direct product, with factor group $\mathbf{G}^F$). Then the subspace of fixed points

$$\big(\mathrm{H}^i_c(\mathscr{L}^{-1}(\mathbf{Y}), \overline{\mathbb{Q}}_\ell) \otimes V\big)^H$$

is still a $\overline{\mathbb{Q}}_\ell\mathbf{G}^F$-module; let $\psi_i^\theta \colon \mathbf{G}^F \to \overline{\mathbb{Q}}_\ell$ be its character. By 2.1.3, we have

$$\psi_i^\theta(g) = \frac{1}{|H|} \sum_{h \in H} \psi_i(g,h)\theta(h) \qquad \text{for all } g \in \mathbf{G}^F.$$

Hence, we see that $R^{\mathbf{G}}_{H,\mathbf{Y}}(\theta) = \sum_i (-1)^i \psi_i^\theta$ is a virtual character of $\mathbf{G}^F$ over $\overline{\mathbb{Q}}_\ell$. Since Lefschetz numbers are integers, the values of $R^{\mathbf{G}}_{H,\mathbf{Y}}(\theta)$ are algebraic integers in $\mathbb{Q}(\mathrm{Trace}(h,V) \mid h \in H) \subseteq \mathbb{K}$. So Remark 2.1.5 shows that $R^{\mathbf{G}}_{H,\mathbf{Y}}(\theta)$ is also a virtual character over $\mathbb{K}$. Thus, we obtain a map $R^{\mathbf{G}}_{H,\mathbf{Y}}$ sending a character of H (over $\mathbb{K}$) to a virtual character of $\mathbf{G}^F$ (over $\mathbb{K}$). Extending this map linearly, we obtain a map from virtual characters of H (over $\mathbb{K}$) to virtual characters of $\mathbf{G}^F$ (over $\mathbb{K}$) and, finally, a map $\mathrm{CF}(H) \to \mathrm{CF}(\mathbf{G}^F)$ as desired. □

We shall assume from now on that $\mathbf{G}$ is connected and reductive[1]. It will also be useful at several places below to fix a setting as in 1.6.1, where $\mathbf{T}_0 \subseteq \mathbf{G}$ is an F-stable maximal torus that is maximally split, that is, contained in an F-stable Borel subgroup $\mathbf{B}_0 \subseteq \mathbf{G}$. Let $\mathbf{W} = N_{\mathbf{G}}(\mathbf{T}_0)/\mathbf{T}_0$ be the Weyl group of $\mathbf{G}$ and q be the positive real number defined by F in Proposition 1.4.19.

Definition 2.2.6 (Deligne–Lusztig [DeLu76], Lusztig [Lu77b]) Let $\mathbf{T} \subseteq \mathbf{G}$ be an F-stable maximal torus and $\theta \in \mathrm{Irr}(\mathbf{T}^F)$. Choose a Borel subgroup $\mathbf{B} \subseteq \mathbf{G}$ such that $\mathbf{T} \subseteq \mathbf{B}$, and let $\mathbf{U} = R_u(\mathbf{B})$ be the unipotent radical of $\mathbf{B}$. (Note that $\mathbf{B}$ is not necessarily F-stable.) Then $\mathbf{Y} := \mathbf{U}$ satisfies condition $(*)$ in 2.2.4 with respect to the finite subgroup $H := \mathbf{T}^F$. Thus, we obtain a virtual character $R^{\mathbf{G}}_{\mathbf{T},\mathbf{U}}(\theta)$ of $\mathbf{G}^F$, which is called a *Deligne–Lusztig character*.

As pointed out in [Lu75], the virtual characters $R^{\mathbf{G}}_{\mathbf{T},\mathbf{U}}(\theta)$ provide a solution to a

[1] There are also extensions of the following constructions to disconnected groups, see [Ma93a], [DiMi94], [Lu12b]; we will briefly discuss this in Section 4.8.

series of conjectures made by Macdonald around 1968, on the basis of the character tables of finite groups of Lie type known at the time, most notably $\mathrm{GL}_n(q)$ (see [Gre55]) and $\mathrm{Sp}_4(q)$ (see [Sr68]). We shall now state some of the basic properties of the virtual characters $R_{\mathbf{T},\mathbf{U}}^{\mathbf{G}}(\theta)$. The original proofs can be found in [DeLu76], [Lu77b]; see also [Ca85], [DiMi20], [Sr79]. The construction of $R_{\mathbf{T},\mathbf{U}}^{\mathbf{G}}(\theta)$ is a kind of 'twisted induction', generalising the usual induction of characters from subgroups, by the following result. (This is a special case of *Harish-Chandra induction* and will be studied in more detail in Section 3.2.)

Proposition 2.2.7 *Let* $\mathbf{T} \subseteq \mathbf{G}$ *be an* F*-stable maximal torus that is contained in an* F*-stable Borel subgroup* $\mathbf{B} \subseteq \mathbf{G}$. *Then* $R_{\mathbf{T},\mathbf{U}}^{\mathbf{G}}(\theta) = \mathrm{Ind}_{\mathbf{B}^F}^{\mathbf{G}^F}(\theta)$ *where* $\theta \in \mathrm{Irr}(\mathbf{T}^F)$ *is regarded by inflation as a character of* $\mathbf{B}^F$.

For the proof, see [Lu77b, 2.6] or [Ca85, 7.2.4]. Now the next major result is a scalar product formula which has a number of important consequences.

Theorem 2.2.8 (Scalar product formula) *Let* $\mathbf{T}, \mathbf{T}' \subseteq \mathbf{G}$ *be* F*-stable maximal tori. Let* $\theta \in \mathrm{Irr}(\mathbf{T}^F)$ *and* $\theta' \in \mathrm{Irr}(\mathbf{T}'^F)$. *Then*

$$\langle R_{\mathbf{T},\mathbf{U}}^{\mathbf{G}}(\theta), R_{\mathbf{T}',\mathbf{U}'}^{\mathbf{G}}(\theta')\rangle = |\{g \in \mathbf{G}^F \mid g\mathbf{T}g^{-1} = \mathbf{T}' \text{ and } {}^g\theta = \theta'\}|/|\mathbf{T}^F|,$$

where $\mathbf{U}$, $\mathbf{U}'$ *are the unipotent radicals of Borel subgroups containing* $\mathbf{T}$, $\mathbf{T}'$, *respectively.*

The original proof in [DeLu76, §6] (see also [Lu75]) has the minor disadvantage that it does not cover certain extreme cases where $q < 2$ (with q as in Proposition 1.4.19). A simpler proof, which works in complete generality, is given in [Lu77b, 2.3]; an exposition of this latter proof can be found in [Ca85, §7.3]. See also [DiMi20, Cor. 9.3.1] where the proof is given via a Mackey formula.

Corollary 2.2.9 *The virtual character* $R_{\mathbf{T},\mathbf{U}}^{\mathbf{G}}(\theta)$ *is independent of* $\mathbf{U}$; *it will henceforth be denoted by* $R_{\mathbf{T}}^{\mathbf{G}}(\theta)$. *Furthermore, the following hold.*

(a) *If* $\theta \in \mathrm{Irr}(\mathbf{T}^F)$ *is in general position (that is, we have* ${}^g\theta \neq \theta$ *for all* $g \in N_{\mathbf{G}}(\mathbf{T})^F \setminus \mathbf{T}^F$*), then either* $R_{\mathbf{T}}^{\mathbf{G}}(\theta)$ *or* $-R_{\mathbf{T}}^{\mathbf{G}}(\theta)$ *is in* $\mathrm{Irr}(\mathbf{G}^F)$.
(b) *For any* $\rho \in \mathrm{Irr}(\mathbf{G}^F)$, *we have* $-|\mathbf{W}|^{1/2} \leqslant \langle R_{\mathbf{T}}^{\mathbf{G}}(\theta), \rho\rangle \leqslant |\mathbf{W}|^{1/2}$. *The number of* $\rho \in \mathrm{Irr}(\mathbf{G}^F)$ *that occur in* $R_{\mathbf{T}}^{\mathbf{G}}(\theta)$ *is bounded above by* $|\mathbf{W}|$.

Proof Let $\mathbf{U}'$ be the unipotent radical of another Borel subgroup containing $\mathbf{T}$. Then Theorem 2.2.8 shows that we have

$$\langle R_{\mathbf{T},\mathbf{U}}^{\mathbf{G}}(\theta), R_{\mathbf{T},\mathbf{U}}^{\mathbf{G}}(\theta)\rangle = \langle R_{\mathbf{T},\mathbf{U}}^{\mathbf{G}}(\theta), R_{\mathbf{T},\mathbf{U}'}^{\mathbf{G}}(\theta)\rangle = \langle R_{\mathbf{T},\mathbf{U}'}^{\mathbf{G}}(\theta), R_{\mathbf{T},\mathbf{U}'}^{\mathbf{G}}(\theta)\rangle.$$

It follows that $R_{\mathbf{T},\mathbf{U}}^{\mathbf{G}}(\theta) - R_{\mathbf{T},\mathbf{U}'}^{\mathbf{G}}(\theta)$ has norm 0 and so $R_{\mathbf{T},\mathbf{U}}^{\mathbf{G}}(\theta) = R_{\mathbf{T},\mathbf{U}'}^{\mathbf{G}}(\theta)$. Thus,

$R^{\mathbf{G}}_{\mathbf{T},\mathbf{U}}(\theta)$ is independent of $\mathbf{U}$. For (a), just note that $\langle R^{\mathbf{G}}_{\mathbf{T}}(\theta), R^{\mathbf{G}}_{\mathbf{T}}(\theta)\rangle = 1$ if θ is in general position. For (b), just note that

$$\langle R^{\mathbf{G}}_{\mathbf{T}}(\theta), \rho\rangle^2 \leqslant \langle R^{\mathbf{G}}_{\mathbf{T}}(\theta), R^{\mathbf{G}}_{\mathbf{T}}(\theta)\rangle \leqslant |N_{\mathbf{G}}(\mathbf{T})^F|/|\mathbf{T}^F| \leqslant |N_{\mathbf{G}}(\mathbf{T})/\mathbf{T}| = |\mathbf{W}|. \qquad \square$$

By similar arguments, we obtain:

Corollary 2.2.10 *Two Deligne–Lusztig characters $R^{\mathbf{G}}_{\mathbf{T}}(\theta)$ and $R^{\mathbf{G}}_{\mathbf{T}'}(\theta')$ are either equal or orthogonal to each other. We have $R^{\mathbf{G}}_{\mathbf{T}}(\theta) = R^{\mathbf{G}}_{\mathbf{T}'}(\theta')$ if and only if there exists some $g \in \mathbf{G}^F$ such that $g\mathbf{T}g^{-1} = \mathbf{T}'$ and ${}^g\theta = \theta'$.*

In order to fix the sign in Corollary 2.2.9(a), we shall need the notion of the *relative F-rank* of $\mathbf{G}$. Following [Ca85, p. 197], this is defined as follows.

Definition 2.2.11 Let $\mathbf{T}$ be any F-stable maximal torus of $\mathbf{G}$. Let φ be the endomorphism of $X = X(\mathbf{T})$ induced by F. Denote by $\varphi_{\mathbb{R}}$ the canonical extension of φ to $X_{\mathbb{R}} = \mathbb{R} \otimes_{\mathbb{Z}} X$. Then the *relative F-rank* of $\mathbf{T}$ is defined as the dimension of the q-eigenspace of $\varphi_{\mathbb{R}}$ on $X_{\mathbb{R}}$ (with q as in Proposition 1.4.19). We also set

$$\varepsilon_{\mathbf{T}} := (-1)^r \qquad \text{where} \qquad r := \text{ relative } F\text{-rank of } \mathbf{T}.$$

If $\mathbf{T}_0 \subseteq \mathbf{G}$ is our fixed maximally split torus, then we set

$$\varepsilon_{\mathbf{G}} := \varepsilon_{\mathbf{T}_0} \qquad \text{and} \qquad \textit{relative } F\textit{-rank of } \mathbf{G} := \text{relative } F\text{-rank of } \mathbf{T}_0.$$

(This does not depend on the particular choice of $\mathbf{T}_0$ since all maximally split tori are $\mathbf{G}^F$-conjugate.) The (absolute) *rank* of $\mathbf{G}$ is defined as $\dim \mathbf{T}_0$.

By [Ca85, 6.5.7], the relative F-rank of $\mathbf{T}_0$ is larger than or equal to the relative F-rank of any F-stable maximal torus of $\mathbf{G}$. Hence, if $\mathbb{G}$ is the complete root datum associated with $\mathbf{G}, F$ (see Example 1.6.11) and $|\mathbb{G}| \in \mathbb{R}[\mathbf{q}]$ is the corresponding order polynomial, then

$$\text{relative } F\text{-rank of } \mathbf{G} = \max\{i \geqslant 0 \mid (\mathbf{q}-1)^i \text{ divides } |\mathbb{G}| \text{ in } \mathbb{R}[\mathbf{q}]\}.$$

In particular, if $\mathbf{G}$ is simple, then the relative F-rank of $\mathbf{G}$ and the sign $\varepsilon_{\mathbf{G}}$ can be easily read off the polynomials in Table 1.3 (p. 73).

Theorem 2.2.12 (Degree formula) *Let $\mathbf{T}$ be an F-stable maximal torus of $\mathbf{G}$ and $\theta \in \mathrm{Irr}(\mathbf{T}^F)$. Then*

$$R^{\mathbf{G}}_{\mathbf{T}}(\theta)(1) = \varepsilon_{\mathbf{G}}\varepsilon_{\mathbf{T}}|\mathbf{G}^F : \mathbf{T}^F|_{p'},$$

where, for any non-zero $m \in \mathbb{Z}$, we denote by $m_{p'}$ the p'-part of m.

For the proof (which uses properties of the Steinberg character of $\mathbf{G}^F$), see [DeLu76, 7.1], [Lu77b, 2.9]; see also [Ca85, 7.5.1, 7.5.2]. Consequently, in Corollary 2.2.9(a), we have $\varepsilon_{\mathbf{G}}\varepsilon_{\mathbf{T}}R^{\mathbf{G}}_{\mathbf{T}}(\theta) \in \mathrm{Irr}(\mathbf{G}^F)$.

Since the definition involves the cohomology spaces $H^i_c(\mathbf{X}, \overline{\mathbb{Q}}_\ell)$ for certain subvarieties $\mathbf{X} \subseteq \mathbf{G}$, it may be expected that the problem of computing the values of $R^{\mathbf{G}}_{\mathbf{T}}(\theta)$ on arbitrary elements of $\mathbf{G}^F$ is difficult, and this is indeed so. A first step consists of using the Jordan decomposition for an element $g \in \mathbf{G}^F$, which allows us to separate the problem according to the semisimple and the unipotent part of g. We begin by recalling some facts about centralisers of semisimple elements.

2.2.13 Let $s \in \mathbf{G}$ be semisimple. Then $C_{\mathbf{G}}(s)$ is a closed subgroup which is, in general, not connected. Let us denote by $\mathbf{H} := C^\circ_{\mathbf{G}}(s)$ the connected component of the identity. Now there always exists some maximal torus $\mathbf{T} \subseteq \mathbf{G}$ such that $s \in \mathbf{T}$ (see, e.g., [Ge03a, Exc. 3.12]). Then $\mathbf{T} \subseteq C_{\mathbf{G}}(s)$ and, hence, $s \in \mathbf{T} \subseteq \mathbf{H}$. It is also known that every unipotent element $u \in C_{\mathbf{G}}(s)$ already belongs to $\mathbf{H}$ (see [Hum95, 1.12]). Furthermore, by [Hum95, 2.2], the connected algebraic group $\mathbf{H}$ is reductive, of the same rank as $\mathbf{G}$ (since $\mathbf{T} \subseteq \mathbf{H}$). In fact, if s belongs to our reference torus $\mathbf{T}_0$, then we have

$$\mathbf{H} = \langle \mathbf{T}_0, \mathbf{U}_\alpha \mid \alpha \in R \text{ such that } \alpha(s) = 1 \rangle$$

and $R_s := \{\alpha \in R \mid \alpha(s) = 1\}$ is the root system of $\mathbf{H}$ with respect to $\mathbf{T}_0$. Note that, since all maximal tori are conjugate in $\mathbf{G}$, every semisimple element of $\mathbf{G}$ is conjugate to an element in $\mathbf{T}_0$.

For later reference we also state the following result, which provides a basic criterion for when $C_{\mathbf{G}}(s)$ is connected.

Theorem 2.2.14 (Steinberg [St68, 8.5, 9.1]) *Let $\pi_{\mathrm{sc}} \colon \mathbf{G}_{\mathrm{sc}} \to \mathbf{G}$ be a simply connected covering of the derived subgroup of $\mathbf{G}$, as in Remark 1.5.13. Let $s \in \mathbf{G}$ be semisimple. Then $C_{\mathbf{G}}(s)/C^\circ_{\mathbf{G}}(s)$ is isomorphic to a subgroup of $\ker(\pi_{\mathrm{sc}})$. In particular, if $\ker(\pi_{\mathrm{sc}}) = \{1\}$, then $C_{\mathbf{G}}(s)$ is connected.*

See [Ca85, §3.5], [DiMi20, §11.2], [Hum95, §2.6], [Bo05], [MaTe11, §14.2], for a further discussion of this important result.

Definition 2.2.15 Let $\mathbf{G}_{\mathrm{uni}}$ be the set of unipotent elements of $\mathbf{G}$. Let $\mathbf{T} \subseteq \mathbf{G}$ be an F-stable maximal torus. Define a function $Q^{\mathbf{G}}_{\mathbf{T}} \colon \mathbf{G}^F_{\mathrm{uni}} \to \mathbb{K}$ by

$$Q^{\mathbf{G}}_{\mathbf{T}}(u) := R^{\mathbf{G}}_{\mathbf{T}}(1_{\mathbf{T}})(u) \qquad \text{for } u \in \mathbf{G}^F_{\mathrm{uni}},$$

where $1_{\mathbf{T}}$ stands for the trivial character of $\mathbf{T}^F$. This function is called a *Green function* (since it already occurred in Green's work [Gre55] on $\mathrm{GL}_n(q)$); see [DeLu76, 4.1]. By Proposition 2.2.5, the values of $Q^{\mathbf{G}}_{\mathbf{T}}$ are integers; hence, we also have $Q^{\mathbf{G}}_{\mathbf{T}}(u) = Q^{\mathbf{G}}_{\mathbf{T}}(u^{-1})$ for all $u \in \mathbf{G}^F_{\mathrm{uni}}$. Furthermore, by Corollary 2.2.10, we have $Q^{\mathbf{G}}_{g\mathbf{T}g^{-1}} = Q^{\mathbf{G}}_{\mathbf{T}}$ for all $g \in \mathbf{G}^F$.

Theorem 2.2.16 (Character formula) *Let $g \in \mathbf{G}^F$ and write $g = su = us$, where s is semisimple and u is unipotent. Let $\mathbf{H} := C^\circ_{\mathbf{G}}(s)$. Then*

$$R^{\mathbf{G}}_{\mathbf{T}}(\theta)(g) = \frac{1}{|\mathbf{H}^F|} \sum_{x \in \mathbf{G}^F \,:\, x^{-1}sx \in \mathbf{T}} Q^{\mathbf{H}}_{x\mathbf{T}x^{-1}}(u)\,\theta(x^{-1}sx).$$

Furthermore, if $s = 1$, then $R^{\mathbf{G}}_{\mathbf{T}}(\theta)(u) = Q^{\mathbf{G}}_{\mathbf{T}}(u) \in \mathbb{Z}$ does not depend on θ.

For the proof, see [DeLu76, 4.2] or [Ca85, 7.2.8]. We observe that the expression $Q^{\mathbf{H}}_{x\mathbf{T}x^{-1}}(u)$ is meaningful: firstly, as already recalled above in 2.2.13, $\mathbf{H}$ is connected and reductive; secondly, if $x^{-1}sx \in \mathbf{T}^F$, then $x\mathbf{T}x^{-1}$ is an F-stable maximal torus contained in $\mathbf{H}$; furthermore, u belongs to $\mathbf{H}$ (as noted in 2.2.13).

Example 2.2.17 (a) The above character formula immediately shows that we have $R^{\mathbf{G}}_{\mathbf{T}}(\theta)(g) = 0$ if s is not conjugate in $\mathbf{G}^F$ to any element of $\mathbf{T}^F$.

(b) Let $g \in \mathbf{G}^F$ and write $g = us = su$, where s is semisimple and u is unipotent. Taking the sum over all $\theta \in \mathrm{Irr}(\mathbf{T}^F)$ in the above character formula, and using the orthogonality relations for the characters of $\mathbf{T}^F$, we obtain the following identity:

$$\sum_{\theta \in \mathrm{Irr}(\mathbf{T}^F)} R^{\mathbf{G}}_{\mathbf{T}}(\theta)(g) = \begin{cases} |\mathbf{T}^F| Q^{\mathbf{G}}_{\mathbf{T}}(u) & \text{if } s = 1, \\ 0 & \text{otherwise.} \end{cases}$$

(c) If $u = 1$, then the character formula reduces to the formula:

$$\mathrm{Ind}^{\mathbf{G}^F}_{\mathbf{T}^F}(\theta)(s) = \varepsilon_{\mathbf{T}}\varepsilon_{\mathbf{H}}|\mathbf{H}^F|_p R^{\mathbf{G}}_{\mathbf{T}}(\theta)(s).$$

Hence, the difficulty of computing the values of $R^{\mathbf{G}}_{\mathbf{T}}(\theta)$ on semisimple elements is equivalent to that of computing the values of $\mathrm{Ind}^{\mathbf{G}^F}_{\mathbf{T}^F}(\theta)$: for this purpose, one has to determine the *class fusion* from $\mathbf{T}^F$ to $\mathbf{G}^F$.

We then also have the following formula for the values of the irreducible characters of $\mathbf{G}^F$ on semisimple elements:

Proposition 2.2.18 (Character values on semisimple elements, [DeLu76, 7.6] or [Ca85, 7.5.5]) *Let $\rho \in \mathrm{Irr}(\mathbf{G}^F)$ and $s \in \mathbf{G}^F$ be semisimple. Then*

$$\rho(s) = \frac{1}{|\mathbf{H}^F|_p} \sum_{(\mathbf{T},\theta)} \varepsilon_{\mathbf{H}}\varepsilon_{\mathbf{T}} \langle R^{\mathbf{G}}_{\mathbf{T}}(\theta), \rho\rangle\, \theta(s),$$

where $\mathbf{H} = C^\circ_{\mathbf{G}}(s)$ and the sum runs over all pairs $(\mathbf{T}, \theta)$ such that $\mathbf{T} \subseteq \mathbf{G}$ is an F-stable maximal torus, $s \in \mathbf{T}$ and $\theta \in \mathrm{Irr}(\mathbf{T}^F)$.

Corollary 2.2.19 *For any irreducible character $\rho \in \mathrm{Irr}(\mathbf{G}^F)$, there exists a pair $(\mathbf{T}, \theta)$ such that $\langle R^{\mathbf{G}}_{\mathbf{T}}(\theta), \rho\rangle \neq 0$.*

Proof Let $s = 1$. Then $\rho(1) \neq 0$ and so some term in the sum in the formula in Proposition 2.2.18 must be non-zero. □

The character formula immediately shows that, if $z \in \mathbf{Z}(\mathbf{G})^F$, then we have $R_{\mathbf{T}}^{\mathbf{G}}(\theta)(z) = \theta(z)R_{\mathbf{T}}^{\mathbf{G}}(\theta)(1)$. Actually, something stronger is true.

Proposition 2.2.20 (See [DeLu76, 1.22]) *Let $\mathbf{T} \subseteq \mathbf{G}$ be an F-stable maximal torus and $\theta \in \mathrm{Irr}(\mathbf{T}^F)$. Let $\rho \in \mathrm{Irr}(\mathbf{G}^F)$ be such that $\langle R_{\mathbf{T}}^{\mathbf{G}}(\theta), \rho\rangle \neq 0$. Then $\rho(z) = \theta(z)\rho(1)$ for all $z \in \mathbf{Z}(\mathbf{G})^F$.*

Proof Since it does not seem to be possible to prove this by purely character-theoretic methods, it may be a good exercise to go through the details. Recall the construction of $R_{\mathbf{T}}^{\mathbf{G}}(\theta)$ from the proof of Proposition 2.2.5. Let $\mathbf{U} \subseteq \mathbf{G}$ be the unipotent radical of a Borel subgroup containing $\mathbf{T}$. Let $\psi_i \colon \mathbf{G}^F \times \mathbf{T}^F \to \overline{\mathbb{Q}}_\ell$ be the character of the $\overline{\mathbb{Q}}_\ell(\mathbf{G}^F \times \mathbf{T}^F)$-module $\mathrm{H}_c^i(\mathscr{L}^{-1}(\mathbf{U}), \overline{\mathbb{Q}}_\ell)$, where the module structure is induced by the action of $\mathbf{G}^F \times \mathbf{T}^F$ on $\mathscr{L}^{-1}(\mathbf{U})$ via $(g,t)\colon x \mapsto gxt^{-1}$ (see 2.2.4). Now assume that $g = z \in \mathbf{Z}(\mathbf{G})^F$. Then (z,t) acts on $\mathscr{L}^{-1}(\mathbf{U})$ in the same way as $(1, z^{-1}t)$. An analogous statement holds for the induced linear action on $\mathrm{H}_c^i(\mathscr{L}^{-1}(\mathbf{U}), \overline{\mathbb{Q}}_\ell)$ and so we have

$$\psi_i(z,t) = \psi_i(1, z^{-1}t) \qquad \text{for all } t \in \mathbf{T}^F.$$

Next, let V_θ be a one-dimensional $\overline{\mathbb{Q}}_\ell\mathbf{T}^F$-module with character θ. Then we have $R_{\mathbf{T}}^{\mathbf{G}}(\theta) = \sum_i (-1)^i \psi_i^\theta$, where $\psi_i^\theta \colon \mathbf{G}^F \to \overline{\mathbb{Q}}_\ell$ is the character of the $\overline{\mathbb{Q}}_\ell\mathbf{G}^F$-module

$$\left(\mathrm{H}_c^i(\mathscr{L}^{-1}(\mathbf{U}), \overline{\mathbb{Q}}_\ell) \otimes V_\theta\right)^{\mathbf{T}^F}.$$

Using the formula in 2.1.3, we obtain that

$$\begin{aligned}\psi_i^\theta(z) &= \frac{1}{|\mathbf{T}^F|}\sum_{t\in\mathbf{T}^F} \psi_i(z,t)\theta(t) = \frac{1}{|\mathbf{T}^F|}\sum_{t\in\mathbf{T}^F} \psi_i(1, z^{-1}t)\theta(t)\\ &= \frac{1}{|\mathbf{T}^F|}\sum_{t\in\mathbf{T}^F} \psi_i(1,t)\theta(zt) = \theta(z)\psi_i^\theta(1).\end{aligned}$$

Since ψ_i^θ is an actual character, we conclude that $z \in \mathbf{Z}(\mathbf{G})^F$ acts by scalar multiplication with $\theta(z)$ on a module affording ψ_i^θ. Since $\langle R_{\mathbf{T}}^{\mathbf{G}}(\theta), \rho\rangle \neq 0$, there is some i such that ρ is a constituent of ψ_i^θ. Hence, z will also act by scalar multiplication with $\theta(z)$ on a module affording ρ. □

The determination of the values of the Green functions is a very hard problem (see Section 2.8 further below). Deligne and Lusztig [DeLu76] were able to determine these values on certain unipotent elements at least, which we now introduce.

2.2.21 The set $\mathbf{G}_{\mathrm{uni}}$ of unipotent elements is a closed irreducible subset of $\mathbf{G}$, invariant under the conjugation action of $\mathbf{G}$ on itself (see [Hum95, §4.2]). In particular, $\mathbf{G}_{\mathrm{uni}}$ is a union of conjugacy classes of $\mathbf{G}$, which are called the *unipotent classes* of $\mathbf{G}$. It is known that the number of unipotent classes is finite (see [Lu76b] and also [DiMi20, §12.1]). Then a general argument shows that there is a unique unipotent class $\mathscr{O}_0$ such that $\mathbf{G}_{\text{uni}}$ is the Zariski closure of $\mathscr{O}_0$ (see [Ca85, 5.1.2], [Hum95, §4.3]). This class $\mathscr{O}_0$ is called the *regular unipotent class* and its elements the *regular unipotent elements* of $\mathbf{G}$. We have $F(\mathscr{O}_0) = \mathscr{O}_0$ and so $\mathscr{O}_0^F \neq \varnothing$. Furthermore, $\dim C_{\mathbf{G}}(u) = \dim \mathbf{T}_0$ for any regular unipotent $u \in \mathbf{G}$. (If one does not want to use the finiteness of the number of unipotent classes, then see Steinberg's existence proof in [Hum95, §4.5, §4.6].)

Theorem 2.2.22 (See [DeLu76, Theorem 9.16]) *Let $\mathbf{T} \subseteq \mathbf{G}$ be an F-stable maximal torus. Then $Q_{\mathbf{T}}^{\mathbf{G}}(u) = 1$ for any regular unipotent $u \in \mathbf{G}^F$.*

Note that, while most of the results in this section so far are fully covered with proofs in [Ca85], [DiMi20], [Sr79], the original article of Deligne and Lusztig appears to be the only source for a proof of the above result. Note also that, in [DeLu76], it is assumed that F is a Frobenius map but the proof applies verbatim for Steinberg maps, as pointed out by [DiMi20, Lemma 12.4.8]. (For the Suzuki and Ree groups, the result is also verified by explicit computations; see Example 2.8.18 below.)

Even if the values of the Green functions are known, the character formula in Theorem 2.2.16 still requires some work as far as the sum over all $x \in \mathbf{G}^F$ such that $x^{-1}sx \in \mathbf{T}^F$ is concerned. The following result is certainly known and much used in the literature (see, e.g., [FoSr82, 1.10], [GePf92, 4.1], [Lue93, Chap. 6], [MaMa18, §II.5.1]) but in order to get all the technicalities right, it is useful to write down a detailed proof.

Lemma 2.2.23 (Simplified character formula) *Assume that we are in the setting of Theorem 2.2.16, where s is conjugate in $\mathbf{G}^F$ to an element in $\mathbf{T}^F$. Let $\mathbf{H} := C_{\mathbf{G}}^{\circ}(s)$ and $\mathbf{T}_1, \ldots, \mathbf{T}_m$ be representatives of the $\mathbf{H}^F$-conjugacy classes of F-stable maximal tori of $\mathbf{H}$ that are conjugate in $\mathbf{G}^F$ to $\mathbf{T}$. For each i let $x_i \in \mathbf{G}^F$ be such that $\mathbf{T}_i = x_i\mathbf{T}x_i^{-1}$. Then $\theta_i := {}^{x_i}\theta \in \operatorname{Irr}(\mathbf{T}_i^F)$ and*

$$R_{\mathbf{T}}^{\mathbf{G}}(\theta)(su) = \sum_{1\leqslant i\leqslant m} Q_{\mathbf{T}_i}^{\mathbf{H}}(u)\frac{1}{|\mathbf{W}(\mathbf{H},\mathbf{T}_i)^F|}\sum_{w\in\mathbf{W}(\mathbf{G},\mathbf{T}_i)^F} \theta_i(\dot{w}^{-1}s\dot{w})$$

where $\mathbf{W}(\mathbf{G},\mathbf{T}_i) = N_{\mathbf{G}}(\mathbf{T}_i)/\mathbf{T}_i$ and $\mathbf{W}(\mathbf{H},\mathbf{T}_i) = N_{\mathbf{H}}(\mathbf{T}_i)/\mathbf{T}_i$.

Proof Let $\mathbf{N} := N_{\mathbf{G}}(\mathbf{T})$. Let D be a set of representatives for the $(\mathbf{H}^F, \mathbf{N}^F)$-double cosets in $\mathbf{G}^F$. Let $x \in \mathbf{G}^F$ and write $x = hdn$ where $h \in \mathbf{H}^F$, $d \in D$ and $n \in \mathbf{N}^F$. We have $x^{-1}sx \in \mathbf{T}$ if and only if $s \in x\mathbf{T}x^{-1}$ if and only if $x\mathbf{T}x^{-1} \subseteq \mathbf{H}$. Using now $x = hdn$, we see that the latter condition is equivalent to $d\mathbf{T}d^{-1} \subseteq \mathbf{H}$ and, hence, to $d^{-1}sd \in \mathbf{T}$. So let D' be the set of all $d \in D$ such that $d^{-1}sd \in \mathbf{T}$. Then the formula in Theorem 2.2.16 can be re-written as follows:

$$R_{\mathbf{T}}^{\mathbf{G}}(\theta)(g) = \frac{1}{|\mathbf{H}^F|} \sum_{d\in D'} \sum_{x\in \mathbf{H}^F d\mathbf{N}^F} Q_{x\mathbf{T}x^{-1}}^{\mathbf{H}}(u)\theta(x^{-1}sx).$$

Now, if $x = hdn$ as above (where $d \in D'$), then $x\mathbf{T}x^{-1} = hd\mathbf{T}d^{-1}h^{-1}$ is an F-stable maximal torus of $\mathbf{H}$ that is conjugate in $\mathbf{H}^F$ to $\mathbf{T}_d := d\mathbf{T}d^{-1}$. Hence, we have $Q_{x\mathbf{T}x^{-1}}^{\mathbf{H}}(u) = Q_{\mathbf{T}_d}^{\mathbf{H}}(u)$. Let $\mathbf{N}_d := N_{\mathbf{G}}(\mathbf{T}_d) = d\mathbf{N}d^{-1}$. Furthermore, let $n_1, \ldots, n_r \in \mathbf{N}_d^F$ be a set of right coset representatives of $\mathbf{H}^F \cap \mathbf{N}_d^F$ in $\mathbf{N}_d^F$, that is, $\mathbf{N}_d^F$ is the disjoint union of $(\mathbf{H}^F \cap \mathbf{N}_d^F)n_i$ for $1 \leqslant i \leqslant r$. Then every $x \in \mathbf{H}^F d\mathbf{N}^F$ has a unique expression as $x = hn_i d$ where $h \in \mathbf{H}^F$ and $1 \leqslant i \leqslant r$. So the above formula can be re-written as

$$\begin{aligned} R_{\mathbf{T}}^{\mathbf{G}}(\theta)(g) &= \sum_{d\in D'} Q_{\mathbf{T}_d}^{\mathbf{H}}(u) \sum_{1\leqslant i\leqslant r} \theta(d^{-1}n_i^{-1}sn_i d) \\ &= \sum_{d\in D'} Q_{\mathbf{T}_d}^{\mathbf{H}}(u) \frac{1}{|\mathbf{H}^F \cap \mathbf{N}_d^F|} \sum_{n\in \mathbf{N}_d^F} {}^d\theta(n^{-1}sn). \end{aligned}$$

Now, $\mathbf{H}^F \cap \mathbf{N}_d^F = (\mathbf{H} \cap N_{\mathbf{G}}(\mathbf{T}_d))^F = N_{\mathbf{H}}(\mathbf{T}_d)^F \supseteq \mathbf{T}_d^F$; we also have $\mathbf{N}_d^F \supseteq \mathbf{T}_d^F$. Since $N_{\mathbf{H}}(\mathbf{T}_d)^F/\mathbf{T}_d^F \cong \mathbf{W}(\mathbf{H}, \mathbf{T}_d)^F$ and $\mathbf{N}_d^F/\mathbf{T}_d^F \cong \mathbf{W}(\mathbf{G}, \mathbf{T}_d)^F$, we obtain:

$$\frac{1}{|\mathbf{H}^F \cap \mathbf{N}_d^F|} \sum_{n\in \mathbf{N}_d^F} {}^d\theta(n^{-1}sn) = \frac{1}{|\mathbf{W}(\mathbf{H}, \mathbf{T}_d)^F|} \sum_{w\in \mathbf{W}(\mathbf{G},\mathbf{T}_d)^F} {}^d\theta(\dot{w}^{-1}s\dot{w}).$$

Finally, it is easy to check that we can take $D' = \{x_1, \ldots, x_m\}$. □

Example 2.2.24 (a) Let $\mathbf{T} \subseteq \mathbf{G}$ be an F-stable torus of type w. Let $g \in \mathbf{G}$ be such that $\mathbf{T} = g\mathbf{T}_0 g^{-1}$ and w is the image of $g^{-1}F(g) \in N_{\mathbf{G}}(\mathbf{T}_0)$. Then the map $C_{\mathbf{W},\sigma}(w) \to \mathbf{W}(\mathbf{G}, \mathbf{T})^F$, $x \mapsto g\dot{x}g^{-1}\mathbf{T}$, is an isomorphism. (See [Ca85, 3.3.6].)

(b) Let $g \in \mathbf{G}^F$ with Jordan decomposition $g = su = us$. Assume that u is regular unipotent in $C_{\mathbf{G}}^\circ(s)$. Let $\mathbf{T} \subseteq \mathbf{G}$ be an F-stable maximal torus such that s is conjugate in $\mathbf{G}^F$ to an element in $\mathbf{T}^F$. Then Theorem 2.2.22 and Lemma 2.2.23 show that

$$R_{\mathbf{T}}^{\mathbf{G}}(1_{\mathbf{T}})(g) = \sum_{1\leqslant i\leqslant m} |\mathbf{W}(\mathbf{G}, \mathbf{T}_i)^F : \mathbf{W}(\mathbf{H}, \mathbf{T}_i)^F|.$$

We can further evaluate this as follows. Assume that $w_1 \in \mathbf{W}$ is such that the F-stable maximal torus $\mathbf{T}_1 \subseteq \mathbf{G}$ is of type w_1. Now $\mathbf{W}(\mathbf{H}, \mathbf{T}_1)$ is a subgroup of $\mathbf{W}_1 := \mathbf{W}(\mathbf{G}, \mathbf{T}_1)$. Furthermore, F induces the automorphism

$$\sigma_1 \colon \mathbf{W}_1 \to \mathbf{W}_1, \qquad y \mapsto w_1\sigma(y)w_1^{-1}.$$

(Indeed, let $g \in \mathbf{G}$ be such that $\mathbf{T}_1 = g\mathbf{T}_0 g^{-1}$ and $g^{-1}F(g) = \dot{w}_1$. Then $N_{\mathbf{G}}(\mathbf{T}_1) =$

$gN_{\mathbf{G}}(\mathbf{T}_0)g^{-1}$. Let $w \in \mathbf{W}$ and $w' := \sigma_1(w)$. Then $F(g\dot{w}g^{-1}) = g\dot{w}_1 F(\dot{w})\dot{w}_1^{-1}g^{-1} = g\dot{w}'g^{-1}$, as required.) Then, using (a), we obtain

$$R_{\mathbf{T}}^{\mathbf{G}}(1_{\mathbf{T}})(g) = |\mathbf{W}(\mathbf{H}, \mathbf{T}_1)|^{-1}|C_{\mathbf{W}_1,\sigma_1}(w)||C_1 \cap \mathbf{W}(\mathbf{H}, \mathbf{T}_1)|,$$

where C_1 denotes the σ_1-conjugacy class of w_1 in $\mathbf{W}_1$.

Definition 2.2.25 Following [Lu77b, 2.15], a class function $f \in \mathrm{CF}(\mathbf{G}^F)$ is called a *uniform function* if f can be written as a $\mathbb{K}$-linear combination of Deligne–Lusztig characters $R_{\mathbf{T}}^{\mathbf{G}}(\theta)$ for various $\mathbf{T}, \theta$.

Example 2.2.26 (a) Let $1_{\mathbf{G}}$ be the trivial character of $\mathbf{G}^F$. Then $\langle R_{\mathbf{T}}^{\mathbf{G}}(1_{\mathbf{T}}), 1_{\mathbf{G}}\rangle = 1$ for all F-stable maximal tori $\mathbf{T} \subseteq \mathbf{G}$; see [Ca85, 7.4.1, 7.4.2], [Lu77b, 2.7]. Furthermore, $1_{\mathbf{G}}$ is uniform:

$$1_{\mathbf{G}} = \frac{1}{|\mathbf{G}^F|}\sum_{\mathbf{T}} |\mathbf{T}^F| R_{\mathbf{T}}^{\mathbf{G}}(1_{\mathbf{T}})$$

where the sum runs over all F-stable maximal tori $\mathbf{T} \subseteq \mathbf{G}$. Evaluating the above identity at 1 and using the degree formula in Theorem 2.2.12, we obtain $|\mathbf{G}^F|_p = \sum_{\mathbf{T}} \varepsilon_{\mathbf{G}}\varepsilon_{\mathbf{T}}$, where the sum runs over all F-stable maximal tori in $\mathbf{G}$.

(b) The character χ_{reg} of the regular representation of $\mathbf{G}^F$ is uniform:

$$\chi_{\mathrm{reg}} = \frac{1}{|\mathbf{G}^F|_p}\sum_{\mathbf{T}}\sum_{\theta\in\mathrm{Irr}(\mathbf{T}^F)} \varepsilon_{\mathbf{G}}\varepsilon_{\mathbf{T}} R_{\mathbf{T}}^{\mathbf{G}}(\theta),$$

where, again, the first sum runs over all F-stable maximal tori $\mathbf{T} \subseteq \mathbf{G}$; see [Ca85, 7.5.6], [Lu77b, 2.11].

Example 2.2.27 Let $\mathrm{St}_{\mathbf{G}}$ be the Steinberg character of $\mathbf{G}^F$. An explicit model of $\mathrm{St}_{\mathbf{G}}$ is constructed in [St57]. In the present context, it is sufficient to know that $\mathrm{St}_{\mathbf{G}}$ is irreducible with the following properties:

$$\mathrm{St}_{\mathbf{G}}(1) = |\mathbf{G}^F|_p \qquad \text{and} \qquad \langle \mathrm{St}_{\mathbf{G}}, \mathrm{Ind}_{\mathbf{B}^F}^{\mathbf{G}^F}(1_{\mathbf{B}})\rangle \neq 0,$$

where $\mathbf{B}$ is an F-stable Borel subgroup of $\mathbf{G}$ and $1_{\mathbf{B}}$ stands for the trivial character of $\mathbf{B}^F$. (See [Ca85, Chap. 6], [DiMi20, §7.4], and also Section 3.4.) This character is also uniform; we have

$$\mathrm{St}_{\mathbf{G}} = \frac{1}{|\mathbf{G}^F|}\sum_{\mathbf{T}} \varepsilon_{\mathbf{G}}\varepsilon_{\mathbf{T}}|\mathbf{T}^F| R_{\mathbf{T}}^{\mathbf{G}}(1_{\mathbf{T}}),$$

where the sum runs over all F-stable maximal tori $\mathbf{T} \subseteq \mathbf{G}$; see [Ca85, 7.6.6].

Remark 2.2.28 Having defined $R_{\mathbf{T}}^{\mathbf{G}} \colon \mathrm{CF}(\mathbf{T}^F) \to \mathrm{CF}(\mathbf{G}^F)$, we obtain by adjunction a unique linear map ${}^*R_{\mathbf{T}}^{\mathbf{G}} \colon \mathrm{CF}(\mathbf{G}^F) \to \mathrm{CF}(\mathbf{T}^F)$ such that

$$\langle {}^*R_{\mathbf{T}}^{\mathbf{G}}(f), f'\rangle = \langle f, R_{\mathbf{T}}^{\mathbf{G}}(f')\rangle \qquad \text{for all } f \in \mathrm{CF}(\mathbf{G}^F) \text{ and } f' \in \mathrm{CF}(\mathbf{T}^F).$$

(Here, on the left-hand side of the equality, the inner product is taken in $\mathrm{CF}(\mathbf{T}^F)$, while on the right-hand side it is taken in $\mathrm{CF}(\mathbf{G}^F)$.) The values of ${}^*R_{\mathbf{T}}^{\mathbf{G}}(f)$ are given as follows (see [DiMi20, §10.1]). Let $t \in \mathbf{T}^F$ and $\mathbf{H} := C_{\mathbf{G}}^\circ(t)$. Then

$$ {}^*R_{\mathbf{T}}^{\mathbf{G}}(f)(t) = |\mathbf{T}^F||\mathbf{H}^F|^{-1} \sum_{u \in \mathbf{H}^F_{\mathrm{uni}}} Q_{\mathbf{T}}^{\mathbf{H}}(u) f(tu). $$

For any $f \in \mathrm{CF}(\mathbf{G}^F)$, we can write uniquely $f = \pi_{\mathrm{un}}^{\mathbf{G}}(f) + f^\perp$ where $\pi_{\mathrm{un}}^{\mathbf{G}}(f)$ is uniform and $f^\perp$ is orthogonal to all uniform class functions. We call $\pi_{\mathrm{un}}^{\mathbf{G}}(f)$ the *uniform projection* of f. Explicitly, we have (see [DiMi20, §10.2]):

$$ \pi_{\mathrm{un}}^{\mathbf{G}}(f) = \frac{1}{|\mathbf{G}^F|} \sum_{\mathbf{T}} |\mathbf{T}^F| R_{\mathbf{T}}^{\mathbf{G}}({}^*R_{\mathbf{T}}^{\mathbf{G}}(f)), $$

where the sum runs over all F-stable maximal tori $\mathbf{T} \subseteq \mathbf{G}$.

Example 2.2.29 Let $\mathbf{G}^F = \mathrm{GL}_2(q)$. In Table 2.5, the most difficult characters to obtain are those of degree $q-1$. Using the notation in Example 2.1.16, consider the cyclic subgroup $H := \langle B_1(1) \rangle$ of index $q(q-1)$. Inducing the linear characters of H to $\mathrm{GL}_2(q)$ yields (reducible) characters $\mu_{q(q-1)}^{(n)}$ (for $0 \leqslant n \leqslant q^2-2$) with the following values.

	$A_1(a)$	$A_2(a)$	$A_3(a,b)$	$B_1(a)$
$\mu_{q(q-1)}^{(n)}$	$q(q-1)\eta^{na(q+1)}$	0	0	$\eta^{na} + \eta^{naq}$

If $n \in E_q$ and $(q+1) \nmid n$, then it turns out that

$$ \chi_{q-1}^{(n)} = \chi_q^{(0)} \chi_{q+1}^{(0,n)} - \chi_{q+1}^{(0,n)} - \mu_{q(q-1)}^{(n)} \in \mathrm{Irr}(\mathrm{GL}_2(q)). $$

(See [St51b, p. 227].) We can now re-interpret this as follows. A maximally split torus and the corresponding Weyl group are given by

$$ \mathbf{T}_0 = \left\{ \begin{pmatrix} \xi & 0 \\ 0 & \xi' \end{pmatrix} \;\middle|\; \xi, \xi' \in k^\times \right\} \quad \text{and} \quad \mathbf{W} = \langle s \rangle \quad \text{where} \quad \dot{s} = \begin{pmatrix} 0 & 1 \\ 1 & 0 \end{pmatrix}. $$

Consider the element $s \in \mathbf{W}$ and let $\mathbf{T}_s \subseteq \mathbf{G}$ be an F-stable maximal torus of type s (see 1.6.4, 1.6.21). By Example 1.6.5, we have

$$ \mathbf{T}_0[s] = \left\{ \begin{pmatrix} \xi & 0 \\ 0 & \xi^q \end{pmatrix} \;\middle|\; \xi \in k^\times,\, \xi^{q^2} = \xi \right\}. $$

For $0 \leqslant a < q^2-1$ let $B_1'(a)$ be the diagonal matrix with diagonal entries τ^a and τ^{aq}, where τ is a generator of $\mathbb{F}_{q^2}^\times$. Thus, $\mathbf{T}_s^F \cong \mathbf{T}_0[s] = \{B_1'(a) \mid 0 \leqslant a < q^2-1\}$. For $0 \leqslant n < q^2-1$ let $\theta_n' \in \mathrm{Irr}(\mathbf{T}_s^F)$ be the linear character that, via $\mathbf{T}_s^F \cong \mathbf{T}_0[s]$, corresponds to the linear character of $\mathbf{T}_0[s]$ sending $B_1'(a)$ to η^{na}. Then we have

$$ R_{\mathbf{T}_s}^{\mathbf{G}}(\theta_n') = -\chi_{q-1}^{(n)} \qquad \text{for all } n \in E_q \text{ such that } (q+1) \nmid n. $$

A completely analogous situation occurs for $\mathrm{GL}_n(q)$ and any $n \geqslant 2$, where the most difficult characters to construct are the so-called 'discrete series' characters, of degree $(q-1)(q^2-1)\ldots(q^{n-1}-1)$; see [Ja86, §3], [Gre99].

Example 2.2.30 Let us re-interpret the character table of $\mathbf{G}^F = \mathrm{SL}_2(q)$ in Table 2.6 (p. 106) in terms of the theory developed so far. In this case, a maximally split torus and the corresponding Weyl group are given by

$$\mathbf{T}_0 = \left\{ \begin{pmatrix} \xi & 0 \\ 0 & \xi^{-1} \end{pmatrix} \;\middle|\; \xi \in k^\times \right\} \quad \text{and} \quad \mathbf{W} = \langle s \rangle \quad \text{where} \quad \dot{s} = \begin{pmatrix} 0 & 1 \\ -1 & 0 \end{pmatrix}.$$

For $a \in \mathbb{Z}$, let $S(a)$ be the diagonal matrix with diagonal entries σ^a and σ^{-a}, where σ is a fixed generator of $\mathbb{F}_q^\times$. Then $\mathbf{T}_0^F = \{S(a) \mid 0 \leqslant a < q-1\}$ is cyclic of order $q-1$. Now consider the element $s \in \mathbf{W}$ and let $\mathbf{T}_s \subseteq \mathbf{G}$ be an F-stable maximal torus of type s (see 1.6.4, 1.6.21). As in Example 1.6.5, one sees that

$$\mathbf{T}_0[s] = \left\{ \begin{pmatrix} \xi & 0 \\ 0 & \xi^q \end{pmatrix} \;\middle|\; \xi \in k^\times,\ \xi^{q^2} = \xi \text{ and } \xi^{q+1} = 1 \right\}$$

is cyclic of order $q+1$. For $b \in \mathbb{Z}$, let $S'(b)$ be the diagonal matrix with diagonal entries τ_0^b and τ_0^{bq}, where $\tau_0 = \tau^{q-1}$ is an element of order $q+1$ in $\mathbb{F}_{q^2}^\times$. (Note that $S'(b)$ has the same eigenvalues as the matrix $T(b)$ defined in Table 2.6.) Thus, $\mathbf{T}_s^F \cong \mathbf{T}_0[s] = \{S'(b) \mid 0 \leqslant b < q+1\}$, where $T(b)$ and $S'(b)$ correspond to each other under this isomorphism.

For $0 \leqslant i < q-1$ let $\theta_i \in \mathrm{Irr}(\mathbf{T}_0^F)$ be the character that sends $S(a)$ to ε^{ai}; for $0 \leqslant j < q+1$ let $\theta_j' \in \mathrm{Irr}(\mathbf{T}_s^F)$ be the character that sends $T(b)$ to η^{bj}. Then we have (see also [Bo11, §5.3]):

$$R_{\mathbf{T}_0}^{\mathbf{G}}(\theta_i) = \begin{cases} \psi_1 + \psi_q & \text{if } i = 0, \\ \psi_{q+1}^{(i)} & \text{if } \varepsilon^{2i} \neq 1, \\ \psi_+' + \psi_+'' & \text{if } i = \frac{q-1}{2}, \end{cases} \qquad R_{\mathbf{T}_s}^{\mathbf{G}}(\theta_j') = \begin{cases} \psi_1 - \psi_q & \text{if } j = 0, \\ -\psi_{q-1}^{(j)} & \text{if } \eta^{2j} \neq 1, \\ -\psi_-' - \psi_-'' & \text{if } j = \frac{q+1}{2}. \end{cases}$$

Thus, we see that ψ_1 and ψ_q, as well as the characters $\psi_{q+1}^{(i)}$ and $\psi_{q-1}^{(j)}$, are uniform, while $\psi_\pm'$, $\psi_\pm''$ are not uniform. (The latter four characters take different values on the regular unipotent elements N, N', while all $R_{\mathbf{T}}^{\mathbf{G}}(\theta)$ have value 1 on N, N' by Theorem 2.2.22.) Assume now that q is odd and let $i_0 = (q-1)/2$, $j_0 = (q+1)/2$. Then we note that the values of the following rational linear combinations take a particularly simple form:

	I	$-I$	N	N'	$-N$	$-N'$	$S(a)$	$T(b)$
$\frac{1}{2}(\psi_+' - \psi_+'' + \psi_-' - \psi_-'')$	.	.	$\sqrt{\delta q}$	$-\sqrt{\delta q}$	.	.	.	.
$\frac{1}{2}(\psi_+' - \psi_+'' - \psi_-' + \psi_-'')$	.	.	.	.	$\delta\sqrt{\delta q}$	$-\delta\sqrt{\delta q}$	.	.

(An interpretation for this special behaviour will be given in Example 2.7.27.) Note that $R_{\mathbf{T}_0}^{\mathbf{G}}(\theta_{i_0})$ and $R_{\mathbf{T}_s}^{\mathbf{G}}(\theta'_{j_0})$ do not have an irreducible constituent in common. However, the pairs $(\mathbf{T}_0, \theta_{i_0})$ and $(\mathbf{T}_s, \theta'_{j_0})$ are still related in a geometric way, which we will discuss in the following section (Example 2.3.6).

It is a good exercise to re-interpret the character tables in Table 2.1 (p. 95) in a similar way. For the Suzuki groups ${}^2B_2(q^2)$, see [Ge03a, §4.6].

2.3 Unipotent Characters and Degree Polynomials

We have seen in the previous section that every $\rho \in \mathrm{Irr}(\mathbf{G}^F)$ occurs in some Deligne–Lusztig character $R_{\mathbf{T}}^{\mathbf{G}}(\theta)$. Since $\pm R_{\mathbf{T}}^{\mathbf{G}}(\theta)$ is not irreducible in general, it makes sense to define a graph $\mathrm{DL}(\mathbf{G}^F)$ as follows. It has vertices in bijection with $\mathrm{Irr}(\mathbf{G}^F)$. Two characters $\rho_1 \neq \rho_2$ in $\mathrm{Irr}(\mathbf{G}^F)$ are joined by an edge if there exists some pair $(\mathbf{T}, \theta)$ such that $\langle R_{\mathbf{T}}^{\mathbf{G}}(\theta), \rho_i \rangle \neq 0$ for $i = 1, 2$. Thus, the connected components of $\mathrm{DL}(\mathbf{G}^F)$ define a partition of $\mathrm{Irr}(\mathbf{G}^F)$. In this section, we will concentrate on a particular connected component of $\mathrm{DL}(\mathbf{G}^F)$: the one containing the trivial character of $\mathbf{G}^F$. We begin with a basic criterion for when two Deligne–Lusztig characters have an irreducible constituent in common, which may happen even if they are orthogonal to each other, since they are in general just virtual characters. This relies on the following auxiliary result.

Lemma 2.3.1 *Let $\mathbf{T} \subseteq \mathbf{G}$ be an F-stable maximal torus. For any integer $d \geqslant 1$, we define a norm map by*

$$N_{F^d/F} : \mathbf{T} \to \mathbf{T}, \qquad t \mapsto tF(t)F^2(t)\ldots F^{d-1}(t).$$

Then $N_{F^d/F}$ is an isogeny of algebraic groups, such that $N_{F^d/F}(\mathbf{T}^{F^d}) = \mathbf{T}^F$.

Proof Since $\mathbf{T}$ is abelian, it is clear that $N_{F^d/F}$ is an abstract group homomorphism. If $t \in \mathbf{T}^{F^d}$, then $F(N_{F^d/F}(t)) = F(t)F^2(t)\ldots F^d(t) = N_{F^d/F}(t)$ and so $N_{F^d/F}(t) \in \mathbf{T}^F$. Since F is a homomorphism of algebraic groups, so is $N_{F^d/F}$. Let us now consider the kernel of $N_{F^d/F}$. Let $t \in \mathbf{T}$. Since $\mathbf{T}$ is F-stable and connected, we can apply the Lang–Steinberg Theorem and write $t = s^{-1}F(s)$ for some $s \in \mathbf{T}$. Then

$$N_{F^d/F}(t) = \big(s^{-1}F(s)\big)F\big(s^{-1}F(s)\big)\ldots F^{d-1}\big(s^{-1}F(s)\big) = s^{-1}F^d(s).$$

Hence, $N_{F^d/F}(t) = 1$ if and only if $s \in \mathbf{T}^{F^d}$. We conclude that

$$\ker(N_{F^d/F}) = \{s^{-1}F(s) \mid s \in \mathbf{T}^{F^d}\} \subseteq \mathbf{T}^{F^d}.$$

So the kernel is finite and, hence, $N_{F^d/F}$ is an isogeny. We have already seen

$N_{F^d/F}(\mathbf{T}^{F^d}) \subseteq \mathbf{T}^F$. It remains to show that equality holds. Let $t \in \mathbf{T}^F$. By Lang–Steinberg, we can write $t = s^{-1}F^d(s)$ where $s \in \mathbf{T}$. Then the above computation shows that $N_{F^d/F}(s^{-1}F(s)) = s^{-1}F^d(s) = t$. Finally,

$$F^d(s^{-1}F(s)) = F^d(s)^{-1}F\big(F^d(s)\big) = (st)^{-1}F(st) = s^{-1}F(s)$$

and so $s^{-1}F(s) \in \mathbf{T}^{F^d}$, as required. □

Using the norm map $N_{F^d/F}$, we obtain an induced map

$$\mathrm{Irr}(\mathbf{T}^F) \to \mathrm{Irr}(\mathbf{T}^{F^d}), \qquad \theta \mapsto \theta \circ N_{F^d/F},$$

which is injective (since $N_{F^d/F}\colon \mathbf{T}^{F^d} \to \mathbf{T}^F$ is surjective).

Theorem 2.3.2 (Exclusion theorem) *Suppose that two Deligne–Lusztig characters $R^{\mathbf{G}}_{\mathbf{T}_1}(\theta_1)$ and $R^{\mathbf{G}}_{\mathbf{T}_2}(\theta_2)$ have an irreducible constituent in common. Then $(\mathbf{T}_1, \theta_1)$ and $(\mathbf{T}_2, \theta_2)$ are 'geometrically conjugate', that is, there exist some $d \geqslant 1$ and some $g \in \mathbf{G}^{F^d}$ such that $\mathbf{T}_2 = g\mathbf{T}_1g^{-1}$ and g conjugates $\theta_1 \circ N_{F^d/F} \in \mathrm{Irr}(\mathbf{T}_1^{F^d})$ to $\theta_2 \circ N_{F^d/F} \in \mathrm{Irr}(\mathbf{T}_2^{F^d})$ (via the isomorphism $\mathbf{T}_1^{F^d} \to \mathbf{T}_2^{F^d}$, $t \mapsto gtg^{-1}$).*

For the proof, see [DeLu76, 5.4, 6.3] or [Ca85, 4.1.1, 7.3.8] or [DiMi20, 11.1.3]; the term 'exclusion theorem' comes from [DeLu76, 7.10].

Remark 2.3.3 Let $\mathbf{T} \subseteq \mathbf{G}$ be an F-stable maximal torus. Then the norm map is transitive, in the sense that

$$N_{F^{de}/F} = N_{(F^d)^e/F^d} \circ N_{F^d/F} \qquad \text{for any integers } d, e \geqslant 1.$$

Consequently, if the pairs $(\mathbf{T}, \theta)$ and $(\mathbf{T}', \theta')$ are geometrically conjugate via the norm map $N_{F^d/F}$, then they will also be geometrically conjugate via the norm map $N_{F^{de}/F}$ for any $e \geqslant 1$.

Definition 2.3.4 (Cf. [Ca85, §12.1]) Let $\rho_1, \rho_2 \in \mathrm{Irr}(\mathbf{G}^F)$. We say that ρ_1, ρ_2 are *geometrically conjugate* if there are Deligne–Lusztig characters $R^{\mathbf{G}}_{\mathbf{T}_i}(\theta_i)$ such that $\langle \rho_i, R^{\mathbf{G}}_{\mathbf{T}_i}(\theta_i)\rangle \neq 0$ for $i = 1, 2$, and the pairs $(\mathbf{T}_1, \theta_1)$, $(\mathbf{T}_2, \theta_2)$ are geometrically conjugate (as in Theorem 2.3.2). This defines an equivalence relation on $\mathrm{Irr}(\mathbf{G}^F)$; the equivalence classes are called *geometric conjugacy classes of characters* or *geometric series of characters*. Clearly, if ρ_1, ρ_2 belong to the same connected component of the graph $\mathrm{DL}(\mathbf{G}^F)$, then ρ_1, ρ_2 belong to the same geometric series of characters.

Lemma 2.3.5 *Let $\rho \in \mathrm{Irr}(\mathbf{G}^F)$ and $\mathbf{T} \subseteq \mathbf{G}$ be an F-stable maximal torus. Then the number of $\theta \in \mathrm{Irr}(\mathbf{T}^F)$ such that $\langle R^{\mathbf{G}}_{\mathbf{T}}(\theta), \rho\rangle \neq 0$ is bounded above by $|\mathbf{W}|$.*

Proof Let us fix some θ such that $\langle R^{\mathbf{G}}_{\mathbf{T}}(\theta), \rho\rangle \neq 0$. If $\langle R^{\mathbf{G}}_{\mathbf{T}}(\theta'), \rho\rangle \neq 0$ for some $\theta' \in \mathrm{Irr}(\mathbf{T}^F)$, then Theorem 2.3.2 implies that there exists some $d \geqslant 1$ and some

$g \in N_{\mathbf{G}}(\mathbf{T})$ such that ${}^g(\theta \circ N_{F^d/F}) = \theta' \circ N_{F^d/F}$. Hence, since $\mathbf{W} \cong N_{\mathbf{G}}(\mathbf{T})/\mathbf{T}$, there are at most $|\mathbf{W}|$ possibilities for $\theta' \circ N_{F^d/F}$. Finally, since the map $\theta' \mapsto \theta' \circ N_{F^d/F}$ is injective, there are at most $|\mathbf{W}|$ possibilities for θ' itself. □

Example 2.3.6 Let $\mathbf{G}^F = \mathrm{SL}_2(q)$ as in Example 2.2.30. Assume that q is odd and consider the non-uniform characters $\psi'_\pm, \psi''_\pm$; we have

$$R^{\mathbf{G}}_{\mathbf{T}_0}(\theta_{i_0}) = \psi'_+ + \psi''_+ \qquad \text{and} \qquad R^{\mathbf{G}}_{\mathbf{T}_s}(\theta_{j'_0}) = -\psi'_- - \psi''_-.$$

Now, although $R^{\mathbf{G}}_{\mathbf{T}_0}(\theta_{i_0})$ and $R^{\mathbf{G}}_{\mathbf{T}_s}(\theta_{j'_0})$ do not have an irreducible constituent in common, the two pairs $(\mathbf{T}_0, \theta_{i_0})$ and $(\mathbf{T}_s, \theta'_{j_0})$ are geometrically conjugate. We leave it as an exercise to check this directly, using the norm map $N_{F^2/F}$. (Or see Example 2.4.7(b).)

Lemma 2.3.7 ([Bo06, 9.11]) *Let $(\mathbf{T}_1, \theta_1)$ and $(\mathbf{T}_2, \theta_2)$ be geometrically conjugate. Then θ_1 and θ_2 have the same restriction to $\mathbf{Z}^\circ(\mathbf{G})^F$.*

Proof Let $d \geqslant 1$ and $g \in \mathbf{G}^{F^d}$ be such that $\mathbf{T}_2 = g\mathbf{T}_1 g^{-1}$, ${}^g(\theta_1 \circ N_{F^d/F}) = \theta_2 \circ N_{F^d/F}$. Let $z \in \mathbf{Z}^\circ(\mathbf{G})^F$. As in the proof of Lemma 2.3.1, we have $z = N_{F^d/F}(z_1^{-1}F(z_1))$ where $z_1 \in \mathbf{Z}^\circ(\mathbf{G})$ is such that $z = z_1^{-1}F^d(z_1)$. This yields

$$\begin{aligned} \theta_2(z) &= \big(\theta_2 \circ N_{F^d/F}\big)(z_1^{-1}F(z_1)) = \theta_1\big(N_{F^d/F}(g^{-1}z_1^{-1}F(z_1)g)\big) \\ &= \theta_1(N_{F^d/F}(z_1^{-1}F(z_1))) = \theta_1(z), \end{aligned}$$

where the third equality holds since $z_1^{-1}F(z_1) \in \mathbf{Z}(\mathbf{G})$. □

Definition 2.3.8 ([DeLu76, 7.8]) A character $\rho \in \mathrm{Irr}(\mathbf{G}^F)$ is called a *unipotent character* if $\langle R^{\mathbf{G}}_{\mathbf{T}}(1_{\mathbf{T}}), \rho\rangle \neq 0$ for some F-stable maximal torus $\mathbf{T} \subseteq \mathbf{G}$ (where $1_{\mathbf{T}}$ stands for the trivial character of $\mathbf{T}^F$). We also denote

$$\mathrm{Uch}(\mathbf{G}^F) := \text{set of all unipotent characters of } \mathbf{G}^F.$$

This set is a geometric series of characters. Indeed, by Example 2.2.26(a), every $\rho \in \mathrm{Uch}(\mathbf{G}^F)$ is geometrically conjugate to $1_{\mathbf{G}}$. Furthermore, if $\mathbf{T} \subseteq \mathbf{G}$ is an F-stable maximal torus and $\theta \in \mathrm{Irr}(\mathbf{T}^F)$, then the Exclusion Theorem 2.3.2 shows that, for $\rho \in \mathrm{Uch}(\mathbf{G}^F)$, we have $\langle R^{\mathbf{G}}_{\mathbf{T}}(\theta), \rho\rangle = 0$ unless $\theta = 1_{\mathbf{T}}$. The same arguments also show that $\mathrm{Uch}(\mathbf{G}^F)$ is precisely the connected component of the graph $\mathrm{DL}(\mathbf{G}^F)$ (introduced in the beginning of this section) to which $1_{\mathbf{G}}$ belongs[2].

Example 2.3.9 Let $\mathbf{B}_0 \subseteq \mathbf{G}$ be an F-stable Borel subgroup such that $\mathbf{T}_0 \subseteq \mathbf{B}_0$ (as in 1.6.1). Then it is a classical part of the character theory of $\mathbf{G}^F$ (well established before the work of Deligne and Lusztig) that there is a bijection

$$\mathrm{Irr}(\mathbf{W}^\sigma) \overset{1-1}{\longleftrightarrow} \big\{\rho \in \mathrm{Irr}(\mathbf{G}^F) \mid \big\langle \mathrm{Ind}_{\mathbf{B}_0^F}^{\mathbf{G}^F}(1_{\mathbf{B}_0}), \rho\big\rangle \neq 0\big\}, \qquad \phi \leftrightarrow \rho_\phi.$$

[2] A characterisation of all connected components of $\mathrm{DL}(\mathbf{G}^F)$ will only be achieved much later, in Remark 2.6.19.

See [CuRe87, §67, §68] and, for example, [GeJa11, §4.3] for a discussion taking into account more recent developments; we will also come back to this in Chapter 3. The above correspondence is canonical once a square root of q has been fixed. (The condition comes from the fact that the values of all characters ρ_ϕ are contained in $\mathbb{Q}(\sqrt{q}) \subseteq \mathbb{K}$; see Remark 3.1.21 and Example 3.2.6 for further details.) By Proposition 2.2.7, we have $\rho_\phi \in \mathrm{Uch}(\mathbf{G}^F)$ for all $\phi \in \mathrm{Irr}(\mathbf{W}^\sigma)$. In particular, the trivial character $1_\mathbf{G}$ and the Steinberg character $\mathrm{St}_\mathbf{G}$ of $\mathbf{G}^F$ are unipotent. (See also Examples 2.2.26 and 2.2.27.)

As we will see in later sections, the unipotent characters are of fundamental importance in the character theory of $\mathbf{G}^F$. It would certainly be interesting and desirable to find an elementary characterisation of these characters, without reference to the theory of ℓ-adic cohomology. For example, given the character table of $\mathbf{G}^F$ (e.g., as in Table 2.1 or the Cambridge ATLAS), how can we identify the unipotent characters in this table? Assuming that q is large enough, such an elementary characterisation is mentioned in the introduction of [Lu77b]. To state it, we need some preparation.

Remark 2.3.10 Let $\mathbf{T}$ be an F-stable maximal torus of $\mathbf{G}$. Then

$$\mathbf{T}_{\mathrm{reg}} := \{t \in \mathbf{T} \mid C^\circ_\mathbf{G}(t) = \mathbf{T}\}$$

is called the set of *regular elements* of $\mathbf{G}$ in $\mathbf{T}$. (See, e.g., [Hum95, Chap. 4] for a further discussion.) Now let $\rho \in \mathrm{Uch}(\mathbf{G}^F)$. As already noted, for $\theta \in \mathrm{Irr}(\mathbf{T}^F)$ we have $\langle R^\mathbf{G}_\mathbf{T}(\theta), \rho\rangle = 0$ unless $\theta = 1_\mathbf{T}$. So the formula in Proposition 2.2.18 takes the following simple form (see [DeLu76, 7.9]):

$$\rho(s) = \langle R^\mathbf{G}_\mathbf{T}(1_\mathbf{T}), \rho\rangle \quad \text{for any } s \in \mathbf{T}^F_{\mathrm{reg}}.$$

(Note that, if $s \in \mathbf{T}^F_{\mathrm{reg}}$, then $\mathbf{T}$ is the only maximal torus of $\mathbf{G}$ that contains s.) Thus, if $\mathbf{T}^F_{\mathrm{reg}} \neq \varnothing$ for all F-stable maximal tori $\mathbf{T} \subseteq \mathbf{G}$, then the knowledge of the multiplicities $\langle R^\mathbf{G}_\mathbf{T}(1_\mathbf{T}), \rho\rangle$ is equivalent to the knowledge of the values of ρ on regular semisimple elements.

Now, it is known that $\mathbf{T}_{\mathrm{reg}}$ is dense in $\mathbf{T}$ which implies that $\mathbf{T}^F_{\mathrm{reg}} \neq \varnothing$ if q is large enough. More precisely, we have:

Lemma 2.3.11 *With $\mathbf{T}$, $\mathbf{T}_{\mathrm{reg}}$ as above, there is a constant $C > 0$ (depending only on the root datum of $\mathbf{G}$) such that $|\mathbf{T}^F_{\mathrm{reg}}|/|\mathbf{T}^F| \geqslant 1 - C/q$.*

Proof We follow the argument in [Ca85, 8.4.2], with minor modifications. First note that the formulae in 1.6.21 show that $|\mathbf{T}^F| = (q - \varepsilon_1)\cdots(q - \varepsilon_r)$ where $r = \dim \mathbf{T}$ and ε_i are roots of unity. Hence,

$$(q-1)^r \leqslant |\mathbf{T}^F| \leqslant (q+1)^r. \tag{a}$$

Let $s \in \mathbf{T}^F$ be non-regular and set $\mathbf{H} := C^\circ_{\mathbf{G}}(s)$. Then $s \in \mathbf{T} \subsetneq \mathbf{H}$, and $\mathbf{H}$ is F-stable, closed, connected and reductive (see 2.2.13). Since s is non-regular, we have $\mathbf{Z}(\mathbf{H}) \subsetneq \mathbf{T}$ and so $\dim \mathbf{Z}(\mathbf{H}) \leqslant r - 1$. Now $\mathbf{Z}^\circ(\mathbf{H})$ is an F-stable torus and so $|\mathbf{Z}^\circ(\mathbf{H})^F| \leqslant (q+1)^{r-1}$ (by the same argument as above for $\mathbf{T}^F$). Furthermore, by 1.3.10(c), $\mathbf{Z}(\mathbf{H})/\mathbf{Z}^\circ(\mathbf{H})$ is isomorphic to the centre of the semisimple group $\mathbf{H}/\mathbf{Z}^\circ(\mathbf{H})$. So, by Proposition 1.5.2 and Table 1.2 (p. 20), we have $|\mathbf{Z}(\mathbf{H})/\mathbf{Z}^\circ(\mathbf{H})| \leqslant r+1$. Thus, we obtain

$$s \in \mathbf{Z}(\mathbf{H})^F \qquad \text{and} \qquad |\mathbf{Z}(\mathbf{H})^F| \leqslant (r+1)(q+1)^{r-1}. \tag{b}$$

Let $R_1 \subseteq X(\mathbf{T})$ be the set of roots of $\mathbf{G}$ with respect to $\mathbf{T}$. Then $\mathbf{H}$ is generated by $\mathbf{T}$ and the root subgroups of $\mathbf{G}$ corresponding to those $\alpha \in R_1$ such that $\alpha(s) = 1$ (see once more 2.2.13). Thus, $\mathbf{H}$ is completely determined by $\mathbf{T}$ and a subset of R_1. So, if we just use (b) and the fact that there are at most $2^{|R_1|}$ subsets of R_1 at all, then the discussion so far shows that

$$|\mathbf{T}^F \setminus \mathbf{T}^F_{\text{reg}}| \leqslant 2^{|R_1|}(r+1)(q+1)^{r-1}. \tag{c}$$

Using (a) and (c), we obtain the estimate

$$\frac{|\mathbf{T}^F_{\text{reg}}|}{|\mathbf{T}^F|} = 1 - \frac{|\mathbf{T}^F \setminus \mathbf{T}^F_{\text{reg}}|}{|\mathbf{T}^F|} \geqslant 1 - \frac{2^{|R_1|}(r+1)(q+1)^{r-1}}{(q-1)^r} \geqslant 1 - \frac{C}{q},$$

where $C > 0$ only depends on the root datum of $\mathbf{G}$. □

One can in fact give precise formulae for the number of regular elements in $\mathbf{T}^F$ but we do not need this here; see, e.g., [Hum95, §8.9] and the references there. The above proof yields a very crude bound but it shows that, if q is large enough, then $\mathbf{T}^F_{\text{reg}} \neq \varnothing$; furthermore, the proportion of elements in $\mathbf{T}^F$ that are regular tends to 1 as $q \to \infty$.

Proposition 2.3.12 (Lusztig, see [Lu77b, Introduction]) *Assume that every F-stable maximal torus of $\mathbf{G}$ contains sufficiently many regular elements fixed by F or, more precisely, that*

$$|\mathbf{T}^F_{\text{reg}}| > \left(1 - 2^{-|\mathbf{W}|}\right)|\mathbf{T}^F| \quad \textit{for any } F\textit{-stable maximal torus } \mathbf{T} \subseteq \mathbf{G}. \tag{$*$}$$

Let $\rho \in \operatorname{Irr}(\mathbf{G}^F)$. Then ρ is unipotent if and only if the restriction of ρ to $\mathbf{T}^F_{\text{reg}}$ is constant, for any F-stable maximal torus $\mathbf{T} \subseteq \mathbf{G}$.

Proof If ρ is unipotent, then we have seen in Remark 2.3.10 that ρ takes the constant value $\langle R^{\mathbf{G}}_{\mathbf{T}}(1_{\mathbf{T}}), \rho\rangle$ on the regular semisimple elements inside a given $\mathbf{T}^F$. Conversely, let $\rho \in \operatorname{Irr}(\mathbf{G}^F)$ be non-unipotent. By Corollary 2.2.19, there is an F-stable maximal torus $\mathbf{T} \subseteq \mathbf{G}$ such that $\langle R^{\mathbf{G}}_{\mathbf{T}}(\theta), \rho\rangle \neq 0$ for some $\theta \in \operatorname{Irr}(\mathbf{T}^F)$ which is not the trivial character. Using our assumption ($*$), we will show that then there exist regular semisimple elements $s, s' \in \mathbf{T}^F$ such that $\rho(s) \neq \rho(s')$. Indeed, for regular semisimple elements, the formula in Proposition 2.2.18 reduces to:

$$\rho(s) = \sum_{1 \leqslant i \leqslant n} m_i \theta_i(s) \qquad \text{for all } s \in \mathbf{T}^F_{\text{reg}}, \tag{$\dagger$}$$

where $\theta_i \in \operatorname{Irr}(\mathbf{T}^F)$ are such that $m_i := \langle R^{\mathbf{G}}_{\mathbf{T}}(\theta_i), \rho \rangle \neq 0$. Note that none of the characters θ_i is the trivial character of $\mathbf{T}^F$. Hence, $\theta_1, \ldots, \theta_n$ together with the trivial character of $\mathbf{T}^F$ are linearly independent functions $\mathbf{T}^F \to \mathbb{K}$, by a classical result due to Dedekind. By a slight generalisation of this result, these functions are even linearly independent when restricted to $\mathbf{T}^F_{\text{reg}}$. Indeed, by [Lu90, 8.1], these $n+1$ restrictions are linearly independent if

$$|\mathbf{T}^F_{\text{reg}}| > \left(1 - 2^{-n}\right)|\mathbf{T}^F|.$$

By Lemma 2.3.5, we have $n \leqslant |\mathbf{W}|$ and so this inequality does hold by our assumption (∗). Hence, on the set $\mathbf{T}^F_{\text{reg}}$, the linear combination $\sum_{1 \leqslant i \leqslant n} m_i\theta_i$ is not a scalar multiple of the trivial character of $\mathbf{T}^F$. Consequently, by (†), there exist $s, s' \in \mathbf{T}^F_{\text{reg}}$ such that $\rho(s) \neq \rho(s')$, as desired. □

The following results show that unipotent characters are much better behaved with respect to normal subgroups of $\mathbf{G}^F$ than arbitrary irreducible characters. Because of its importance for the classification of unipotent characters (see Chapter 4), we discuss this in some detail. This is also a good illustration of some of the methods developed so far.

Lemma 2.3.13 *Let $\eta \in \operatorname{Irr}(\mathbf{G}^F)$ be a linear character such that $\eta(u) = 1$ for all unipotent elements $u \in \mathbf{G}^F$. Then $\eta \cdot R^{\mathbf{G}}_{\mathbf{T}}(\theta) = R^{\mathbf{G}}_{\mathbf{T}}(\eta|_{\mathbf{T}^F} \cdot \theta)$ for any F-stable maximal torus $\mathbf{T} \subseteq \mathbf{G}$ and $\theta \in \operatorname{Irr}(\mathbf{T}^F)$.*

Proof This easily follows from the character formula in Theorem 2.2.16, applied to $R^{\mathbf{G}}_{\mathbf{T}}(\eta|_{\mathbf{T}^F} \cdot \theta)$. Just note that, if $g = su = us$ is the Jordan decomposition of $g \in \mathbf{G}^F$ and if $x \in \mathbf{G}^F$ is an element such that $x^{-1}sx \in \mathbf{T}$, then we have $\left(\eta|_{\mathbf{T}^F} \cdot \theta\right)(x^{-1}sx) = \eta(s)\theta(x^{-1}sx) = \eta(g)\theta(x^{-1}sx)$. □

Lemma 2.3.14 *Let $\mathbf{G}^F_u = \langle u \in \mathbf{G}^F \mid u \text{ unipotent} \rangle \subseteq \mathbf{G}^F_{\text{der}}$ (see Remark 1.5.13). If $\rho \in \operatorname{Uch}(\mathbf{G}^F)$, then the restriction of ρ to $\mathbf{G}^F_u$ is irreducible. Furthermore, distinct unipotent characters of $\mathbf{G}^F$ have distinct restrictions to $\mathbf{G}^F_u$.*

Proof First note that $\mathbf{G}^F_u$ is a normal subgroup of $\mathbf{G}^F$ and $\mathbf{G}^F/\mathbf{G}^F_u$ is abelian (see Remark 1.5.13). Hence, by Remark 2.1.4, we must show that $\eta \cdot \rho \neq \rho$ for every non-trivial linear character $\eta \in \operatorname{Irr}(\mathbf{G}^F)$ with $\mathbf{G}^F_u \subseteq \ker(\eta)$. For this purpose, let $\mathbf{T} \subseteq \mathbf{G}$ be an F-stable maximal torus such that $\langle R^{\mathbf{G}}_{\mathbf{T}}(1_{\mathbf{T}}), \rho \rangle \neq 0$. Let $\eta \in \operatorname{Irr}(\mathbf{G}^F)$ be a linear character such that $\mathbf{G}^F_u \subseteq \ker(\eta)$. Using Lemma 2.3.13, we deduce that

$\langle R_{\mathbf{T}}^{\mathbf{G}}(\eta|_{\mathbf{T}^F}), \eta \cdot \rho\rangle \neq 0$. Hence, if $\rho = \eta \cdot \rho$, then the Exclusion Theorem 2.3.2 implies that $\eta|_{\mathbf{T}^F}$ must be trivial. But, the two statements (b) and (c) in Remark 1.5.13 imply that $\mathbf{G}^F = \mathbf{G}_u^F.\mathbf{T}^F$ and so η itself must be trivial, as desired. □

Proposition 2.3.15 (See [DeLu76, 7.10]) *Let $\mathbf{G}'$ be another connected reductive algebraic group over k and $F'\colon \mathbf{G}' \to \mathbf{G}'$ be a Steinberg map. Let $\pi\colon \mathbf{G} \to \mathbf{G}'$ be a surjective homomorphism of algebraic groups such that $F' \circ \pi = \pi \circ F$ and $\ker(\pi) \subseteq \mathbf{Z}(\mathbf{G})$. Then the following hold.*

(a) *If $\mathbf{T} \subseteq \mathbf{G}$ is an F-stable maximal torus, then $\mathbf{T}' := \pi(\mathbf{T}) \subseteq \mathbf{G}'$ is an F'-stable maximal torus and $R_{\mathbf{T}}^{\mathbf{G}}(1_{\mathbf{T}}) = R_{\mathbf{T}'}^{\mathbf{G}'}(1_{\mathbf{T}'}) \circ \pi|_{\mathbf{G}^F}$.*
(b) *We have a bijection $\mathrm{Uch}(\mathbf{G}'^{F'}) \to \mathrm{Uch}(\mathbf{G}^F)$, $\rho' \mapsto \rho' \circ \pi|_{\mathbf{G}^F}$.*

Proof (a) This is contained in [DeLu76, 7.10]; see also [DiMi20, 11.3.8]. Here, one has to use once more the definition of $R_{\mathbf{T}}^{\mathbf{G}}(1_{\mathbf{T}})$ as in Proposition 2.2.5.

(b) By Proposition 1.4.13, $\pi(\mathbf{G}^F)$ is a normal subgroup of $\mathbf{G}'^{F'}$, with abelian factor group of order prime to p. Thus, all unipotent elements of $\mathbf{G}'^{F'}$ are contained in $\pi(\mathbf{G}^F)$. By Lemma 2.3.14, the restriction of any $\rho' \in \mathrm{Uch}(\mathbf{G}'^{F'})$ to $\pi(\mathbf{G}^F)$ is irreducible. Hence, $\rho := \rho' \circ \pi|_{\mathbf{G}^F} \in \mathrm{Irr}(\mathbf{G}^F)$ for any $\rho' \in \mathrm{Uch}(\mathbf{G}'^{F'})$; furthermore, the map $\rho' \mapsto \rho$ is injective. Now let $\mathbf{T}' \subseteq \mathbf{G}'$ be an F'-stable maximal torus and write $R_{\mathbf{T}'}^{\mathbf{G}'}(1_{\mathbf{T}'}) = \sum_i n_i \rho_i'$ where $\rho_i' \in \mathrm{Uch}(\mathbf{G}'^{F'})$ and $n_i \in \mathbb{Z}$. Then (a) shows that $R_{\mathbf{T}}^{\mathbf{G}}(1_{\mathbf{T}}) = \sum_i n_i \rho_i$. Thus, we conclude that $\mathrm{Uch}(\mathbf{G}^F) = \{\rho \mid \rho' \in \mathrm{Uch}(\mathbf{G}'^{F'})\}$. □

Remark 2.3.16 Let $\mathbf{G}'$ be another connected reductive algebraic group over k and $F'\colon \mathbf{G}' \to \mathbf{G}'$ be a Steinberg map. Let $f\colon \mathbf{G} \to \mathbf{G}'$ be an isotypy (see 1.3.21), that is, f is a homomorphism of algebraic groups such that $\ker(f) \subseteq \mathbf{Z}(\mathbf{G})$ and $\mathbf{G}'_{\mathrm{der}} \subseteq f(\mathbf{G})$; also assume that $f \circ F = F' \circ f$. Let $\mathbf{T}' \subseteq \mathbf{G}'$ be an F'-stable maximal torus and consider the F-stable maximal torus $\mathbf{T} := f^{-1}(\mathbf{T}') \subseteq \mathbf{G}$. Let $\theta' \in \mathrm{Irr}(\mathbf{T}'^{F'})$ and $\theta = \theta' \circ f|_{\mathbf{T}^F} \in \mathrm{Irr}(\mathbf{T}^F)$. If $\ker(f)$ is connected, then

$$R_{\mathbf{T}}^{\mathbf{G}}(\theta) = R_{\mathbf{T}'}^{\mathbf{G}'}(\theta') \circ f|_{\mathbf{G}^F}.$$

See [Bo00, Cor. 2.1.3] and [DiMi20, Prop. 11.3.10]. If one is only interested in $\theta = 1_{\mathbf{T}}$ and $\theta' = 1_{\mathbf{T}'}$, then the assumption on $\ker(f)$ can be dropped, as in Proposition 2.3.15(a) above.

Example 2.3.17 Let $\alpha\colon \mathbf{G} \to \mathbf{G}$ be a bijective homomorphism of algebraic groups such that $\alpha \circ F = F \circ \alpha$. Then α restricts to an automorphism of $\mathbf{G}^F$ which we denote by the same symbol. Let $\mathbf{T} \subseteq \mathbf{G}$ be an F-stable maximal torus. Then the formula in Remark 2.3.16 shows that

$$R_{\mathbf{T}}^{\mathbf{G}}(1_{\mathbf{T}}) = R_{\alpha(\mathbf{T})}^{\mathbf{G}}(1_{\alpha(\mathbf{T})}) \circ \alpha.$$

It follows that, if $\rho \in \mathrm{Uch}(\mathbf{G}^F)$, then $\rho \circ \alpha \in \mathrm{Uch}(\mathbf{G}^F)$. This will be used, for example, in the situation where $\mathbf{G}$ is a closed subgroup of some larger group and α is given by conjugation with an element in the normaliser of $\mathbf{G}$.

As noted in Definition 2.3.8, the unipotent characters form a geometric series of characters of $\mathbf{G}^F$. As preparation for the discussion of the remaining geometric series, we introduce an alternative model for $R_{\mathbf{T}}^{\mathbf{G}}(\theta)$, which appears already in [DeLu76] (see [Lu77b, 3.3]) and which is interesting and useful in its own right. One advantage of this model is that we do not have to deal with arbitrary maximal tori in $\mathbf{G}$, but that everything is done in terms of a fixed maximally split torus $\mathbf{T}_0 \subseteq \mathbf{G}$ as in 1.6.1. We will treat this in some detail, as this is also a good illustration for working with Lefschetz numbers.

2.3.18 Recall the setting of 1.6.1. Thus, $\mathbf{T}_0 \subseteq \mathbf{G}$ is an F-stable maximal torus that is maximally split, that is, it is contained in an F-stable Borel subgroup $\mathbf{B}_0 \subseteq \mathbf{G}$. Let $\mathbf{W} = N_{\mathbf{G}}(\mathbf{T}_0)/\mathbf{T}_0$, identified with the Weyl group of the corresponding root datum $\mathscr{R} = (X, R, Y, R^\vee)$ (relative to $\mathbf{T}_0$). For any $w \in \mathbf{W}$, we consider the finite subgroup

$$\mathbf{T}_0[w] := \{t \in \mathbf{T}_0 \mid F(t) = \dot{w}^{-1}t\dot{w}\} \subseteq \mathbf{T}_0 \qquad \text{(see 1.6.4)},$$

where $\dot{w}$ is a representative of w in $N_{\mathbf{G}}(\mathbf{T}_0)$. Then, as discussed in 1.6.21, every subgroup of $\mathbf{G}^F$ of the form $\mathbf{T}^F$ (where $\mathbf{T} \subseteq \mathbf{G}$ is an F-stable maximal torus) can be realised as $\mathbf{T}_0[w]$ for some $w \in \mathbf{W}$. More precisely, given $\mathbf{T}$, there exists some $g \in \mathbf{G}$ such that $\mathbf{T} = g\mathbf{T}_0g^{-1}$ and conjugation with g defines an isomorphism $\mathbf{T}^F \cong \mathbf{T}_0[w]$, where w is the image of $g^{-1}F(g)$ in $\mathbf{W}$; thus, $\mathbf{T}$ is a *torus of type* w.

Let $\mathbf{U}_0 = R_u(\mathbf{B}_0)$ be the unipotent radical of $\mathbf{B}_0$. Then $\dot{w}\mathbf{U}_0$ satisfies condition (∗) in 2.2.4 with respect to the finite subgroup $\mathbf{T}_0[w]$. Hence, by Proposition 2.2.5, we obtain virtual characters of $\mathbf{G}^F$ by setting

$$R_w^\theta := R_{\mathbf{T}_0[w],\dot{w}\mathbf{U}_0}^{\mathbf{G}}(\theta) \qquad \text{for any } \theta \in \mathrm{Irr}(\mathbf{T}_0[w]).$$

We now identify these characters with the $R_{\mathbf{T}}^{\mathbf{G}}(\theta)$ introduced earlier.

Lemma 2.3.19 *The virtual character R_w^θ does not depend on the choice of the representative $\dot{w}$. Furthermore, let $g \in \mathbf{G}$ be such that $g^{-1}F(g) = \dot{w}$ and consider the corresponding torus $\mathbf{T} = g\mathbf{T}_0g^{-1}$ of type w, as in* 2.3.18. *Then we have*

$$R_w^\theta = R_{\mathbf{T}}^{\mathbf{G}}({}^g\theta) \quad \textit{where} \quad {}^g\theta(t) := \theta(g^{-1}tg) \textit{ for } t \in \mathbf{T}^F.$$

Proof First we show that R_w^θ only depends on w. An arbitrary representative of w in $N_{\mathbf{G}}(\mathbf{T}_0)$ is of the form $h_0\dot{w}$ where $h_0 \in \mathbf{T}_0$. Let $F' \colon \mathbf{G} \to \mathbf{G}$ be defined by $F'(x) = \dot{w}F(x)\dot{w}^{-1}$ for all $x \in \mathbf{G}$. By Lemma 1.4.14, F' also is a Steinberg map on $\mathbf{G}$, and we have $F'(\mathbf{T}_0) \subseteq \mathbf{T}_0$. Hence, by Theorem 1.4.8 (Lang–Steinberg), we can write $h_0 = h^{-1}F'(h)$ for some $h \in \mathbf{T}_0$. We claim that we have a bijection

$$f \colon \mathscr{L}^{-1}(\dot{w}\mathbf{U}_0) \to \mathscr{L}^{-1}(h_0\dot{w}\mathbf{U}_0), \qquad x \mapsto xh.$$

Indeed, let $x \in \mathbf{G}$ be such that $x^{-1}F(x) \in \dot{w}\mathbf{U}_0$. Since $F(h) \in \mathbf{T}_0$ normalises $\mathbf{U}_0$, we have $(xh)^{-1}F(xh) \in h^{-1}\dot{w}F(h)\mathbf{U}_0$. We have $h^{-1}\dot{w}F(h) = h_0\dot{w}$ and so $(xh)^{-1}F(xh) \in h_0\dot{w}\mathbf{U}_0$. Thus, f is a bijection. Furthermore, f commutes with the action of an element $(g', t) \in \mathbf{G}^F \times \mathbf{T}_0[w]$ on $\mathscr{L}^{-1}(\dot{w}\mathbf{U}_0)$ and on $\mathscr{L}^{-1}(h_0\dot{w}\mathbf{U}_0)$. Hence, the property of Lefschetz numbers in Example 2.2.3(b) implies that, for all $g' \in \mathbf{G}^F$ and $t \in \mathbf{T}_0[w]$, we have

$$\mathfrak{L}\big((g', t), \mathscr{L}^{-1}(\dot{w}\mathbf{U}_0)\big) = \mathfrak{L}\big((g', t), \mathscr{L}^{-1}(h_0\dot{w}\mathbf{U}_0)\big).$$

Thus, R_w^θ only depends on w but not on the choice of the representative $\dot{w} \in N_{\mathbf{G}}(\mathbf{T}_0)$. Next we establish the identity $R_w^\theta = R_{\mathbf{T}}^{\mathbf{G}}({}^g\theta)$. Let $\mathbf{T} = g\mathbf{T}_0g^{-1}$ as above. Then we also have $\mathbf{T} = F(g)\mathbf{T}_0F(g)^{-1}$ and so $F(g)\mathbf{B}_0F(g)^{-1}$ is a Borel subgroup containing $\mathbf{T}$. Consequently, by Corollary 2.2.9, we have

$$\begin{aligned} R_{\mathbf{T}}^{\mathbf{G}}({}^g\theta)(g') &= \frac{1}{|\mathbf{T}^F|} \sum_{t\in\mathbf{T}^F} \mathfrak{L}\big((g', t), \mathscr{L}^{-1}(F(g)\mathbf{U}_0F(g)^{-1})\big)\, \theta(g^{-1}tg) \\ &= \frac{1}{|\mathbf{T}_0[w]|} \sum_{t\in\mathbf{T}_0[w]} \mathfrak{L}\big((g', gtg^{-1}), \mathscr{L}^{-1}(F(g)\mathbf{U}_0F(g)^{-1})\big)\, \theta(t) \end{aligned}$$

for all $g' \in \mathbf{G}^F$. So it remains to show that the Lefschetz number appearing in the above sum equals $\mathfrak{L}((g', t), \mathscr{L}^{-1}(\dot{w}\mathbf{U}_0))$. For this purpose, note that the morphism

$$f\colon \mathscr{L}^{-1}(F(g)\mathbf{U}_0F(g)^{-1}) \to \mathscr{L}^{-1}(\dot{w}\mathbf{U}_0), \qquad y \mapsto yg,$$

is bijective. (Indeed, for $y \in \mathbf{G}$ we have $\mathscr{L}(y) \in F(g)\mathbf{U}_0F(g)^{-1} = g\dot{w}\mathbf{U}_0\dot{w}^{-1}g^{-1}$ if and only if $\mathscr{L}(yg) = g^{-1}\mathscr{L}(y)F(g) \in g^{-1}F(g)\mathbf{U}_0 = \dot{w}\mathbf{U}_0$.) It is also straightforward to check that f commutes with the action of $\mathbf{G}^F$ (by left multiplication on both sides) and transforms the action of $\mathbf{T}^F$ on $\mathscr{L}^{-1}(F(g)\mathbf{U}_0F(g)^{-1})$ into the action of $\mathbf{T}_0[w]$ on $\mathscr{L}^{-1}(\dot{w}\mathbf{U}_0)$. Hence, a further application of Example 2.2.3(b) yields the desired equality. □

2.3.20 We define $\mathfrak{X}(\mathbf{G}, F)$ to be the set consisting of all pairs $(\mathbf{T}, \theta)$, where $\mathbf{T}$ is an F-stable maximal torus of $\mathbf{G}$ and $\theta \in \mathrm{Irr}(\mathbf{T}^F)$. If $(\mathbf{T}, \theta) \in \mathfrak{X}(\mathbf{G}, F)$ and $g \in \mathbf{G}^F$, then $\mathbf{T}' := g\mathbf{T}g^{-1}$ is an F-stable maximal torus of $\mathbf{G}$ and we obtain an irreducible character ${}^g\theta \in \mathrm{Irr}(\mathbf{T}'^F)$ by setting ${}^g\theta(t') := \theta(g^{-1}t'g)$ for all $t' \in \mathbf{T}'^F$. Thus, $\mathbf{G}^F$ acts on $\mathfrak{X}(\mathbf{G}, F)$ via

$$\mathbf{G}^F \times \mathfrak{X}(\mathbf{G}, F) \to \mathfrak{X}(\mathbf{G}, F), \qquad \big(g, (\mathbf{T}, \theta)\big) \mapsto (g\mathbf{T}g^{-1}, {}^g\theta). \tag{a}$$

We denote by $\overline{\mathfrak{X}}(\mathbf{G}, F)$ the set of $\mathbf{G}^F$-orbits for this action. On the other hand, we define $\mathfrak{X}(\mathbf{W}, \sigma) := \{(w, \theta) \mid w \in \mathbf{W},\ \theta \in \mathrm{Irr}(\mathbf{T}_0[w])\}$. If $w, w', x \in \mathbf{W}$ are such that $w' = xw\sigma(x)^{-1}$, then a straightforward computation shows that we have a group isomorphism $\mathbf{T}_0[w] \to \mathbf{T}_0[w']$, $t \mapsto \dot{x}t\dot{x}^{-1}$, where $\dot{x}$ is a fixed representative of x in $N_{\mathbf{G}}(\mathbf{T}_0)$; furthermore, if $\theta \in \mathrm{Irr}(\mathbf{T}_0[w])$, then we obtain an irreducible character

${}^{x}\theta \in \mathrm{Irr}(\mathbf{T}_0[w'])$ by setting ${}^{x}\theta(t) := \theta(\dot{x}^{-1}t\dot{x})$ for all $t \in \mathbf{T}_0[w']$. Thus, $\mathbf{W}$ acts on $\mathfrak{X}(\mathbf{W},\sigma)$ via

$$\mathbf{W} \times \mathfrak{X}(\mathbf{W},\sigma) \to \mathfrak{X}(\mathbf{W},\sigma), \qquad \big(x,(w,\theta)\big) \mapsto (xw\sigma(x)^{-1}, {}^{x}\theta). \tag{b}$$

We denote by $\overline{\mathfrak{X}}(\mathbf{W},\sigma)$ the set of $\mathbf{W}$-orbits for this action. Then it is straightforward to verify that we have a natural bijection

$$\overline{\mathfrak{X}}(\mathbf{W},\sigma) \quad \longleftrightarrow \quad \overline{\mathfrak{X}}(\mathbf{G},F), \tag{c}$$

defined as follows: The $\mathbf{W}$-orbit of a pair $(w,\theta) \in \mathfrak{X}(\mathbf{W},\sigma)$ corresponds to the $\mathbf{G}^F$-orbit of a pair $(\mathbf{T},\theta') \in \mathfrak{X}(\mathbf{G},F)$ if there exists some $g \in \mathbf{G}$ such that $\mathbf{T} = g\mathbf{T}_0g^{-1}$, $g^{-1}F(g) \in N_{\mathbf{G}}(\mathbf{T}_0)$ is a representative of w, and $\theta'(gtg^{-1}) = \theta(t)$ for all $t \in \mathbf{T}_0[w]$.

Remark 2.3.21 Via 2.3.20(c), the $\mathbf{W}$-orbits of pairs $(w,1)$ correspond to the $\mathbf{G}^F$-orbits of pairs $(\mathbf{T},1)$ (where, in both cases, 1 stands for the trivial character). Hence, the correspondence in 2.3.20(c) is a refinement of the well-known correspondence between σ-conjugacy classes of $\mathbf{W}$ and $\mathbf{G}^F$-conjugacy classes of F-stable maximal tori of $\mathbf{G}$. For further details see, e.g., [Ca85, 3.3.3], [Ge03a, 4.3.5].

Using R_w^θ instead of $R_{\mathbf{T}}^{\mathbf{G}}(\theta)$, certain identities involving Deligne–Lusztig characters simplify considerably. We give a few examples.

Example 2.3.22 (a) The scalar product formula in Theorem 2.2.8 can be re-expressed as follows. If $(w_i,\theta_i) \in \mathfrak{X}(\mathbf{W},\sigma)$ for $i = 1,2$, then

$$\langle R_{w_1}^{\theta_1}, R_{w_2}^{\theta_2}\rangle = |\{x \in \mathbf{W} \mid x.(w_1,\theta_1) = (w_2,\theta_2)\}|.$$

In particular, the scalar product is zero unless w_1, w_2 are σ-conjugate in $\mathbf{W}$. (See [Ge18, 2.6] for further details about the translation of the previous formula into the new one.)

(b) Let $(w,\theta) \in \mathfrak{X}(\mathbf{W},\sigma)$ correspond to the pair $(\mathbf{T},\theta') \in \mathfrak{X}(\mathbf{G},F)$ under the bijection in 2.3.20(c). Then $R_w^\theta = R_{\mathbf{T}}^{\mathbf{G}}(\theta')$; furthermore, $\varepsilon_{\mathbf{G}}\varepsilon_{\mathbf{T}} = (-1)^{l(w)}$ where $l(w)$ denotes the usual length of $w \in \mathbf{W}$; see [Ca85, 7.5.2]. Consequently, the degree formula in Theorem 2.2.12 translates to the formula

$$R_w^\theta(1) = (-1)^{l(w)}\frac{|\mathbf{G}^F|_{p'}}{|\mathbf{T}_0[w]|} = (-1)^{l(w)}\frac{|\mathbf{G}^F|_{p'}}{\det(q\,\mathrm{id}_{X_\mathbb{R}} - \varphi_0 \circ w^{-1})},$$

where the second equality holds by Lemma 1.6.6.

Example 2.3.23 If $\theta = 1$ is the trivial character of $\mathbf{T}_0[w]$, then we shall simply write R_w instead of R_w^1. Then the trivial character $1_{\mathbf{G}}$ and the Steinberg character $\mathrm{St}_{\mathbf{G}}$ of $\mathbf{G}^F$ (see Example 2.2.27) can be expressed as follows:

$$1_{\mathbf{G}} = \frac{1}{|\mathbf{W}|}\sum_{w\in\mathbf{W}} R_w \qquad \text{and} \qquad \mathrm{St}_{\mathbf{G}} = \frac{1}{|\mathbf{W}|}\sum_{w\in\mathbf{W}} (-1)^{l(w)} R_w.$$

We have $\langle R_w, 1_{\mathbf{G}}\rangle = 1$ and $\langle R_w, \mathrm{St}_{\mathbf{G}}\rangle = (-1)^{l(w)}$ for all $w \in \mathbf{W}$. The character of the regular representation of $\mathbf{G}^F$ is given by

$$\chi_{\mathrm{reg}} = \frac{1}{|\mathbf{W}|} \sum_{(w,\theta)\in\mathfrak{X}(\mathbf{W},\sigma)} R_w^\theta(1)\, R_w^\theta.$$

Compare with the formulae in Examples 2.2.26 and 2.2.27! The above formulations are taken from [DiMi20, 10.2.5 and 10.2.6]. Finally, let $\rho \in \mathrm{Irr}(\mathbf{G}^F)$. Taking the scalar product of ρ with the above expression for χ_{reg}, we obtain

$$\rho(1) = \frac{1}{|\mathbf{W}|} \sum_{(w,\theta)\in\mathfrak{X}(\mathbf{W},\sigma)} \langle R_w^\theta, \rho\rangle R_w^\theta(1),$$

which simplifies to $\rho(1) = \frac{1}{|\mathbf{W}|}\sum_{w\in\mathbf{W}}\langle R_w, \rho\rangle R_w(1)$ if ρ is unipotent.

Proposition 2.3.24 (See [Lu77b, 3.12]) *We have*

$$\sum_{\rho\in\mathrm{Uch}(\mathbf{G}^F)} \rho(1)\,\rho = \frac{1}{|\mathbf{W}|}\sum_{w\in\mathbf{W}} R_w(1)\, R_w.$$

Proof This follows from the identity

$$\sum_{\rho\in\mathrm{Irr}(\mathbf{G}^F)} \rho(1)\,\rho = \chi_{\mathrm{reg}} = \frac{1}{|\mathbf{W}|} \sum_{(w,\theta)\in\mathfrak{X}(\mathbf{W},\sigma)} R_w^\theta(1)\, R_w^\theta$$

and the Exclusion Theorem 2.3.2. (See also [Lu77a, Lemma 7.7].) □

Combining the formula for $\rho(1)$ in Example 2.3.23 with the degree formula for R_w^θ in Example 2.3.22(b), we are led to the following definition which, in this form, appeared in [Ge18, 3.3].

Definition 2.3.25 Let $\mathbb{G} = (\mathscr{R}, \varphi_0\mathbf{W})$ be the complete root datum associated with $\mathbf{G}, F$ (see Example 1.6.11). For any $\rho \in \mathrm{Irr}(\mathbf{G}^F)$, the corresponding *degree polynomial* is defined as

$$\mathbb{D}_\rho := \frac{1}{|\mathbf{W}|} \sum_{(w,\theta)\in\mathfrak{X}(\mathbf{W},\sigma)} (-1)^{l(w)}\langle R_w^\theta, \rho\rangle\, \mathbf{q}^{-N}\frac{|\mathbb{G}|}{|\mathbb{T}_w|} \in \mathbb{R}[\mathbf{q}],$$

where N is the number of positive roots, $|\mathbb{G}| \in \mathbb{R}[\mathbf{q}]$ is the order polynomial of $\mathbf{G}^F$ and $|\mathbb{T}_w| \in \mathbb{R}[\mathbf{q}]$ is the order polynomial of $\mathbf{T}_0[w]$; note that $\mathbf{q}^N|\mathbb{T}_w|$ divides $|\mathbb{G}|$ in $\mathbb{R}[\mathbf{q}]$ (see 1.6.21) and, hence, $\mathbb{D}_\rho$ is indeed a polynomial (and not just a rational function). Using the formula in Example 2.3.23, we see that the actual character degree $\rho(1)$ is obtained by evaluating the polynomial $\mathbb{D}_\rho$ at q; in particular, $\mathbb{D}_\rho \neq 0$. Analogous to the notation $|\mathbf{G}^F|_{p'}$ (for the p'-part of $|\mathbf{G}^F|$), we will also write $|\mathbb{G}|_{\mathbf{q}'}$ instead of $\mathbf{q}^{-N}|\mathbb{G}|$.

Remark 2.3.26 Let $\rho \in \mathrm{Irr}(\mathbf{G}^F)$. Using the degree polynomial $\mathbb{D}_\rho \in \mathbb{R}[\mathbf{q}]$, we can define numerical invariants of ρ as follows.

$$A_\rho := \text{degree of } \mathbb{D}_\rho,$$
$$a_\rho := \text{largest non-negative integer such that } \mathbf{q}^{a_\rho} \text{ divides } \mathbb{D}_\rho.$$

We obtain a further invariant $n_\rho \in \mathbb{R}_{>0}$ by the condition that

$$\mathbb{D}_\rho = \pm\frac{1}{n_\rho}\mathbf{q}^{a_\rho} + \text{combination of powers } \mathbf{q}^i \text{ where } i > a_\rho.$$

All we can say at this stage is that $0 \leqslant a_\rho \leqslant A_\rho \leqslant N$ (since $\mathbf{q}^{-N}|\mathbb{G}|/|\mathbb{T}_w| \in \mathbb{R}[\mathbf{q}]$ has degree N). The above invariants first appeared in [Lu79a, §8]; see also [Lu84a, 4.26]. They are an important tool for organising the irreducible characters into 'series' and 'families' (see Section 4.2 for a further discussion).

Remark 2.3.27 One can show that the degree polynomials behave in many ways like true character degrees. For example, by analogy with results known for character degrees, the following are true:

(a) $\mathbb{D}_\rho$ divides $|\mathbb{G}|$ in $\mathbb{R}[\mathbf{q}]$.
(b) If $\langle R_w^\theta, \rho\rangle \neq 0$, then $\mathbb{D}_\rho$ divides $|\mathbb{G}|/|\mathbb{T}_w|$ in $\mathbb{R}[\mathbf{q}]$ (cf. [Ge92, 2.5]).

See [Ge18, Remark 3.11] for some hints concerning the proofs. Furthermore, let $\mathbf{G} \subseteq \tilde{\mathbf{G}}$ be a regular embedding (see Section 1.7). Let $\tilde{\rho} \in \mathrm{Irr}(\tilde{\mathbf{G}}^F)$. By Theorem 1.7.15, we have $\tilde{\rho}|_{\mathbf{G}^F} = \rho_1 + \cdots + \rho_r$ with distinct irreducible characters ρ_i of $\mathbf{G}^F$. Since the ρ_i are conjugate under $\tilde{\mathbf{G}}^F$, we have $\rho_i(1) = \frac{1}{r}\tilde{\rho}(1)$ for all i. The analogous relation also holds for the degree polynomials:

(c) $\mathbb{D}_{\rho_i} = \frac{1}{r}\mathbb{D}_{\tilde{\rho}}$ for $i = 1, \ldots, r$ (see [Ge18, Lemma 6.5]).

Remark 2.3.28 Once the degree polynomials of all unipotent characters are determined, and the Jordan decomposition of characters is establised, one observes that $\mathbb{D}_\rho$, for any $\rho \in \mathrm{Irr}(\mathbf{G}^F)$, is of the form

$$\mathbb{D}_\rho = \frac{1}{n_\rho}\left(\pm\mathbf{q}^{a_\rho} + \cdots + \mathbf{q}^{A_\rho}\right), \qquad \text{where} \qquad n_\rho\mathbb{D}_\rho \in \mathbb{Z}[\mathbf{q}].$$

This will follow by a combination of Remark 2.3.27(c) (reduction to the case where $\mathbf{Z}(\mathbf{G})$ is connected), Proposition 2.5.11 (reduction to the case where ρ is unipotent), Remark 4.2.1 (further reduction to the case where $\mathbf{G}$ is simple) and, finally, by inspection of the tables of unipotent character degrees in the appendix of [Lu84a] (see also [Ca85, §13.8, §13.9]).

Remark 2.3.29 Let $\mathbb{G}$ be the complete root datum associated with $\mathbf{G}, F$, as above. We have a corresponding (infinite) series of finite groups of Lie type $\{\mathbb{G}(q') \mid q' \in$

$\mathscr{P}_{\mathbb{G}}\}$ as in Remark 1.6.12, and our given group $\mathbf{G}^F$ is isomorphic to the member $\mathbb{G}(q)$ of this series. Then it easily follows from Remark 2.3.28 that the set of all possible degree polynomials

$$\{\mathbb{D}_\rho \mid \rho \in \mathrm{Irr}(\mathbb{G}(q')) \text{ for some } q' \in \mathscr{P}_{\mathbb{G}}\}$$

is finite. By [Ge12a, 4.2], this can also be proved directly by a general argument.

Table 2.8 *The unipotent characters of* $\mathbf{G}^F = \mathrm{Sp}_4(q)$

	$\mathbf{T}_1$	$\mathbf{T}_2$	$\mathbf{T}_3$	$\mathbf{T}_4$	$\mathbf{T}_5$
$\mathbf{W}$-class	$\{s_1s_2, s_2s_1\}$	$\{s_2, s_1s_2s_1\}$	$\{1\}$	$\{s_1s_2s_1s_2\}$	$\{s_1, s_2s_1s_2\}$
$\lvert\mathbf{T}_i^F\rvert$	q^2+1	q^2-1	$(q-1)^2$	$(q+1)^2$	q^2-1

$\mathbb{D}_{\theta_0} = 1$	$R_{s_1s_2}$	$= \theta_0 - \theta_9 + \theta_{10} + \theta_{13},$
$\mathbb{D}_{\theta_9} = \frac{1}{2}\mathbf{q}(\mathbf{q}+1)^2$	R_{s_2}	$= \theta_0 + \theta_{11} - \theta_{12} - \theta_{13},$
$\mathbb{D}_{\theta_{10}} = \frac{1}{2}\mathbf{q}(\mathbf{q}-1)^2$	R_1	$= \theta_0 + 2\theta_9 + \theta_{11} + \theta_{12} + \theta_{13},$
$\mathbb{D}_{\theta_{11}} = \mathbb{D}_{\theta_{12}} = \frac{1}{2}\mathbf{q}(\mathbf{q}^2+1)$	$R_{s_1s_2s_1s_2}$	$= \theta_0 - 2\theta_{10} - \theta_{11} - \theta_{12} + \theta_{13},$
$\mathbb{D}_{\theta_{13}} = \mathbf{q}^4$	R_{s_1}	$= \theta_0 - \theta_{11} + \theta_{12} - \theta_{13}.$

Example 2.3.30 Let $\mathbf{G}^F = \mathrm{Sp}_4(q)$ where $|\mathbf{G}^F| = q^4(q^2-1)^2(q^2+1)$. The complete character table (for q odd) was determined in [Sr68] (see [Pr82] for the correction of some minor errors). The unipotent characters are explicitly described in [Sr91, A.1] in terms of the Deligne–Lusztig characters $R_{\mathbf{T}}^{\mathbf{G}}(1_{\mathbf{T}})$; see [Eno72] for q a power of 2. The Weyl group $\mathbf{W} = \langle s_1, s_2\rangle$ is dihedral of order 8 (with F acting trivially on $\mathbf{W}$), where we fix the notation so that s_1 corresponds to a long root and s_2 to a short root. Corresponding to the five conjugacy classes of $\mathbf{W}$, there are five $\mathbf{G}^F$-conjugacy classes of F-stable maximal tori in $\mathbf{G}$. As in [Sr91], we denote representatives by $\mathbf{T}_i$ ($1 \leqslant i \leqslant 5$). There are six unipotent characters, denoted by $\theta_0 = 1_{\mathbf{G}}, \theta_9, \theta_{10}, \theta_{11}, \theta_{12}$ and $\theta_{13} = \mathrm{St}_{\mathbf{G}}$ in [Sr68]. This information, plus the decomposition of the $R_{\mathbf{T}_i}^{\mathbf{G}}(1_{\mathbf{T}_i})$ into unipotent characters, is contained in Table 2.8. Under the correspondence in Example 2.3.9, and using the notation for the characters of $\mathbf{W}$ in Example 2.1.12, we have:

$$\theta_0 \leftrightarrow 1_{\mathbf{W}}, \qquad \theta_9 \leftrightarrow \phi_1, \qquad \theta_{11} \leftrightarrow \varepsilon', \qquad \theta_{12} \leftrightarrow \varepsilon'', \qquad \theta_{13} \leftrightarrow \varepsilon.$$

Finally, the notation is different in [Eno72], but the information in Table 2.8 remains valid for q even. See also [Shi82] where the character table of the conformal symplectic group $\mathrm{CSp}_4(q)$ (for q odd) is determined.

Example 2.3.31 Let $\mathbf{G}^F = {}^2B_2(q^2)$ be a Suzuki group, where $q = \sqrt{2}^{2m+1}$ for some $m \in \mathbb{Z}_{\geqslant 0}$; see Example 1.4.22. We have $|\mathbf{G}^F| = q^4(q^2-1)(q^4+1)$. The known

Table 2.9 *Values of unipotent characters of* $\mathbf{G}^F = {}^2B_2(q^2)$

g	1	u'	u_0	u_0^{-1}	π_0^a	π_1^b	π_2^c
$\lvert C_{\mathbf{G}}(g)^F\rvert$	$\lvert\mathbf{G}^F\rvert$	q^4	$2q^2$	$2q^2$	$\lvert\mathbf{T}_0^F\rvert$	$\lvert\mathbf{T}_1^F\rvert$	$\lvert\mathbf{T}_2^F\rvert$
$1_{\mathbf{G}}$	1	1	1	1	1	1	1
ϖ	$\frac{1}{2}\sqrt{2}\,q(q^2-1)$	$-\frac{1}{2}\sqrt{2}\,q$	$\frac{1}{2}\sqrt{-2}\,q$	$-\frac{1}{2}\sqrt{-2}\,q$	.	1	-1
ϖ'	$\frac{1}{2}\sqrt{2}\,q(q^2-1)$	$-\frac{1}{2}\sqrt{2}\,q$	$-\frac{1}{2}\sqrt{-2}\,q$	$\frac{1}{2}\sqrt{-2}\,q$	.	1	-1
$\mathrm{St}_{\mathbf{G}}$	q^4	.	.	.	1	-1	-1

$\mathbf{T}_0$ is of type $1 \in \mathbf{W}$ and $\lvert\mathbf{T}_0^F\rvert = q^2-1$; $\quad R_1 = 1_{\mathbf{G}} + \mathrm{St}_{\mathbf{G}}$.
$\mathbf{T}_1$ is of type $s_1s_2s_1$ and $\lvert\mathbf{T}_1^F\rvert = q^2+\sqrt{2}q+1$; $\quad R_{s_1s_2s_1} = 1_{\mathbf{G}} + \varpi + \varpi' - \mathrm{St}_{\mathbf{G}}$.
$\mathbf{T}_2$ is of type s_1 and $\lvert\mathbf{T}_2^F\rvert = q^2-\sqrt{2}q+1$; $\quad R_{s_1} = 1_{\mathbf{G}} - \varpi - \varpi' - \mathrm{St}_{\mathbf{G}}$.

character table in [Suz62] has been re-interpreted in [Ge03a, §4.6] in terms of the Deligne–Lusztig characters $R_{\mathbf{T}}^{\mathbf{G}}(\theta)$. The Weyl group $\mathbf{W} = \langle s_1, s_2\rangle$ is again dihedral of order 8 but now F induces an automorphism $\sigma\colon \mathbf{W} \to \mathbf{W}$ such that $\sigma(s_1) = s_2$ and $\sigma(s_2) = s_1$. Corresponding to the three σ-conjugacy classes of $\mathbf{W}$ in Example 2.1.12, there are three $\mathbf{G}^F$-conjugacy classes of F-stable maximal tori in $\mathbf{G}$; representatives will be denoted by $\mathbf{T}_0, \mathbf{T}_1, \mathbf{T}_2$. We have $\mathrm{Uch}(\mathbf{G}^F) = \{1_{\mathbf{G}}, \mathrm{St}_{\mathbf{G}}, \varpi, \varpi'\}$, where ϖ, ϖ' are complex conjugate to each other. The values of the unipotent characters are given in Table 2.9, where u', u_0 are unipotent elements; furthermore, $\pi_0^a, \pi_1^b, \pi_2^c$ are regular semisimple and a, b, c run over certain index ranges that we do not specify here. (There are q^2+3 conjugacy classes.) The degree polynomials are given by the polynomial expressions in Table 2.9. We have

$$a_\varpi = a_{\varpi'} = 1, \qquad A_\varpi = A_{\varpi'} = 3 \qquad \text{and} \qquad n_\varpi = n_{\varpi'} = \sqrt{2}.$$

Thus, the invariant n_ρ is not always an integer. Finally, as in Example 2.2.30, we note that the values of the linear combination $\frac{1}{2}\sqrt{2}(\varpi - \varpi')$ take a particularly simple form:

	1	u'	u_0	u_0^{-1}	$\{\pi_0^a\}$	$\{\pi_1^b\}$	$\{\pi_2^c\}$
$\frac{1}{2}\sqrt{2}(\varpi - \varpi')$	.	.	$\sqrt{-1}\,q$	$-\sqrt{-1}\,q$	.	.	.

If we consider the example where $q = \sqrt{2}^3$, then the complete character table is printed in Table 2.1 (p. 95). By just looking at character degrees, we see that $\{\mathtt{X.2}, \mathtt{X.3}\} = \{\varpi, \varpi'\}$. In order to match these characters exactly, one would need to specify actual representatives of the classes labelled by 4a, 4b.

Example 2.3.32 Let $\mathbf{G}^F = {}^2F_4(q^2)$ be a Ree group of type F_4, where $q = \sqrt{2}^{2m+1}$ for some $m \in \mathbb{Z}_{\geqslant 0}$; see Example 1.4.22. By [Lu84a, §14.2], there is a unique unipotent character $\rho \in \mathrm{Uch}(\mathbf{G}^F)$ such that

$$\mathbb{D}_\rho = \tfrac{1}{3}\mathbf{q}^4(\mathbf{q}^2-1)^2(\mathbf{q}^4-\mathbf{q}^2+1)(\mathbf{q}^8-\mathbf{q}^4+1) \quad \text{and} \quad \langle R_{\mathbf{T}}^{\mathbf{G}}(1_{\mathbf{T}}), \rho\rangle \in 2\mathbb{Z}$$

for any F-stable maximal torus $\mathbf{T} \subseteq \mathbf{G}$. In particular, we have $a_\rho = 4$, $A_\rho = 20$ and $n_\rho = 3$. Somewhat related to this particular behaviour is the fact that ρ is rational valued but its Frobenius–Schur indicator is -1 and, hence, it can not be realised by a representation over $\mathbb{Q}$ (see [Ge03b, §7]).

2.4 Towards Lusztig's Main Theorem 4.23

In the previous section, we introduced the partition of $\mathrm{Irr}(\mathbf{G}^F)$ into geometric series, where the unipotent characters $\mathrm{Uch}(\mathbf{G}^F)$ appear to play a special role. The determination of that partition essentially relies on knowing the scalar products $\langle R_w^\theta, \rho\rangle$ for all $\rho \in \mathrm{Irr}(\mathbf{G}^F)$, all $w \in \mathbf{W}$ and all $\theta \in \mathrm{Irr}(\mathbf{T}_0[w])$. One may hope that this information—once available—would then lead to:

- a parametrisation of the characters inside a geometric series and
- explicit formulae for character degrees and character values, at least on semi-simple elements (via Proposition 2.2.18).

This hope has been fully realised in Lusztig's work, where the book [Lu84a] develops the main results assuming that $\mathbf{Z}(\mathbf{G})$ is connected and [Lu88] extends this to the general case (via regular embeddings and the Multiplicity-Freeness Theorem 1.7.15). It is the purpose of this section to introduce the basic formalism of [Lu84a] (with complete proofs for a number of intermediate results), and to explain the 'Main Theorem 4.23' in [Lu84a, p. 131]. (See also the survey [Ca95].)

In non-technical terms, that main result yields explicitly computable formulae for the multiplicities $\langle R_w^\theta, \rho\rangle$ and a *Jordan decomposition of characters* by which the characters in an arbitrary series of $\mathrm{Irr}(\mathbf{G}^F)$ are put in bijection with the unipotent characters of a (usually) smaller group. As far as the unipotent characters themselves are concerned, there is a purely combinatorial classification which is 'independent of q', as follows.

Theorem 2.4.1 (Lusztig [Lu84a, 4.23]) *There exist a finite set $\bar{X}(\mathbf{W}, \sigma)$ and a collection of integers $\{m(w, \bar{x}) \mid w \in \mathbf{W}, \bar{x} \in \bar{X}(\mathbf{W}, \sigma)\} \subseteq \mathbb{Z}$ (both depending only on $\mathbf{W}$ and σ) such that the following holds. There is a bijection*

$$\mathrm{Uch}(\mathbf{G}^F) \overset{1-1}{\longleftrightarrow} \bar{X}(\mathbf{W}, \sigma), \qquad \rho \leftrightarrow \bar{x}_\rho,$$

such that $\langle R_w, \rho\rangle = m(w, \bar{x}_\rho)$ for all $w \in \mathbf{W}$ and $\rho \in \mathrm{Uch}(\mathbf{G}^F)$. In particular, for *each $\rho \in \mathrm{Uch}(\mathbf{G}^F)$, the corresponding degree polynomial is given by*

$$\mathbb{D}_\rho = \frac{1}{|\mathbf{W}|} \sum_{w \in \mathbf{W}} (-1)^{l(w)} m(w, \bar{x}_\rho) \frac{|\mathbb{G}|_{\mathbf{q}'}}{|\mathbb{T}_w|}.$$

Remark 2.4.2 (a) Using Proposition 2.3.15 and the arguments in [Lu76c, 1.18] (see also Remark 4.2.1 in Chapter 4), the proof of Theorem 2.4.1 can be reduced to the case where $\mathbf{G}$ is simple of adjoint type. In this case, and assuming that q is sufficiently large, the above result had already been obtained in earlier papers by Lusztig; see [Lu80a], [Lu81c], [Lu82b] (and [Lu84a, Appendix] for the Suzuki and Ree groups). Then [Lu84a, 4.23] shows that these results remain valid for any q. A more conceptual explanation for the fact that $\mathrm{Uch}(\mathbf{G}^F)$ is classified 'independently of q' is provided by [Lu14b]. It even makes sense to define 'exotic' parameter sets and corresponding degree polynomials for any (not necessarily crystallographic) finite Coxeter group (see [Lu93], [Lu94]). This is the starting point for the 'spetses' philosophy in [BMM93], [BMM99], [BMM14].

(b) Let $\bar{x}_1 \in \bar{X}(\mathbf{W}, \sigma)$ correspond to the trivial character of $\mathbf{G}^F$. Then, by Example 2.2.26, we have $m(w, \bar{x}_1) = 1$ for all $w \in \mathbf{W}$ and one easily sees that $\bar{x}_1$ is uniquely determined by this condition (see, e.g., [Ge18, 4.9]). In general, the correspondence $\rho \leftrightarrow \bar{x}_\rho$ is not uniquely determined by the above properties. For example, the two unipotent characters ϖ, ϖ' of $\mathbf{G}^F = {}^2B_2(q^2)$ in Example 2.3.31 cannot be distinguished by their multiplicities in the various R_w. (A similar thing necessarily happens whenever there exist unipotent characters that are not rational valued, which is the case in almost all groups of exceptional type.) These uniqueness issues will also be discussed in further detail in Chapter 4.

Remark 2.4.3 The explicit description of the sets $\bar{X}(\mathbf{W}, \sigma)$ occupies almost all of [Lu84a, Chap. 4]. This is a formidable piece of technical and combinatorial machinery. It proceeds in various stages starting with the case where $\mathbf{W}$ is irreducible and $\sigma = \mathrm{id}_{\mathbf{W}}$, then going on to the case where σ is non-trivial but ordinary, and finally presenting reduction arguments for the general case. (For the case where σ is not ordinary, see [Lu84a, §14.2].) For a somewhat different parametrisation, see [Lu14a, §3], [Lu14c, §18]. (This is also described in [Ge18, §4].) The integers $m(w, \bar{x})$ are not directly described in [Lu84a] but, assuming that σ is ordinary, in terms of the σ-character table of $\mathbf{W}$ and a certain 'non-abelian Fourier matrix', with rows indexed by $\bar{X}(\mathbf{W}, \sigma)$ and columns indexed by a second set $X(\mathbf{W}, \sigma)$; see Section 4.2 for further details. (In the case where $\mathbf{W}$ is irreducible and σ is not ordinary, the integers $m(w, \bar{x})$ are printed in [Lu84a, Appendix].)

Thus, given any $\mathbf{G}$, F with corresponding $\mathbf{W}$, σ, the set $\bar{X}(\mathbf{W}, \sigma)$ and the integers $m(w, \bar{x})$ can be worked out explicitly, by an entirely combinatorial procedure. This is electronically available in Michel's version of CHEVIE [MiChv], through the functions

`UnipotentCharacters` and `DeligneLusztigCharacter`

which take as input an arbitrary finite Weyl group and an automorphism. The first function displays the parameter set $\bar{X}(\mathbf{W}, \sigma)$ (including the corresponding degree polynomials) and the second function returns, for any $w \in \mathbf{W}$, the decomposition of the Deligne–Lusztig character R_w into unipotent characters; see the online help for further information and examples. (E.g., one can immediately recover the formulae in Examples 2.3.30, 2.3.31, 2.3.32 in this way. One can even use as input for the function `UnipotentCharacters` a finite Coxeter group of non-crystallographic type $I_2(m)$, H_3 or H_4, in which case one obtains the 'exotic' parameter sets and degree polynomials in [Lu93].) Given the amount of complicated mathematics involved in them, the above two functions may be regarded as a highlight of modern computer algebra techniques!

Our next task is to explain, following [Lu84a], how the characters in an arbitrary geometric series of $\mathrm{Irr}(\mathbf{G}^F)$ can be described in terms of the information available for unipotent characters in Theorem 2.4.1 (for $\mathbf{G}$ itself and further, usually smaller groups).

2.4.4 As in [DeLu76, §5], we fix some choices of a purely number-theoretic nature that allow us to connect the multiplicative group of $k = \overline{\mathbb{F}}_p$ with roots of unity in $\mathbb{C}$. It is known that $k^\times$ is (non-canonically) isomorphic to $(\mathbb{Q}/\mathbb{Z})_{p'}$, the group of all elements of $\mathbb{Q}/\mathbb{Z}$ of order prime to p; see, e.g., [Ca85, 3.1.3], [Hum95, §2.7]. We fix once and for all a group isomorphism

$$\iota\colon k^\times \xrightarrow{\sim} (\mathbb{Q}/\mathbb{Z})_{p'}.$$

Furthermore, the exponential map induces a group isomorphism

$$\exp\colon (\mathbb{Q}/\mathbb{Z})_{p'} \xrightarrow{\sim} \mu_{p'}, \qquad x + \mathbb{Z} \mapsto \exp(2\pi i x),$$

where $\mu_{p'} := \{z \in \mathbb{C} \mid z^n = 1 \text{ for some } n \in \mathbb{Z}_{\geqslant 1} \text{ such that } p \nmid n\}$ is the group of all roots of unity in $\mathbb{C}$ of order prime to p. Composing these two isomorphisms, we obtain a group isomorphism

$$\psi\colon k^\times \xrightarrow{\sim} \mu_{p'} \subseteq \mathbb{C}^\times \qquad (\psi = \exp \circ \iota).$$

The following constructions will depend on these choices[3].

2.4.5 We now introduce the basic set-up of [Lu84a, 2.1][4]. Let us consider a pair (λ, n) where $\lambda \in X = X(\mathbf{T}_0)$ and $n \geqslant 1$ is an integer prime to p. (Here, $\mathbf{T}_0$ is our

[3] Isomorphisms ι, ψ as above can be obtained from a realisation of k as $k = \mathbb{A}/\mathfrak{p}$ where $\mathbb{A}$ is the ring of algebraic integers in $\mathbb{C}$ and $\mathfrak{p}$ is a maximal ideal of $\mathbb{A}$ such that $p \in \mathfrak{p}$. Definite choices would arise from a construction of k via Conway polynomials; see [LuPa10, §4.2].

[4] Lusztig actually works with line bundles L over $\mathbf{G}/\mathbf{B}_0$ instead of characters $\lambda \in X(\mathbf{T}_0)$, but one can pass from one to the other as described in [Lu84a, 1.3.2].

fixed maximally split torus of $\mathbf{G}$, as in 2.3.18.) Let $\mathscr{Z}_{\lambda,n}$ be the set of all $w \in \mathbf{W}$ for which there exists some $\lambda_w \in X$ such that

$$\lambda_w(t^n) = \lambda\big(F(t)\dot{w}^{-1}t^{-1}\dot{w}\big) \qquad \text{for all } t \in \mathbf{T}_0. \tag{$\spadesuit$}$$

Note that the character λ_w, if it exists, is uniquely determined by w (since $k^\times = \{\xi^n \mid \xi \in k^\times\}$ and, hence, also $\mathbf{T}_0 = \{t^n \mid t \in \mathbf{T}_0\}$). Also note that, using additive notation in X, we can re-phrase ($\spadesuit$) as $n\lambda_w = \lambda \circ F - w.\lambda$. Assume now that $\mathscr{Z}_{\lambda,n} \neq \varnothing$. Then, for any $w \in \mathscr{Z}_{\lambda,n}$, the restriction of λ_w to $\mathbf{T}_0[w]$ is a group homomorphism

$$\bar{\lambda}_w \colon \mathbf{T}_0[w] \to k^\times \qquad \text{such that} \qquad \bar{\lambda}_w(t)^n = 1 \quad (t \in \mathbf{T}_0[w]).$$

Using our chosen isomorphism $\psi \colon k^\times \xrightarrow{\sim} \mu_{p'} \subseteq \mathbb{C}^\times$ from 2.4.4, we obtain a linear character $\theta_w := \psi \circ \bar{\lambda}_w \in \mathrm{Irr}(\mathbf{T}_0[w])$, such that $\theta_w(t)^n = 1$ for all $t \in \mathbf{T}_0[w]$. If $w \in \mathscr{Z}_{\lambda,n}$ and $\theta = \theta_w \in \mathrm{Irr}(\mathbf{T}_0[w])$, then we will simply write $(\lambda, n) \leftrightsquigarrow (w, \theta)$ in the following. We shall see that this relation provides an efficient tool for dealing with geometric conjugacy in a way that does not involve the norm map.

Definition 2.4.6 (Lusztig) Let (λ, n) be a pair as in 2.4.5, with $\mathscr{Z}_{\lambda,n} \neq \varnothing$. We define a corresponding subset of $\mathrm{Irr}(\mathbf{G}^F)$ by

$$\mathscr{E}_{\lambda,n} := \{\rho \in \mathrm{Irr}(\mathbf{G}^F) \mid \langle R_w^{\theta_w}, \rho\rangle \neq 0 \text{ for some } w \in \mathscr{Z}_{\lambda,n}\}.$$

(Note that the $\mathscr{E}_{\lambda,n}$ are originally introduced in [Lu84a, 2.19], assuming that $\mathbf{Z}(\mathbf{G})$ is connected, and using individual cohomology spaces instead of the characters $R_w^{\theta_w}$. Then it is shown in [Lu84a, 6.5] that the original definition is equivalent to the one above. Here, the above definition in terms of virtual characters will work for all our purposes[5], without any assumption on $\mathbf{Z}(\mathbf{G})$.)

Example 2.4.7 (a) Let $n = 1$. Let $\lambda_0 \in X$ be the neutral element, that is, $\lambda_0(t) = 1$ for all $t \in \mathbf{T}_0$. For $w \in \mathbf{W}$, we can just set $\lambda_w := \lambda_0$ and then ($\spadesuit$) in 2.4.5 holds. Hence, $\mathscr{Z}_{\lambda_0,1} = \mathbf{W}$. For any $w \in \mathbf{W}$, the restriction of λ_w to $\mathbf{T}_0[w]$ is trivial and so $\theta_w = 1$ is the trivial character of $\mathbf{T}_0[w]$. Hence, we have $(\lambda_0, 1) \leftrightsquigarrow (w, 1)$ for all $w \in \mathbf{W}$. This shows that $\mathscr{E}_{\lambda_0,1} = \mathrm{Uch}(\mathbf{G}^F)$ is the set of unipotent characters.

(b) Let $\mathbf{G}^F = \mathrm{SL}_2(q)$ where q is odd. As in Example 2.2.30, we write $\mathbf{W} = \{1, s\}$ and $\mathbf{T}_0 = \{S(\xi) \mid \xi \in k^\times\}$ where, for any $\xi \in k^\times$, we denote by $S(\xi)$ the diagonal matrix with diagonal entries ξ, ξ^{-1}. Let $n = 2$ and define $\lambda \in X$ by $\lambda(S(\xi)) = \xi$ for all $\xi \in k^\times$. Now note that, if $t \in \mathbf{T}_0$, then $t^{q-1} = F(t)t^{-1}$ and $t^{q+1} = F(t)\dot{s}^{-1}t^{-1}\dot{s}$. Thus, if we define $\lambda_1, \lambda_s \in X$ by

$$\lambda_1(S(\xi)) = \xi^{(q-1)/2} \quad \text{and} \quad \lambda_s(S(\xi)) = \xi^{(q+1)/2} \qquad \text{for all } \xi \in k^\times,$$

then ($\spadesuit$) holds for $w = 1$ and $w = s$, respectively. So $\mathscr{Z}_{\lambda,2} = \mathbf{W}$. We have $(\lambda, 2) \leftrightsquigarrow$

[5] We will see in Corollary 2.4.29 below that the sets $\mathscr{E}_{\lambda,n}$ are precisely the geometric series of characters as introduced in the previous section.

$(1, \theta_1)$ and $(\lambda, 2) \leftrightsquigarrow (s, \theta_s)$ where $\theta_1 \in \mathrm{Irr}(\mathbf{T}_0[1])$ and $\theta_s \in \mathrm{Irr}(\mathbf{T}_0[s])$ are the unique non-trivial characters of order 2. Thus, we obtain

$$\mathscr{E}_{\lambda,2} = \{\psi'_+, \psi''_+, \psi'_-, \psi''_-\}.$$

(We will see in Example 2.4.25 that $(1, \theta_1)$ and (s, θ_s) are geometrically conjugate and, hence, $\psi'_+, \psi''_+, \psi'_-, \psi''_-$ are geometrically conjugate.)

Lemma 2.4.8 (Cf. [Lu84a, 6.2(iii)]) *Let $w \in \mathbf{W}$ and $\theta \in \mathrm{Irr}(\mathbf{T}_0[w])$. Let $n \geqslant 1$ be any integer prime to p such that $\theta(t)^n = 1$ for all $t \in \mathbf{T}_0[w]$. Then there exists some $\lambda \in X$ such that $w \in \mathscr{Z}_{\lambda,n}$ and $(\lambda, n) \leftrightsquigarrow (w, \theta)$. If $\mu \in X$ also satisfies these conditions, then $\mu = \lambda + n\nu$ for some $\nu \in X$.*

Proof Define $F' \colon \mathbf{T}_0 \to \mathbf{T}_0$ by $F'(t) = \dot{w}F(t)\dot{w}^{-1}$ for all $t \in \mathbf{T}_0$. Then $\mathbf{T}_0[w] = \mathbf{T}_0^{F'}$ and, as in the proof of Lemma 1.6.6, we have a surjective map

$$X(\mathbf{T}_0) \to \mathrm{Hom}(\mathbf{T}_0^{F'}, k^\times), \qquad \lambda \mapsto \lambda|_{\mathbf{T}_0^{F'}}, \tag{$*$}$$

with kernel $\{\lambda \circ F' - \lambda \mid \lambda \in X\}$. Now, the values of θ are nth roots of unity in $\mathbb{C}^\times$. Hence, using the isomorphism $\psi \colon k^\times \xrightarrow{\sim} \mu_{p'}$, we can write uniquely $\theta = \psi \circ \bar{\theta}$ where $\bar{\theta} \colon \mathbf{T}_0^{F'} \to k^\times$ is a group homomorphism. By $(*)$, we know that $\bar{\theta}$ is the restriction of some $\lambda_1 \in X$. Now θ^n is the trivial character and so $n\lambda_1$ is in the kernel of the map in $(*)$. Hence, there exists some $\lambda' \in X$ such that $\lambda' \circ F' - \lambda' = n\lambda_1$. Setting $\lambda := w^{-1}.\lambda' \in X$, we obtain

$$\lambda_1(t^n) = (n\lambda_1)(t) = \lambda'(F'(t)t^{-1}) = \lambda(F(t)\dot{w}^{-1}t^{-1}\dot{w}) \qquad \text{for all } t \in \mathbf{T}_0,$$

that is, we have $w \in \mathscr{Z}_{\lambda,n}$ and $(\lambda, n) \leftrightsquigarrow (w, \theta)$, as desired. Now assume that $\mu \in X$ also satisfies these conditions, that is, we have

$$\mu_1(t^n) = \mu(F(t)\dot{w}^{-1}t^{-1}\dot{w}) \qquad \text{for all } t \in \mathbf{T}_0,$$

where $\mu_1 \in X$ is such that $\bar{\theta}$ is the restriction of μ_1. Then the restriction of $\lambda_1 - \mu_1$ to $\mathbf{T}_0^{F'}$ is trivial and so $\mu_1 - \lambda_1 = \nu' \circ F' - \nu'$ for some $\nu' \in X$. Setting $\nu := w^{-1}.\nu' \in X$, it follows that

$$\begin{aligned}\mu\big(F(t)\dot{w}^{-1}t^{-1}\dot{w}\big) &= \mu_1(t^n) = \lambda_1(t^n)\nu'\big(F'(t^n)(t^n)^{-1}\big)\\ &= \lambda\big(F(t)\dot{w}^{-1}t^{-1}\dot{w}\big)\nu'\big(F'(t)t^{-1})\big)^n = \lambda\big(F(t)\dot{w}^{-1}t^{-1}\dot{w}\big)\nu\big(F(t)\dot{w}^{-1}t^{-1}\dot{w}\big)^n\end{aligned}$$

for $t \in \mathbf{T}_0$. But the map $\mathbf{T}_0 \to \mathbf{T}_0$, $t \mapsto F(t)\dot{w}^{-1}t^{-1}\dot{w}$, is surjective by the Lang–Steinberg theorem. Hence, we conclude that $\mu(t) = \lambda(t)\nu(t)^n$ for all $t \in \mathbf{T}_0$, that is, $\mu = \lambda + n\nu$. □

Remark 2.4.9 Let (λ, n) be a pair as in 2.4.5, with $\mathscr{Z}_{\lambda,n} \neq \varnothing$. Let $n' \geqslant 1$ also be an integer prime to p. If $w \in \mathscr{Z}_{\lambda,n}$ and $t \in \mathbf{T}_0$, then

$$\lambda_w(t^{nn'}) = \lambda_w\big((t^{n'})^n\big) = \lambda\big(F(t^{n'})\dot{w}^{-1}(t^{n'})^{-1}\dot{w}\big) = \lambda\big(F(t)\dot{w}^{-1}t^{-1}\dot{w}\big)^{n'}.$$

So $w \in \mathscr{Z}_{n'\lambda,nn'}$ and the two pairs (λ, n) and $(n'\lambda, nn')$ give rise to the same $\lambda_w \in X$ and, hence, to the same character $\theta_w \in \mathrm{Irr}(\mathbf{T}_0[w])$. Thus,

$$(\lambda, n) \leftrightsquigarrow (w, \theta) \quad \Rightarrow \quad (n'\lambda, nn') \leftrightsquigarrow (w, \theta).$$

One checks that the reverse implication also holds and so $\mathscr{Z}_{\lambda,n} = \mathscr{Z}_{n'\lambda,nn'}$. As in [Lu84a, 6.1], we say that (λ, n) is *indivisible* if it is impossible to write $\lambda = n_1'\lambda_1$ where $n = n_1 n_1'$ with $n_1 \geqslant 1$ and $n_1' \geqslant 2$. Thus, in the definition of $\mathscr{E}_{\lambda,n}$ we may assume without loss of generality that (λ, n) is indivisible.

Remark 2.4.10 Assume that $(\lambda, n) \leftrightsquigarrow (w, \theta)$. Then (λ, n) is indivisible if and only if $n \geqslant 1$ is the smallest integer such that $\theta(t)^n = 1$ for all $t \in \mathbf{T}_0[w]$.

Indeed, the assumption $(\lambda, n) \leftrightsquigarrow (w, \theta)$ implies, in particular, that $\theta(t)^n = 1$ for all $t \in \mathbf{T}_0[w]$. So, if $m \geqslant 1$ is the smallest integer such that $\theta(t)^m = 1$ for all $t \in \mathbf{T}_0[w]$, then $n = mn'$ for some integer $n' \geqslant 1$. By Lemma 2.4.8, there exists some $\mu \in X$ such that $(\mu, m) \leftrightsquigarrow (w, \theta)$. Then we also have $(n'\mu, n) \leftrightsquigarrow (w, \theta)$; see Remark 2.4.9. Again by Lemma 2.4.8, we have $n'\mu - \lambda = mn'\nu$ for some $\nu \in X$. Hence, (λ, n) is indivisible if and only if $n' = 1$.

Corollary 2.4.11 *If $\rho \in \mathrm{Irr}(\mathbf{G}^F)$, then there exists an indivisible pair (λ, n) as in 2.4.5 such that $\mathscr{Z}_{\lambda,n} \neq \varnothing$ and $\rho \in \mathscr{E}_{\lambda,n}$.*

Proof By Corollary 2.2.19 and Lemma 2.3.19, there exists some pair $(w, \theta) \in \mathfrak{X}(\mathbf{W}, \sigma)$ such that $\langle R_w^\theta, \rho\rangle \neq 0$. Let $n \geqslant 1$ be the smallest integer such that $\theta(t)^n = 1$ for all $t \in \mathbf{T}_0[w]$; then n is prime to p. By Lemma 2.4.8, there exists a pair (λ, n) as in 2.4.5 such that $(\lambda, n) \leftrightsquigarrow (w, \theta)$ and so $\rho \in \mathscr{E}_{\lambda,n}$. By Remark 2.4.10, the pair (λ, n) is indivisible. □

Lemma 2.4.12 *Let (λ, n) be a pair as in 2.4.5, with $\mathscr{Z}_{\lambda,n} \neq \varnothing$. Then $\mathscr{Z}_{\lambda,n}$ is a coset with respect to the subgroup*

$$\hat{\mathbf{W}}_{\lambda,n} := \{w \in \mathbf{W} \mid w.\lambda - \lambda \in nX\} \subseteq \mathbf{W}.$$

Proof Let $w, w' \in \mathscr{Z}_{\lambda,n}$. Using additive notation in X, we have $n\lambda_w = \lambda \circ F - w.\lambda$ and $n\lambda_{w'} = \lambda \circ F - w'.\lambda$. Setting $x := w^{-1}w'$, it follows that

$$x.\lambda - \lambda = w^{-1}.\big(w'.\lambda - w.\lambda\big) = n(w^{-1}.\lambda_{w'} - w^{-1}.\lambda_w) \in nX.$$

Thus, $x \in \hat{\mathbf{W}}_{\lambda,n}$ and so $\mathscr{Z}_{\lambda,n}$ is contained in the coset $w\hat{\mathbf{W}}_{\lambda,n}$. Now consider any $y \in \hat{\mathbf{W}}_{\lambda,n}$. Then $\lambda(F(t)\dot{y}^{-1}\dot{w}^{-1}t^{-1}\dot{w}\dot{y}) = \lambda(F(t))\big((wy).\lambda\big)(t^{-1})$. Writing $y.\lambda = \lambda + n\nu$ where $\nu \in X$, we also have

$$(wy).\lambda = w.(y.\lambda) = w.(\lambda + n\nu) = w.\lambda + nw.\nu = \lambda \circ F - n\lambda_w + nw.\nu$$

and so $\lambda \circ F - (wy).\lambda = n(\lambda_w - w, \nu) \in nX$. Thus, $wy \in \mathscr{Z}_{\lambda,n}$. □

2.4.13 Let (λ, n) be a pair as in 2.4.5, with $\mathscr{Z}_{\lambda,n} \neq \varnothing$. In general, $\hat{\mathbf{W}}_{\lambda,n}$ is not a reflection subgroup of $\mathbf{W}$. But one can always find a reflection subgroup inside $\hat{\mathbf{W}}_{\lambda,n}$, as follows. As in [Lu84a, 1.8], we define a subgroup by

$$\mathbf{W}_{\lambda,n} := \{w \in \mathbf{W} \mid w.\lambda - \lambda \in n\mathbb{Z}R\} \subseteq \hat{\mathbf{W}}_{\lambda,n},$$

where $R \subseteq X$ is the set of roots of $\mathbf{G}$ with respect to $\mathbf{T}_0$. We also set

$$R_{\lambda,n} := \{\alpha \in R \mid \langle \lambda, \alpha^\vee \rangle \in \mathbb{Z} \text{ is divisible by } n\}.$$

Note that, for $\alpha \in R$, we have $w_\alpha.\lambda - \lambda = -\langle \lambda, \alpha^\vee \rangle \alpha$. Hence, one easily sees that $w_\alpha \in \mathbf{W}_{\lambda,n}$ if and only if $\alpha \in R_{\lambda,n}$. Thus, $\{w_\alpha \mid \alpha \in R_{\lambda,n}\}$ is precisely the set of all reflections of $\mathbf{W}$ that are contained in $\mathbf{W}_{\lambda,n}$. Now, it is known that $\mathbf{W}_{\lambda,n}$ is generated by $\{w_\alpha \mid \alpha \in R_{\lambda,n}\}$; see [Bou68, Chap. VI, Exc. 1 of §2] and [Hum95, 2.10]. So $\mathbf{W}_{\lambda,n}$ is a Weyl group with root system $R_{\lambda,n}$; furthermore, $R^+ \cap R_{\lambda,n}$ is a positive system in $R_{\lambda,n}$ and so there is a unique base $\Pi_{\lambda,n}$ for $R_{\lambda,n}$ such that $\Pi_{\lambda,n} \subseteq R^+ \cap R_{\lambda,n}$. Thus,

$$\big(\mathbf{W}_{\lambda,n}, S_{\lambda,n}\big) \text{ is a Coxeter system, where } S_{\lambda,n} := \{w_\alpha \mid \alpha \in \Pi_{\lambda,n}\}. \qquad \text{(a)}$$

We also note that, by [Lu84a, 1.9], every coset $w\mathbf{W}_{\lambda,n}$ ($w \in \mathbf{W}$) contains a unique element w_1 of minimal length (for the usual length function of $\mathbf{W}$) and we have

$$l(w_1 y) > l(w_1) \qquad \text{for all } y \in \mathbf{W}_{\lambda,n},\ y \neq 1. \qquad \text{(b)}$$

When $\mathbf{Z}(\mathbf{G})$ is connected, we obtain a more precise statement, as follows.

Lemma 2.4.14 (Cf. [Lu84a, 2.15, 2.19]) *In the above setting, assume that $\mathbf{Z}(\mathbf{G})$ is connected. Then $\mathbf{W}_{\lambda,n} = \hat{\mathbf{W}}_{\lambda,n}$ and the following hold.*

(a) *There is a unique element $w_1 \in \mathscr{Z}_{\lambda,n}$ of minimal length (for the usual length function of $\mathbf{W}$). We have $\mathscr{Z}_{\lambda,n} = w_1\mathbf{W}_{\lambda,n}$.*
(b) *There is a well-defined group automorphism $\gamma\colon \mathbf{W}_{\lambda,n} \to \mathbf{W}_{\lambda,n}$ such that $\gamma(y) = \sigma(w_1 y w_1^{-1})$ for all $y \in \mathbf{W}_{\lambda,n}$. We have $\gamma(S_{\lambda,n}) = S_{\lambda,n}$.*

Proof The equality $\mathbf{W}_{\lambda,n} = \hat{\mathbf{W}}_{\lambda,n} = \{w \in \mathbf{W} \mid w.\lambda - \lambda \in nX\}$ follows rather directly from the fact that $\mathbf{Z}(\mathbf{G})$ is connected, as shown in [DiMi20, 11.2.1]. Combining this with Lemma 2.4.12, we see that $\mathscr{Z}_{\lambda,n}$ is a coset of $\mathbf{W}_{\lambda,n}$. As already mentioned above, such a coset contains a unique element w_1 of minimal length; by [Lu84a, 1.9], this element is characterised by the property that $w_1.\alpha \in R^+$ for any $\alpha \in R^+ \cap R_{\lambda,n}$. Thus, (a) holds.

For (b), let $y \in \mathbf{W}_{\lambda,n}$ and $t \in \mathbf{T}_0$. Setting $x := \sigma(w_1 y w_1^{-1}) \in \mathbf{W}$, we must show that $x.\lambda - \lambda \in nX$. Let $t \in \mathbf{T}_0$ and consider $t_1 := \dot{w}_1 \dot{y}^{-1} \dot{w}_1^{-1} t \dot{w}_1 \dot{y} \dot{w}_1^{-1} \in \mathbf{T}_0$. Since

$\dot{x} \equiv F(\dot{w}_1)F(\dot{y})F(\dot{w}_1)^{-1}$ (modulo $\mathbf{T}_0$), we obtain

$$\begin{aligned}(x.\lambda)\big(F(t)\big) &= \lambda\big(\dot{x}^{-1}F(t)\dot{x}\big) = (\lambda \circ F)(t_1) = (w_1.\lambda)(t_1) \cdot \lambda_{w_1}(t_1)^n \\ &= \lambda(\dot{y}^{-1}\dot{w}_1^{-1}t\dot{w}_1\dot{y}) \cdot \lambda_{w_1}(t_1)^n = \big(w_1y.\lambda\big)(t) \cdot \lambda_{w_1}(t_1)^n \\ &= (\lambda \circ F)(t) \cdot \lambda_{w_1y}(t^{-1})^n \cdot \lambda_{w_1}(t_1)^n \\ &= \lambda(F(t)) \cdot \big(\lambda_{w_1y}(t^{-1}) \cdot \lambda_{w_1}(t_1)\big)^n,\end{aligned}$$

where we used that $w_1 \in \mathscr{Z}_{\lambda,n}$ (third equality) and $w_1y \in \mathscr{Z}_{\lambda,n}$ (sixth equality). Now, the map $t \mapsto \lambda_{w_1y}(t^{-1}) \cdot \lambda_{w_1}(t_1)$ is a character of $\mathbf{T}_0$ (where t_1 is defined in terms of t as above). Thus, we see that $(x.\lambda - \lambda) \circ F \in nX$. We must show that this implies that $x.\lambda - \lambda \in nX$. Now, by Proposition 1.4.18, there exist integers $d, m \geqslant 1$ such that the map induced by F^d on X is given by scalar multiplication with p^m. Since $(x.\lambda - \lambda) \circ F \in nX$, we also have $p^m(x.\lambda - \lambda) = (x.\lambda - \lambda) \circ F^d \in nX$. As n is prime to p, this easily implies that $x.\lambda - \lambda \in nX$.

It remains to show that $\gamma(S_{\lambda,n}) = S_{\lambda,n}$. Recall from 1.6.1 that there is a permutation $\alpha \mapsto \alpha^\dagger$ of R such that $\alpha^\dagger \circ F = q_\alpha \alpha$; we have $\alpha^\dagger \in R^+$ for all $\alpha \in R^+$. We claim that this induces a permutation

$$R^+ \cap R_{\lambda,n} \to R^+ \cap R_{\lambda,n}, \qquad \alpha \mapsto (w_1.\alpha)^\dagger.$$

Indeed, let $\alpha \in R^+ \cap R_{\lambda,n}$. By Remark 1.2.10, we have $\sigma(w_\alpha) = w_{\alpha^\dagger}$ and so

$$w_{(w_1.\alpha)^\dagger} = \sigma(w_{w_1.\alpha}) = \sigma(w_1 w_\alpha w_1^{-1}) = \gamma(w_\alpha) \in \mathbf{W}_{\lambda,n}.$$

Hence, $(w_1.\alpha)^\dagger \in R_{\lambda,n}$. On the other hand, we also have $w_1.\alpha \in R^+$ (as noted in the proof of (a)) and so $(w_1.\alpha)^\dagger \in R^+$, as required. But then the above permutation also preserves the unique base $\Pi_{\lambda,n}$ contained in $R^+ \cap R_{\lambda,n}$. □

We obtain the following extension of Theorem 2.4.1 to non-unipotent characters of $\mathbf{G}^F$. Recall from Corollary 2.4.11 that $\mathrm{Irr}(\mathbf{G}^F) = \bigcup_{(\lambda,n)} \mathscr{E}_{\lambda,n}$ where the union runs over all pairs (λ, n) as in 2.4.5, with $\mathscr{Z}_{\lambda,n} \neq \varnothing$.

Theorem 2.4.15 (Lusztig [Lu84a, 4.23]) *Assume that $\mathbf{Z}(\mathbf{G})$ is connected. Let (λ, n) be a pair as in 2.4.5, with $\mathscr{Z}_{\lambda,n} \neq \varnothing$. Let $\mathbf{W}_{\lambda,n}$, w_1, γ be as in Lemma 2.4.14, such that $\mathscr{Z}_{\lambda,n} = w_1\mathbf{W}_{\lambda,n}$. Then there is a bijection*

$$\mathscr{E}_{\lambda,n} \overset{1-1}{\longleftrightarrow} \bar{X}(\mathbf{W}_{\lambda,n}, \gamma), \qquad \rho \leftrightarrow \bar{x}_\rho,$$

such that $\langle R_{w_1y}^{\theta_{w_1y}}, \rho\rangle = (-1)^{l(w_1)} m(y, \bar{x}_\rho)$ for all $y \in \mathbf{W}_{\lambda,n}$ and $\rho \in \mathscr{E}_{\lambda,n}$.

Remark 2.4.16 (a) In the above statement, the set $\bar{X}(\mathbf{W}_{\lambda,n}, \gamma)$ and the integers $m(y, \bar{x})$ are determined as in Remark 2.4.3, by an entirely combinatorial procedure. We shall see in the next section that there exists a connected reductive algebraic group $\hat{\mathbf{H}}$ over k and a Steinberg map $\hat{F}\colon \hat{\mathbf{H}} \to \hat{\mathbf{H}}$ such that $\mathbf{W}_{\lambda,n}$ can be identified with the Weyl group of $\hat{\mathbf{H}}$ (relative to a maximally split torus of $\hat{\mathbf{H}}$) and γ is the automorphism induced by the action of $\hat{F}$ (as in 1.6.1). Thus, in combination with Theorem 2.4.1, we obtain bijections

$$\mathscr{E}_{\lambda,n} \xleftrightarrow{1-1} \bar{X}(\mathbf{W}_{\lambda,n}, \gamma) \xleftrightarrow{1-1} \mathrm{Uch}(\hat{\mathbf{H}}^{\hat{F}})$$

which are almost what is meant by the Jordan decomposition of characters.

(b) The above result is established in [Lu84a] assuming that $F\colon \mathbf{G} \to \mathbf{G}$ is a Frobenius map. By the arguments in [Lu84a, 8.8], the proof can be reduced to the case where the derived subgroup $\mathbf{G}_{\mathrm{der}}$ is simple of simply connected type. As indicated in [Lu84a, 14.2], it can be shown that the above theorem also holds for the Suzuki and Ree groups and, hence, for any Steinberg map $F\colon \mathbf{G} \to \mathbf{G}$. (For an extension of the reduction arguments to the more general situation involving Steinberg maps, see also [Ta19].) There is a certain amount of case-by-case arguments involved in the proof of Theorem 2.4.15. More conceptual arguments are nowadays available via Lusztig's work on 'categorical centres'; see [Lu14b], [Lu16], [LuYu19].

Remark 2.4.17 In the setting of Theorem 2.4.15, let $y, y' \in \mathbf{W}_{\lambda,n}$. By [Lu84a, Prop. 3.9] (or, rather, its proof), we have

$$\begin{aligned}\langle R_{w_1y}^{\theta_{w_1y}}, R_{w_1y'}^{\theta_{w_1y'}} \rangle &= |\{x \in \mathbf{W}_{\lambda,n} \mid y'x = \gamma^{-1}(x)y\}| \\ &= |\{x' \in \mathbf{W}_{\lambda,n} \mid y'\gamma(x') = x'y\}|.\end{aligned}$$

(Compare with the formula in Example 2.3.22(a).) Let us now consider the γ-character table of $\mathbf{W}_{\lambda,n}$ (as in Section 2.1), where we fix a γ-extension $\tilde{\phi}$ for each $\phi \in \mathrm{Irr}(\mathbf{W}_{\lambda,n})^\gamma$ as in Proposition 2.1.14. Following [Lu77b, 3.17], [Lu84a, 3.7], we define the corresponding *uniform almost character*[6] by

$$R_{\tilde{\phi}} := \frac{1}{|\mathbf{W}_{\lambda,n}|} \sum_{y \in \mathbf{W}_{\lambda,n}} \tilde{\phi}(y) R_{w_1y}^{\theta_{w_1y}} \in \mathrm{CF}(\mathbf{G}^F).$$

Note that this is, in general, an $\mathbb{R}$-linear combination of irreducible characters of $\mathbf{G}^F$. If we replace the chosen σ-extension $\tilde{\phi}$ by $-\tilde{\phi}$, then this will result in replacing $R_{\tilde{\phi}}$ by $-R_{\tilde{\phi}}$. By [Lu84a, Prop. 3.9], we have

$$\langle R_{\tilde{\phi}}, R_{\tilde{\phi}'} \rangle = \begin{cases} 1 & \text{if } \phi = \phi', \\ 0 & \text{if } \phi \neq \phi' \end{cases}$$

(see also [Lu77b, 3.19]). Consequently, we also have

$$R_{w_1y}^{\theta_{w_1y}} = \sum_{\phi \in \mathrm{Irr}(\mathbf{W}_{\lambda,n})^\gamma} \tilde{\phi}(y)\, R_{\tilde{\phi}} \qquad \text{for all } y \in \mathbf{W}_{\lambda,n}.$$

[6] This is a special case of the more general definitions in [Lu84a, 4.24], [Lu19].

Thus, a knowledge of the integers $m(y, \bar{x})$ is equivalent to a knowledge of the scalar products $\langle R_{\tilde{\phi}}, \rho\rangle$ for $\phi \in \mathrm{Irr}(\mathbf{W}_{\lambda,n})^{\gamma}$ and $\rho \in \mathscr{E}_{\lambda,n}$. Furthermore, by [Lu84a, 4.26.1], we have

$$\rho(1) = \sum_{\phi \in \mathrm{Irr}(\mathbf{W}_{\lambda,n})^{\gamma}} \langle R_{\tilde{\phi}}, \rho\rangle R_{\tilde{\phi}}(1) \qquad \text{for } \rho \in \mathscr{E}_{\lambda,n}.$$

In fact, it turns out that the matrix of scalar products $\langle R_{\tilde{\phi}}, \rho\rangle$ has a much simpler shape than the matrix of integers $m(y, \bar{x})$; this will be further discussed in Section 4.2.

Example 2.4.18 Assume that $\mathbf{W}_{\lambda,n}$ is a direct product of Weyl groups of type A_{m_i} (for various $m_i \geqslant 0$). Then one can extract the following information from [Lu84a, 4.4, 4.19] and the general reduction arguments in [Lu84a, 4.21]. First, we can identify $\bar{X}(\mathbf{W}_{\lambda,n}, \gamma)$ and $\mathrm{Irr}(\mathbf{W}_{\lambda,n})^{\gamma}$. If $\bar{x} \in \bar{X}(\mathbf{W}_{\lambda,n}, \gamma)$ corresponds to $\phi \in \mathrm{Irr}(\mathbf{W}_{\lambda,n})^{\gamma}$ under this identification, then $m(y, \bar{x}) = \pm\tilde{\phi}(y)$ for all $y \in \mathbf{W}_{\lambda,n}$, where the sign only depends on the choice of the (real-valued) γ-extension $\tilde{\phi}$ of ϕ. Furthermore, we have

$$\mathscr{E}_{\lambda,n} = \{\rho \in \mathrm{Irr}(\mathbf{G}^F) \mid \rho = \pm R_{\tilde{\phi}} \text{ for some } \phi \in \mathrm{Irr}(\mathbf{W}_{\lambda,n})^{\gamma}\}.$$

Indeed, by Theorem 2.4.15, the subspace of $\mathrm{CF}(\mathbf{G}^F)$ spanned by $\mathscr{E}_{\lambda,n}$ has dimension $|\bar{X}(\mathbf{W}_{\lambda,n}, \gamma)| = |\mathrm{Irr}(\mathbf{W}_{\lambda,n})^{\gamma}|$. So the orthogonality relations in Remark 2.4.17 imply that this subspace is also spanned by $\{R_{\tilde{\phi}} \mid \phi \in \mathrm{Irr}(\mathbf{W}_{\lambda,n})^{\gamma}\}$. Consequently, every $\rho \in \mathscr{E}_{\lambda,n}$ can be written as a linear combination of the virtual characters $R_{w_1 y}^{\theta_{w_1 y}}$ for $y \in \mathbf{W}_{\lambda,n}$. Since $m(y, \bar{x}) = \pm\tilde{\phi}(y)$ for $y \in \mathbf{W}_{\lambda,n}$ (where $\bar{x}$ and ϕ correspond to each as above), we conclude that $\rho = \pm R_{\tilde{\phi}}$ if $\bar{x} = \bar{x}_\rho$, as required. In order to fix the sign, one has to choose the γ-extension $\tilde{\phi}$ so that $R_{\tilde{\phi}}(1) > 0$ (and there is a unique such choice).

We can now state the following result that shows why the groups $\mathrm{GL}_n(q)$ and $\mathrm{GU}_n(q)$ are so much easier to deal with than groups of other types. (The situation is already more complicated for $\mathrm{SL}_n(q)$; see Example 2.2.30 and also [Leh73], [Leh78], [Bo06].)

Corollary 2.4.19 (Cf. [LuSr77]) *Assume that* $\mathbf{G} = \mathrm{GL}_n(k)$ *or, more generally, that* $\mathbf{Z}(\mathbf{G})$ *is connected and* $\mathbf{W}$ *is a direct product of Weyl groups of type* A_{m_i} *(for various* m_i*). Then every* $\rho \in \mathrm{Irr}(\mathbf{G}^F)$ *is a uniform almost character, that is,* $\rho = \pm R_{\tilde{\phi}}$ *where* $\phi \in \mathrm{Irr}(\mathbf{W}_{\lambda,n})$ *for some pair* (λ, n) *as in 2.4.5 such that* $\mathscr{L}_{\lambda,n} \neq \varnothing$.

Proof Since $\mathbf{W}$ is a direct product of Weyl groups of type A_{m_i} (for various m_i), an analogous statement holds for the reflection subgroups $\mathbf{W}_{\lambda,n}$. So the assertion is an immediate consequence of Corollary 2.4.11 and the discussion in Example 2.4.18. Note that [LuSr77, 2.2] provides a direct argument showing that $\mathrm{Uch}(\mathbf{G}^F) = \{\pm R_{\tilde{\phi}} \mid \phi \in \mathrm{Irr}(\mathbf{W})^{\sigma}\}$. See also [Lu77a, 7.14], [DiMi20, §11.7]. □

Example 2.4.20 Assume that $\mathbf{G} = \mathrm{GL}_n(k)$ and $F\colon \mathbf{G} \to \mathbf{G}$ is the standard Frobenius map (see Example 1.4.21) such that $\mathbf{G}^F = \mathrm{GL}_n(q)$. We have $\mathbf{W} \cong \mathfrak{S}_n$ (the symmetric group of degree n) and $\sigma\colon \mathbf{W} \to \mathbf{W}$ is the identity. There is a natural labelling of the irreducible characters of $\mathbf{W}$ by the partitions $\nu \vdash n$; see Example 4.1.2. We shall write

$$\mathrm{Irr}(\mathbf{W}) = \{\phi^\nu \mid \nu \vdash n\}.$$

For example, $\phi^{(n)}$ is the trivial character and $\phi^{(1^n)}$ is the sign character. For each $\nu \vdash n$, we let $\tilde{\phi}^\nu := \phi^\nu$ be the trivial extension as in Example 2.1.8(a) and form the corresponding unipotent uniform almost character

$$R_\nu := \frac{1}{|\mathbf{W}|} \sum_{w \in \mathbf{W}} \phi^\nu(w) R_w^1 \qquad \text{(cf. Remark 2.4.17).}$$

Then $\mathrm{Uch}(\mathbf{G}^F) = \{R_\nu \mid \nu \vdash n\}$; see Example 2.4.18 (and [DiMi20, 11.7.2] where it is shown how the signs in Example 2.4.18 are fixed). On the other hand, we can apply the discussion in Example 2.3.9. To each $\phi^\nu \in \mathrm{Irr}(\mathbf{W})$ corresponds a character $\rho_\nu \in \mathrm{Uch}(\mathbf{G}^F)$, such that ρ_ν occurs in the permutation character of $\mathbf{G}^F$ on the cosets of $\mathbf{B}_0^F$. Then $\rho_\nu = R_\nu$ for all $\nu \vdash n$. (This is a very special case of [Lu84a, 12.6, 12.14.2].) Hence, the restrictions of the functions R_w^1 to $\mathbf{G}_{\mathrm{uni}}^F$ are indeed the Green functions first investigated in [Gre55]. (The discussion of this case will be continued in Example 2.8.7.)

Our next aim is to establish, following [Lu84a, Chap. 6], an important fact related to Theorem 2.4.15: the various sets $\mathscr{E}_{\lambda,n}$ are precisely the geometric series of characters. This will work without any assumptions on the centre $\mathbf{Z}(\mathbf{G})$. We begin with a version of Theorem 2.3.2 for the characters R_w^θ.

2.4.21 Let $w \in \mathbf{W}$ and recall from 1.6.4 that $\mathbf{T}_0[w] = \mathbf{T}_0^{F'}$ where $F'\colon \mathbf{T}_0 \to \mathbf{T}_0$ is defined by $F'(t) = \dot{w}F(t)\dot{w}^{-1}$ for $t \in \mathbf{T}_0$. To indicate the dependence on w, we will denote F' by wF (short-hand for: F followed by conjugation with $\dot{w}$); thus, $\mathbf{T}_0[w] = \mathbf{T}_0^{wF}$. Now, for any integer $d \geqslant 1$ and $t \in \mathbf{T}_0$, we have

$$(wF)^d(t) = F'^d(t) = \dot{w}F(\dot{w})\ldots F^{d-1}(\dot{w})F^d(t)F^{d-1}(\dot{w})^{-1}\ldots F(\dot{w})^{-1}\dot{w}^{-1}.$$

Hence, setting $y := w\sigma(w)\ldots\sigma^{d-1}(w) \in \mathbf{W}$, we have $(wF)^d(t) = \dot{y}F^d(t)\dot{y}^{-1}$ for $t \in \mathbf{T}_0$. Let $d_0 \geqslant 1$ be the order of σ. If d is a multiple of d_0, then

$$y = \big(w\sigma(w)\ldots\sigma^{d_0-1}(w)\big)^{d/d_0}.$$

Hence, if d is a sufficiently large multiple of d_0, then $y = 1$ and so $(wF)^d(t) = F^d(t)$ for all $t \in \mathbf{T}_0$. Since $\mathbf{W}$ is finite, we can even find $d \geqslant 1$ such that

$$d_0 \mid d \qquad \text{and} \qquad (wF)^d(t) = F^d(t) \quad \text{for all } t \in \mathbf{T}_0 \text{ and all } w \in \mathbf{W}.$$

Any such d will be called *admissible*. If d is admissible, then $\mathbf{T}_0^{(wF)^d} = \mathbf{T}_0^{F^d}$ does not depend on w any more; let $N_d^{(w)} := N_{(wF)^d/wF} \colon \mathbf{T}_0 \to \mathbf{T}_0$ be the corresponding norm map. By Lemma 2.3.1, we have

$$N_d^{(w)}(\mathbf{T}_0^{F^d}) = \mathbf{T}_0^{wF} = \mathbf{T}_0[w] \qquad \text{for all } w \in \mathbf{W}.$$

As in Remark 2.3.3, one sees that $N_d^{(w)}$ is transitive.

Definition 2.4.22 (Cf. [BoRo93, §4.4]) Recall from 2.3.20 the definition of the set $\mathfrak{X}(\mathbf{W}, \sigma)$. We say that two pairs (w_1, θ_1) and (w_2, θ_2) in $\mathfrak{X}(\mathbf{W}, \sigma)$ are *geometrically conjugate* if, in the above setting, there exists an admissible integer $d \geqslant 1$ such that

$$\theta_2 \circ N_d^{(w_2)} = {}^{\dot{y}}(\theta_1 \circ N_d^{(w_1)}) \in \operatorname{Irr}(\mathbf{T}_0^{F^d}) \qquad \text{for some } y \in \mathbf{W}.$$

Here, we tacitly assume that $\dot{y} \in N_{\mathbf{G}}(\mathbf{T}_0)^{F^d}$, which we may since $\sigma^d = \mathrm{id}_{\mathbf{W}}$ (see Example 1.4.11(b)).

Lemma 2.4.23 *The correspondence $\overline{\mathfrak{X}}(\mathbf{G}, F) \longleftrightarrow \overline{\mathfrak{X}}(\mathbf{W}, \sigma)$ in 2.3.20 induces a bijection between geometric conjugacy classes of pairs $(\mathbf{T}, \theta)$ (see Theorem 2.3.2) and geometric conjugacy classes of pairs (w, θ) as defined above.*

Proof For $i = 1, 2$ let $w_i \in \mathbf{W}$ and $\theta_i \in \operatorname{Irr}(\mathbf{T}_0[w_i])$. Let $\mathbf{T}_i := g_i \mathbf{T}_0 g_i^{-1}$ where $g_i \in \mathbf{G}$ is such that $\dot{w}_i = g_i^{-1} F(g_i)$. Assume first that $(\mathbf{T}_1, {}^{g_1}\theta_1)$ and $(\mathbf{T}_2, {}^{g_2}\theta_2)$ are geometrically conjugate. So there exist an integer $d \geqslant 1$ and an element $x \in \mathbf{G}^{F^d}$ such that $\mathbf{T}_2 = x\mathbf{T}_1 x^{-1}$ and ${}^x\hat{\theta}_1 = \hat{\theta}_2$, where

$$\hat{\theta}_i := ({}^{g_i}\theta_i) \circ N_{F^d/F} \in \operatorname{Irr}(\mathbf{T}_i^{F^d}) \qquad \text{for } i = 1, 2.$$

Replacing d by a multiple if necessary we may assume, by Remark 2.3.3 and the discussion in 2.4.21, that d is admissible and also that $F^d(g_i) = g_i$ for $i = 1, 2$. Now define $\theta_i' \in \operatorname{Irr}(\mathbf{T}_0^{F^d})$ by $\hat{\theta}_i = {}^{g_i}\theta_i'$ for $i = 1, 2$. We note that $y := g_2^{-1} x g_1 \in N_{\mathbf{G}}(\mathbf{T}_0)^{F^d}$; furthermore, $\theta_2' = {}^y\theta_1'$. Hence, it remains to show that $\theta_i' = \theta_i \circ N_d^{(w_i)}$ for $i = 1, 2$. For this purpose, let $t \in \mathbf{T}$. Then $g_i^{-1} t g_i \in \mathbf{T}_0$ and we compute:

$$(w_i F)(g_i^{-1} t g_i) = \dot{w}_i F(g_i)^{-1} F(t) F(g_i) \dot{w}_i^{-1} = g^{-1} F(t) g_i;$$

hence, $(w_i F)^m (g_i^{-1} t g_i) = g_i^{-1} F^m(t) g_i$ for all $m \geqslant 1$. We obtain that

$$\begin{aligned} N_d^{(w_i)}(g_i^{-1} t g_i) &= (g_i^{-1} t g_i)(w_i F)(g_i^{-1} t g_i) \cdots (w_i F)^{d-1}(g_i^{-1} t g_i) \\ &= g_i^{-1} t F(t) \cdots F^{d-1}(t) g_i = g_i^{-1} N_{F^d/F}(t) g_i. \end{aligned}$$

Now let $s \in \mathbf{T}_0^{F^d}$. Then $t := g_i s g_i^{-1} \in \mathbf{T}^{F^d}$ and so

$$\theta_i'(s) = \hat{\theta}_i(t) = \theta_i\big(g_i^{-1} N_{F^d/F}(t) g_i\big) = \theta_i\big(N_d^{(w_i)}(g_i^{-1} t g_i)\big) = \theta_i(N_d^{(w_i)}(s)).$$

Thus, $\theta_i' = \theta_i \circ N_d^{(w_i)}$ for $i = 1, 2$. So, (w_1, θ_1) and (w_2, θ_2) are geometrically conjugate, as required. One can also run this argument backwards, which proves the reverse implication. □

Corollary 2.4.24 (Exclusion Theorem; 2nd version) *Assume that two Deligne–Lusztig characters $R_{w_1}^{\theta_1}$ and $R_{w_2}^{\theta_2}$ (as in 2.3.18) have an irreducible constituent in common. Then the pairs (w_1, θ_1) and (w_2, θ_2) are geometrically conjugate.*

Proof Immediate by Theorem 2.3.2, Lemma 2.3.19 and Lemma 2.4.23. □

Example 2.4.25 Let (λ, n) be a pair as in 2.4.5, with $\mathscr{Z}_{\lambda,n} \neq \varnothing$. For each $w \in \mathscr{Z}_{\lambda,n}$, we have a corresponding character $\theta_w \in \mathrm{Irr}(\mathbf{T}_0[w])$. We claim that all pairs $\{(w, \theta_w) \mid w \in \mathscr{Z}_{\lambda,n}\}$ are geometrically conjugate.

Indeed, let $w \in \mathscr{Z}_{\lambda,n}$. Then $\theta_w = \psi \circ \bar{\lambda}_w$ and $\theta_w(t)^n = 1$ for all $t \in \mathbf{T}_0[w]$. Recall from 2.4.5 that $\lambda_w \in X$ is defined by the condition:

$$\lambda_w(t^n) = \lambda(F(t)\dot{w}^{-1}t^{-1}\dot{w}) = (w.\lambda)\big(F'(t)t^{-1}\big) \qquad \text{for all } t \in \mathbf{T}_0,$$

where we set again $F'(t) := \dot{w}F(t)\dot{w}^{-1}$ for $t \in \mathbf{T}_0$. Now consider the norm map $N_d^{(w)} \colon \mathbf{T}_0^{F^d} \to \mathbf{T}_0[w]$ and the linear character $\theta_w \circ N_d^{(w)} \in \mathrm{Irr}(\mathbf{T}_0^{F^d})$, where $d \geqslant 1$ is an admissible integer as in 2.4.21. Let $t \in \mathbf{T}_0$ and set $t' := N_d^{(w)}(t)$. Then

$$\begin{aligned} F'(t') = F'\big(N_d^{(w)}(t)\big) &= F'\big(tF'(t)\dots F'^{d-1}(t)\big) \\ &= t^{-1}N_d^{(w)}(t)F'^{d}(t) = t^{-1}t'F^d(t) \end{aligned}$$

where the last equality holds since d is admissible. Hence, we obtain:

$$\lambda_w(t'^n) = (w.\lambda)\big(F'(t')t'^{-1}\big) = (w.\lambda)\big(t^{-1}t'F^d(t)t'^{-1}\big) = (w.\lambda)\big(F^d(t)t^{-1}\big).$$

On the other hand, the left-hand side equals

$$\lambda_w(t'^n) = \lambda_w\big((N_d^{(w)}(t))^n\big) = \lambda_w\big(N_d^{(w)}(t^n)\big) = \big(\lambda_w \circ N_d^{(w)}\big)(t^n).$$

Thus, we obtain that $\big(\lambda_w \circ N_d^{(w)}\big)(t^n) = (w.\lambda)\big(F^d(t)t^{-1}\big)$ for all $t \in \mathbf{T}_0$. This means that $1 \in \mathscr{Z}_{w.\lambda,n}^{(d)}$ (where the super-script (d) indicates that F^d is used instead of F) and that (♠) holds with $(w.\lambda)_1^{(d)} := \lambda_w \circ N_d^{(w)} \in X$. Hence

$$\theta_w \circ N_d^{(w)} = (\psi \circ \bar{\lambda}_w) \circ N_d^{(w)} = \psi \circ \overline{(w.\lambda)}_1^{(d)} = \theta_1^{(d)} \in \mathrm{Irr}(\mathbf{T}_0^{F^d})$$

and so $(w.\lambda, n) \leftrightsquigarrow (1, \theta_1^{(d)}) = (1, \theta \circ N_d^{(w)})$ (where $\leftrightsquigarrow$ is defined with respect to F^d). Now let also $w' \in \mathscr{Z}_{\lambda,n}$; then $\theta_{w'} = \psi \circ \bar{\lambda}_{w'}$. Setting $x := w'w^{-1} \in \mathbf{W}$, we have

$$\big(\lambda_w \circ N_d^{(w)}\big)(\dot{x}^{-1}t^n\dot{x}) = \big(\lambda_w \circ N_d^{(w)}\big)\big((\dot{x}^{-1}t\dot{x})^n\big) = (w.\lambda)\big(\dot{x}^{-1}t^{-1}\dot{x}F^d(\dot{x}^{-1}t\dot{x})\big)$$

for all $t \in \mathbf{T}_0$. Since $F^d(\dot{x}) = \dot{x}$ and $xw = w'$, the right-hand side equals

$$(w.\lambda)\big(\dot{x}^{-1}t^{-1}F^d(t)\dot{x}\big) = (w'.\lambda)\big(t^{-1}F^d(t)\big) = \big(\lambda_{w'} \circ N_d^{(w')}\big)(t^n).$$

Thus, $\lambda_{w'} \circ N_d^{(w')} = {}^{\dot{x}}(\lambda_w \circ N_d^{(w)})$ and so $\theta_{w'} \circ N_d^{(w')} = {}^{\dot{x}}(\theta_w \circ N_d^{(w)})$. Hence, the pairs (w, θ_w) and $(w', \theta_{w'})$ are geometrically conjugate, as claimed.

Lemma 2.4.26 (Cf. [Lu84a, 6.2(i)]) *Let (λ, n) be a pair as in 2.4.5. Let $w \in \mathscr{Z}_{\lambda,n}$ and assume that $(\lambda, n) \leftrightsquigarrow (w, \theta)$ where $\theta \in \mathrm{Irr}(\mathbf{T}_0[w])$.*

(a) *Let $\mu := \lambda + n\nu$ for some $\nu \in X$. Then $w \in \mathscr{Z}_{\mu,n}$ and $(\mu, n) \leftrightsquigarrow (w, \theta)$.*
(b) *For any $y \in \mathbf{W}$, we have $y.w \in \mathscr{Z}_{\sigma(y).\lambda,n}$ and $(\sigma(y).\lambda, n) \leftrightsquigarrow y.(w, \theta)$ (where the action of y on $\mathfrak{X}(\mathbf{W}, \sigma)$ is defined in 2.3.20(b)).*

Proof By assumption, we have $\theta = \psi \circ \bar{\lambda}_w$ and $\theta(t)^n = 1$ for all $t \in \mathbf{T}_0[w]$. Recall from 2.4.5 that $\lambda_w \in X$ is defined by the condition:

$$\lambda_w(t^n) = \lambda(F(t)\dot{w}^{-1}t^{-1}\dot{w}) \qquad \text{for all } t \in \mathbf{T}_0.$$

(a) Define $\nu_w \colon \mathbf{T}_0 \to k^\times$ by $\nu_w(t) := \nu(F(t)\dot{w}^{-1}t^{-1}\dot{w})$ for $t \in \mathbf{T}_0$. Since $\mathbf{T}_0$ is abelian and $\dot{w} \in N_{\mathbf{G}}(\mathbf{T}_0)$, we have $\nu_w \in X$. Now set $\mu_w := \lambda_w + \nu_w \in X$. Then we obtain

$$\begin{aligned} \mu_w(t)^n &= \lambda_w(t)^n \nu_w(t)^n = \lambda(F(t)\dot{w}^{-1}t^{-1}\dot{w})\nu(F(t)\dot{w}^{-1}t^{-1}\dot{w})^n \\ &= (\lambda + n\nu)(F(t)\dot{w}^{-1}t^{-1}\dot{w}) = \mu(F(t)\dot{w}^{-1}t^{-1}\dot{w}) \end{aligned}$$

for all $t \in \mathbf{T}_0$ and so $w \in \mathscr{Z}_{\mu,n}$. Since the restriction of ν_w to $\mathbf{T}_0[w]$ is trivial, both λ_w and μ_w have the same restriction and so $(\mu, n) \leftrightsquigarrow (w, \theta)$, as desired.

(b) A straightforward computation shows that, for any $t \in \mathbf{T}_0$, we have

$$(y.\lambda_w)(t^n) = \lambda_w((\dot{y}^{-1}t\dot{y})^n) = \cdots = (\sigma(y).\lambda)(F(t)\dot{w}'^{-1}t^{-1}\dot{w}')$$

where $w' := y.w = yw\sigma(y)^{-1}$. Thus, $w' \in \mathscr{Z}_{\sigma(y).\lambda,n}$ and $(\sigma(y).\lambda)_{w'} = y.\lambda_w$. Now let $\theta' \in \mathrm{Irr}(\mathbf{T}_0[w'])$ be the linear character such that $(\sigma(y).\lambda, n) \leftrightsquigarrow (w', \theta')$. For $t \in \mathbf{T}_0$, we have $\dot{y}t\dot{y}^{-1} \in \mathbf{T}_0[w']$ and

$$\theta'(\dot{y}t\dot{y}^{-1}) = \psi((y.\lambda_w)(\dot{y}t\dot{y}^{-1})) = \psi(\lambda_w(t)) = \theta(t),$$

which shows that $\theta' = {}^y\theta$, as required. □

Corollary 2.4.27 *Let (λ, n) be a pair as in 2.4.5, with $\mathscr{Z}_{\lambda,n} \neq \varnothing$. Let $x \in \mathbf{W}$ and $\nu \in X$. Then the following hold.*

(a) *We have $\mathscr{Z}_{x.\lambda+n\nu,n} = \mathscr{Z}_{x.\lambda,n} \neq \varnothing$ and $\mathscr{E}_{\lambda,n} = \mathscr{E}_{x.\lambda+n\nu,n}$.*
(b) *If $w \in \mathscr{Z}_{\lambda,n}$, then $w' := yw\sigma(y)^{-1} \in \mathscr{Z}_{x.\lambda,n}$ and $R_w^{\theta_w} = R_{w'}^{\theta_{w'}}$, where $y \in \mathbf{W}$ is such that $x = \sigma(y)$.*

(These statements appear in [Lu84a, 6.5(ii)] but the proof there is more complicated, because Lusztig's original definition of $\mathscr{E}_{\lambda,n}$ is different; see the comments in Definition 2.4.6.)

Proof Firstly, Lemma 2.4.26(a) immediately shows that $\mathscr{Z}_{\lambda+n\nu,n} = \mathscr{Z}_{\lambda,n}$ and $\mathscr{E}_{\lambda+n\nu,n} = \mathscr{E}_{\lambda,n}$ for all $\nu \in X$. So we may assume now that $\nu = 0$. Let $x \in \mathbf{W}$ and set $y := \sigma^{-1}(x) \in \mathbf{W}$. Then Lemma 2.4.26(b) shows that, for any $w \in \mathscr{Z}_{\lambda,n}$, we have $w' := y.w = yw\sigma(y)^{-1} \in \mathscr{Z}_{x.\lambda,n}$ and $(x.\lambda, n) \leftrightsquigarrow y.(w, \theta_w) = (w', \theta_{w'})$. Furthermore, the scalar product formula in Example 2.3.22(a) implies that $R_w^{\theta_w} = R_{y.w}^{y.\theta_w} = R_{w'}^{\theta_{w'}}$. In particular, we conclude that $\{y.w \mid w \in \mathscr{Z}_{\lambda,n}\} \subseteq \mathscr{Z}_{x.\lambda,n}$ and $\mathscr{E}_{\lambda,n} \subseteq \mathscr{E}_{x.\lambda,n}$. By symmetry, we also get the reverse inclusions. □

We can now establish the following strengthening of Example 2.4.25. First, some notation. Let $\Lambda(\mathbf{G}, F)$ be the set of all indivisible pairs (λ, n) as in 2.4.5, with $\mathscr{Z}_{\lambda,n} \neq \varnothing$. For $i = 1, 2$ let $(\lambda_i, n_i) \in \Lambda(\mathbf{G}, F)$. Then write $(\lambda_1, n_1) \sim (\lambda_2, n_2)$ if $n_1 = n_2$ and there exists some $x \in \mathbf{W}$ and $\nu \in X$ such that $\lambda_2 = x.\lambda_1 + n_1\nu$. This defines an equivalence relation on $\Lambda(\mathbf{G}, F)$.

Proposition 2.4.28 (Cf. [DeLu76, 5.7], [Lu84a, 6.5]) *The relation $(\lambda, n) \leftrightsquigarrow (w, \theta)$ induces a bijection between equivalence classes of pairs $(\lambda, n) \in \Lambda(\mathbf{G}, F)$ and geometric conjugacy classes of pairs $(w, \theta) \in \mathfrak{X}(\mathbf{W}, \sigma)$ (as in Definition 2.4.22), which in turn are in bijection with geometric conjugacy classes of pairs $(\mathbf{T}, \theta) \in \mathfrak{X}(\mathbf{G}, F)$ (see Lemma 2.4.23).*

Proof By Lemma 2.4.8 and Remark 2.4.10, every pair in $\mathfrak{X}(\mathbf{W}, \sigma)$ arises from a pair in $\Lambda(\mathbf{G}, F)$ via the relation $\leftrightsquigarrow$. Now let $(\lambda_i, n_i) \in \Lambda(\mathbf{G}, F)$ for $i = 1, 2$. Let $w_i \in \mathscr{Z}_{\lambda_i,n_i}$ and set $\theta_i := \theta_{w_i}$. We must show that $(\lambda_1, n_1) \sim (\lambda_2, n_2)$ if and only if (w_1, θ_1) and (w_2, θ_2) are geometrically conjugate. First note that, by Remark 2.4.10, $n_i \geqslant 1$ is the smallest integer such that $\theta_i(t)^{n_i} = 1$ for all $t \in \mathbf{T}_0[w_i]$. Now consider the norm map $N_d^{(w_i)} \colon \mathbf{T}_0^{F^d} \to \mathbf{T}_0[w_i]$ where $d \geqslant 1$ is an admissible integer. We have seen in Example 2.4.25 that

$$(w_i.\lambda_i, n_i) \leftrightsquigarrow (1, \theta_i \circ N_d^{(w_i)}) \qquad \text{for } i = 1, 2, \qquad (*)$$

where $\leftrightsquigarrow$ is defined using F^d. Assume now that (w_1, θ_1) and (w_2, θ_2) are geometrically conjugate, that is, there exists some $x \in \mathbf{W}$ such that

$$^{\dot{x}}(\theta_1 \circ N_d^{(w_1)}) = \theta_2 \circ N_d^{(w_2)} \in \mathrm{Irr}(\mathbf{T}_0^{F^d}) \qquad \text{where} \qquad \dot{x} \in N_{\mathbf{G}}(\mathbf{T}_0)^{F^d}.$$

First we note that this implies that $n_1 = n_2$. We now apply Lemma 2.4.26 to $(w_1.\lambda_1, n_1) \leftrightsquigarrow (1, \theta_1 \circ N_d^{(w_1)})$. Since, here, $\leftrightsquigarrow$ is defined with respect to F^d and since $\sigma^d = \mathrm{id}_{\mathbf{W}}$, we obtain

$$(xw_1.\lambda_1, n_1) \leftrightsquigarrow x.\big(1, \theta_1 \circ N_d^{(w_1)}\big) = \big(1, {}^x(\theta_1 \circ N_d^{(w_1)})\big) = \big(1, \theta_2 \circ N_d^{(w_2)}\big).$$

Finally, since both $(w_2.\lambda_2, n_2) \leftrightsquigarrow (1, \theta_2 \circ N_d^{(w_2)})$ and $(xw_1.\lambda_1, n_1) \leftrightsquigarrow (1, \theta_2 \circ N_d^{(w_2)})$, we can apply Lemma 2.4.8 which yields that $w_2.\lambda_2 = xw_1.\lambda_1 + n_1\nu$ for some $\nu \in X$. Hence, $(\lambda_1, n_1) \sim (\lambda_2, n_2)$. Conversely, assume that $(\lambda_1, n_1) \sim (\lambda_2, n_2)$.

Hence, $n_1 = n_2$ and there exists some $x \in \mathbf{W}$ such that $\lambda_2 = x.\lambda_1 + n_1 \nu$ for some $\nu \in X$.

We set $y := w_2 x w_1^{-1} \in \mathbf{W}$. Then $w_2.\lambda_2 = w_2 x.\lambda_1 + n_1 w_2.\nu = y w_1.\lambda_1 + n_1 w_2.\nu$. By $(*)$, Lemmas 2.4.8 and 2.4.26 (applied with F^d), we have

$$(y w_1.\lambda_1 + n_1 w_2.\nu, n_1) \leftrightarrow \big(1, {}^y(\theta_1 \circ N_d^{(w_1)})\big).$$

Since also $(w_2.\lambda_2, n_2) \leftrightarrow (1, \theta_2 \circ N_d^{(w_2)})$, we conclude that $\theta_2 \circ N_d^{(w_2)} = {}^y(\theta_1 \circ N_d^{(w_1)})$, that is, (w_1, θ_1) and (w_2, θ_2) are geometrically conjugate. □

Corollary 2.4.29 (Cf. [DeLu76, 10.1], [Lu84a, 6.5]) *We have a partition*

$$\operatorname{Irr}(\mathbf{G}^F) = \bigsqcup_{(\lambda,n)} \mathscr{E}_{\lambda,n} \qquad (\mathscr{E}_{\lambda,n} \textit{ as in Definition 2.4.6})$$

where (λ, n) runs over a set of representatives for the classes of $\Lambda(\mathbf{G}, F)$ under the equivalence relation $\sim$. Two characters $\rho_1, \rho_2 \in \operatorname{Irr}(\mathbf{G}^F)$ belong to the same piece in the above partition if and only if ρ_1, ρ_2 belong to the same geometric series of characters (as in Definition 2.3.4).

Proof First, by Corollary 2.4.11, we have $\operatorname{Irr}(\mathbf{G}^F) = \bigcup_{(\lambda,n)} \mathscr{E}_{\lambda,n}$ where the union runs over all $(\lambda, n) \in \Lambda(\mathbf{G}, F)$. Now let (λ_1, n_1) and (λ_2, n_2) be pairs in $\Lambda(\mathbf{G}, F)$ and assume that there exists some $\rho \in \mathscr{E}_{\lambda_1,n_1} \cap \mathscr{E}_{\lambda_2,n_2}$. For $i = 1, 2$ let $w_i \in \mathscr{Z}_{\lambda_i,n_i}$ be such that $\langle \rho, R_{w_i}^{\theta_i} \rangle \neq 0$ where we set $\theta_i := \theta_{w_i}$. Then, by Corollary 2.4.24, the pairs (w_1, θ_1) and (w_2, θ_2) are geometrically conjugate. Consequently, by Proposition 2.4.28, we have $(\lambda_1, n_1) \sim (\lambda_2, n_2)$. So Corollary 2.4.27 shows that $\mathscr{E}_{\lambda_1,n_1} = \mathscr{E}_{\lambda_2,n_2}$. Thus, we have a disjoint union as stated above.

By Proposition 2.4.28, it is now sufficient to show that, if $\rho_1, \rho_2 \in \mathscr{E}_{\lambda,n}$, then ρ_1, ρ_2 belong to the same geometric series of characters. There exist $w_i \in \mathscr{Z}_{\lambda,n}$ such that $\langle R_{w_i}^{\theta_i}, \rho_i \rangle \neq 0$ where we set $\theta_i := \theta_{w_i}$ for $i = 1, 2$. By Example 2.4.25, (w_1, θ_1) and (w_2, θ_2) are geometrically conjugate. Hence, ρ_1, ρ_2 belong to the same geometric series of characters by Definition 2.3.4 and Lemma 2.4.23. □

Lemma 2.4.30 *Let (λ, n) be a pair as in 2.4.5, with $\mathscr{Z}_{\lambda,n} \neq \varnothing$. Let $\rho \in \mathscr{E}_{\lambda,n}$ and $(w', \theta') \in \mathfrak{X}(\mathbf{W}, \sigma)$ be such that $\langle R_{w'}^{\theta'}, \rho \rangle \neq 0$. Then there exists some $w \in \mathscr{Z}_{\lambda,n}$ such that $R_{w'}^{\theta'} = R_w^{\theta_w}$.*

Proof Let $m \geqslant 1$ be the smallest integer such that $\theta'(t)^m = 1$ for all $t \in \mathbf{T}_0[w']$. By Lemma 2.4.8, there exists some $\mu \in X$ such that $(\mu, m) \leftrightarrow (w', \theta')$; furthermore, (μ, m) is indivisible by Remark 2.4.10. Now $\rho \in \mathscr{E}_{\lambda,n} \cap \mathscr{E}_{\mu,m}$ and so $(\lambda, n) \sim (\mu, m)$ by Corollary 2.4.29. Write $\mu = x.\lambda + n\nu$ where $x \in \mathbf{W}$ and $\nu \in X$. If we set $y := \sigma^{-1}(x)$, then Corollary 2.4.27 shows that $R_w^{\theta_w} = R_{w'}^{\theta_{w'}} = R_{w'}^{\theta'}$ where $w \in \mathscr{Z}_{\lambda,n}$ is defined by the condition that $w' = y.w = yw\sigma(y)^{-1}$. □

Finally, we re-interpret the degree polynomials in Definition 2.3.25 in terms of the formalism of Theorem 2.4.15. First, some preparations.

Remark 2.4.31 In the setting of Remark 2.4.17, we have $R_{\tilde{\phi}}(1) = \mathbb{D}_{\tilde{\phi}}(q)$, where the polynomial $\mathbb{D}_{\tilde{\phi}} \in \mathbb{R}[\mathbf{q}]$ is defined by

$$\mathbb{D}_{\tilde{\phi}} := \frac{1}{|\mathbf{W}_{\lambda,n}|} \sum_{y \in \mathbf{W}_{\lambda,n}} (-1)^{l(w_1 y)} \tilde{\phi}(y) \frac{|\mathbb{G}|_{\mathbf{q}'}}{|\mathbb{T}_{w_1 y}|}.$$

(As in Definition 2.3.25, one sees that $\mathbb{D}_{\tilde{\phi}}$ indeed is a polynomial, and not just a rational function in $\mathbf{q}$.) These polynomials first appeared in [St51a], [Lu77a, §2], [Lu77b, 3.16], [BeLu78, §1]; following [AlLu82, §3], $\mathbb{D}_{\tilde{\phi}}$ is called the (twisted) *fake degree* of ϕ. Note that $\mathbb{D}_{\tilde{\phi}}$ only depends on the complete root datum associated with $\mathbf{G}, F$ and the choice of the γ-extension $\tilde{\phi}$ of ϕ.

Now assume that $n = 1$ and $\lambda = \lambda_0$ is the trivial character. Then $\mathbf{W}_{\lambda_0,1} = \mathbf{W}$ and so $w_1 = 1$; furthermore, the automorphism γ is just the automorphism $\sigma \colon \mathbf{W} \to \mathbf{W}$ induced by F, and $\mathscr{E}_{\lambda_0,1} = \mathrm{Uch}(\mathbf{G}^F)$ is the set of unipotent characters of $\mathbf{G}^F$; see Example 2.4.7. In this case, the class functions $R_{\tilde{\phi}}$ (for $\phi \in \mathrm{Irr}(\mathbf{W})^\sigma$) are the *unipotent uniform almost characters*; the above formula now reads:

$$\mathbb{D}_{\tilde{\phi}} = \frac{1}{|\mathbf{W}|} \sum_{w \in \mathbf{W}} (-1)^{l(w)} \tilde{\phi}(w) \frac{|\mathbb{G}|_{\mathbf{q}'}}{|\mathbb{T}_w|} \qquad (\phi \in \mathrm{Irr}(\mathbf{W})^\sigma).$$

We will see alternative interpretations of $\mathbb{D}_{\tilde{\phi}}$ in 4.1.6, 4.1.26, and Proposition 4.2.5. See also Proposition 2.4.32 at the end of this section.

Proposition 2.4.32 (Simplified degree polynomials) *Assume that we are in the setting of Theorem 2.4.15, and let $\rho \in \mathscr{E}_{\lambda,n}$. Then the degree polynomial $\mathbb{D}_\rho \in \mathbb{R}[\mathbf{q}]$ (see Definition 2.3.25) can be re-written as follows.*

$$\begin{aligned}\mathbb{D}_\rho &= \frac{1}{|\mathbf{W}_{\lambda,n}|} \sum_{y \in \mathbf{W}_{\lambda,n}} (-1)^{l(y)} m(y, \bar{x}_\rho) \frac{|\mathbb{G}|_{\mathbf{q}'}}{|\mathbb{T}_{w_1 y}|} \\ &= \sum_{\phi \in \mathrm{Irr}(\mathbf{W}_{\lambda,n})^\gamma} \langle R_{\tilde{\phi}}, \rho \rangle \, \mathbb{D}_{\tilde{\phi}},\end{aligned}$$

where the fake degree polynomials $\mathbb{D}_{\tilde{\phi}}$ are defined as in Remark 2.4.31.

Proof By Corollary 2.4.24, the sum over all pairs $(w, \theta) \in \mathfrak{X}(\mathbf{W}, \sigma)$ in the definition of $\mathbb{D}_\rho$ can be restricted to those pairs that lie in a fixed geometric conjugacy class of $\mathfrak{X}(\mathbf{W}, \sigma)$. Furthermore, that geometric conjugacy class is a union of orbits under the action of $\mathbf{W}$ on $\mathfrak{X}(\mathbf{W}, \sigma)$; see 2.3.20. Let $\{(w_a, \theta_a) \mid a \in I\}$ be a complete set of representatives for those orbits (where I is some finite index set). By the scalar product formula in Example 2.3.22(a), we have $\langle R_{w_a}^{\theta_a}, R_{w_b}^{\theta_b} \rangle = 0$ for $a, b \in I$ with $a \neq b$; furthermore, we have $R_w^\theta = R_{w_a}^{\theta_a}$ if $(w, \theta) \in \mathfrak{X}(\mathbf{W}, \sigma)$ is in the same $\mathbf{W}$-orbit as

(w_a, θ_a). Finally, $\langle R_{w_a}^{\theta_a}, R_{w_a}^{\theta_a}\rangle$ is precisely the size of the stabiliser of (w_a, θ_a) under the action of $\mathbf{W}$. Thus, the defining formula for $\mathbb{D}_\rho$ can be re-written as follows.

$$\mathbb{D}_\rho = \sum_{a\in I}(-1)^{l(w_a)}\frac{\langle R_{w_a}^{\theta_a}, \rho\rangle}{\langle R_{w_a}^{\theta_a}, R_{w_a}^{\theta_a}\rangle}\frac{|\mathbb{G}|_{\mathbf{q}'}}{|\mathbb{T}_{w_a}|}.$$

By Lemma 2.4.30, we may assume that $w_a \in \mathscr{X}_{\lambda,n}$ and $\theta_a = \theta_{w_a}$ for $a \in I$. (Note that, if $R_{w'}^{\theta'} = R_w^{\theta_w}$ as in Lemma 2.4.30, then Example 2.3.22(a) also shows that the pairs (w', θ') and (w, θ_w) are in the same $\mathbf{W}$-orbit of $\mathfrak{X}(\mathbf{W}, \sigma)$.) So we can write $w_a = w_1 y_a$, where $y_a \in \mathbf{W}_{\lambda,n}$ for $a \in I$. The scalar product formula in Remark 2.4.17 shows that

$$\langle R_{w_a}^{\theta_a}, R_{w_a}^{\theta_a}\rangle = \langle R_{w_1 y_a}^{\theta_{w_1 y_a}}, R_{w_1 y_a}^{\theta_{w_1 y_a}}\rangle = |C_{\mathbf{W}_{\lambda,n},\gamma}(y_a)|$$

(where the γ-centraliser is defined as in 2.1.6). Hence, we obtain

$$\mathbb{D}_\rho = \sum_{a\in I}(-1)^{l(w_1 y_a)}\frac{\langle R_{w_1 y_a}^{\theta_{w_1 y_a}}, \rho\rangle}{|C_{\mathbf{W}_{\lambda,n},\gamma}(y_a)|}\frac{|\mathbb{G}|_{\mathbf{q}'}}{|\mathbb{T}_{w_1 y_a}|}.$$

We now claim that $\{y_a \mid a \in I\}$ is a complete set of representatives of the γ-conjugacy classes of $\mathbf{W}_{\lambda,n}$. This is seen as follows. Let $a, b \in I$, $a \neq b$. Then $\langle R_{w_a}^{\theta_a}, R_{w_b}^{\theta_b}\rangle = 0$ and so the formula in Remark 2.4.17 shows that y_a, y_b are not γ-conjugate in $\mathbf{W}_{\lambda,n}$. On the other hand, by Example 2.4.25, all pairs $(w_1 y, \theta_{w_1 y})$ (for $y \in \mathbf{W}_{\lambda,n}$) are in the same geometric conjugacy class of $\mathfrak{X}(\mathbf{W}, \sigma)$. Hence, each such pair $(w_1 y, \theta_{w_1 y})$ must be in the same $\mathbf{W}$-orbit as $(w_a, \theta_a) = (w_1 y_a, \theta_{w_1 y_a})$, for some $a \in I$. But then

$$|\{x' \in \mathbf{W}_{\lambda,n} \mid y_a\gamma(x') = x'y\}| = \langle R_{w_1 y}^{\theta_{w_1 y}}, R_{w_1 y_a}^{\theta_{w_1 y_a}}\rangle \neq 0$$

(see again Remark 2.4.17 and Example 2.3.22(a)), and we conclude that y, y_a are γ-conjugate. Thus, the above claim is proved. Hence, the formula for $\mathbb{D}_\rho$ can be re-written as

$$\mathbb{D}_\rho = \frac{1}{|\mathbf{W}_{\lambda,n}|}\sum_{y\in\mathbf{W}_{\lambda,n}}(-1)^{l(w_1 y)}\langle R_{w_1 y}^{\theta_{w_1 y}}, \rho\rangle\frac{|\mathbb{G}|_{\mathbf{q}'}}{|\mathbb{T}_{w_1 y}|}.$$

It remains to use the formula for $\langle R_{w_1 y}^{\theta_{w_1 y}}, \rho\rangle$ in Theorem 2.4.15 and the identities concerning $R_{\tilde{\phi}}$ in Remark 2.4.17. □

Remark 2.4.33 Finally, we remark that the proofs of the main results of [Lu84a] involve a further ingredient that we did not discuss here at all: the theory of 'cells' of finite Coxeter groups, as introduced by Kazhdan and Lusztig [KaLu79], and further developed by Lusztig; see [Lu03b] and the survey [Cu88]. At the time of writing [Lu84a], some crucial properties of cells (see, e.g., [Lu81a]) were established using the theory of primitive ideals in enveloping algebras of semisimple Lie algebras.

As pointed out in [Lu87b, p. 253] (see also [Lu18b]), the further development of the cell theory made it possible to avoid that dependence on the theory of primitive ideals.

2.5 Geometric Conjugacy and the Dual Group

The basic assumptions of the previous section remain in force; in particular, we assume that an isomorphism $\iota\colon k^\times \xrightarrow{\sim} (\mathbb{Q}/\mathbb{Z})_{p'}$ as in 2.4.4 has been fixed. We will now reformulate the partition of $\mathrm{Irr}(\mathbf{G}^F)$ into geometric series in terms of 'dual groups', already introduced in Section 1.5. Following [Lu84a, 8.4], this can be done in a rather straightforward way by using the formalism of pairs (λ, n) and the sets $\mathscr{E}_{\lambda,n}$ from the previous section. Since this set-up is somewhat different from that in [Ca85], [DiMi20], we will give detailed proofs whenever appropriate.

2.5.1 Let $\mathbf{G}^*$ be a connected reductive algebraic group over k. Let $F^*\colon \mathbf{G}^* \to \mathbf{G}^*$ be a Steinberg map such that the pairs $(\mathbf{G}, F)$ and $(\mathbf{G}^*, F^*)$ are in duality; see Definition 1.5.17 and Example 1.6.19. Such a duality is defined with respect to maximally split tori $\mathbf{T}_0 \subseteq \mathbf{G}$ and $\mathbf{T}_0^* \subseteq \mathbf{G}^*$. Thus, we are given an isomorphism $\delta\colon X(\mathbf{T}_0) \xrightarrow{\sim} Y(\mathbf{T}_0^*)$ satisfying the conditions in Definition 1.5.17. Now, by [Ca85, 3.1.2], the map $k^\times \times Y(\mathbf{T}_0^*) \to \mathbf{T}_0^*$, $(\xi, \nu) \mapsto \nu(\xi)$, induces an isomorphism of abelian groups $k^\times \otimes_{\mathbb{Z}} Y(\mathbf{T}_0^*) \xrightarrow{\sim} \mathbf{T}_0^*$. (We have already used this in Example 1.5.6.) Combining this with our chosen isomorphism $\iota\colon k^\times \xrightarrow{\sim} (\mathbb{Q}/\mathbb{Z})_{p'}$ and the map $\delta\colon X(\mathbf{T}_0) \xrightarrow{\sim} \mathbf{Y}(\mathbf{T}_0^*)$, we obtain an isomorphism of abelian groups

$$(\mathbb{Q}/\mathbb{Z})_{p'} \otimes_{\mathbb{Z}} X(\mathbf{T}_0) \xrightarrow{\sim} \mathbf{T}_0^*, \qquad (x+\mathbb{Z}) \otimes \lambda \mapsto \delta(\lambda)\big(\iota^{-1}(x+\mathbb{Z})\big). \tag{a}$$

Next, recall that $\mathbf{W} = N_{\mathbf{G}}(\mathbf{T}_0)/\mathbf{T}_0$ is the Weyl group of $\mathbf{G}$ and $\sigma\colon \mathbf{W} \to \mathbf{W}$ is the automorphism induced by F; see 1.6.1. Similarly, $\mathbf{W}^* = N_{\mathbf{G}^*}(\mathbf{T}_0^*)/\mathbf{T}_0^*$ is the Weyl group of $\mathbf{G}^*$ and $\sigma^*\colon \mathbf{W}^* \to \mathbf{W}^*$ is the automorphism induced by F^*. By [Ca85, 4.2.3, 4.3.2], there is a group isomorphism $\mathbf{W} \xrightarrow{\sim} \mathbf{W}^*$, $w \mapsto w^*$, such that

$$\big(\sigma(w)\big)^* = (\sigma^*)^{-1}(w^*) \quad\text{and}\quad \delta(w.\lambda) = w^*.\delta(\lambda) \quad\text{for all } \lambda \in X(\mathbf{T}_0). \tag{b}$$

Furthermore, $l(w) = l^*(w^*)$ for all $w \in \mathbf{W}$, where $l\colon \mathbf{W} \to \mathbb{Z}_{\geqslant 0}$ and $l^*\colon \mathbf{W}^* \to \mathbb{Z}_{\geqslant 0}$ are the length functions as in 1.6.1. (See also Remark 1.5.19.) Using the isomorphism $\mathbf{W} \xrightarrow{\sim} \mathbf{W}^*$, $w \mapsto w^*$, we obtain a natural bijection

$$\left\{\begin{array}{c}\mathbf{G}^F\text{-classes of } F\text{-stable}\\ \text{maximal tori } \mathbf{T} \subseteq \mathbf{G}\end{array}\right\} \longleftrightarrow \left\{\begin{array}{c}\mathbf{G}^{*F^*}\text{-classes of } F^*\text{-stable}\\ \text{maximal tori } \mathbf{T}^* \subseteq \mathbf{G}^*\end{array}\right\}. \tag{c}$$

Here, the $\mathbf{G}^F$-conjugacy class of $\mathbf{T} \subseteq \mathbf{G}$ corresponds to the $\mathbf{G}^{*F^*}$-conjugacy class of $\mathbf{T}^* \subseteq \mathbf{G}^*$ if there exists some $w \in \mathbf{W}$ such that $\mathbf{T}$ is of type w and $\mathbf{T}^*$ is of type $(w^*)^{-1}$; see [Ca85, 4.3.4].

Lemma 2.5.2 *In the above setting, assume that $\mathbf{T} \subseteq \mathbf{G}$ and $\mathbf{T}^* \subseteq \mathbf{G}^*$ correspond to each other via 2.5.1(c). Then we have $|\mathbf{T}^F| = |\mathbf{T}^{*F^*}|$; furthermore, the order polynomials of $(\mathbf{T}, F)$ and $(\mathbf{T}^*, F^*)$ are equal.*

Proof By [DeLu76, 5.2] or [Ca85, 4.3.4], the two pairs $(\mathbf{T}, F)$ and $(\mathbf{T}^*, F^*)$ are in duality, that is, there exists an isomorphism $\delta\colon X(\mathbf{T}) \to Y(\mathbf{T}^*)$ satisfying the conditions in Definition 1.5.17. Using the formula in 1.6.21, it follows that $(\mathbf{T}, F)$ and $(\mathbf{T}^*, F^*)$ have the same order polynomials. Alternatively, one can argue as follows. By [Ca85, 4.4.2], we have $|\mathbf{T}^F| = |\mathbf{T}^{*F^*}|$. By [DeLu76, 5.3], this equality remains valid when we replace F,F^* by F^n, $(F^*)^n$, for any $n \geqslant 1$. But then not only the individual orders are the same, but also the corresponding order polynomials. □

Remark 2.5.3 Let $X = X(\mathbf{T}_0)$. For working with $(\mathbb{Q}/\mathbb{Z})_{p'} \otimes_{\mathbb{Z}} X$, we note:

(a) Every element of $(\mathbb{Q}/\mathbb{Z})_{p'} \otimes_{\mathbb{Z}} X$ is a tensor of the form $(\frac{1}{n}+\mathbb{Z}) \otimes \lambda$ where $\lambda \in X$ and $n \geqslant 1$ is an integer prime to p.
(b) Let $\lambda, \mu \in X$ and $n, m \geqslant 1$ be integers prime to p. Then $(\frac{1}{n}+\mathbb{Z}) \otimes \lambda = (\frac{1}{m}+\mathbb{Z}) \otimes \mu$ if and only if there exist integers $d, d' \geqslant 1$, both prime to p, such that $dn = d'm$ and $d'\mu = d\lambda + dn\nu$ for some $\nu \in X$.
(c) Let $m \geqslant 1$ be an integer prime to p. Then every element of $(\mathbb{Q}/\mathbb{Z})_{p'} \otimes_{\mathbb{Z}} X$ of order m can be written as $(\frac{1}{m}+\mathbb{Z}) \otimes \mu$ where $\mu \in X$.

This immediately follows from standard properties of tensor products and the fact that X is a free $\mathbb{Z}$-module of finite rank. (We omit further details.)

The following results contain a basic feature of a duality as above: it replaces characters of a torus by elements in a dual torus. We set

$$t_{\lambda,n} := \delta(\lambda)\big(\iota^{-1}(\tfrac{1}{n}+\mathbb{Z})\big) \in \mathbf{T}_0^* \qquad \text{for any pair } (\lambda, n) \text{ as in 2.4.5.}$$

Thus, $t_{\lambda,n}$ is the image of $(\frac{1}{n}+\mathbb{Z}) \otimes \lambda$ under the isomorphism in 2.5.1(a).

Lemma 2.5.4 *Let $\lambda \in X$ and $n \geqslant 1$ be an integer prime to p, as in 2.4.5.*

(a) *We have $F^*(t_{\lambda,n}) = t_{\lambda\circ F,n}$ and $t_{w.\lambda,n} = \dot{w}^* t_{\lambda,n} (\dot{w}^*)^{-1}$ for any $w \in \mathbf{W}$.*
(b) *The $\mathbf{G}^*$-conjugacy class of $t_{\lambda,n}$ is F^*-stable if and only if $\mathscr{Z}_{\lambda,n} \neq \varnothing$. If $\mathscr{Z}_{\lambda,n} \neq \varnothing$, then $F^*(t_{\lambda,n}) = \dot{w}^* t_{\lambda,n} (\dot{w}^*)^{-1}$ for all $w \in \mathscr{Z}_{\lambda,n}$.*
(c) *(λ, n) is indivisible if and only if n is the order of the element $t_{\lambda,n}$.*

Proof Let $\xi := \iota^{-1}(\frac{1}{n}+\mathbb{Z}) \in k^\times$; then $t_{\lambda,n} = \delta(\lambda)(\xi)$.

(a) By Definition 1.5.17(b), we have $F^* \circ \delta(\lambda) = \delta(\lambda \circ F)$. This yields that

$$F^*(t_{\lambda,n}) = F^*\big(\delta(\lambda)(\xi)\big) = \big(F^* \circ \delta(\lambda)\big)(\xi) = \delta\big(\lambda \circ F\big)(\xi) = t_{\lambda\circ F,n}.$$

Now let $w \in \mathbf{W}$. Then

$$t_{w.\lambda,n} = \delta(w.\lambda)(\xi) = \big(w^*.\delta(\lambda)\big)(\xi) = \dot{w}^*\delta(\lambda)(\xi)(\dot{w}^*)^{-1} = \dot{w}^* t_{\lambda,n}(\dot{w}^*)^{-1}$$

where the second equality holds by 2.5.1(b) and the third equality just expresses the natural action of $\mathbf{W}^*$ on $Y(\mathbf{T}_0^*)$; see 1.3.1.

(b) First assume that $F^*(t_{\lambda,n})$ is $\mathbf{G}^*$-conjugate to $t_{\lambda,n}$. It is well known (see, e,g., [Ca85, 3.7.1]) that then we can already find some $w \in \mathbf{W}$ such that $F^*(t_{\lambda,n}) = \dot{w}^* t_{\lambda,n}(\dot{w}^*)^{-1}$. Now, by (a), the left-hand side equals $t_{\lambda\circ F,n}$ and the right-hand side equals $t_{w.\lambda,n}$. So we have $(\frac{1}{n}+\mathbb{Z})\otimes(\lambda\circ F) = (\frac{1}{n}+\mathbb{Z})\otimes(w.\lambda)$. Using Remark 2.5.3(b), it follows that $\lambda \circ F = w.\lambda + n\mu$ for some $\mu \in X$ and so $\mathscr{Z}_{\lambda,n} \neq \varnothing$. Conversely, assume that $w \in \mathscr{Z}_{\lambda,n}$, that is, $\lambda \circ F = w.\lambda + n\lambda_w$. Using (a), we obtain

$$F^*(t_{\lambda,n}) = t_{\lambda\circ F,n} = \delta(w.\lambda + n\lambda_w)(\xi) = \delta(w.\lambda)(\xi)\cdot\delta(\lambda_w)(\xi^n).$$

We certainly have $\delta(\lambda_w)(\xi^n) = 1$ since $\xi^n = \iota^{-1}(1+\mathbb{Z}) = 1 \in k^\times$. Hence, using (a) again, the right-hand side of the above equation equals $\dot{w}^* t_{\lambda,n}(\dot{w}^*)^{-1}$ and so the $\mathbf{G}^*$-conjugacy class of $t_{\lambda,n}$ is F^*-stable.

(c) Let $m \geqslant 1$ be the order of $t_{\lambda,n}$. Since $\xi^n = 1$, we have $t_{\lambda,n}^n = 1$ and so $m \mid n$. Under the isomorphism in 2.5.1(a), $t_{\lambda,n}$ corresponds to an element of order m in $(\mathbb{Q}/\mathbb{Z})_{p'} \otimes_{\mathbb{Z}} X$. By Remark 2.5.3(c), there exists some $\mu \in X$ such that $t_{\lambda,n} = \delta(\mu)\big(\iota^{-1}(\frac{1}{m}+\mathbb{Z})\big)$ and so $(\frac{1}{n}+\mathbb{Z})\otimes\lambda = (\frac{1}{m}+\mathbb{Z})\otimes\mu$. Using Remark 2.5.3(b), it follows that (λ, n) is indivisible if and only if $n = m$. □

Proposition 2.5.5 (Cf. [DeLu76, 5.7, 5.22]) *The map $(\lambda, n) \mapsto t_{\lambda,n}$ induces a bijection between equivalence classes of pairs $(\lambda, n) \in \Lambda(\mathbf{G}, F)$ (see Proposition 2.4.28) and F^*-stable conjugacy classes of semisimple elements in $\mathbf{G}^*$.*

Proof Let $(\lambda_i, n_i) \in \Lambda(\mathbf{G}, F)$ for $i = 1, 2$. By Lemma 2.5.4(b), the $\mathbf{G}^*$-conjugacy class of t_{λ_i,n_i} is F^*-stable. We must show that $(\lambda_1, n_1) \sim (\lambda_2, n_2)$ if and only if t_{λ_1,n_1} and t_{λ_2,n_2} are conjugate in $\mathbf{G}^*$. Assume first that $(\lambda_1, n_1) \sim (\lambda_2, n_2)$, that is, $n_1 = n_2$ and $\lambda_2 = w.\lambda_1 + n_1\nu$ for some $w \in \mathbf{W}$ and $\nu \in X$. Let $\xi := \iota^{-1}(\frac{1}{n_1}+\mathbb{Z}) \in k^\times$. Then $t_{n_1,\lambda_1} = \delta(\lambda_1)(\xi)$ and

$$t_{\lambda_2,n_2} = t_{\lambda_2,n_1} = \delta(\lambda_2)(\xi) = \delta(w.\lambda_1 + n_1\nu)(\xi) = \delta(w.\lambda_1)(\xi)\cdot\delta(\nu)(\xi^{n_1}).$$

Now, $\delta(\nu)(\xi^{n_1}) = 1$ since $\xi^{n_1} = \iota^{-1}(1+\mathbb{Z}) = 1 \in k^\times$. Furthermore, $\delta(w.\lambda_1)(\xi) = \dot{w}^* t_{\lambda_1,n_1}(\dot{w}^*)^{-1}$ by Lemma 2.5.4(a). Hence, t_{λ_1,n_1} and t_{λ_2,n_2} are conjugate in $\mathbf{G}^*$. Conversely, assume that t_{λ_1,n_1} and t_{λ_2,n_2} are conjugate in $\mathbf{G}^*$. Then t_{λ_1,n_1} and t_{λ_2,n_2} have the same order and so $n := n_1 = n_2$; see Lemma 2.5.4(c). As in the proof of Lemma 2.5.4(b), there exists some $w \in \mathbf{W}$ such that $t_{\lambda_2,n_2} = \dot{w}^* t_{\lambda_1,n_1}(\dot{w}^*)^{-1}$. But $t_{w.\lambda_1,n_1} = \dot{w}^* t_{\lambda_1,n_1}(\dot{w}^*)^{-1}$ and so $(\frac{1}{n}+\mathbb{Z})\otimes(w.\lambda_1) = (\frac{1}{n}+\mathbb{Z})\otimes\lambda_2$. Using Remark 2.5.3(b), it follows that $(\lambda_1, n_1) \sim (\lambda_2, n_2)$, as required. Finally, we show that every F^*-stable conjugacy class of semisimple elements in $\mathbf{G}^*$ corresponds to an equivalence class of pairs in $\Lambda(\mathbf{G}, F)$ by the above procedure. So let C be such a conjugacy class in $\mathbf{G}^*$. By [Ca85, 3.7.1], we can find an element $s \in C \cap \mathbf{T}_0^*$. Furthermore, since C is F^*-stable, there exists some $w \in \mathbf{W}$ such that $F^*(s) = \dot{w}^* s (\dot{w}^*)^{-1}$. Using the isomorphism in 2.5.1(a) and Remark 2.5.3, we can write $s = t_{\lambda,n}$ where (λ, n) is a pair as in 2.4.5 and n is the order of s. By Lemma 2.5.4, we have $\mathscr{Z}_{\lambda,n} \neq \varnothing$ and so $(\lambda, n) \in \Lambda(\mathbf{G}, F)$. □

Combining the above result with Proposition 2.4.28, we obtain a bijection

$$\{\text{geometric conjugacy classes of pairs } (\mathbf{T}, \theta) \in \mathfrak{X}(\mathbf{G}, F)\}$$
$$\xleftrightarrow{1-1} \quad \{F^*\text{-stable conjugacy classes of semisimple elements in } \mathbf{G}^*\}.$$

By Corollary 2.4.29, we have a partition $\operatorname{Irr}(\mathbf{G}^F) = \bigsqcup_{(\lambda,n)} \mathscr{E}_{\lambda,n}$ where the union runs over a set of representatives for the equivalence classes of $\Lambda(\mathbf{G}, F)$ (as above), and each piece $\mathscr{E}_{\lambda,n}$ is a geometric series of characters.

Proposition 2.5.6 *The number of geometric conjugacy classes of pairs in $\mathfrak{X}(\mathbf{G}, F)$ is $|\mathbf{Z}^\circ(\mathbf{G})^F|q^l$ where $l = \dim \mathbf{T}_0 - \dim \mathbf{Z}^\circ(\mathbf{G})$ is the semisimple rank of $\mathbf{G}$.*

Proof By [St68, 14.8] (or [Ca85, 3.7.6]), it is known that the number of F-stable conjugacy classes of semisimple elements of $\mathbf{G}$ is given by $|\mathbf{Z}^\circ(\mathbf{G})^F|q^l$. By [Ca85, 4.4.5], we have $|\mathbf{Z}^\circ(\mathbf{G}^*)^{F^*}| = |\mathbf{Z}^\circ(\mathbf{G})^F|$. Furthermore, l also is the semisimple rank of $\mathbf{G}^*$. Hence, we conclude that $|\mathbf{Z}^0(\mathbf{G})^F|q^l$ is the number of F^*-stable conjugacy classes of semisimple elements of $\mathbf{G}^*$. By Proposition 2.5.5, the latter number is the number of geometric conjugacy classes of pairs in $\mathfrak{X}(\mathbf{G}, F)$. □

The map $(\lambda, n) \mapsto t_{\lambda,n}$ gives rise to a number of further constructions.

Lemma 2.5.7 (Cf. [DeLu76, 5.2], [Lu84a, 8.4]) *For each $w \in \mathbf{W}$, there is a unique isomorphism of abelian groups*

$$\operatorname{Irr}(\mathbf{T}_0[w]) \xrightarrow{\sim} \mathbf{T}_0^*[(w^*)^{-1}], \qquad \theta \mapsto s_\theta,$$

satisfying the following condition. If $\theta \in \operatorname{Irr}(\mathbf{T}_0[w])$ and (λ, n) is any pair as in Lemma 2.4.8 such that $(\lambda, n) \leftrightsquigarrow (w, \theta)$, then $s_\theta = t_{\lambda,n} \in \mathbf{T}_0^*$.

Proof Let $X = X(\mathbf{T}_0)$. Let $\theta \in \operatorname{Irr}(\mathbf{T}_0[w])$ and $n \geqslant 1$ be the smallest integer such that $\theta(t)^n = 1$ for all $t \in \mathbf{T}_0[w]$. By Lemma 2.4.8, there exists some $\lambda \in X$ such that $w \in \mathscr{Z}_{\lambda,n}$ and $\theta = \psi \circ \bar{\lambda}_w$ (see 2.4.5). Then we set $s_\theta := t_{\lambda,n}$. We must show that this is well defined. So let (μ, m) be any pair as in 2.4.5 such that $(\mu, m) \leftrightsquigarrow (w, \theta)$. Then $n \mid m$. Writing $m = dn$ where $d \geqslant 1$, we also have $(d\lambda, dn) \leftrightsquigarrow (w, \theta)$; see Remark 2.4.9. Now Lemma 2.4.8 implies that $\mu = d\lambda + dn\nu$ for some $\nu \in X$. Hence, by Remark 2.5.3(b), the pairs (μ, m) and (λ, n) define the same element of $(\mathbb{Q}/\mathbb{Z})_{p'} \otimes_{\mathbb{Z}} X$ and so $s_\theta = t_{\lambda,n} = t_{\mu,m}$, as desired.

Next, since $(\lambda, n) \leftrightsquigarrow (w, \theta)$, we have $w \in \mathscr{Z}_{\lambda,n}$ and so we obtain $F^*(t_{\lambda,n}) = \dot{w}^* t_{\lambda,n}(\dot{w}^*)^{-1}$; see Lemma 2.5.4(b). Thus, $s_\theta = t_{\lambda,n} \in \mathbf{T}_0^*[(w^*)^{-1}]$, as desired. So we have a well-defined map $\theta \mapsto s_\theta$. To see that this is a group homomorphism, we set $m := |\mathrm{Irr}(\mathbf{T}_0[w])|$; note that this is prime to p. For each $\theta \in \mathrm{Irr}(\mathbf{T}_0[w])$, we have $s_\theta = t_{\mu,m}$ where $\mu \in X$ is such that $(\mu, m) \leftrightsquigarrow (w, \theta)$. Working with this fixed m, it is then easy to see that $\theta \mapsto s_\theta$ is a group homomorphism.

Finally, we show that the map $\theta \mapsto s_\theta$ is bijective. First, let $\theta \in \mathrm{Irr}(\mathbf{T}_0[w])$ and write again $s_\theta = t_{\lambda,n}$ where, as above, $n \geqslant 1$ is the smallest integer such that $\theta(t)^n = 1$ for all $t \in \mathbf{T}_0[w]$. So (λ, n) is indivisible by Remark 2.4.10; consequently, n is the order of $s_\theta = t_{\lambda,n}$ by Lemma 2.5.4(c). Hence, if $\theta \neq 1$, then $n > 1$ and so $s_\theta \neq 1$. Thus, $\theta \mapsto s_\theta$ is injective. On the other hand, by Lemma 2.5.2, we have

$$|\mathrm{Irr}(\mathbf{T}_0[w])| = |\mathbf{T}_0[w]| = |\mathbf{T}_0^*[(w^*)^{-1}]|.$$

Hence, the map $\theta \mapsto s_\theta$ is also surjective. □

Since we systematically work in the set-up of [Lu84a] (as in 2.4.5), the above construction of an isomorphism $\mathrm{Irr}(\mathbf{T}_0[w]) \cong \mathbf{T}_0^*[(w^*)^{-1}]$ is somewhat different from that in [Ca85, 4.4.1] or [DiMi20, 11.1.7, 11.1.14] (although, of course, the two constructions can be seen to be equivalent by following the exact sequences in [DeLu76, 5.2]; see also [DiMi20, 11.1.7]). We can now state the following version of the dictionary between geometric conjugacy classes and the dual group.

Corollary 2.5.8 *For $i = 1, 2$ let $(w_i, \theta_i) \in \mathfrak{X}(\mathbf{W}, \sigma)$ and consider the corresponding semisimple elements $s_{\theta_i} \in \mathbf{T}_0^*[(w_i^*)^{-1}] \subseteq \mathbf{T}_0^*$, obtained via the isomorphisms in Lemma 2.5.7. Then (w_1, θ_1) and (w_2, θ_2) are geometrically conjugate if and only if $s_{\theta_2} = \dot{y}\, s_{\theta_1}\, \dot{y}^{-1}$ for some $y \in \mathbf{W}^*$.*

Proof For $i = 1, 2$ let $(\lambda_i, n_i) \in \Lambda(\mathbf{G}, F)$ be such that $(\lambda_i, n_i) \leftrightsquigarrow (w_i, \theta_i)$. By Proposition 2.4.28, the pairs (λ_1, n_1) and (λ_2, n_2) are equivalent if and only if (w_1, θ_1), (w_2, θ_2) are geometrically conjugate. Now $s_{\theta_i} = t_{\lambda_i,n_i}$ for $i = 1, 2$. It remains to use Proposition 2.5.5 and to recall that two elements in $\mathbf{T}_0^*$ are conjugate in $\mathbf{G}^*$ if and only if they are conjugate in $N_{\mathbf{G}^*}(\mathbf{T}_0^*)$. □

Remark 2.5.9 For future reference, we state a compatibility property (see (∗) below) of the isomorphism in Lemma 2.5.7. Assume that $(\mathbf{G}', F')$ and $(\mathbf{G}'^*, F'^*)$ also is a pair of connected reductive groups in duality, with respect to maximally split tori $\mathbf{T}_0' \subseteq \mathbf{G}'$ and $\mathbf{T}_0'^* \subseteq \mathbf{G}'^*$. Let $f\colon \mathbf{G} \to \mathbf{G}'$ and $f^*\colon \mathbf{G}'^* \to \mathbf{G}^*$ be central isotypies that correspond to each other by duality, exactly as in 1.7.11; in particular, we have $f(\mathbf{T}_0) \subseteq \mathbf{T}_0'$ and $f^*(\mathbf{T}_0'^*) \subseteq \mathbf{T}_0^*$. Using f and f^*, we may identify $\mathbf{W}$ with the Weyl group of $\mathbf{G}'$ (with respect to $\mathbf{T}_0'$) and $\mathbf{W}^*$ with the Weyl group of $\mathbf{G}'^*$ (with respect to $\mathbf{T}_0'^*$). Thus, for each $w \in \mathbf{W}$, we also have an isomorphism

$$\mathrm{Irr}(\mathbf{T}_0'[w]) \xrightarrow{\sim} \mathbf{T}_0'^*[(w^*)^{-1}], \qquad \theta' \mapsto s_{\theta'}.$$

Using $f \circ F = F' \circ f$ and $f^* \circ F'^* = F^* \circ f^*$, one immediately checks that

$$f\big(\mathbf{T}_0[w]\big) \subseteq \mathbf{T}_0'[w] \qquad \text{and} \qquad f^*\big(\mathbf{T}_0'^*[(w^*)^{-1}]\big) \subseteq \mathbf{T}_0^*[(w^*)^{-1}].$$

Now let $\theta' \in \mathrm{Irr}(\mathbf{T}_0'[w])$. Then we claim that

$$s_\theta = f^*(s_{\theta'}) \qquad \text{where} \qquad \theta := \theta' \circ f|_{\mathbf{T}_0[w]} \in \mathrm{Irr}(\mathbf{T}_0[w]). \tag{$*$}$$

To see this, let $(\lambda', n) \in \Lambda(\mathbf{G}', F')$ be such that $(\lambda', n) \leftrightsquigarrow (w, \theta')$. Then $s_{\theta'} = \delta'(\lambda')(\xi)$ where $\xi := \iota^{-1}(\frac{1}{n} + \mathbb{Z}) \in k^\times$ and $\delta' \colon X(\mathbf{T}_0') \to Y(\mathbf{T}_0'^*)$ is the isomorphism that defines the duality between $(\mathbf{G}', F')$ and $(\mathbf{G}'^*, F'^*)$. Now set $\lambda := \lambda' \circ f \in X(\mathbf{T}_0)$ and $\lambda_w := \lambda_w' \circ f \in X(\mathbf{T}_0)$. A straightforward computation shows that $n\lambda_w = \lambda \circ F - w.\lambda$ and so $w \in \mathscr{Z}_{\lambda,n}$; furthermore, we have $(\lambda, n) \leftrightsquigarrow (w, \theta)$ and

$$s_\theta = \delta(\lambda)(\xi) = \big(\delta(\lambda' \circ f)\big)(\xi) = \big(f^* \circ (\delta'(\lambda'))\big)(\xi) = f^*\big(\delta'(\lambda')(\xi)\big) = f^*(s_{\theta'}),$$

where the third equality holds by the compatibility relation in 1.7.11.

2.5.10 Let us fix a pair $(\lambda, n) \in \Lambda(\mathbf{G}, F)$ and set $s := t_{\lambda,n} \in \mathbf{T}_0^*$. Since $\mathscr{Z}_{\lambda,n} \neq \varnothing$, the $\mathbf{G}^*$-conjugacy class of s is F^*-stable (see Lemma 2.5.4). Let $R^* \subseteq X(\mathbf{T}_0^*)$ be the set of roots of $\mathbf{G}^*$ with respect to $\mathbf{T}_0^*$. Recall from Definition 1.5.17 that $R^* = \{\alpha^* \mid \alpha \in R\}$, where α^* is determined by the condition that $\delta(\alpha) = \alpha^{*\vee} \in Y(\mathbf{T}_0^*)$. Then, by 2.2.13, we have

$$C^\circ_{\mathbf{G}^*}(s) = \langle \mathbf{T}_0^*, \mathbf{U}_{\alpha^*} \mid \alpha^* \in R_s^* \rangle$$

where $R_s^* := \{\alpha^* \in R^* \mid \alpha^*(s) = 1\}$ is the root system of $C^\circ_{\mathbf{G}^*}(s)$ with respect to $\mathbf{T}_0^*$. Let $\xi = \iota^{-1}(\frac{1}{n} + \mathbb{Z}) \in k^\times$. Then

$$\alpha^*(s) = \alpha^*(t_{\lambda,n}) = \alpha^*\big(\delta(\lambda)(\xi)\big) = \xi^{\langle \alpha^*, \delta(\lambda)\rangle} = \xi^{\langle \lambda, \alpha^\vee\rangle},$$

where we used the relations in 1.1.11 and Definition 1.5.17(a). Thus, since ξ has order n, we have $\alpha^* \in R_s^*$ if and only if $\langle \lambda, \alpha^\vee \rangle \in \mathbb{Z}$ is divisible by n. Hence, we obtain that

$$R_s^* = \{\alpha^* \mid \alpha \in R_{\lambda,n}\}, \qquad \text{with } R_{\lambda,n} \subseteq R \text{ as in 2.4.13}.$$

So the Weyl group of $C^\circ_{\mathbf{G}^*}(s)$ (with respect to $\mathbf{T}_0^*$) is given by the subgroup

$$\mathbf{W}_{\lambda,n}^* := \{w^* \mid w \in \mathbf{W}_{\lambda,n}\} \subseteq \mathbf{W}^*.$$

Assume now that $\mathbf{Z}(\mathbf{G})$ is connected. By Lemma 1.5.22, this implies that the fundamental group of $\mathbf{G}^*$ is trivial. Hence, Theorem 2.2.14 shows that $C_{\mathbf{G}^*}(s) = C^\circ_{\mathbf{G}^*}(s)$ is connected. Let w_1 be the unique element of minimal length in $\mathscr{Z}_{\lambda,n}$ (as in

Lemma 2.4.14) and define $F' : \mathbf{G}^* \to \mathbf{G}^*$ by $F'(g) = (\dot{w}_1^*)^{-1} F^*(g)\dot{w}_1^*$ for $g \in \mathbf{G}^*$. Then F' is a Steinberg map by Lemma 1.4.14. We have seen in Lemma 2.5.4(b) that $F^*(t_{\lambda,n}) = \dot{w}_1^* t_{\lambda,n}(\dot{w}_1^*)^{-1}$. Hence, we obtain that $F'(s) = s$ and so $C_{\mathbf{G}^*}(s)$ is F'-stable. The map induced by F' on $\mathbf{W}_{\lambda,n}^*$ is given by

$$\gamma'(w^*) = (w_1^*)^{-1}\sigma^*(w^*)w_1^* \qquad \text{for all } w \in \mathbf{W}_{\lambda,n}.$$

Recall from Lemma 2.4.14 that we have an automorphism $\gamma : \mathbf{W}_{\lambda,n} \to \mathbf{W}_{\lambda,n}$ such that $\gamma(w) = \sigma(w_1 w w_1^{-1})$ for all $w \in \mathbf{W}_{\lambda,n}$. By 2.5.1(b), we have $\sigma^*(w^*) = \left(\sigma^{-1}(w)\right)^*$ for all $w \in \mathbf{W}$. This yields

$$\gamma'(w^*) = (w_1^*)^{-1}\left(\sigma^{-1}(w)\right)^* w_1^* = \left(w_1^{-1}\sigma^{-1}(w)w_1\right)^* = \left(\gamma^{-1}(w)\right)^*$$

for all $w \in \mathbf{W}_{\lambda,n}$. Thus, under the isomorphism $\mathbf{W}_{\lambda,n} \xrightarrow{\sim} \mathbf{W}_{\lambda,n}^*$, $w \mapsto w^*$,

(a) $\gamma' \in \mathrm{Aut}(\mathbf{W}_{\lambda,n}^*)$ corresponds to $\gamma^{-1} \in \mathrm{Aut}(\mathbf{W}_{\lambda,n})$.

It follows that the set of simple reflections $S_{\lambda,n}^* = \{w^* \mid w \in S_{\lambda,n}\}$ of $\mathbf{W}_{\lambda,n}^*$ is preserved by γ'. Finally, this also shows that the torus $\mathbf{T}_0^* \subseteq C_{\mathbf{G}^*}(s)$ is maximally split with respect to F'. Note that, by [Ca85, 3.5.5],

$$\mathbf{B}_s^* := \langle \mathbf{T}_0^*, \mathbf{U}_{\alpha^*} \mid \alpha \in R^+ \cap R_{\lambda,n}\rangle$$

is a Borel subgroup of $C_{\mathbf{G}^*}(s)$, and this Borel subgroup is F'-stable (see the proof of Lemma 2.4.14(b)). We conclude that

(b) the Weyl group $\mathbf{W}_{\lambda,n}^*$ and the automorphism γ' arise from the connected reductive algebraic group $C_{\mathbf{G}^*}(s)$ and the Steinberg map F' by the same procedure by which $\mathbf{W}, \sigma$ arise from $\mathbf{G}, F$ as in 1.6.1.

Now let $\mathbf{H} := C_{\mathbf{G}^*}(s)$. We can apply the construction of dual groups to the pair $(\mathbf{H}, F')$. Thus, we obtain a connected reductive algebraic group $\mathbf{H}^*$ and a Steinberg map $F'^* : \mathbf{H}^* \to \mathbf{H}^*$ such that $(\mathbf{H}^*, F'^*)$ is in duality with $(\mathbf{H}, F')$ with respect to suitable maximally split tori in $\mathbf{H}$ and in $\mathbf{H}^*$. Then we can identify the Weyl group of $\mathbf{H}^*$ with $\mathbf{W}_{\lambda,n}$; by (a) and 2.5.1(b), the automorphism induced by F'^* on $\mathbf{W}_{\lambda,n}$ will then be γ. Thus, as already announced in Remark 2.4.16(a), we can now state:

(c) There is a bijection $\mathrm{Uch}(\mathbf{H}^{*F'^*}) \xleftrightarrow{1-1} \bar{X}(\mathbf{W}_{\lambda,n}, \gamma)$, $\psi \leftrightarrow \bar{x}_\psi$, satisfying the conditions in Theorem 2.4.1.

Now, it is even known that $(\mathbf{H}^*, F'^*)$ can be replaced by $(\mathbf{H}, F')$ in (c). This will be further discussed in the next section (see Remark 2.6.5).

Proposition 2.5.11 *In the setting of Theorem 2.4.15, let $\rho \in \mathscr{E}_{\lambda,n}$. As in 2.5.10(c),* *let* $\psi \in \mathrm{Uch}(\mathbf{H}^{*F'^*})$ *be the unipotent character such that* $\bar{x}_\psi = \bar{x}_\rho \in \bar{X}(\mathbf{W}_{\lambda,n}, \gamma)$. *Then*

$$\mathbb{D}_\rho = \frac{|\mathbb{G}|_{\mathbf{q}'}}{|\mathbb{H}'^*|_{\mathbf{q}'}} \mathbb{D}_\psi,$$

where $\mathbb{G}$ *and* $\mathbb{H}'^*$ *are the complete root data associated with* $\mathbf{G}, F$ *and with* $\mathbf{H}^*, F'^*$, *respectively; furthermore,* $\mathbb{D}_\rho \in \mathbb{R}[\mathbf{q}]$ *and* $\mathbb{D}_\psi \in \mathbb{R}[\mathbf{q}]$ *are the degree polynomials of* ρ *and* ψ, *respectively (see Definition 2.3.25).*

Proof By Theorems 2.4.1 and 2.4.15, and Proposition 2.4.32, we have

$$\mathbb{D}_\rho = \frac{1}{|\mathbf{W}_{\lambda,n}|} \sum_{y \in \mathbf{W}_{\lambda,n}} (-1)^{l(y)} m(y, \bar{x}_\rho) \frac{|\mathbb{G}|_{\mathbf{q}'}}{|\mathbb{T}_{w_1 y}|},$$

$$\mathbb{D}_\psi = \frac{1}{|\mathbf{W}_{\lambda,n}|} \sum_{y \in \mathbf{W}_{\lambda,n}} (-1)^{l(y)} m(y, \bar{x}_\psi) \frac{|\mathbb{H}'^*|_{\mathbf{q}'}}{|\mathbb{T}_y'^*|}.$$

Here, $|\mathbb{T}_{w_1 y}|$ is the order polynomial of an F-stable maximal torus of $\mathbf{G}$ of type $w_1 y$, while $|\mathbb{T}_y'^*|$ is the order polynomial of an F'^*-stable maximal torus of $\mathbf{H}^*$ of type y. Since $\bar{x}_\psi = \bar{x}_\rho \in \bar{X}(\mathbf{W}_{\lambda,n}, \gamma)$, it only remains to show that

$$|\mathbb{T}_{w_1 y}| = |\mathbb{T}_y'^*| \qquad \text{for all } y \in \mathbf{W}_{\lambda,n}.$$

This is seen as follows. By duality (see 2.5.1), an F-stable maximal torus $\mathbf{T}$ of $\mathbf{G}$ of type $w_1 y$ corresponds to an F^*-stable maximal torus $\mathbf{T}^*$ of $\mathbf{G}^*$ of type $(w_1^* y^*)^{-1}$; furthermore, $(\mathbf{T}, F)$ and $(\mathbf{T}^*, F^*)$ have the same order polynomial (see Lemma 2.5.2). Similarly, an F'^*-stable maximal torus of $\mathbf{H}^*$ of type y corresponds to an F'-stable maximal torus of $\mathbf{H}$ of type $(y^*)^{-1}$, and these two tori have the same order polynomial. Now we use the order formula in 1.6.21. Let $X^* = X(\mathbf{T}_0^*)$ and $\varphi^*\colon X^* \to X^*$ be the map induced by F^*. Write $\varphi^* = q\varphi_0^*$ where $\varphi_0^*\colon X_\mathbb{R}^* \to X_\mathbb{R}^*$ has finite order. Then, by 1.6.21 and the above discussion, we have

$$|\mathbb{T}_{w_1 y}| = \det\big(\mathbf{q}\,\mathrm{id}_{X_\mathbb{R}^*} - \varphi_0^* \circ (w_1^* y^*)\big).$$

Similarly, let $\varphi'\colon X^* \to X^*$ be the map induced by F'. Then

$$|\mathbb{T}_y'^*| = \det\big(\mathbf{q}\,\mathrm{id}_{X_\mathbb{R}^*} - \varphi_0' \circ y^*\big),$$

where $\varphi' = q\varphi_0'$ and φ_0' has finite order. By a computation as in 1.6.4, one checks that $\varphi' = \varphi^* \circ w_1^*$ and, hence, also $\varphi_0' = \varphi_0^* \circ w_1^*$. Comparing the above two formulae, we conclude that $|\mathbb{T}_{w_1 y}| = |\mathbb{T}_y'^*|$, as desired. □

We introduce the following notation, dual to that in 2.3.20. (We first formulate this in terms of $\mathbf{G}$, in order to avoid cumbersome notation.)

2.5.12 Define $\mathfrak{Y}(\mathbf{G}, F)$ as the set of all pairs $(\mathbf{T}, s)$, where $\mathbf{T}$ is an F-stable maximal torus of $\mathbf{G}$ and $s \in \mathbf{T}^F$. Then $\mathbf{G}^F$ acts on $\mathfrak{Y}(\mathbf{G}, F)$ via

$$\mathbf{G}^F \times \mathfrak{Y}(\mathbf{G}, F) \to \mathfrak{Y}(\mathbf{G}, F), \quad \big(g, (\mathbf{T}, s)\big) \mapsto (g\mathbf{T}g^{-1}, gsg^{-1}). \tag{a}$$

We denote by $\overline{\mathfrak{Y}}(\mathbf{G}, F)$ the set of $\mathbf{G}^F$-orbits for this action. On the other hand, the group $\mathbf{W}$ does not only act on itself by σ-conjugation but also on the set $\mathfrak{Y}(\mathbf{W}, \sigma) := \{(w, t) \mid w \in \mathbf{W}, t \in \mathbf{T}_0[w]\}$, via

$$\mathbf{W} \times \mathfrak{Y}(\mathbf{W}, \sigma) \to \mathfrak{Y}(\mathbf{W}, \sigma), \quad \big(x, (w, t)\big) \mapsto (xw\sigma(x)^{-1}, \dot{x}t\dot{x}^{-1}) \tag{b}$$

(where $\dot{x}$ denotes a representative of x in $N_{\mathbf{G}}(\mathbf{T}_0)$). We denote by $\overline{\mathfrak{Y}}(\mathbf{W}, \sigma)$ the set of $\mathbf{W}$-orbits for this action. Then it is straightforward to verify that there is a natural bijection

$$\overline{\mathfrak{Y}}(\mathbf{W}, \sigma) \quad \longleftrightarrow \quad \overline{\mathfrak{Y}}(\mathbf{G}, F) \tag{c}$$

defined as follows: The $\mathbf{W}$-orbit of a pair $(w, t) \in \mathfrak{Y}(\mathbf{W}, \sigma)$ corresponds to the $\mathbf{G}^F$-orbit of a pair $(\mathbf{T}, s) \in \mathfrak{Y}(\mathbf{G}, F)$ if there exists some $g \in \mathbf{G}$ such that $\mathbf{T} = g\mathbf{T}_0 g^{-1}$, $g^{-1}F(g) \in N_{\mathbf{G}}(\mathbf{T}_0)$ is a representative of w, and $s = gtg^{-1}$.

Remark 2.5.13 The isomorphism in Lemma 2.5.7 has the following compatibility property. Let (w_1, θ_1) and (w_2, θ_2) be two pairs in $\mathfrak{X}(\mathbf{W}, \sigma)$ that are in the same $\mathbf{W}$-orbit, that is, there exists some $x \in \mathbf{W}$ such that $w_2 = xw_1\sigma(x)^{-1}$ and $\theta_2 = {}^x\theta_1$. Then 2.5.1(b) shows that

$$(w_2^*)^{-1} = y^*(w_1^*)^{-1}\sigma^*(y^*)^{-1} \qquad \text{where} \qquad y := \sigma(x) \in \mathbf{W}^*.$$

Thus, $(w_1^*)^{-1}$ and $(w_2^*)^{-1}$ are σ^*-conjugate in $\mathbf{W}^*$. Now let (λ, n) be a pair as in 2.4.5 such that $(\lambda, n) \leftrightsquigarrow (w_1, \theta_1)$. Then $s_{\theta_1} = t_{\lambda,n} \in \mathbf{T}_0^*[(w_1^*)^{-1}]$. By Lemma 2.4.26, we have $(y.\lambda, n) \leftrightsquigarrow (w_2, \theta_2)$. So Lemma 2.5.4 shows that

$$\dot{y}^*\, s_{\theta_1}\, (\dot{y}^*)^{-1} = \dot{y}^*\, t_{\lambda,n}\, (\dot{y}^*)^{-1} = t_{y.\lambda,n} = s_{\theta_2} \in \mathbf{T}_0^*[(w_2^*)^{-1}].$$

In particular, the pairs $\big((w_1^*)^{-1}, s_{\theta_1}\big)$ and $\big((w_2^*)^{-1}, s_{\theta_2}\big)$ in $\mathfrak{Y}(\mathbf{W}^*, \sigma^*)$ are in the same $\mathbf{W}^*$-orbit.

Corollary 2.5.14 (Cf. [DeLu76, 5.21]) *There are natural bijections*

$$\overline{\mathfrak{X}}(\mathbf{G}, F) \overset{1-1}{\longleftrightarrow} \overline{\mathfrak{X}}(\mathbf{W}, \sigma) \overset{1-1}{\longleftrightarrow} \overline{\mathfrak{Y}}(\mathbf{W}^*, \sigma^*) \overset{1-1}{\longleftrightarrow} \overline{\mathfrak{Y}}(\mathbf{G}^*, F^*),$$

where the first one is given by 2.3.20, the third one by 2.5.12, and the middle one is induced by the isomorphisms $\mathrm{Irr}(\mathbf{T}_0[w]) \xrightarrow{\sim} \mathbf{T}_0^*[(w^*)^{-1}]$ *($w \in \mathbf{W}$) in Lemma 2.5.7.*

Proof Let $(w_1, \theta_1) \in \mathfrak{X}(\mathbf{W}, \sigma)$. Using Lemma 2.5.7, we obtain an element $s_{\theta_1} \in \mathbf{T}_0^*[(w_1^*)^{-1}]$ and, hence, a pair $\big((w_1^*)^{-1}, s_{\theta_1}\big) \in \mathfrak{Y}(\mathbf{W}^*, \sigma^*)$. Let also $(w_2, \theta_2) \in \mathfrak{X}(\mathbf{W}, \sigma)$, with corresponding $\big((w_2^*)^{-1}, s_{\theta_2}\big) \in \mathfrak{Y}(\mathbf{W}^*, \sigma^*)$. Now, if (w_1, θ_1) and

(w_2, θ_2) are in the same $\mathbf{W}$-orbit, then Remark 2.5.13 shows that $((w_1^*)^{-1}, s_{\theta_1})$ and $((w_2^*)^{-1}, s_{\theta_2})$ are in the same $\mathbf{W}^*$-orbit. Conversely, assume that $((w_1^*)^{-1}, s_{\theta_1})$ and $((w_2^*)^{-1}, s_{\theta_2})$ are in the same $\mathbf{W}^*$-orbit. So there exists some $y \in \mathbf{W}$ such that $(w_2^*)^{-1} = y^*(w_1^*)^{-1}(y^*)^{-1}$ and $s_{\theta_2} = y^* s_{\theta_1}(y^*)^{-1}$. We write $y = \sigma(x)$ where $x \in \mathbf{W}$. Then $w_2 = xw_1\sigma(x)^{-1}$. Let $\theta' := {}^x\theta_1 \in \mathrm{Irr}(\mathbf{T}_0[w_2])$. Then Remark 2.5.13 shows that $s_{\theta_2} = y^* s_{\theta_1}(y^*)^{-1} = y^* t_{\lambda,n}(y^*)^{-1} = t_{y.\lambda,n} = s_{\theta'}$. Since the map $\mathrm{Irr}(\mathbf{T}_0[w_2]) \xrightarrow{\sim} \mathbf{T}_0^*[(w_2^*)^{-1}]$ in Lemma 2.5.7 is an isomorphism, it follows that $\theta_2 = \theta' = {}^x\theta_1$. Thus, we obtain a bijection between $\mathbf{W}$-orbits of $\mathfrak{X}(\mathbf{W}, \sigma)$ and $\mathbf{W}^*$-orbits of $\mathfrak{Y}(\mathbf{W}^*, \sigma^*)$, giving rise to the middle correspondence. □

Remark 2.5.15 Let $(\mathbf{T}, \theta) \in \mathfrak{X}(\mathbf{G}, F)$ and $(\mathbf{T}^*, s) \in \mathfrak{Y}(\mathbf{G}^*, F^*)$ be such that the $\mathbf{G}^F$-orbit of $(\mathbf{T}, \theta)$ corresponds to the $\mathbf{G}^{*F^*}$-orbit of $(\mathbf{T}^*, s)$ via the bijections in Corollary 2.5.14. Then, for any integer $r \in \mathbb{Z}$, the $\mathbf{G}^F$-orbit of $(\mathbf{T}, \theta^r)$ also corresponds to the $\mathbf{G}^{*F^*}$-orbit of $(\mathbf{T}^*, s^r)$ via those bijections.

To see this, one first passes from $(\mathbf{T}, \theta)$ to a pair $(w, \theta') \in \mathfrak{X}(\mathbf{G}, F)$ as in 2.3.20; then $(\mathbf{T}, \theta^r)$ passes to the pair (w, θ'^r). Similarly, one passes from $(\mathbf{T}^*, s)$ to a pair $((w^*)^{-1}, t) \in \mathfrak{Y}(\mathbf{G}^*, F^*)$ as in 2.5.12; then $(\mathbf{T}^*, s^r)$ passes to $((w^*)^{-1}, t^r)$. Now $\theta' \in \mathrm{Irr}(\mathbf{T}_0[w])$ and $t \in \mathbf{T}_0^*[(w^*)^{-1}]$ correspond to each other under the above group isomorphism $\mathrm{Irr}(\mathbf{T}_0[w]) \xrightarrow{\sim} \mathbf{T}_0^*[(w^*)^{-1}]$. Hence, θ'^r and t^r also correspond to each other in this way. (See also [Hi90a, §2].)

Remark 2.5.16 We note that $|\mathfrak{X}(\mathbf{G}, F)| = |\mathfrak{Y}(\mathbf{G}^*, F^*)|$. Indeed, by [St67, 14.14] (see also [Ca85, 3.4.1]), the number of F-stable maximal tori in $\mathbf{G}$ is given by $|\mathbf{G}^F|_p^2$. Furthermore, we have $|\mathbf{G}^F| = |\mathbf{G}^{*F^*}|$; see Example 1.6.19. So the number of F^*-stable maximal tori in $\mathbf{G}^*$ is given by the same number. It remains to group the F-stable maximal tori of $\mathbf{G}$ and the F^*-stable maximal tori of $\mathbf{G}^*$ into $\mathbf{G}^F$-classes and $\mathbf{G}^{*F^*}$-classes, respectively, and to use the bijection in 2.5.1(c).

Definition 2.5.17 (See [Lu77a, 7.5]) Let $\mathbf{T}^* \subseteq \mathbf{G}^*$ be an F^*-stable maximal torus and $s \in \mathbf{T}^{*F^*}$. Via Corollary 2.5.14, the $\mathbf{G}^{*F^*}$-orbit of $(\mathbf{T}^*, s)$ corresponds to the $\mathbf{G}^F$-orbit of some $(\mathbf{T}, \theta) \in \mathfrak{X}(\mathbf{G}, F)$. Then we denote[7]

$$R_{\mathbf{T}^*}^{\mathbf{G}}(s) := R_{\mathbf{T}}^{\mathbf{G}}(\theta).$$

By Corollary 2.2.10, this is well defined, that is, $R_{\mathbf{T}^*}^{\mathbf{G}}(s)$ does not depend on the choice of the pair $(\mathbf{T}, \theta) \in \mathfrak{X}(\mathbf{G}, F)$ in its $\mathbf{G}^F$-orbit.

Following [Lu77a, §7], we will now re-express some properties of the virtual characters $R_{\mathbf{T}}^{\mathbf{G}}(\theta)$ in terms of the new notation $R_{\mathbf{T}^*}^{\mathbf{G}}(s)$.

[7] We note explicitly that there is no known direct way of constructing a virtual character of $\mathbf{G}^F$ from a pair $(\mathbf{T}^*, s) \in \mathfrak{Y}(\mathbf{G}^*, F^*)$. Whenever we want to prove something about $R_{\mathbf{T}^*}^{\mathbf{G}}(s)$, we have to translate the desired statement back to the original definition of $R_{\mathbf{T}}^{\mathbf{G}}(\theta)$ or R_w^θ.

Proposition 2.5.18 ([Lu77a, 7.5.1]) *For $i = 1, 2$ let $(\mathbf{T}_i^*, s_i) \in \mathfrak{Y}(\mathbf{G}^*, F^*)$. Then we have the following scalar product formula:*

$$\langle R_{\mathbf{T}_1^*}^{\mathbf{G}}(s_1), R_{\mathbf{T}_2^*}^{\mathbf{G}}(s_2)\rangle = |\{x \in \mathbf{G}^{*F^*} \mid x\mathbf{T}_1^*x^{-1} = \mathbf{T}_2^*, xs_1x^{-1} = s_2\}|/|\mathbf{T}_1^{*F^*}|.$$

In particular, if $(\mathbf{T}_1^*, s_1)$ and $(\mathbf{T}_2^*, s_2)$ are not in the same $\mathbf{G}^{*F^*}$-orbit, then $R_{\mathbf{T}_1^*}^{\mathbf{G}}(s_1)$ and $R_{\mathbf{T}_2^*}^{\mathbf{G}}(s_2)$ are orthogonal.

Proof For $i = 1, 2$ let $(\mathbf{T}_i^*, s_i)$ correspond to $(\mathbf{T}_i, \theta_i) \in \mathfrak{X}(\mathbf{G}, F)$ via the bijections in Corollary 2.5.14. If $(\mathbf{T}_1, \theta_1)$ and $(\mathbf{T}_2, \theta_2)$ are not in the same $\mathbf{G}^F$-orbit, then we have $\langle R_{\mathbf{T}_1}^{\mathbf{G}}(\theta_1), R_{\mathbf{T}_2}^{\mathbf{G}}(\theta_2)\rangle = 0$ by Corollary 2.2.10. But then $(\mathbf{T}_1^*, s_1)$ and $(\mathbf{T}_2^*, s_2)$ are not in the same $\mathbf{G}^{*F^*}$-orbit either, so the right-hand side of the desired scalar product formula is zero. Now assume that $(\mathbf{T}_1, \theta_1)$, $(\mathbf{T}_2, \theta_2)$ are in the same $\mathbf{G}^F$-orbit, and, hence, that $(\mathbf{T}_1^*, s_1)$, $(\mathbf{T}_2^*, s_2)$ are in the same $\mathbf{G}^{*F^*}$-orbit. For $i = 1, 2$ let $(\mathbf{T}_i, \theta_i)$ correspond to $(w_i, \theta_i') \in \overline{\mathfrak{X}}(\mathbf{W}, \sigma)$ (as in 2.3.20). Then $\langle R_{\mathbf{T}_1}^{\mathbf{G}}(\theta_1), R_{\mathbf{T}_2}^{\mathbf{G}}(\theta_2)\rangle$ equals

$$\langle R_{w_1}^{\theta_1'}, R_{w_2}^{\theta_2'}\rangle = |\{x \in \mathbf{W} \mid x.(w_1, \theta_1') = (w_2, \theta_2')\}|; \tag{1}$$

see Lemma 2.3.19 and Example 2.3.22(a). Similarly, for $i = 1, 2$ let $(\mathbf{T}_i^*, s_i)$ correspond to $(w_i, s_i') \in \overline{\mathfrak{Y}}(\mathbf{W}^*, \sigma^*)$ (as in 2.5.12). Then one easily sees that the right-hand side of the desired formula $\langle R_{\mathbf{T}_1}^{\mathbf{G}}(\theta_1), R_{\mathbf{T}_2}^{\mathbf{G}}(\theta_2)\rangle$ equals

$$|\{y \in \mathbf{W}^* \mid y.(w_1, s_1') = (w_2, s_2')\}|. \tag{2}$$

But, we have $s_i' = s_{\theta_i'}$ for $i = 1, 2$ (as in Lemma 2.5.7) and so the compatibility of the isomorphisms in Lemma 2.5.7 with the actions of $\mathbf{W}$ and of $\mathbf{W}^*$ (see Remark 2.5.13) shows that the right-hand side of (1) equals (2). □

Example 2.5.19 We give a few examples where a relation in Section 2.2 involving the virtual characters $R_{\mathbf{T}}^{\mathbf{G}}(\theta)$ is reformulated in terms of $R_{\mathbf{T}^*}^{\mathbf{G}}(s)$.

(a) For $i = 1, 2$ let $(\mathbf{T}_i^*, s_i) \in \mathfrak{Y}(\mathbf{G}^*, F^*)$. If $R_{\mathbf{T}_1^*}^{\mathbf{G}}(s_1)$ and $R_{\mathbf{T}_2^*}^{\mathbf{G}}(s_2)$ have an irreducible constituent in common, then s_1, s_2 are conjugate in $\mathbf{G}^*$.

Indeed, for $i = 1, 2$ let $(\mathbf{T}_i^*, s_i)$ correspond to $(\mathbf{T}_i, \theta_i)$ via Corollary 2.5.14. So, if $R_{\mathbf{T}_1^*}^{\mathbf{G}}(s_1)$ and $R_{\mathbf{T}_2^*}^{\mathbf{G}}(s_2)$ have an irreducible constituent in common, then $(\mathbf{T}_1, \theta_1)$ and $(\mathbf{T}_2, \theta_2)$ are geometrically conjugate by Theorem 2.3.2. Via Propositions 2.4.28 and 2.5.5, this translates into the condition that s_1, s_2 are conjugate in $\mathbf{G}^*$.

(b) Using Remark 2.5.16 (see also [Lu77a, 7.5.3]), the character of the regular representation of $\mathbf{G}^F$ in Example 2.2.26(b) can be re-written as follows:

$$\chi_{\text{reg}} = \frac{1}{|\mathbf{G}^F|_p} \sum_{(\mathbf{T}^*, s) \in \mathfrak{Y}(\mathbf{G}^*, F^*)} \varepsilon_{\mathbf{G}}\varepsilon_{\mathbf{T}^*} R_{\mathbf{T}^*}^{\mathbf{G}}(s).$$

If $s_1, \ldots, s_n$ is a set of representatives of the conjugacy classes of semisimple elements of $\mathbf{G}^{*F^*}$ and $\mathbf{H}_i := C_{\mathbf{G}^*}(s_i)$ for $1 \leqslant i \leqslant n$, then this also equals:

$$\chi_{\text{reg}} = \sum_{i=1}^{n} \frac{|\mathbf{G}^F|_{p'}}{|\mathbf{H}_i^{F^*}|} \sum_{\mathbf{T}^*} \varepsilon_{\mathbf{G}}\varepsilon_{\mathbf{T}^*} R_{\mathbf{T}^*}^{\mathbf{G}}(s_i);$$

where the second sum runs over all F^*-stable maximal tori $\mathbf{T}^* \subseteq \mathbf{H}$; see [Lu77a, 7.7]. Note that $|\mathbf{G}^F| = |\mathbf{G}^{*F^*}|$, $\varepsilon_{\mathbf{G}} = \varepsilon_{\mathbf{G}^*}$ and $\varepsilon_{\mathbf{T}} = \varepsilon_{\mathbf{T}^*}$, if $\mathbf{T}$ and $\mathbf{T}^*$ correspond to each other as in 2.5.1(c). Indeed, as already mentioned in Example 1.6.19, $(\mathbf{G}, F)$ and $(\mathbf{G}^*, F^*)$ have the same order polynomial and so $|\mathbf{G}^F| = |\mathbf{G}^{*F^*}|$. For the other two identities, one can then use the characterisation of the relative rank in terms of order polynomials (see Definition 2.2.11).

The linear characters of $\mathbf{G}^F$ have an elegant description via the above dictionary between $\mathbf{G}$ and $\mathbf{G}^*$. As in Remark 1.5.13, let

$$\mathbf{G}_u^F = \langle u \in \mathbf{G}^F \mid u \text{ unipotent}\rangle \subseteq \mathbf{G}_{\text{der}}^F.$$

We denote by $\Theta_u \subseteq \text{Irr}(\mathbf{G}^F)$ the group of all linear characters $\eta\colon \mathbf{G}^F \to \mathbb{K}^\times$ such that $\mathbf{G}_u^F \subseteq \ker(\eta)$. We can construct a canonical group homomorphism

$$\Theta_u \to \mathbf{T}_0^{*F^*}, \qquad \eta \mapsto z_\eta,$$

as follows. If $\eta \in \Theta_u$, then we apply the isomorphism $\text{Irr}(\mathbf{T}_0^F) \xrightarrow{\sim} \mathbf{T}_0^{*F^*}$ in Lemma 2.5.7 to $\eta|_{\mathbf{T}_0^F} \in \text{Irr}(\mathbf{T}_0^F)$ and obtain an element $z_\eta \in \mathbf{T}_0^{*F^*}$.

Proposition 2.5.20 ([Lu77a, 7.4.2]) *We have $z_\eta \in \mathbf{Z}(\mathbf{G}^*)^{F^*}$ for any $\eta \in \Theta_u$. The resulting map $\Theta_u \to \mathbf{Z}(\mathbf{G}^*)^{F^*}$, $\eta \mapsto z_\eta$, is an isomorphism.*

If $z \in \mathbf{Z}(\mathbf{G}^*)^{F^*}$ and $\eta \in \Theta_u$ is such that $z_\eta = z$, then it will also be convenient to denote η by $\hat{z}$.

Proof We follow the hint given in [Lu77a, 7.4.2]. By Remark 1.5.13(b), we have $\mathbf{G}_u^F = \pi_{\text{sc}}(\mathbf{G}_{\text{sc}}^F)$ where $\pi_{\text{sc}}\colon \mathbf{G}_{\text{sc}} \to \mathbf{G}$ is a simply connected covering of the derived subgroup of $\mathbf{G}$. Let $\tilde{\mathbf{T}}_0 := \pi_{\text{sc}}^{-1}(\mathbf{T}_0) \subseteq \mathbf{G}_{\text{sc}}$. Then, by Remark 1.5.13(c) the inclusion $\mathbf{T}_0 \subseteq \mathbf{G}$ induces an isomorphism $\mathbf{T}_0^F/\pi_{\text{sc}}(\tilde{\mathbf{T}}_0^F) \cong \mathbf{G}^F/\pi_{\text{sc}}(\mathbf{G}_{\text{sc}}^F) = \mathbf{G}^F/\mathbf{G}_u^F$. Let

$$\text{Irr}'(\mathbf{T}_0^F) := \{\theta \in \text{Irr}(\mathbf{T}_0^F) \mid \pi_{\text{sc}}(\tilde{\mathbf{T}}_0^F) \subseteq \ker(\theta)\}.$$

Then we have an isomorphism $\Theta_u \to \text{Irr}'(\mathbf{T}_0^F)$, $\eta \mapsto \eta|_{\mathbf{T}_0^F}$, and z_η is obtained by applying the isomorphism $\text{Irr}(\mathbf{T}_0^F) \xrightarrow{\sim} \mathbf{T}_0^{*F^*}$, $\theta \mapsto s_\theta$, in Lemma 2.5.7 to $\eta|_{\mathbf{T}_0^F} \in \text{Irr}'(\mathbf{T}_0^F)$. Hence, we must show that the isomorphism $\theta \mapsto s_\theta$ in Lemma 2.5.7 maps the subset $\text{Irr}'(\mathbf{T}_0^F) \subseteq \text{Irr}(\mathbf{T}_0^F)$ onto $\mathbf{Z}(\mathbf{G}^*)^{F^*} \subseteq \mathbf{T}_0^{*F^*}$.

For this purpose, as in 1.7.11, we consider a dual central isotypy $\pi_{\text{sc}}^*\colon \mathbf{G}^* \to \mathbf{G}_{\text{sc}}^*$, with respect to maximally split tori $\mathbf{T}_0^* \subseteq \mathbf{G}^*$ and $\tilde{\mathbf{T}}_0^* \subseteq \mathbf{G}_{\text{sc}}^*$. There are Steinberg maps $F^*\colon \mathbf{G}^* \to \mathbf{G}^*$ and $\tilde{F}^*\colon \mathbf{G}_{\text{sc}}^* \to \mathbf{G}_{\text{sc}}^*$ such that $\pi_{\text{sc}}^* \circ F^* = \tilde{F}^* \circ \pi_{\text{sc}}^*$. Furthermore, there are isomorphisms $\delta\colon X(\mathbf{T}_0) \xrightarrow{\sim} Y(\mathbf{T}_0^*)$ and $\tilde{\delta}\colon X(\tilde{\mathbf{T}}_0) \xrightarrow{\sim} Y(\tilde{\mathbf{T}}_0^*)$ satisfying the appropriate conditions. As in Remark 2.5.9, we also have an isomorphism

$$\mathrm{Irr}(\tilde{\mathbf{T}}_0^F) \xrightarrow{\sim} (\tilde{\mathbf{T}}_0^*)^{\tilde{F}^*}, \qquad \tilde{\theta} \mapsto s_{\tilde{\theta}},$$

which is compatible with $\mathrm{Irr}(\mathbf{T}_0^F) \xrightarrow{\sim} \mathbf{T}_0^{*F^*}$, $\theta \mapsto s_\theta$, in the following sense. If $\theta \in \mathrm{Irr}(\mathbf{T}_0^F)$ and $\tilde{\theta} := \theta \circ \pi_{\mathrm{sc}}|_{\tilde{\mathbf{T}}_0^F} \in \mathrm{Irr}(\tilde{\mathbf{T}}_0^{\tilde{F}})$, then $s_{\tilde{\theta}} = \pi_{\mathrm{sc}}^*(s_\theta)$.

Now we can argue as follows. Let $\theta \in \mathrm{Irr}'(\mathbf{T}_0^F)$. By definition, this means that $\tilde{\theta} := \theta \circ \pi_{\mathrm{sc}}|_{\tilde{\mathbf{T}}_0^F}$ is the trivial character of $\tilde{\mathbf{T}}_0^F$ and so $s_{\tilde{\theta}} = 1$. But then $\pi_{\mathrm{sc}}^*(s_\theta) = s_{\tilde{\theta}} = 1$ and so $s_\theta \in \ker(\pi_{\mathrm{sc}}^*) \subseteq \mathbf{Z}(\mathbf{G}^*)^{F^*}$. Conversely, let $z \in \mathbf{Z}(\mathbf{G}^*)^{F^*} \subseteq \mathbf{T}_0^{*F^*}$ and $\theta \in \mathrm{Irr}(\mathbf{T}_0^F)$ be such that $z = s_\theta$. Since $\mathbf{G}_{\mathrm{sc}}^*$ is semisimple of adjoint type (see Example 1.5.20(a)), we have $\mathbf{Z}(\mathbf{G}_{\mathrm{sc}}^*) = \{1\}$ (see Example 1.5.3) and so $\pi_{\mathrm{sc}}^*(s_\theta) = \pi_{\mathrm{sc}}^*(z) = 1$. Hence, if we set $\tilde{\theta} := \theta \circ \pi_{\mathrm{sc}}|_{\tilde{\mathbf{T}}_0^F} \in \mathrm{Irr}(\tilde{\mathbf{T}}_0^F)$ as above, then $s_{\tilde{\theta}} = \pi_{\mathrm{sc}}^*(s_\theta) = 1$ and so $\tilde{\theta}$ is the trivial character, that is, $\theta \in \mathrm{Irr}'(\mathbf{T}_0^F)$. □

Proposition 2.5.21 ([Lu77a, 7.5.5]) *We have $\eta \cdot R_{\mathbf{T}^*}^{\mathbf{G}}(s) = R_{\mathbf{T}^*}^{\mathbf{G}}(z_\eta s)$ for any $\eta \in \Theta_u$ and any pair $(\mathbf{T}^*, s) \in \mathfrak{Y}(\mathbf{G}^*, F^*)$.*

Proof Let $(\mathbf{T}, \theta) \in \mathfrak{X}(\mathbf{G}, F)$ correspond to $(\mathbf{T}^*, s)$ via Corollary 2.5.14. Then we already know from Lemma 2.3.13 that $\eta \cdot R_{\mathbf{T}}^{\mathbf{G}}(\theta) = R_{\mathbf{T}}^{\mathbf{G}}(\eta|_{\mathbf{T}^F} \cdot \theta)$. So we must show that the pairs $(\mathbf{T}, \eta|_{\mathbf{T}^F} \cdot \theta) \in \mathfrak{X}(\mathbf{G}, F)$ and $(\mathbf{T}^*, z_\eta s) \in \mathfrak{Y}(\mathbf{G}^*, F^*)$ also correspond to each other via Corollary 2.5.14. For this purpose, we go explicitly through the correspondences in Corollary 2.5.14. Let $w \in \mathbf{W}$ be such that $\mathbf{T}$ is of type w. Let $g \in \mathbf{G}$ be such that $\dot{w} = g^{-1}F(g)$ and $\mathbf{T} = g\mathbf{T}_0 g^{-1}$. Define $\theta' \in \mathrm{Irr}(\mathbf{T}_0[w])$ by $\theta'(t) = \theta(gtg^{-1})$ for all $t \in \mathbf{T}_0[w]$ (see 2.3.20). Now $\mathbf{T}^*$ is of type $(w^*)^{-1}$. Let $h \in \mathbf{G}^*$ be such that $(\dot{w}^*)^{-1} = h^{-1}F^*(h)$ and $\mathbf{T}^* = h\mathbf{T}_0^* h^{-1}$. Define $s' \in \mathbf{T}_0^*[(w^*)^{-1}]$ by $s = hs'h^{-1}$ (see 2.5.12). Thus, we have

$$(\mathbf{T}, \theta) \longleftrightarrow (w, \theta') \longleftrightarrow \big((w^*)^{-1}, s'\big) \longleftrightarrow (\mathbf{T}^*, s),$$

where $s' = s_{\theta'}$ by the isomorphism $\mathrm{Irr}(\mathbf{T}_0[w]) \xrightarrow{\sim} \mathbf{T}_0^*[(w^*)^{-1}]$ in Lemma 2.5.7. Since $z_\eta \in \mathbf{Z}(\mathbf{G}^*)^{F^*}$, we have $z_\eta s = z_\eta hs'h^{-1} = hz_\eta s'h^{-1}$ and so $(\mathbf{T}^*, z_\eta s) \in \mathfrak{Y}(\mathbf{G}^*, F^*)$ corresponds to $\big((w^*)^{-1}, z_\eta s'\big) \in \mathfrak{Y}(\mathbf{W}^*, \sigma^*)$ as in 2.5.12. Define $\eta' \in \mathrm{Irr}(\mathbf{T}_0[w])$ by $\eta'(t) = \eta(gtg^{-1})$ for all $t \in \mathbf{T}_0[w]$. Then $(\mathbf{T}, \eta|_{\mathbf{T}^F} \cdot \theta) \in \mathfrak{X}(\mathbf{G}, F)$ corresponds to $(w, \eta' \cdot \theta') \in \mathfrak{X}(\mathbf{W}, \sigma)$ via 2.3.20. Hence, it remains to show that $s_{\eta'} = z_\eta$ via the isomorphism in Lemma 2.5.7. Now, by Example 2.2.25 and Lemma 2.3.13, we have

$$\langle R_{\mathbf{T}}^{\mathbf{G}}(\eta|_{\mathbf{T}^F}), \eta\rangle = \langle \eta \cdot R_{\mathbf{T}}^{\mathbf{G}}(1_{\mathbf{T}}), \eta\rangle = \langle R_{\mathbf{T}}^{\mathbf{G}}(1_{\mathbf{T}}), 1_{\mathbf{G}}\rangle = 1$$

and, similarly, $\langle R_{\mathbf{T}_0}^{\mathbf{G}}(\eta|_{\mathbf{T}_0^F}), \eta\rangle = 1$. Thus, by Theorem 2.3.2, the two pairs $(\mathbf{T}, \eta|_{\mathbf{T}^F})$ and $(\mathbf{T}_0, \eta|_{\mathbf{T}_0^F})$ in $\mathfrak{X}(\mathbf{G}, F)$ are geometrically conjugate and, hence, the corresponding two pairs (w, η') and $(1, \eta|_{\mathbf{T}_0^F})$ in $\mathfrak{X}(\mathbf{W}, \sigma)$ are geometrically conjugate (see Lemma 2.4.23). Via Lemma 2.5.7, the first of these pairs gives rise to the element

$s_{\eta'} \in \mathbf{T}_0^*[(w^*)^{-1}]$, while the second pair gives rise to $z_\eta \in \mathbf{T}_0^{*F^*}$ (by the definition of the isomorphism in Proposition 2.5.20). These two elements are conjugate in $\mathbf{G}^*$ by Corollary 2.5.8. Since $z_\eta \in \mathbf{Z}(\mathbf{G}^*)$, we conclude that $s_{\eta'} = z_\eta$, as desired. □

Next, it will be useful to know how the virtual characters $R_{\mathbf{T}^*}^{\mathbf{G}}(s)$ behave with respect to a regular embedding $i\colon \mathbf{G} \to \tilde{\mathbf{G}}$ as in Section 1.7. We use the convention in Remark 1.7.6 where we identify $\mathbf{G}$ with $i(\mathbf{G}) \subseteq \tilde{\mathbf{G}}$. Let $(\tilde{\mathbf{G}}^*, \tilde{F}^*)$ be in duality with $(\tilde{\mathbf{G}}, \tilde{F})$. We have a corresponding central isotypy $i^*\colon \tilde{\mathbf{G}}^* \to \mathbf{G}^*$ such that $i^* \circ \tilde{F}^* = F^* \circ i^*$; see 1.7.11. Since $i\colon \mathbf{G} \hookrightarrow \tilde{\mathbf{G}}$ is a closed embedding, we have:

- $\mathbf{K} := \ker(i^*)$ is an $\tilde{F}^*$-stable torus contained in $\mathbf{Z}(\tilde{\mathbf{G}}^*)$,
- $i^*(\tilde{\mathbf{G}}^*) = \mathbf{G}^*$ and $i^*\big((\tilde{\mathbf{G}}^*)^{\tilde{F}^*}\big) = \mathbf{G}^{*F^*}$;

see Lemma 1.7.12. In this setting, we can now state:

Proposition 2.5.22 (Cf. [Lu88, p. 164]) *Let $\mathbf{T}^* \subseteq \mathbf{G}^*$ be an F^*-stable maximal torus and $s \in \mathbf{T}^{*F^*}$. Let $\tilde{\mathbf{T}}^* := i^{*-1}(\mathbf{T}^*) \subseteq \tilde{\mathbf{G}}^*$. Then there exists a semisimple element $\tilde{s} \in (\tilde{\mathbf{T}}^*)^{\tilde{F}^*}$ such that $i^*(\tilde{s}) = s$. For any such $\tilde{s}$ we have $R_{\mathbf{T}^*}^{\mathbf{G}}(s) = R_{\tilde{\mathbf{T}}^*}^{\tilde{\mathbf{G}}}(\tilde{s})|_{\mathbf{G}^F}$.*

Proof First note that $\tilde{\mathbf{T}}^* := i^{*-1}(\mathbf{T}^*)$ is an $\tilde{F}^*$-stable maximal torus of $\tilde{\mathbf{G}}^*$. Hence, $i^{*-1}(s) \subseteq \tilde{\mathbf{T}}^*$ consists of semisimple elements. Now note that $i^{*-1}(s)$ is an $\tilde{F}^*$-stable coset of $\mathbf{K} = \ker(i^*)$. The connected group $\mathbf{K}$ acts transitively on this coset by multiplication and so Proposition 1.4.9 shows that $i^{*-1}(s)^{\tilde{F}^*} \neq \varnothing$. Thus, there exists a semisimple element $\tilde{s} \in (\tilde{\mathbf{G}}^*)^{\tilde{F}^*}$ such that $i^*(\tilde{s}) = s$.

Next note that we are in a setting as in Remark 2.5.9. Let $\tilde{\mathbf{T}}_0 \subseteq \tilde{\mathbf{G}}$ be the maximally split torus such that $\mathbf{T}_0 = \mathbf{G} \cap \tilde{\mathbf{T}}_0$. (We have $\tilde{\mathbf{T}}_0 = \mathbf{T}_0.\mathbf{Z}(\tilde{\mathbf{G}})$.) Let $\tilde{\mathbf{T}}_0^* := i^{*-1}(\mathbf{T}_0^*) \subseteq \tilde{\mathbf{G}}^*$ be the corresponding dual maximally split torus. Using i and i^*, we can identify $\mathbf{W}$ with the Weyl group of $\tilde{\mathbf{G}}$ (with respect to $\tilde{\mathbf{T}}_0$) and $\mathbf{W}^*$ with the Weyl group of $\tilde{\mathbf{G}}^*$ (with respect to $\tilde{\mathbf{T}}^*$). Let $w \in \mathbf{W}$ be such that $\mathbf{T}$ is of type w. Then $\tilde{\mathbf{T}}$ also is of type w. Furthermore, one immediately checks that $\mathbf{T}_0[w] \subseteq \tilde{\mathbf{T}}_0[w]$. Let $\tilde{\theta} \in \mathrm{Irr}(\tilde{\mathbf{T}}_0[w])$ be such that $\tilde{s} = s_{\tilde{\theta}}$ via the isomorphism in Lemma 2.5.7. Let

$$\theta := \tilde{\theta}|_{\mathbf{T}_0[w]} = \tilde{\theta} \circ i|_{\mathbf{T}_0[w]} \in \mathrm{Irr}(\mathbf{T}_0[w]).$$

Let $g \in \mathbf{G}$ be such that $\dot{w} = g^{-1}F(g)$ and $\mathbf{T} = g\mathbf{T}_0 g^{-1}$. Then we also have $\tilde{\mathbf{T}} = g\tilde{\mathbf{T}}_0 g^{-1}$. Define $\theta' \in \mathrm{Irr}(\mathbf{T}^F)$ by $\theta'(t) = \theta(g^{-1}tg)$ for all $t \in \mathbf{T}^F$ and $\tilde{\theta}' \in \mathrm{Irr}(\tilde{\mathbf{T}}^{\tilde{F}})$ by $\tilde{\theta}'(\tilde{t}) = \tilde{\theta}(g^{-1}\tilde{t}g)$ for all $\tilde{t} \in \tilde{\mathbf{T}}^{\tilde{F}}$ (see 2.3.20). Then

$$R_{\tilde{\mathbf{T}}^*}^{\tilde{\mathbf{G}}}(\tilde{s}) = R_{\tilde{\mathbf{T}}^*}^{\tilde{\mathbf{G}}}(s_{\tilde{\theta}}) = R_{\tilde{\mathbf{T}}}^{\tilde{\mathbf{G}}}(\tilde{\theta}') \qquad \text{and} \qquad R_{\mathbf{T}^*}^{\mathbf{G}}(s_\theta) = R_{\mathbf{T}}^{\mathbf{G}}(\theta');$$

furthermore, it is clear that $\theta' = \tilde{\theta}'|_{\mathbf{T}^F}$. Hence, we have $R_{\mathbf{T}}^{\mathbf{G}}(\theta') = R_{\tilde{\mathbf{T}}}^{\tilde{\mathbf{G}}}(\tilde{\theta}')\big|_{\mathbf{G}^F}$. (This is just a very special case of the formula in Remark 2.3.16.) Combining this with the previous equalities, we deduce that $R_{\mathbf{T}^*}^{\mathbf{G}}(s_\theta) = R_{\tilde{\mathbf{T}}^*}^{\tilde{\mathbf{G}}}(\tilde{s})|_{\mathbf{G}^F}$. It remains to use that $s_\theta = i^*(\tilde{s}) = s$, which holds by Remark 2.5.9(∗). □

Now, Proposition 2.5.20 also holds for $\tilde{\mathbf{G}}^{\tilde{F}}$. So we have an isomorphism

$$\tilde{\Theta}_u \xrightarrow{\sim} \mathbf{Z}(\tilde{\mathbf{G}}^*)^{\tilde{F}^*}, \qquad \tilde{\eta} \mapsto z_{\tilde{\eta}},$$

where $\tilde{\Theta}_u$ is the group of all linear characters $\tilde{\eta}\colon \tilde{\mathbf{G}}^F \to \mathbb{K}^\times$ such that $\tilde{\mathbf{G}}_u^{\tilde{F}} \subseteq \ker(\tilde{\eta})$; also note that $\mathbf{G}_u^F = \tilde{\mathbf{G}}_u^{\tilde{F}}$.

Lemma 2.5.23 *We have* $\ker(i^*) = \{z_{\tilde{\eta}} \mid \tilde{\eta} \in \tilde{\Theta}_u$ *such that* $\mathbf{G}^F \subseteq \ker(\tilde{\eta})\}$.

Proof Let Θ be the group of all $\tilde{\eta} \in \tilde{\Theta}_u$ such that $\mathbf{G}^F \subseteq \ker(\tilde{\eta})$. Let $\mathrm{Irr}'(\tilde{\mathbf{T}}_0^{\tilde{F}})$ be the group of all $\tilde{\theta} \in \mathrm{Irr}(\tilde{\mathbf{T}}_0^{\tilde{F}})$ such that $\mathbf{T}_0^F \subseteq \ker(\tilde{\theta})$. Now note that $\mathbf{T}_0^F = \mathbf{G}^F \cap \tilde{\mathbf{T}}_0^{\tilde{F}}$ and $\tilde{\mathbf{G}}^{\tilde{F}} = \mathbf{G}^F.\tilde{\mathbf{T}}_0^{\tilde{F}}$; see Lemma 1.7.7. Hence, we have an isomorphism

$$\Theta \to \mathrm{Irr}'(\tilde{\mathbf{T}}_0^{\tilde{F}}), \qquad \tilde{\eta} \mapsto \tilde{\eta}|_{\tilde{\mathbf{T}}_0^{\tilde{F}}}.$$

We can now argue exactly as in the proof of Proposition 2.5.22, where the role of $\pi_{\mathrm{sc}}\colon \mathbf{G}_{\mathrm{sc}} \to \mathbf{G}$ is replaced by that of $i\colon \mathbf{G} \to \tilde{\mathbf{G}}$ and, hence, the role of $\mathbf{Z}(\mathbf{G}_{\mathrm{sc}}^*) = \ker(\pi_{\mathrm{sc}}^*)$ is replaced by that of $\mathbf{K} = \ker(i^*)$. □

We have the following refined version of the Exclusion Theorem 2.3.2, which shows an advantage of the new notation $R_{\mathbf{T}^*}^{\mathbf{G}}(s)$: it is not at all obvious how to even state the result in terms of the original notation $R_{\mathbf{T}}^{\mathbf{G}}(\theta)$!

Theorem 2.5.24 (Refined Exclusion Theorem; see [Lu77a, 7.5.2]) *Assume that $R_{\mathbf{T}_1^*}^{\mathbf{G}}(s_1)$ and $R_{\mathbf{T}_2^*}^{\mathbf{G}}(s_2)$ (as in Definition 2.5.17) have an irreducible constituent in common. Then s_1, s_2 are conjugate not only in $\mathbf{G}^*$ (see Example 2.5.19) but in $\mathbf{G}^{*F^*}$.*

Proof Assume that $\rho \in \mathrm{Irr}(\mathbf{G}^F)$ is a common irreducible constituent of $R_{\mathbf{T}_1^*}^{\mathbf{G}}(s_1)$ and $R_{\mathbf{T}_2^*}^{\mathbf{G}}(s_2)$. Following the hint in [Lu77a, 7.5.2], we consider a regular embedding $i\colon \mathbf{G} \to \tilde{\mathbf{G}}$ as above, and let $i^*\colon \tilde{\mathbf{G}}^* \to \mathbf{G}^*$ be a corresponding dual central isotypy. Let $i \in \{1,2\}$ and $\tilde{\mathbf{T}}_i^* := i^{*-1}(\mathbf{T}_i^*) \subseteq \tilde{\mathbf{G}}^*$. By Frobenius reciprocity and Proposition 2.5.22, we have

$$\langle R_{\tilde{\mathbf{T}}_i^*}^{\tilde{\mathbf{G}}}(\tilde{s}_i), \mathrm{Ind}_{\mathbf{G}^F}^{\tilde{\mathbf{G}}^{\tilde{F}}}(\rho)\rangle = \langle R_{\mathbf{T}_i^*}^{\mathbf{G}}(s_i), \rho\rangle \neq 0,$$

where $\tilde{s}_i \in (\tilde{\mathbf{T}}_i^*)^{\tilde{F}^*}$ is such that $i^*(\tilde{s}_i) = s_i$. Hence, there is some $\tilde{\rho}_i \in \mathrm{Irr}(\tilde{\mathbf{G}}^{\tilde{F}})$ such that $\langle R_{\tilde{\mathbf{T}}_i^*}^{\tilde{\mathbf{G}}}(\tilde{s}_i), \tilde{\rho}_i\rangle \neq 0$ and ρ occurs in the restriction of $\tilde{\rho}_i$ to $\mathbf{G}^F$. Since this holds for $i = 1$ and $i = 2$, we see by Remark 2.1.4 that $\tilde{\rho}_2 = \tilde{\eta} \cdot \tilde{\rho}_1$ for some linear character $\tilde{\eta} \in \mathrm{Irr}(\tilde{\mathbf{G}}^{\tilde{F}})$ such that $\mathbf{G}^F \subseteq \ker(\tilde{\eta})$. Using Proposition 2.5.21, we obtain

$$\langle R_{\tilde{\mathbf{T}}_1^*}^{\tilde{\mathbf{G}}}(z_{\tilde{\eta}}\,\tilde{s}_1), \tilde{\rho}_2\rangle = \langle \tilde{\eta} \cdot R_{\tilde{\mathbf{T}}_1^*}^{\tilde{\mathbf{G}}}(\tilde{s}_1), \tilde{\eta} \cdot \tilde{\rho}_1\rangle = \langle R_{\tilde{\mathbf{T}}_1^*}^{\tilde{\mathbf{G}}}(\tilde{s}_1), \tilde{\rho}_1\rangle \neq 0.$$

Hence, $\tilde{\rho}_2$ is a common irreducible constituent of $R_{\tilde{\mathbf{T}}_1^*}^{\tilde{\mathbf{G}}}(z_{\tilde{\eta}}\,\tilde{s}_1)$ and $R_{\tilde{\mathbf{T}}_2^*}^{\tilde{\mathbf{G}}}(\tilde{s}_2)$. By Example 2.5.19(a), the elements $z_{\tilde{\eta}}\,\tilde{s}_1$ and $\tilde{s}_2$ are conjugate in $\tilde{\mathbf{G}}^*$. But, as already noted in 2.5.10, since $\mathbf{Z}(\tilde{\mathbf{G}})$ is connected, the centraliser $C_{\tilde{\mathbf{G}}^*}(\tilde{s}_2)$ is connected and so $z_{\tilde{\eta}}\,\tilde{s}_1$ and $\tilde{s}_2$ are already conjugate in $(\tilde{\mathbf{G}}^*)^{\tilde{F}^*}$ (see Example 1.4.10). Hence, $i^*(z_{\tilde{\eta}})s_1$ and s_2 are conjugate in $\mathbf{G}^{*F^*}$. Finally, since $\mathbf{G}^F \subseteq \ker(\tilde{\eta})$, we have $i^*(z_{\tilde{\eta}}) = 1$ by Lemma 2.5.23. Thus, s_1 and s_2 are conjugate in $\mathbf{G}^{*F^*}$. □

For somewhat different proofs, see [DiMi20, Prop. 12.4.4] and [Bo06, 11.8].

2.6 The Jordan Decomposition of Characters

Let $\mathbf{G}$ be a connected reductive algebraic group over k and $F\colon \mathbf{G} \to \mathbf{G}$ be a Steinberg map. Based on the preparations from the previous sections, we are now ready to formulate and to discuss the fundamental *Jordan decomposition of characters* of the finite group $\mathbf{G}^F$, both in the original case where $\mathbf{Z}(\mathbf{G})$ is connected (see [Lu84a]), and in the general case (see [Lu88], [Lu08a]).

Let $\mathbf{G}^*$ be a connected reductive algebraic group over k and $F^*\colon \mathbf{G}^* \to \mathbf{G}^*$ be a Steinberg map such that $(\mathbf{G}, F)$ and $(\mathbf{G}^*, F^*)$ are in duality as in Definition 1.5.17; we also assume that the choices in 2.4.4 have been made.

Recall from the previous section that we have then a canonical bijection between $\mathbf{G}^F$-orbits of pairs $(\mathbf{T}, \theta)$ (where $\mathbf{T} \subseteq \mathbf{G}$ is an F-stable maximal torus and $\theta \in \mathrm{Irr}(\mathbf{T}^F)$) and $\mathbf{G}^{*F^*}$-orbits of pairs $(\mathbf{T}^*, s)$ (where $\mathbf{T}^* \subseteq \mathbf{G}^*$ is an F^*-stable maximal torus and $s \in \mathbf{T}^{*F^*}$). If the pairs $(\mathbf{T}, \theta)$ and $(\mathbf{T}^*, s)$ correspond to each other in this way, then we write $R_{\mathbf{T}^*}^{\mathbf{G}}(s) = R_{\mathbf{T}}^{\mathbf{G}}(\theta)$.

Definition 2.6.1 (Cf. [Lu77a, 7.6]) Let $s \in \mathbf{G}^{*F^*}$ be semisimple. We define $\mathscr{E}(\mathbf{G}^F, s)$ to be the set of all $\rho \in \mathrm{Irr}(\mathbf{G}^F)$ such that $\langle R_{\mathbf{T}^*}^{\mathbf{G}}(s), \rho\rangle \neq 0$ for some F^*-stable maximal torus $\mathbf{T}^* \subseteq \mathbf{G}^*$ with $s \in \mathbf{T}^*$. This set is called a *rational series of characters* of $\mathbf{G}^F$, or *Lusztig series of characters*.

For example, we have $\mathscr{E}(\mathbf{G}^F, 1) = \mathrm{Uch}(\mathbf{G}^F)$; see Definition 2.3.8. Indeed, just note that the pairs $(\mathbf{T}^*, 1) \in \mathfrak{Y}(\mathbf{G}^*, F^*)$ correspond to the pairs $(\mathbf{T}, 1_{\mathbf{T}}) \in \mathfrak{X}(\mathbf{G}, F)$. Thus, the notions of 'rational series of characters' and 'geometric series of characters' coincide as far as $\mathrm{Uch}(\mathbf{G}^F)$ is concerned. (We will see that this is not the case in general; see Remark 2.6.19.)

Theorem 2.6.2 ([Lu77a, 7.6]) *If $s_1, s_2 \in \mathbf{G}^{*F^*}$ are semisimple and conjugate in $\mathbf{G}^{*F^*}$, then $\mathscr{E}(\mathbf{G}^F, s_1) = \mathscr{E}(\mathbf{G}^F, s_2)$. We have a partition*

$$\mathrm{Irr}(\mathbf{G}^F) = \bigsqcup_{s} \mathscr{E}(\mathbf{G}^F, s)$$

where s runs over a set of representatives of the conjugacy classes of semisimple *elements in* $\mathbf{G}^{*F^*}$. *If* $\rho \in \mathscr{E}(\mathbf{G}^F, s)$, *then*

$$\rho(1) = \frac{|\mathbf{G}^F|_{p'}}{|\mathbf{H}^{F^*}|} \sum_{\mathbf{T}^*} \varepsilon_{\mathbf{G}} \varepsilon_{\mathbf{T}^*} \langle R_{\mathbf{T}^*}^{\mathbf{G}}(s), \rho \rangle$$

where $\mathbf{H} := C_{\mathbf{G}^*}(s)$ *and the sum runs over all* F^**-stable maximal tori* $\mathbf{T}^* \subseteq \mathbf{H}^\circ$. *(Note that* $\mathbf{H}$ *need not be connected here!)*

Proof Since every $\rho \in \mathrm{Irr}(\mathbf{G}^F)$ occurs in some Deligne–Lusztig character $R_{\mathbf{T}}^{\mathbf{G}}(\theta)$, we have $\mathrm{Irr}(\mathbf{G}^F) = \bigcup_s \mathscr{E}(\mathbf{G}^F, s)$ where s runs over all semisimple elements of $\mathbf{G}^{*F^*}$. Now let s_1, s_2 be semisimple elements in $\mathbf{G}^{*F^*}$. We claim that $\mathscr{E}(\mathbf{G}^F, s_1) = \mathscr{E}(\mathbf{G}^F, s_2)$ if s_1, s_2 are conjugate in $\mathbf{G}^{*F^*}$. Indeed, let $x \in \mathbf{G}^{*F^*}$ be such that $s_2 = xs_1x^{-1}$. Let $\mathbf{T}_1^* \subseteq \mathbf{G}^*$ be an F^*-stable maximal torus in $\mathbf{G}^*$ such that $s_1 \in \mathbf{T}_1^*$. Setting $\mathbf{T}_2^* := x\mathbf{T}_1^*x^{-1}$, we have $s_2 \in \mathbf{T}_2^*$ and $(\mathbf{T}_2^*, s_2) \in \mathfrak{Y}(\mathbf{G}^*, F^*)$ is in the same $\mathbf{G}^{*F^*}$-orbit as $(\mathbf{T}_1^*, s_1)$. So $R_{\mathbf{T}_2^*}^{\mathbf{G}}(s_2) = R_{\mathbf{T}_1^*}^{\mathbf{G}}(s_1)$. It follows that $\mathscr{E}(\mathbf{G}^F, s_1) \subseteq \mathscr{E}(\mathbf{G}^F, s_2)$. The reverse inclusion holds by symmetry and so $\mathscr{E}(\mathbf{G}^F, s_1) = \mathscr{E}(\mathbf{G}^F, s_2)$, as claimed.

On the other hand, Theorem 2.5.24 shows that, if $\mathscr{E}(\mathbf{G}^F, s_1) \cap \mathscr{E}(\mathbf{G}^F, s_2) \neq \varnothing$, then s_1, s_2 are conjugate in $\mathbf{G}^{*F^*}$ and so $\mathscr{E}(\mathbf{G}^F, s_1) = \mathscr{E}(\mathbf{G}^F, s_2)$. Consequently, we obtain a partition as stated above. The formula for $\rho(1)$ follows by using the formula for χ_{reg} in Example 2.5.19(b) and the fact that we have a partition of $\mathrm{Irr}(\mathbf{G}^F)$ as above. □

Remark 2.6.3 Let us fix a semisimple element $s \in \mathbf{G}^{*F^*}$. Let $\mathscr{C}$ be the $\mathbf{G}^*$-conjugacy class of s, which is F-stable. By Proposition 2.5.5, $\mathscr{C}$ corresponds to the equivalence class of a pair $(\lambda, n) \in \Lambda(\mathbf{G}, F)$. Corresponding to (λ, n), we have a subset $\mathscr{E}_{\lambda,n} \subseteq \mathrm{Irr}(\mathbf{G}^F)$ that is a geometric series of characters; see Corollary 2.4.29. Then the definitions immediately imply that

$$\mathscr{E}_{\lambda,n} = \bigcup_{s' \in \mathscr{C}, F^*(s')=s'} \mathscr{E}(\mathbf{G}^F, s'). \tag{a}$$

(Just follow once more the correspondences in Propositions 2.4.28 and 2.5.5, and Corollary 2.5.14.) In particular, this shows that every geometric series of characters is a union of rational series of characters. Now assume that $\mathbf{Z}(\mathbf{G})$ is connected. Then $C_{\mathbf{G}^*}(s)$ is connected (see Theorem 2.2.14) and so the set $\{s' \in \mathscr{C} \mid F^*(s') = s'\}$ is a single $\mathbf{G}^{*F^*}$-conjugacy class (see Example 1.4.10). Thus, we conclude that

$$\mathscr{E}_{\lambda,n} = \mathscr{E}(\mathbf{G}^F, s) \qquad \text{(if } \mathbf{Z}(\mathbf{G}) \text{ is connected);} \tag{b}$$

that is, every geometric series of characters is just a rational series of characters in this case. We can now state the *Jordan decomposition of characters* for the connected centre case:

Theorem 2.6.4 ([Lu84a, 4.23], [Lu88, Cor. 6.1]) *Assume that* $\mathbf{Z}(\mathbf{G})$ *is connected.*

Let $s \in \mathbf{G}^*$ be a semisimple element such that $F^*(s) = s$. Let $\mathbf{H} := C_{\mathbf{G}^*}(s)$; note that $\mathbf{H}$ is connected. There is a bijection

$$\mathscr{E}(\mathbf{G}^F, s) \overset{1-1}{\longleftrightarrow} \mathrm{Uch}(\mathbf{H}^{F^*}), \qquad \rho \leftrightarrow \rho_u,$$

such that, for any F^*-stable maximal torus $\mathbf{T}^* \subseteq \mathbf{H}$, we have

$$\langle R_{\mathbf{T}^*}^{\mathbf{G}}(s), \rho\rangle = \varepsilon_{\mathbf{G}}\varepsilon_{\mathbf{H}}\langle R_{\mathbf{T}^*}^{\mathbf{H}}(1_{\mathbf{T}^*}), \rho_u\rangle.$$

(Here $R_{\mathbf{T}^*}^{\mathbf{H}}(1_{\mathbf{T}^*})$ is for the Deligne–Lusztig character of $\mathbf{H}^{F^*}$ corresponding to the trivial character of $\mathbf{T}^{*F^*}$, as in the original set-up of Section 2.2.)

A bijection as above will *not* in general be unique; we will discuss the uniqueness question under additional assumptions in Section 4.7. If we speak about Jordan decomposition later, we will mean *any* bijection satisfying the conditions in Theorem 2.6.4.

Theorem 2.6.4 is obtained from Theorem 2.4.15 and the discussion in 2.5.10, via the various correspondences established in the previous sections, by which one can pass from equivalence classes of pairs (λ, n) as in 2.4.5 to F^*-stable conjugacy classes of semisimple elements in $\mathbf{G}^*$ and, hence, to conjugacy classes of semisimple elements in $\mathbf{G}^{*F^*}$. It also involves the following fact (already mentioned in 2.5.10):

Remark 2.6.5 If we take $s = 1$ in Theorem 2.6.4, then we obtain a bijection

$$\mathrm{Uch}(\mathbf{G}^F) \overset{1-1}{\longleftrightarrow} \mathrm{Uch}(\mathbf{G}^{*F^*}), \qquad \rho \leftrightarrow \rho^*,$$

such that $\langle R_{\mathbf{T}}^{\mathbf{G}}(1_{\mathbf{T}}), \rho\rangle = \langle R_{\mathbf{T}^*}^{\mathbf{G}^*}(1_{\mathbf{T}^*}), \rho^*\rangle$ for corresponding maximal tori $\mathbf{T} \subseteq \mathbf{G}$ and $\mathbf{T}^* \subseteq \mathbf{G}^*$ as in 2.5.1(c). This was first stated explicitly in [Lu88, 6.1]. Note, however, that this bijection is not really a consequence of Theorem 2.6.4 but one needs to prove it beforehand, in order to be able at all to reformulate Theorem 2.4.15 as above. A proof of the above bijection can be obtained as follows. Using Proposition 2.3.15 and the reduction arguments in [Lu76c, 1.18] (see also Remark 4.2.1 in Chapter 4), it is sufficient to consider the case where $\mathbf{G}$ is simple. Then the only case which requires a special argument is when $\mathbf{G}$ is of type B_n or C_n, where the desired statement is known by the results in [Lu81c], [Lu84a, Chap. 9] (see also Remark 4.2.18 below). A more conceptual proof can nowadays be obtained by [Lu14b, §7].

While Theorem 2.4.15 already contains all the essential information, the above formulation is particularly useful in many applications; it leads to elegant formulations of certain identities that would be hard to state otherwise.

Corollary 2.6.6 *In the setting of Theorem 2.6.4, we have*

$$\rho(1) = |\mathbf{G}^{*F^*} : \mathbf{H}^{F^*}|_{p'}\rho_u(1) \qquad \text{for all } \rho \in \mathscr{E}(\mathbf{G}^F, s).$$

In particular, $\rho(1)$ is divisible by $|\mathbf{G}^{*F^*} : \mathbf{H}^{F^*}|_{p'}$ for any $\rho \in \mathscr{E}(\mathbf{G}^F, s)$.

Proof Using the degree formula in Theorem 2.6.2, we have

$$\rho(1) = \frac{|\mathbf{G}^F|_{p'}}{|\mathbf{H}^{F^*}|} \sum_{\mathbf{T}^*} \varepsilon_{\mathbf{G}}\varepsilon_{\mathbf{T}^*}\langle R_{\mathbf{T}^*}^{\mathbf{G}}(s), \rho\rangle$$

where the sum runs over all F^*-stable maximal tori in $\mathbf{H}$. On the other hand, by Proposition 2.2.18 (applied to $\mathbf{H}$ and $s = 1$) and Theorem 2.3.2, we obtain

$$\rho_u(1) = \frac{1}{|\mathbf{H}^{F^*}|_p} \sum_{\mathbf{T}^*} \varepsilon_{\mathbf{H}}\varepsilon_{\mathbf{T}^*}\langle R_{\mathbf{T}^*}^{\mathbf{H}}(1_{\mathbf{T}^*}), \rho_u\rangle$$

where, again, the sum runs over all F^*-stable maximal tori in $\mathbf{H}$. So Theorem 2.6.4 immediately yields the desired formula. □

Example 2.6.7 In the setting of Theorem 2.6.4, assume that s is a regular element as in Remark 2.3.10, that is, $\mathbf{H} = C_{\mathbf{G}^*}(s) \subseteq \mathbf{G}^*$ is a maximal torus. Then $\mathscr{E}(\mathbf{G}^F, s) = \{\varepsilon_{\mathbf{G}}\varepsilon_{\mathbf{H}}R_{\mathbf{H}}^{\mathbf{G}}(s)\}$. Now, we have seen in Lemma 2.3.11 that, if q is large, then most elements in a torus are regular. Hence, as already noted in the introduction of [Lu77b], 'almost all' irreducible characters of $\mathbf{G}^F$ are of the form $\pm R_{\mathbf{T}^*}^{\mathbf{G}}(s)$ where $s \in \mathbf{T}^{*F^*}$ is regular.

2.6.8 It had already been shown in [DeLu76, §10] (that is, long before the full version of Theorem 2.6.4 became available) that every geometric series $\mathscr{E}(\mathbf{G}^F, s)$ (where $\mathbf{Z}(\mathbf{G})$ is connected) contains certain distinguished characters which – in the language of Theorem 2.6.4 – correspond to the trivial and the Steinberg character of $\mathbf{H}^{F^*}$; thus, their degrees are given by

$$|\mathbf{G}^{*F^*} : \mathbf{H}^{F^*}|_{p'} \qquad \text{and} \qquad |\mathbf{G}^{*F^*} : \mathbf{H}^{F^*}|_{p'}|\mathbf{H}^{F^*}|_p,$$

respectively (see Example 2.2.27 for the degree of the Steinberg character; also note that, in the situation of Example 2.6.7, there is only one character in $\mathscr{E}(\mathbf{G}^F, s)$). We will now explain how the character corresponding to the trivial character of $\mathbf{H}^{F^*}$ is obtained, following (and streamlining) the exposition in [Ca85, Chap. 8]. In the beginning of the following discussion, we make no assumption on $\mathbf{Z}(\mathbf{G})$. Let $\mathscr{O}_0 \subseteq \mathbf{G}$ be the conjugacy class of regular unipotent elements; see 2.2.21. Let $l := \dim \mathbf{T}_0 - \dim \mathbf{Z}^\circ(\mathbf{G})$ be the *semisimple rank* of $\mathbf{G}$. By [Ca85, 5.1.9], we have

$$|\mathscr{O}_0^F||\mathbf{Z}^\circ(\mathbf{G})^F|q^l = |\mathbf{G}^F|.$$

Following [DeLu76, 10.2], we define a class function $\Delta_{\mathbf{G}} \in \mathrm{CF}(\mathbf{G}^F)$ by

$$\Delta_{\mathbf{G}}(g) = \begin{cases} |\mathbf{Z}^\circ(\mathbf{G})^F|q^l & \text{if } g \in \mathscr{O}_0^F, \\ 0 & \text{otherwise.} \end{cases}$$

(In [Ca85, §8.3], this class function is denoted by Ξ.) Then, for $f \in \mathrm{CF}(\mathbf{G}^F)$, we have $\langle f, \Delta_{\mathbf{G}}\rangle \neq 0$ if and only if the average value of f on $\mathscr{O}_0^F$ is non-zero.

Definition 2.6.9 (Cf. [Ca85, p. 280]) Let $\rho \in \mathrm{Irr}(\mathbf{G}^F)$. Using the above notation, we say that ρ is a *semisimple character* if $\langle \rho, \Delta_\mathbf{G} \rangle \neq 0$. Thus, ρ is semisimple if and only if the average value of ρ on $\mathscr{O}_0^F$ is non-zero. Let $\mathscr{S}_0(\mathbf{G}^F)$ denote the set of all semisimple characters of $\mathbf{G}^F$.

Clearly, the trivial character of $\mathbf{G}^F$ is semisimple, and so are all those linear characters with $\mathbf{G}_{\mathrm{uni}}^F$ in their kernel.

The reason why the values of $\Delta_\mathbf{G}$ are taken to be $|\mathbf{Z}^\circ(\mathbf{G})^F|q^l$ on elements in $\mathscr{O}_0^F$ (and not just to be the constant value 1) is the following:

(♣) *$\Delta_\mathbf{G}$ is a virtual character of $\mathbf{G}^F$ if $\mathbf{Z}(\mathbf{G}) = \mathbf{Z}^\circ(\mathbf{G})$ is connected.*

See [Ca85, 8.1.7] or [DeLu76, 10.3] for a proof, which uses the Gelfand–Graev character of $\mathbf{G}^F$. (The proof is elementary in the sense that it does not use the theory of ℓ-adic cohomology; note that, for the following discussion, we do not require the multiplicity-freeness of Gelfand–Graev characters in [Ca85, Thm. 8.1.3].)

2.6.10 By Theorem 2.2.22, any Deligne–Lusztig character takes value 1 on $\mathscr{O}_0^F$. Hence, for any pair $(\mathbf{T}^*, s) \in \mathfrak{Y}(\mathbf{G}^*, F^*)$, we obtain that

$$\langle R_{\mathbf{T}^*}^\mathbf{G}(s), \Delta_\mathbf{G} \rangle = \frac{1}{|\mathbf{G}^F|} \sum_{u \in \mathscr{O}_0^F} |\mathbf{Z}^\circ(\mathbf{G})^F|q^l = 1. \qquad \text{(a)}$$

It follows that there exists at least one $\rho \in \mathscr{S}_0(\mathbf{G}^F)$ such that $\langle R_{\mathbf{T}^*}^\mathbf{G}(s), \rho \rangle \neq 0$. In particular, the rational series of characters $\mathscr{E}(\mathbf{G}^F, s)$ contains at least one semisimple character. Thus,

$$\mathscr{E}(\mathbf{G}^F, s) \cap \mathscr{S}_0(\mathbf{G}^F) \neq \varnothing \quad \text{for every semisimple element } s \in \mathbf{G}^{*F^*}. \qquad \text{(b)}$$

Now assume that $\mathbf{Z}(\mathbf{G}) = \mathbf{Z}^\circ(\mathbf{G})$ is connected. Then the rational series of characters are precisely the geometric series of characters; see Remark 2.6.3. Furthermore, by (♣), the class function $\Delta_\mathbf{G}$ is an integral linear combination of the semisimple characters of $\mathbf{G}^F$. Thus, $|\mathscr{S}_0(\mathbf{G}^F)| \leqslant \langle \Delta_\mathbf{G}, \Delta_\mathbf{G} \rangle = |\mathbf{Z}(\mathbf{G})^F|q^l$, where the last equality is clear by the definition of $\Delta_\mathbf{G}$. On the other hand, by Proposition 2.5.6, $|\mathbf{Z}(\mathbf{G})^F|q^l$ also is the number of geometric conjugacy classes of characters. Taking into account (b), we conclude that

$$|\mathscr{E}(\mathbf{G}^F, s) \cap \mathscr{S}_0(\mathbf{G}^F)| = 1 \quad \text{for every semisimple element } s \in \mathbf{G}^{*F^*}. \qquad \text{(c)}$$

Thus, if $\mathbf{Z}(\mathbf{G})$ is connected, then $\mathscr{E}(\mathbf{G}^F, s)$ contains exactly one semisimple character, which we will denote by $\rho_s \in \mathscr{S}_0(\mathbf{G}^F)$. Hence $\Delta_\mathbf{G}$ has at least $|\mathbf{Z}(\mathbf{G})^F|q^l$ irreducible constituents. Since $\langle \Delta_\mathbf{G}, \Delta_\mathbf{G} \rangle = |\mathbf{Z}(\mathbf{G})^F|q^l$, we conclude that

$$\Delta_\mathbf{G} = \sum_s \epsilon_s \rho_s \qquad (\epsilon_s \in \{\pm 1\}) \qquad \text{(d)}$$

where s runs over a set of representatives of the conjugacy classes of semisimple elements of $\mathbf{G}^{*F^*}$.

Now we can state the following result, which summarises a big part of [DeLu76, §10] and interprets it in terms of the Jordan decomposition.

Theorem 2.6.11 *Assume that $\mathbf{Z}(\mathbf{G})$ is connected and let $s \in \mathbf{G}^{*F^*}$ be a semisimple element. As above, let $\rho_s \in \mathscr{S}_0(\mathbf{G}^F)$ be the unique semisimple character that belongs to $\mathscr{E}(\mathbf{G}^F, s)$. Then the following hold.*

(a) *For any F^*-stable maximal torus $\mathbf{T}^* \subseteq \mathbf{G}^*$ with $s \in \mathbf{T}^*$, we have $\langle R^{\mathbf{G}}_{\mathbf{T}^*}(s), \rho_s\rangle = \epsilon_s$, where $\epsilon_s = \pm 1$ is as in 2.6.10(d).*
(b) *Let $\mathbf{H} := C_{\mathbf{G}^*}(s)$. Then $\rho_s(1) = |\mathbf{G}^{*F^*} : \mathbf{H}^{F^*}|_{p'}$ and $\epsilon_s = \varepsilon_{\mathbf{G}}\varepsilon_{\mathbf{H}}$.*
(c) *Via the Jordan decomposition of characters in Theorem 2.6.4, the character $\rho_s \in \mathscr{E}(\mathbf{G}^F, s)$ corresponds to the trivial character of $\mathbf{H}^{F^*}$.*

Proof (a) We have $1 = \langle R^{\mathbf{G}}_{\mathbf{T}^*}(s), \Delta_{\mathbf{G}}\rangle = \sum_{s'} \epsilon_{s'}\langle R^{\mathbf{G}}_{\mathbf{T}^*}(s), \rho_{s'}\rangle$ where s' runs over a set of representatives of the conjugacy classes of semisimple elements of $\mathbf{G}^{*F^*}$. Using the partition in Theorem 2.6.2, we conclude that all terms in the sum are zero, except for one term corresponding to the unique s' that is conjugate to s. Thus, $1 = \epsilon_s\langle R^{\mathbf{G}}_{\mathbf{T}^*}(s), \rho_s\rangle$, as desired.

(b) Using the degree formula in Theorem 2.6.2, we have

$$\rho_s(1) = \frac{|\mathbf{G}^F|_{p'}}{|\mathbf{H}^{F^*}|}\sum_{\mathbf{T}^*} \varepsilon_{\mathbf{G}}\varepsilon_{\mathbf{T}^*}\langle R^{\mathbf{G}}_{\mathbf{T}^*}(s), \rho_s\rangle = \epsilon_s \frac{|\mathbf{G}^F|_{p'}}{|\mathbf{H}^{F^*}|}\sum_{\mathbf{T}^*} \varepsilon_{\mathbf{G}}\varepsilon_{\mathbf{T}^*},$$

where the sums run over all F^*-stable maximal tori $\mathbf{T}^* \subseteq \mathbf{H}$; the second equality holds by (a). Now, by Example 2.2.26(a), we have $\sum_{\mathbf{T}^*} \varepsilon_{\mathbf{G}}\varepsilon_{\mathbf{T}^*} = \varepsilon_{\mathbf{G}}\varepsilon_{\mathbf{H}}|\mathbf{H}^{F^*}|_p$. This yields the formula for $\rho_s(1)$ and the identity $\epsilon_s = \varepsilon_{\mathbf{G}}\varepsilon_{\mathbf{H}}$.

(c) Let $\psi \in \mathrm{Uch}(\mathbf{H}^{F^*})$ correspond to ρ_s via a bijection as in Theorem 2.6.4. Let $\mathbf{T}^* \subseteq \mathbf{H}$ be an F^*-stable maximal torus. Using (a) and (b), we obtain

$$\langle R^{\mathbf{H}}_{\mathbf{T}^*}(1_{\mathbf{T}^*}), \psi\rangle = \varepsilon_{\mathbf{G}}\varepsilon_{\mathbf{H}}\langle R^{\mathbf{G}}_{\mathbf{T}^*}(s), \rho_s\rangle = \varepsilon_{\mathbf{G}}\varepsilon_{\mathbf{H}}\epsilon_s = 1.$$

So Remark 2.4.2(b) shows that ψ must be the trivial character of $\mathbf{H}^{F^*}$. □

Remark 2.6.12 Assume that $\mathbf{Z}(\mathbf{G})$ is connected. By (♣), the class function $\Delta_{\mathbf{G}}$ is a virtual character of $\mathbf{G}^F$. We have seen in 2.6.10 that $\langle \Delta_{\mathbf{G}}, \Delta_{\mathbf{G}}\rangle = |\mathbf{Z}(\mathbf{G})^F|q^l$ and that the scalar product of $\Delta_{\mathbf{G}}$ with any Deligne–Lusztig character equals 1. Then there is a general argument which implies that $\Delta_{\mathbf{G}}$ and all irreducible constituents of $\Delta_{\mathbf{G}}$ are uniform; see [Ca85, 8.4.4] or [DeLu76, 10.6]. In particular, each ρ_s is uniform. An explicit expression of $\Delta_{\mathbf{G}}$ as a linear combination of Deligne–Lusztig characters is given by

$$\Delta_{\mathbf{G}} = \frac{1}{|\mathbf{G}^F|} \sum_{(\mathbf{T}^*,s)\in\mathfrak{Y}(\mathbf{G}^*,F^*)} |\mathbf{T}^{*F^*}| R_{\mathbf{T}^*}^{\mathbf{G}}(s).$$

This immediately follows by re-writing [DeLu76, 10.7.4] in terms of $R_{\mathbf{T}^*}^{\mathbf{G}}(s)$.

Remark 2.6.13 Assume that $\mathbf{Z}(\mathbf{G})$ is connected.

(a) If $(\lambda, n) \in \Lambda(\mathbf{G}, F)$ corresponds to s as in Remark 2.6.3, then one easily sees that $\pm\rho_s$ is equal to the uniform almost character $R_{\tilde{\phi}}$ (see Remark 2.4.17), where ϕ is the trivial character of $\mathbf{W}_{\lambda,n}$ and $\tilde{\phi}$ is its canonical γ-extension as in Example 2.1.8(c).

(b) We have mentioned above that there is a further distinguished member of $\mathscr{E}(\mathbf{G}^F, s)$, corresponding to the Steinberg character of $\mathbf{H}^{F^*} = C_{\mathbf{G}^*}(s)^{F^*}$. The most natural and elegant way to obtain this character is by using a duality operation on $\mathrm{CF}(\mathbf{G}^F)$ which we will discuss in Section 3.4.

Our final aim in this section is to present Lusztig's extension of Theorem 2.6.4 to the case where $\mathbf{Z}(\mathbf{G})$ is not necessarily connected. This passes by an intermediate version involving regular embeddings as in Section 1.7. The following discussion is a worked-out version of [Lu84a, §14.1].

Let us fix a regular embedding $i\colon \mathbf{G} \to \tilde{\mathbf{G}}$ and use the notational convention in Remark 1.7.6, where we identify $\mathbf{G}$ with $i(\mathbf{G}) \subseteq \tilde{\mathbf{G}}$. Let $(\tilde{\mathbf{G}}^*, \tilde{F}^*)$ be in duality with $(\tilde{\mathbf{G}}, \tilde{F})$. As in 1.7.11, we have a corresponding central isotypy $i^*\colon \tilde{\mathbf{G}}^* \to \mathbf{G}^*$ such that $i^* \circ \tilde{F}^* = F^* \circ i^*$. Recall from Lemma 1.7.12 that

- $\mathbf{K} := \ker(i^*)$ is an $\tilde{F}^*$-stable torus contained in $\mathbf{Z}(\tilde{\mathbf{G}}^*)$,
- $i^*(\tilde{\mathbf{G}}^*) = \mathbf{G}^*$ and $i^*((\tilde{\mathbf{G}}^*)^{\tilde{F}^*}) = \mathbf{G}^{*F^*}$.

Also recall that $\tilde{\mathbf{G}}^{\tilde{F}}/\mathbf{G}^F$ is abelian of order prime to p. In particular, every unipotent element of $\tilde{\mathbf{G}}^{\tilde{F}}$ already belongs to $\mathbf{G}^F$.

2.6.14 In order to simplify the notation, let us write

$$G := \mathbf{G}^F, \qquad \tilde{G} := \tilde{\mathbf{G}}^{\tilde{F}}, \qquad G^* := \mathbf{G}^{*F^*}, \qquad \tilde{G}^* := (\tilde{\mathbf{G}}^*)^{\tilde{F}^*}.$$

We denote the restriction of i^* to $\tilde{G}^*$ again by the same symbol. Now $\tilde{G}$ acts by conjugation on G and, hence, on $\mathrm{Irr}(G)$. Every $g \in G$ acts trivially on $\mathrm{Irr}(G)$ and so we have an action of $\tilde{G}/G$ on $\mathrm{Irr}(G)$. Let Θ be the group of all linear characters $\eta \in \mathrm{Irr}(\tilde{G})$ such that $G \subseteq \ker(\eta)$. Then Θ acts on $\mathrm{Irr}(\tilde{G})$ via the usual tensor product of class functions. Given $\tilde{\rho} \in \mathrm{Irr}(\tilde{G})$, let

$$\Theta(\tilde{\rho}) := \{\tilde{\eta} \in \Theta \mid \tilde{\eta} \cdot \tilde{\rho} = \tilde{\rho}\}$$

be the stabiliser of $\tilde{\rho}$ in Θ. By Theorem 1.7.15, we have $\tilde{\rho}|_G = \rho_1 + \cdots + \rho_r$, where

$\rho_1, \ldots, \rho_r$ are distinct irreducible characters of G. We have

$$r = \langle \tilde{\rho}|_G, \tilde{\rho}|_G \rangle = |\Theta(\tilde{\rho})|; \qquad \text{see Remark 2.1.4.}$$

Furthermore, let $\mathbf{Z} = \mathbf{Z}(\mathbf{G})$, $Z := \mathbf{Z}^F$ and $\tilde{Z} := \mathbf{Z}(\tilde{\mathbf{G}})^{\tilde{F}}$. Then

$$\Theta(\tilde{\rho}) \subseteq \{\tilde{\eta} \in \Theta \mid \tilde{Z} \subseteq \ker(\tilde{\eta})\},$$

since any element of $\tilde{Z}$ acts as a scalar in a representation affording $\tilde{\rho}$. Consequently, $r = |\Theta(\tilde{\rho})|$ divides $|\tilde{G} : G.\tilde{Z}|$ and, hence, the order of $(\mathbf{Z}/\mathbf{Z}^\circ)_F$, where we use the isomorphism in Remark 1.7.6(b). In particular, if $\mathbf{G}$ itself has a connected centre, then the restriction of $\tilde{\rho} \in \mathrm{Irr}(\tilde{G})$ to G is irreducible.

Remark 2.6.15 As in the previous section, let $\tilde{\Theta}_u \subseteq \mathrm{Irr}(\tilde{G})$ be the group of all linear characters $\tilde{\eta} \colon \tilde{G} \to \mathbb{K}^\times$ such that $\tilde{\eta}(u) = 1$ for $u \in \tilde{G}$ unipotent. Since all unipotent elements of $\tilde{G}$ already belong to G, we conclude that Θ (as defined above) is contained in $\tilde{\Theta}_u$. So, by Lemma 2.5.23, we have

$$\mathbf{K}^{\tilde{F}^*} = \{z_{\tilde{\eta}} \mid \tilde{\eta} \in \Theta\}. \tag{a}$$

Using Proposition 2.5.21, one immediately sees that

$$\tilde{\eta} \cdot \mathscr{E}(\tilde{G}, \tilde{s}) := \{\tilde{\eta} \cdot \tilde{\rho} \mid \tilde{\rho} \in \mathscr{E}(\tilde{G}, \tilde{s})\} = \mathscr{E}(\tilde{G}, z_{\tilde{\eta}}\tilde{s}) \tag{b}$$

for any semisimple element $\tilde{s} \in \tilde{G}^*$. Thus, the action of Θ on $\mathrm{Irr}(\tilde{G})$ permutes the geometric series $\mathscr{E}(\tilde{G}, \tilde{s})$. By Theorem 2.6.2, the stabiliser of $\mathscr{E}(\tilde{G}, \tilde{s})$ under this action is the subgroup

$$\Theta(\tilde{s}) := \{\tilde{\eta} \in \Theta \mid z_{\tilde{\eta}} \text{ and } z_{\tilde{\eta}}\tilde{s} \text{ are } \tilde{G}^*\text{-conjugate}\}. \tag{c}$$

Thus, the subgroup $\Theta(\tilde{s}) \subseteq \Theta$ permutes the characters in $\mathscr{E}(\tilde{G}, \tilde{s})$. Clearly, if $\tilde{\rho} \in \mathscr{E}(\tilde{G}, \tilde{s})$, then $\Theta(\tilde{\rho}) \subseteq \Theta(\tilde{s})$.

Proposition 2.6.16 *Let $s \in G^*$ be semisimple and $\tilde{s} \in \tilde{G}^*$ be any semisimple element such that $i^*(\tilde{s}) = s$. Then*

$$\mathscr{E}(G, s) = \big\{\rho \in \mathrm{Irr}(G) \mid \langle \tilde{\rho}|_G, \rho \rangle \neq 0 \textit{ for some } \tilde{\rho} \in \mathscr{E}(\tilde{G}, \tilde{s})\big\}.$$

Proof First we show the inclusion '$\subseteq$'. Let $\rho \in \mathscr{E}(G, s)$. There exists an F^*-stable maximal torus $\mathbf{T}^* \subseteq \mathbf{G}^*$ such that $s \in \mathbf{T}^*$ and $\langle R_{\mathbf{T}^*}^{\mathbf{G}}(s), \rho \rangle \neq 0$. Then $\tilde{s} \in \tilde{\mathbf{T}}^* := i^{*-1}(\mathbf{T}^*) \subseteq \tilde{\mathbf{G}}^*$. Using Proposition 2.5.22 and Frobenius reciprocity, we obtain

$$\big\langle R_{\tilde{\mathbf{T}}^*}^{\tilde{\mathbf{G}}}(\tilde{s}), \mathrm{Ind}_G^{\tilde{G}}(\rho) \big\rangle = \big\langle R_{\mathbf{T}^*}^{\mathbf{G}}(s), \rho \big\rangle \neq 0.$$

So there exists some $\tilde{\rho} \in \mathrm{Irr}(\tilde{G})$ such that $\langle R_{\tilde{\mathbf{T}}^*}^{\tilde{\mathbf{G}}}(\tilde{s}), \tilde{\rho} \rangle \neq 0$ and $\tilde{\rho}$ occurs in $\mathrm{Ind}_G^{\tilde{G}}(\rho)$. But then $\tilde{\rho} \in \mathscr{E}(\tilde{G}, \tilde{s})$ and ρ occurs in $\tilde{\rho}|_G$ (by Frobenius reciprocity), as required. To prove the reverse inclusion we use an indirect argument, as follows. Let $s_1, \ldots, s_n \in$

G^* be a set of representatives of the conjugacy classes of semisimple elements of G^*. For each i, we choose a semisimple element $\tilde{s}_i \in \tilde{G}^*$ such that $i^*(\tilde{s}_i) = s_i$. By the previous part of the proof, we have for all i:

$$\mathscr{E}(G, s_i) \subseteq \tilde{\mathscr{E}}_i := \{\rho \in \mathrm{Irr}(G) \mid \langle \tilde{\rho}|_G, \rho\rangle \neq 0 \text{ for some } \tilde{\rho} \in \mathscr{E}(\tilde{G}, \tilde{s}_i)\}.$$

Assume that there exists some $\rho \in \tilde{\mathscr{E}}_i \cap \tilde{\mathscr{E}}_j$. Thus, we have $\langle \tilde{\rho}|_G, \rho\rangle \neq 0$ for some $\tilde{\rho} \in \mathscr{E}(\tilde{G}, \tilde{s}_i)$ and $\langle \tilde{\rho}'|_G, \rho\rangle \neq 0$ for some $\tilde{\rho}' \in \mathscr{E}(\tilde{G}, \tilde{s}_j)$. By Remark 2.1.4, there exists some $\tilde{\eta} \in \Theta$ such that $\tilde{\rho}' = \tilde{\eta} \cdot \tilde{\rho}$. By Remark 2.6.15, we have

$$z_{\tilde{\eta}} \in \mathbf{K}^{\tilde{F}^*} \qquad \text{and} \qquad \tilde{\rho}' \in \tilde{\eta} \cdot \mathscr{E}(\tilde{G}, \tilde{s}_i) = \mathscr{E}(\tilde{G}, z_{\tilde{\eta}}\tilde{s}_i).$$

By Theorem 2.6.2, $\tilde{s}_j$ and $z_{\tilde{\eta}}\tilde{s}_i$ are conjugate in $\tilde{G}^*$. Hence s_j and $i^*(z_{\tilde{\eta}})s_i$ are conjugate in G^*. But $i^*(z_{\tilde{\eta}}) = 1$ since $z_{\tilde{\eta}} \in \mathbf{K}^{\tilde{F}^*}$ and so $i = j$. Hence, $\{\tilde{\mathscr{E}}_i \mid 1 \leqslant i \leqslant n\}$ is a family of disjoint subsets of $\mathrm{Irr}(G)$. But we also have $\mathrm{Irr}(G) = \bigcup_{1 \leqslant i \leqslant n} \mathscr{E}(G, s_i)$ and so we must have $\mathscr{E}(G, s_i) = \tilde{\mathscr{E}}_i$ for all i. □

For a somewhat different proof of the above proposition, see [Bo06, 11.7].

Proposition 2.6.17 ([Lu84a, §14.1], [Lu88, §11]) *Let $\tilde{s} \in \tilde{G}^*$ be semisimple and $\tilde{\rho} \in \mathscr{E}(\tilde{G}, \tilde{s})$. By Theorem 1.7.15, we can write*

$$\tilde{\rho}|_{\mathbf{G}^F} = \rho_1 + \cdots + \rho_r \quad \textit{where} \quad \rho_1, \ldots, \rho_r \in \mathrm{Irr}(\mathbf{G}^F)$$

and $\rho_i \neq \rho_j$ for $i \neq j$. Let $s := i^*(\tilde{s}) \in G^*$ and $\mathbf{T}^* \subseteq \mathbf{G}^*$ be an F^*-stable maximal torus with $s \in \mathbf{T}^*$. Then, setting $\tilde{\mathbf{T}}^* = i^{*-1}(\mathbf{T}^*) \subseteq \tilde{\mathbf{G}}^*$, we have

$$\rho_i \in \mathscr{E}(G, s) \quad \textit{and} \quad \langle R^{\mathbf{G}}_{\mathbf{T}^*}(s), \rho_i\rangle = \sum_{\tilde{\rho}' \in O} \langle R^{\tilde{\mathbf{G}}}_{\tilde{\mathbf{T}}^*}(\tilde{s}), \tilde{\rho}'\rangle \quad \textit{for } 1 \leqslant i \leqslant r,$$

where $O \subseteq \mathscr{E}(\tilde{G}, \tilde{s})$ is the orbit of $\tilde{\rho}$ under the action of $\Theta(\tilde{s})$.

Since every $\rho \in \mathrm{Irr}(\mathbf{G}^F)$ occurs as some ρ_i as above, this result in combination with Theorem 2.6.2 reduces the problem of computing the multiplicities $\langle R^{\mathbf{G}}_{\mathbf{T}^*}(s), \rho\rangle$ for all $(\mathbf{T}^*, s) \in \mathfrak{Y}(\mathbf{G}^*, F^*)$ to an analogous problem for $\mathrm{Irr}(\tilde{G})$.

Proof The fact that $\rho_i \in \mathscr{E}(G, s)$ is clear by Proposition 2.6.16. Now, by Remark 2.1.4 and Theorem 1.7.15, we have

$$\mathrm{Ind}_G^{\tilde{G}}(\rho_i) = \sum_{\tilde{\rho}' \in O'} \tilde{\rho}'$$

where O' is the full orbit of $\tilde{\rho}$ under the action of Θ. Hence, by Proposition 2.5.22 and Frobenius reciprocity, we obtain:

$$\langle R^{\mathbf{G}}_{\mathbf{T}^*}(s), \rho_i\rangle = \langle R^{\tilde{\mathbf{G}}}_{\tilde{\mathbf{T}}^*}(\tilde{s})|_{\mathbf{G}^F}, \rho_i\rangle = \langle R^{\tilde{\mathbf{G}}}_{\tilde{\mathbf{T}}^*}(\tilde{s}), \mathrm{Ind}_{\mathbf{G}^F}^{\tilde{\mathbf{G}}^{\tilde{F}}}(\rho_i)\rangle = \sum_{\tilde{\rho}' \in O'} \langle R^{\tilde{\mathbf{G}}}_{\tilde{\mathbf{T}}^*}(\tilde{s}), \tilde{\rho}'\rangle.$$

Let $\tilde{\rho}' \in O'$ be such that the corresponding term in the above sum is non-zero. Then

$\tilde{\rho}' \in \mathscr{E}(\tilde{G}, \tilde{s})$. Writing $\tilde{\rho}' = \tilde{\eta} \cdot \tilde{\rho}$ and using Remark 2.6.15, we see that $\tilde{\eta}$ preserves $\mathscr{E}(\tilde{G}, \tilde{s})$ and, hence, $\tilde{\eta} \in \Theta(\tilde{s})$. So $\tilde{\rho}' \in O$ as required. □

The following result completely describes the semisimple characters of G. (See [DiMi20, §12.4] for a slightly different discussion.)

Corollary 2.6.18 *Let $s \in G^*$ be semisimple. Let $\tilde{s} \in \tilde{G}^*$ be semisimple such that $i^*(\tilde{s}) = s$. Let $\tilde{\rho} \in \mathrm{Irr}(\tilde{G})$ be the unique semisimple character that belongs to $\mathscr{E}(\tilde{G}, \tilde{s})$; see 2.6.10(c).*

(a) *The semisimple characters in $\mathscr{E}(G, s)$ are precisely the irreducible constituents of $\tilde{\rho}|_G$.*
(b) *Let $\rho \in \mathscr{E}(G, s)$ be a semisimple character. Then $\langle R_{\mathbf{T}^*}^{\mathbf{G}}(s), \rho\rangle = \pm 1$ for any F^*-stable maximal torus $\mathbf{T}^* \subseteq \mathbf{G}^*$ such that $s \in \mathbf{T}^*$.*

Proof (a) Take any $\tilde{\psi} \in \mathscr{E}(\tilde{G}, \tilde{s})$ and let $\rho \in \mathrm{Irr}(G)$ be an irreducible constituent of $\tilde{\psi}|_G$; by Proposition 2.6.16, we have $\rho \in \mathscr{E}(G, s)$. By Clifford's theorem, $\tilde{\psi}|_G$ is a sum of conjugates of ρ, that is, characters of the form ${}^{\tilde{g}}\rho$ where $\tilde{g} \in \tilde{G}$. Since $\mathscr{O}_0^F \subseteq G$ is invariant under conjugation in $\tilde{G}$, all conjugates of ρ have the same average value on $\mathscr{O}_0^F$. So $\tilde{\psi}$ has a non-zero average value on $\mathscr{O}_0^F$ if and only if ρ has a non-zero average value on $\mathscr{O}_0^F$. Thus, if $\tilde{\psi} = \tilde{\rho}$, then ρ is semisimple and all semisimple characters in $\mathscr{E}(G, s)$ are constituents of $\tilde{\rho}|_G$.

(b) Note that $\tilde{\eta} \cdot \tilde{\rho}$ is also semisimple, for every $\tilde{\eta} \in \Theta$. Hence, since $\tilde{\rho}$ is the unique semisimple character in $\mathscr{E}(\tilde{G}, \tilde{s})$, the orbit of $\tilde{\rho}$ under the action of $\Theta(\tilde{s})$ is $O = \{\tilde{\rho}\}$. So Proposition 2.6.17 yields that

$$\langle R_{\mathbf{T}^*}^{\mathbf{G}}(s), \rho\rangle = \langle R_{\tilde{\mathbf{T}}^*}^{\tilde{\mathbf{G}}}(\tilde{s}), \tilde{\rho}\rangle = \pm 1$$

where the last equality holds by Theorem 2.6.11(a). □

Remark 2.6.19 Recall from the beginning of Section 2.3 the definition of the graph $\mathrm{DL}(G)$. As discussed in [Ge18, 6.14], it now also follows that the partition of $\mathrm{Irr}(G)$ defined by the connected components of this graph is precisely the partition into rational series of characters as in Theorem 2.6.2.

Indeed, by Corollary 2.6.18, each character in $\mathscr{E}(G, s)$ is directly linked to a fixed semisimple character in $\mathscr{E}(G, s)$. Hence, all characters in $\mathscr{E}(G, s)$ belong to the same connected component of $\mathrm{DL}(G)$. Assume, if possible, that $\mathscr{E}(G, s)$ is strictly contained in a connected component of $\mathrm{DL}(G)$. Then there exists some $\rho_1 \in \mathscr{E}(G, s)$ and some $\rho_2 \in \mathrm{Irr}(G)$ such that $\rho_2 \notin \mathscr{E}(G, s)$ and ρ_1, ρ_2 are directly linked in $\mathrm{DL}(G)$, that is, there is some pair $(\mathbf{T}'^*, s') \in \mathfrak{Y}(\mathbf{G}^*, F^*)$ such that $\langle R_{\mathbf{T}'^*}^{\mathbf{G}}(s'), \rho_i\rangle \neq 0$ for $i = 1, 2$. But then s, s' are conjugate in G^* by Theorem 2.5.24. So we have $\mathscr{E}(G, s) = \mathscr{E}(G, s')$ by Theorem 2.6.2 and, hence, $\rho_2 \in \mathscr{E}(G, s)$, contradiction. (For a slightly different argument, see [DiMi20, Thm. 12.4.13].)

To illustrate this, consider again the example where $G = \mathrm{SL}_2(q)$, with q odd. By Example 2.2.30, $\{\psi'_+, \psi''_+\}$ and $\{\psi'_-, \psi''_-\}$ are connected components of DL(G) and so these are also rational series of characters; in fact, we have

$$\mathscr{E}(\mathbf{G}^F, s) = \{\psi'_+, \psi''_+\} \qquad \text{and} \qquad \mathscr{E}(\mathbf{G}^F, s') = \{\psi'_-, \psi''_-\}$$

where s is the unique element of order 2 in a maximally split torus of $\mathbf{G}^* = \mathrm{PGL}_2(k)$ and s' is the unique element of order 2 in an F^*-stable maximal torus of $\mathbf{G}^*$ that is not maximally split. On the other hand, by Example 2.4.7(b), the union $\mathscr{E}(\mathbf{G}^F, s) \cup \mathscr{E}(\mathbf{G}^F, s') = \{\psi'_+, \psi''_+, \psi'_-, \psi''_-\}$ is a geometric series of characters of $\mathbf{G}^F$.

In particular, this shows that rational series of characters and geometric series of characters are not the same in general — and this already happens in the smallest possible case where $\mathbf{Z}(\mathbf{G})$ is not connected!

In order to make an effective use of Proposition 2.6.17, one needs to know the action of Θ on $\mathrm{Irr}(\tilde{G})$. Ideally, it should be possible to describe this action via the bijections in Theorem 2.6.4 and, hence, reduce the problem to a question about unipotent characters. This reduction can indeed be done, but requires more work.

Lemma 2.6.20 ([Lu88, §8]) *Let $\tilde{s} \in \tilde{G}^*$ be semisimple and $s := i^*(\tilde{s}) \in G^*$. Let $A(s) := C_{\mathbf{G}^*}(s)/C^\circ_{\mathbf{G}^*}(s)$. Then F^* induces an automorphism of $A(s)$, which we denote by the same symbol. There is a canonical isomorphism*

$$A(s)^{F^*} \xrightarrow{\sim} \Theta(\tilde{s})$$

defined as follows. Let $x \in C_{\mathbf{G}^*}(s)^{F^*}$ and $\dot{x} \in \tilde{G}^*$ be such that $i^*(\dot{x}) = x$; then the coset of x in $A(s)$ is sent to the unique $\tilde{\eta} \in \Theta(\tilde{s})$ such that $z_{\tilde{\eta}} = \dot{x}^{-1}\tilde{s}^{-1}\dot{x}\tilde{s}$. Hence, via this isomorphism, we obtain an action of $A(s)^{\tilde{F}^*}$ on $\mathscr{E}(\tilde{G}, \tilde{s})$.

Proof We apply Lemma 1.1.9 with $A = \tilde{\mathbf{G}}^*$, $B = \mathbf{G}^*$, $f = i^*$, $\sigma(\tilde{x}) = \tilde{s}^{-1}\tilde{x}\tilde{s}$ ($\tilde{x} \in \tilde{\mathbf{G}}^*$), $\tau(x) = s^{-1}xs$ ($x \in \mathbf{G}^*$). Then $A^\sigma = C_{\tilde{\mathbf{G}}^*}(\tilde{s})$ and $B^\tau = C_{\mathbf{G}^*}(s)$. Since $C_{\tilde{\mathbf{G}}^*}(\tilde{s})$ is connected, we have $i^*(A^\sigma) = C^\circ_{\mathbf{G}^*}(s)$ by 1.3.10(e); furthermore, $C = \{a^{-1}\sigma(a) \mid a \in \ker(f)\} = \{1\}$ and we obtain a canonical isomorphism

$$\delta\colon C_{\mathbf{G}^*}(s)/C^\circ_{\mathbf{G}^*}(s) \xrightarrow{\sim} \{\tilde{x}^{-1}\tilde{s}^{-1}\tilde{x}\tilde{s} \mid \tilde{x} \in \tilde{\mathbf{G}}^*\} \cap \mathbf{K}.$$

The group on the right-hand side equals $\{\tilde{z} \in \mathbf{K} \mid \tilde{s}, \tilde{z}\tilde{s} \text{ are conjugate in } \tilde{\mathbf{G}}^*\}$. Now F^* is compatible with δ and we obtain an isomorphism

$$\delta\colon A(s)^{F^*} \xrightarrow{\sim} \{\tilde{z} \in \mathbf{K}^{F^*} \mid \tilde{s}, \tilde{z}\tilde{s} \text{ are conjugate in } \tilde{\mathbf{G}}^*\}.$$

Note also that, by Example 1.4.11(b), the inclusion $C_{G^*}(s) \subseteq C_{\mathbf{G}^*}(s)$ induces an isomorphism $C_{\mathbf{G}^*}(s)^{F^*}/C^\circ_{\mathbf{G}^*}(s)^{F^*} \cong A(s)^{F^*}$. Finally, since $C_{\tilde{\mathbf{G}}^*}(\tilde{s})$ is connected, the group on the right side of the above isomorphism equals

$$\{\tilde{z} \in \mathbf{K}^{F^*} \mid \tilde{s}, \tilde{z}\tilde{s} \text{ are conjugate in } \tilde{G}^*\}$$

and the latter group is isomorphic to $\Theta(\tilde{s})$ via Remark 2.6.15(c). The above description of the map $A(s)^{F^*} \to \Theta(\tilde{s})$ follows from the description of δ in Lemma 1.1.9. $\square$

In the setting of Lemma 2.6.20, let $\tilde{\mathbf{H}} := C_{\tilde{\mathbf{G}}^*}(\tilde{s})$ and $\mathbf{H} := C_{\mathbf{G}^*}(s)$; we also set $\tilde{H} := \tilde{\mathbf{H}}^{\tilde{F}^*}$, $H := \mathbf{H}^{F^*}$ and $H^\circ := \mathbf{H}^{\circ F^*}$. Then we obtain a natural action of $A(s)^{\tilde{F}^*}$ on $\mathrm{Uch}(\tilde{H})$, as follows. In the above proof we already saw that $i^*(\tilde{\mathbf{H}}) = \mathbf{H}^\circ$. Hence, by Proposition 2.3.15, we obtain a canonical bijection

$$\mathrm{Uch}(H^\circ) \xrightarrow{1-1} \mathrm{Uch}(\tilde{H}), \qquad \psi \mapsto \psi \circ i^*|_{\tilde{H}}.$$

Now the conjugation action of H on H° induces an action of H on $\mathrm{Uch}(H^\circ)$ (see Example 2.3.17). Clearly, H° is in the kernel of this action and so we obtain a natural action of $A(s)^{F^*}$ on $\mathrm{Uch}(H^\circ)$. Via the above bijection, this becomes an action of $A(s)^{F^*}$ on $\mathrm{Uch}(\tilde{H})$. Now we can state:

Theorem 2.6.21 ([Lu88, 8.1]) *Let $\tilde{s} \in \tilde{G}^*$ be semisimple and $s := i^*(\tilde{s}) \in G^*$. Let $\tilde{\mathbf{H}} := C_{\tilde{\mathbf{G}}^*}(\tilde{s})$ and $\tilde{H} := \tilde{\mathbf{H}}^{\tilde{F}^*}$. Then we can choose the bijection*

$$\mathscr{E}(\tilde{G}, \tilde{s}) \xleftrightarrow{1-1} \mathrm{Uch}(\tilde{H}), \qquad \tilde{\rho} \leftrightarrow \tilde{\rho}_u, \qquad \textit{(see Theorem 2.6.4)}$$

so that $\tilde{\rho} \leftrightarrow \tilde{\rho}_u$ commutes with the action of $A(s)^{F^*}$ on $\mathscr{E}(\tilde{G}, \tilde{s})$ (see Lemma 2.6.20) and the action of $A(s)^{F^*}$ on $\mathrm{Uch}(\tilde{H})$ just defined.

The proof of this result is quite a tour de force. It proceeds by a reduction to the case where $\mathbf{G}$ is simple of simply connected type. Then it uses the full power of Theorem 2.4.15 (that is, the Main Theorem 4.23 of [Lu84a]), including the explicit description of the parameter sets $\bar{X}(\mathbf{W}_{\lambda,n})$ and the multiplicities $m(y, \bar{x})$. Various special situations have to be considered case-by-case, including one where $\mathbf{G}$ is of type E_7; see also [Lu84a, §14.1].

We can now state the general version of the *Jordan decomposition of characters*, where the formulation does not refer to a regular embedding of $\mathbf{G}$.

Theorem 2.6.22 ([Lu88, 5.1], [Lu08a, 5.2]) *Let $s \in \mathbf{G}^*$ be semisimple such that $F^*(s) = s$. Let $\mathbf{H} := C_{\mathbf{G}^*}(s)$ and $\mathfrak{U}_{F^*}(\mathbf{H})$ be the set of all pairs (O, a) where O is an orbit of $A(s)^{F^*}$ on $\mathrm{Uch}(\mathbf{H}^{\circ F^*})$ and $a \in A(s)^{F^*}$ fixes some (or, equivalently, any) character in O. (Since $A(s)^{F^*}$ is abelian, all characters in an orbit O have the same stabiliser.) Then there is a bijection*

$$\mathscr{E}(\mathbf{G}^F, s) \xleftrightarrow{1-1} \mathfrak{U}_{F^*}(\mathbf{H}), \qquad \rho \leftrightarrow (O_\rho, a_\rho),$$

such that, for any F^*-stable maximal torus $\mathbf{T}^* \subseteq \mathbf{H}^\circ$, we have

$$\langle R_{\mathbf{T}^*}^{\mathbf{G}}(s), \rho \rangle = \varepsilon_{\mathbf{G}} \varepsilon_{\mathbf{H}} \sum_{1 \leqslant i \leqslant r} \langle R_{\mathbf{T}^*}^{\mathbf{H}^\circ}(1), \rho_i \rangle \quad \textit{where} \quad O_\rho = \{\rho_1, \ldots, \rho_r\}.$$

This is now a formal consequence of Theorems 2.6.4, 2.6.21, Proposition 2.6.17 and general Clifford theory; see [Lu88, §11] and [CE04, 15.14] for further details.

Remark 2.6.23 If $A(s)^{F^*} = \{1\}$, then Theorem 2.6.22 essentially reduces to the statement of Theorem 2.6.4, even if $\mathbf{H}$ is not connected. Indeed, in this case, every orbit O consists of one unipotent character of $\mathbf{H}^{\circ F^*}$ and so

$$\mathfrak{U}_{F^*}(\mathbf{H}) = \left\{(\psi, 1) \mid \psi \in \mathbf{H}^{\circ F^*}\right\} \xleftrightarrow{1-1} \mathrm{Uch}(\mathbf{H}^{\circ F^*}), \qquad (\psi, 1) \leftrightarrow \psi.$$

Hence, we obtain a bijection $\mathscr{E}(\mathbf{G}^F, s) \xleftrightarrow{1-1} \mathrm{Uch}(\mathbf{H}^{\circ F^*})$ satisfying the scalar product formula in Theorem 2.6.4. In general, the following facts are known.

(a) By Lemma 2.6.20 and the discussion in 2.6.14, the order of $A(s)^{F^*}$ divides the order of $(\mathbf{Z}/\mathbf{Z}^\circ)_F$.
(b) The exponent of $A(s)^{F^*}$ divides the order of the element s or, more precisely, the order of the image of s in $\mathbf{G}_{\mathrm{ad}}$ where $\pi_{\mathrm{ad}}\colon \mathbf{G} \to \mathbf{G}_{\mathrm{ad}}$ denotes an adjoint quotient as in Remark 1.5.12 (see [Bo05, Cor. 2.9], [BrMi89, Lemma 2.1] and [Bor70, Part E, II, 4.4, 4.6]).

So, if the order of s is prime to the order of $(\mathbf{Z}/\mathbf{Z}^\circ)_F$, then $A(s)^{F^*} = \{1\}$. For example, if $\mathbf{G}$ is a simple algebraic group of classical type B_n, C_n or D_n and s has odd order, then $A(s)^{F^*} = \{1\}$.

Remark 2.6.24 In the setting of Theorem 2.6.22, let us fix an orbit O of $A(s)^{F^*}$ on $\mathrm{Uch}(\mathbf{H}^{\circ F^*})$ and consider the set of characters

$$\{\rho \in \mathscr{E}(\mathbf{G}^F, s) \mid O_\rho = O\}.$$

In [Lu88, §3], it is explained how one can define a natural action of the group $(\mathbf{Z}/\mathbf{Z}^\circ)_F$ on $\mathscr{E}(\mathbf{G}^F, s)$. Then each set of characters as above is precisely an orbit under this action of $(\mathbf{Z}/\mathbf{Z}^\circ)_F$; see [Lu88, 5.1].

Example 2.6.25 Let $G = \mathbf{G}^F = \mathrm{SL}_4(q)$ where q is odd; then $Z = \mathbf{Z}(\mathbf{G})^F$ has order 2 (if $4 \mid q+1$) or order 4 (if $4 \mid q-1$). Since $\mathbf{G}$ is simple of simply connected type, we can take $\mathbf{G}^* = \mathrm{PGL}_4(k)$ and $G^* = \mathrm{PGL}_4(q)$ (see 1.5.18). Consider the following two matrices in $\mathrm{GL}_4(k)$:

$$s = \begin{pmatrix} 1 & 0 & 0 & 0 \\ 0 & 1 & 0 & 0 \\ 0 & 0 & -1 & 0 \\ 0 & 0 & 0 & -1 \end{pmatrix} \quad \text{and} \quad a = \begin{pmatrix} 0 & 0 & 0 & 1 \\ 0 & 0 & 1 & 0 \\ 0 & 1 & 0 & 0 \\ 1 & 0 & 0 & 0 \end{pmatrix}.$$

Let $\bar{s}$ be the image of s in G^*. Let $\mathbf{H} := C_{\mathbf{G}^*}(\bar{s})$. Then $\mathbf{H}^\circ$ has type $A_1 \times A_1$, the centre of $\mathbf{H}^\circ$ has dimension 1, and $\mathbf{H}/\mathbf{H}^\circ$ is cyclic of order 2, generated by $\bar{a}$ (the image of a). We have $\mathrm{Uch}(H^\circ) = \{\psi_1, \psi_2, \psi_2', \psi_3\}$ where ψ_1 is the trivial character,

ψ_3 is the Steinberg character (of degree q^2) and ψ_2, ψ_2' have degree q. There are four H°-conjugacy classes of F^*-stable maximal tori in $\mathbf{H}^\circ$; we denote representatives by $\mathbf{T}_1^*, \mathbf{T}_2^*, \mathbf{T}_2'^*, \mathbf{T}_3^*$ where

$$|\mathbf{T}_1^{*F^*}| = (q-1)^3,\ |\mathbf{T}_2^{*F^*}| = |\mathbf{T}_2'^{*F^*}| = (q-1)^2(q+1),\ |\mathbf{T}_3^{*F^*}| = (q-1)(q+1)^2.$$

The Deligne–Lusztig characters have the following decompositions:

$$\begin{aligned}
R_{\mathbf{T}_1^*}^{\mathbf{H}^\circ}(1_{\mathbf{T}_1^*}) &= \psi_1 + \psi_2 + \psi_2' + \psi_3,\\
R_{\mathbf{T}_2^*}^{\mathbf{H}^\circ}(1_{\mathbf{T}_2^*}) &= \psi_1 + \psi_2 - \psi_2' - \psi_3,\\
R_{\mathbf{T}_2'^*}^{\mathbf{H}^\circ}(1_{\mathbf{T}_2'^*}) &= \psi_1 - \psi_2 + \psi_2' - \psi_3,\\
R_{\mathbf{T}_3^*}^{\mathbf{H}^\circ}(1_{\mathbf{T}_3^*}) &= \psi_1 - \psi_2 - \psi_2' + \psi_3.
\end{aligned}$$

Now, the action of $\bar{a}$ on $\mathbf{H}^\circ$ exchanges the two components of type A_1. This action fixes $\mathbf{T}_1^*$ and $\mathbf{T}_3^*$ but exchanges $\mathbf{T}_2^*, \mathbf{T}_2'^*$ (up to conjugation in $\mathbf{H}^\circ$). Similarly, it fixes ψ_1 and ψ_3 but exchanges ψ_2, ψ_2'. Thus, the orbits of $A(\bar{s})^{F^*}$ on $\mathrm{Uch}(H^\circ)$ are

$$O_1 = \{\psi_1\}, \qquad O_2 = \{\psi_2, \psi_2'\}, \qquad O_3 = \{\psi_3\}.$$

The stabilisers of ψ_1, ψ_3 in $A(\bar{s})^{F^*}$ have order 2, while the stabilisers of ψ_2, ψ_2' in $A(\bar{s})^{F^*}$ are trivial. Hence, we obtain

$$\mathfrak{U}_{F^*}(\mathbf{H}) = \{(O_1, 1), (O_1, \bar{a}), (O_2, 1), (O_3, 1), (O_3, \bar{a})\}.$$

Thus, $\mathscr{E}(G, \bar{s})$ contains exactly five irreducible characters. Let $\rho \in \mathscr{E}(G, \bar{s})$ correspond to $(O_2, 1)$ under a bijection as in Theorem 2.6.22. Then we obtain:

$$\begin{aligned}
\langle R_{\mathbf{T}_1^*}^{\mathbf{H}}(\bar{s}), \rho\rangle &= \langle R_{\mathbf{T}_1^*}^{\mathbf{H}^\circ}(1_{\mathbf{T}_1^*}), \psi_2\rangle + \langle R_{\mathbf{T}_1^*}^{\mathbf{H}^\circ}(1_{\mathbf{T}_1^*}), \psi_2'\rangle &&= 1+1 = 2,\\
\langle R_{\mathbf{T}_2^*}^{\mathbf{H}}(\bar{s}), \rho\rangle &= \langle R_{\mathbf{T}_2^*}^{\mathbf{H}^\circ}(1_{\mathbf{T}_2^*}), \psi_2\rangle + \langle R_{\mathbf{T}_2^*}^{\mathbf{H}^\circ}(1_{\mathbf{T}_2^*}), \psi_2'\rangle &&= 1-1 = 0,\\
\langle R_{\mathbf{T}_2'^*}^{\mathbf{H}}(\bar{s}), \rho\rangle &= \langle R_{\mathbf{T}_2'^*}^{\mathbf{H}^\circ}(1_{\mathbf{T}_2'^*}), \psi_2\rangle + \langle R_{\mathbf{T}_2'^*}^{\mathbf{H}^\circ}(1_{\mathbf{T}_2'^*}), \psi_2'\rangle &&= -1+1 = 0,\\
\langle R_{\mathbf{T}_3^*}^{\mathbf{H}}(\bar{s}), \rho\rangle &= \langle R_{\mathbf{T}_3^*}^{\mathbf{H}^\circ}(1_{\mathbf{T}_1^*}), \psi_2\rangle + \langle R_{\mathbf{T}_3^*}^{\mathbf{H}^\circ}(1_{\mathbf{T}_3^*}), \psi_2'\rangle &&= -1-1 = -2.
\end{aligned}$$

(Note that $\varepsilon_{\mathbf{G}}\varepsilon_{\mathbf{H}} = 1$ since a maximally split torus of $\mathbf{G}$ is contained in $\mathbf{H}^\circ$.)

See [Ge18, 6.11] for the discussion of a similar example in $G = \mathrm{Sp}_4(q)$.

Remark 2.6.26 Let $s \in G^*$ be semisimple; as above, we write $\mathbf{H} = C_{\mathbf{G}^*}(s)$, $H := \mathbf{H}^{F^*}$ and $H^\circ := \mathbf{H}^{\circ F^*}$. As in [Lu88, §12], we define $\mathrm{Uch}(H)$ to be the set of all $\rho \in \mathrm{Irr}(H)$ such that the restriction of ρ to H° has an irreducible constituent that is unipotent. (See Proposition 4.8.19 for a different characterisation.) Note that, if $\rho \in \mathrm{Uch}(H)$, then all irreducible constituents of the restriction of ρ to H° are unipotent. (Indeed, the irreducible constituents of $\rho|_{H^\circ}$ form an orbit under the natural action of H on $\mathrm{Irr}(H^\circ)$ by conjugation; so, if one of them is unipotent, then all are unipotent by Example 2.3.17.) Now, it is known that the restriction of any

$\rho \in \mathrm{Uch}(H)$ to H° is multiplicity-free (see [DiMi20, Prop. 11.5.3]). Hence, as in [Lu88, §12], one can conclude that there is a bijection $\mathscr{E}(G, s) \overset{1-1}{\longleftrightarrow} \mathrm{Uch}(H)$; see [DiMi20, Chap. 11] for a further discussion.

2.7 Average Values and Unipotent Support

Let $\mathbf{G}$ be a connected reductive algebraic group over $k = \overline{\mathbb{F}}_p$ and $F\colon \mathbf{G} \to \mathbf{G}$ be a Steinberg map. The results exposited in the previous sections show that there is an efficient classification of the irreducible characters of $\mathbf{G}^F$. This classification also yields all character values on semisimple elements of $\mathbf{G}^F$ (see Proposition 2.2.18), which includes the character degrees, but does not tell us much about the remaining values. In order to attack this problem in general, Lusztig developed the theory of *character sheaves* [LuCS], which tries to produce some geometric objects over the algebraic group $\mathbf{G}$ from which the irreducible characters of $\mathbf{G}^F$, for any F as above, could be deduced in a uniform manner (as exemplified by the *generic character tables* in Section 2.1). In this and the following section, we give but a brief introduction into this theory. Our focus will be on uniform functions and the determination of the values of the virtual characters $R_{\mathbf{T}}^{\mathbf{G}}(\theta)$, which already is a highly complex and technically intricate story.

We begin with some elementary remarks relating conjugacy classes in the finite group $\mathbf{G}^F$ to F-stable conjugacy classes of $\mathbf{G}$.

2.7.1 Let $\mathscr{C}$ be an F-stable conjugacy class of $\mathbf{G}$. Then $\mathscr{C}^F$ is non-empty (see Example 1.4.10) and we pick an element $g \in \mathscr{C}^F$. Let

$$A(g) := C_{\mathbf{G}}(g)/C_{\mathbf{G}}^\circ(g)$$

be the finite group of components of the centraliser of g. (If it is necessary to indicate the underlying algebraic group, then we write $A_{\mathbf{G}}(g)$ instead of just $A(g)$.) Since $F(g) = g$, the groups $C_{\mathbf{G}}(g)$ and $C_{\mathbf{G}}^\circ(g)$ are F-stable and so F induces an automorphism of $A(g)$ which we denote by the same symbol. By Example 1.4.10, we have:

$$\mathscr{C}^F \text{ is a single } \mathbf{G}^F\text{-conjugacy class if } A(g) = \{1\}. \tag{a}$$

In general, the following happens. Let $a \in A(g)$ and $\dot{a} \in C_{\mathbf{G}}(g)$ be a representative of a. By the Lang–Steinberg theorem, we can write $\dot{a} = x^{-1}F(x)$ for some $x \in \mathbf{G}$. One immediately checks that $g_a := xgx^{-1} \in \mathscr{C}^F$. Thus, we have associated with any element $a \in A(g)$ an element $g_a \in \mathscr{C}^F$; this depends on the choices of $\dot{a}$ and x as above, but the $\mathbf{G}^F$-conjugacy class of g_a does not depend on these choices. If we denote this $\mathbf{G}^F$-conjugacy class by C_a, then we obtain a canonical bijection

$$H^1(F, A(g)) \overset{1-1}{\longleftrightarrow} \{\mathbf{G}^F\text{-conjugacy classes contained in } \mathscr{C}^F\}, \tag{b}$$

where the F-conjugacy class of the element $a \in A(g)$ corresponds to the $\mathbf{G}^F$-conjugacy class C_a. (See, e.g., [Ge03a, 4.3.5] for a detailed proof; also recall from 2.1.6 that $H^1(F, A(g))$ denotes the set of F-conjugacy classes of $A(g)$.) Now let us consider what happens if we replace the chosen element $g \in \mathscr{C}^F$ by another element $g' \in \mathscr{C}^F$. Then $g' \in C_a$ for some $a \in A(g)$. We can reverse the above procedure and find some $x \in \mathbf{G}$ such that $g' = xgx^{-1}$ and $\dot{a} = x^{-1}F(x)$. Then $C_{\mathbf{G}}(g') = xC_{\mathbf{G}}(g)x^{-1}$ and $C_{\mathbf{G}}^{\circ}(g') = xC_{\mathbf{G}}^{\circ}(g)x^{-1}$. So conjugation by x induces an isomorphism $A(g) \xrightarrow{\sim} A(g')$ and it is straightforward to check that this restricts to an isomorphism

$$C_{A(g),F}(a) \xrightarrow{\sim} A(g')^F; \tag{c}$$

here, $C_{A(g),F}(a)$ denotes the F-centraliser of a (as in 2.1.6).

2.7.2 In the setting of 2.7.1, consider the irreducible characters of the finite group $A(g)$. For $\varsigma \in \mathrm{Irr}(A(g))^F$, we fix an F-extension $\tilde{\varsigma} \in \mathrm{CF}_F(A(g))$ of ς (as in 2.1.7). Then define a class function $Y_{(g,\tilde{\varsigma})} \in \mathrm{CF}(\mathbf{G}^F)$ by

$$Y_{(g,\tilde{\varsigma})}(g') = \begin{cases} \tilde{\varsigma}(a) & \text{if } g' = g_a \in \mathscr{C}^F \text{ for some } a \in A(g), \\ 0 & \text{if } g' \notin \mathscr{C}^F. \end{cases} \tag{a}$$

As noted in the remarks following Definition 2.1.11, the F-character table of $A(g)$ is invertible. In particular, it follows that

$$\text{the functions } \{Y_{(g,\tilde{\varsigma})} \mid \varsigma \in \mathrm{Irr}(A(g))^F\} \text{ are linearly independent.} \tag{b}$$

Now let $\mathrm{Cl}(\mathbf{G})^F$ be the set of all F-stable conjugacy classes of $\mathbf{G}$. For each $\mathscr{C} \in \mathrm{Cl}(\mathbf{G})^F$, let us pick an element $g_{\mathscr{C}} \in \mathscr{C}^F$; for each $\varsigma \in \mathrm{Irr}(A(g_{\mathscr{C}}))^F$, we choose a particular F-extension $\tilde{\varsigma}$ of ς. Then (a), (b) show that the functions

$$\mathscr{B} := \{Y_{(g_{\mathscr{C}},\tilde{\varsigma})} \mid \mathscr{C} \in \mathrm{Cl}(\mathbf{G})^F,\ \varsigma \in \mathrm{Irr}(A(g_{\mathscr{C}}))^F\} \tag{c}$$

form a vector space basis of $\mathrm{CF}(\mathbf{G}^F)$. Thus, the problem of computing the character table of $\mathbf{G}^F$ can be reformulated as the problem of determining the base change from the basis $\mathrm{Irr}(\mathbf{G}^F)$ of $\mathrm{CF}(\mathbf{G}^F)$ to the basis $\mathscr{B}$. Note that this involves the issue of making choices of the representatives $g_{\mathscr{C}} \in \mathscr{C}^F$ and of F-extensions of the F-invariant irreducible characters of $A(g_{\mathscr{C}})$.

Remark 2.7.3 The advantage of using the functions in 2.7.2(c) (rather than just the indicator functions of the conjugacy classes of $\mathbf{G}^F$) is that these have a topological interpretation: each $Y_{(g_{\mathscr{C}},\tilde{\varsigma})}$ is a '*characteristic function*' of an F-stable, irreducible, $\mathbf{G}$-equivariant $\overline{\mathbb{Q}}_\ell$-local system on $\mathscr{C}$. (This is a special case of a more general geometric principle; see [Lu04b, 19.7] and also Example 2.7.27 below.) It opens the possibility of using the powerful machinery of 'intersection cohomology', which is the underlying theme in the theory of character sheaves; see [Lu84c], [Lu06] and the references there. Beyond its application to the character theory of $\mathbf{G}^F$, this machinery has turned out to be extremely helpful in attacking a broad range of problems in representation theory; see [Spr82] and the surveys [Lu91], [Lu14c].

Remark 2.7.4 Let $\mathscr{C}$ be an F-stable conjugacy class of $\mathbf{G}$. A priori, we may choose any $g \in \mathscr{C}^F$ and then perform the constructions in 2.7.2. So the question arises whether there are natural choices for g. We certainly have a favourable situation when we can find some $g \in \mathscr{C}^F$ such that F acts trivially on $A(g)$. For example, this clearly happens when $|A(g)| \leqslant 2$ for $g \in \mathscr{C}$. Further conditions are described in [Sho86, §5], [Ta13, §2]. However, one should keep in mind that this is not always the case. We give two examples.

(a) Let $\mathbf{G}^F = \mathrm{SL}_3(q)$ where $q \equiv -1 \bmod 3$. Let $u \in \mathbf{G}^F$ be a regular unipotent element (the Jordan normal form of u consists of one block with 1 on the diagonal). Now, we have $\mathbf{Z}(\mathbf{G}) = \{\zeta I_3 \mid \zeta^3 = 1\}$ and one easily checks that $A(u)$ is cyclic of order 3, equal to the image of $\mathbf{Z}(\mathbf{G})$ in $A(u)$. Since $q \equiv -1 \bmod 3$, the elements of $\mathbf{Z}(\mathbf{G})$ are inverted by F. Hence, F acts non-trivially on $A(u)$ and so $A(u)^F = \{1\}$. (This example can be easily generalised to other types of groups; see [Ta13, §2].)

(b) Consider the Ree group $\mathbf{G}^F = {}^2F_4(q^2)$. There exists an F-stable unipotent class $\mathscr{C}$ of $\mathbf{G}$ (denoted by $F_4(a_2)$ in [Spa85, p. 330]) such that $A(g)$ is dihedral of order 8 for any $g \in \mathscr{C}$, but there is no $g \in \mathscr{C}^F$ such that F acts trivially on $A(g)$. (By [Shi75, Table II], the class $\mathscr{C}$ splits into only three classes in $\mathbf{G}^F$, with representatives u_{10}, u_{11}, u_{12}.)

Definition 2.7.5 Let $\mathscr{C}$ be an F-stable conjugacy class of $\mathbf{G}$. For any class function $f \in \mathrm{CF}(\mathbf{G}^F)$ we define the *average value* of f on $\mathscr{C}^F$ by

$$\mathrm{AV}(f, \mathscr{C}) := \sum_{1 \leqslant j \leqslant r} |A(g_j) : A(g_j)^F| \, f(g_j),$$

where $g_1, \ldots, g_r \in \mathscr{C}^F$ are representatives of the $\mathbf{G}^F$-conjugacy classes contained in $\mathscr{C}^F$ and, as above, we denote by $A(g) := C_{\mathbf{G}}(g)/C_{\mathbf{G}}^\circ(g)$ the finite group of components of the centraliser of an element $g \in \mathscr{C}^F$.

(Note that $\mathrm{AV}(f, \mathscr{C})$ does not depend on the choice of $g_1, \ldots, g_r$; also note that $|A(g_j)|$ does not depend on j, but $|A(g_j)^F|$ may well depend on j.)

It will turn out that the above definition of an average value is more efficient than just averaging the values of f over the whole set $\mathscr{C}^F$.

Remark 2.7.6 Let $\mathscr{C}$ be an F-stable conjugacy class of $\mathbf{G}$. Let us fix an element $g \in \mathscr{C}^F$. Let $1 = a_1, a_2, \ldots, a_r \in A(g)$ be a set of representatives of the F-conjugacy classes of $A(g)$. For $1 \leqslant j \leqslant r$, let $g_j \in C_{a_j}$ (where C_{a_j} is defined as in 2.7.1). Then $A(g_j) \cong C_{A(g),F}(a_j)$ and so

$$|A(g_j) : A(g_j)^F| = \text{size of the } F\text{-conjugacy class of } a_j \text{ in } A(g).$$

Hence, we may also write $\mathrm{AV}(f, \mathscr{C}) = \sum_{a \in A(g)} f(g_a)$, where $g_a \in \mathscr{C}^F$ is associated with $a \in A(g)$ as in 2.7.1.

Example 2.7.7 Let $\mathscr{C}$ be an F-stable conjugacy class of $\mathbf{G}$ and $Y_{(g_{\mathscr{C}}, \tilde{\varsigma})} \in \mathrm{CF}(\mathbf{G}^F)$ be one of the basis functions in 2.7.2, where $g_{\mathscr{C}} \in \mathscr{C}^F$ and $\tilde{\varsigma}$ is an F-extension of $\varsigma \in \mathrm{Irr}(A(g_{\mathscr{C}}))$. Then we have

$$\mathrm{AV}(Y_{(g_{\mathscr{C}}, \tilde{\varsigma})}, \mathscr{C}) = \begin{cases} \tilde{\varsigma}(1)|A(g_{\mathscr{C}})| & \text{if } \varsigma \text{ is the trivial character of } A(g_{\mathscr{C}}), \\ 0 & \text{otherwise.} \end{cases}$$

Indeed, using the notation in Remark 2.7.6, we have

$$\mathrm{AV}(Y_{(g_{\mathscr{C}}, \tilde{\varsigma})}, \mathscr{C}) = \sum_{a \in A(g_{\mathscr{C}})} Y_{(g_{\mathscr{C}}, \tilde{\varsigma})}(g_a) = \sum_{a \in A(g_{\mathscr{C}})} \tilde{\varsigma}(a)$$

and the right-hand side equals $|A(g_{\mathscr{C}})|\langle \tilde{\varsigma}, \tilde{1} \rangle_F$ where the inner product is defined in 2.1.7 and $\tilde{1}$ denotes the trivial extension of the trivial character of $A(g_{\mathscr{C}})$. So the orthogonality relations in 2.1.10 yield the desired identity.

Example 2.7.8 Let $\mathscr{C}$ be an F-stable conjugacy class of $\mathbf{G}$.

(a) Let $g \in \mathscr{C}^F$ and assume that $C_{\mathbf{G}}(g)$ is connected. Then $\mathscr{C}^F$ is a single conjugacy class of $\mathbf{G}^F$ (see Proposition 1.4.9) and so $\mathrm{AV}(f, \mathscr{C}) = f(g)$.

(b) Let $g \in \mathscr{C}^F$ and assume that $A(g)$ is abelian and F acts trivially on $A(g)$. By Example 2.1.8(c), we then have $|H^1(F, A(g))| = |A(g)|$ and so $A(g')^F = A(g')$ for any $g' \in \mathscr{C}^F$. Hence, in this case, we have

$$\mathrm{AV}(f, \mathscr{C}) = \sum_{1 \leqslant j \leqslant r} f(g_j),$$

where $r = |A(g)|$ and $g_1, \ldots, g_r \in \mathscr{C}^F$ are as in Definition 2.7.5.

(c) Let $g \in \mathscr{C}^F$ and assume that $A(g)$ is isomorphic to the symmetric group $\mathfrak{S}_3$. Then every (abstract) group automorphism of $A(g)$ is inner and we can choose g so that F acts trivially on $A(g)$ (see, e.g., [Ta13, Lemma 2.3]). Let $1 = a_1, a_2, a_3 \in A(g)$ be representatives of the conjugacy classes of $A(g)$, where a_2 corresponds to a 2-cycle and a_3 corresponds to a 3-cycle in $\mathfrak{S}_3 \cong A(g)$. Thus, we have

$$\mathrm{AV}(f, \mathscr{C}) = f(g_1) + 3f(g_2) + 2f(g_3),$$

with $g_j \in C_{a_j}$ for $j = 1, 2, 3$; note that $|A(g_j)^F| = |C_{\mathfrak{S}_3}(a_j)|$.

Remark 2.7.9 Let $\mathscr{C} \subseteq \mathbf{G}$ be as in Definition 2.7.5. As in [Ge96, §1], we have $\mathrm{AV}(f, \mathscr{C}) = \langle f, \alpha_{\mathscr{C}} \rangle$ where the function $\alpha_{\mathscr{C}} \colon \mathbf{G}^F \to \mathbb{K}$ is defined by

$$\alpha_{\mathscr{C}}(g) := \begin{cases} |A(g)||C^\circ_{\mathbf{G}}(g)^F| & \text{if } g \in \mathscr{C}^F, \\ 0 & \text{otherwise.} \end{cases}$$

One easily sees that $\alpha_{\mathscr{C}}$ is a class function. The formula $\mathrm{AV}(f, \mathscr{C}) = \langle f, \alpha_{\mathscr{C}} \rangle$ immediately follows from the fact that $|C_{\mathbf{G}}(g)^F| = |A(g)^F||C^\circ_{\mathbf{G}}(g)^F|$ for $g \in \mathbf{G}^F$ (see Example 1.4.11(b)).

The following result (mentioned in the proof of [GeMa00, Theorem 3.7]) shows that $\mathrm{AV}(\rho, \mathscr{C})$ behaves well with respect to regular embeddings.

Lemma 2.7.10 *Let $\mathbf{G} \subseteq \tilde{\mathbf{G}}$ be a regular embedding (see Section 1.7). Let $\mathscr{C}$ be an F-stable conjugacy class of $\mathbf{G}$. Then $\mathscr{C}$ is also an $\tilde{F}$-stable conjugacy class in $\tilde{\mathbf{G}}$. Let $\tilde{\rho} \in \mathrm{Irr}(\tilde{\mathbf{G}}^{\tilde{F}})$ and $\rho \in \mathrm{Irr}(\mathbf{G}^F)$ be a constituent of the restriction of $\tilde{\rho}$ to $\mathbf{G}^F$. Then*

$$|A(g)|\mathrm{AV}(\tilde{\rho}, \mathscr{C}) = r|\tilde{A}(g)|\mathrm{AV}(\rho, \mathscr{C}),$$

where $\tilde{A}(g) = C_{\tilde{\mathbf{G}}}(g)/C^\circ_{\tilde{\mathbf{G}}}(g)$ (for $g \in \mathscr{C}$) and $r \geqslant 1$ is the number of irreducible constituents of the restriction of $\tilde{\rho}$ to $\mathbf{G}^F$.

Proof Let us denote $\mathbf{Z} := \mathbf{Z}(\mathbf{G})$ and $\tilde{\mathbf{Z}} := \mathbf{Z}(\tilde{\mathbf{G}})$. Let $x \in \tilde{\mathbf{G}}^{\tilde{F}}$. Since $\tilde{\mathbf{G}} = \mathbf{G}.\tilde{\mathbf{Z}}$, we can write $x = yz$ where $y \in \mathbf{G}$ and $z \in \tilde{\mathbf{Z}}$ and so $x\mathscr{C}x^{-1} = y\mathscr{C}y^{-1} = \mathscr{C}$. Thus, $\mathscr{C}$ is also an $\tilde{F}$-stable conjugacy class in $\tilde{\mathbf{G}}$. By Remark 2.7.9, we have $\mathrm{AV}(\tilde{\rho}, \mathscr{C}) = \langle \tilde{\rho}, \tilde{\alpha}_{\mathscr{C}} \rangle$ where $\tilde{\alpha}_{\mathscr{C}} \colon \tilde{\mathbf{G}}^{\tilde{F}} \to \mathbb{K}$ is defined by

$$\tilde{\alpha}_{\mathscr{C}}(g) := \begin{cases} |\tilde{A}(g)||C^\circ_{\tilde{\mathbf{G}}}(g)^{\tilde{F}}| & \text{if } g \in \mathscr{C}^F, \\ 0 & \text{otherwise.} \end{cases}$$

Let $g \in \mathscr{C}^F$. Let $a = |A(g)|$ and $\tilde{a} = |\tilde{A}(g)|$. (Note that these two numbers do not depend on g.) Since $\tilde{\mathbf{G}} = \mathbf{G}.\tilde{\mathbf{Z}}$, it is clear that $C_{\tilde{\mathbf{G}}}(g) = C_{\mathbf{G}}(g).\tilde{\mathbf{Z}}$. Since $\tilde{\mathbf{Z}}$ is connected, this implies that $C^\circ_{\tilde{\mathbf{G}}}(g) = C^\circ_{\mathbf{G}}(g).\tilde{\mathbf{Z}}$. Since $\mathbf{Z}^\circ \subseteq C^\circ_{\mathbf{G}}(g)$, we can use Lemma 1.7.8 and obtain that

$$|C^\circ_{\tilde{\mathbf{G}}}(g)^{\tilde{F}}| = |C^\circ_{\mathbf{G}}(g)^F||\tilde{\mathbf{Z}}^{\tilde{F}}|/|\mathbf{Z}^{\circ F}|.$$

Now Lemma 1.7.8 also shows that $|\tilde{\mathbf{G}}^{\tilde{F}}| = |\mathbf{G}^F||\tilde{\mathbf{Z}}^{\tilde{F}}|/|\mathbf{Z}^{\circ F}|$ and so

$$a|\mathbf{G}^F|\tilde{\alpha}_{\mathscr{C}}(g) = \tilde{a}|\tilde{\mathbf{G}}^{\tilde{F}}|\alpha_{\mathscr{C}}(g) \qquad \text{for all } g \in \mathscr{C}^F.$$

Thus, the restriction of $\tilde{\alpha}_{\mathscr{C}}$ to $\mathbf{G}^F$ is a scalar multiple of $\alpha_{\mathscr{C}}$, and we obtain

$$a\mathrm{AV}(\tilde{\rho}, \mathscr{C}) = a\langle \tilde{\rho}, \tilde{\alpha}_{\mathscr{C}} \rangle = \tilde{a}\langle \tilde{\rho}|_{\mathbf{G}^F}, \alpha_{\mathscr{C}} \rangle = \tilde{a}\mathrm{AV}(\tilde{\rho}|_{\mathbf{G}^F}, \mathscr{C})$$

(where the left-hand average value is taken with respect to $\tilde{\mathbf{G}}^{\tilde{F}}$ and the right-hand average value with respect to $\mathbf{G}^F$). It remains to show that $\mathrm{AV}(\tilde{\rho}|_{\mathbf{G}^F}, \mathscr{C}) =$

$r\mathrm{AV}(\rho, \mathscr{C})$. To see this, we write $\tilde{\rho}|_{\mathbf{G}^F} = \rho_1 + \cdots + \rho_r$ where $\rho_i \in \mathrm{Irr}(\mathbf{G}^F)$ and $\rho_1 = \rho$. By Clifford's theorem, each ρ_i is conjugate to ρ via some element $x_i \in \tilde{\mathbf{G}}^{\tilde{F}}$. Now, the fact that $\alpha_{\mathscr{C}}$ is the restriction of a class function on $\tilde{\mathbf{G}}^{\tilde{F}}$ to $\mathbf{G}^F$ implies that $\alpha_{\mathscr{C}}$ is invariant under the action of $\tilde{\mathbf{G}}^{\tilde{F}}$ on $\mathrm{CF}(\mathbf{G}^F)$ by conjugation. Hence, we have $\langle \rho_i, \alpha_{\mathscr{C}}\rangle = \langle \rho, \alpha_{\mathscr{C}}\rangle$ for all i and so $\mathrm{AV}(\tilde{\rho}|_{\mathbf{G}^F}, \mathscr{C}) = \langle \tilde{\rho}|_{\mathbf{G}^F}, \alpha_{\mathscr{C}}\rangle = r\langle \rho, \alpha_{\mathscr{C}}\rangle = r\mathrm{AV}(\rho, \mathscr{C})$, as required. □

The following result is obtained as a combination of [Ge96, Prop. 1.3] and [DiMi15, Cor. 6.8]. An analogous result concerning the indicator function on $\mathscr{C}^F$ is established in [Ge18, §8], confirming a conjecture in [Lu77b, 2.16].

Theorem 2.7.11 *Let* $\mathbf{G}$ *be connected reductive and* F *be a Frobenius map. Let* $\mathscr{C}$ *be an* F*-stable conjugacy class of* $\mathbf{G}$*. Then the function* $\alpha_{\mathscr{C}} \in \mathrm{CF}(\mathbf{G}^F)$ *is uniform (see Definition 2.2.25). Thus, we have*

$$\mathrm{AV}(\rho, \mathscr{C}) = \mathrm{AV}\big(\pi_{\mathrm{un}}^{\mathbf{G}}(\rho), \mathscr{C}\big) \qquad \textit{for any } \rho \in \mathrm{Irr}(\mathbf{G}^F),$$

where $\pi_{\mathrm{un}}^{\mathbf{G}}(\rho)$ *denotes the uniform projection as in Remark 2.2.28.*

Proof By [Ge96, Prop. 1.3], the statement holds in the special case where $\mathscr{C}$ is unipotent. The proof uses the results on Green functions in the following section, most notably the full power of Theorem 2.8.3 and 2.8.4 below. In order to deal with the general case, we follow the reduction argument in [DiMi15, §6] and introduce the following notation. Let $s \in \mathbf{G}^F$ be semisimple and $\mathbf{H}_s := C_{\mathbf{G}}(s)$. For any $f \in \mathrm{CF}(\mathbf{G}^F)$, we define $d_s(f) \in \mathrm{CF}(\mathbf{H}_s^{\circ F})$ by

$$d_s(f)(x) := \begin{cases} f(sx) & \text{if } x \in \mathbf{H}_s^{\circ F} \text{ is unipotent,} \\ 0 & \text{otherwise.} \end{cases}$$

Now let $\mathscr{C}$ be an arbitrary F-stable conjugacy class of $\mathbf{G}$ and consider the class function $f = \alpha_{\mathscr{C}} \in \mathrm{CF}(\mathbf{G}^F)$. Let

$$\mathscr{C}' := \{x \in \mathbf{H}_s \mid x \text{ unipotent and } sx \in \mathscr{C}\}.$$

If $\mathscr{C}' = \varnothing$, then $d_s(f)$ is identically zero. Now assume that $\mathscr{C}' \neq \varnothing$. Since $|\mathbf{H}_s : \mathbf{H}_s^\circ| < \infty$ and since every unipotent element of $\mathbf{H}_s$ belongs to $\mathbf{H}_s^\circ$ (see 2.2.13), we conclude that $\mathscr{C}'$ is a finite union of conjugacy classes of $\mathbf{H}_s^\circ$; we denote these classes by $\mathscr{C}_1, \ldots, \mathscr{C}_m$ where $m \geqslant 1$. For $1 \leqslant i \leqslant m$ we set $e_i := |C_{\mathbf{H}_s}(x) : C_{\mathbf{H}_s^\circ}(x)|$ (where $x \in \mathscr{C}_i$).

Let us now work out the class function $d_s(\alpha_{\mathscr{C}}) \in \mathrm{CF}(\mathbf{H}_s^{\circ F})$. Let $x \in \mathbf{H}_s^{\circ F}$ be unipotent and set $g := sx = xs$. If $g \notin \mathscr{C}$, then $d_s(\alpha_{\mathscr{C}})(x) = \alpha_{\mathscr{C}}(g) = 0$. Now assume that $g \in \mathscr{C}$. Then $x \in \mathscr{C}'$ and so $x \in \mathscr{C}_i$ for a unique $i \in \{1, \ldots, m\}$; in particular, $\mathscr{C}_i$ is F-stable. We claim that, in this case, we have

$$d_s(\alpha_{\mathscr{C}})(x) = e_i \alpha_{\mathscr{C}_i}(x) \qquad (\text{where } x \in \mathscr{C}_i).$$

Indeed, since $g = sx = xs$, were s is semisimple and x is unipotent, we certainly have $C_{\mathbf{G}}(g) = C_{\mathbf{H}_s}(x)$. This also implies that $C_{\mathbf{G}}^\circ(g) = C_{\mathbf{H}_s}^\circ(x)$. Since $C_{\mathbf{H}_s^\circ}^\circ(x)$ is connected and contained in $C_{\mathbf{G}}(g)$, we have $C_{\mathbf{H}_s^\circ}^\circ(x) \subseteq C_{\mathbf{G}}^\circ(g)$. On the other hand, $C_{\mathbf{H}_s^\circ}^\circ(x)$ has finite index in $C_{\mathbf{H}_s}(x)$ and so $C_{\mathbf{H}_s^\circ}^\circ(x) \supseteq C_{\mathbf{H}_s}^\circ(x) = C_{\mathbf{G}}^\circ(g)$. Thus, we have $C_{\mathbf{G}}^\circ(g) = C_{\mathbf{H}_s}^\circ(x) = C_{\mathbf{H}_s^\circ}^\circ(x)$. It follows that

$$A_{\mathbf{H}_s^\circ}(x) = C_{\mathbf{H}_s^\circ}(x)/C_{\mathbf{H}_s^\circ}^\circ(x) \subseteq C_{\mathbf{H}_s}(x)/C_{\mathbf{H}_s^\circ}^\circ(x) = C_{\mathbf{G}}(g)/C_{\mathbf{G}}^\circ(g) = A_{\mathbf{G}}(g)$$

and $|A_{\mathbf{G}}(g)| = e_i|A_{\mathbf{H}_s^\circ}(x)|$. Thus, we conclude that

$$\alpha_{\mathscr{C}}(g) = |A_{\mathbf{G}}(g)||C_{\mathbf{G}}^\circ(g)^F| = e_i|A_{\mathbf{H}_s^\circ}(x)||C_{\mathbf{H}_s^\circ}^\circ(x)^F| = e_i\alpha_{\mathscr{C}_i}(x).$$

Since the left-hand side equals $d_s(\alpha_{\mathscr{C}})(x)$ by definition, the above claim is proved. It follows that

$$d_s(\alpha_{\mathscr{C}}) = \sum_i e_i\alpha_{\mathscr{C}_i} \in \mathrm{CF}(\mathbf{H}_s^{\circ F}),$$

where the sum runs over all $i \in \{1, \ldots, m\}$ such that $\mathscr{C}_i$ is F-stable. Now each $\alpha_{\mathscr{C}_i}$ in the above sum is a uniform function by the special case dealt with in [Ge96, Prop. 1.3]. Hence, $d_s(\alpha_{\mathscr{C}})$ is a uniform function. We have seen that this holds for every semisimple element $s \in \mathbf{G}^F$. So [DiMi15, Cor. 6.3] implies that $\alpha_{\mathscr{C}} \in \mathrm{CF}(\mathbf{G}^F)$ is uniform. □

Remark 2.7.12 In the above result, we assumed that F is a Frobenius map. This was used when we referred to Theorems 2.8.3 and 2.8.4 in the proof. We will verify in Example 2.8.18 below by an explicit computation that the statements of these two theorems continue to hold for the Suzuki and Ree groups. Hence, Theorem 2.7.11 will also hold for these groups.

Corollary 2.7.13 ([Lu84a, Introduction, p. xx]) *Assume that F is a Frobenius map. Then, for any $\rho \in \mathrm{Irr}(\mathbf{G}^F)$ and any F-stable conjugacy class $\mathscr{C}$ of $\mathbf{G}$, the average value $\mathrm{AV}(\rho, \mathscr{C})$ can be explicitly determined (in the form of an algorithm).*

In fact, at the time of writing [Lu84a], Lusztig had to assume that the characteristic p is large enough, but this condition can now be removed.

Sketch of proof By Lemma 2.7.10, it is sufficient to determine $\mathrm{AV}(\rho, \mathscr{C})$ in the case where $\mathbf{Z}(\mathbf{G})$ is connected. Let us now assume that this is the case. Then we can determine $\pi_{\mathrm{un}}^{\mathbf{G}}(\rho)$ using Theorem 2.4.15 (the 'Main Theorem 4.23' of [Lu84a]). So the next step will be to work out the average values $\mathrm{AV}(R_{\mathbf{T}}^{\mathbf{G}}(\theta), \mathscr{C})$, for any F-stable maximal torus $\mathbf{T} \subseteq \mathbf{G}$ and any $\theta \in \mathrm{Irr}(\mathbf{T}^F)$. Using the character formula in Theorem 2.2.16 (or Lemma 2.2.23), we can reduce to the case where $\mathscr{C}$ is unipotent. Thus, finally, we need to determine the average values $\mathrm{AV}(Q_{\mathbf{T}}^{\mathbf{G}}, \mathscr{C})$, where $Q_{\mathbf{T}}^{\mathbf{G}}$ is the Green function corresponding to an F-stable maximal torus $\mathbf{T} \subseteq \mathbf{G}$. For this purpose, we express $Q_{\mathbf{T}}^{\mathbf{G}}$ in the basis $\mathscr{B}$ of $\mathrm{CF}(\mathbf{G}^F)$ defined in 2.7.2. By Example 2.7.7, $\mathrm{AV}(Q_{\mathbf{T}}^{\mathbf{G}}, \mathscr{C})$ only depends on the coefficient of the basis function $Y_{(g_{\mathscr{C}},\varsigma)} \in \mathscr{B}$ where ς is the trivial character of $A(g_{\mathscr{C}})$. An algorithm for the determination of that coefficient will be explained in the following section; see 2.8.11. □

We now focus on unipotent classes of **G**; these play a special role in the theory of reductive algebraic groups. It is shown in [Lu76b] (by a general argument) that the number of unipotent classes of **G** is always finite. An exposition of this argument can be found in [DiMi20, §12.1]; it uses Lusztig induction, to be discussed in Section 3.3. For **G** simple, the unipotent classes have been explicitly determined in all cases; see, e.g., [Miz80] where this is done for **G** of type E_7, E_8, which is a truly amazing achievement!

In the study of unipotent classes, one typically encounters exceptional situations when the characteristic p is 'small'. More precisely, the distinction is between 'good' and 'bad' primes for **G**. Before we continue with average values, this is now a good place to introduce these notions, which we could completely avoid so far.

2.7.14 Good and bad primes. Recall from [SpSt70, I, §4], [Ca85, §1.14] that p is a *good prime* for **G**, if p is good for each simple factor of **G**, and that the conditions for the various simple types are as follows.

$$\begin{aligned} A_n &: \quad \text{no condition,} \\ B_n, C_n, D_n &: \quad p \neq 2, \\ G_2, F_4, E_6, E_7 &: \quad p \neq 2, 3, \\ E_8 &: \quad p \neq 2, 3, 5. \end{aligned}$$

It turns out that the classification of unipotent classes of **G** is independent of p, as long as p is a good prime for **G**. (If p is a bad prime, then usually there are more unipotent classes than in good characteristic.) See [Spa82a], [Hum95] for further details, and [Lu05], [ClPr13] for more recent developments regarding the bad prime case.

The following result (or, rather, a slightly different version where the average values $\mathrm{AV}(\rho, \mathscr{C})$ are replaced by the global sums $\sum_{g \in \mathscr{C}^F} \rho(g)$) was first conjectured by Lusztig [Lu80b, §1]. Quite remarkably, it is uniformly true in all characteristics although, as mentioned above, unipotent classes behave differently in bad or good characteristic.

Theorem 2.7.15 ([Lu92a], [GeMa00]) *Let* **G** *be connected reductive and F be a Frobenius map. Let $\rho \in \mathrm{Irr}(\mathbf{G}^F)$. Then there exists a unique F-stable unipotent class $\mathscr{O}$ of* **G** *such that $\mathrm{AV}(\rho, \mathscr{O}) \neq 0$, and such that $\mathrm{AV}(\rho, \mathscr{O}') = 0$ for any F-stable unipotent class $\mathscr{O}'$ of* **G**, *unless $\mathscr{O}' = \mathscr{O}$ or $\dim \mathscr{O}' < \dim \mathscr{O}$.*

Given $\rho \in \mathrm{Irr}(\mathbf{G}^F)$, the unique unipotent class $\mathscr{O}$ attached to ρ as above is called the *unipotent support* of ρ and denoted by $\mathscr{O}_\rho$.

For example, the unipotent support of the trivial character is the class of regular unipotent elements; at the other extreme, the unipotent support of the Steinberg character is the class of the identity element.

The proof of the above result uses the full power of the whole theory developed so far (e.g., Theorem 2.7.11 and the main results of the book [Lu84a]), as well as some further ingredients that we did not discuss here at all, most notably the theory of '*generalised Gelfand–Graev representations*' (see [Kaw86], [Kaw87] and also [Lu92a], [GeHe08], [Ta16] for more recent developments concerning these representations). In [Lu92a, §11], Theorem 2.7.15 is established assuming that p, q are sufficiently large; subsequently, it is shown in [GeMa00] that these assumptions can be removed.

The following complementary result shows that the invariants a_ρ and n_ρ, which were defined in Remark 2.3.26 using the degree polynomial of the character ρ, can be recovered from $\mathscr{O}_\rho$ and the average value of ρ on $\mathscr{O}_\rho$. (As far as a_ρ is concerned, this was also conjectured in [Lu80b, §1]; see also [Lu09a] for further interpretations of these invariants.)

Proposition 2.7.16 ([Lu92a], [GeMa00]) *Assume that F is a Frobenius map. Let $\rho \in \mathrm{Irr}(\mathbf{G}^F)$. Then $a_\rho = \dim \mathbf{G} - \mathrm{rank}(\mathbf{G}) - \dim \mathscr{O}_\rho$ and*

$$\mathrm{AV}(\rho, \mathscr{O}_\rho) = \pm n_\rho^{-1} q^{a_\rho} |A(u)| \qquad (u \in \mathscr{O}_\rho).$$

Remark 2.7.17 If $\mathbf{G}^F$ is a Suzuki or Ree group (where F is not a Frobenius map), then the complete character table of $\mathbf{G}^F$ is known by [Suz62], [War66], [Ma90]. As mentioned in [GeMa00, §5], the statements of Theorem 2.7.15 and Proposition 2.7.16 continue to hold in these cases as well. (In [GeMa00, §5], this was stated somewhat incorrectly, because we did not use the correct definition of n_ρ.)

Example 2.7.18 Let $\mathbf{G} \subseteq \tilde{\mathbf{G}}$ be a regular embedding (see Section 1.7). Let $\tilde{\rho} \in \mathrm{Irr}(\tilde{\mathbf{G}}^F)$ and write $\tilde{\rho}|_{\mathbf{G}^F} = \rho_1 + \cdots + \rho_r$ where the ρ_i are distinct irreducible characters of $\mathbf{G}^F$. Then all ρ_i have the same unipotent support, and this is the unipotent support of $\tilde{\rho}$. This immediately follows from Lemma 2.7.10.

Example 2.7.19 Let $\mathscr{O}_0$ be the class of regular unipotent elements of $\mathbf{G}$ (see 2.2.21). Let $\rho \in \mathrm{Irr}(\mathbf{G}^F)$. Then we claim that

$$\mathscr{O}_\rho = \mathscr{O}_0 \qquad \Longleftrightarrow \qquad \rho \text{ is semisimple (see Definition 2.6.9).}$$

To see this, assume first that $\mathbf{Z}(\mathbf{G})$ is connected. Then it easily follows from the discussion in [Ge18, Remark 3.7] that $\alpha_{\mathscr{O}_0} = |A(u)|\Delta_{\mathbf{G}}$, where $u \in \mathscr{O}_0$ and the function

$\Delta_{\mathbf{G}}$ is defined in 2.6.8. Hence, we have $\mathrm{AV}(\rho, \mathcal{O}_0) = \langle \rho, \alpha_{\mathcal{O}_0} \rangle = |A(u)| \langle \rho, \Delta_{\mathbf{G}} \rangle$. Now ρ is semisimple if and only if $\langle \rho, \Delta_{\mathbf{G}} \rangle \neq 0$, in which case we have $\langle \rho, \Delta_{\mathbf{G}} \rangle = \pm 1$; see 2.6.10(d). Hence, the desired equivalence holds; furthermore, we obtain that

$$\mathrm{AV}(\rho, \mathcal{O}_0) = \pm |A(u)| \qquad \text{if } \rho \text{ is semisimple (and } \mathbf{Z}(\mathbf{G}) \text{ is connected).}$$

Now assume that $\mathbf{Z}(\mathbf{G})$ is not connected. Then consider a regular embedding $\mathbf{G} \subseteq \tilde{\mathbf{G}}$ as in Section 1.7. Let $\tilde{\rho} \in \mathrm{Irr}(\tilde{\mathbf{G}}^F)$ be such that ρ occurs in the restriction of $\tilde{\rho}$ to $\mathbf{G}^F$. By Example 2.7.18, the characters ρ and $\tilde{\rho}$ have the same unipotent support. Furthermore, by Corollary 2.6.18, ρ is semisimple if and only if $\tilde{\rho}$ is semisimple. Hence, the desired equivalence holds in general.

As far as the interaction of the unipotent support and individual character values are concerned, we have the following result.

Theorem 2.7.20 ([Lu92a, §11]) *Assume that F is a Frobenius map and p, q are sufficiently large. Let $\rho \in \mathrm{Irr}(\mathbf{G}^F)$ and $\mathcal{O}_\rho$ be the unipotent support of ρ. Let $g \in \mathbf{G}^F$ be such that $\rho(g) \neq 0$. Let u be the unipotent part of g and $\mathcal{O}$ be the conjugacy class of u in $\mathbf{G}$. Then either* $\dim \mathcal{O} < \dim \mathcal{O}_\rho$ *or* $\mathcal{O} = \mathcal{O}_\rho$.

Remark 2.7.21 Here, the assumption that p, q are sufficiently large means that one can operate with the Lie algebra of $\mathbf{G}$ as if we were in characteristic 0. (In particular, the variety of nilpotent elements in the Lie algebra may be identified with the variety of unipotent elements in $\mathbf{G}$, via an exponential map; see [Lu92a, 1.3].) It is shown in [Ta19, §9] that the conclusion of Theorem 2.7.20 continues to hold if we only assume that p is a good prime for $\mathbf{G}$ and $\mathbf{Z}(\mathbf{G})$ is connected. It is likely that the assumption on $\mathbf{Z}(\mathbf{G})$ is unnecessary. However, the following examples will show that some assumptions on the characteristic p are necessary.

Example 2.7.22 Let $\mathbf{G}^F = \mathrm{PCSp}_4(q)$ where $q = p^f$ ($f \geqslant 1$). The unipotent characters of $\mathbf{G}^F$ are denoted as in Table 2.8 (p. 132); the classification and the degree polynomials do not depend on p or q. However, we will see a difference when we look at the values of the unipotent characters; here, we just focus on unipotent elements. The values are printed in Table 2.10; they can be extracted from the tables in [Sr68] and [Eno72]. (We work with the simple group $\mathbf{G} = \mathrm{PCSp}_4(k)$ of adjoint type, because the unipotent classes are easier to describe in this case.)

If $p \neq 2$, then there are four unipotent classes which we denote by $\mathcal{O}_\mu$, where the subscript μ specifies the Jordan type of the elements in the class. (There is an isogeny $\mathrm{Sp}_4(k) \to \mathbf{G}$ which induces a bijection between the unipotent classes of $\mathrm{Sp}_4(k)$ and of $\mathbf{G}$; thus, we may also speak of the Jordan type of unipotent elements in $\mathbf{G}$.) For example, the elements in $\mathcal{O}_{(211)}$ have one Jordan block of size 2 and two blocks of size 1. The set $\mathcal{O}_{(22)}^F$ splits into two classes in $\mathbf{G}^F$ which we denote by

Table 2.10 *Unipotent characters of* $\mathrm{PCSp}_4(q)$ *on unipotent elements*

$p \neq 2$	$O_{(1111)}$	$O_{(211)}$	$O_{(22)}$	$O'_{(22)}$	$O_{(4)}$
$\lvert C_{\mathbf{G}}(u)^F \rvert$	$\lvert \mathbf{G}^F \rvert$	$q^4(q^2-1)$	$2q^3(q-1)$	$2q^3(q+1)$	q^2
θ_0	1	1	1	1	1
θ_9	$\frac{1}{2}q(q+1)^2$	$\frac{1}{2}q(q+1)$	q	.	.
θ_{10}	$\frac{1}{2}q(q-1)^2$	$-\frac{1}{2}q(q-1)$	.	q	.
θ_{11}	$\frac{1}{2}q(q^2+1)$	$-\frac{1}{2}q(q-1)$	q	.	.
θ_{12}	$\frac{1}{2}q(q^2+1)$	$\frac{1}{2}q(q+1)$	.	q	.
θ_{13}	q^4	.	.	.	.

$p = 2$	$O_{(1111)}$	$O_{(211)}$	$O^*_{(22)}$	$O_{(22)}$	$O_{(4)}$	$O'_{(4)}$
$\lvert C_{\mathbf{G}}(u)^F \rvert$	$\lvert \mathbf{G}^F \rvert$	$q^4(q^2-1)$	$q^4(q^2-1)$	q^4	$2q^2$	$2q^2$
θ_0	1	1	1	1	1	1
θ_9	$\frac{1}{2}q(q+1)^2$	$\frac{1}{2}q(q+1)$	$\frac{1}{2}q(q+1)$	$\frac{q}{2}$	$\frac{q}{2}$	$-\frac{q}{2}$
θ_{10}	$\frac{1}{2}q(q-1)^2$	$-\frac{1}{2}q(q-1)$	$-\frac{1}{2}q(q-1)$	$\frac{q}{2}$	$\frac{q}{2}$	$-\frac{q}{2}$
θ_{11}	$\frac{1}{2}q(q^2+1)$	$-\frac{1}{2}q(q-1)$	$\frac{1}{2}q(q+1)$	$\frac{q}{2}$	$-\frac{q}{2}$	$\frac{q}{2}$
θ_{12}	$\frac{1}{2}q(q^2+1)$	$\frac{1}{2}q(q+1)$	$-\frac{1}{2}q(q-1)$	$\frac{q}{2}$	$-\frac{q}{2}$	$\frac{q}{2}$
θ_{13}	q^4	.	.	.	.	.

$O_{(22)}$ and $O'_{(22)}$; each of the remaining classes $\mathscr{O}_\mu$ gives rise to exactly one class in $\mathbf{G}^F$, which we denote by O_μ.

If $p = 2$, then we use a similar convention for denoting unipotent classes as above; just note that, now, there are two unipotent classes of $\mathbf{G}$ with elements of Jordan type (22), which we denote by $\mathscr{O}_{(22)}$ and $\mathscr{O}^*_{(22)}$.

By inspection of Table 2.10, we obtain the unipotent support of all $\rho \in \mathrm{Uch}(\mathbf{G}^F)$:

ρ	a_ρ	n_ρ	$\mathscr{O}_\rho$	$\mathrm{AV}(\rho, \mathscr{O}_\rho)$	
θ_0	0	1	$\mathscr{O}_{(4)}$	$1\,(p{\neq}2)$,	$2\,(p{=}2)$
$\theta_9, \theta_{10}, \theta_{11}, \theta_{12}$	1	2	$\mathscr{O}_{(22)}$	$q\,(p{\neq}2)$,	$\frac{q}{2}\,(p{=}2)$
θ_{13}	4	1	$\mathscr{O}_{(1111)}$	$q^4\,(p{\neq}2)$,	$q^4\,(p{=}2)$

We also see in Table 2.10 that, for $p = 2$, the conclusion of Theorem 2.7.20 fails: the four characters $\theta_9, \theta_{10}, \theta_{11}, \theta_{12}$ have non-zero values on $\mathscr{O}_{(4)}$ but the unipotent support is a class of strictly smaller dimension. (It may also be instructive to work with the individual character tables for $p = q = 3$ and $p = q = 2$ printed in Table 2.1, p. 95.)

Finally, by considering the whole character table of $\mathbf{G}^F$ for $p = 2$, one sees that $\mathscr{O}^*_{(22)}$ is not the unipotent support of any irreducible character of $\mathbf{G}^F$. (This is not an isolated phenomenon, see [GeMa00, Remark 3.9].)

Example 2.7.23 Let $\mathbf{G}^F = {}^2B_2(q^2)$ be a Suzuki group, where $q = \sqrt{2}^{2m+1}$ for some $m \geqslant 0$. The values of the unipotent characters of $\mathbf{G}^F$ are printed in Table 2.9 (p. 133). Let $\mathscr{O}'$ be the unipotent class of the element denoted by u' in that table. We see that $\mathscr{O}'$ is the unipotent support of the characters ϖ, ϖ'; furthermore, we have

$$\mathrm{AV}(\varpi, \mathscr{O}') = \mathrm{AV}(\varpi', \mathscr{O}') = -\tfrac{1}{2}\sqrt{2}q,$$

which is consistent with the formula in Proposition 2.7.16 since $n_\varpi = n_{\varpi'} = \sqrt{2}$ and $a_\rho = 1$. Again, we observe that ϖ, ϖ' take non-zero values on a class of strictly bigger dimension than that of $\mathscr{O}'$.

Finally, we briefly indicate how Lusztig's theory of character sheaves enters the picture. By the basic construction of [DeLu76] (as explained in the previous sections), we obtain representations of $\mathbf{G}^F$ on cohomology spaces $\mathrm{H}_c^i(\mathbf{X}, \overline{\mathbb{Q}}_\ell)$. In the theory of character sheaves, a completely different approach is used. One obtains class functions on $\mathbf{G}^F$ by taking the trace of the action of the Frobenius map F on certain cohomology spaces associated with $\mathbf{G}$. A priori, it is not at all clear that these class functions have anything to do with actual (or even just virtual) representations of $\mathbf{G}^F$. But, by this theory, Lusztig obtains a new basis of $\mathrm{CF}(\mathbf{G}^F)$ which is, at least in principle, computable (see also the next section). Hence, in this picture, the whole problem of computing the character table of $\mathbf{G}^F$ amounts to finding the base change from this new basis of $\mathrm{CF}(\mathbf{G}^F)$ to the basis $\mathrm{Irr}(\mathbf{G}^F)$.

The new basis comes about by working with perverse sheaves in the bounded derived category $\mathscr{D}\mathbf{G}$ of constructible $\overline{\mathbb{Q}}_\ell$-sheaves on $\mathbf{G}$, in the sense of Beilinson, Bernstein, Deligne [BBD82]. The objects of this category are extremely complicated, and we will not even try to attempt to explain this. (See [Lu87a], [Lau89], [MaSp89], [Sho88] for further expositions and introductions.)

2.7.24 The '*character sheaves*' on $\mathbf{G}$, introduced and studied by Lusztig [LuCS], are certain irreducible perverse sheaves in $\mathscr{D}\mathbf{G}$ that are equivariant for the action of $\mathbf{G}$ on itself by conjugation. What is remarkable is that in many situations we do not need to know exactly how these objects are defined, but only how to manipulate them (assuming some general familiarity with sheaf theory). Let $K \in \mathscr{D}\mathbf{G}$. Then K is represented by a complex of $\overline{\mathbb{Q}}_\ell$-sheaves

$$K: \qquad \ldots \to K_{i-1} \to K_i \to K_{i+1} \to \ldots,$$

such that $K_i = \{0\}$ if $|i|$ is large. For any $i \in \mathbb{Z}$, we have the ith cohomology sheaf, denoted by $\mathscr{H}^i(K)$. Given $g \in \mathbf{G}$, the stalks $\mathscr{H}^i_g(K)$ are finite-dimensional $\overline{\mathbb{Q}}_\ell$-vector spaces. Using the Frobenius map $F\colon \mathbf{G} \to \mathbf{G}$, we can form the inverse image $F^*K \in \mathscr{D}\mathbf{G}$, which has the property that

$$\mathscr{H}^i_g(F^*K) = \mathscr{H}^i_{F(g)}(K) \qquad \text{for } i \in \mathbb{Z},\, g \in \mathbf{G}.$$

Let us assume now that K is isomorphic to F^*K in $\mathscr{D}\mathbf{G}$. An isomorphism $\phi\colon F^*K \xrightarrow{\sim} K$ induces linear maps $\phi_{i,g}\colon \mathscr{H}^i_g(F^*K) \to \mathscr{H}^i_g(K)$ for each $i \in \mathbb{Z}$ and $g \in \mathbf{G}$. If $g \in \mathbf{G}^F$, then $\mathscr{H}^i_g(F^*K) = \mathscr{H}^i_{F(g)}(K) = \mathscr{H}^i_g(K)$ and so we obtain endomorphisms $\phi_{i,g} \in \operatorname{End}\big(\mathscr{H}^i_g(K)\big)$ for any $i \in \mathbb{Z}$. Following [LuCS, II, 8.4.1], the function

$$\chi_{K,\phi}\colon \mathbf{G}^F \to \overline{\mathbb{Q}}_\ell, \qquad g \mapsto \sum_i (-1)^i \operatorname{Trace}(\phi_{i,g}, \mathscr{H}^i_g(K)),$$

is called the *characteristic function* of K (with respect to ϕ). If K is irreducible, then, by a version of Schur's Lemma, ϕ is unique up to a non-zero scalar; hence, $\chi_{K,\phi}$ is unique up to a non-zero scalar.

Theorem 2.7.25 ([LuCS, V, §25], [Lu12a])

(1) *Let $A \in \mathscr{D}\mathbf{G}$ be a character sheaf with $F^*A \cong A$. Then there is an isomorphism $\phi\colon F^*A \xrightarrow{\sim} A$ such that the values of $\chi_{A,\phi}$ are cyclotomic numbers (so we can assume $\chi_{A,\phi}(g) \in \mathbb{K}$ for all $g \in \mathbf{G}^F$) and we have $\langle \chi_{A,\phi}, \chi_{A,\phi}\rangle = 1$.*
(2) *If A, ϕ are as in (1), then the values of $\chi_{A,\phi}$ on all elements of $\mathbf{G}^F$ can be computed 'in principle'.*
(3) *The characteristic functions $\{\chi_{A,\phi} \mid A, \phi$ as in (1), up to isomorphism$\}$ form an orthonormal basis of the space of class functions on $\mathbf{G}^F$.*

The existence of such a basis was already conjectured in [Lu84a, 13.7]. In [LuCS, Part V], the above theorem is proved under some mild conditions on p. These conditions were subsequently removed in [Lu12a]. We just mention that a part of the proof in [Lu12a] relies on explicit computations using data and programs in CHEVIE, concerning a canonical map

$$\{\text{conjugacy classes of } W\} \twoheadrightarrow \{\text{unipotent classes of } \mathbf{G}\} \quad \text{(see [Lu11])}.$$

See [Ge11], [MiChv] for some further explanations about the computations.

Remark 2.7.26 The problem of computing the character table of $\mathbf{G}^F$ is now reduced to the problem of finding the base change from $\operatorname{Irr}(\mathbf{G}^F)$ to the basis of class functions in Theorem 2.7.25(c). That is, we need to express each characteristic function $\chi_{A,\phi}$ explicitly as a linear combination of the irreducible characters of $\mathbf{G}^F$. Conjecturally (see [LuCS, II, p. 226]), these linear combinations should be given by the *almost characters* defined by [Lu84a, 4.24.1] (in the case where $\mathbf{Z}(\mathbf{G})$ is connected) and by [Lu18a] (in general); see also Remark 2.4.17 where the uniform almost characters were defined. The conjecture is known in many cases, but not in general; solving this problem involves, in particular, the tricky issue of specifying a particular isomorphism $\phi\colon F^*A \xrightarrow{\sim} A$ for any F-stable character sheaf A on $\mathbf{G}$.

The first successful realization of this whole program was carried out in [Lu86a], where character values on unipotent elements are determined. For further cases see [Lu92b], [Sho95], [Sho97], [Sho06a], [Sho09], [Bo06], [Wal04] and the references there. (See also the surveys [Sho98], [Ge18].)

Example 2.7.27 There is a notion of 'cuspidal' character sheaves; see [Lu84c, Def. 2.4], [LuCS, I, Def. 30.10; II, §7], or [Lu92b, 1.1]. Assume now that $\mathbf{G}$ is simple. Then there are only finitely many cuspidal character sheaves on $\mathbf{G}$ (up to isomorphism), and these have been classified in all cases; see [Lu84c, 2.10], [Lu84c, §10–§15], [LuSp85]; see also [Sho95, I, §7; II, §5]. Assume that $A \in \mathscr{D}\mathbf{G}$ is a cuspidal character sheaf such that $F^*A \cong A$; let $\phi\colon F^*A \xrightarrow{\sim} A$ be an isomorphism as in Theorem 2.7.25. Then, by [Lu04b, 19.7] and the 'cleanness' result of [Lu12a], there is an F-stable conjugacy class $\mathscr{C}$ of $\mathbf{G}$ (which has some very specific properties) such that

$$\chi_{A,\phi} = q^{(\dim \mathbf{G} - \dim \mathscr{C})/2} Y_{(g,\tilde{\varsigma})}, \qquad (\diamond)$$

where $g \in \mathscr{C}^F$ and $\tilde{\varsigma}$ is a suitable F-extension of some $\varsigma \in \mathrm{Irr}(A(g))^F$, as in 2.7.2. Here, the choice of ϕ determines $\tilde{\varsigma}$ and vice versa.

In [Lu92b, Theorem 0.8], the decomposition of $\chi_{A,\phi}$ as a linear combination of $\mathrm{Irr}(\mathbf{G}^F)$ is determined in each case, assuming that p is sufficiently large, and up to specifying the exact choice of $\tilde{\varsigma}$.

We have in fact already encountered some instances of this problem. For example, let $\mathbf{G}^F = \mathrm{SL}_2(q)$ where q is odd. Then the two class functions

$$\tfrac{1}{2}(\psi'_+ - \psi''_+ + \psi'_- - \psi''_-) \qquad \text{and} \qquad \tfrac{1}{2}(\psi'_+ - \psi''_+ - \psi'_- + \psi''_-)$$

shown in Example 2.2.30 are characteristic functions of F-stable cuspidal character sheaves. Similarly, the class function $\frac{1}{2}\sqrt{2}(\varpi - \varpi')$ of $\mathbf{G}^F = {}^2B_2(q^2)$ in Example 2.3.31 is such a characteristic function. Finally, if $\mathbf{G}^F = \mathrm{PCSp}_4(q)$, then the class function

$$\Gamma_1 := \tfrac{1}{2}(\theta_9 + \theta_{10} - \theta_{11} - \theta_{12}) \qquad \text{(notation as in Table 2.10)}$$

is the characteristic function of an F-stable cuspidal character sheaf. If $p \neq 2$, then Γ_1 takes values $\pm q$ on the conjugacy class of elements $su \in \mathbf{G}^F$ where s is a certain involution and u is regular unipotent in $C^\circ_{\mathbf{G}}(s)$; see [Sr91, A.1, p. 192]. If $p = 2$, then we can read off Table 2.10 that Γ_1 takes values $\pm q$ on the two unipotent classes $O_{(4)}, O'_{(4)}$ (and the whole character table in [Eno72] shows that Γ_1 is 0 everywhere else).

It is an on-going project to extend the above-mentioned [Lu92b, Theorem 0.8] to the cases where p is small, and to determine the appropriate choice of $\tilde{\varsigma}$ in each case; see [Sho06a], [Sho09], [Ta14b], [Ge19a], [He19].

2.8 On the Values of Green Functions

In this section, we consider in more detail the problem of computing the values of a Deligne–Lusztig character $R_{\mathbf{T}}^{\mathbf{G}}(\theta)$ at an arbitrary element $g \in \mathbf{G}^F$. The character formula in Theorem 2.2.16 essentially reduces this problem to the case where $g = u$ is unipotent and, hence, to Green functions. All this is part of the more general program sketched at the end of the previous section.

2.8.1 Let $\mathbf{T} \subseteq \mathbf{G}$ be an F-stable maximal torus. Recall from Definition 2.2.15 that the Green function $Q_{\mathbf{T}}^{\mathbf{G}}$ is defined by

$$Q_{\mathbf{T}}^{\mathbf{G}}(u) := R_{\mathbf{T}}^{\mathbf{G}}(1_{\mathbf{T}})(u) \qquad \text{for } u \in \mathbf{G}_{\text{uni}}^F.$$

Also recall that $Q_{\mathbf{T}}^{\mathbf{G}}(u) = Q_{\mathbf{T}}^{\mathbf{G}}(u^{-1}) \in \mathbb{Z}$ for all unipotent $u \in \mathbf{G}^F$. It will be convenient to regard $Q_{\mathbf{T}}^{\mathbf{G}}$ as a function on all of $\mathbf{G}^F$, which takes value 0 outside $\mathbf{G}_{\text{uni}}^F$. Let $g \in \mathbf{G}^F$ and write $g = us = su$ where $s \in \mathbf{G}^F$ is semisimple and $u \in \mathbf{G}^F$ is unipotent. By Example 2.2.17(b), we have

$$\sum_{\theta \in \operatorname{Irr}(\mathbf{T}^F)} R_{\mathbf{T}}^{\mathbf{G}}(\theta)(g) = \begin{cases} |\mathbf{T}^F| Q_{\mathbf{T}}^{\mathbf{G}}(u) & \text{if } s = 1, \\ 0 & \text{otherwise.} \end{cases} \tag{a}$$

Combining (a) with the orthogonality relations for $R_{\mathbf{T}}^{\mathbf{G}}(\theta)$ in Theorem 2.2.8, we obtain the following *orthogonality relations for Green functions*:

$$\langle Q_{\mathbf{T}}^{\mathbf{G}}, Q_{\mathbf{T}'}^{\mathbf{G}} \rangle = \frac{|\{g \in \mathbf{G}^F \mid \mathbf{T}' = g\mathbf{T}g^{-1}\}|}{|\mathbf{T}^F||\mathbf{T}'^F|} \tag{b}$$

where $\mathbf{T}' \subseteq \mathbf{G}$ is another F-stable maximal torus (see also [Ca85, 7.6.2]).

2.8.2 Let $\mathbf{T} \subseteq \mathbf{G}$ be an F-stable maximal torus and let $w \in \mathbf{W}$ be such that $\mathbf{T}$ is of type w (see 2.3.18). By Lemma 2.3.19, we have $R_{\mathbf{T}}^{\mathbf{G}}(1_{\mathbf{T}}) = R_w^1$ (where the superscript 1 stands for the trivial character of $\mathbf{T}_0[w]$). We set $Q_w(u) := R_w^1(u)$ for $u \in \mathbf{G}_{\text{uni}}^F$ (and also regard Q_w as a function on all of $\mathbf{G}^F$, which takes value 0 outside $\mathbf{G}_{\text{uni}}^F$). Then $Q_{\mathbf{T}}^{\mathbf{G}} = Q_w$. As in Example 2.3.22, we may re-write the above orthogonality relations as follows:

$$\langle Q_w, Q_{w'} \rangle = \frac{|\{x \in \mathbf{W} \mid w' = xw\sigma(x)^{-1}\}|}{|\mathbb{T}_w|(q)} \qquad (w, w' \in \mathbf{W})$$

where $\sigma \colon \mathbf{W} \to \mathbf{W}$ is the automorphism induced by F and

$$|\mathbb{T}_w| = \det(\mathbf{q}\,\mathrm{id}_{X_{\mathbb{R}}} - \varphi_0 \circ w^{-1}) \qquad \text{for all } w \in \mathbf{W} \text{ (see 1.6.21).}$$

Note that $Q_w = Q_{w'}$ if $w, w' \in \mathbf{W}$ are σ-conjugate; moreover, $\langle Q_w, Q_{w'}\rangle = 0$ if w, w' are not σ-conjugate in $\mathbf{W}$. We now perform a transformation in analogy to that in Remark 2.4.17. For each $\phi \in \mathrm{Irr}(\mathbf{W})^\sigma$, let us fix a real-valued σ-extension $\tilde{\phi}$, which exists by Proposition 2.1.14. Then we set

$$Q_{\tilde{\phi}} := \frac{1}{|\mathbf{W}|} \sum_{w \in \mathbf{W}} \tilde{\phi}(w) Q_w.$$

We can invert these relations and also write $Q_w = \sum_{\phi \in \mathrm{Irr}(\mathbf{W})^\sigma} \tilde{\phi}(w) Q_{\tilde{\phi}}$ for all $w \in \mathbf{W}$. Thus, the problem of computing the values of the Green functions Q_w is equivalent to that of computing the values of the functions $Q_{\tilde{\phi}}$.

For each unipotent class $\mathcal{O}$ of $\mathbf{G}$, we fix once and for all a representative $u_{\mathcal{O}} \in \mathcal{O}$; here, we tacitly assume that $F(u_{\mathcal{O}}) = u_{\mathcal{O}}$ if $\mathcal{O}$ is F-stable. Let $\mathscr{I}_{\mathbf{G}}^F$ be the set of all pairs $(\mathcal{O}, \varsigma)$ where $\mathcal{O}$ is an F-stable unipotent class of $\mathbf{G}$ and $\varsigma \in \mathrm{Irr}(A(u_{\mathcal{O}}))^F$. If $(\mathcal{O}, \varsigma) \in \mathscr{I}_{\mathbf{G}}^F$ and $\tilde{\varsigma}$ is an F-extension of ς, then we have a corresponding function $Y_{(u_{\mathcal{O}}, \tilde{\varsigma})} \in \mathrm{CF}(\mathbf{G}^F)$ as in 2.7.2.

Theorem 2.8.3 (Springer, Kazhdan, Lusztig, Shoji, ...) *Assume that* $F\colon \mathbf{G} \to \mathbf{G}$ *is a Frobenius map. For each* $\phi \in \mathrm{Irr}(\mathbf{W})^\sigma$, *there exists a unique pair* $(\mathcal{O}, \varsigma) \in \mathscr{I}_{\mathbf{G}}^F$ *such that* $\{u \in \mathbf{G}_{\mathrm{uni}}^F \mid Q_{\tilde{\phi}}(u) \neq 0\} \subseteq \overline{\mathcal{O}}$ *and*

$$Q_{\tilde{\phi}}|_{\mathcal{O}^F} = q^{(\dim \mathbf{G} - \dim \mathbf{T}_0 - \dim \mathcal{O})/2}\, Y_{(u_{\mathcal{O}}, \tilde{\varsigma})}$$

where $\tilde{\varsigma}$ *is a suitable* F*-extension of* ς. *(See Remark 2.8.8 below for further comments about* $\tilde{\varsigma}$.*) The resulting map* $\mathrm{Irr}(\mathbf{W})^\sigma \to \mathscr{I}_{\mathbf{G}}^F$ *is injective.*

The map $\mathrm{Irr}(\mathbf{W})^\sigma \hookrightarrow \mathscr{I}_{\mathbf{G}}^F$ is called the *Springer correspondence*. The pairs in the image of this map will be called the *uniform pairs* in $\mathscr{I}_{\mathbf{G}}^F$. These pairs and the correspondence itself are explicitly known for all $\mathbf{G}$; see the summaries in [Ca85, §13.3] (with some restrictions on p) and [Lu84c], [LuSp85], [Spa85]. See also [Hum95, Chap. 9] for further comments and references.

Theorem 2.8.4 (Springer, Kazhdan, Lusztig, Shoji, ...) *With the assumptions of Theorem 2.8.3, let* $\Upsilon \subseteq \mathrm{CF}(\mathbf{G}^F)$ *be the subspace spanned by* $\{Q_{\tilde{\phi}} \mid \phi \in \mathrm{Irr}(\mathbf{W})^\sigma\}$.

(a) Υ *is equal to the subspace spanned by the functions* $Y_{(u_{\mathcal{O}}, \tilde{\varsigma})}$ *where* $(\mathcal{O}, \varsigma) \in \mathscr{I}_{\mathbf{G}}^F$ *is a uniform pair and* $\tilde{\varsigma}$ *is any* F*-extension of* ς. *In particular, each such* $Y_{(u_{\mathcal{O}}, \tilde{\varsigma})}$ *is a uniform function (cf. Definition 2.2.25).*

(b) *If* $(\mathcal{O}, \varsigma) \in \mathscr{I}_{\mathbf{G}}^F$ *is not a uniform pair, then* $Y_{(u_{\mathcal{O}}, \tilde{\varsigma})}$ *is orthogonal to the subspace* Υ, *for any* F*-extension* $\tilde{\varsigma}$ *of* ς.

It is almost impossible to trace here the exact history of the proof of these important results, which stretches from [DeLu76] up until quite recently. Based on the fundamental papers of Springer [Spr76], [Spr78] and Kazhdan [Kaz77], first instances of the above results were established by Shoji [Sho82], [Sho83] and Beynon–Spaltenstein [BeSp84], assuming that p, q are sufficiently large. It turned out to be a formidable task to remove the assumptions on p and q. The state of knowledge up until around 1986, incorporating further results by Borho–MacPherson, Lusztig and others, is summarised in Shoji's survey articles [Sho86], [Sho88]. At that stage the general plan was laid out, with Lusztig's theory of character sheaves [LuCS] being the new essential ingredient. In this picture one had to show that the Green functions Q_w coincide with another type of Green functions defined in terms of character sheaves, for which the properties in Theorem 2.8.3 are known to hold by [LuCS, 24.1, 24.2] and those in Theorem 2.8.4 would hold by [LuCS, 24.4], assuming certain 'cleanness' assumptions. In [Spa82b], [Ma93b], Green functions were explicitly computed for some groups of exceptional type and the 'bad' prime $p = 2$; thus establishing Theorems 2.8.3, 2.8.4 in these cases. A decisive step in the general direction was taken by [Lu90], which was then completed by [Sho95, Part II, 5.5]. As a result, the two types of Green functions are indeed known to coincide without any assumption on p, q; this immediately yields Theorem 2.8.3. The final step is provided by [Lu12a] which establishes the cleanness assumption of [LuCS, §24] in complete generality and immediately implies Theorem 2.8.4. (Actually, the full version of cleanness is not needed to get Theorem 2.8.4; see [Ge96, §3] for a further discussion. On the other hand, there are also 'generalised' Green functions in [LuCS] for which the full version of cleanness is required.)

Remark 2.8.5 The interpretation in terms of character sheaves (as mentioned above) yields the following compatibility property of the Springer correspondence $\mathrm{Irr}(\mathbf{W})^\sigma \hookrightarrow \mathscr{I}_{\mathbf{G}}^F$. Let $n \geqslant 1$. Then $F^n\colon \mathbf{G} \to \mathbf{G}$ also is a Frobenius map; it induces the automorphism $\sigma^n\colon \mathbf{W} \to \mathbf{W}$. Clearly, we have $\mathrm{Irr}(\mathbf{W})^\sigma \subseteq \mathrm{Irr}(\mathbf{W})^{\sigma^n}$ and $\mathscr{I}_{\mathbf{G}}^F \subseteq \mathscr{I}_{\mathbf{G}}^{F^n}$. Now [LuCS, 24.2] yields a commutative diagram

$$\begin{array}{ccc} \mathrm{Irr}(\mathbf{W})^{\sigma^n} & \hookrightarrow & \mathscr{I}_{\mathbf{G}}^{F^n} \\ \cup\!\!| & & \cup\!\!| \\ \mathrm{Irr}(\mathbf{W})^{\sigma} & \hookrightarrow & \mathscr{I}_{\mathbf{G}}^{F} \end{array}$$

where the top arrow is the correspondence in Theorem 2.8.3 with respect to $\mathbf{G}, F^n$.

Example 2.8.6 Let us consider the trivial character $1_\mathbf{W}$ and the sign character ε of $\mathbf{W}$. We choose the trivial σ-extensions $\tilde{1}_\mathbf{W} = 1_\mathbf{W}$ and $\tilde{\varepsilon} = \varepsilon$ (see Example 2.1.8(c)). Now let $u \in \mathbf{G}_{\mathrm{uni}}^F$. By Example 2.3.23, we have

$$Q_{\tilde{1}_\mathbf{W}}(u) = 1_\mathbf{G}(u) = 1 \quad \text{and} \quad Q_{\tilde{\varepsilon}}(u) = \mathrm{St}_\mathbf{G}(u) = \begin{cases} |\mathbf{G}^F|_p & \text{if } u = 1, \\ 0 & \text{if } u \neq 1. \end{cases}$$

(Note that an irreducible character of a finite group with p-defect 0 has value 0 on all non-trivial p-elements.) Hence, the defining conditions in Theorem 2.8.3

immediately show that $1_{\mathbf{W}}$ corresponds to the uniform pair $(\mathscr{O}_0, 1) \in \mathscr{I}_{\mathbf{G}}^F$ where $\mathscr{O}_0$ is the *regular unipotent class* and 1 stands for the trivial character of $A(u_{\mathscr{O}_0})$. Similarly, ε corresponds to the uniform pair $(\{1\}, 1) \in \mathscr{I}_{\mathbf{G}}^F$ where, again, 1 stands for the trivial character of $A(1)$.

Example 2.8.7 Let $\mathbf{G} = \mathrm{GL}_n(k)$ and $F\colon \mathbf{G} \to \mathbf{G}$ be the standard Frobenius map (see Example 1.4.21) such that $\mathbf{G}^F = \mathrm{GL}_n(q)$. We have $\mathbf{W} \cong \mathfrak{S}_n$ (the symmetric group of degree n) and $\sigma\colon \mathbf{W} \to \mathbf{W}$ is the identity. As above, let Q_w (for $w \in \mathbf{W}$) denote the restriction of R_w^1 to $\mathbf{G}_{\mathrm{uni}}^F$. We have noted in Example 2.4.20 that the functions Q_w are the same as the Green functions originally introduced and investigated in [Gre55]. Let us write again $\mathrm{Irr}(\mathbf{W}) = \{\phi^\nu \mid \nu \vdash n\}$ and set

$$Q_\nu := \frac{1}{|\mathbf{W}|} \sum_{w \in \mathbf{W}} \phi^\nu(w) Q_w \qquad \text{for any } \nu \vdash n.$$

Now the unipotent classes of $\mathbf{G}$ are also naturally parametrised by the partitions of n. For $\mu \vdash n$, let $u_\mu \in \mathbf{G}$ be the block-diagonal matrix formed by the Jordan blocks (with eigenvalue 1) of sizes given by the non-zero parts of μ. Let $\mathscr{O}_\mu$ be the $\mathbf{G}$-conjugacy class of u_μ. Then $\{\mathscr{O}_\mu \mid \mu \vdash n\}$ is the set of unipotent classes of $\mathbf{G}$ and, clearly, we have $F(u_\mu) = u_\mu$ for all $\mu \vdash n$. Further note that $C_{\mathbf{G}}(g)$ is connected for any $g \in \mathbf{G}$, and so $A(u_\mu) = \{1\}$ for all $\mu \vdash n$. Finally, if we set

$$n(\mu) := \sum_{1 \leqslant i \leqslant r} (i-1)\mu_i \qquad \text{where } \mu = (\mu_1 \geqslant \mu_2 \geqslant \cdots \geqslant \mu_r) \vdash n,$$

then $(\dim \mathbf{G} - \dim \mathbf{T}_0 - \dim \mathscr{O}_\mu)/2 = n(\mu)$; see, e.g., [Ge03a, 2.6.1]. Based on the fundamental results of [Gre55], and extensions of these results in [Mor63], it is shown in [Ohm77, 2.14] that, for any $\mu, \nu \vdash n$, we have:

$$Q_\nu(u_\nu) = q^{n(\nu)} \qquad \text{and} \qquad Q_\nu(u_\mu) = 0 \text{ unless } \mu \leqslant \nu.$$

Here, we write $\mu \leqslant \nu$ if μ is less than or equal to ν in the lexicographical ordering. Now, given $\mu, \mu' \vdash n$, it is well known that $\mathscr{O}_\mu \subseteq \overline{\mathscr{O}}_{\mu'} \Leftrightarrow \mu \trianglelefteq \mu'$, where $\trianglelefteq$ denotes the dominance order on partitions; see, e.g., [Ge03a, 2.6.5]. It was noticed in [Kaw85, 3.2.19] that, in the above assertion about $Q_\nu(u_\mu)$, we can replace the lexicographical ordering by the dominance order. Hence, we conclude that

$$\{u \in \mathbf{G}_{\mathrm{uni}}^F \mid Q_\nu(u) \neq 0\} \subseteq \overline{\mathscr{O}}_\nu \qquad \text{and} \qquad Q_\nu|_{\mathscr{O}_\nu^F} = q^{n(\nu)} Y_{(u_\nu, \tilde{\varsigma})}$$

where $Y_{(u_\nu,\tilde{\varsigma})}(u) = 1$ for $u \in \mathscr{O}_\nu^F$ (that is, $\tilde{\varsigma} = \varsigma$ is the trivial id-extension of the trivial character of $A(u_\nu) = \{1\}$). Thus, we have explicitly verified Theorems 2.8.3 and 2.8.4 for $\mathbf{G}^F = \mathrm{GL}_n(q)$, where the correspondence $\mathrm{Irr}(\mathbf{W})^\sigma \hookrightarrow \mathscr{I}_{\mathbf{G}}^F$ is a bijection: it sends the character ϕ^ν of $\mathbf{W} \cong \mathfrak{S}_n$ to the pair $(\mathscr{O}_\nu, 1)$ (where 1 stands for the trivial character of $A(u_\nu) = \{1\}$). For a further detailed discussion of the Green functions in this case, including both combinatorial and cohomological aspects, see [HoSh79] and [Mac95].

Remark 2.8.8 Let $\phi \in \mathrm{Irr}(\mathbf{W})^\sigma$ and $(\mathcal{O}, \varsigma) \in \mathscr{I}_{\mathbf{G}}^F$ as in Theorem 2.8.3. Let $\tilde{\varsigma}$ be the F-extension of ς such that

$$Q_{\tilde{\phi}}|_{\mathcal{O}^F} = q^{(\dim \mathbf{G} - \dim \mathbf{T}_0 - \dim \mathcal{O})/2}\, Y_{(u_{\mathcal{O}}, \tilde{\varsigma})}.$$

In general, what can we say about the F-extension $\tilde{\varsigma}$? This is quite a delicate matter, which is solved in almost all cases, but there are still open cases for type E_8 and $p = 2, 3, 5$. (See the references below.) To begin with, the values of the Green functions Q_w are rational integers and so are the values of the σ-extensions of the characters in $\mathrm{Irr}(\mathbf{W})^\sigma$ (since F is a Frobenius map and so σ is ordinary; see Proposition 2.1.14). It follows that the values of the function $\tilde{\varsigma}$ are in $\mathbb{Q}$. On the other hand, by [LuCS, 24.2.4], [Lu04b, 19.7], the F-extension $\tilde{\varsigma}$ is indeed formed using a matrix E of finite order (cf. 2.1.9) and so the values of $\tilde{\varsigma}$ are cyclotomic integers in $\mathbb{K}$. Hence,

$$\tilde{\varsigma}(a) \in \mathbb{Z} \qquad \text{for all } a \in A(u_{\mathcal{O}}).$$

Finally, assume that F acts trivially on $A(u_{\mathcal{O}})$, or that ς is a linear character. (This will cover most cases in practice.) Then the above discussion shows that there is a root of unity δ such that $\tilde{\varsigma}(a) = \delta\varsigma(a)$ for $a \in A(u_{\mathcal{O}})$. Since the values of $\tilde{\varsigma}$ are in $\mathbb{Q}$, it follows that $\delta = \pm 1$ and $\varsigma(a) \in \mathbb{Z}$ for all $a \in A(u_{\mathcal{O}})$. Furthermore, in this case, $\delta = \pm 1$ is determined by the equality

$$Y_{(u_{\mathcal{O}}, \tilde{\varsigma})}(u_{\mathcal{O}}) = \delta\varsigma(1). \qquad (\heartsuit)$$

See the remarks in [Sho86, 5.1], and also [BeSp84], [Sho06b], [Sho07], [Ge19b] for further results and details. It is important to keep in mind that, even if F acts trivially on $A(u_{\mathcal{O}})$ and if ς is a linear character, then there are examples where $\delta = -1$; see [BeSp84, §3, Case V], [Ge19c, §9].

2.8.9 The Lusztig–Shoji algorithm. We can now describe the fundamental algorithm in [LuCS, 24.4] which yields explicit expressions of the functions $Q_{\tilde{\phi}}$ as linear combinations of the functions $Y_{(u_{\mathcal{O}}, \tilde{\varsigma})}$. (It modifies and simplifies an algorithm described earlier by Shoji; see [Sho86, §5].) First, we need to prepare some notation. Let $\phi \in \mathrm{Irr}(\mathbf{W})^\sigma$ and $(\mathcal{O}, \varsigma) \in \mathscr{I}_{\mathbf{G}}^F$ be the corresponding uniform pair as in Theorem 2.8.3, such that

$$Q_{\tilde{\phi}}|_{\mathcal{O}^F} = q^{(\dim \mathbf{G} - \dim \mathbf{T}_0 - \dim \mathcal{O})/2}\, Y_{(u_{\mathcal{O}}, \tilde{\varsigma})}.$$

Then we denote the function $Y_{(u_{\mathcal{O}}, \tilde{\varsigma})}$ by $Y_{\tilde{\phi}}$. Furthermore, we set

$$X_{\tilde{\phi}} := q^{-d_{\mathcal{O}}} Q_{\tilde{\phi}} \qquad \text{where} \qquad d_{\mathcal{O}} := (\dim \mathbf{G} - \dim \mathbf{T}_0 - \dim \mathcal{O})/2;$$

it will also be convenient to set $d_\phi := d_{\mathcal{O}}$ in this situation. The functions $X_{\tilde{\phi}}$ will be called *modified Green functions*. By 2.7.2(b) and Theorem 2.8.4, the functions $\{Y_{\tilde{\phi}} \mid \phi \in \mathrm{Irr}(\mathbf{W})^\sigma\}$ form a basis of the vector space Υ; in particular, we obtain a system of equations

$$X_{\tilde{\phi}} = \sum_{\phi' \in \mathrm{Irr}(\mathbf{W})^\sigma} p_{\phi',\phi} Y_{\tilde{\phi}'}$$

where the coefficients $p_{\phi',\phi} \in \mathbb{K}$ are uniquely determined. Using the relations in 2.1.10 and 2.8.2, we obtain $|\mathbf{G}^F|\langle X_{\tilde{\phi}}, X_{\tilde{\phi}'}\rangle = \tilde{\omega}_{\phi,\phi'}(q)$ where

$$\tilde{\omega}_{\phi,\phi'} := \frac{1}{|\mathbf{W}|}\mathbf{q}^{-d_\phi - d_{\phi'}} \sum_{w \in \mathbf{W}} \frac{|\mathbb{G}|}{|\mathbb{T}_w|}\tilde{\phi}(w)\tilde{\phi}'(w) \in \mathbb{R}[\mathbf{q}] \quad \text{for } \phi, \phi' \in \mathrm{Irr}(\mathbf{W})^\sigma.$$

Thus, the scalar products $\langle X_{\tilde{\phi}}, X_{\tilde{\phi}'}\rangle$ are obtained from a symmetric matrix

$$\tilde{\Omega} := \big(\tilde{\omega}_{\phi,\phi'}\big)_{\phi,\phi' \in \mathrm{Irr}(\mathbf{W})^\sigma}$$

with entries in $\mathbb{R}[\mathbf{q}]$, which can be computed explicitly using the σ-character table of $\mathbf{W}$ and the order polynomials $|\mathbb{T}_w| \in \mathbb{R}[\mathbf{q}]$. Now we obtain:

$$\tilde{\omega}_{\phi,\phi'}(q) = |\mathbf{G}^F|\langle X_{\tilde{\phi}}, X_{\tilde{\phi}'}\rangle = \sum_{\psi,\psi' \in \mathrm{Irr}(\mathbf{W})^\sigma} p_{\psi,\phi}\overline{p}_{\psi',\phi'}|\mathbf{G}^F|\langle Y_{\tilde{\psi}}, Y_{\tilde{\psi}'}\rangle.$$

Let us set $\tilde{\lambda}_{\phi,\phi'} := |\mathbf{G}^F|\langle Y_{\tilde{\phi}}, Y_{\tilde{\phi}'}\rangle \in \mathbb{K}$ for all $\phi, \phi' \in \mathrm{Irr}(\mathbf{W})^\sigma$. Since the values of each $Q_{\tilde{\phi}}$ and of each $Y_{\tilde{\phi}}$ are in $\mathbb{Q}$, it follows that

$$p_{\phi',\phi} \in \mathbb{Q} \qquad \text{and} \qquad \tilde{\lambda}_{\phi,\phi'} = \tilde{\lambda}_{\phi',\phi} \in \mathbb{Q} \qquad \text{for all } \phi, \phi' \in \mathrm{Irr}(\mathbf{W})^\sigma.$$

We can now write the above equations as a single matrix equation

$$P^{\mathrm{tr}} \cdot \tilde{\Lambda} \cdot P = \tilde{\Omega}(q) \qquad \text{where} \qquad \begin{cases} P := \big(p_{\psi,\phi}\big)_{\psi,\phi \in \mathrm{Irr}(\mathbf{W})^\sigma}, \\ \tilde{\Lambda} := \big(\tilde{\lambda}_{\phi,\phi'}\big)_{\phi,\phi' \in \mathrm{Irr}(\mathbf{W})^\sigma}. \end{cases}$$

In general, this system of equations will not have a unique solution for $P, \tilde{\Lambda}$. But if we take into account the additional information in Theorem 2.8.3, then it does have a unique solution. Indeed, let $\mathcal{O}_1, \ldots, \mathcal{O}_m$ be the F-stable unipotent classes of $\mathbf{G}$; let us also denote $u_i := u_{\mathcal{O}_i} \in \mathcal{O}_i^F$. Here, we choose the labelling such that

$$\dim \mathcal{O}_1 \leqslant \dim \mathcal{O}_2 \leqslant \cdots \leqslant \dim \mathcal{O}_m.$$

This gives rise to a partition

$$\mathrm{Irr}(\mathbf{W})^\sigma = \mathscr{I}_1 \sqcup \cdots \sqcup \mathscr{I}_m,$$

where each subset $\mathscr{I}_j$ consists of those $\phi \in \mathrm{Irr}(\mathbf{W})^\sigma$ that correspond (via Theorem 2.8.3) to a uniform pair of the form $(\mathcal{O}_j, \varsigma) \in \mathscr{I}_{\mathbf{G}}^F$ for some $\varsigma \in \mathrm{Irr}(A(u_j))$. We now enumerate the characters in $\mathrm{Irr}(\mathbf{W})^\sigma$ in a way that is compatible with the above partition of $\mathrm{Irr}(\mathbf{W})^\sigma$. Then it is clear that $\tilde{\Lambda}$ has a block diagonal shape, where the blocks correspond to the sets $\mathscr{I}_1, \ldots, \mathscr{I}_m$. Furthermore, Theorem 2.8.3 implies that P has an upper block triangular shape with identity matrices on the diagonal. More precisely, we can write:

$$P = \begin{bmatrix} I_{e_1} & P_{1,2} & \cdots & P_{1,m} \\ 0 & I_{e_2} & & \vdots \\ \vdots & & \ddots & P_{m-1,m} \\ 0 & \cdots & 0 & I_{e_m} \end{bmatrix} \quad \text{and} \quad \tilde{\Lambda} = \begin{bmatrix} \tilde{\Lambda}_1 & 0 & \cdots & 0 \\ 0 & \tilde{\Lambda}_2 & & \vdots \\ \vdots & & \ddots & 0 \\ 0 & \cdots & 0 & \tilde{\Lambda}_m \end{bmatrix};$$

here, $e_j = |\mathscr{I}_j|$ and I_{e_j} denotes the identity matrix of size e_j.

For $1 \leqslant i < j \leqslant m$, the block $P_{i,j}$ has size $e_i \times e_j$ and entries $p_{\phi,\phi'} \in \mathbb{Q}$ for $\phi \in \mathscr{I}_i$ and $\phi' \in \mathscr{I}_j$; similarly, the block $\tilde{\Lambda}_j$ has size $e_j \times e_j$ and entries $\tilde{\lambda}_{\phi,\phi'} \in \mathbb{Q}$ for $\phi, \phi' \in \mathscr{I}_j$.

It is then a standard linear algebra problem to compute P and $\tilde{\Lambda}$ from the above matrix equation; see [LuCS, 24.4] for further details, and [LaSr90], [Ge03d, §5] for the discussion of some examples. Thus, given $\tilde{\Omega}$ and the Springter correspondence $\mathrm{Irr}(\mathbf{W})^\sigma \hookrightarrow \mathscr{I}_{\mathbf{G}}^F$, we have a purely automatic procedure for the computation of P and $\tilde{\Lambda}$. (See also [GeMa99] for a variation of that algorithm.) This is implemented in Michel's version of CHEVIE [MiChv], through the functions

`UnipotentClasses` and `ICCTable`

which take as input a finite Weyl group (or, more generally, a root datum) and a prime p (the characteristic of the field over which $\mathbf{G}$ is defined). The functions use extensive databases within CHEVIE which hold the explicitly known information about the Springer correspondence $\mathrm{Irr}(\mathbf{W})^\sigma \hookrightarrow \mathscr{I}_{\mathbf{G}}^F$.

Remark 2.8.10 The matrix $\tilde{\Omega}$ has the property that every principal minor is non-zero. (This follows from the fact that $\tilde{\Omega}(q)$ is the matrix of mutual scalar products of linearly independent class functions on $\mathbf{G}^F$.) Hence, the system of equations $P^{\mathrm{tr}} \cdot \tilde{\Lambda} \cdot P = \tilde{\Omega}(q)$ can be solved at the level of rational functions in $\mathbf{q}$. We will not formalise this here, but it means that the entries of P and $\tilde{\Lambda}$ can be expressed as values at q of well-defined rational functions in $\mathbf{q}$. By letting q vary over a suitable infinite set of prime powers, and using [LuCS, 24.5.2], one can even show that those rational functions are polynomials in $\mathbf{q}$. (See the argument in the first part of the proof of [LuCS, 24.8].)

Using the output of the above algorithm, we can now also solve the issue concerning the F-extensions $\tilde{\varsigma}$ in Remark 2.8.8, at least in some special cases.

2.8.11 Let $\mathrm{Irr}(\mathbf{W})_1^\sigma$ be the set of all $\phi \in \mathrm{Irr}(\mathbf{W})^\sigma$ such that ϕ corresponds to a pair $(\mathscr{O}, 1) \in \mathscr{I}_{\mathbf{G}}^F$ (as in Theorem 2.8.3) where 1 stands for the trivial character of

$A(u_{\mathscr{O}})$. Thus, given $\phi \in \mathrm{Irr}(\mathbf{W})_1^\sigma$ with corresponding pair $(\mathscr{O}, 1) \in \mathscr{I}_{\mathbf{G}}^F$, there is a sign $\delta = \pm 1$ such that

$$Y_{\tilde{\phi}}(u) = Y_{(u_{\mathscr{O}}, \tilde{1})}(u) = \delta \qquad \text{for all } u \in \mathscr{O}^F$$

(see Remark 2.8.8). Let $\tilde{1}_{\mathbf{W}} = 1_{\mathbf{W}}$ be the trivial σ-extension of the trivial character of $\mathbf{W}$. Then, as in [BeSp84, p. 587], we claim that

$$\delta = p_{\phi, 1_{\mathbf{W}}} \qquad \text{(where } p_{\phi,1_{\mathbf{W}}} \text{ is an entry of the matrix } P \text{ in 2.8.9).}$$

Indeed, by Example 2.8.6, we have $Q_{\tilde{1}_{\mathbf{W}}}(u) = 1$ for all $u \in \mathbf{G}_{\mathrm{uni}}^F$; the uniform pair corresponding to $1_{\mathbf{W}}$ is given by $(\mathscr{O}_0, 1) \in \mathscr{I}_{\mathbf{G}}^F$ where $\mathscr{O}_0$ is the regular unipotent class. By 2.2.21, we have $d_{\mathscr{O}_0} = 0$. As in 2.8.9, we have

$$1 = Q_{\tilde{1}_{\mathbf{W}}}(u) = q^{d_{\mathscr{O}_0}} X_{\tilde{1}_{\mathbf{W}}}(u) = \sum_{\psi \in \mathrm{Irr}(\mathbf{W})^\sigma} p_{\psi,1_{\mathbf{W}}} Y_{\tilde{\psi}}(u)$$

for all $u \in \mathbf{G}_{\mathrm{uni}}^F$. By 2.7.2(b), the functions $\{Y_{\tilde{\psi}} \mid \psi \in \mathrm{Irr}(\mathbf{W})^\sigma\}$ are linearly independent. Hence, we deduce that $p_{\psi,1_{\mathbf{W}}} = 0$ if $\psi \notin \mathrm{Irr}(\mathbf{W})_1^\sigma$; furthermore, if $\psi = \phi \in \mathrm{Irr}(\mathbf{W})_1^\sigma$, then the non-zero values of $p_{\phi,1_{\mathbf{W}}} Y_{\tilde{\phi}}$ must be equal to 1, and so $p_{\phi,1_{\mathbf{W}}}\delta = 1$, as desired. This also shows that, if $\phi \in \mathrm{Irr}(\mathbf{W})_1^\sigma$, then the corresponding pair $(\mathscr{O}, 1) \in \mathscr{I}_{\mathbf{G}}^F$ is a uniform pair.

Remark 2.8.12 Explicit descriptions of the subsets $\mathrm{Irr}(\mathbf{W})_1^\sigma \subseteq \mathrm{Irr}(\mathbf{W})^\sigma$ can be found in [Lu05, 1.3] (completing earlier work, assuming that p is large enough, in [Lu79b]). In particular, we obtain an explicit bijection

$$\mathrm{Irr}(\mathbf{W})_1^\sigma \xrightarrow{\sim} \{F\text{-stable unipotent classes of } \mathbf{G}\}, \qquad \phi \mapsto \mathscr{O}_\phi,$$

where $\mathscr{O}_\phi$ is the F-stable unipotent class such that ϕ corresponds to the uniform pair $(\mathscr{O}_\phi, 1) \in \mathscr{I}_{\mathbf{G}}^F$. Thus, the number of F-stable unipotent classes of $\mathbf{G}$ is seen to be bounded above by the cardinality of $\mathrm{Irr}(\mathbf{W})^\sigma$ and, hence, by the number of σ-conjugacy classes of $\mathbf{W}$.

Example 2.8.13 Let $\mathbf{G}^F = \mathrm{GL}_n(q)$ as in Example 2.8.7. In this case, the characters of $\mathbf{W} \cong \mathfrak{S}_n$ and the F-stable unipotent classes of $\mathbf{G}$ are both parametrised by the partitions of n. Furthermore, each piece $\mathscr{I}_j$ in the partition of $\mathrm{Irr}(\mathbf{W})^\sigma$ in 2.8.9 is just a singleton set. The requirement to order the unipotent classes by increasing dimension means that we have to order the partitions $\nu \vdash n$ according to decreasing value of $n(\nu)$. Then P is an upper triangular matrix with 1 on the diagonal and $\tilde{\Lambda}$ is a diagonal matrix. Let us write

$$P = (p_{\mu\nu})_{\mu,\nu \vdash n} \qquad \text{where} \qquad p_{\mu\nu} := p_{\phi_\mu,\phi_\nu}.$$

We already saw in Example 2.8.7 that each function $Y_{(u_\mu,\tilde{\varsigma})}$ takes the constant value 1

Table 2.11 *Modified Green functions for* $\mathrm{GL}_n(q)$, $n = 2, 3, 4$

n=2	$n(\nu)$	(11)	(2)
(11)	1	1	.
(2)	0	1	1

n=3	$n(\nu)$	(111)	(21)	(3)
(111)	3	1	.	.
(21)	1	$q+1$	1	.
(3)	0	1	1	1

n=4	$n(\nu)$	(1111)	(211)	(22)	(31)	(4)
(1111)	6	1	.	.	.	.
(211)	3	q^2+q+1	1	.	.	.
(22)	2	q^2+1	1	1	.	.
(31)	1	q^2+q+1	$q+1$	1	1	.
(4)	0	1	1	1	1	1

on $\mathscr{O}_\mu^F$. Thus, we conclude that

$$Q_\nu(u_\mu) = q^{n(\nu)} p_{\mu\nu} \qquad \text{for all } \mu, \nu \vdash n.$$

Finally, by [Gre55] (see also [LuCS, 24.8] for a more general argument), it is known that there exist polynomials $\pi_{\mu\nu} \in \mathbb{Q}[\mathbf{q}]$ (depending only on n, and defined in terms of 'Hall polynomials') such that $p_{\mu\nu} = \pi_{\mu\nu}(q)$ for all prime powers q. By the algorithm in 2.8.9, we obtain the example matrices P^{tr} in Table 2.11. Note that the entries of the diagonal matrix $\tilde{\Lambda}$ are given by

$$\tilde{\lambda}_{\mu\mu} = |\mathbf{G}^F : C_{\mathbf{G}}(u_\mu)^F| \qquad \text{for all } \mu \vdash n$$

in this case, and these numbers are also obtained by evaluating certain well-defined polynomials at q. Thus, if we denote these polynomials by $c_\mu \in \mathbb{Q}[\mathbf{q}]$ ($\mu \vdash n$), then we actually have a system of polynomial equations

$$\tilde{\omega}_{\mu\nu} = \sum_{\nu' \vdash n} \pi_{\nu'\mu} c_{\nu'} \pi_{\nu'\nu} \qquad \text{for } \mu, \nu \vdash n,$$

where the matrix $(\pi_{\mu\nu})$ is upper triangular with 1 on the diagonal.

Example 2.8.14 Let $\mathbf{G}^F = G_2(q)$ where $|\mathbf{G}^F| = q^6(q^2-1)(q^6-1)$ and q is a power of a prime $p \neq 2, 3$. The Green functions in this case were first determined in [Spr76, 7.16], before the above machinery was available. The Weyl group $\mathbf{W} = \langle s_1, s_2 \rangle$ is a dihedral group of order 12, with F acting trivially on $\mathbf{W}$. We fix the notation so that s_1 corresponds to a simple long root (which we denote by β) and s_2 corresponds to a simple short root (which we denote by α).

There are five unipotent classes in $\mathbf{G}$, which are all F-stable. The relevant information about these classes, as well as the Springer correspondence $\mathrm{Irr}(\mathbf{W})^\sigma \hookrightarrow \mathscr{I}_{\mathbf{G}}^F$, are given in Table 2.12; here, the characters of $\mathbf{W}$ are denoted as in Example 2.1.13. (For the information about the classes $\mathscr{O}_i$, see [Cha68] and Remark 2.8.15 below;

Table 2.12 *The Springer correspondence for* G_2, $p \neq 2, 3$

$\mathscr{O}_i$	$d_{\mathscr{O}_i}$	u_i	$\lvert C_{\mathbf{G}}(u_i)^F \rvert$	$A(u_i)$	ς	$\phi \in \mathrm{Irr}(\mathbf{W})$
$\mathscr{O}_1 = \varnothing$	6	1	$\lvert \mathbf{G}^F \rvert$	$\{1\}$	1	ε
$\mathscr{O}_2 = A_1$	3	$x_{3\alpha+2\beta}(1)$	$q^6(q^2-1)$	$\{1\}$	1	ε''
$\mathscr{O}_3 = \tilde{A}_1$	2	$x_{2\alpha+\beta}(1)$	$q^4(q^2-1)$	$\{1\}$	1	ϕ_2
$\mathscr{O}_4 = G_2(a_1)$	1	$x_\beta(1)x_{2\alpha+\beta}(-3)$	$6q^4$	$\mathfrak{S}_3$	$\varsigma_{(3)}$	ϕ_1
					$\varsigma_{(21)}$	ε'
					$\varsigma_{(111)}$	non-uniform
$\mathscr{O}_5 = G_2$	0	$x_\alpha(1)x_\beta(1)$	q^2	$\{1\}$	1	$1_{\mathbf{W}}$

P^{tr}	$\varnothing$	A_1	$\tilde{A}_1$	$G_2(a_1)$	G_2
ε	1	.	.	.	.
ε''	1	1	.	.	.
ϕ_2	q^2+1	1	1	.	.
ϕ_1	q^4+1	1	1	1	.
ε'	q^2	.	1	. 1	.
$1_{\mathbf{W}}$	1	1	1	1 .	1

for the Springer correspondence, see [Ca85, p. 427] or [Spa85, p. 329].) By the Lusztig–Shoji algorithm in 2.8.9, we obtain the matrix P^{tr} printed in Table 2.12.

Now let $\phi \in \mathrm{Irr}(\mathbf{W})^\sigma$ and $(\mathscr{O}, \varsigma) \in \mathscr{I}_{\mathbf{G}}^F$ be the corresponding uniform pair. It remains to determine the F-extension $\tilde{\varsigma}$ that enters in the definition of $Y_{\tilde{\phi}} = Y_{(u_{\mathscr{O}}, \tilde{\varsigma})}$; see Remark 2.8.8. First of all, since σ is the identity, we can take $\tilde{\phi} = \phi$ for each $\phi \in \mathrm{Irr}(\mathbf{W})^\sigma$. Since F acts trivially on $A(u_{\mathscr{O}})$, we can apply Remark 2.8.8($\heartsuit$). So there exists a sign $\delta = \pm 1$ such that $\tilde{\varsigma} = \delta\varsigma$ and

$$Y_{\tilde{\phi}}(u_{\mathscr{O}}) = Y_{(u_{\mathscr{O}}, \tilde{\varsigma})}(u_{\mathscr{O}}) = \delta\varsigma(1).$$

We claim that $\delta = 1$ in all cases. If $\varsigma = 1$ is the trivial character of $A(u_{\mathscr{O}})$, then this immediately follows from the entries in the last row of the above matrix P^{tr} and 2.8.11. It remains to consider the case where $\mathscr{O} = \mathscr{O}_4$ and $\varsigma = \varsigma_{(21)}$ is the character of degree 2 of $A(u_4) \cong \mathfrak{S}_3$. Now one either has to enter a study of the geometry of the variety of Borel subgroups containing u_4 (e.g., as in [BeSp84, §3]), or one has to use ad hoc information that is available in some other way. In the present situation, we can argue as follows. Let $\mathbf{B}_0 \subseteq \mathbf{G}$ be an F-stable Borel subgroup containing our reference torus $\mathbf{T}_0$ such that $u_4 \in \mathbf{B}_0^F$. Since $A(u_4) \cong \mathfrak{S}_3$ is non-abelian and of order prime to p, we certainly have $C_{\mathbf{G}}(u_4)^F \not\subseteq \mathbf{B}_0^F$ and so

$$\mathrm{Ind}_{\mathbf{B}_0^F}^{\mathbf{G}^F}(1_{\mathbf{B}_0})(u_4) > 1.$$

On the other hand, by Proposition 2.2.7, the above induced character equals $R_1^1 =$

$R_{\mathbf{T}_0}^{\mathbf{G}}(1_{\mathbf{T}_0})$. We now compute:

$$\begin{aligned} R_1^1(u_4) = Q_1(u_4) &= \sum_{\phi \in \mathrm{Irr}(\mathbf{W})^\sigma} \phi(1) Q_{\tilde{\phi}}(u_4) = \sum_{\psi,\phi \in \mathrm{Irr}(\mathbf{W})^\sigma} \phi(1) p_{\psi,\phi} q^{d_\phi} Y_{\tilde{\psi}}(u_4) \\ &= \phi_1(1) q Y_{\tilde{\phi}_1}(u_4) + \varepsilon'(1) q Y_{\tilde{\varepsilon}'}(u_4) + 1_{\mathbf{W}}(1) Y_{\tilde{\phi}_1}(u_4) \\ &= (2q+1) Y_{\tilde{\phi}_1}(u_4) + q Y_{\tilde{\varepsilon}'}(u_4) = (2q+1) + 2q\delta. \end{aligned}$$

Since $R_1^1(u_4) > 1$, we must have $\delta = 1$ in this case as well. Hence, for every uniform pair $(\mathscr{O}, \varsigma) \in \mathscr{I}_{\mathbf{G}}^F$, the function $Y_{(u_{\mathscr{O}}, \tilde{\varsigma})}$ is formed using $\tilde{\varsigma} = \varsigma$.

(Using the character tables in [Eno76], [EnYa86], one also recovers easily the Green functions for $G_2(q)$ where q is a power of 2 or 3.)

Remark 2.8.15 Consider the unipotent class $\mathscr{O}_4 = G_2(a_1)$ in Table 2.12. According to [Cha68] (see also [Hum95, 8.16]), we have $A(u) \cong \mathfrak{S}_3$ for $u \in \mathscr{O}_4$, and $\mathscr{O}_4^F$ splits into three classes in the finite group $\mathbf{G}^F$, with centraliser orders $6q^4, 2q^4, 3q^4$. Hence, these three classes into which $\mathscr{O}_4^F$ splits can be distinguished just by looking at centraliser orders (or class sizes). So it is natural to fix a representative $u_4 \in \mathscr{O}_4^F$ such that $|C_{\mathbf{G}}(u_4)^F| = 6q^4$, in which case F acts trivially on $A(u_4)$. By [Cha68, 3.18], such a representative is given by

$$u_4 = \begin{cases} x_\beta(1) x_{2\alpha+\beta}(1) & \text{if } q \equiv 1 \bmod 3, \\ x_\beta(1) x_{2\alpha+\beta}(\xi) & \text{if } q \equiv -1 \bmod 3, \end{cases}$$

where $0 \neq \xi \in \mathbb{F}_q$ is a non-square. In fact, the formulae in [Cha68, §2] for the action of elements of $\mathbf{T}_0^F$ on unipotent elements show that we can just take

$$u_4 = x_\beta(1) x_{2\alpha+\beta}(\xi')$$

where $0 \neq \xi' \in \mathbb{F}_q$ is a square if $q \equiv 1 \bmod 3$, and a non-square if $q \equiv -1 \bmod 3$. Finally, it would be desirable to fix ξ' in some way, in order to obtain a uniform description of u_4, valid for all q (assuming that $p \neq 2, 3$). It is an easy application of quadratic reciprocity to see that we can always take $\xi' = -3$.

Example 2.8.16 Let again $\mathbf{G} = \mathrm{GL}_n(k)$ and $F \colon \mathbf{G} \to \mathbf{G}$ be the standard Frobenius map (see Example 1.4.21) such that $\mathbf{G}^F = \mathrm{GL}_n(q)$. Let us now consider the automorphism $\gamma \colon \mathbf{G} \to \mathbf{G}$ defined by

$$\gamma(g) = J_n (g^{\mathrm{tr}})^{-1} J_n \qquad \text{with } J_n \in \mathbf{G} \text{ as in Example 1.3.19.}$$

Then $F' := F \circ \gamma$ is a Frobenius map such that $\mathbf{G}^{F'} = \mathrm{GU}_n(q)$ is the finite general unitary group (see Example 1.4.21). The automorphism $\sigma' \colon \mathbf{W} \to \mathbf{W}$ induced by F' is given by conjugation with the longest element $w_0 \in \mathbf{W}$; note that J_n is the permutation matrix associated with w_0. Write again $\mathrm{Irr}(\mathbf{W}) = \{\phi^\nu \mid \nu \vdash n\}$. Each ϕ^ν is invariant under σ' and we could define a corresponding σ'-extension as in

Table 2.13 *Modified Green functions for* $\mathrm{GU}_n(q)$, $n = 2, 3, 4$

n=2	$n(\nu)$	(11)	(2)
(11)	1	1	.
(2)	0	1	1

n=3	$n(\nu)$	(111)	(21)	(3)
(111)	3	1	.	.
(21)	1	$1-q$	1	.
(3)	0	1	1	1

n=4	$n(\nu)$	(1111)	(211)	(22)	(31)	(4)
(1111)	6	1	.	.	.	.
(211)	3	q^2-q+1	1	.	.	.
(22)	2	q^2+1	1	1	.	.
(31)	1	q^2-q+1	$1-q$	1	1	.
(4)	0	1	1	1	1	1

Example 2.1.8(c). However, following [LuCS, 17.2] (see also Remark 4.1.30), it will be more convenient to define a σ'-extension by

$$\tilde{\phi}^{\nu}(w) := (-1)^{n(\nu)}\phi^{\nu}(ww_0) \qquad \text{for all } w \in \mathbf{W}.$$

As in Example 2.8.7, we denote the unipotent classes of $\mathbf{G}$ by $\{\mathscr{O}_\mu \mid \mu \vdash n\}$. Since these are characterised by the Jordan normal form of matrices, it is clear that $F'(\mathscr{O}_\mu) = \mathscr{O}_\mu$ for all $\mu \vdash n$; let us fix a representative $u'_\mu \in \mathscr{O}_\mu^{F'}$ for each $\mu \vdash n$. (It will not matter which representative we choose since $C_{\mathbf{G}}(u'_\mu)$ is connected and so $\mathscr{O}_\mu^{F'}$ is a single conjugacy class in $\mathbf{G}^{F'}$.) Using the above σ'-extensions of the characters of $\mathbf{W}$, we obtain

$$Q'_\nu = \frac{1}{|\mathbf{W}|}(-1)^{n(\nu)} \sum_{w\in\mathbf{W}} \phi^{\nu}(w_0 w) Q'_w \qquad \text{and} \qquad X'_\nu = q^{-n(\nu)} Q'_\nu,$$

where Q'_w are the Green functions of $\mathbf{G}^{F'}$. By the algorithm in 2.8.9, we obtain the matrices in Table 2.13. We notice that these matrices are obtained from those in Table 2.11 by simply 'changing q to $-q$'. The general theory developed so far now yields a rather straightforward proof of this fact.

Theorem 2.8.17 (*Ennola duality*; see [Enn63], [HoSp77, 3.1], [Kaw85, 4.1]) *In the setting of Example 2.8.16, we have*

$$Q'_\nu(u'_\mu) = q^{n(\nu)}\pi_{\mu\nu}(-q) \qquad \textit{for all } \mu, \nu \vdash n,$$

where $\pi_{\mu\nu} \in \mathbb{Q}[\mathbf{q}]$ *are the polynomials in Example 2.8.13.*

Proof We consider the algorithm in 2.8.9 for $\mathbf{G}, F'$ and compare it with the analogous algorithm for $\mathbf{G}, F$. Thus, we have to consider two systems of matrix equations

$$P'^{\mathrm{tr}} \cdot \tilde{\Lambda}' \cdot P' = \tilde{\Omega}'(q) \qquad \text{and} \qquad P^{\mathrm{tr}} \cdot \tilde{\Lambda} \cdot P = \tilde{\Omega}(q),$$

where the first one refers to $\mathbf{G}, F'$ and the second to $\mathbf{G}, F$; all of the above matrices have rows and columns indexed by the partitions of n. In order to obtain unique solutions to the above systems of equations, we have to consider the unipotent classes $\{\mathscr{O}_\mu \mid \mu \vdash n\}$ and order them by increasing dimension. Via Theorem 2.8.3, this gives rise to two partitions

$$\mathrm{Irr}(\mathbf{W})^{\sigma'} = \bigsqcup_{\mu \vdash n} \mathscr{I}'_\mu \qquad \text{and} \qquad \mathrm{Irr}(\mathbf{W})^{\sigma} = \bigsqcup_{\mu \vdash n} \mathscr{I}_\mu$$

where, again, the first one refers to $\mathbf{G}, F'$ and the second one to $\mathbf{G}, F$. (These partitions define the required block-triangular shape of the matrices P, P'.) Now, since γ has order 2 and commutes with F, we have $F'^2 = F^2$ and so $\mathbf{G}^{F'^2} = \mathbf{G}^{F^2} = \mathrm{GL}_n(q^2)$. Since $\mathrm{Irr}(\mathbf{W}) = \mathrm{Irr}(\mathbf{W})^{\sigma'} = \mathrm{Irr}(\mathbf{W})^{\sigma}$, we conclude using Remark 2.8.5 that the correspondence $\mathrm{Irr}(\mathbf{W})^{\sigma'} \hookrightarrow \mathscr{I}_{\mathbf{G}}^{F'}$ is actually the same as the correspondence $\mathrm{Irr}(\mathbf{W})^{\sigma} \hookrightarrow \mathscr{I}_{\mathbf{G}}^{F}$. Thus, $\mathscr{I}'_\mu = \mathscr{I}_\mu = \{\phi_\mu\}$ for all $\mu \vdash n$ and the matrices P, P' have the same triangular shape with 1 on the diagonal. In particular, the rows and columns of $\tilde{\Omega}$ and $\tilde{\Omega}'$ are indexed and ordered in the same way. Now let us compare these two matrices. Let $|\mathbb{G}'| \in \mathbb{R}[\mathbf{q}]$ be the order polynomial associated to $\mathbf{G}, F'$ and $|\mathbb{G}| \in \mathbb{R}[\mathbf{q}]$ be the order polynomial associated to $\mathbf{G}, F$. Let $\mathbf{T}_0 \subseteq \mathbf{G}$ be the maximal torus consisting of the diagonal matrices; note that $\mathbf{T}_0$ is maximally split both for F and for F'. For $w \in \mathbf{W}$, let $|\mathbb{T}'_w| \in \mathbb{R}[\mathbf{q}]$ be the order polynomial of an F'-stable maximal torus in $\mathbf{G}$ of type w. Similarly, let $|\mathbb{T}_w| \in \mathbb{R}[\mathbf{q}]$ be the order polynomial of an F-stable maximal torus in $\mathbf{G}$ of type w. Then the entries of $\tilde{\Omega}' = (\tilde{\omega}'_{\mu\nu})_{\mu,\nu \vdash n}$ and $\tilde{\Omega} = (\tilde{\omega}_{\mu\nu})_{\mu,\nu \vdash n}$ are the polynomials given by

$$\tilde{\omega}'_{\mu\nu} = \frac{1}{|\mathbf{W}|}\mathbf{q}^{-n(\mu)-n(\nu)} \sum_{w \in \mathbf{W}} \frac{|\mathbb{G}'|}{|\mathbb{T}'_w|} \tilde{\phi}^\mu(w)\tilde{\phi}^\nu(w) \in \mathbb{R}[\mathbf{q}],$$

$$\tilde{\omega}_{\mu\nu} = \frac{1}{|\mathbf{W}|}\mathbf{q}^{-n(\mu)-n(\nu)} \sum_{w \in \mathbf{W}} \frac{|\mathbb{G}|}{|\mathbb{T}_w|} \phi^\mu(w)\phi^\nu(w) \in \mathbb{R}[\mathbf{q}].$$

By Example 1.6.18(a), we have $|\mathbb{G}'| = (-1)^n|\mathbb{G}|(-\mathbf{q})$. Next we show that

$$|\mathbb{T}'_w| = (-1)^n|\mathbb{T}_{ww_0}|(-\mathbf{q}) \qquad \text{for any } w \in \mathbf{W}. \tag{$*$}$$

This is seen as follows. Let $X = X(\mathbf{T}_0)$ be the character group of $\mathbf{T}_0$. Let $\varphi\colon X \to X$ be the map induced by F; then φ is just given by multiplication with q and so $|\mathbb{T}_{ww_0}| = \det(\mathbf{q}\,\mathrm{id}_X - (ww_0)^{-1})$; see 1.6.21. Let $\varphi'\colon X \to X$ be the map induced by F'; for $t \in \mathbf{T}_0$, we have

$$F'(t) = \gamma(F(t)) = \gamma(t)^q = (J_n t^{-1} J_n)^q = (\dot{w}_0^{-1} t \dot{w}_0)^{-q}.$$

Hence, φ' is given by $-qw_0$ and so

$$|\mathbb{T}'_w| = \det(\mathbf{q}\,\mathrm{id}_X - (-w_0) \circ w^{-1}) = \det(\mathbf{q}\,\mathrm{id}_X + (ww_0)^{-1}),$$

where the right-hand side equals $(-1)^n|\mathbb{T}_{ww_0}|(-\mathbf{q})$, as desired. Thus, (∗) is proved. Combining this with the definition of $\tilde{\phi}_\nu$ in Example 2.8.16, we conclude that

$$\tilde{\omega}'_{\mu\nu} = \tilde{\omega}_{\mu\nu}(-\mathbf{q}) \qquad \text{for all } \mu, \nu \vdash n.$$

As noted in Example 2.8.13, the entries of $P, \tilde{\Lambda}$ are obtained by evaluating certain well-defined polynomials at q and the equation $P^{\text{tr}} \cdot \tilde{\Lambda} \cdot P = \tilde{\Omega}(q)$ can actually be solved at the polynomial level. Since $\tilde{\Omega}' = \tilde{\Omega}(-\mathbf{q})$, the polynomial equation in Example 2.8.13 yields that

$$\tilde{\omega}'_{\mu\nu}(q) = \tilde{\omega}_{\mu\nu}(-q) = \sum_{\nu'\vdash n} \pi_{\nu'\mu}(-q)d_{\nu'}(-q)\pi_{\nu'\nu}(-q) \qquad \text{for } \mu, \nu \vdash n.$$

On the other hand, P' and $\tilde{\Lambda}'$ are determined by the system of equations

$$\tilde{\omega}'_{\mu\nu}(q) = \sum_{\nu'\vdash n} p'_{\nu'\mu}\tilde{\lambda}'_{\nu'\nu'}p'_{\nu'\nu} \qquad \text{for } \mu, \nu \vdash n.$$

Since the matrices $(\pi_{\mu\nu})$ and $(p'_{\mu\nu})$ have the same upper triangular shape with 1 on the diagonal for the given ordering of the partitions of n, the above two systems of equations have the same solution. So we conclude that

$$p'_{\mu\nu} = \pi_{\mu\nu}(-q) \quad \text{and} \quad \tilde{\lambda}'_{\mu\mu} = d_\mu(-q) \quad \text{for all } \mu, \nu \vdash n.$$

Using 2.8.11, we also see that $Y_{\tilde{\phi}_\nu}$ is equal to 1 on $\mathcal{O}_\nu^{F'}$. (From the results for $\mathbf{G}^F = \mathrm{GL}_n(q)$, we know that $\pi_{\nu,(n)} = 1$ for all $\nu \vdash n$.) Hence, we conclude that $Q'_\nu(u'_\mu) = q^{n(\nu)}p'_{\mu\nu}$ for all $\mu, \nu \vdash n$. □

An analogous argument also works for the Green functions of the groups $\mathbf{G}^{F'} = {}^2E_6(q)$ and ${}^2D_{2n+1}(q)$ where, again, the automorphism induced on $\mathbf{W}$ is given by conjugation with the longest element; see [Sho86, §6(B)].

Example 2.8.18 Let $\mathbf{G}$ be simple of type B_2, G_2 or F_4, and $F\colon \mathbf{G} \to \mathbf{G}$ be a Steinberg map such that $\mathbf{G}^F$ is Suzuki or Ree group. Although F is not a Frobenius map, we can simply run the algorithm in 2.8.9 in these cases as well and see what we get; the output is displayed in Table 2.14. By comparison with the known tables of unipotent characters in [Suz62], [War66], [Ma90], one verifies that Theorems 2.8.3 and 2.8.4 still hold.

For ${}^2B_2$, we use the σ-extensions specified in Example 2.1.12. There are three F-stable unipotent classes of $\mathbf{G}$, specified by indicating the Jordan normal form; the Springer correspondence in this case can be found in [LuSp85, 6.1]. The Green functions can be extracted from the complete character table in [Suz62, Thm. 13] and the multiplicity formulae in [Lu84a, p. 373]; see also [Ge03a, Prop. 4.6.8].

For ${}^2G_2$, we use the σ-extensions specified in Example 2.1.13. There are four F-stable unipotent classes of $\mathbf{G}$, denoted as in [Spa85, p. 329] where one can also find

Table 2.14 *Values of modified Green functions for* $^2B_2$, $^2G_2$, $^2F_4$

$^2B_2$		$\varnothing$	$\mathcal{O}_{(22)}$	$\mathcal{O}_{(4)}$	
ϕ	d_ϕ	1	σ	ρ	ρ^{-1}
$\tilde{\varepsilon}$	4	1	.	.	.
$\tilde{\phi}_1$	1	$-q^2+1$	1	.	.
$\tilde{1}$	0	1	1	1	1

$^2G_2$		$\varnothing$	$\tilde{A}_1$	$G_2(a_1)$		G_2		
ϕ	d_ϕ	1	X	T	T^{-1}	Y	YT	YT^{-1}
$\tilde{\varepsilon}$	6	1	.	.	.	.	.	.
$\tilde{\phi}_2$	2	$-q^2+1$	1	.	.	.	.	.
$\tilde{\phi}_1$	1	$-q^4+1$	1	1	1	.	.	.
$\tilde{1}$	0	1	1	1	1	1	1	1

$^2F_4$		$\varnothing$	$\tilde{A}_1$	$A_1+\tilde{A}_1$	$(B_2)_2$	
ϕ	d_ϕ	$u_0 = 1$	u_1	u_2	u_3	u_4
$\tilde{\phi}_{1,24}$	24	1	.	.	.	.
$\tilde{\phi}_{4,13}$	13	$-q^{10}+q^6-q^4+1$	1	.	.	.
$\tilde{\phi}_{9,10}$	10	$q^{12}-q^{10}+q^6-q^2+1$	$-q^2+1$	1	.	.
$\tilde{\phi}_{4,8}$	8	q^8+1	1	1	1	1
$\tilde{\phi}'_{6,6}$	6	$q^{12}-q^{10}-q^2+1$	$-q^2+1$	$-q^2+1$	.	.
$\tilde{\phi}_{16,5}$	5	$q^{14}-2q^{10}+q^8+q^6-2q^4+1$	q^6-2q^4+1	$-q^4+1$	$-q^2+1$	$-q^2+1$
$\tilde{\phi}_{12,4}$	4	$q^{16}+1$	1	1	1	1
$\tilde{\phi}''_{6,6}$	4	$q^{14}-q^{12}+2q^8-q^4+q^2$	$q^8-q^4+q^2$	q^2	1	1
$\tilde{\phi}_{9,2}$	2	$q^{12}-q^{10}+q^6-q^2+1$	q^6-q^2+1	$-q^2+1$	q^4-q^2+1	q^4-q^2+1
$\tilde{\phi}_{4,1}$	1	$-q^{10}+q^6-q^4+1$	q^6-q^4+1	$-q^4+1$	$-q^2+1$	$-q^2+1$
$\tilde{\phi}_{1,0}$	0	1	1	1	1	1

cont'd		$\tilde{A}_2+A_1$	$C_3(a_1)$	$F_4(a_3)$			$F_4(a_2)$			$F_4(a_1)$		F_4			
ϕ	d_ϕ	u_5	u_6	u_7	u_8	u_9	u_{10}	u_{11}	u_{12}	u_{13}	u_{14}	u_{15}	u_{16}	u_{17}	u_{18}
$\tilde{\phi}_{1,24}$	24	.	.	.	.	.	.	.	.	.	.	.	.	.	.
$\tilde{\phi}_{4,13}$	13	.	.	.	.	.	.	.	.	.	.	.	.	.	.
$\tilde{\phi}_{9,10}$	10	.	.	.	.	.	.	.	.	.	.	.	.	.	.
$\tilde{\phi}_{4,8}$	8	.	.	.	.	.	.	.	.	.	.	.	.	.	.
$\tilde{\phi}'_{6,6}$	6	1	.	.	.	.	.	.	.	.	.	.	.	.	.
$\tilde{\phi}_{16,5}$	5	1	1	.	.	.	.	.	.	.	.	.	.	.	.
$\tilde{\phi}_{12,4}$	4	1	1	1	1	1	.	.	.	.	.	.	.	.	.
$\tilde{\phi}''_{6,6}$	4	.	1	2	.	-1	.	.	.	.	.	.	.	.	.
$\tilde{\phi}_{9,2}$	2	$-q^2+1$	$-q^2+1$	1	1	1	1	1	1	.	.	.	.	.	.
$\tilde{\phi}_{4,1}$	1	1	$-q^2+1$	$-2q^2+1$	1	q^2+1	1	1	1	1	1	.	.	.	.
$\tilde{\phi}_{1,0}$	0	1	1	1	1	1	1	1	1	1	1	1	1	1	1

the Springer correspondence. Representatives of the unipotent classes are denoted as in [War66, Table III]. The Green functions can be extracted from the complete character table in [War66, p. 87] and the multiplicity formulae in [Lu84a, p. 376]; see also [Ge03a, Example 4.5.12].

For $^2F_4$, we use the σ-extensions specified in Example 2.1.15. There are ten F-stable unipotent classes of $\mathbf{G}$, denoted as in [Spa85, p. 330] where one can also find the Springer correspondence. Representatives of the unipotent classes are denoted as in [Shi75, Table II]. The Green functions have been first determined in [Ma90, §5].

Remark 2.8.19 Let $\mathbf{G}, F$ be as in Example 2.8.18 above. Let $\phi \in \mathrm{Irr}(\mathbf{W})^\sigma$ and consider the corresponding pair $(\mathscr{O}, \varsigma) \in \mathscr{I}_{\mathbf{G}}^F$ as in Theorem 2.8.3. Then we observe that, in all cases, ς is the trivial character or there exists a representative $u_{\mathscr{O}} \in \mathscr{O}^F$ such that F acts trivially on $A(u_{\mathscr{O}})$. Hence, as in Remark 2.8.8, there is a sign $\delta = \pm 1$ such that $Y_{(u_{\mathscr{O}}, \tilde{\varsigma})}(u_{\mathscr{O}}) = \delta\varsigma(1)$. Clearly, the sign δ depends on the choice of the σ-extension of ϕ that is used to define $Q_{\tilde{\phi}}$. Hence, replacing $\tilde{\phi}$ by $-\tilde{\phi}$ if necessary, we can achieve that $\delta = 1$. This defines a particular set of σ-extensions of the characters in $\mathrm{Irr}(\mathbf{W})^\sigma$, and these are exactly those chosen in Examples 2.1.12, 2.1.13, 2.1.15.

3

Harish-Chandra Theories

In this chapter we introduce the concepts and main results of Harish-Chandra theory for the finite groups of Lie type. Harish-Chandra theories provide an inductive approach to the classification of irreducible characters or modules in whole families of groups of Lie type.

The first of these, ordinary Harish-Chandra theory, can be developed solely on the basis of the fact that a group of Lie type G is in a natural way a group with a BN-pair: here B and N are the group of F-fixed points of an F-stable Borel subgroup and of the normaliser of a maximally split maximal torus, respectively, in an underlying algebraic group $\mathbf{G}$ in characteristic p with a Steinberg endomorphism F. Attached to such a BN-pair is its collection of standard Levi subgroups. It induces a subdivision of the set $\mathrm{Irr}_{\mathbb{K}}(G)$ of irreducible characters of G over a field $\mathbb{K}$ of characteristic different from p into (ordinary) Harish-Chandra series. These are indexed by so-called cuspidal characters of standard Levi subgroups (up to G-conjugation), and the Harish-Chandra series above a given cuspidal character is parametrised by the set of irreducible characters of a suitable endomorphism algebra. The latter turns out to be closely related to an Iwahori–Hecke algebra associated to a finite Coxeter group.

This construction was first proposed by Harish-Chandra [HaCh] and then studied in more detail by Springer [Spr70] and Howlett–Lehrer [HoLe80]. The main drawback of this inductive approach to the classification of irreducible characters is the fact that it does not yield any information about the cuspidal characters of G itself, which have to be determined in some other way. We discuss the concepts and main results in this area in Section 3.1. We then specialise this approach to the case of finite groups of Lie type in Section 3.2 where stronger results in particular on the attached Iwahori–Hecke algebras and on character degrees are available.

The second instance of Harish-Chandra theories we present here is the so-called d-Harish-Chandra theories introduced by Fong–Srinivasan [FoSr86] and then in full generality by Broué–Malle–Michel [BMM93] which have become a fundamental tool in the block theory of finite reductive groups. Their definition and the investigation of their properties is in terms of Lusztig's induction functor which is based on ℓ-adic cohomology theory and thus on the Weil conjectures; no elementary construction for the corresponding generalised induction and restriction is known.

Lusztig induction and its basic properties are discussed in Section 3.3. We then present the Alvis–Curtis–Kawanaka–Lusztig duality functor and use it to define the important Steinberg character of a finite reductive group in Section 3.4. Finally, in Section 3.5 we introduce d-tori and d-split Levi subgroups for arbitrary $d \geqslant 1$ and lay out the foundations of d-Harish-Chandra theory.

3.1 Harish-Chandra Theory for BN-Pairs

The purpose of this section is to introduce a setting in which the usual Harish-Chandra theory for finite groups with a BN-pair can be formulated. For a more general approach allowing for different collections of subquotients, so-called Mackey systems, see for example [CE04, §1] and [DD93].

3.1.1 (Finite BN-pairs) We fix a prime p. Let G be a finite group with an algebraic BN-pair in characteristic p satisfying the commutator relations (see [Ca85, §2]) with associated Weyl group W generated by the set of simple reflections S. The cardinality $|S|$ is called the *rank* of G. We write $T := B \cap N$, and for any $w \in W = N/T$ we choose once and for all a representative $\dot{w} \in N$.

For $I \subseteq S$ let $P_I := \langle B, \dot{s} \mid s \in I \rangle$ be the associated *standard parabolic subgroup* of G and $U_I := O_p(P_I)$ the largest normal p-subgroup of P_I. It can be shown that the extension P_I of U_I splits and there exists a natural complement to U_I in P_I, the *standard Levi subgroup* L_I, uniquely determined by the requirement that $T \leqslant L_I$ (see [Ca85, §2.6]). The semidirect product $P_I = U_I \rtimes L_I$ is called the *(standard) Levi decomposition* of P_I. Here, L_I is again a group with an algebraic BN-pair (B_I, N_I) in characteristic p, where $B_I := B \cap L_I$ and $N_I = \langle T, \dot{s} \mid s \in I \rangle$, with Weyl group $W_I = \langle I \rangle$ and set of simple reflections I (see [Ca85, Prop. 2.6.3]).

If $P_I = U_I.L_I$, for $I \subseteq S$, is a standard parabolic subgroup of G, then any conjugate $P = P_I^g$ (with $g \in G$) is called a *parabolic subgroup* of G, and $L = L_I^g$ is called a *Levi complement* of P, or *Levi subgroup* of G.

We will mainly be interested in the following important special case:

Example 3.1.2 Let $\mathbf{G}$ be a connected reductive linear algebraic group in characteristic $p > 0$ and $F : \mathbf{G} \to \mathbf{G}$ a Steinberg endomorphism. We have seen in Theorem 1.3.2 that $\mathbf{G}$ has an algebraic BN-pair $(\mathbf{B}_0, \mathbf{N}_0)$, where $\mathbf{B}_0$ is a Borel subgroup of $\mathbf{G}$ and $\mathbf{N}_0$ is the normaliser in $\mathbf{G}$ of a maximal torus $\mathbf{T}_0$ of $\mathbf{B}_0$.

Now assume that $\mathbf{T}_0$ and $\mathbf{B}_0$ are chosen to be F-stable. Then the finite group of fixed points $G = \mathbf{G}^F$ has an algebraic BN-pair (B, N), where $B := \mathbf{B}_0^F$ and $N := N_{\mathbf{G}}(\mathbf{T}_0)^F$ are the groups of F-fixed points of $\mathbf{B}_0, \mathbf{N}_0$, respectively, see [MaTe11, Thm. 24.10]. Let $\mathbf{W}$ be the Weyl group of $\mathbf{G}$ with respect to $\mathbf{T}_0$ with set of simple reflections $\tilde{S}$. Then $W = \mathbf{W}^F$ is again a Coxeter group, with set of simple reflections S in bijection with the set of F-orbits on $\tilde{S}$ ([MaTe11, Thm. C.5]). Let $\tilde{I} \subseteq \tilde{S}$ be an F-stable subset and denote by $I \subseteq S$ the subset corresponding to the F-orbits in $\tilde{I}$. Then for a parabolic subgroup $\mathbf{P}_{\tilde{I}}$ of $\mathbf{G}$ with F-stable Levi decomposition $\mathbf{P}_{\tilde{I}} = \mathbf{U}_{\tilde{I}}.\mathbf{L}_{\tilde{I}}$ we have $P_I = \mathbf{P}_{\tilde{I}}^F = \mathbf{U}_{\tilde{I}}^F.\mathbf{L}_{\tilde{I}}^F$. The Levi subgroups of the BN-pair G are thus exactly the G-conjugates of F-fixed points $L_I = \mathbf{L}_{\tilde{I}}^F$, hence of F-stable Levi subgroups of F-stable parabolic subgroups of $\mathbf{G}$. We will call such F-stable Levi subgroups of F-stable parabolic subgroups of $\mathbf{G}$ *split* or *1-split* Levi subgroups for short. These can be characterised as follows (this will be the motivation for our definition of d-split Levi subgroups in 3.5.1):

Lemma 3.1.3 *In the situation of Example 3.1.2 the split Levi subgroups of* $\mathbf{G}$ *are up to G-conjugation exactly the centralisers of split subtori of* $\mathbf{T}_0$.

Here, an F-stable torus $\mathbf{S} \leqslant \mathbf{G}$ is called *(1-)split* if $F(t) = t^q$ for all $t \in \mathbf{S}$. In particular, $\mathbf{S}^F \cong \mathbb{F}_q^\times \times \cdots \times \mathbb{F}_q^\times$ ($\dim \mathbf{S}$ factors), and the order polynomial of $\mathbf{S}$ is given by $(\mathbf{q}-1)^{\dim \mathbf{S}}$. Observe that any F-stable torus $\mathbf{T}$ has a uniquely determined maximal F-stable split subtorus $\mathbf{T}_1$. Indeed, if $\mathbf{T}$ has complete root datum $((X, \varnothing, Y, \varnothing), \varphi)$, then the root datum of $\mathbf{T}_1$ is given by $((X', \varnothing, Y', \varnothing), \varphi')$ where X' is the largest quotient of X on which the characteristic polynomial of φ is a power of $\mathbf{q} - 1$, Y' is the kernel of $\varphi - 1$ on Y, and φ' the map on X' induced by φ (see also 3.5.1).

Proof Let $\mathbf{S} \leqslant \mathbf{T}_0$ be a split subtorus. Then its centraliser $\mathbf{L} := C_{\mathbf{G}}(\mathbf{S})$ is the F-stable Levi subgroup generated by $\mathbf{T}_0$ together with the root subgroups $\{\mathbf{U}_\alpha \mid \alpha \in R,\ \alpha|_{\mathbf{S}} = 0\}$, where R is the root system of $\mathbf{G}$ with respect to $\mathbf{T}_0$ (see [DiMi20, Thms. 1.3.3(iii) and 2.3.1(iv)]). In particular the Weyl group $\mathbf{W}_\mathbf{L}$ of $\mathbf{L}$ is generated by the reflections s_α with $\alpha|_{\mathbf{S}} = 1$, that is, the reflections that centralise $Y(\mathbf{S})$. Thus $\mathbf{W}_\mathbf{L}$ is a parabolic subgroup of $\mathbf{W}$ (see e.g. [MaTe11, Cor. A.29]), F-stable as $\mathbf{L}$ is. Then a set of simple roots for $\mathbf{W}_\mathbf{L}$ is $\mathbf{W}^F$-conjugate to a set of simple roots corresponding to an F-stable subset $\tilde{I} \subseteq \tilde{S}$ (see [MaTe11, Thm. C.5]) and hence $\mathbf{W}_\mathbf{L}$ is $\mathbf{W}^F$-conjugate to $\mathbf{W}_{\tilde{I}}$. But then $\mathbf{L}$ is G-conjugate to $\mathbf{L}_{\tilde{I}}$, so contained in the F-stable parabolic subgroup $\mathbf{P}_{\tilde{I}}$ of $\mathbf{G}$ as claimed.

Conversely, if $\mathbf{L}$ is 1-split, then after conjugation in G we may assume that $\mathbf{L}$ is a standard Levi subgroup. In particular it contains $\mathbf{T}_0$ and so $\mathbf{Z}^\circ(\mathbf{L}) \leqslant \mathbf{T}_0$. By construction of the BN-pair for $G = \mathbf{G}^F$ (see e.g. [MaTe11, Thm. C.5]) the Weyl group $W = \mathbf{W}^F$ of G acts faithfully on $Y(\mathbf{T}_1)$ for the maximal split subtorus $\mathbf{T}_1$

of $\mathbf{T}_0$. Reversing the above arguments we then see that $\mathbf{L} = C_{\mathbf{G}}(\mathbf{S})$, where $\mathbf{S}$ is the maximal split subtorus of $\mathbf{Z}^\circ(\mathbf{L})$. □

Example 3.1.4 (BN-pair in $\mathrm{GL}_n(q)$ and $\mathrm{GU}_n(q)$) In $\mathbf{G} = \mathrm{GL}_n$ by Example 1.3.7 the subgroup $\mathbf{B}_0$ of all upper triangular matrices of GL_n together with the group $\mathbf{N}_0$ of monomial matrices form an algebraic BN-pair in $\mathbf{G}$. Both of these are F-stable with respect to the standard Frobenius map F raising matrix entries to their qth power. So, by Example 3.1.2 the subgroup B of upper triangular matrices in $\mathbf{G}^F = \mathrm{GL}_n(q)$ together with the group $N = \mathbf{N}_0^F$ of monomial matrices over $\mathbb{F}_q$ form a finite algebraic BN-pair for $\mathrm{GL}_n(q)$, with Weyl group $W = N/(B \cap N) \cong \mathfrak{S}_n$ the symmetric group of degree n. The standard parabolic subgroups are the subgroups consisting of block upper triangular matrices of a fixed shape, and the corresponding Levi subgroups are the block diagonal matrices in $\mathrm{GL}_n(q)$ of the same fixed shape.

Now consider GL_n with the Frobenius map F' from Example 1.3.19 with fixed point group $G' := \mathbf{G}^{F'} = \mathrm{GU}_n(q)$. Again, both $\mathbf{B}_0$ and $\mathbf{N}_0$ are F'-stable, so $B' = \mathbf{B}_0^{F'}$ and $N' = \mathbf{N}_0^{F'}$ form a finite algebraic BN-pair in $\mathrm{GU}_n(q)$. But here F acts as the non-trivial graph automorphism on the Dynkin diagram of the Weyl group $\mathbf{W}$ of $\mathbf{G}$, and the Weyl group $W' = N'/(B \cap N')$ is a proper subgroup of the Weyl group of $\mathbf{G}$, namely a Weyl group of type B_m with $m = \lfloor n/2 \rfloor$. The F-orbits on $\tilde{S}$ are the subsets $s_i := \{\tilde{s}_i, \tilde{s}_{n-i}\}$, for $1 \leqslant i \leqslant m$. The standard parabolic subgroup of $\mathrm{GU}_n(q)$ corresponding to a subset $S \setminus \{s_{i_1}, \ldots, s_{i_r}\}$ of the orbit set S, with $i_1 < \cdots < i_r$, consists of the block upper triangular matrices in G' with fixed symmetric vector of block lengths $i_1, i_2 - i_1, \ldots, i_r - i_{r-1}, i_r - i_{r-1}, \ldots, i_1$. Its standard Levi subgroup consists of the block diagonal matrices of the same block shape.

We will specialise to this setting of finite reductive groups from Section 3.2 onwards. Let us for the moment return to an arbitrary finite algebraic BN-pair G in characteristic p. Throughout let $\mathbb{K}$ be a field of characteristic different from p.

Definition 3.1.5 Attached to any parabolic subgroup $P \leqslant G$ with Levi decomposition $P = U.L$ there are two natural functors

$$R_{L \leqslant P}^G \colon \mathbb{K}L\text{-mod} \longrightarrow \mathbb{K}G\text{-mod},$$
$${}^*R_{L \leqslant P}^G \colon \mathbb{K}G\text{-mod} \longrightarrow \mathbb{K}L\text{-mod},$$

called *Harish-Chandra induction* and *Harish-Chandra restriction* respectively, defined as follows: via the identification $L \cong P/U$, any $\mathbb{K}L$-module X can be considered as a $\mathbb{K}P$-module with U acting trivially, and then

$$R_{L \leqslant P}^G(X) := \mathrm{Ind}_P^G(\mathrm{Infl}_L^P(X))$$

is the induction to G of this inflated module; conversely, if Y is a $\mathbb{K}G$-module, then L acts on the U-fixed points U^Y and the resulting $\mathbb{K}L$-module is denoted by ${}^*R_{L \leqslant P}^G(Y)$.

Thus, $R^G_{L\leqslant P}(X) = \mathbb{K}G \otimes_{\mathbb{K}P} \mathrm{Infl}^P_L(X)$. It is immediate from this definition that

$$\dim(R^G_{L\leqslant P}(X)) = |G:P| \dim X = |G:L|_{p'} \dim X.$$

There's another interpretation of Harish-Chandra induction and restriction due to Broué which is sometimes more useful. Consider the $(\mathbb{K}G, \mathbb{K}L)$-bimodule $\mathbb{K}[G/U]$, where the actions are given by left and right multiplication respectively. Then the decompositions $gP = \coprod_{l\in L} gUl$ for g in G induce a canonical isomorphism

$$R^G_{L\leqslant P}(X) \xrightarrow{\sim} \mathbb{K}[G/U] \otimes_{\mathbb{K}L} X, \quad g\otimes x \mapsto gU \otimes x,$$

and similarly we have

$$^*R^G_{L\leqslant P}(Y) \xrightarrow{\sim} Y \otimes_{\mathbb{K}G} \mathbb{K}[G/U]$$

(see [DiMi20, Prop. 5.1.8]).

As both functors are thus given by tensor products with projective right modules, the following is immediate (see [Ge01, Lemma 3.4]):

Corollary 3.1.6 *The functors $R^G_{L\leqslant P}$ and $^*R^G_{L\leqslant P}$ are exact, and they preserve direct sums and projectives.*

Example 3.1.7 In the set-up of Example 3.1.2, the Borel subgroup $B = \mathbf{B}^F$ is a parabolic subgroup of G with Levi complement $T = \mathbf{T}^F$. Let $\theta \in \mathrm{Irr}(T)$. Then Proposition 2.2.7 shows that in this case $R^G_{T\leqslant B}(\theta)$ is the same as the Deligne–Lusztig character $R^{\mathbf{G}}_{\mathbf{T},\mathbf{U}}(\theta)$. We choose a slightly different notation here to conform with the existing literature.

Example 3.1.8 Assume that Y is a $\mathbb{K}G$-module with character χ and $P = U.L$ is a parabolic subgroup of G. Then the character of $^*R^G_{L\leqslant P}(Y)$ is given by

$$^*R^G_{L\leqslant P}(\chi)(l) = \frac{1}{|U|}\sum_{u\in U} \chi(lu) \qquad \text{for } l \in L.$$

This follows easily from the definition of $^*R^G_{L\leqslant P}(Y)$ as $C_Y(U)$. In particular this shows that for $\mathbb{K}_G$ the trivial $\mathbb{K}G$-module, $^*R^G_{L\leqslant P}(\mathbb{K}_G) = \mathbb{K}_L$ is the trivial $\mathbb{K}L$-module.

Note that if $J \subseteq I \subseteq S$ then $U_I \leqslant U_J \leqslant P_J \leqslant P_I$ and L_J is a standard Levi subgroup of the parabolic subgroup $V_J.L_J$ of L_I corresponding to the subset $J \subseteq I$, where $V_J = U_J/U_I$. Thus, if $Q = V.M \leqslant P = U.L$ are parabolic subgroups such that $M \leqslant L$ then M is a Levi subgroup of the parabolic subgroup $L\cap Q$ of L with Levi decomposition $L \cap Q = (V\cap L).M$. With this notation the Harish-Chandra functors satisfy natural transitivity properties:

Proposition 3.1.9 (Transitivity) *Let $Q \leqslant P$ be parabolic subgroups of G with Levi complements M, L respectively such that $M \leqslant L$. Then*

$$
\begin{aligned}
R^G_{L\leqslant P} \circ R^L_{M\leqslant L\cap Q}(X) &\cong R^G_{M\leqslant Q}(X) && \text{for all } X \in \mathbb{K}M\text{-mod},\\
{}^*R^L_{M\leqslant L\cap Q} \circ {}^*R^G_{L\leqslant P}(Y) &\cong {}^*R^G_{M\leqslant Q}(Y) && \text{for all } Y \in \mathbb{K}G\text{-mod}.
\end{aligned}
$$

See Dipper–Fleischmann [DF92, Lemma 1.12], or [DiMi20, Prop. 5.1.11] for the case of $\mathbb{K}$ of characteristic 0. Furthermore, the two functors are adjoint to one another:

Proposition 3.1.10 (Adjointness) *Let $P = U.L$ be a parabolic subgroup of G, $X \in \mathbb{K}L$-mod and $Y \in \mathbb{K}G$-mod. Then*

$$
\begin{aligned}
\mathrm{Hom}_{\mathbb{K}G}(R^G_{L\leqslant P}(X), Y) &\cong \mathrm{Hom}_{\mathbb{K}L}(X, {}^*R^G_{L\leqslant P}(Y)),\\
\mathrm{Hom}_{\mathbb{K}G}(Y, R^G_{L\leqslant P}(X)) &\cong \mathrm{Hom}_{\mathbb{K}L}({}^*R^G_{L\leqslant P}(Y), X),
\end{aligned}
$$

as $\mathbb{K}$-vector spaces.

See [DF92, 1.8]. It is a crucial property of Harish-Chandra induction and restriction that like ordinary induction and restriction they are intertwined by a Mackey type formula:

Theorem 3.1.11 (Mackey formula) *Let P, Q be parabolic subgroups of G with Levi complements L, M respectively. Then for all $X \in \mathbb{K}M$-mod we have*

$$
{}^*R^G_{L\leqslant P} \circ R^G_{M\leqslant Q}(X) \cong \bigoplus_{w\in P\backslash G/Q} R^L_{L\cap {}^wM\leqslant L\cap {}^wQ} \circ \mathrm{ad}(w) \circ {}^*R^M_{L^w\cap M\leqslant P^w\cap M}(X),
$$

where w runs over a system of P–Q double coset representatives in G.

See [DF92, Thm. 1.4], or [DiMi20, Thm. 5.2.1] for characters. The set $P\backslash G/Q$ can be expressed in terms of the Weyl groups: Let $I, J \subseteq S$ and P_I, P_J the corresponding standard parabolic subgroups of G. Then there is a natural system D_{IJ} of double coset representatives for $W_I\backslash W/W_J$ consisting of elements of minimal length in their respective double cosets. With this, $\{\dot{w} \mid w \in D_{IJ}\}$ is a system of P_I–P_J double coset representatives in G, see [CuRe87, §64C], or [DiMi20, Prop. 3.2.3(ii)] in the case of finite reductive groups.

Example 3.1.12 Observe that a subgroup L of G may be a Levi subgroup of two non-conjugate parabolic subgroups: Let q be a prime power and $G = \mathrm{GL}_3(q)$, with the BN-pair consisting of the upper triangular invertible matrices B and the monomial matrices N in G (see Example 3.1.4). Then $W = \langle s_1, s_2\rangle \cong \mathfrak{S}_3$. Now it is easily seen that the two standard parabolic subgroups P_i corresponding to $\{s_i\} \subset S$, $i = 1, 2$, are not conjugate in G (in fact, one of them is a point stabiliser while the other is the stabiliser of a hyperplane; modulo G-conjugation they are interchanged by the transpose-inverse automorphism of G), but their standard Levi subgroups L_1, L_2 are G conjugate.

Nevertheless it can be shown that Harish-Chandra induction and restriction are independent of the parabolic subgroup P containing a given Levi subgroup L (see [DD93, 5.2] or [HoLe94, Thm. 1.1], as well as [DiMi20, Thm. 5.3.1] for characters):

Theorem 3.1.13 *Let P, P' be parabolic subgroups of G containing the same Levi subgroup L. Then*

$$\begin{aligned} R^G_{L\leqslant P}(X) &\cong R^G_{L\leqslant P'}(X) && \text{for all } X \in \mathbb{K}L\text{-mod},\\ {}^*R^G_{L\leqslant P}(Y) &\cong {}^*R^G_{L\leqslant P'}(Y) && \text{for all } Y \in \mathbb{K}G\text{-mod}. \end{aligned}$$

Proof for characters (Deligne, see [LuSp79, Thm. 2.4]) We proceed by induction on the rank of G. Note that we may assume that L is strictly smaller than G. Now for χ a character of L the Mackey formula 3.1.11 shows that the scalar product

$$\langle R^G_{L\leqslant P}(\chi), R^G_{L\leqslant P'}(\chi)\rangle = \langle \chi, {}^*R^G_{L\leqslant P}(R^G_{L\leqslant P'}(\chi))\rangle$$

only involves Harish-Chandra induction and restriction in groups of smaller rank than G, so does not depend on the chosen parabolic subgroups. Thus, writing $\psi_P = R^G_{L\leqslant P}(\chi)$ we obtain

$$\langle \psi_P, \psi_P\rangle = \langle \psi_P, \psi_{P'}\rangle = \langle \psi_{P'}, \psi_{P'}\rangle,$$

which implies that $\psi_P - \psi_{P'}$ has norm zero and so $\psi_P = \psi_{P'}$. The claim for Harish-Chandra restriction follows by adjunction. □

Thus, from now on we will write R^G_L and ${}^*R^G_L$ in place of $R^G_{L\leqslant P}$ and ${}^*R^G_{L\leqslant P}$, as these constructions do not depend on the choice of parabolic subgroup P containing L as a Levi subgroup.

We define an order relation $\leqslant_1$ on the set of pairs

$$\mathscr{L}(G) := \{(L, X) \mid L \leqslant G \text{ Levi subgroup, } X \in \mathbb{K}L\text{-mod simple}\}$$

as follows: $(L, X) \leqslant_1 (M, Y)$ if and only if $L \leqslant M$ and X is isomorphic to a composition factor of the socle of ${}^*R^M_L(Y)$. Observe that due to Proposition 3.1.9 this relation is transitive.

Definition 3.1.14 A pair $(L, X) \in \mathscr{L}(G)$ is called a *cuspidal pair* in G if it is minimal with respect to the partial order $\leqslant_1$. Then X is called a *cuspidal* $\mathbb{K}L$-module. Thus, $X \in \mathbb{K}G$-mod is cuspidal if and only if ${}^*R^G_L(X) = 0$ for every proper Levi subgroup $L < G$.

For a cuspidal pair $(L, X) \in \mathscr{L}(G)$ the corresponding *Harish-Chandra series* $\mathrm{Irr}_{\mathbb{K}}(G, (L, X))$ is defined to be the set of all simple $\mathbb{K}G$-modules Y (up to isomorphism) such that

(1) L is minimal such that ${}^*R_L^G(Y) \neq 0$, and
(2) X is a composition factor of ${}^*R_L^G(Y)$ (see [Hi93, p. 224]).

Observe that $\mathrm{Irr}_{\mathbb{K}}(G, (L, X))$ only depends on the G-conjugacy class of (L, X). But note that at this point it is not clear that Harish-Chandra series are disjoint, nor even that they are non-empty.

Example 3.1.15 Assume that G has a BN-pair with trivial Weyl group, that is, with $N = B = G$. Then the only parabolic subgroup of G, as well as the only Levi subgroup of G, is G itself. Thus, every simple $\mathbb{K}G$-module is cuspidal.

More generally, if G has an algebraic BN-pair in characteristic p (see 1.1.14) then $B = O_p(B).T$, where $T = B \cap N$, is the smallest parabolic subgroup, and by the previous observation, all of its simple modules are cuspidal. Thus, any (T, X) with X a simple $\mathbb{K}T$-module, is a cuspidal pair. Moreover, according to Example 3.1.8 the trivial $\mathbb{K}G$-module $\mathbb{K}_G$ lies in the Harish-Chandra series of $(T, \mathbb{K}_T)$. The union of the Harish-Chandra series $\mathrm{Irr}_{\mathbb{K}}(G, (T, X))$ with $X \in \mathbb{K}T$-mod simple is called the *principal series* of G.

Proposition 3.1.16 *Let Y be a simple $\mathbb{K}G$-module and (L, X) be a cuspidal pair for G. Then the following are equivalent:*

(i) $Y \in \mathrm{Irr}_{\mathbb{K}}(G, (L, X))$*;*
(ii) *Y is contained in the socle of $R_L^G(X)$;*
(iii) *Y is contained in the head of $R_L^G(X)$.*

See [Hi93, Thm. 5.8]. The following is an easy consequence:

Corollary 3.1.17 *The Harish-Chandra series partition* $\mathrm{Irr}_{\mathbb{K}}(G)$. *More precisely,* $\mathrm{Irr}_{\mathbb{K}}(G, (L, X))$ *is non-empty for every cuspidal pair (L, X) in G, and*

$$\mathrm{Irr}_{\mathbb{K}}(G, (L, X)) \cap \mathrm{Irr}_{\mathbb{K}}(G, (M, Y)) = \varnothing$$

for all cuspidal pairs (M, Y) of G not G-conjugate to (L, X).

For the further study of Harish-Chandra series we need the following important object. Let (L, X) be a cuspidal pair in G and set $Y := R_L^G(X)$. Then $\mathscr{H}_G(L, X) := \mathrm{End}_{\mathbb{K}G}(Y)^{\mathrm{opp}}$ is called the *Hecke algebra* associated with (L, X). Note that by Theorem 3.1.13 we obtain an isomorphic algebra if we choose another parabolic subgroup containing L for the definition of R_L^G. The corresponding *Hom-functor* is now defined by

$$\mathscr{F}_{\mathbb{K}} : \mathbb{K}G\text{-mod} \longrightarrow \mathscr{H}_G(L, X)\text{-mod}, \qquad Z \mapsto \mathscr{F}_{\mathbb{K}}(Z) := \mathrm{Hom}_{\mathbb{K}G}(Y, Z),$$

where $\mathscr{H}_G(L, X)$ acts on $\mathscr{F}_{\mathbb{K}}(Z)$ via $h.f := f \circ h$ for $h \in \mathscr{H}_G(L, X)$ and $f \in \mathscr{F}_{\mathbb{K}}(Z)$, and a $\mathbb{K}G$-homomorphism $\varphi : Z \to Z'$ is sent to $\mathscr{F}_{\mathbb{K}}(\varphi) = \varphi_*$ with $\varphi_*(f) := \varphi \circ f$ for $f \in \mathscr{F}_{\mathbb{K}}(Z)$.

The theory of endomorphism algebras now yields (see Harish-Chandra [HaCh] for $\mathbb{K}$ of characteristic 0, and [GHM96, Thm. 2.4] for the general case):

Theorem 3.1.18 *For any cuspidal pair (L, X) in G the Hom-functor $\mathscr{F}_{\mathbb{K}}$ induces a bijection*

$$\mathrm{Irr}_{\mathbb{K}}(G, (L, X)) \longrightarrow \mathrm{Irr}_{\mathbb{K}}(\mathscr{H}_G(L, X)), \qquad Z \mapsto \mathrm{Hom}_{\mathbb{K}G}(Y, Z),$$

from the Harish-Chandra series above (L, X) to the set of simple $\mathscr{H}_G(L, X)$-modules up to isomorphism.

This induces a bijection between the set of isomorphism classes of simple $\mathbb{K}G$-modules and the set of equivalence classes of triples (L, X, Ψ) where L is a Levi subgroup of G, X is a simple cuspidal $\mathbb{K}L$-module and Ψ a simple $\mathscr{H}_G(L, X)$-module (where (L, X) has to be taken modulo G-conjugation and Ψ up to isomorphism).

Thus, Harish-Chandra theory provides the following inductive approach to a classification of the simple $\mathbb{K}G$-modules: first determine the cuspidal pairs (L, X) in G, then for each such cuspidal pair (up to conjugation) parametrise the members of the corresponding Harish-Chandra series $\mathrm{Irr}_{\mathbb{K}}(G, (L, X))$. While the first problem cannot be solved purely in the framework of Harish-Chandra theory, there are powerful results about the structure of Harish-Chandra series, which we will describe now.

3.1.19 In order to formulate the relevant result, we need to recall the construction and some basic facts about Iwahori–Hecke algebras for finite Coxeter groups. Let (W, S) be a Coxeter system with a finite Coxeter group W and a distinguished set S of involutive Coxeter generators of W (see e.g. [Bou68, IV,§1] or [GePf00, §1]). For $s, t \in S$ we write m_{st} for the order of st. We denote the corresponding length function on W by l, so $l(s) = 1$ for all $s \in S$, and more generally $l(w)$ is the length of the shortest expression of $w \in W$ as a product of elements from S. Let $\underline{\mathbf{v}} = (\mathbf{v}_s \mid s \in S)$ be a set of indeterminates such that $\mathbf{v}_s = \mathbf{v}_t$ whenever $s, t \in S$ are conjugate in W. We will write $\mathbf{x}_s := \mathbf{v}_s^2$ and $\underline{\mathbf{x}} = (\mathbf{x}_s \mid s \in S)$. The *generic Iwahori–Hecke algebra* $\mathscr{H}(W, \underline{\mathbf{x}})$ of W over the ring $A := \mathbb{Z}[\underline{\mathbf{v}}^{\pm 1}] := \mathbb{Z}[\mathbf{v}_s, \mathbf{v}_s^{-1} \mid s \in S]$ is the associative A-algebra with identity generated by elements T_s, $s \in S$, subject to the relations

$$(T_s - \mathbf{x}_s)(T_s + 1) = 0 \qquad \text{for } s \in S,$$

$$T_s T_t T_s \cdots = T_t T_s T_t \cdots \ (\, m_{st} \text{ terms each}) \qquad \text{for } s, t \in S.$$

For $w \in W$ with an expression $w = s_1 \cdots s_r$ $(s_i \in S)$ of shortest possible length $r = l(w)$ (a *reduced expression* for w) we set $T_w := T_{s_1} \cdots T_{s_r} \in \mathscr{H}(W, \underline{\mathbf{x}})$. It is an important fact that this does not depend on the choice of reduced expression for w (Matsumoto's Lemma, see [GePf00, Thm. 1.2.2]). Then $\mathscr{H}(W, \underline{\mathbf{x}})$ is a free

A-module of rank $|W|$ with basis $\{T_w \mid w \in W\}$, and the multiplication in $\mathscr{H}(W, \underline{\mathbf{x}})$ is described by the rules

$$T_w T_s = \begin{cases} T_{ws} & \text{if } l(ws) = l(w) + 1, \\ \mathbf{x}_s T_{ws} + (\mathbf{x}_s - 1) T_w & \text{if } l(ws) = l(w) - 1, \end{cases}$$

for $w \in W$, $s \in S$ (see [GePf00, Lemma 4.4.3]).

It follows easily from this that the generic Iwahori–Hecke algebra for a product of Coxeter groups is just the product of the Iwahori–Hecke algebras of the factors (see [GePf00, Ex. 8.4]).

3.1.20 For any homomorphism $\varphi : A \longrightarrow R$ from A into a commutative ring R we denote by

$$\mathscr{H}_R(W, \varphi(\underline{\mathbf{x}})) := \mathscr{H}(W, \underline{\mathbf{x}}) \otimes_A R$$

the *specialisation of* $\mathscr{H}(W, \underline{\mathbf{x}})$ *along* φ, and we also write φ for the induced map on $\mathscr{H}(W, \underline{\mathbf{x}})$. Obviously, such a specialisation is uniquely determined by specifying the images $\varphi(\mathbf{v}_s)$, $s \in S$. By a slight abuse of notation we will sometimes also denote the corresponding specialisation by $\mathscr{H}(W, \varphi(\underline{\mathbf{x}}))$. Any such specialisation is called an *Iwahori–Hecke algebra* associated to the Coxeter group W. An important example is the so-called *one-parameter Hecke algebra* $\mathscr{H}(W, \mathbf{x})$ obtained by the specialisation

$$\varphi_{\mathbf{x}} : A \longrightarrow \mathbb{Z}[\sqrt{\mathbf{x}}^{\pm 1}], \qquad \mathbf{v}_s \mapsto \sqrt{\mathbf{x}} \text{ for all } s \in S$$

(so $\varphi_{\mathbf{x}}(\mathbf{x}_s) = \mathbf{x}$), for an indeterminate $\sqrt{\mathbf{x}}$. Another example is given by the specialisation $\varphi_1 : A \longrightarrow \mathbb{Q}$, $\mathbf{v}_s \mapsto 1$ for all $s \in S$. The images $\varphi_1(T_s)$, $s \in S$, then satisfy the relations

$$\begin{aligned} &\varphi_1(T_s)^2 = 1 && \text{for } s \in S, \\ &\varphi_1(T_s)\varphi_1(T_t)\cdots = \varphi_1(T_t)\varphi_1(T_s)\cdots \ (\, m_{st} \text{ terms each}) && \text{for } s, t \in S, \end{aligned}$$

so $\mathscr{H}(W, \varphi_1(\underline{\mathbf{x}}))$ is the rational group algebra $\mathbb{Q}[W]$ of W. Thus, $\mathscr{H}(W, \underline{\mathbf{x}})$ has a semisimple specialisation and so is itself a semisimple algebra over $\mathrm{Frac}(A) = \mathbb{Q}(\underline{\mathbf{v}})$. Let $\mathbb{Q}_W \leqslant \mathbb{C}$ denote the character field of the reflection representation of W. (So $\mathbb{Q}_W = \mathbb{Q}$ if W is a Weyl group.) It can be shown moreover that $\mathscr{H}(W, \underline{\mathbf{x}})$ is split over $\mathbb{Q}_W(\underline{\mathbf{v}})$ ([GePf00, Thm. 9.3.5]). Thus, by a fundamental result of Tits (Tits' deformation theorem, see [CuRe87, Thm. 68.17] or [GePf00, Thm. 7.4.6]), we have that

$$\mathscr{H}_{\mathbb{Q}_W(\underline{\mathbf{v}})}(W, \underline{\mathbf{x}}) = \mathscr{H}(W, \underline{\mathbf{x}}) \otimes_A \mathbb{Q}_W(\underline{\mathbf{v}}) \cong \mathbb{Q}_W(\underline{\mathbf{v}})[W].$$

In particular, the specialisation φ_1 induces a natural one-to-one correspondence

$$\mathrm{Irr}(\mathscr{H}(W, \underline{\mathbf{x}})) \longrightarrow \mathrm{Irr}(W), \qquad \phi \mapsto \phi_1,$$

between the irreducible characters of $\mathscr{H}_{\mathbb{Q}_W(\underline{\mathbf{v}})}(W, \underline{\mathbf{x}})$ and $\mathrm{Irr}(W)$. This can be described as follows: the values of the irreducible characters of $\mathscr{H}_{\mathbb{Q}_W(\underline{\mathbf{v}})}(W, \underline{\mathbf{x}})$ on the basis $\{T_w \mid w \in W\}$ are all contained in $\mathbb{Q}_W[\underline{\mathbf{v}}]$ (see [GePf00, Prop. 7.3.8]) and so can be specialised according to φ_1. The image will then be the character of an irreducible representation of $\mathbb{Q}_W[W]$ which we denote by ϕ_1.

Remark 3.1.21 Benson and Curtis [BeCu72] showed that all irreducible representations of the one-parameter Iwahori–Hecke algebra $\mathscr{H}(W, \mathbf{x})$ of an irreducible Weyl group W can in fact already be realised over $\mathbb{Q}(\mathbf{x})$, except for the two 512-dimensional irreducible representations for W of type E_7 and the four 4096-dimensional irreducible representations for W of type E_8, for which a square root of the parameters is needed (see [Lu81a] and also [GePf00, Thm. 9.3.5, Ex. 9.3.4(a) and 9.2.3] and the further references given there). These so-called *exceptional characters* will also play a special role in the decomposition of unipotent characters, see for example Theorem 4.2.16.

This rationality assertion remains true for the generic Iwahori-Hecke algebra $\mathscr{H}(W, \underline{\mathbf{x}})$, except for the two 2-dimensional irreducible representations for W the Weyl group of type G_2 that are non-rational. Thus, except for these eight irreducible representations, we can work with the set of parameters $\underline{\mathbf{x}}$ and do not need to worry about their square roots $\underline{\mathbf{v}}$.

Example 3.1.22 (Index and sign representation) The irreducible representation of $\mathscr{H}(W, \underline{\mathbf{x}})$ specialising to the trivial character of W is the *index representation*

$$\mathrm{ind} : \mathscr{H}(W, \underline{\mathbf{x}}) \longrightarrow A \quad \text{defined by} \quad \mathrm{ind}(T_s) = \mathbf{x}_s,$$

so $\mathrm{ind}(T_w) = \prod_{i=1}^r \mathbf{x}_{s_i}$ if $w = s_1 \cdots s_r$, with $s_i \in S$, is any reduced expression of $w \in W$. The sign representation of the reflection group W is the specialisation under φ_1 of the *sign representation*

$$\varepsilon : \mathscr{H}(W, \underline{\mathbf{x}}) \longrightarrow A, \qquad \varepsilon(T_w) = (-1)^{l(w)}.$$

More generally, let $\varphi : A \longrightarrow R$ be any specialisation to a field R such that $\mathscr{H}_R(W, \varphi(\underline{\mathbf{x}}))$ is split semisimple. Then application of the procedure explained above yields natural bijections

$$\mathrm{Irr}(\mathscr{H}_R(W, \varphi(\underline{\mathbf{x}}))) \xleftrightarrow{1-1} \mathrm{Irr}(\mathscr{H}(W, \underline{\mathbf{x}})) \xleftrightarrow{1-1} \mathrm{Irr}(W) \qquad (*)$$

by specialisation of character values.

Example 3.1.23 The specialisations of $\mathscr{H}(W, \underline{\mathbf{x}})$ of importance to us here are all of the following form: Let q be a prime power with positive square root $\sqrt{q}$ and define $\varphi : A \longrightarrow \mathbb{R}$ by $\varphi(\mathbf{v}_s) = \sqrt{q}^{a_s}$ for integers a_s $(s \in S)$, so $\varphi(\mathbf{x}_s) = q^{a_s}$. Then from (*) we obtain natural bijections

$$\phi_q \longleftarrow \phi \longmapsto \phi_1.$$

These will become relevant in the description of the decomposition of Harish-Chandra induction (see e.g. Example 3.2.6).

3.1.24 The setting of Iwahori–Hecke algebras is not quite general enough to describe the endomorphism algebras of induced cuspidal representations. Let W_1 be a finite Coxeter group with distinguished set of Coxeter generators S_1 and Ω a finite group with a homomorphism $\Omega \longrightarrow \mathrm{Aut}(W_1, S_1)$, where $\mathrm{Aut}(W_1, S_1)$ denotes the subgroup of automorphisms of W_1 that stabilise S_1. Denote by $W = W_1 \rtimes \Omega$ the corresponding semidirect product. This is called an *extended Coxeter group*. The length function l on W_1 extends to W by decreeing that $l(w) = 0$ for all $w \in \Omega$. Let $\underline{\mathbf{v}} = (\mathbf{v}_s \mid s \in S_1)$ be a set of indeterminates such that $\mathbf{v}_s = \mathbf{v}_t$ whenever $s, t \in S_1$ are conjugate in W. We set $\mathbf{x}_s := \mathbf{v}_s^2$ and $\underline{\mathbf{x}} = (\mathbf{x}_s \mid s \in S_1)$. Let $\mu : W \times W \longrightarrow \mathbb{K}^\times$ be a 2-cocycle that is trivial on $W_1 \times W_1$. Let $\mathcal{O} \subseteq \mathbb{K}$ be the integral closure of the subring generated by the values of μ. The *generic twisted extended Iwahori–Hecke algebra* $\mathcal{H}^\mu(W_1 \rtimes \Omega, \underline{\mathbf{x}})$ of W with respect to μ over the ring $A^\mu := \mathcal{O}[\underline{\mathbf{v}}^{\pm 1}] := \mathcal{O}[\mathbf{v}_s, \mathbf{v}_s^{-1} \mid s \in S_1]$ is the free A-module with basis $\{T_w \mid w \in W\}$ and multiplication given by

$$T_w T_\omega = \mu(w, \omega) T_{w\omega} \quad \text{and} \quad T_\omega T_w = \mu(\omega, w) T_{\omega w}$$

for all $w \in W$ and $\omega \in \Omega$, and

$$T_w T_s = \begin{cases} T_{ws} & \text{if } l(ws) = l(w) + 1, \\ \mathbf{x}_s T_{ws} + (\mathbf{x}_s - 1) T_w & \text{if } l(ws) = l(w) - 1, \end{cases}$$

for all $w \in W_1, s \in S_1$ (see [Ca85, §10.8]). In particular, the subalgebra $\bigoplus_{w \in W_1} \mathcal{O} T_w$ of $\mathcal{H}^\mu(W_1 \rtimes \Omega, \underline{\mathbf{x}})$ is isomorphic to the Iwahori–Hecke algebra $\mathcal{H}_{\mathcal{O}}(W_1, \underline{\mathbf{x}})$ for the Coxeter group W_1 with parameters $\underline{\mathbf{x}}$.

Under the specialisation $\varphi_1 : A \longrightarrow \mathbb{K}$, $v_s \mapsto 1$ for all $s \in S_1$, $\mathcal{H}^\mu(W_1 \rtimes \Omega, \underline{\mathbf{x}})$ maps to the twisted group ring $\mathbb{K}_\mu[W]$. Thus, if $\mathbb{K}_\mu[W]$ is semisimple, which is the case for example if $\mathbb{K}$ has characteristic 0, then so is the generic twisted extended Iwahori–Hecke algebra $\mathcal{H}^\mu(W_1 \rtimes \Omega, \underline{\mathbf{x}})$. Let $\mathbb{K}'$ be a finite extension of $\mathbb{K}(\underline{\mathbf{v}})$ over which $\mathcal{H}^\mu(W_1 \rtimes \Omega, \underline{\mathbf{x}})$ splits. Then again by Tits' deformation theorem (see [CuRe87, Thm. 68.17] or [GePf00, Thm. 7.4.6]) we have that

$$\mathcal{H}^\mu_{\mathbb{K}'}(W_1 \rtimes \Omega, \underline{\mathbf{x}}) = \mathcal{H}^\mu(W_1 \rtimes \Omega, \underline{\mathbf{x}}) \otimes_A \mathbb{K}' \cong \mathbb{K}'_\mu[W],$$

and so the specialisation φ_1 naturally induces a one-to-one correspondence

$$\mathrm{Irr}(\mathcal{H}^\mu(W_1 \rtimes \Omega, \underline{\mathbf{x}})) \longrightarrow \mathrm{Irr}(\mathbb{K}'_\mu[W]), \qquad \phi \mapsto \phi_1.$$

3.1.25 Let (L, X) be a cuspidal pair in G. Then the endomorphism algebra $\mathrm{End}_{\mathbb{K}G}(R_L^G(X))^{\mathrm{opp}}$ is also called the *Hecke algebra* of X in G. It is closely related to an Iwahori–Hecke algebra of a finite Coxeter group. To describe it more precisely, we need to introduce the relative Weyl group of a cuspidal pair. Assume that $I \subseteq S$. Then by [Ca85, Lemma 9.2.1] the normaliser of the parabolic subgroup W_I in W has a semidirect product decomposition

$$N_W(W_I) = W_I \rtimes C_I \qquad \text{with } C_I := \{w \in W \mid w(\Delta_I) = \Delta_I\},$$

with Δ_I the simple roots corresponding to I, and the map $N_G(L_I) \cap N \to N_W(W_I)$, $n \mapsto nT$, induces an isomorphism

$$W_G(L_I) := (N_G(L_I) \cap N)L_I/L_I \cong N_W(W_I)/W_I \cong C_I.$$

The group $W_G(L_I)$ is called the *relative Weyl group* of L_I in G.

Example 3.1.26 The relative Weyl group of L_I in G is in general different from $N_G(L_I)/L_I$; for example assume that $G = \mathrm{GL}_n(2)$ with the BN-pair as in Example 3.1.4, then the Levi complement $L_\varnothing$ of the smallest parabolic subgroup $B = U.T$ of G satisfies $L_\varnothing = T = 1$, so $N_G(L_\varnothing)/L_\varnothing = G$, while $W_G(L_\varnothing) = (G \cap N)T/T = N/T = W \cong \mathfrak{S}_n$.

If $G = \mathbf{G}^F$ as in Example 3.1.2 then the relative Weyl group of $L = \mathbf{L}^F$ is also given by $W_G(L) = N_{\mathbf{G}}(\mathbf{L})^F/\mathbf{L}^F = N_G(\mathbf{L})/L$, see also Definition 3.5.8.

The groups $N_W(W_I)$ are described in detail in [Ho80].

Definition 3.1.27 The *relative Weyl group of a cuspidal pair* (L_I, X) in G is defined as

$$\begin{aligned} W_G(L_I, X) :=& \{g \in (N_G(L_I) \cap N)L_I \mid \mathrm{ad}(g)(X) \cong X \text{ as } \mathbb{K}L_I\text{-module}\}/L_I \\ \cong& \{w \in C_I \mid \mathrm{ad}(\dot{w})(X) \cong X \text{ as } \mathbb{K}L_I\text{-module}\}, \end{aligned}$$

a subgroup of $W_G(L_I)$. If L is an arbitrary Levi subgroup of G, then there is $g \in G$ and $I \subseteq S$ such that $L = L_I^g$, and we define the relative Weyl groups $W_G(L)$ and $W_G(L, X)$ by transport of structure. Note that this is well defined, since it can be shown that if $I, J \subseteq S$ are such that $L_I^g = L_J$ then there is $w \in W$ with $I^w = J$ (see [Ca85, Prop. 9.2.2]).

The endomorphism algebra $\mathscr{H}_G(L, X) = \mathrm{End}_{\mathbb{K}G}(R_L^G(X))^{\mathrm{opp}}$ of a Harish-Chandra induced cuspidal $\mathbb{K}L$-module X can now be described as follows:

Theorem 3.1.28 *Let (L, X) be a cuspidal pair in G. Then there is a natural semidirect product decomposition $W_G(L, X) = W_1 \rtimes \Omega$ with (W_1, S_1) a Coxeter group, and a 2-cocycle $\mu : W_G(L, X) \times W_G(L, X) \to \mathbb{K}^\times$ such that $\mathscr{H}_G(L, X)$*

is isomorphic to the twisted extended Iwahori–Hecke algebra $\mathcal{H}^{\mu}(W_1 \rtimes \Omega, \underline{q})$ *for suitable parameters* $\underline{q} = (q_s)_{s\in S_1} \subseteq \mathbb{K}^{\times}$, *with* $q_s \neq 1$ *for all* $s \in S_1$. *In particular*

$$\dim_{\mathbb{K}} \mathcal{H}_G(L, X) = |W_G(L, X)| = |W_1| \cdot |\Omega|$$

and the subalgebra $\bigoplus_{w\in W_1} \mathbb{K}T_w$ *is isomorphic to the Iwahori–Hecke algebra* $\mathcal{H}_{\mathbb{K}}(W_1, \underline{q})$ *for* W_1 *with parameters* $\underline{q}$.

Moreover $\mathcal{H}_G(L, X)$ *is a symmetric algebra.*

Special cases of this for $\mathbb{K}$ of characteristic 0 were first proved by Iwahori, Harish-Chandra, Lusztig and Kilmoyer. The general case of characteristic 0 was established by Howlett and Lehrer [HoLe80], and building on their work, by Geck, Hiss and Malle [GHM96, §3] in arbitrary characteristic prime to p; the above formulation is due to Ackermann [Ac02]. The fact that $\mathcal{H}_G(L, X)$ is symmetric was shown in [GeHi97].

For a complete description of the Hecke algebra in Theorem 3.1.28 one needs to know its parameters. Remarkably, it turns out that they are determined locally, as we will now explain. Assume, as we may, that $L = L_I$ is a standard Levi subgroup for some subset $I \subseteq S$ and that X is a cuspidal $\mathbb{K}L_I$-module such that $W_G(L_I, X) \neq 1$. The construction of $W_G(L_I, X)$ yields a distinguished Coxeter system (W_1, S_1) for the Coxeter group W_1 such that for every standard generator $s \in S_1$ there exists a minimal subset $J \subseteq S$ containing I with $W_{L_J}(L_I, X) = \langle s\rangle$, the Weyl group of type A_1. Then the parameter q_s of the Hecke algebra $\mathcal{H}_G(L_I, X)$ at s is determined already by the situation when $G = L_J$ (see [GHM96, (3.14)]). Here we have by [GHM96, Lemmas 3.15–3.17]:

Proposition 3.1.29 (Parameters of the Hecke algebra) *Let* (L, X) *be a cuspidal pair in* G *such that* $W_G(L, X)$ *is the Coxeter group of type* A_1, *with Hecke algebra* $\mathcal{H}_G(L, X) = \mathcal{H}(W(A_1), q)$. *Then:*

(a) $R_L^G(X)$ *is indecomposable if and only if* $q = -1$. *In this case the characteristic of* $\mathbb{K}$ *divides* $\dim_{\mathbb{K}}(R_L^G(X))$.
(b) *Else,* $R_L^G(X)$ *is the direct sum of two non-isomorphic simple* $\mathbb{K}G$*-modules* Y_1, Y_2 *and (up to possibly interchanging* Y_1, Y_2*) we have*

$$\dim_{\mathbb{K}}(Y_2) = q \dim_{\mathbb{K}}(Y_1).$$

In particular, if $\mathbb{K}$ has characteristic 0, or more generally, if the characteristic of $\mathbb{K}$ does not divide $\dim_{\mathbb{K}}(R_L^G(X)) = |G : L|_{p'} \dim(X)$, then the parameter q of $\mathcal{H}_G(L, X)$ is given by $q = \dim_{\mathbb{K}}(Y_2)/\dim_{\mathbb{K}}(Y_1)$.

Example 3.1.30 Let G be a finite algebraic BN-pair of rank 1. The BN-pair axioms show that G then acts doubly transitively on the cosets of B, with N the setwise stabiliser of a pair of points, and $T = B \cap N$ the 2-point stabiliser. Let $X = \mathbb{K}_T$ be the trivial $\mathbb{K}T$-module. Then by definition $Y := R_T^G(X)$ is the permutation module on the cosets of B. It is well known that for $\mathbb{K}$ of characteristic ℓ dividing $|G : B|$, this permutation module Y has (at least) two trivial constituents, one in the head and one in the socle. So then Y is not semisimple and we are in case (a) of Proposition 3.1.29. On the other hand, if $\dim_{\mathbb{K}}(Y) = |G : B|$ is prime to ℓ, then according to Proposition 3.1.29, the $\mathbb{K}G$-module Y has two simple summands. (In characteristic 0 this is true for any 2-transitive group, but need not hold in positive characteristic.) In particular, the non-trivial constituent of the permutation character remains irreducible modulo ℓ. In this case by Proposition 3.1.29(b) and the subsequent remark, the parameter of the Hecke algebra $\mathscr{H}_G(T, X)$ is given by

$$(\dim_{\mathbb{K}}(R_T^G(X)) - 1)/1 = |G : B| - 1 = |U|,$$

where $U = O_p(B)$. Indeed, as G is doubly transitive with point stabiliser B and 2-point stabiliser T, we have $|G : B| = |B : T| + 1 = |U| + 1$. For example, if $G = \mathrm{SL}_2(q)$ then $|G : B| = q + 1$, whence the parameter equals q in this case.

Remark 3.1.31 The extended Iwahori–Hecke algebras in 3.1.24 do not only occur as endomorphism algebras of induced cuspidal modules in finite reductive groups with non-connected centre as we will see in the next section, but also for disconnected groups of Lie type, see e.g. Digne–Michel [DiMi85] or Malle [Ma91, §1].

3.2 Harish-Chandra Theory for Groups of Lie Type

3.2.1 The most complete results on Harish-Chandra series are available in the case of representations in characteristic 0, and when G is a finite group of Lie type. We will therefore assume for the rest of this chapter that we are in the setting of Example 3.1.2, that is, $\mathbb{K}$ is a field of characteristic 0 and G is obtained as the group of fixed points $G = \mathbf{G}^F$ of a connected reductive algebraic group $\mathbf{G}$ over $\overline{\mathbb{F}}_p$ under a Steinberg map $F : \mathbf{G} \to \mathbf{G}$, with its natural algebraic BN-pair in characteristic p coming from the one of $\mathbf{G}$. Recall that this means that the Weyl group of $\mathbf{G}^F$ is $W = \mathbf{W}^F$, the group of F-fixed points of the Weyl group of a maximally split torus of $\mathbf{G}$, with Coxeter generators S in bijection with the F-orbits on the set of Coxeter generators $\tilde{S}$ of $\mathbf{W}$ (see e.g. [MaTe11, §23]). We write q for the absolute value of the eigenvalues of F on the character group of an F-stable maximal torus of $\mathbf{G}$.

Moreover let us assume for simplicity that the field $\mathbb{K}$ of characteristic 0 is a splitting field for G and all of its subgroups. Then $\mathbb{K}G$ is split semisimple, and we can work with $\mathbb{K}$-characters of G instead of $\mathbb{K}G$-modules. In analogy to Deligne–Lusztig induction and in view of Example 3.1.7 we will employ the notation $R_{\mathbf{L}}^{\mathbf{G}}$

and ${}^*R_{\mathbf{L}}^{\mathbf{G}}$ in place of R_L^G and ${}^*R_L^G$ if $L = \mathbf{L}^F$ is the group of F-fixed points of a split Levi subgroup $\mathbf{L} \leqslant \mathbf{G}$, and also use these to denote the induced linear maps on the corresponding character rings. We will also write $\mathscr{E}(G, (L, X))$ in place of $\mathrm{Irr}_{\mathbb{K}}(G, (L, X))$ for the Harish-Chandra series of a cuspidal pair (L, X) in G, as this is closer to Lusztig's notation for characters of finite reductive groups.

A first connection to the Deligne–Lusztig theory presented in Section 2.2 is given by the fact that in the case of groups of Lie type, there is a criterion for cuspidality that can be phrased purely in terms of uniform functions, that is, of Deligne–Lusztig characters, see [Lu77b, Prop. 2.18]:

Proposition 3.2.2 (Uniform criterion for cuspidality) *Let $G = \mathbf{G}^F$ be a finite group of Lie type and $\rho \in \mathrm{Irr}(G)$. Then ρ is cuspidal if and only if ${}^*R_{\mathbf{T}}^{\mathbf{G}}(\rho) = 0$ for all F-stable maximal tori $\mathbf{T}$ contained in some proper (1-)split Levi subgroup of $\mathbf{G}$.*

Proof Assume that ${}^*R_{\mathbf{T}}^{\mathbf{G}}(\rho) \neq 0$ for some F-stable maximal torus $\mathbf{T}$ of a proper 1-split Levi subgroup $\mathbf{L} < \mathbf{G}$. Then we have ${}^*R_{\mathbf{T}}^{\mathbf{G}}(\rho) = {}^*R_{\mathbf{T}}^{\mathbf{L}}({}^*R_{\mathbf{L}}^{\mathbf{G}}(\rho))$ by the transitivity statement in Proposition 2.2.7. This implies ${}^*R_{\mathbf{L}}^{\mathbf{G}}(\rho) \neq 0$, whence ρ is not cuspidal. Conversely, if ρ is not cuspidal, then there is some proper 1-split Levi subgroup $\mathbf{L} < \mathbf{G}$ such that ${}^*R_{\mathbf{L}}^{\mathbf{G}}(\rho) \neq 0$, and thus as ${}^*R_{\mathbf{L}}^{\mathbf{G}}(\rho)$ is a character of L, ${}^*R_{\mathbf{L}}^{\mathbf{G}}(\rho)(1) \neq 0$. Since the characteristic function of the identity element is uniform by Proposition 2.3.23, there must be some F-stable maximal torus $\mathbf{T} \leqslant \mathbf{L}$ and some $\theta \in \mathrm{Irr}(\mathbf{T}^F)$ with

$$\langle \theta, {}^*R_{\mathbf{T}}^{\mathbf{G}}(\rho) \rangle = \langle \theta, {}^*R_{\mathbf{T}}^{\mathbf{L}}({}^*R_{\mathbf{L}}^{\mathbf{G}}(\rho)) \rangle = \langle R_{\mathbf{T}}^{\mathbf{L}}(\theta), {}^*R_{\mathbf{L}}^{\mathbf{G}}(\rho) \rangle \neq 0,$$

and so ${}^*R_{\mathbf{T}}^{\mathbf{G}}(\rho) \neq 0$. □

According to Lemma 3.1.3 the 1-split Levi subgroups are just the centralisers in $\mathbf{G}$ of split tori. So another way to phrase the condition on $\mathbf{T}$ in the preceding statement is as follows: an F-stable maximal torus $\mathbf{T}$ does not lie in a proper 1-split Levi subgroup of $\mathbf{G}$ if and only if its maximal split subtorus is central in $\mathbf{G}$.

3.2.3 The uniform criterion provides strong information on the degree polynomials of cuspidal characters. For this, we need to define the *relative rank* of a Levi subgroup of G: If $L = L_I^g$ for some $g \in G$ and some standard Levi subgroup of G associated to a subset $I \subseteq S$, then the *relative rank* of L is defined as $r(L) := |I|$. This is closely related to the notion of relative F-rank from Definition 2.2.11. Indeed, by definition the relative F-rank $r_F(\mathbf{G})$ of $\mathbf{G}$ is the dimension of the q-eigenspace of $\varphi_{\mathbb{R}}$ on $X_{\mathbb{R}} = X \otimes_{\mathbb{Z}} \mathbb{R}$, where X is the character group of a maximally split torus of $\mathbf{G}$, hence $r_F(\mathbf{G})$ is equal to $\dim(X_{\mathbb{R}}^{\sigma})$ where $\sigma := \frac{1}{q}\varphi_{\mathbb{R}}$ is the automorphism of finite order associated to F. Now assume that $\mathbf{G}$ is semisimple. Then according to [MaTe11, Thm. C.5] we have that $\dim(X_{\mathbb{R}}^{\sigma}) = |S|$ equals the number of simple reflections for the Coxeter group $W = \mathbf{W}^F = \mathbf{W}^{\sigma}$. Thus, if $\mathbf{L} = \mathbf{L}_{\tilde{I}} \leqslant \mathbf{G}$ is a 1-split Levi subgroup, corresponding to the F-stable set $\tilde{I} \subseteq \tilde{S}$ of simple reflections of $\tilde{S}$, then $\mathbf{L}^F$ is the standard Levi subgroup L_I of G corresponding to the subset $I \subseteq S$ in bijection with the set of σ-orbits $\tilde{I}$, and by the above we similarly have $r_F([\mathbf{L},\mathbf{L}]) = |\tilde{I}^\sigma| = |I|$. Thus $r(\mathbf{L}^F) = r_F([\mathbf{L},\mathbf{L}])$ and so in particular $(-1)^{r(L)} = \varepsilon_{[\mathbf{L},\mathbf{L}]}$.

Corollary 3.2.4 *Let $\lambda \in \mathrm{Irr}(G)$ be cuspidal. Then the degree polynomial of λ has the form*

$$\mathbb{D}_\lambda = (\mathbf{q}-1)^{r(G)} f(\mathbf{q})$$

where $f \in \mathbb{Q}[\mathbf{q}]$ is not divisible by $\mathbf{q}-1$.

Proof By Definition 2.3.25 the degree polynomial of λ is given by

$$\mathbb{D}_\lambda = \frac{1}{|\mathbf{W}|} \sum_{(w,\theta)\in\mathfrak{X}(\mathbf{W},\sigma)} (-1)^{l(w)} \langle R_w^\theta, \lambda\rangle\, \mathbf{q}^{-N} \frac{|\mathbb{G}|}{|\mathbb{T}_w|}.$$

As λ is cuspidal, Proposition 3.2.2 shows that we only need to sum over $w \in \mathbf{W}$ such that $\mathbf{T}_w$ is not contained in any proper 1-split Levi subgroup of $\mathbf{G}$. Since the centraliser of a split subtorus of $\mathbf{T}_w$ is a 1-split Levi subgroup by definition and contains $\mathbf{T}_w$, the maximal split subtorus of $\mathbf{T}_w$ must be central. In particular, the power of $\mathbf{q}-1$ dividing $|\mathbb{T}_w|$ is the same as for the order polynomial of $\mathbf{Z}^\circ(\mathbf{G})$. As $\mathbf{G} = [\mathbf{G},\mathbf{G}]\mathbf{Z}^\circ(\mathbf{G})$ this shows that the polynomial $|\mathbb{G}|/|\mathbb{T}_w|$ must be divisible by $(\mathbf{q}-1)^{r(G)}$. So $\mathbb{D}_\lambda$ is divisible at least by this power of $\mathbf{q}-1$. On the other hand, the degree polynomial of λ divides the order polynomial of $[\mathbf{G},\mathbf{G}]$, and so it cannot be divisible by a higher power of $\mathbf{q}-1$. □

It will become apparent in Corollary 3.2.21 that the converse of this statement also holds.

In characteristic 0, a very explicit description of the endomorphism algebra of a Harish-Chandra induced cuspidal module is available, considerably sharpening Theorem 3.1.28:

Theorem 3.2.5 *Assume that $G = \mathbf{G}^F$ is a finite group of Lie type and (L, X) is a cuspidal pair of G. Then*

$$\mathscr{H}_G(L,X) := \mathrm{End}_{\mathbb{K}G}(R_{\mathbf{L}}^{\mathbf{G}}(X))^{\mathrm{opp}} \cong \mathbb{K}[W_G(L,X)],$$

the group ring of $W_G(L,X)$ over $\mathbb{K}$. In particular, there is a bijection

$$I_{L,X}^G : \mathrm{Irr}_{\mathbb{K}}(W_G(L,X)) \xleftrightarrow{1-1} \mathscr{E}(G,(L,X)).$$

If $\mathbf{G}$ has connected centre, or if X is unipotent, then $W_G(L,X)$ is a Coxeter group and $\mathscr{H}_G(L,X)$ is the associated Iwahori–Hecke algebra for suitable parameters $\underline{q}$ all of which are positive integral powers of q.

Lusztig showed that $\mathscr{H}_G(L, X)$ is an Iwahori–Hecke algebra if X is unipotent [Lu76c, Cor. 5.12] or if $\mathbf{G}$ has connected centre [Lu84a, Thm. 8.6], with parameters of the stated form. Geck [Ge93b, Cor. 2] settled the general case by showing that in the situation of the theorem, $R_{\mathbf{L}}^{\mathbf{G}}(X)$ always has a constituent with multiplicity 1, so $\mathscr{H}_G(L, X)$ has a 1-dimensional representation, and thus the 2-cocycle from Theorem 3.1.28 is trivial. The statement about the parameters can be found in [Lu84a, §8], see also [Ca85, §10.5].

Thus, from Theorem 3.1.18 we obtain a parametrisation of the irreducible $\mathbb{K}$-characters of G by triples (L, λ, ϕ), with L a Levi subgroup of G, $\lambda \in \operatorname{Irr}(L)$ cuspidal, and $\phi \in \operatorname{Irr}(W_G(L, \lambda))$, where (L, λ) is taken modulo G-conjugation. This parametrisation is effective as the cuspidal irreducible characters of all finite reductive groups are known: they can be obtained by applying the uniform criterion from Proposition 3.2.2 to Lusztig's decomposition of $R_{\mathbf{T}}^{\mathbf{G}}$ ([Lu84a, Thm. 4.23]; see Theorem 4.2.16).

Example 3.2.6 (The unipotent principal series) Let $G = \mathbf{G}^F$ with natural BN-pair $B = \mathbf{B}^F$ and $N = N_{\mathbf{G}}(\mathbf{T})^F$, where $\mathbf{T} \leqslant \mathbf{B}$ is an F-stable maximal torus in an F-stable Borel subgroup of $\mathbf{G}$, and set $T := \mathbf{T}^F = \mathbf{B}^F \cap N$. According to Example 3.1.15 all irreducible $\mathbb{K}T$-modules are cuspidal, hence in particular so is the trivial module with character $1_{\mathbf{T}}$. By Example 3.1.7, $R_{\mathbf{T}}^{\mathbf{G}}(1_{\mathbf{T}})$ is a Deligne–Lusztig character, hence by Definition 2.3.8 all constituents of $R_{\mathbf{T}}^{\mathbf{G}}(1_{\mathbf{T}})$ are unipotent. The elements of the Harish-Chandra series $\mathscr{E}(G, (T, 1))$ are called the *unipotent principal series characters* of G. In this way Harish-Chandra theory provides an approach to at least a part of the unipotent characters of finite reductive groups.

We now describe the parameters of the corresponding endomorphism algebra, which by Theorem 3.2.5 is an Iwahori–Hecke algebra. These are determined locally and can be read off from the result of Example 3.1.30: The relative Weyl group is $W_G(T, 1_{\mathbf{T}}) = W_G(T) = \mathbf{W}^\sigma = W$ with Coxeter system (W, S), and the parameter of $\mathscr{H}(T, 1_{\mathbf{T}})$ at $s \in S$ is given by $|U_s|$, where $U_s = O_p(B_s)$ for a Borel subgroup B_s of the standard Levi subgroup $L_{\{s\}}$ of rank 1 corresponding to the subset $\{s\} \subseteq S$. These orders can be computed from the underlying root system of G as explained for example in [MaTe11, §23.2]. For example, for G a group of split type obtained from a standard Frobenius map with respect to an $\mathbb{F}_q$-structure, all parameters are equal to the size q of the underlying field of G. More generally, the parameters are equal to $q^{l(w_s)}$, where w_s is the longest element of the parabolic subgroup $\mathbf{W}_I$ of $\mathbf{W}$, where I is the F-orbit on the set of Coxeter generators of $\mathbf{W}$ corresponding to s.

The choice of a square root of q now establishes a canonical bijection between the unipotent principal series $\mathscr{E}(G, (T, 1))$ and $\operatorname{Irr}(\mathbf{W}^\sigma)$ by combining the bijection from Theorem 3.2.5 induced by the Hom-functor (see Theorem 3.1.18) with the canonical bijections in 3.1.23.

The values for the various twisted types are collected in Table 3.1 (see [MaTe11, Table 23.2]). Observe that we may easily reduce to the case when **G** is simple: all characters in the unipotent principal series are trivial on the centre, so we may pass to a group of adjoint type, and this has a direct decomposition as explained in Corollary 1.5.16. On the other hand, the corresponding Iwahori–Hecke algebra is the product of the Iwahori–Hecke algebras corresponding to the simple factors (see 3.1.19).

Table 3.1 *Parameters for the principal series in twisted types*

G	$W_G(T)$	parameters	G	$W_G(T)$	parameters
${}^2A_{2n-1}(q)$	B_n	$q; q^2, \ldots, q^2$	${}^2E_6(q)$	F_4	$q^2, q^2; q, q$
${}^2A_{2n}(q)$	B_n	$q^3; q^2, \ldots, q^2$	${}^2B_2(q^2)$	A_1	q^4
${}^2D_n(q)$	B_{n-1}	$q^2; q, \ldots, q$	${}^2G_2(q^2)$	A_1	q^6
${}^3D_4(q)$	G_2	$q^3; q, \ldots, q$	${}^2F_4(q^2)$	$I_2(8)$	$q^4; q^2$

The Ree groups ${}^2F_4(q^2)$ are the only instance in which a relative Weyl group for the principal series is a real reflection group which is not a Weyl group, namely the dihedral group $I_2(8)$ of order 16. The parameters are given in the order induced by the labelling on the Dynkin diagrams in Table 1.1.

Theorem 3.2.5 can be refined to relate the multiplicities in Harish-Chandra induced characters to the decomposition of induced characters in the corresponding relative Weyl groups:

Theorem 3.2.7 (Howlett–Lehrer Comparison Theorem) *Let (L, λ) be a cuspidal pair in G. Then the collection of bijections*

$$I_{L,\lambda}^M : \mathrm{Irr}_{\mathbb{K}}(W_M(L,\lambda)) \longrightarrow \mathscr{E}(M,(L,\lambda))$$

from Theorem 3.2.5, where M runs over split Levi subgroups $L \leqslant M \leqslant G$, can be chosen such that the diagrams

$$\begin{array}{ccc} \mathbb{Z}\mathrm{Irr}_{\mathbb{K}}(W_G(L,\lambda)) & \xrightarrow{I_{L,\lambda}^G} & \mathbb{Z}\mathscr{E}(G,(L,\lambda)) \\ \uparrow{\scriptstyle \mathrm{Ind}} & & \uparrow{\scriptstyle R_{\mathbf{M}}^{\mathbf{G}}} \\ \mathbb{Z}\mathrm{Irr}_{\mathbb{K}}(W_M(L,\lambda)) & \xrightarrow{I_{L,\lambda}^M} & \mathbb{Z}\mathscr{E}(M,(L,\lambda)) \end{array}$$

commute for all M, where Ind *denotes ordinary induction.*

See [HoLe83, Thm. 5.9], and [GHM96, Prop. 2.7] for a version in arbitrary non-defining characteristic. Thus the decomposition of Harish-Chandra induction (and restriction) can be computed purely inside relative Weyl groups. We will see a far-reaching generalisation, for unipotent characters, in Theorem 4.6.21.

3.2.8 We next relate Harish-Chandra theory to regular embeddings. Let $\mathbf{G} \hookrightarrow \tilde{\mathbf{G}}$ be a regular embedding of $\mathbf{G}$ into a group $\tilde{\mathbf{G}}$ with connected centre and same derived subgroup as $\mathbf{G}$ and denote by F an extension to $\tilde{\mathbf{G}}$ of the Steinberg map on $\mathbf{G}$ (see Definition 1.7.1). Then $\tilde{\mathbf{G}} = \mathbf{G}\mathbf{Z}(\tilde{\mathbf{G}})$ and $\tilde{\mathbf{G}}^F/\mathbf{G}^F$ is abelian. Furthermore, $\tilde{\mathbf{T}} = \mathbf{T}\mathbf{Z}(\tilde{\mathbf{G}})$ is an F-stable maximal torus of $\tilde{\mathbf{G}}$ in the F-stable Borel subgroup $\tilde{\mathbf{B}} = \mathbf{B}\mathbf{Z}(\tilde{\mathbf{G}})$, and the natural map $N_{\mathbf{G}}(\mathbf{T})^F \longrightarrow N_{\tilde{\mathbf{G}}}(\tilde{\mathbf{T}})^F$ induces an isomorphism $\mathbf{W}^F \longrightarrow \tilde{\mathbf{W}}^F$ between the Weyl groups of $G = \mathbf{G}^F$ and $\tilde{G} = \tilde{\mathbf{G}}^F$ sending simple reflections to simple reflections. Now if $\mathbf{P}$ is an F-stable parabolic subgroup of $\mathbf{G}$, then $\tilde{\mathbf{P}} := \mathbf{P}\mathbf{Z}(\tilde{\mathbf{G}})$ is an F-stable parabolic subgroup of $\tilde{\mathbf{G}}$, with the same unipotent radical as $\mathbf{P}$, and for $\mathbf{L}$ an F-stable Levi subgroup of $\mathbf{P}$, $\tilde{\mathbf{L}} := \mathbf{L}\mathbf{Z}(\tilde{\mathbf{G}})$ is an F-stable Levi subgroup of $\tilde{\mathbf{P}}$. This defines a natural bijection between the sets of split Levi subgroups of $\mathbf{G}$ and of $\tilde{\mathbf{G}}$.

Proposition 3.2.9 *Let $\mathbf{G} \hookrightarrow \tilde{\mathbf{G}}$ be a regular embedding. Let $\mathbf{L}$ be a split Levi subgroup of $\mathbf{G}$ and $\tilde{\mathbf{L}}$ the corresponding split Levi subgroup of $\tilde{\mathbf{G}}$. Then we have*

$$\begin{aligned}
\mathrm{Ind}_G^{\tilde{G}} \circ R_{\mathbf{L}}^{\mathbf{G}} &= R_{\tilde{\mathbf{L}}}^{\tilde{\mathbf{G}}} \circ \mathrm{Ind}_L^{\tilde{L}}, &\qquad \mathrm{Ind}_L^{\tilde{L}} \circ {}^*R_{\mathbf{L}}^{\mathbf{G}} &= {}^*R_{\tilde{\mathbf{L}}}^{\tilde{\mathbf{G}}} \circ \mathrm{Ind}_G^{\tilde{G}},\\
\mathrm{Res}_L^{\tilde{L}} \circ {}^*R_{\tilde{\mathbf{L}}}^{\tilde{\mathbf{G}}} &= {}^*R_{\mathbf{L}}^{\mathbf{G}} \circ \mathrm{Res}_G^{\tilde{G}}, &\qquad \mathrm{Res}_G^{\tilde{G}} \circ R_{\tilde{\mathbf{L}}}^{\tilde{\mathbf{G}}} &= R_{\mathbf{L}}^{\mathbf{G}} \circ \mathrm{Res}_L^{\tilde{L}}.
\end{aligned}$$

Proof The first formula is immediate from the definition of $R_{\mathbf{L}}^{\mathbf{G}}$ as induction and inflation over the same kernel commute, and the second follows as taking fixed points with respect to the same normal subgroup commutes with induction. The third and fourth are obtained by adjunction. □

The following statement is now immediate from the definitions and Proposition 3.2.9 (see [Bo06, Prop. 12.1]):

Corollary 3.2.10 *In the situation of Proposition 3.2.9, let $\rho \in \mathrm{Irr}(G)$ and suppose that $\tilde{\rho} \in \mathrm{Irr}(\tilde{G})$ lies above ρ. Then ρ is cuspidal if and only if $\tilde{\rho}$ is.*

3.2.11 Let (L, X) be a cuspidal pair in G. Since the endomorphism algebra $\mathrm{End}_{\mathbb{K}G}(R_{\mathbf{L}}^{\mathbf{G}}(X))$ is semisimple, the degrees of the irreducible constituents of $R_{\mathbf{L}}^{\mathbf{G}}(X)$ can be computed from the Schur elements of the natural symmetrising form on the symmetric algebra $\mathrm{End}_{\mathbb{K}G}(R_{\mathbf{L}}^{\mathbf{G}}(X))^{\mathrm{opp}} = \mathcal{H}_G(L, X)$.

Now assume that we are in the setting of 3.1.24; that is, W_1 is a finite Coxeter group with distinguished set of Coxeter generators S_1, Ω a finite group with a homomorphism $\Omega \longrightarrow \mathrm{Aut}(W_1, S_1)$, and $W = W_1 \rtimes \Omega$ is the corresponding semidirect product. Let $\mathcal{H}(W_1 \rtimes \Omega, \underline{\mathbf{x}})$ be the associated generic extended Iwahori–Hecke algebra over $A := \mathbb{Z}[\underline{\mathbf{v}}^{\pm 1}]$ with respect to the trivial 2-cocycle, as introduced in 3.1.24.

Then $\mathscr{H}(W_1 \rtimes \Omega, \underline{\mathbf{x}})$ is symmetric; more precisely, it carries a natural non-degenerate symmetrising form $\tau : \mathscr{H}(W_1 \rtimes \Omega, \underline{\mathbf{x}}) \to A$ defined by

$$\tau(T_w) = \begin{cases} 1 & \text{if } w = 1, \\ 0 & \text{else,} \end{cases} \tag{3.1}$$

on the A-basis $\{T_w \mid w \in W\}$ of $\mathscr{H}(W_1 \rtimes \Omega, \underline{\mathbf{x}})$ from 3.1.24 (see [Ca85, §10.9], or also [GePf00, Prop. 8.1.1] for the case $\Omega = 1$). Any such trace form can be written uniquely as a linear combination of irreducible characters of $\mathscr{H}(W_1 \rtimes \Omega, \underline{\mathbf{x}})$ with non-zero coefficients (see [GePf00, Thm. 7.2.6]), say

$$\tau = \sum_{\phi \in \mathrm{Irr}(\mathscr{H}(W_1 \rtimes \Omega, \underline{\mathbf{x}}))} \frac{1}{c_\phi} \phi,$$

where $\mathrm{Irr}(\mathscr{H}(W_1 \rtimes \Omega, \underline{\mathbf{x}})$ denotes the set of characters of irreducible representations of $\mathscr{H}(W_1 \rtimes \Omega, \underline{\mathbf{x}})$ over a splitting field $\mathbb{K}'$ as introduced in 3.1.24. The non-zero elements $c_\phi \in A_W$, where A_W is the integral closure of A in $\mathbb{K}'$, are called the *Schur elements* of $\mathscr{H}(W_1 \rtimes \Omega, \underline{\mathbf{x}})$ (with respect to τ).

In the case $\Omega = 1$, when $W = W_1$ is a Coxeter group, we saw in 3.1.20 that we may choose $\mathbb{K}' = \mathbb{Q}_W(\underline{\mathbf{v}})$ as a splitting field, in which case we have $A_W = \mathbb{Z}_W[\underline{\mathbf{v}}^{\pm 1}]$, with $\mathbb{Z}_W$ the ring of integers in $\mathbb{Q}_W$.

Example 3.2.12 Under the specialisation φ_1 to the group algebra of W, the trace form on $\mathscr{H}(W, \underline{\mathbf{x}})$ becomes the regular character of W divided by $|W|$, so the Schur elements of the specialisation $\mathbb{Z}[W] = \mathscr{H}(W, \varphi_1(\underline{\mathbf{x}}))$ are given by $\varphi_1(c_\phi) = |W|/\phi(1)$.

Definition 3.2.13 The 1-parameter specialisation $\varphi_{\mathbf{q}}(c_{\mathrm{ind}})$ of the Schur element c_{ind} for the index representation from Example 3.1.22 is called the *Poincaré polynomial* of W. By the previous example it specialises to $|W|$ under φ_1, so can be thought of as a 'quantisation' of the order of W. In view of Example 3.2.12 the quotient $D_\phi := c_{\mathrm{ind}}/c_\phi$ is called the *generic degree* of $\phi \in \mathrm{Irr}(W)$. We thus have $\varphi_1(D_\phi) = \phi(1)$, so D_ϕ is a 'quantisation' of the character degree $\phi(1)$. Note that by definition the trivial character of W always has generic degree equal to 1.

Clifford theory provides a simple way to relate the Schur elements of an extended Iwahori–Hecke algebra $\mathscr{H}(W_1 \rtimes \Omega, \underline{\mathbf{x}})$ to those of the Iwahori–Hecke algebra $\mathscr{H}(W_1, \underline{\mathbf{x}})$ of the underlying Coxeter group W_1:

Proposition 3.2.14 *Let $\phi \in \mathrm{Irr}(\mathscr{H}(W_1 \rtimes \Omega, \underline{\mathbf{x}}))$. Then we have*

$$c_\phi\, \phi(1) = |\Omega| c_{\phi_1}\, \phi_1(1)$$

for all $\phi_1 \in \mathrm{Irr}(\mathscr{H}(W_1, \underline{\mathbf{x}}))$ which occur in the restriction of ϕ to $\mathscr{H}(W_1, \underline{\mathbf{x}})$.

This is shown in [Ge00, Prop. 4.6] for the equal-parameter case, but the proof remains valid in our more general setting, see also [Ma95, Lemma 5.11]. It is therefore sufficient to know Schur elements for Iwahori–Hecke algebras.

Example 3.2.15 Assume that $W = W_1 \times W_2$ is a direct product of two Coxeter groups W_1, W_2. Write τ_i for the natural symmetrising form on $\mathscr{H}(W_i, \underline{\mathbf{x}}_i)$, $i = 1, 2$. Then the natural symmetrising form on $\mathscr{H}(W, \underline{\mathbf{x}}) = \mathscr{H}(W_1, \underline{\mathbf{x}}_1) \otimes \mathscr{H}(W_2, \underline{\mathbf{x}}_2)$, $\underline{\mathbf{x}} = (\underline{\mathbf{x}}_1, \underline{\mathbf{x}}_2)$, is given by $\tau = \tau_1 \otimes \tau_2$ and the above definitions show that the corresponding Schur elements and generic degrees are multiplicative:

$$c_{\phi_1 \otimes \phi_2}(\underline{\mathbf{x}}) = c_{\phi_1}(\underline{\mathbf{x}}_1)\, c_{\phi_2}(\underline{\mathbf{x}}_2) \qquad \text{and} \qquad D_{\phi_1 \otimes \phi_2} = D_{\phi_1}\, D_{\phi_2}$$

for all $\phi_i \in \mathrm{Irr}(W_i)$, $i = 1, 2$.

3.2.16 The Schur elements of all finite irreducible Coxeter groups are known explicitly (see e.g. [Ca85, §13.5] for Weyl groups, where the generic degrees of $\mathscr{H}(W, \underline{\mathbf{x}})$ are given). (The general case reduces to the one of irreducible groups by Example 3.2.15.) They were calculated by Hoefsmit [Ho74] for the classical types, and by Kilmoyer, Surowski [Su77, Su78], Lusztig [Lu79a], and Benson [Be79] for the exceptional types. Lusztig [Lu77a, 9.6] observed that the Schur elements for type B_n (and thus for type D_n) could also be obtained by interpolation from the known unipotent character degrees of the general unitary groups, using Theorem 3.2.18 together with the fact that the Hecke algebra of type B_n occurs as endomorphism algebra of Harish-Chandra induced cuspidal modules for infinitely many different choices of parameters. See also Iancu [Ia01] for a different proof for type B_n using Markov traces, and Geck–Iancu–Malle [GIM00], Mathas [Mat04] and Rui [Ru01] for the generalisation to imprimitive complex reflection groups.

Observe that the Schur elements are determined for example as the unique solution of the linear system of equations obtained by evaluating the defining equation 3.2.11(3.1) of τ on some basis of $\mathscr{H}(W, \underline{\mathbf{x}})$, hence from a knowledge of the character table of $\mathscr{H}(W, \underline{\mathbf{x}})$.

It is a remarkable fact, first explicitly stated by Chlouveraki [Chl09], that all Schur elements of Iwahori–Hecke algebras have the following form: they are products of cyclotomic polynomials evaluated at monomials in the parameters of the Iwahori–Hecke algebra, times a monomial in the parameters, possibly times an integer. While this phenomenon can be explained rigorously for example for type A_{n-1} by the fact that the specialisations at q of the Schur elements divide the order of the finite groups $\mathrm{GL}_n(q)$, for an infinitude of prime powers q (see [CuRe87, Thm. 68.31]), no such explanation is available for example for the 2-parameter Iwahori–Hecke algebra of type F_4. A general proof of this observation using the theory of rational Cherednik algebras has been given by Rouquier [Rou08, Thm. 3.5].

Example 3.2.17 Let $W = \mathfrak{S}_n$, the symmetric group of degree n. The irreducible characters of W are labelled by the partitions $\alpha \vdash n$. Here all reflections are conjugate so the Iwahori–Hecke algebra of W has just a single parameter $\mathbf{x}$. Let $\alpha = (\alpha_1 \geqslant \alpha_2 \ldots \geqslant \alpha_r) \vdash n$ be a partition. The Schur element c_α of $\phi^\alpha \in \mathrm{Irr}(\mathfrak{S}_n)$ is then given by

$$c_\alpha = \mathbf{x}^{-a(\alpha)} \prod_{i=1}^{n} \frac{\mathbf{x}^{l_i} - 1}{\mathbf{x} - 1},$$

where l_i denotes the hook length at the ith box of the Young diagram of α, and $a(\alpha) = \sum_{i<j}(\alpha_i - \alpha_j)$ is the *a-invariant* of α (see Definition 4.1.10, and also [Ca85, §13.5] and [Ma95, Bem. 3.12]). For example, for the trivial and the sign character of $\mathfrak{S}_n$, which are labelled by the partitions (n) and $(1)^n$ respectively, we obtain the Schur elements

$$c_{(n)} = \prod_{i=1}^{n} \frac{\mathbf{x}^{i} - 1}{\mathbf{x} - 1} \quad \text{and} \quad c_{(1)^n} = \mathbf{x}^{-\binom{n}{2}} \prod_{i=1}^{n} \frac{\mathbf{x}^{i} - 1}{\mathbf{x} - 1}$$

and thus the generic degree $D_{(1)^n} = c_{(n)}/c_{(1)^n} = \mathbf{x}^{\binom{n}{2}}$ for the sign character.

We can now give a degree formula for Harish-Chandra induction in terms of the bijection $I_{L,\lambda}^G : \mathrm{Irr}(W_G(L,\lambda)) \to \mathscr{E}(G,(L,\lambda))$ from Theorem 3.2.5 (depending on a choice of specialisation of the parameters $\mathbf{x}$ of the generic Hecke algebra of $W_G(L,\lambda)$ and thus on a choice of isomorphism $\mathscr{H}_G(L,\lambda) \cong \mathbb{K}[W_G(L,\lambda)]$).

Theorem 3.2.18 (Degree formula) *Let (L,λ) be a cuspidal pair in G. Then for $\phi \in \mathrm{Irr}(W_G(L,\lambda))$ the degree of $\rho = I_{L,\lambda}^G(\phi) \in \mathscr{E}(G,(L,\lambda))$ is given by*

$$\rho(1) = R_{\mathbf{L}}^{\mathbf{G}}(\lambda)(1)\, c_\phi(\underline{q})^{-1} = \lambda(1)\frac{|G|_{p'}}{|L|_{p'}}\, c_\phi(\underline{q})^{-1},$$

where $\underline{q}$ are the parameters of the relative Hecke algebra $\mathscr{H}_G(L,\lambda)$.

See [Ca85, Thm. 10.11.5], or [CuRe87, Prop. 68.30(iii)] for the characters in the principal series.

Let us point out an immediate consequence:

Corollary 3.2.19 *In the situation of Theorem 3.2.18 if ρ lies in the Harish-Chandra series above (L,λ) then $\rho(1)_{p'}$ divides $R_{\mathbf{L}}^{\mathbf{G}}(\lambda)(1)$. In particular $\rho(1)$ divides $|G : \mathbf{Z}(L)|$.*

Proof The first claim follows from the formula for $\rho(1)$ in Theorem 3.2.18 together with the fact that the Schur elements lie in A_W (see 3.2.11), so their specialisations at powers of p are integral up to powers of p. As $|\mathbf{Z}(L)|$ is prime to p the second part then is an immediate consequence of the first and the well-known property of irreducible characters $\lambda \in \mathrm{Irr}(L)$ that $\lambda(1)$ divides $|L : \mathbf{Z}(L)|$. □

In order to be able to apply Theorem 3.2.18, the parameters of the Hecke algebra $\mathcal{H}_G(L,\lambda)$ have to be known. They can be determined locally, as described in Proposition 3.1.29. Note that over a field of characteristic 0 we are always in the case (b). The parameters are known explicitly in all cases by the work of Lusztig [Lu84a, Thm. 8.6]. For unipotent characters they will be given in Table 4.8.

We now relate the degree formula to regular embeddings. Choose a regular embedding $\mathbf{G} \hookrightarrow \tilde{\mathbf{G}}$. Then for $\mathbf{L} \leqslant \mathbf{G}$ an F-stable Levi subgroup and $\tilde{\mathbf{L}} = \mathbf{LZ}(\tilde{\mathbf{G}})$ the corresponding Levi subgroup of $\tilde{\mathbf{G}}$, observe that $|\tilde{\mathbf{G}}^F : \mathbf{G}^F| = |\tilde{\mathbf{L}}^F : \mathbf{L}^F|$. Let $\lambda \in \mathrm{Irr}(\mathbf{L}^F)$ be cuspidal and $\rho \in \mathcal{E}(\mathbf{G}^F, (\mathbf{L},\lambda))$. By Theorem 1.7.15 induction of characters from $\mathbf{G}^F$ to $\tilde{\mathbf{G}}^F$ and from $\mathbf{L}^F$ to $\tilde{\mathbf{L}}^F$ is multiplicity free, so we have

$$\mathrm{Ind}_L^{\tilde{L}}(\lambda) = \sum_{i=1}^{a} \tilde{\lambda}_i, \quad \mathrm{Ind}_G^{\tilde{G}}(\rho) = \sum_{j=1}^{b} \tilde{\rho}_j,$$

for distinct irreducible characters $\tilde{\lambda}_i \in \mathrm{Irr}(\tilde{\mathbf{L}}^F)$ and $\tilde{\rho}_j \in \mathrm{Irr}(\tilde{\mathbf{G}}^F)$. Setting $\tilde{\lambda} := \tilde{\lambda}_1$ we have $\tilde{\lambda}$ is cuspidal by Corollary 3.2.10 and we may choose notation such that $\tilde{\rho} := \rho_1 \in \mathcal{E}(\tilde{\mathbf{G}}^F, (\tilde{\mathbf{L}}, \tilde{\lambda}))$. Then $a\rho(1)/\lambda(1) = b\tilde{\rho}(1)/\tilde{\lambda}(1)$. So we obtain:

Corollary 3.2.20 *In the situation of Theorem 3.2.18 and with the above notation, for $\phi \in \mathrm{Irr}(W_G(L,\lambda))$ the degree of $\rho = I_{L,\lambda}^G(\phi) \in \mathcal{E}(G,(L,\lambda))$ is given by*

$$\rho(1) = d\, R_{\mathbf{L}}^{\mathbf{G}}(\lambda)(1)\, c_{\tilde{\phi}}(\underline{q})^{-1} = d\,\lambda(1)\frac{|G|_{p'}}{|L|_{p'}}\, c_{\tilde{\phi}}(\underline{q})^{-1},$$

where $d = a/b = \langle \mathrm{Ind}_G^{\tilde{G}}(\lambda), \mathrm{Ind}_G^{\tilde{G}}(\lambda)\rangle / \langle \mathrm{Ind}_L^{\tilde{L}}(\rho), \mathrm{Ind}_L^{\tilde{L}}(\rho)\rangle$ and $c_{\tilde{\phi}}$ denotes the Schur element of an extension $\tilde{\phi}$ of ϕ to the Hecke algebra of $(\tilde{L},\tilde{\lambda})$ in $\tilde{G}$, with parameters $\underline{q}$.

Observe that this formula is generic in the sense that it relates the degree polynomials $\mathbb{D}_\lambda$ of λ and $\mathbb{D}_\rho$ of ρ by

$$\mathbb{D}_\rho = d\,\mathbb{D}_\lambda \frac{|\mathbb{G}|_{\mathbf{q}'}}{|\mathbb{L}|_{\mathbf{q}'}} c_{\tilde{\phi}}(\mathbf{q})^{-1},$$

where $c_{\tilde{\phi}}(\mathbf{q})$ denotes the 1-parameter specialisation of $c_{\tilde{\phi}}$ through which the specialisation to $\underline{q}$ factors (see Example 3.1.23).

We obtain the following generalisation and converse of Corollary 3.2.4:

Corollary 3.2.21 *Let $\rho \in \mathrm{Irr}(G)$ lie in the Harish-Chandra series of the cuspidal pair (L,λ). Then the degree polynomial of ρ has the form*

$$\mathbb{D}_\rho = (\mathbf{q}-1)^{r(\mathbf{L}^F)} f(\mathbf{q}),$$

where $f \in \mathbb{Q}[\mathbf{q}]$ is not divisible by $\mathbf{q}-1$.

In particular ρ is cuspidal if and only if $(\mathbf{q}-1)^{r(\mathbf{G}^F)}$ is the precise power of $\mathbf{q}-1$ dividing its degree polynomial.

Proof First assume that $\mathbf{G}$ has connected centre. Then by Theorem 3.2.5 the parameters $\underline{q}$ of $\mathscr{H}_G(L,\lambda)$ are integral powers of q, say, $\underline{q} = (q_s \mid s \in S)$ with $q_s = q^{a_s}$. Thus the corresponding specialisation $\varphi_{\underline{q}} : \mathbb{Z}[\underline{\mathbf{v}}^{\pm 1}] \to \mathbb{Z}$, $\mathbf{x}_s \mapsto q_s = q^{a_s}$, factors through a specialisation

$$\varphi_{\mathbf{q}} : \mathbb{Z}[\underline{\mathbf{v}}^{\pm 1}] \to \mathbb{Z}[\mathbf{q}^{\pm 1}], \quad \mathbf{x}_s \mapsto \mathbf{q}^{a_s}.$$

Let $c_{\phi,\mathbf{q}} = \varphi_{\mathbf{q}}(c_\phi)$, $\phi \in \mathrm{Irr}(W_G(L,\lambda))$, denote the specialised Schur elements. These lie in $\mathbb{Z}[\mathbf{q}^{\pm 1}]$, so have no pole at $\mathbf{q} = 1$. But the specialisation φ_1 to the group algebra of the relative Weyl group $W_G(L,\lambda)$ also factorises through $\varphi_{\mathbf{q}}$, and by Example 3.2.12 it sends the Schur elements to the non-zero elements $|W_G(L,\lambda)|/\phi(1)$, so the $c_{\phi,\mathbf{q}}$ do not have a zero at $\mathbf{q} = 1$ either.

Since $\mathbf{L}$ is a 1-split Levi subgroup of $\mathbf{G}$, it contains a maximal split subtorus of $\mathbf{G}$ by Lemma 3.1.3 and so the order polynomials of $\mathbf{L}$ and $\mathbf{G}$ are divisible by the same power of $\mathbf{q}-1$. Since the Schur elements $c_{\phi,\mathbf{q}}$ have neither a zero nor a pole at $\mathbf{q} = 1$, the form of $\mathbb{D}_\rho$ now follows by Corollary 3.2.4 combined with Theorem 3.2.18.

The last claim follows with Corollary 3.2.4 from the observation that $r(\mathbf{L}^F) = |I| < |S| = r(\mathbf{G}^F)$ for any proper split Levi subgroup $\mathbf{L}$ of $\mathbf{G}$ (if $\mathbf{L}^F$ is conjugate to the standard Levi subgroup L_I for $I \subsetneq S$).

In the general case, let $\mathbf{G} \hookrightarrow \tilde{\mathbf{G}}$ be a regular embedding and be $\tilde{\mathbf{L}} \leqslant \mathbf{G}$ the F-stable Levi subgroup with $\mathbf{L} = \tilde{\mathbf{L}} \cap \mathbf{G}$, $\tilde{\lambda} \in \mathrm{Irr}(\tilde{\mathbf{L}}^F)$ a cuspidal character lying above λ and $\tilde{\rho} \in \mathrm{Irr}(\tilde{\mathbf{G}}^F)$ in $\mathscr{E}(\tilde{\mathbf{G}}^F, \tilde{\lambda})$ above ρ. Then the degree polynomials of ρ and of $\tilde{\rho}$ differ by an integer factor by Remark 2.3.27(c) and hence the claim for ρ follows from the one for $\tilde{\rho}$. □

A formula for values of $R_{\mathbf{L}}^{\mathbf{G}}$ and ${}^*R_{\mathbf{L}}^{\mathbf{G}}$ on arbitrary elements will be given in Theorem 3.3.12 in the more general setting of Lusztig induction.

We now clarify the relation between cuspidality and Jordan decomposition. For this let $(\mathbf{G}^*, F)$ be dual to $(\mathbf{G}, F)$ (see Definition 1.5.17). (For simplicity we denote the Steinberg map on the dual group $\mathbf{G}^*$ again by F.) The following result (which first appeared in special cases in [Lu77a, 7.8]) shows that in order to understand the cuspidal characters of all connected reductive groups it is enough to classify unipotent cuspidal characters:

Theorem 3.2.22 *Let $s \in \mathbf{G}^{*F}$ be semisimple. Let $\rho \in \mathscr{E}(\mathbf{G}^F, s)$ and let ψ be in a $C_{\mathbf{G}^*}(s)^F$-orbit of unipotent characters of $C^\circ_{\mathbf{G}^*}(s)^F$ corresponding to ρ under Jordan decomposition (see Theorem 2.6.22). Then ρ is cuspidal if and only if*

(1) *ψ is cuspidal, and*
(2) *$\mathbf{Z}^\circ(\mathbf{G}^*)$ and $\mathbf{Z}^\circ(C^\circ_{\mathbf{G}^*}(s))$ have the same $\mathbb{F}_q$-rank, that is, the maximal split subtorus of $\mathbf{Z}(C^\circ_{\mathbf{G}^*}(s))$ is contained in $\mathbf{Z}^\circ(\mathbf{G}^*)$.*

Proof (See [CE99, Prop. 1.10].) Let $i : \mathbf{G} \hookrightarrow \tilde{\mathbf{G}}$ be a regular embedding. Then by

Corollary 3.2.10 cuspidality of ρ and ψ is equivalent to cuspidality of corresponding characters of $\tilde{\mathbf{G}}^F$ and $C_{\tilde{\mathbf{G}}^*}(\tilde{s})^F$ respectively, where $\tilde{s} \in \tilde{\mathbf{G}}^{*F}$ is a preimage of s under the epimorphism $i^* : \tilde{\mathbf{G}}^* \twoheadrightarrow \mathbf{G}^*$ dual to i. Clearly the condition (2) in the statement is also preserved by regular embeddings, so we may and will assume for the proof that $\mathbf{G}$ has connected centre and thus $C_{\mathbf{G}^*}(s)$ is connected.

As the regular character reg_G is uniform by Proposition 2.3.23, there exists an F-stable maximal torus $\mathbf{T} \leqslant \mathbf{G}$ with $\langle R_{\mathbf{T}}^{\mathbf{G}}(\theta), \rho\rangle \neq 0$ for some $\theta \in \mathrm{Irr}(\mathbf{T}^F)$. Moreover, as $\rho \in \mathscr{E}(G, s)$, the pair $(\mathbf{T}, \theta)$ must lie in the geometric conjugacy class of $(\mathbf{T}^*, s)$, where $\mathbf{T}^*$ is an F-stable maximal torus of $\mathbf{G}^*$ dual to $\mathbf{T}$ (see Corollary 2.5.14 and Definition 2.5.17). If ρ is cuspidal, then by the uniform criterion in Proposition 3.2.2 the maximal split subtorus of $\mathbf{T}$ must be contained in $\mathbf{Z}(\mathbf{G})$, and thus the maximal split subtorus of $\mathbf{T}^*$ lies in $\mathbf{Z}(\mathbf{G}^*)$. Since $s \in \mathbf{T}^*$ we also have $\mathbf{Z}^\circ(C_{\mathbf{G}^*}(s)) \leqslant \mathbf{T}^*$, so we obtain condition (2). Conversely, if (2) is satisfied, then the maximal split subtorus of $\mathbf{T}$ lies in $\mathbf{Z}(\mathbf{G})$ if and only if the maximal split subtorus of $\mathbf{T}^*$ lies in $\mathbf{Z}(C_{\mathbf{G}^*}(s))$. But then ρ satisfies the uniform criterion for cuspidality if and only if ψ does, since

$$\left\langle R_{\mathbf{T}}^{\mathbf{G}}(\theta), \rho\right\rangle = \pm\left\langle R_{\mathbf{T}^*}^{C_{\mathbf{G}^*}(s)}(1), \psi\right\rangle$$

by the fundamental property of Jordan decomposition for groups with connected centre (see Theorem 2.6.4). □

A special case of this was already shown in [Lu77a, Prop. 7.9].

3.3 Lusztig Induction and Restriction

We introduce a far-reaching common generalisation of both Harish-Chandra induction and restriction and of Deligne–Lusztig characters introduced by Lusztig [Lu76b]. It provides a generalised induction from a family of subgroups encompassing the split Levi subgroups used in Harish-Chandra induction as well as the maximal tori featuring in Deligne–Lusztig induction: the F-fixed points of arbitrary F-stable Levi subgroups. Throughout this section, let $\mathbf{G}$ denote a connected reductive algebraic group over an algebraic closure of a finite field of characteristic p, with a Steinberg endomorphism $F : \mathbf{G} \to \mathbf{G}$. We let q denote the absolute value of the eigenvalues of F on the character group of an F-stable maximal torus of $\mathbf{G}$. As before $\mathbb{K}$ denotes a sufficiently large field of characteristic 0.

3.3.1 We start by classifying F-stable Levi subgroups of $\mathbf{G}$ in terms of combinatorial data by generalising the corresponding classification of F-stable maximal tori from 1.6.4. Fix a maximally split torus $\mathbf{T}_0$ of $\mathbf{G}$ contained in an F-stable Borel subgroup of $\mathbf{G}$, with corresponding set of simple reflections $\tilde{S}$. Let $\mathbf{L} \leqslant \mathbf{G}$ be an F-stable Levi subgroup of a (not necessarily F-stable) parabolic subgroup $\mathbf{P}$ of $\mathbf{G}$. Choose an F-stable maximal torus $\mathbf{T} \leqslant \mathbf{L}$. Then there is $x \in \mathbf{G}$ such that $\mathbf{L} = x\mathbf{L}_I x^{-1}$ for the standard Levi subgroup $\mathbf{L}_I$ corresponding to a subset $I \subseteq \tilde{S}$ and moreover $\mathbf{T} = x\mathbf{T}_0 x^{-1}$. Then we have

$$F(x)\mathbf{T}_0 F(x)^{-1} = F(\mathbf{T}) = \mathbf{T} = x\mathbf{T}_0 x^{-1} \quad \text{and}$$
$$F(x)F(\mathbf{L}_I)F(x)^{-1} = F(\mathbf{L}) = \mathbf{L} = x\mathbf{L}_I x^{-1}.$$

The first equation shows that $x^{-1}F(x) \in N_{\mathbf{G}}(\mathbf{T}_0)$ and so $x^{-1}F(x) = \dot{w}$ for some $w \in \mathbf{W}$. In the second equation, we have $F(\mathbf{L}_I) = \mathbf{L}_{F(I)}$, so $\dot{w}\mathbf{L}_{F(I)}\dot{w}^{-1} = \mathbf{L}_I$ and hence $wF(I) = I$. Thus, $\mathbf{L}$ determines a pair (I, w) with $I \subseteq S$ and $w \in \mathbf{W}$ with $wF(I) = I$. We then say that $\mathbf{L}$ is a *Levi subgroup of type* (I, w). Conversely, any such pair (I, w) defines an F-stable Levi subgroup $x\mathbf{L}_I x^{-1}$ of $\mathbf{G}$, where $x \in \mathbf{G}$ is such that $x^{-1}F(x) = \dot{w}$. The complete root datum for a Levi subgroup of type (I, w) is of the form $\big((X, R_I, Y, R_I^\vee), w\mathbf{W}_I\big)$, where R_I is the parabolic subsystem of R determined by I with Weyl group $\mathbf{W}_I = \langle I \rangle$.

It is easily seen that this construction sets up a bijection between $\mathbf{G}^F$-conjugacy classes of F-stable Levi subgroups of $\mathbf{G}$ and equivalence classes of pairs (I, w) as before, with $(I, w), (I, w')$ equivalent if and only if there is $v \in \mathbf{W}$ with $vI = I'$ and $vw = w'F(v)$; see [Lu77a, 7.2]. Assume that $\mathbf{L}$ is 1-split. Then it contains a maximally split torus, hence up to conjugation we have $\mathbf{T}_0 \leqslant \mathbf{L}$ and thus we can choose $w = 1$ in our construction. Thus, the 1-split Levi subgroups are parametrised by the equivalence classes of pairs of the form $(I, 1)$, as is also clear from the BN-pair structure of G.

In the Chevie-system [MiChv] the various rational forms of a standard Levi subgroup can be obtained with the command `Twistings`.

Recall the construction of Deligne–Lusztig characters in Definition 2.2.6 starting from an F-stable maximal torus $\mathbf{T}$ of $\mathbf{G}$ contained in a not necessarily F-stable Borel subgroup $\mathbf{B}$. Then $R_{\mathbf{T}}^{\mathbf{G}}$ was defined from the ℓ-adic cohomology of the preimage of the unipotent radical $\mathbf{U}$ of $\mathbf{B}$ under the Lang–Steinberg map $\mathscr{L}$.

Definition 3.3.2 (Lusztig [Lu76b]) Let $\mathbf{L}$ be an F-stable Levi subgroup of a (not necessarily F-stable) parabolic subgroup $\mathbf{P}$ of $\mathbf{G}$. Then the unipotent radical $\mathbf{Y} := R_u(\mathbf{P})$ of $\mathbf{P}$ satisfies condition $(*)$ in 2.2.4 with respect to the finite subgroup $H := \mathbf{L}^F$. Thus, for any $\lambda \in \mathrm{Irr}(\mathbf{L}^F)$ we obtain a virtual character of $\mathbf{G}^F$, called the *Lusztig induced character* of λ, defined by

$$R_{\mathbf{L}\leqslant\mathbf{P}}^{\mathbf{G}}(\lambda)(g) := \frac{1}{|\mathbf{L}^F|} \sum_{l \in \mathbf{L}^F} \mathfrak{L}\big((g, l), \mathscr{L}^{-1}(\mathbf{Y})\big)\, \lambda(l)$$
$$= \sum_{i \geqslant 0} (-1)^i \,\mathrm{Trace}\big(g^*, \mathrm{H}_c^i(\mathscr{L}^{-1}(\mathbf{Y}), \overline{\mathbb{Q}}_\ell)_\lambda\big)$$

for $g \in \mathbf{G}^F$, where $\mathrm{H}^i_c(\mathscr{L}^{-1}(\mathbf{Y}), \overline{\mathbb{Q}}_\ell)_\lambda$ denotes the λ-isotypic part of the $\overline{\mathbb{Q}}_\ell \mathbf{L}^F$-module $\mathrm{H}^i_c(\mathscr{L}^{-1}(\mathbf{Y}), \overline{\mathbb{Q}}_\ell)$ and g^* is the automorphism induced by g on it (see 2.2.1). We denote by ${}^*R^{\mathbf{G}}_{\mathbf{L} \leqslant \mathbf{P}}$ the adjoint map, *Lusztig restriction*, which sends characters of $\mathbf{G}^F$ to virtual characters of $\mathbf{L}^F$, given by

$$ {}^*R_{\mathbf{L}\leqslant\mathbf{P}}(\rho)(l) = \sum_{i \geqslant 0} (-1)^i \operatorname{Trace}\big(l^*, {}_\rho\mathrm{H}^i_c(\mathscr{L}^{-1}(\mathbf{Y}), \overline{\mathbb{Q}}_\ell)\big) $$

for $\rho \in \operatorname{Irr}(\mathbf{G}^F)$, $l \in \mathbf{L}^F$, where ${}_\rho\mathrm{H}^i_c(\mathscr{L}^{-1}(\mathbf{Y}), \overline{\mathbb{Q}}_\ell)$ denotes the ρ-isotypic part of the $\mathbb{K}\mathbf{G}^F$-module $\mathrm{H}^i_c(\mathscr{L}^{-1}(\mathbf{Y}), \overline{\mathbb{Q}}_\ell)$. Thus we have

$$ \big\langle R^{\mathbf{G}}_{\mathbf{L}\leqslant\mathbf{P}}(\lambda), \rho\big\rangle = \big\langle \lambda, {}^*R^{\mathbf{G}}_{\mathbf{L}\leqslant\mathbf{P}}(\rho)\big\rangle \quad \text{for all } \lambda \in \operatorname{Irr}(\mathbf{L}^F),\ \rho \in \operatorname{Irr}(\mathbf{G}^F). $$

Lusztig originally called $R^{\mathbf{G}}_{\mathbf{L}\leqslant\mathbf{P}}$ *twisted induction*, a term which is sometimes still used.

Note that the values of the generalised characters $R^{\mathbf{G}}_{\mathbf{L}\leqslant\mathbf{P}}(\lambda)$ and ${}^*R_{\mathbf{L}\leqslant\mathbf{P}}(\rho)$ only involve roots of unity in $\overline{\mathbb{Q}}_\ell$, so (after a suitable identification) can be considered as elements of our chosen large enough field $\mathbb{K}$ of characteristic 0.

It is clear from the definition that $R^{\mathbf{G}}_{\mathbf{L}\leqslant\mathbf{P}}$ is a generalisation of the Deligne–Lusztig characters, which arise in the special situation that the parabolic subgroup $\mathbf{P}$ is minimal possible, that is, $\mathbf{P}$ is a Borel subgroup and $\mathbf{L}$ an F-stable maximal torus, and ${}^*R^{\mathbf{G}}_{\mathbf{L}\leqslant\mathbf{P}}$ is a generalisation of ${}^*R^{\mathbf{G}}_{\mathbf{T}}$. But Lusztig induction also generalises Harish-Chandra induction:

Proposition 3.3.3 *Assume that $\mathbf{L}$ is an F-stable Levi subgroup of an F-stable parabolic subgroup $\mathbf{P}$ of $\mathbf{G}$, that is, $\mathbf{L}$ is a split Levi subgroup of $\mathbf{G}$. Then $R^{\mathbf{G}}_{\mathbf{L}\leqslant\mathbf{P}}$ is just Harish-Chandra induction from $\mathbf{L}^F$ to $\mathbf{G}^F$ and ${}^*R^{\mathbf{G}}_{\mathbf{L}\leqslant\mathbf{P}}$ is Harish-Chandra restriction.*

See the remarks after [DiMi20, Prop. 9.1.4]. This result justifies our use of the same symbol for both.

Example 3.3.4 In generalisation of Example 3.1.8 we have ${}^*R^{\mathbf{G}}_{\mathbf{L}}(1_{\mathbf{G}}) = 1_{\mathbf{L}}$ for any F-stable Levi subgroup $\mathbf{L} \leqslant \mathbf{G}$, see [DiMi20, proof of Cor. 10.1.7].

3.3.5 We briefly discuss another realisation of Lusztig induction from [Lu84a, p. 215] (see also [BoRo93, §5]), generalising the alternative model for Deligne–Lusztig characters introduced in 2.3.18. Fix a maximally split torus $\mathbf{T}_0$ of $\mathbf{G}$ and an F-stable Borel subgroup $\mathbf{B}$ containing $\mathbf{T}_0$. Let $\mathbf{L} \leqslant \mathbf{G}$ be an F-stable Levi subgroup, of type (I, w) for some $I \subseteq \tilde{S}$ and $w \in \mathbf{W}$ with $wF(I) = I$ (see 3.3.1). As discussed in 2.3.18,

$$ F' \colon \mathbf{G} \to \mathbf{G}, \qquad g \mapsto \dot{w}F(g)\dot{w}^{-1}, $$

is again a Steinberg map on $\mathbf{G}$ and conjugation with x defines an isomorphism

$\mathbf{G}^F \cong \mathbf{G}^{F'}$, where $x \in \mathbf{G}$ is such that $x^{-1}F(x) = \dot{w}$. Let $\mathbf{L}_I$ be the standard Levi subgroup of $\mathbf{G}$ corresponding to I. Then $\mathbf{L}_I$ is F'-stable, and we set

$$\mathbf{L}_I[w] := \mathbf{L}_I^{F'} = \{l \in \mathbf{L}_I \mid F(l) = \dot{w}^{-1} l \dot{w}\}.$$

Note that $\mathbf{L}_I[w]$ is a finite subgroup of $\mathbf{L}_I$ that depends on w, but not on the representative $\dot{w}$. (Another common notation for this subgroup is $\mathbf{L}_I^{wF}$.) Then conjugation by x^{-1} defines an isomorphism $\mathbf{L}^F \cong \mathbf{L}_I[w]$.

Let $\mathbf{P}_I$ be the standard parabolic subgroup of $\mathbf{G}$ corresponding to I and hence containing $\mathbf{L}_I$, and $\mathbf{Y}_I := R_u(\mathbf{P}_I)$ its unipotent radical. Then $\mathbf{Y}_I\dot{w}$ satisfies the condition (*) in 2.2.4 with respect to the finite subgroup $\mathbf{L}_I[w]$. So as before we obtain virtual characters of $\mathbf{G}^F$ by setting, for any $\lambda \in \operatorname{Irr}(\mathbf{L}_I[w])$,

$$R_{I,w}^{\lambda}(g) := \sum_{i \geqslant 0} (-1)^i \operatorname{Trace}\big(g^*, \mathrm{H}_c^i(\mathscr{L}^{-1}(\mathbf{Y}_I), \overline{\mathbb{Q}}_\ell)_\lambda\big) \quad \text{for } g \in \mathbf{G}^F.$$

With an identical proof as in Lemma 2.3.19 for the virtual characters R_w^θ (the case when $I = \varnothing$, so $\mathbf{L}_I = \mathbf{L}_\varnothing = \mathbf{T}_0$) we then have that $R_{I,w}^{\lambda}$ does not depend on the choice of a representative $\dot{w}$, and furthermore, with the parabolic subgroup $\mathbf{P} = x\mathbf{P}_I x^{-1}$ containing $\mathbf{L}$,

$$R_{I,w}^{\lambda} = R_{\mathbf{L} \leqslant \mathbf{P}}^{\mathbf{G}}({}^x\lambda) \quad \text{where} \quad {}^x\lambda(l) := \lambda(x^{-1}lx) \text{ for } l \in \mathbf{L}^F.$$

As for Harish-Chandra induction (see Proposition 3.1.9), Lusztig induction and restriction are transitive:

Theorem 3.3.6 (Transitivity) *Let $\mathbf{Q} \leqslant \mathbf{P}$ be parabolic subgroups of $\mathbf{G}$ with F-stable Levi subgroups $\mathbf{M}, \mathbf{L}$ respectively, such that $\mathbf{M} \leqslant \mathbf{L}$. Then*

$$\begin{aligned} R_{\mathbf{L} \leqslant \mathbf{P}}^{\mathbf{G}} \circ R_{\mathbf{M} \leqslant \mathbf{L} \cap \mathbf{Q}}^{\mathbf{L}}(\psi) &= R_{\mathbf{M} \leqslant \mathbf{Q}}^{\mathbf{G}}(\psi) && \textit{for all } \psi \in \operatorname{Irr}(\mathbf{M}^F), \\ {}^*R_{\mathbf{M} \leqslant \mathbf{L} \cap \mathbf{Q}}^{\mathbf{L}} \circ {}^*R_{\mathbf{L} \leqslant \mathbf{P}}^{\mathbf{G}}(\rho) &= {}^*R_{\mathbf{M} \leqslant \mathbf{Q}}^{\mathbf{G}}(\rho) && \textit{for all } \rho \in \operatorname{Irr}(\mathbf{G}^F). \end{aligned}$$

See [DiMi20, Prop. 9.1.8].

It is generally expected that Lusztig induction and restriction do also satisfy a Mackey formula, but this has not yet been proved in full generality. It is true in the case that both parabolic subgroups are F-stable by Theorem 3.1.11, it follows from the work of Deligne and Lusztig if one of the two Levi subgroups is a maximal torus (see [DeLu83, Thm. 7] or [DiMi20, Thm. 9.2.6]), and it has also been shown if q is large enough [Bo98]. Results of Shoji can then be used to deduce that the Mackey formula always holds for unipotent characters, see [BMM93, Thm. 1.35]. (Observe that Shoji's proofs [Sho87] need the assumption that $q \not\equiv -1 \pmod 3$ for $\mathbf{G}$ of type E_8. But in fact by [Sho87, Rem. 3.3] this assumption is not necessary for the case of unipotent characters.) The most far-reaching result to date was obtained by Bonnafé–Michel [BoMi11, Thm.] using sophisticated computational techniques, and building upon that by Taylor [Ta18a]:

Theorem 3.3.7 (Mackey formula) *Let* $\mathbf{P}, \mathbf{Q}$ *be parabolic subgroups of* $\mathbf{G}$ *with* F*-stable Levi complements* $\mathbf{L}, \mathbf{M}$ *respectively. Let* $\psi \in \mathrm{Irr}(\mathbf{M}^F)$ *and assume that one of the following conditions is satisfied:*

(1) *either* $\mathbf{L}$ *or* $\mathbf{M}$ *is a maximal torus;*
(2) ψ *is unipotent;*
(3) F *is a Frobenius map and* $q > 2$*;*
(4) F *is a Frobenius map and* $\mathbf{G}^F$ *does not possess a component of type* ${}^2E_6$, E_7 *or* E_8*; or*
(5) F *is a Frobenius map,* $\mathbf{Z}(\mathbf{G})$ *is connected and* $\mathbf{G}$ *does not have a factor of type* E_8.

Then

$$ {}^*R^{\mathbf{G}}_{\mathbf{L}\leqslant\mathbf{P}} \circ R^{\mathbf{G}}_{\mathbf{M}\leqslant\mathbf{Q}}(\psi) = \sum_{w \in \mathbf{L}^F \backslash \mathscr{S}(\mathbf{L},\mathbf{M})^F / \mathbf{M}^F} R^{\mathbf{L}}_{\mathbf{L}\cap {}^w\mathbf{M} \leqslant \mathbf{L}\cap {}^w\mathbf{Q}} \circ \mathrm{ad}(w) \circ {}^*R^{\mathbf{M}}_{\mathbf{L}^w \cap \mathbf{M} \leqslant \mathbf{P}^w \cap \mathbf{M}}(\psi), $$

where w *runs over a system of* $\mathbf{L}^F$*–*$\mathbf{M}^F$ *double coset representatives in*

$$ \mathscr{S}(\mathbf{L},\mathbf{M})^F := \{g \in \mathbf{G}^F \mid \mathbf{L} \cap {}^g\mathbf{M} \text{ contains a maximal torus of } \mathbf{G}\}. $$

As in the Harish-Chandra case, this result immediately implies that $R^{\mathbf{G}}_{\mathbf{L}\leqslant\mathbf{P}}$ and ${}^*R^{\mathbf{G}}_{\mathbf{L}\leqslant\mathbf{P}}$ are independent of the particular parabolic subgroup $\mathbf{P}$ containing the F-stable Levi subgroup $\mathbf{L}$, unless possibly when we are in one of the excluded cases in Theorem 3.3.7:

Theorem 3.3.8 *In the notation of Theorem 3.3.7, assume that any one of the assumptions (1)–(5) of that theorem are satisfied, or that the centre of* $\mathbf{G}$ *is connected. Then* $R^{\mathbf{G}}_{\mathbf{L}\leqslant\mathbf{P}}$ *and* ${}^*R^{\mathbf{G}}_{\mathbf{L}\leqslant\mathbf{P}}$ *are independent of the parabolic subgroup* $\mathbf{P}$ *containing* $\mathbf{L}$.

The statement for connected centre had been shown by Shoji [Sho96, Thm. 4.2] in the case that the underlying characteristic is almost good for $\mathbf{G}$. This latter assumption has now been removed by Lusztig by showing the 'cleanness' of character sheaves, see e.g., [Ge18, §7].

We will from now on write $R^{\mathbf{G}}_{\mathbf{L}}$ and ${}^*R^{\mathbf{G}}_{\mathbf{L}}$ in place of $R^{\mathbf{G}}_{\mathbf{L}\leqslant\mathbf{P}}$ and ${}^*R^{\mathbf{G}}_{\mathbf{L}\leqslant\mathbf{P}}$ respectively whenever the result will not depend on the choice of $\mathbf{P}$, for example whenever the Mackey formula is known to hold.

Remark 3.3.9 Since the Mackey formula holds if one Levi subgroup is a maximal torus, it follows by transitivity of Lusztig induction that the Mackey formula, and thus also the independence of Lusztig induction from the chosen parabolic subgroup, holds on the subspace of uniform functions for arbitrary $\mathbf{L}$ and $\mathbf{M}$.

By transitivity Lusztig induction sends uniform functions to uniform functions. The Mackey formula implies the following stronger property:

Proposition 3.3.10 *Let* **G** *be connected reductive. For all F-stable Levi subgroups* $\mathbf{L} \leqslant \mathbf{G}$ *we have*

$$\pi_{un}^{\mathbf{G}} \circ R_{\mathbf{L}}^{\mathbf{G}} = R_{\mathbf{L}}^{\mathbf{G}} \circ \pi_{un}^{\mathbf{L}} \quad \textit{and} \quad \pi_{un}^{\mathbf{L}} \circ {}^*R_{\mathbf{L}}^{\mathbf{G}} = {}^*R_{\mathbf{L}}^{\mathbf{G}} \circ \pi_{un}^{\mathbf{G}},$$

that is, Lusztig induction and restriction commute with uniform projection.

Proof For a class function f on $\mathbf{L}^F$ let $f^\perp := f - \pi_{\mathrm{un}}^{\mathbf{L}}(f)$ be the part orthogonal to the space of uniform functions. By transitivity of Lusztig induction (see Theorem 3.3.6) we have that $R_{\mathbf{L}}^{\mathbf{G}}(\pi_{\mathrm{un}}^{\mathbf{L}}(f))$ is uniform. We claim that $R_{\mathbf{L}}^{\mathbf{G}}(f^\perp)$ is orthogonal to all uniform functions. Indeed, for any F-stable maximal torus $\mathbf{T} \leqslant \mathbf{G}$ and any $\theta \in \mathrm{Irr}(\mathbf{T}^F)$ we have

$$\big\langle R_{\mathbf{T}}^{\mathbf{G}}(\theta), R_{\mathbf{L}}^{\mathbf{G}}(f^\perp)\big\rangle = \big\langle {}^*R_{\mathbf{L}}^{\mathbf{G}}(R_{\mathbf{T}}^{\mathbf{G}}(\theta)), f^\perp\big\rangle.$$

By the Mackey formula, which holds as $\mathbf{T}$ is a maximal torus, ${}^*R_{\mathbf{L}}^{\mathbf{G}}(R_{\mathbf{T}}^{\mathbf{G}}(\theta))$ is a linear combination of Deligne–Lusztig characters $R_{\mathbf{T}'}^{\mathbf{L}}(\theta')$ of $\mathbf{L}^F$ and hence orthogonal to $f^\perp$. So $R_{\mathbf{L}}^{\mathbf{G}}(f) = R_{\mathbf{L}}^{\mathbf{G}}(\pi_{\mathrm{un}}^{\mathbf{L}}(f) + f^\perp) = R_{\mathbf{L}}^{\mathbf{G}}(\pi_{\mathrm{un}}^{\mathbf{L}}(f)) + R_{\mathbf{L}}^{\mathbf{G}}(f^\perp)$ is the decomposition into the uniform part and the part orthogonal to it. The second formula follows by adjunction. □

In contrast to the situation for Deligne–Lusztig characters and for Harish-Chandra induction, the decomposition of Lusztig induction $R_{\mathbf{L}}^{\mathbf{G}}$ is not yet known in general. Nevertheless, quite substantial partial results are known in important special cases, for example in the case of unipotent characters, or when **G** has connected centre, see Section 4.6.

Lusztig induction satisfies a character formula generalising Theorem 2.2.16 for $R_{\mathbf{T}}^{\mathbf{G}}$. For this we need a generalisation of the Green functions introduced there:

Definition 3.3.11 Let **L** be an F-stable Levi subgroup of a parabolic subgroup **P** of **G** with Levi decomposition $\mathbf{P} = \mathbf{UL}$. Then

$$Q_{\mathbf{L}\leqslant\mathbf{P}}^{\mathbf{G}} : \mathbf{G}_{\mathrm{uni}}^F \times \mathbf{L}_{\mathrm{uni}}^F \to \mathbb{Q},\ (u, v) \mapsto \frac{1}{|\mathbf{L}^F|}\sum_{i\geqslant 0}(-1)^i \mathrm{Trace}\big((u, v) \mid \mathrm{H}_c^i(\mathscr{L}^{-1}(\mathbf{U}))\big),$$

is called the associated *2-parameter Green function*. Here $\mathbf{G}_{\mathrm{uni}}$ denotes the set of unipotent elements of **G**.

If $\mathbf{L} = \mathbf{T}$ is a maximal torus, so $\mathbf{P} = \mathbf{B}$ is a Borel subgroup, then $\mathbf{L}_{\mathrm{uni}}^F = \{1\}$ and the defining formula shows that $Q_{\mathbf{T}\leqslant\mathbf{B}}^{\mathbf{G}}(u, 1) = R_{\mathbf{T}}^{\mathbf{G}}(1)(u)$ ($u \in \mathbf{G}_{\mathrm{uni}}^F$), which is the (1-parameter) Green function introduced in Definition 2.2.15. See Lusztig [Lu90, Thm. 1.14] for a more general result, and also Corollary 3.3.19.

As for Lusztig induction, we will write $Q_{\mathbf{L}}^{\mathbf{G}}$ in place of $Q_{\mathbf{L}\leqslant\mathbf{P}}^{\mathbf{G}}$ whenever the result does not depend on $\mathbf{P}$.

Theorem 3.3.12 (Character formula) *Let $\mathbf{L}$ be an F-stable Levi subgroup of $\mathbf{G}$. Then*

$$R_{\mathbf{L}}^{\mathbf{G}}(\psi)(g) =$$
$$\frac{1}{|\mathbf{L}^F|\,|C_{\mathbf{G}}^\circ(s)^F|}\sum_{h\in\mathbf{G}^F\,|\,s^h\in\mathbf{L}} |C_{^h\mathbf{L}}^\circ(s)^F| \sum_{v\in C_{^h\mathbf{L}}^\circ(s)^F_{\mathrm{uni}}} Q_{C_{^h\mathbf{L}}^\circ(s)}^{C_{\mathbf{G}}^\circ(s)}(u,v^{-1})\,{}^h\psi(sv),$$

for $\psi\in\mathrm{Irr}(\mathbf{L}^F)$ and $g\in\mathbf{G}^F$ with Jordan decomposition $g=su$, and

$$^*R_{\mathbf{L}}^{\mathbf{G}}(\rho)(l) = \frac{|C_{\mathbf{L}}^\circ(s)^F|}{|C_{\mathbf{G}}^\circ(s)^F|}\sum_{u\in C_{\mathbf{G}}^\circ(s)^F_{\mathrm{uni}}} Q_{C_{\mathbf{L}}^\circ(s)}^{C_{\mathbf{G}}^\circ(s)}(u,v^{-1})\,\rho(su),$$

for $\rho\in\mathrm{Irr}(\mathbf{G}^F)$ and $l\in\mathbf{L}^F$ with Jordan decomposition $l=sv$.

For proofs, see [DiMi83, Thm. 2.2], [Lu86a, Prop. 6.2], [Lu90, 1.7(b)] or [DiMi20, Prop. 10.1.2]. The right-hand side of the character formula for $^*R_{\mathbf{L}}^{\mathbf{G}}$ does not change when replacing $\mathbf{G}$ by $C_{\mathbf{G}}^\circ(s)$ and $\mathbf{L}$ by $C_{\mathbf{L}}^\circ(s)$, so this immediately implies:

Corollary 3.3.13 *Let $\mathbf{L}$ be an F-stable Levi subgroup of $\mathbf{G}$ and $\rho\in\mathrm{Irr}(\mathbf{G}^F)$. Let $s\in\mathbf{G}^F$ be semisimple. Then*

$$(\mathrm{Res}^{\mathbf{L}^F}_{C_{\mathbf{L}}^\circ(s)^F}\circ{}^*R_{\mathbf{L}}^{\mathbf{G}})(\rho)(g) = ({}^*R_{C_{\mathbf{L}}^\circ(s)}^{C_{\mathbf{G}}^\circ(s)}\circ\mathrm{Res}^{\mathbf{G}^F}_{C_{\mathbf{G}}^\circ(s)^F})(\rho)(g)$$

for any $g\in\mathbf{L}^F$ with semisimple part s. In particular, if $C_{\mathbf{G}}^\circ(s)\leqslant\mathbf{L}$ then

$$^*R_{\mathbf{L}}^{\mathbf{G}}(\rho)(g) = \rho(g).$$

Observe that by [DiMi20, Prop. 3.5.3], $C_{\mathbf{L}}(s)/C_{\mathbf{L}}^\circ(s)$ consists of semisimple elements, so any $g\in\mathbf{L}^F$ with semisimple part s lies in the connected component $C_{\mathbf{L}}^\circ(s)$ of the centraliser of s, hence the first formula makes sense. The second part follows from the first as then $C_{\mathbf{G}}^\circ(s)=C_{\mathbf{L}}^\circ(s)$.

Let $\mathbf{L}\leqslant\mathbf{G}$ be an F-stable Levi subgroup. Then for $\rho\in\mathrm{Irr}(\mathbf{G}^F)$ we may write

$$^*R_{\mathbf{L}}^{\mathbf{G}}(\rho) = \sum_{\lambda\in\mathrm{Irr}(\mathbf{L}^F)}\langle{}^*R_{\mathbf{L}}^{\mathbf{G}}(\rho),\lambda\rangle\,\lambda = \sum_{\lambda\in\mathrm{Irr}(\mathbf{L}^F)}\langle\rho,R_{\mathbf{L}}^{\mathbf{G}}(\lambda)\rangle\,\lambda.$$

Thus, from Corollary 3.3.13 we obtain Schewe's formula [Sche85, Satz 1.3]

$$\rho(g) = \sum_{\lambda\in\mathrm{Irr}(\mathbf{L}^F)}\langle\rho,R_{\mathbf{L}}^{\mathbf{G}}(\lambda)\rangle\,\lambda(g)$$

for any $g\in\mathbf{G}^F$ with Jordan decomposition $g=su$ with $C_{\mathbf{G}}^\circ(s)\leqslant\mathbf{L}$.

The character formula, together with the fact that the Green functions take values in $\mathbb{Q}$, also implies compatibility with Galois automorphisms:

Corollary 3.3.14 *Let* $\mathbf{L} \leqslant \mathbf{G}$ *be an* F*-stable Levi subgroup and* σ *be a field automorphism of* $\mathbb{K}$. *Then we have*

$$\sigma(R_{\mathbf{L}}^{\mathbf{G}}(\lambda)) = R_{\mathbf{L}}^{\mathbf{G}}(\sigma(\lambda)) \qquad \textit{and} \quad \sigma({}^*R_{\mathbf{L}}^{\mathbf{G}}(\rho)) = {}^*R_{\mathbf{L}}^{\mathbf{G}}(\sigma(\rho))$$

for all $\lambda \in \mathrm{Irr}(\mathbf{L}^F)$, $\rho \in \mathrm{Irr}(\mathbf{G}^F)$.

This allows us to deduce the following action of Galois automorphisms on Lusztig series:

Proposition 3.3.15 *Let* $s \in \mathbf{G}^{*F}$ *be semisimple and* σ *be a field automorphism of* $\mathbb{K}$. *Then* $\sigma(\mathscr{E}(\mathbf{G}^F, s)) = \mathscr{E}(\mathbf{G}^F, s^r)$ *where* $r \in \mathbb{N}$ *is such that* $\sigma(\zeta) = \zeta^r$ *for all* $|\mathbf{G}^F|$*th roots of unity* ζ.

Proof By definition $\rho \in \mathscr{E}(\mathbf{G}^F, s)$ if there is a pair $(\mathbf{T}, \theta) \in \mathfrak{X}(\mathbf{G}, F)$, with $\mathbf{T} \leqslant \mathbf{G}$ an F-stable maximal torus and $\theta \in \mathrm{Irr}(\mathbf{T}^F)$, corresponding to $(\mathbf{T}^*, s) \in \mathfrak{Y}(\mathbf{G}^*, F)$ via Corollary 2.5.14 such that $\langle \rho, R_{\mathbf{T}}^{\mathbf{G}}(\theta)\rangle \neq 0$. Now by assumption we have $\sigma(\theta) = \theta^r$, and according to Remark 2.5.15 the pair $(\mathbf{T}, \theta^r)$ corresponds to $(\mathbf{T}^*, s^r)$, so we obtain

$$\begin{aligned}\big\langle \sigma(\rho), R_{\mathbf{T}^*}^{\mathbf{G}}(s^r)\big\rangle = \big\langle \sigma(\rho), R_{\mathbf{T}}^{\mathbf{G}}(\theta^r)\big\rangle &= \big\langle \sigma(\rho), R_{\mathbf{T}}^{\mathbf{G}}(\sigma(\theta))\big\rangle \\ &= \big\langle \sigma(\rho), \sigma(R_{\mathbf{T}}^{\mathbf{G}}(\theta))\big\rangle = \big\langle \rho, R_{\mathbf{T}}^{\mathbf{G}}(\theta)\big\rangle \neq 0,\end{aligned}$$

where the third equality holds by Corollary 3.3.14. Hence $\sigma(\rho) \in \mathscr{E}(\mathbf{G}^F, s^r)$. □

For a finite group H denote by $\mathrm{CF}(H)_{p'}$ the set of *p-constant* class functions on H, that is, class functions f such that $f(h) = f(h_{p'})$ for any $h \in H$. The following result, which is strongly reminiscent of a similar assertion for ordinary induction and restriction, is easily deduced from the character formula (see [DiMi20, Prop. 10.1.6]):

Proposition 3.3.16 *Let* $f \in \mathrm{CF}(\mathbf{G}^F)_{p'}$ *and* $\mathbf{L}$ *be an* F*-stable Levi subgroup of* $\mathbf{G}$. *Then*

$$\begin{aligned} R_{\mathbf{L}}^{\mathbf{G}}\big(\pi \otimes \mathrm{Res}_{\mathbf{L}^F}^{\mathbf{G}^F}(f)\big) &= R_{\mathbf{L}}^{\mathbf{G}}(\pi) \otimes f && \textit{for all } \pi \in \mathrm{CF}(\mathbf{L}^F), \\ {}^*R_{\mathbf{L}}^{\mathbf{G}}(\eta) \otimes \mathrm{Res}_{\mathbf{L}^F}^{\mathbf{G}^F}(f) &= {}^*R_{\mathbf{L}}^{\mathbf{G}}(\eta \otimes f) && \textit{for all } \eta \in \mathrm{CF}(\mathbf{G}^F). \end{aligned}$$

In particular

$${}^*R_{\mathbf{L}}^{\mathbf{G}}(f) = \mathrm{Res}_{\mathbf{L}^F}^{\mathbf{G}^F}(f).$$

Proof The first claim is immediate from the character formula in Theorem 3.3.12; from this the second is obtained by adjunction.

The last assertion follows by choosing $\eta = 1_{\mathbf{G}}$ in the preceding formula and using that ${}^*R_{\mathbf{L}}^{\mathbf{G}}(1_{\mathbf{G}}) = 1_{\mathbf{L}}$ by Example 3.3.4. □

In particular the previous result shows that for $f \in \mathrm{CF}(\mathbf{G}^F)_{p'}$,

$$R_{\mathbf{L}}^{\mathbf{G}}(\pi \otimes {}^*R_{\mathbf{L}}^{\mathbf{G}}(f)) = R_{\mathbf{L}}^{\mathbf{G}}(\pi) \otimes f \quad \text{and} \quad {}^*R_{\mathbf{L}}^{\mathbf{G}}(\eta) \otimes {}^*R_{\mathbf{L}}^{\mathbf{G}}(f) = {}^*R_{\mathbf{L}}^{\mathbf{G}}(\eta \otimes f)$$

for all $\pi \in \mathrm{CF}(\mathbf{L}^F)$ and $\eta \in \mathrm{CF}(\mathbf{G}^F)$.

For applications to the block theory of finite reductive groups, the following consequence of the character formula is fundamental; here for an ℓ-element $g \in G$ and a prime ℓ,

$$d_\ell^g : \mathrm{CF}(\mathbf{G}^F) \to \mathrm{CF}(C_{\mathbf{G}}(g)^F)$$

denotes the *generalised decomposition map* defined by

$$d_\ell^g(f)(h) = \begin{cases} f(gh) & \text{if } h \in C_{\mathbf{G}}(g)^F_{\ell'}, \\ 0 & \text{else} \end{cases}$$

(see e.g. [CE04, Def. 5.7]).

Proposition 3.3.17 *Let ℓ be a prime different from p and $\mathbf{L} \leqslant \mathbf{G}$ be an F-stable Levi subgroup. Then*

$$d_\ell^s \circ {}^*R_{\mathbf{L}}^{\mathbf{G}} = {}^*R_{C_{\mathbf{L}}^\circ(s)}^{C_{\mathbf{G}}^\circ(s)} \circ d_\ell^s$$

for all (semisimple) ℓ-elements $s \in \mathbf{L}^F$.

Observe that the image of d_ℓ^s consists of class functions on $C_{\mathbf{G}}(s)^F$ vanishing outside $C_{\mathbf{G}}(s)^F_{\ell'}$, and the latter is contained in $C_{\mathbf{G}}^\circ(s)^F$ (see e.g. [DiMi20, Prop. 3.5.3]), so the composition on the right-hand side of the formula does make sense. The claim now readily follows from the character formula for ${}^*R_{\mathbf{L}}^{\mathbf{G}}$ as given in Corollary 3.3.13.

The degree of Lusztig induced characters is given by the same formula as in the Harish-Chandra case, but this is somewhat harder to show (see [Lu76b, Prop. 12]):

Proposition 3.3.18 *Let $\mathbf{L}$ be an F-stable Levi subgroup of $\mathbf{G}$. Then*

$$R_{\mathbf{L}}^{\mathbf{G}}(\lambda)(1) = \varepsilon_{\mathbf{G}}\varepsilon_{\mathbf{L}}|\mathbf{G}^F : \mathbf{L}^F|_{p'}\, \lambda(1) \qquad \textit{for all } \lambda \in \mathrm{Irr}(\mathbf{L}^F).$$

Thus in particular

$${}^*R_{\mathbf{L}}^{\mathbf{G}}(\mathrm{reg}_G) = \varepsilon_{\mathbf{G}}\varepsilon_{\mathbf{L}}|\mathbf{G}^F : \mathbf{L}^F|_{p'}\, \mathrm{reg}_L.$$

The proof only uses the degree formula in terms of Deligne–Lusztig characters in Proposition 2.3.23 and the Mackey formula in the case (1) of Theorem 3.3.7, see [DiMi20, Prop. 10.2.9]. The second statement follows by applying the first to obtain

$$\langle \lambda, {}^*R_{\mathbf{L}}^{\mathbf{G}}(\mathrm{reg}_G)\rangle = \langle R_{\mathbf{L}}^{\mathbf{G}}(\lambda), \mathrm{reg}_G\rangle = \varepsilon_{\mathbf{G}}\varepsilon_{\mathbf{L}}|\mathbf{G}^F : \mathbf{L}^F|_{p'}\langle \lambda, \mathrm{reg}_L\rangle$$

for all $\lambda \in \mathrm{Irr}(\mathbf{L}^F)$. In contrast there is no simple formula for the degree of a Lusztig restricted character!

The previous formula allows us to determine certain values of the 2-parameter Green functions introduced in 3.3.11:

Corollary 3.3.19 *Let* $\mathbf{L}$ *be an* F*-stable Levi subgroup of* $\mathbf{G}$ *and* $v \in \mathbf{L}^F_{\mathrm{uni}}$ *be unipotent. The 2-parameter Green function satisfies*

$$Q^{\mathbf{G}}_{\mathbf{L}}(1, v) = \begin{cases} \varepsilon_{\mathbf{G}}\varepsilon_{\mathbf{L}}|\mathbf{G}^F : \mathbf{L}^F|_{p'} & \text{if } v = 1, \\ 0 & \text{otherwise.} \end{cases}$$

Proof Apply the character formula in Theorem 3.3.12 to the formula for ${}^*R^{\mathbf{G}}_{\mathbf{L}}(\mathrm{reg}_G)$ in Proposition 3.3.18 at the element v. □

We now relate Lusztig induction and restriction to the rational Lusztig series introduced in Definition 2.6.1. For this, let $(\mathbf{G}^*, F)$ be in duality with $(\mathbf{G}, F)$ as in Definition 1.5.17. Let $\mathbf{L} \leqslant \mathbf{G}$ be an F-stable Levi subgroup of $\mathbf{G}$ of type (I, w) (see 3.3.1), and let $\mathbf{L}'$ be an F-stable Levi subgroup of $\mathbf{G}^*$ of type $(I^*, (w^*)^{-1})$, where $I^* = \{s^* \mid s \in I\}$. (Observe that the map $(I, w) \mapsto (I^*, (w^*)^{-1})$ induces a bijection between $\mathbf{G}^F$-conjugacy classes of F-stable Levi subgroups of $\mathbf{G}$ and $\mathbf{G}^{*F}$-conjugacy classes of F-stable Levi subgroups of $\mathbf{G}^*$.) Then the complete root data of $\mathbf{L}$ and $\mathbf{L}'$ are given by $\big((X, R_I, Y, R_I^\vee), w\mathbf{W}_I\big)$ and $\big((Y, R_I^\vee, X, R_I), (w^*)^{-1}\mathbf{W}_I^*\big)$ respectively, with R_I the parabolic subsystem of R determined by I. Thus, $\mathbf{L}'$ has the same complete root datum as a group $(\mathbf{L}^*, F)$ dual to $(\mathbf{L}, F)$ and hence $\mathbf{L}' \cong \mathbf{L}^*$. In this way we can and will identify the dual $\mathbf{L}^*$ of $\mathbf{L}$ with an F-stable Levi subgroup of $\mathbf{G}^*$ (up to conjugation).

Proposition 3.3.20 *Let* $\mathbf{L}$ *be an* F*-stable Levi subgroup of* $\mathbf{G}$ *and* $s \in \mathbf{L}^{*F}$ *be semisimple. Then*

$$R^{\mathbf{G}}_{\mathbf{L}}(\lambda) \in \mathbb{Z}\mathscr{E}(\mathbf{G}^F, s) \qquad \text{for all } \lambda \in \mathscr{E}(\mathbf{L}^F, s),$$

that is, Lusztig induction preserves rational Lusztig series. Furthermore, for all $\rho \in \mathscr{E}(\mathbf{G}^F, s)$*, any constituent of* ${}^*R^{\mathbf{G}}_{\mathbf{L}}(\rho)$ *lies in a Lusztig series* $\mathscr{E}(\mathbf{L}^F, t)$ *for some* $t \in \mathbf{L}^{*F}$ *in the* $\mathbf{G}^{*F}$*-conjugacy class of* s*.*

This is shown in [Lu76b, Cor. 6], see also [Bo06, Thm. 11.10, Cor. 11.11], for F a Frobenius endomorphism; see [BoRo93, Thm. 10.3] for the case of Steinberg maps. It requires consideration of the individual cohomology groups, not just the alternating trace. Note that the second statement follows from the first by adjunction.

Corollary 3.3.21 *Lusztig series are unions of Harish-Chandra series.*

Proof Let $(\mathbf{L}^F, \lambda)$ be a cuspidal pair, with $\lambda \in \mathscr{E}(\mathbf{L}^F, s)$ for some semisimple element $s \in \mathbf{L}^{*F}$. Now if $\rho \in \mathrm{Irr}(\mathbf{G}^F)$ lies in the Harish-Chandra series of $(\mathbf{L}^F, \lambda)$, then by definition $\langle R^{\mathbf{G}}_{\mathbf{L}}(\lambda), \rho\rangle \neq 0$ and hence $\rho \in \mathscr{E}(\mathbf{G}^F, s)$ by Proposition 3.3.20. Thus indeed $\mathscr{E}(\mathbf{G}^F, (\mathbf{L}^F, \lambda)) \subseteq \mathscr{E}(\mathbf{G}^F, s)$. □

We now return to Lusztig's important Jordan decomposition of characters stated in Section 2.6, in particular concerning its connection and interrelation with Lusztig induction. The next result gives a conceptual construction of the Jordan decomposition in most cases, that is, for Lusztig series parametrised by semisimple elements of the dual group whose centraliser lies inside a proper F-stable Levi subgroup. In fact, it can be shown that for a semisimple algebraic group $\mathbf{G}$ there is only a finite number of (F-stable) semisimple conjugacy classes of $\mathbf{G}^*$ whose centraliser is not contained in a proper (F-stable) Levi subgroup of $\mathbf{G}^*$; they are called *quasi-isolated*. See Bonnafé [Bo05] for a classification of such elements.

Theorem 3.3.22 (Lusztig [Lu76b, Prop. 10], [Lu77a, (7.9)]) *Let $\mathbf{L}$ be an F-stable Levi subgroup of $\mathbf{G}$ and $s \in \mathbf{G}^{*F}$ be a semisimple element such that $C_{\mathbf{G}^*}(s)^F \leqslant \mathbf{L}^{*F}$. Then $R_{\mathbf{L}}^{\mathbf{G}}$ induces a bijection of rational Lusztig series*

$$\mathscr{E}(\mathbf{L}^F, s) \longrightarrow \mathscr{E}(\mathbf{G}^F, s), \qquad \lambda \mapsto \varepsilon_{\mathbf{L}}\varepsilon_{\mathbf{G}} R_{\mathbf{L}}^{\mathbf{G}}(\lambda).$$

If moreover we have that $C_{\mathbf{G}^*}(s)^F = \mathbf{L}^{*F}$ then the composition

$$J_s^{\mathbf{G}} := (\_ \otimes \hat{s}^{-1}) \circ \varepsilon_{\mathbf{L}}\varepsilon_{\mathbf{G}}(R_{\mathbf{L}}^{\mathbf{G}})^{-1} : \mathscr{E}(\mathbf{G}^F, s) \to \mathscr{E}(\mathbf{L}^F, s) \to \mathrm{Uch}(\mathbf{L}^F)$$

(where $(\_ \otimes \hat{s}^{-1})$ denotes tensoring with the linear character $\hat{s}^{-1}$) is a Jordan decomposition as in Theorem 2.6.22.

Note that we don't assume $C_{\mathbf{G}^*}(s) = \mathbf{L}^*$ in the second part; so $C_{\mathbf{G}^*}(s)$ might in fact be disconnected, as long as $C_{\mathbf{G}^*}(s)^F = C^\circ_{\mathbf{G}^*}(s)^F$.

In fact, as first conjectured by Broué, even more is true in this situation: for any prime $\ell \neq p$ such that s is of ℓ'-order, $R_{\mathbf{L}}^{\mathbf{G}}$ induces a Morita equivalence between the union of ℓ-blocks of $\mathbf{L}^F$ containing characters from $\mathscr{E}(\mathbf{L}^F, s)$ and the union of ℓ-blocks of $\mathbf{G}^F$ containing characters from $\mathscr{E}(\mathbf{G}^F, s)$ (see Bonnafé–Rouquier [BoRo93, Thm. 10.7], from which the above result follows for arbitrary Steinberg endomorphisms, as well as [CE04, §12], and Bonnafé–Dat–Rouquier [BDR15] for an even more far-reaching result).

We obtain the following partial compatibility between Jordan decomposition and Harish-Chandra induction; see also Corollary 4.7.6 for a stronger statement for groups with connected centre.

Corollary 3.3.23 *Let $(\mathbf{L}, \lambda)$ be a cuspidal pair in $\mathbf{G}$ with $\lambda \in \mathscr{E}(\mathbf{L}^F, s)$ for some $s \in \mathbf{L}^{*F}$ and let $\rho \in \mathrm{Irr}(\mathbf{G}^F)$ lie in the Harish-Chandra series above $(\mathbf{L}, \lambda)$. Assume that $C_{\mathbf{G}^*}(s) \leqslant \mathbf{G}_s^*$ for some F-stable Levi subgroup $\mathbf{G}_s$ of $\mathbf{G}$ and set $\mathbf{L}_s^* := C_{\mathbf{L}^*}(s) = \mathbf{L}^* \cap \mathbf{G}_s^*$, a Levi subgroup of $\mathbf{G}_s^*$. Let $\rho_s \in \mathscr{E}(\mathbf{G}_s^F, s)$ and $\lambda_s \in \mathscr{E}(\mathbf{L}_s^F, s)$ with*

$$\rho = \varepsilon_{\mathbf{G}_s}\varepsilon_{\mathbf{G}} R_{\mathbf{G}_s}^{\mathbf{G}}(\rho_s) \quad \text{and} \quad \lambda = \varepsilon_{\mathbf{L}_s}\varepsilon_{\mathbf{L}} R_{\mathbf{L}_s}^{\mathbf{L}}(\lambda_s)$$

as in Theorem 3.3.22. Then $(\mathbf{L}_s, \lambda_s)$ is a cuspidal pair, ρ_s lies in the Harish-Chandra *series above* $(\mathbf{L}_s, \lambda_s)$ *and*

$$\langle R_{\mathbf{L}}^{\mathbf{G}}(\lambda), \rho\rangle = \langle R_{\mathbf{L}_s}^{\mathbf{G}_s}(\lambda_s), \rho_s\rangle.$$

That is, in the situation of Theorem 3.3.22, Jordan decomposition preserves Harish-Chandra series.

Proof First of all, $\rho \in \mathscr{E}(\mathbf{G}^F, s)$, $\rho_s \in \mathscr{E}(\mathbf{G}_s^F, s)$ and $\lambda_s \in \mathscr{E}(\mathbf{L}_s^F, s)$ by Proposition 3.3.20. Since $\mathbf{L}^*$ is split in $\mathbf{G}^*$ we have $\mathbf{L}^* = C_{\mathbf{G}^*}(Z^\circ(\mathbf{L}^*)_1)$ and thus $\mathbf{L}_s^* = \mathbf{G}_s^* \cap \mathbf{L}^* = C_{\mathbf{G}_s^*}(Z^\circ(\mathbf{L}^*)_1)$ is also a split Levi subgroup of $\mathbf{G}_s^*$ by Lemma 3.1.3. Moreover, as λ is cuspidal, so is λ_s by Theorem 3.2.22. Now we calculate

$$\begin{aligned}\varepsilon_{\mathbf{G}_s}\varepsilon_{\mathbf{G}}\langle R_{\mathbf{L}_s}^{\mathbf{G}_s}(\lambda_s), \rho_s\rangle &= \langle R_{\mathbf{L}_s}^{\mathbf{G}_s}(\lambda_s), {}^*R_{\mathbf{G}_s}^{\mathbf{G}}(\rho)\rangle = \langle R_{\mathbf{G}_s}^{\mathbf{G}}(R_{\mathbf{L}_s}^{\mathbf{G}_s}(\lambda_s)), \rho\rangle \\ &= \langle R_{\mathbf{L}}^{\mathbf{G}}(R_{\mathbf{L}_s}^{\mathbf{L}}(\lambda_s)), \rho\rangle = \varepsilon_{\mathbf{L}_s}\varepsilon_{\mathbf{L}}\langle R_{\mathbf{L}}^{\mathbf{G}}(\lambda), \rho\rangle \neq 0.\end{aligned}$$

In particular, this shows that ρ_s lies in the Harish-Chandra series above $(\mathbf{L}_s, \lambda_s)$. Moreover, as both multiplicities are non-negative, the signs must cancel. □

As a consequence of Proposition 3.3.20 and Theorem 3.3.22 the problem of decomposing Lusztig induction reduces to cases of characters lying in quasi-isolated Lusztig series, of various F-stable Levi subgroups of $\mathbf{G}$. For unipotent characters this decomposition has been determined completely up to one small indeterminacy, see Section 4.6. Further partial results were obtained by Kessar–Malle [KeMa13] and Hollenbach [Ho19] for certain d-split Levi subgroups (to be introduced in Section 3.5) and quasi-isolated series in exceptional groups of Lie type.

The problem of decomposing $R_{\mathbf{L}}^{\mathbf{G}}$ for arbitrary characters would reduce to the unipotent case if it were known in general that Lusztig induction and restriction commute with Jordan decomposition of characters, or more precisely, that a Jordan decomposition can always be chosen having this property. At least for groups with connected centre this is now known except for very few situations in exceptional groups. This will be explained in Section 4.7 once we have collected detailed information on unipotent characters.

We next discuss the behaviour of Lusztig induction with respect to regular embeddings. For this, as in Remark 2.3.16, let us first consider a slightly more general situation. Let $\mathbf{G}'$ also be connected reductive and $F' : \mathbf{G}' \to \mathbf{G}'$ be a Steinberg map. Let $f : \mathbf{G} \to \mathbf{G}'$ be an isotypy (see 1.3.21), that is, f is a homomorphism of algebraic groups such that $\ker(f) \leqslant \mathbf{Z}(\mathbf{G})$ and $\mathbf{G}'_{\mathrm{der}} \leqslant f(\mathbf{G})$, and assume further that $f \circ F = F' \circ f$. Then the map $\mathbf{L} \mapsto \mathbf{L}' := f(\mathbf{L})\mathbf{Z}(\mathbf{G}')$ sets up a bijection between the F-stable Levi subgroups of $\mathbf{G}$ and the F'-stable Levi subgroups of $\mathbf{G}'$, with inverse given by $\mathbf{L} := f^{-1}(\mathbf{L}') \leqslant \mathbf{G}$ for $\mathbf{L}'$ an F'-stable Levi subgroup of $\mathbf{G}'$.

Proposition 3.3.24 *In the above setting, let* $\mathbf{L}$ *be an* F*-stable Levi subgroup of*

$\mathbf{G}$ and $\mathbf{L}' := f(\mathbf{L})\mathbf{Z}(\mathbf{G}') \leqslant \mathbf{G}'$, an F'-stable Levi subgroup of $\mathbf{G}'$. If $\ker(f)$ *is connected then we have*

$$\begin{aligned} f^*_{\mathbf{G}} \circ R^{\mathbf{G}'}_{\mathbf{L}'} &= R^{\mathbf{G}}_{\mathbf{L}} \circ f^*_{\mathbf{L}}, \\ f^*_{\mathbf{L}} \circ {}^*R^{\mathbf{G}'}_{\mathbf{L}'} &= {}^*R^{\mathbf{G}}_{\mathbf{L}} \circ f^*_{\mathbf{G}}, \end{aligned}$$

where $f^*_{\mathbf{G}}(\psi) = \psi \circ f$ for class functions ψ on $\mathbf{G}'^F$, and $f^*_{\mathbf{L}}(\eta) = \eta \circ f$ for class functions η on $\mathbf{L}'^F$.

See [Bo00, Cor. 2.1.3] and [DiMi90, Cor. 9.2] for two different proofs.

Now assume that $i : \mathbf{G} \hookrightarrow \tilde{\mathbf{G}}$ is a regular embedding (see Definition 1.7.1). This is, in particular, an isotypy, and we have $i^*_{\mathbf{G}} = \mathrm{Res}^{\tilde{G}}_{G}$ and $i^*_{\mathbf{L}} = \mathrm{Res}^{\tilde{L}}_{L}$. Thus the above yields the following generalisation of Proposition 3.2.9:

Corollary 3.3.25 *Let $\mathbf{L}$ be an F-stable Levi subgroup of $\mathbf{G}$ with corresponding Levi subgroup $\tilde{\mathbf{L}} = \mathbf{L}\mathbf{Z}(\tilde{\mathbf{G}}) \leqslant \tilde{\mathbf{G}}$. Then we have*

$$\begin{aligned} \mathrm{Ind}^{\tilde{G}}_{G} \circ R^{\mathbf{G}}_{\mathbf{L}} &= R^{\tilde{\mathbf{G}}}_{\tilde{\mathbf{L}}} \circ \mathrm{Ind}^{\tilde{L}}_{L}, & \mathrm{Ind}^{\tilde{L}}_{L} \circ {}^*R^{\mathbf{G}}_{\mathbf{L}} &= {}^*R^{\tilde{\mathbf{G}}}_{\tilde{\mathbf{L}}} \circ \mathrm{Ind}^{\tilde{G}}_{G}, \\ \mathrm{Res}^{\tilde{L}}_{L} \circ {}^*R^{\tilde{\mathbf{G}}}_{\tilde{\mathbf{L}}} &= {}^*R^{\mathbf{G}}_{\mathbf{L}} \circ \mathrm{Res}^{\tilde{G}}_{G}, & \mathrm{Res}^{\tilde{G}}_{G} \circ R^{\tilde{\mathbf{G}}}_{\tilde{\mathbf{L}}} &= R^{\mathbf{G}}_{\mathbf{L}} \circ \mathrm{Res}^{\tilde{L}}_{L}. \end{aligned}$$

Here, the first two formulas follow from the second two by adjunction, see [Bo00, (2.1.4)–(2.1.7)].

Another basic and useful property of Lusztig induction (and restriction) is its commutation with restriction of scalars, which is sometimes useful in reduction statements. For this let $\mathbf{G} = \mathbf{G}_1 \times \cdots \times \mathbf{G}_r$ be a direct product of isomorphic simple groups $\mathbf{G}_i$ and $F : \mathbf{G} \to \mathbf{G}$ a Steinberg endomorphism such that $F(\mathbf{G}_i) = \mathbf{G}_{i+1}$ (with the indices taken modulo r). Thus $\mathbf{G}$ is F-simple in the sense of 1.5.14. Then F^r stabilises each $\mathbf{G}_1$ and induces a Steinberg endomorphism on it. Let $\mathbf{L}_1 \leqslant \mathbf{G}_1$ be an F^r-stable Levi subgroup and $\mathbf{L}_i := F^{i-1}(\mathbf{L}_1) \leqslant \mathbf{G}_i$. Denote by π_1^* the inflation of class functions from $\mathbf{G}_1^{F^r}$ to $\mathbf{G}^F$.

Proposition 3.3.26 *Lusztig induction commutes with restriction of scalars, that is, in the above situation we have $\pi_1^* \circ R^{\mathbf{G}_1}_{\mathbf{L}_1} = R^{\mathbf{G}}_{\mathbf{L}} \circ \pi_1^*$.*

This is shown in [Ta19, Prop. 8.3], see also [DiMi14, Prop. 7.17].

3.4 Duality and the Steinberg Character

The Harish-Chandra induction and restriction functors can be used to define an important duality operation on the space of $\mathbb{K}$-valued class functions on a finite reductive group, see [Al79, Cu80, DeLu82, Kaw82] and also [Lu17]. We continue the set-up of the previous section, that is, $\mathbf{G}$ is a connected reductive algebraic group over an algebraically closed field of characteristic p with a Steinberg endomorphism

$F : \mathbf{G} \to \mathbf{G}$, and we set $G := \mathbf{G}^F$. Furthermore, S denotes the set of Coxeter generators of the Weyl group $W = \mathbf{W}^F$ of $\mathbf{G}^F$ with respect to some maximally split torus of $\mathbf{G}$.

Definition 3.4.1 The *Alvis–Curtis–Kawanaka–Lusztig duality* operator on the space of class functions $\mathrm{CF}(G)$ on G is defined as

$$\mathrm{D}_{\mathbf{G}} := \sum_{I \subseteq S} (-1)^{|I|} R_{\mathbf{L}_I}^{\mathbf{G}} \circ {}^*R_{\mathbf{L}_I}^{\mathbf{G}}.$$

As the functors $R_{\mathbf{L}}^{\mathbf{G}}$ and ${}^*R_{\mathbf{L}}^{\mathbf{G}}$ are adjoint to one another by Proposition 3.1.10 it easily follows that $\mathrm{D}_{\mathbf{G}}$ is self-adjoint:

Proposition 3.4.2 *We have* $\langle \rho, \mathrm{D}_{\mathbf{G}}(\psi)\rangle = \langle \mathrm{D}_{\mathbf{G}}(\rho), \psi\rangle$ *for all class functions* $\rho, \psi \in \mathrm{CF}(G)$*, that is,* $\mathrm{D}_{\mathbf{G}}$ *is self-adjoint.*

The third and fourth formulae in Proposition 3.2.9 show that duality commutes with regular embeddings:

Proposition 3.4.3 *Let* $\mathbf{G} \hookrightarrow \tilde{\mathbf{G}}$ *be a regular embedding. Then*

$$\mathrm{Res}_G^{\tilde{G}} \circ \mathrm{D}_{\tilde{\mathbf{G}}} = \mathrm{D}_{\mathbf{G}} \circ \mathrm{Res}_G^{\tilde{G}}.$$

The Mackey formula can be used to show the following commutation with Lusztig induction:

Theorem 3.4.4 *Let* $\mathbf{L}$ *be an* F*-stable Levi subgroup of* $\mathbf{G}$ *such that the Mackey formula 3.3.7 holds for* $R_{\mathbf{L}}^{\mathbf{G}}$*. Then*

$$\varepsilon_{\mathbf{G}}\, \mathrm{D}_{\mathbf{G}} \circ R_{\mathbf{L}}^{\mathbf{G}} = \varepsilon_{\mathbf{L}} R_{\mathbf{L}}^{\mathbf{G}} \circ \mathrm{D}_{\mathbf{L}}$$

and

$$\varepsilon_{\mathbf{G}}\, {}^*R_{\mathbf{L}}^{\mathbf{G}} \circ \mathrm{D}_{\mathbf{G}} = \varepsilon_{\mathbf{L}}\, \mathrm{D}_{\mathbf{L}} \circ {}^*R_{\mathbf{L}}^{\mathbf{G}}.$$

In particular for all F*-stable maximal tori* $\mathbf{T}$ *of* $\mathbf{G}$ *we have*

$$\varepsilon_{\mathbf{G}}\, \mathrm{D}_{\mathbf{G}} \circ R_{\mathbf{T}}^{\mathbf{G}} = \varepsilon_{\mathbf{T}} R_{\mathbf{T}}^{\mathbf{G}}.$$

See [DiMi20, Thms. 7.2.4 and 10.2.1]. The proof of the first assertion uses only transitivity of $R_{\mathbf{L}}^{\mathbf{G}}$ and the Mackey formula, the second claim follows from the first by adjunction, and the third by using that the Mackey formula holds if one of the Levi subgroups is a torus and that by definition duality is the identity on the characters of a torus.

We conclude from this:

Corollary 3.4.5 *The operator* $\mathrm{D}_{\mathbf{G}} \circ \mathrm{D}_{\mathbf{G}}$ *is the identity on* $\mathrm{CF}(G)$*.*

Proof (See [DiMi20, Cor. 7.2.8].) From the definition and Theorem 3.4.4 we have

$$\begin{aligned}
D_{\mathbf{G}} \circ D_{\mathbf{G}} &= \sum_{I \subseteq S} (-1)^{|I|} R_{\mathbf{L}_I}^{\mathbf{G}} \circ {}^*R_{\mathbf{L}_I}^{\mathbf{G}} \circ D_{\mathbf{G}} \\
&= \sum_{I \subseteq S} (-1)^{|I|} R_{\mathbf{L}_I}^{\mathbf{G}} \circ D_{\mathbf{L}_I} \circ {}^*R_{\mathbf{L}_I}^{\mathbf{G}} \\
&= \sum_{I \subseteq S} (-1)^{|I|} \sum_{J \subseteq I} (-1)^{|J|} R_{\mathbf{L}_I}^{\mathbf{G}} \circ R_{\mathbf{L}_J}^{\mathbf{L}_I} \circ {}^*R_{\mathbf{L}_J}^{\mathbf{L}_I} \circ {}^*R_{\mathbf{L}_I}^{\mathbf{G}} \\
&= \sum_{J \subseteq S} (-1)^{|J|} \sum_{I \supseteq J} (-1)^{|I|} R_{\mathbf{L}_J}^{\mathbf{G}} \circ {}^*R_{\mathbf{L}_J}^{\mathbf{G}},
\end{aligned}$$

using transitivity and the Mackey formula for Harish-Chandra induction. The claim now follows as clearly $\sum_{I \supseteq J} (-1)^{|I|} = 0$ for all proper subsets J of S. □

Since Lusztig series are defined in terms of multiplicities in Deligne–Lusztig characters (see Definition 2.6.1) the previous theorem also immediately implies the preservation of rational Lusztig series:

Corollary 3.4.6 *If* $\rho \in \mathscr{E}(G, s)$ *then* $\pm D_{\mathbf{G}}(\rho) \in \mathscr{E}(G, s)$.

A fundamental property of the duality operation is that it sends irreducible characters to irreducible characters up to sign; recall from 3.2.3 that the relative rank $r(L)$ of a Levi subgroup $L = \mathbf{L}^F$ of G satisfies $(-1)^{r(L)} = \varepsilon_{[\mathbf{L},\mathbf{L}]}$.

Proposition 3.4.7 *The duality* $D_{\mathbf{G}}$ *permutes the irreducible characters of* G *up to sign. More precisely, if* $\rho \in \mathscr{E}(G, (L, \lambda))$ *for a cuspidal pair* (L, λ) *in* G *then* $(-1)^{r(L)} D_{\mathbf{G}}(\rho)$ *is again irreducible and also lies in* $\mathscr{E}(G, (L, \lambda))$.

The irreducibility up to sign was first shown in [Al79, Thm. 4.2], see also [DiMi20, Cor. 7.2.9]. It is immediate from the definition that $D_{\mathbf{L}}(\lambda) = (-1)^{r(L)}\lambda$ for a cuspidal character λ of L. It then follows by adjointness (Proposition 3.1.10) and commutation with $D_{\mathbf{G}}$ (Theorem 3.4.4) that

$$\begin{aligned}
\langle D_{\mathbf{G}}(\rho), R_{\mathbf{L}}^{\mathbf{G}}(\lambda) \rangle &= \langle {}^*R_{\mathbf{L}}^{\mathbf{G}}(D_{\mathbf{G}}(\rho)), \lambda \rangle = \langle D_{\mathbf{L}}({}^*R_{\mathbf{L}}^{\mathbf{G}}(\rho)), \lambda \rangle \\
&= \langle {}^*R_{\mathbf{L}}^{\mathbf{G}}(\rho), D_{\mathbf{L}}(\lambda) \rangle = \pm \langle {}^*R_{\mathbf{L}}^{\mathbf{G}}(\rho), \lambda \rangle \neq 0,
\end{aligned}$$

so $D_{\mathbf{G}}(\rho)$ lies in the Harish-Chandra series of (L, λ) as claimed.

This can be made even more precise. For this, let ε denote the sign character of the Coxeter group $W_G(L, \lambda)$ and recall the bijection $I_{L,\lambda}^G : \mathrm{Irr}_{\mathbb{K}}(W_G(L, \lambda)) \to \mathscr{E}(G, (L, \lambda))$ from Theorem 3.2.5.

Theorem 3.4.8 *Let* (L, λ) *be a cuspidal pair in* G. *We have*

$$D_{\mathbf{G}}(I_{L,\lambda}^G(\phi)) = (-1)^{r(L)} I_{L,\lambda}^G(\varepsilon \otimes \phi)$$

for all $\phi \in \mathrm{Irr}(W_G(L, \lambda))$.

See [HoLe83, Thm. 7.5], or [CuRe87, Thm. 71.14] for the case of the principal series. We will have more to say about the degree of $\mathrm{D}_{\mathbf{G}}(\rho)$ in Propositions 3.4.20 and 3.4.21 below.

Duality gives a particularly elegant approach, due to Curtis [Cu66], to the Steinberg character, originally introduced by Steinberg [St57] in the case of groups of Lie type, of a finite group with an algebraic BN-pair. By Example 3.1.8 we have that ${}^*R_{\mathbf{L}}^{\mathbf{G}}(1_{\mathbf{G}}) = 1_{\mathbf{L}}$ for any Levi subgroup $L \leqslant G$. Applying this to $L = T$ we see that $1_G \in \mathscr{E}(G, (T, 1_{\mathbf{T}}))$. Since $r(T) = 0$ it follows from Proposition 3.4.7 that $\mathrm{D}_{\mathbf{G}}(1_{\mathbf{G}})$ is an irreducible character.

Definition 3.4.9 The irreducible character $\mathrm{St}_{\mathbf{G}} := \mathrm{D}_{\mathbf{G}}(1_{\mathbf{G}})$, where $1_{\mathbf{G}}$ is the trivial character of G, is called the *Steinberg character* of G.

According to Theorem 3.4.8 the Steinberg character $\mathrm{St}_{\mathbf{G}}$ lies in the principal series $\mathscr{E}(G, (T, 1_{\mathbf{T}}))$ of G and is labelled by the sign character of the Weyl group $W_G(T, 1_{\mathbf{G}}) = W$. Moreover, as $1_{\mathbf{G}}$ is a uniform function, the same holds for $\mathrm{St}_{\mathbf{G}}$ by construction; more explicitly we have

$$\mathrm{St}_{\mathbf{G}} = \frac{1}{|\mathbf{W}|} \sum_{w \in \mathbf{W}} (-1)^{l(w)} R_w$$

(see Example 2.3.23). Since duality commutes with Lusztig induction by Theorem 3.4.4 we conclude via Remark 3.3.9 that

$$\begin{aligned} {}^*R_{\mathbf{L}}^{\mathbf{G}}(\mathrm{St}_{\mathbf{G}}) = {}^*R_{\mathbf{L}}^{\mathbf{G}}(\mathrm{D}_{\mathbf{G}}(1_{\mathbf{G}})) &= \varepsilon_{\mathbf{L}}\varepsilon_{\mathbf{G}}\, \mathrm{D}_{\mathbf{L}}({}^*R_{\mathbf{L}}^{\mathbf{G}}(1_{\mathbf{G}})) \\ &= \varepsilon_{\mathbf{L}}\varepsilon_{\mathbf{G}}\, \mathrm{D}_{\mathbf{L}}(1_{\mathbf{L}}) = \varepsilon_{\mathbf{L}}\varepsilon_{\mathbf{G}} \mathrm{St}_{\mathbf{L}} \end{aligned}$$

for any F-stable Levi subgroup $\mathbf{L} \leqslant \mathbf{G}$. The values of the Steinberg character are easy to describe:

Proposition 3.4.10 *Let $G = \mathbf{G}^F$ be a finite group of Lie type and $g \in G$. Then*

$$\mathrm{St}_{\mathbf{G}}(g) = \begin{cases} \varepsilon_{\mathbf{G}}\varepsilon_{C_{\mathbf{G}}^\circ(g)} |C_{\mathbf{G}}^\circ(g)^F|_p & \textit{if } g \textit{ is semisimple,} \\ 0 & \textit{else.} \end{cases}$$

See [DiMi20, Cor. 7.4.4]. In particular the Steinberg character $\mathrm{St}_{\mathbf{G}}$ has degree $\mathrm{St}_{\mathbf{G}}(1) = |G|_p$, so it is an irreducible character of G of p-defect zero. Now assume that G is perfect (which is often the case for $\mathbf{G}$ of simply connected type, see e.g. [MaTe11, Thm. 24.17]). Then it can be shown using the representation theory of algebraic groups that all other irreducible characters of $G = \mathbf{G}^F$ have positive p-defect. (This also follows from Lusztig's classification of irreducible characters.) Indeed, Weyl's character formula shows that all other p-modular irreducible representations of G have degree less than $|G|_p$. Since any complex irreducible character of G of degree divisible by $|G|_p$ remains irreducible under p-modular reduction, no other complex irreducible character of G can have p-defect zero.

The Steinberg character relates Lusztig induction to ordinary induction:

Corollary 3.4.11 *Let* $\mathbf{L}$ *be an* F*-stable Levi subgroup of* $\mathbf{G}$. *Then*

$$\varepsilon_{\mathbf{G}}\, R_{\mathbf{L}}^{\mathbf{G}}(\lambda) \otimes \mathrm{St}_{\mathbf{G}} = \varepsilon_{\mathbf{L}}\, \mathrm{Ind}_{\mathbf{L}^F}^{\mathbf{G}^F}(\lambda \otimes \mathrm{St}_{\mathbf{L}}) \qquad \text{for all } \lambda \in \mathrm{Irr}(\mathbf{L}^F),$$
$$\varepsilon_{\mathbf{G}}\, {}^*R_{\mathbf{L}}^{\mathbf{G}}(\rho \otimes \mathrm{St}_{\mathbf{G}}) = \varepsilon_{\mathbf{L}}\, \mathrm{Res}_{\mathbf{L}^F}^{\mathbf{G}^F}(\rho) \otimes \mathrm{St}_{\mathbf{L}} \qquad \text{for all } \rho \in \mathrm{Irr}(\mathbf{G}^F).$$

Proof Application of the character formula 3.3.12 and Proposition 3.4.10 show that in order to verify the second formula we need to prove that

$$\varepsilon_{C^\circ_{\mathbf{G}}(s)} \frac{|C^\circ_{\mathbf{L}}(s)^F|}{|C^\circ_{\mathbf{G}}(s)^F|_{p'}} Q^{C^\circ_{\mathbf{G}}(s)}_{C^\circ_{\mathbf{L}}(s)}(1, v^{-1})\, \rho(s) = \begin{cases} \varepsilon_{C^\circ_{\mathbf{L}}(s)} |C^\circ_{\mathbf{L}}(s)^F|_p\, \rho(s) & \text{if } v = 1, \\ 0 & \text{else,} \end{cases}$$

for any semisimple element $s \in \mathbf{L}^F$ and unipotent $v \in \mathbf{L}^F_{\mathrm{uni}}$ centralising s. But this is just Corollary 3.3.19 applied to the F-stable Levi subgroup $C^\circ_{\mathbf{L}}(s)$ of $C^\circ_{\mathbf{G}}(s)$. The first formula follows from the second by adjunction. □

3.4.12 Recall the semisimple characters $\rho \in \mathscr{S}_0(\mathbf{G}^F)$ of $\mathbf{G}^F$ introduced in Definition 2.6.9, characterised by the property that $\langle \rho, \Delta_{\mathbf{G}} \rangle \neq 0$, where $\Delta_{\mathbf{G}}$ is the class function taking values

$$\Delta_{\mathbf{G}}(g) = \begin{cases} |\mathbf{Z}^\circ(\mathbf{G})^F| q^l & \text{if } g \in \mathscr{O}_0^F, \\ 0 & \text{otherwise,} \end{cases}$$

with l the semisimple rank of $\mathbf{G}$ and $\mathscr{O}_0 \subseteq \mathbf{G}$ the class of regular unipotent elements.

Definition 3.4.13 Let $\rho \in \mathrm{Irr}(\mathbf{G}^F)$. Then ρ is a *regular character* of $\mathbf{G}^F$ if $\pm D_{\mathbf{G}}(\rho)$ is a semisimple character. We write $\mathscr{S}^0(\mathbf{G}^F)$ for the set of all regular characters of $\mathbf{G}^F$.

This notion should not be confused with the (in general reducible) regular character of an abstract finite group.

Example 3.4.14 According to 2.6.9 the trivial character is semisimple and thus the Steinberg character $\mathrm{St}_{\mathbf{G}}$ is an example of a regular character.

3.4.15 Assume that $\mathbf{Z}(\mathbf{G})$ is connected. Let $(\mathbf{G}^*, F)$ be dual to $(\mathbf{G}, F)$ (see Definition 1.5.17). Then the (rational) Lusztig series of any semisimple element $s \in \mathbf{G}^{*F}$ contains a unique semisimple character ρ_s (see 2.6.10). By Theorems 2.6.11 and 3.4.4 we obtain

$$\left\langle R^{\mathbf{G}}_{\mathbf{T}^*}(s), D_{\mathbf{G}}(\rho_s) \right\rangle = \left\langle D_{\mathbf{G}}(R^{\mathbf{G}}_{\mathbf{T}^*}(s)), \rho_s \right\rangle = \varepsilon_{\mathbf{G}}\varepsilon_{\mathbf{T}^*} \left\langle R^{\mathbf{G}}_{\mathbf{T}^*}(s), \rho_s \right\rangle = \varepsilon_{\mathbf{H}}\varepsilon_{\mathbf{T}^*},$$

where $\mathbf{H} = C_{\mathbf{G}^*}(s)$. The degree formula in Theorem 2.6.2 then gives

$$D_{\mathbf{G}}(\rho_s)(1) = \frac{|\mathbf{G}^F|}{|\mathbf{H}^F|} \sum_{\mathbf{T}^*} \varepsilon_{\mathbf{G}}\varepsilon_{\mathbf{T}^*} \langle R_{\mathbf{T}^*}^{\mathbf{G}}(s), D_{\mathbf{G}}(\rho_s)\rangle = \frac{|\mathbf{G}^F|}{|\mathbf{H}^F|} \sum_{\mathbf{T}^* \leqslant \mathbf{H}} \varepsilon_{\mathbf{G}}\varepsilon_{\mathbf{H}},$$

where the sum runs over all F-stable maximal tori of $\mathbf{H}$; thus the regular character corresponding to ρ_s is $\rho^s := \varepsilon_{\mathbf{G}}\varepsilon_{\mathbf{H}} D_{\mathbf{G}}(\rho_s) \in \mathrm{Irr}(\mathbf{G}^F)$. It lies in the same (rational) Lusztig series $\mathscr{E}(\mathbf{G}^F, s)$ as ρ_s. We have the following analogue of Theorem 2.6.11:

Theorem 3.4.16 (See [DeLu76, Thm. 7]) *Assume that $\mathbf{Z}(\mathbf{G})$ is connected and let $s \in \mathbf{G}^{*F}$ be semisimple. Let $\rho^s \in \mathscr{S}^0(\mathbf{G}^F)$ be the unique regular character that belongs to $\mathscr{E}(\mathbf{G}^F, s)$. Then the following hold.*

(a) *For any F-stable maximal torus $\mathbf{T}^* \leqslant \mathbf{G}^*$ with $s \in \mathbf{T}^*$, $\langle R_{\mathbf{T}^*}^{\mathbf{G}}(s), \rho^s\rangle = \varepsilon_{\mathbf{G}}\varepsilon_{\mathbf{T}^*}$.*
(b) *Let $\mathbf{H} := C_{\mathbf{G}^*}(s)$. Then $\rho^s(1) = |\mathbf{G}^{*F} : \mathbf{H}^F|_{p'} \cdot |\mathbf{H}^F|_p$.*
(c) *Via the Jordan decomposition of characters in Theorem 2.6.4, the character $\rho^s \in \mathscr{E}(\mathbf{G}^F, s)$ corresponds to the Steinberg character of $\mathbf{H}^F$.*

Proof (a) was already shown above, and (b) follows from the above formula for $D_{\mathbf{G}}(\rho_s)(1)$ with Remark 2.5.16.

(c) Let $\psi \in \mathrm{Uch}(\mathbf{H}^F)$ correspond to ρ^s via a bijection as in Theorem 2.6.4. Let $\mathbf{T}^* \subseteq \mathbf{H}$ be an F-stable maximal torus. Using (a) we obtain

$$\langle R_{\mathbf{T}^*}^{\mathbf{H}}(1_{\mathbf{T}^*}), \psi\rangle = \varepsilon_{\mathbf{G}}\varepsilon_{\mathbf{H}}\langle R_{\mathbf{T}^*}^{\mathbf{G}}(s), \rho^s\rangle = \varepsilon_{\mathbf{H}}\varepsilon_{\mathbf{T}^*}.$$

Hence, Example 2.2.27 shows that ψ must be the Steinberg character of $\mathbf{H}^F$. □

Regular characters also satisfy the analogue of Corollary 2.6.18. For this let $i : \mathbf{G} \hookrightarrow \tilde{\mathbf{G}}$ be a regular embedding with dual epimorphism $i^* : \tilde{\mathbf{G}}^* \to \mathbf{G}^*$.

Corollary 3.4.17 *Let $s \in G^*$ be semisimple and let $\tilde{s} \in \tilde{G}^*$ be such that $i^*(\tilde{s}) = s$. Let $\tilde{\rho} \in \mathrm{Irr}(\tilde{G})$ be the unique regular character in $\mathscr{E}(\tilde{G}, \tilde{s})$.*

(a) *The regular characters in $\mathscr{E}(G, s)$ are precisely the irreducible constituents of $\tilde{\rho}|_G$.*
(b) *Let $\rho \in \mathscr{E}(G, s)$ be a regular character. Then $\langle R_{\mathbf{T}^*}^{\mathbf{G}}(s), \rho\rangle = \pm 1$ for any F-stable maximal torus $\mathbf{T}^* \subseteq \mathbf{G}^*$ such that $s \in \mathbf{T}^*$.*

Proof Let $\rho \in \mathscr{E}(G, s)$ be regular. Then there exists a semisimple character $\rho' \in \mathscr{E}(G, s)$ with $\rho = \pm D_{\mathbf{G}}(\rho')$. By Corollary 2.6.18 this is a constituent of $\tilde{\rho}'|_G$ for the (unique) semisimple character $\tilde{\rho}' \in \mathscr{E}(\tilde{G}, \tilde{s})$. As $\tilde{\mathbf{G}}$ has connected centre, $\mathscr{E}(\tilde{G}, \tilde{s})$ also contains a unique regular character, so that $\tilde{\rho} = \pm D_{\tilde{\mathbf{G}}}(\tilde{\rho}')$. Thus we have

$$\begin{aligned}\langle \mathrm{Res}_G^{\tilde{G}}(\tilde{\rho}), \rho\rangle &= \pm\langle \mathrm{Res}_G^{\tilde{G}}(D_{\tilde{\mathbf{G}}}(\tilde{\rho}')), \rho\rangle = \pm\langle D_{\mathbf{G}}(\mathrm{Res}_G^{\tilde{G}}(\tilde{\rho}')), \rho\rangle \\ &= \pm\langle \mathrm{Res}_G^{\tilde{G}}(\tilde{\rho}'), D_{\mathbf{G}}(\rho)\rangle = \pm\langle \mathrm{Res}_G^{\tilde{G}}(\tilde{\rho}'), \rho'\rangle \neq 0,\end{aligned}$$

where the second equality holds by Proposition 3.2.9; so ρ is a constituent of $\tilde{\rho}|_G$. The same sequence of equations also shows that any constituent of $\mathrm{Res}_G^{\tilde{G}}(\tilde{\rho})$ is in fact regular.

In (b), as $\pm\,\mathrm{D}_{\mathbf{G}}(\rho)$ is semisimple by definition, we have

$$\langle R_{\mathbf{T}^*}^{\mathbf{G}}(s), \rho\rangle = \langle \mathrm{D}_{\mathbf{G}}(R_{\mathbf{T}^*}^{\mathbf{G}}(s)), \mathrm{D}_{\mathbf{G}}(\rho)\rangle = \pm\langle R_{\mathbf{T}^*}^{\mathbf{G}}(s), \mathrm{D}_{\mathbf{G}}(\rho)\rangle = \pm 1$$

by Corollary 2.6.18(b). □

Remark 3.4.18 Assume $\mathbf{G}$ has connected centre. By Theorems 2.6.11 and 3.4.16 the dual

$$\Gamma_{\mathbf{G}} := \mathrm{D}_{\mathbf{G}}(\Delta_{\mathbf{G}}) = \sum_{s \in \mathbf{G}^{*F}/\sim} \rho^s$$

of $\Delta_{\mathbf{G}}$ is an actual character (where the sum runs over a system of representatives of the semisimple conjugacy classes in $\mathbf{G}^{*F}$). It is the so-called *Gelfand–Graev character* of $\mathbf{G}^F$. It can be constructed as an induced character from a suitable unipotent subgroup of $\mathbf{G}^F$. Applying $\mathrm{D}_{\mathbf{G}}$ to the formula for $\Delta_{\mathbf{G}}$ in Remark 2.6.12 we obtain the expression

$$\Gamma_{\mathbf{G}} = \frac{1}{|\mathbf{G}^F|} \sum_{(\mathbf{T}^*, s) \in \mathfrak{Y}(\mathbf{G}^*, F)} \varepsilon_{\mathbf{G}}\varepsilon_{\mathbf{T}^*} |\mathbf{T}^{*F}| R_{\mathbf{T}^*}^{\mathbf{G}}(s)$$

for the Gelfand–Graev character of $\mathbf{G}^F$.

The Gelfand–Graev character, together with its generalisations, the so-called *generalised Gelfand–Graev characters*, plays a crucial role in the study of decomposition matrices of finite groups of Lie type for non-defining primes, see e.g. [GeHi97], [DuMa18]. See also [Ca85, §8.1]; [DiMi20, §12.3] discuss Gelfand–Graev characters in the non-connected centre case, leading to a different notion of generalised Gelfand–Graev characters.

Let $G_{\mathrm{uni}} \subset G$ denote the set of unipotent elements of G (i.e., the set of p-elements), and δ_u the characteristic function of G_{uni}, taking value 1 on elements of G_{uni} and zero elsewhere. Then the dual of the regular character reg_G of G is given as follows (see [DiMi20, Cor. 7.4.5]):

Corollary 3.4.19 *We have* $\mathrm{D}_{\mathbf{G}}(\mathrm{reg}_G) = |G|_{p'}\, \delta_u$.

Proof The function δ_u has the property that $\delta_u(g) = \delta_u(g_{p'})$ for all $g \in G$, where $g_{p'}$ denotes the p'-part of g. It then follows from Proposition 3.3.16 (with $f = \delta_u$ and $\pi = 1_L$) that $R_{\mathbf{L}}^{\mathbf{G}}(\mathrm{Res}_L^G(\delta_u)) = R_{\mathbf{L}}^{\mathbf{G}}(1_{\mathbf{L}}) \otimes \delta_u$ and ${}^*R_{\mathbf{L}}^{\mathbf{G}}(\delta_u) = \mathrm{Res}_L^G(\delta_u)$ for any F-stable Levi subgroup $\mathbf{L} \leqslant \mathbf{G}$. But then $\mathrm{D}_{\mathbf{G}}(\delta_u) = \mathrm{D}_{\mathbf{G}}(1_{\mathbf{G}}) \otimes \delta_u$ and so

$$\mathrm{D}_{\mathbf{G}}(|G|_{p'}\, \delta_u) = |G|_{p'}\, \mathrm{D}_{\mathbf{G}}(1_{\mathbf{G}}) \otimes \delta_u = |G|_{p'}\, \mathrm{St}_{\mathbf{G}} \otimes \delta_u = \mathrm{reg}_G$$

by Proposition 3.4.10. □

From this, we find with Proposition 3.4.2 and Corollary 3.4.5 that

$$|G| = \langle \mathrm{reg}_G, \mathrm{reg}_G\rangle = \big\langle \mathrm{D}_{\mathbf{G}}(\mathrm{reg}_G), \mathrm{D}_{\mathbf{G}}(\mathrm{reg}_G)\big\rangle$$
$$= |G|_{p'}^2 \,\langle \delta_u, \delta_u\rangle = |G|_{p'}^2\, |G_{\mathrm{uni}}|/|G|.$$

This shows that G has exactly $|G_{\mathrm{uni}}| = |G|_p^2$ unipotent elements, a result originally due to Steinberg [St68, Thm. 15.1]. As a further consequence we obtain (see [CuRe87, Thm. 71.11], and also [Lu84a, (8.5.12)]):

Proposition 3.4.20 *Let* $\rho \in \mathrm{Irr}(G)$. *Then*

$$\mathrm{D}_{\mathbf{G}}(\rho)(1)_{p'} = \pm\rho(1)_{p'},$$

that is, the degrees of ρ *and* $\pm\,\mathrm{D}_{\mathbf{G}}(\rho)$ *differ only by a power of* p.

Proof From Corollary 3.4.19 and Proposition 3.4.2 we get

$$\sum_{u\in G_{\mathrm{uni}}} \rho(u) = |G|\langle \rho, \delta_u\rangle = |G|\big\langle \mathrm{D}_{\mathbf{G}}(\rho), \mathrm{D}_{\mathbf{G}}(\delta_u)\big\rangle$$
$$= |G|_p\big\langle \mathrm{D}_{\mathbf{G}}(\rho), \mathrm{reg}_G\big\rangle = |G|_p\, \mathrm{D}_{\mathbf{G}}(\rho)(1).$$

Now $\rho(1)^{-1}\sum_{u\in G_{\mathrm{uni}}}\rho(u)$ is a sum of values of central characters on the p-singular classes of G, hence an algebraic integer. Thus $|G|_p\,\mathrm{D}_{\mathbf{G}}(\rho)(1)/\rho(1) \in \mathbb{Z}$, and, replacing ρ by $\pm\,\mathrm{D}_{\mathbf{G}}(\rho)$ in this argument and using that $\mathrm{D}_{\mathbf{G}}^2 = \mathrm{id}$ by Corollary 3.4.5, we also have $|G|_p\rho(1)/\mathrm{D}_{\mathbf{G}}(\rho)(1) \in \mathbb{Z}$. The claim follows from this. □

This has the following nice explanation due to Alvis [Al82, Cor. 3.6]:

Proposition 3.4.21 *Let* $\rho \in \mathrm{Irr}(G)$. *Then the degree polynomials of* ρ *and* $\mathrm{D}_{\mathbf{G}}(\rho)$ *are related by*

$$\mathbb{D}_{\mathrm{D}_{\mathbf{G}}(\rho)}(\mathbf{q}) = \mathbf{q}^N\, \mathbb{D}_\rho(\mathbf{q}^{-1}).$$

Proof Recall that the order polynomial $|\mathbb{G}|_{p'}$ is a product of cyclotomic polynomials of degree $N+r$, where $N = |R^+|$ and $r = \mathrm{rnk}(\mathbf{G})$. Similarly, the order $|\mathbb{T}|$ for any maximal torus $\mathbb{T}$ of $\mathbb{G}$ is a product of cyclotomic polynomials of degree r dividing $|\mathbb{G}|_{p'}$, and moreover the number of factors $\mathbf{q}-1$ in this polynomial equals $r(\mathbb{T})$ by Corollary 3.2.21. Now if Φ is a cyclotomic polynomial, then $\Phi(\mathbf{q}^{-1}) = \pm\mathbf{q}^{-d}\Phi(\mathbf{q})$, where $d = \deg(\Phi)$ and the sign is positive unless $\Phi = \mathbf{q}-1$. Thus if we let $f_{\mathbb{T}} := |\mathbb{G}|_{p'}/|\mathbb{T}| \in \mathbb{Q}[\mathbf{q}]$ then $\mathbf{q}^N f_{\mathbb{T}}(\mathbf{q}^{-1}) = \varepsilon_{\mathbf{T}}\varepsilon_{\mathbf{G}} f_{\mathbb{T}}(\mathbf{q})$.

Now by Remark 2.2.28 the uniform projection of ρ can be written as a sum over conjugacy classes of pairs $(\mathbf{T},\theta) \in \mathfrak{X}(\mathbf{G}, F)$

$$\sum_{(\mathbf{T},\theta)} \varepsilon_{\mathbf{T}}\varepsilon_{\mathbf{G}}\, a(\mathbf{T},\theta)\, R_{\mathbf{T}}^{\mathbf{G}}(\theta),$$

with rational coefficients $a(\mathbf{T},\theta) = \langle\rho, R_{\mathbf{T}}^{\mathbf{G}}(\theta)\rangle / \langle R_{\mathbf{T}}^{\mathbf{G}}(\theta), R_{\mathbf{T}}^{\mathbf{G}}(\theta)\rangle$. Similarly, the uniform projection of $\mathrm{D}_{\mathbf{G}}(\rho)$ is given by

$$\sum_{(\mathbf{T},\theta)} \varepsilon_{\mathbf{T}}\varepsilon_{\mathbf{G}}\, b(\mathbf{T},\theta)\, R_{\mathbf{T}}^{\mathbf{G}}(\theta)$$

with $\langle R_{\mathbf{T}}^{\mathbf{G}}(\theta), R_{\mathbf{T}}^{\mathbf{G}}(\theta)\rangle\, b(\mathbf{T},\theta) = \langle \mathrm{D}_{\mathbf{G}}(\rho), R_{\mathbf{T}}^{\mathbf{G}}(\theta)\rangle$. By Proposition 3.4.2 and the commutation property of $\mathrm{D}_{\mathbf{G}}$ with $R_{\mathbf{T}}^{\mathbf{G}}$ from Theorem 3.4.4 this latter quantity equals

$$\langle \rho, \mathrm{D}_{\mathbf{G}}(R_{\mathbf{T}}^{\mathbf{G}}(\theta))\rangle = \varepsilon_{\mathbf{T}}\varepsilon_{\mathbf{G}}\langle\rho, R_{\mathbf{T}}^{\mathbf{G}}(\theta)\rangle = \varepsilon_{\mathbf{T}}\varepsilon_{\mathbf{G}}\langle R_{\mathbf{T}}^{\mathbf{G}}(\theta), R_{\mathbf{T}}^{\mathbf{G}}(\theta)\rangle\, a(\mathbf{T},\theta),$$

whence $b(\mathbf{T},\theta) = \varepsilon_{\mathbf{T}}\varepsilon_{\mathbf{G}} a(\mathbf{T},\theta)$. Evaluation at the identity element thus gives

$$\begin{aligned}\mathrm{D}_{\mathbf{G}}(\rho)(1) &= \varepsilon_{\mathbf{T}}\varepsilon_{\mathbf{G}} \sum_{(\mathbf{T},\theta)} a(\mathbf{T},\theta)\, R_{\mathbf{T}}^{\mathbf{G}}(\theta)(1)\\ &= \varepsilon_{\mathbf{T}}\varepsilon_{\mathbf{G}} \sum_{(\mathbf{T},\theta)} a(\mathbf{T},\theta)\, f_{\mathbf{T}}(q) = q^N \sum_{(\mathbf{T},\theta)} a(\mathbf{T},\theta)\, f_{\mathbf{T}}(q^{-1}),\end{aligned}$$

the degree polynomial of ρ evaluated at q^{-1}. □

Let $\mathbf{q}^{a_\rho}$ denote the precise power of $\mathbf{q}$ dividing the degree polynomial of ρ, and A_ρ its degree. Then the preceding result shows that

$$a_{\mathrm{D}_{\mathbf{G}}(\rho)} = N - A_\rho.$$

We will have more to say about the invariants a_ρ and A_ρ in Section 4.1.

Proposition 3.4.22 *Assume that* $\mathbf{G}$ *has only components of classical type,* F *is a Frobenius map and that* $\mathbf{Z}(\mathbf{G})$ *is connected. Then the Jordan decomposition for* $\mathbf{G}$ *commutes with* $\mathrm{D}_{\mathbf{G}}$.

Proof Let $s \in G^*$ be semisimple and $\mathbf{H} := C_{\mathbf{G}^*}(s)$. By our assumption on $\mathbf{Z}(\mathbf{G})$, $\mathbf{H}$ is connected. We write $J_s : \mathscr{E}(\mathbf{G}^F, s) \to \mathrm{Uch}(\mathbf{H}^F)$ for a Jordan decomposition as in Theorem 2.6.4. Then for all $\rho \in \mathscr{E}(\mathbf{G}^F, s)$, all F-stable maximal tori $\mathbf{T} \leqslant \mathbf{H}^*$ and all $\theta \in \mathrm{Irr}(\mathbf{T}^F)$ we have

$$\begin{aligned}\langle \mathrm{D}_{\mathbf{H}}(J_s(\rho)), R_{\mathbf{T}}^{\mathbf{H}}(\theta)\rangle &= \langle J_s(\rho), \mathrm{D}_{\mathbf{H}}(R_{\mathbf{T}}^{\mathbf{H}}(\theta))\rangle = \langle J_s(\rho), R_{\mathbf{T}}^{\mathbf{H}}(\mathrm{D}_{\mathbf{T}}(\theta))\rangle\\ &= \varepsilon_{\mathbf{G}}\varepsilon_{\mathbf{H}}\langle\rho, R_{\mathbf{T}}^{\mathbf{G}}(\mathrm{D}_{\mathbf{T}}(\theta))\rangle = \varepsilon_{\mathbf{G}}\varepsilon_{\mathbf{H}}\langle\rho, \mathrm{D}_{\mathbf{G}}(R_{\mathbf{T}}^{\mathbf{G}}(\theta))\rangle\\ &= \varepsilon_{\mathbf{G}}\varepsilon_{\mathbf{H}}\langle \mathrm{D}_{\mathbf{G}}(\rho), R_{\mathbf{T}}^{\mathbf{G}}(\theta)\rangle = \langle J_s(\mathrm{D}_{\mathbf{G}}(\rho)), R_{\mathbf{T}}^{\mathbf{H}}(\theta)\rangle.\end{aligned}$$

So $\mathrm{D}_{\mathbf{H}}(J_s(\rho))$ and $J_s(\mathrm{D}_{\mathbf{G}}(\rho))$ have the same multiplicities in all Deligne–Lusztig characters of $\mathbf{H}$. Now note that $\mathbf{H}$ has only components of classical type. But the unipotent characters of groups of classical type are uniquely determined by their multiplicities in Deligne–Lusztig characters, see Theorem 4.4.23. Thus $\mathrm{D}_{\mathbf{H}}(J_s(\rho)) = J_s(\mathrm{D}_{\mathbf{G}}(\rho))$ as claimed. □

In fact the proof shows that the assertion holds for any Lusztig series $\mathscr{E}(\mathbf{G}^F, s)$ for $\mathbf{G}$ of classical type without assumption on $\mathbf{Z}(\mathbf{G})$ as long as $C_{\mathbf{G}^*}(s)$ is connected. Note that Proposition 3.4.22 also follows immediately from the (deeper) result of Enguehard (see Theorem 4.7.2) on the commutation of Jordan decomposition with Lusztig induction.

It can be shown that the statement continues to hold for groups with connected centre and with no components of type E_7 or E_8 by using that $\mathrm{D}_{\mathbf{G}}$ preserves the so-called eigenvalues of Frobenius (see 4.2.21) and appealing to Corollary 4.5.4.

3.5 d-Harish-Chandra Theories

As a consequence of the investigation of Brauer ℓ-blocks of finite groups of Lie type it became apparent that Harish-Chandra theory has an important generalisation in terms of Lusztig induction, which is obtained by replacing the set of 1-split Levi subgroups by d-split Levi subgroups and Harish-Chandra induction by Lusztig induction. This has proved to be of fundamental importance for example in the understanding and the description of the block theory of the finite groups of Lie type. While it is expected that these d-Harish-Chandra theories share many of the properties of ordinary Harish-Chandra theory, this has at present only been proved in particular circumstances.

We start by defining the relevant subgroups, which were first introduced in [BrMa92].

3.5.1 As before let $\mathbf{G}$ be connected reductive with a Steinberg map F. First assume that F is not very twisted. An F-stable torus $\mathbf{T}$ of $\mathbf{G}$ is called a *d-torus* (of $(\mathbf{G}, F)$) if its complete root datum $((X, \varnothing, Y, \varnothing), \varphi)$ (see Definition 1.6.10) is such that the characteristic polynomial of φ on Y is a power of the dth cyclotomic polynomial Φ_d. So, $\mathbf{T}$ is a d-torus if and only if its order polynomial is a power of the dth cyclotomic polynomial Φ_d: there is $a \geqslant 0$ such that $|\mathbf{T}^{F^k}| = \Phi_d(q^k)^a$ for all $k \equiv 1 \pmod d$, with $a = \dim \mathbf{T}/\phi(d)$ where here ϕ denotes the Euler totient function. Alternatively, the torus $\mathbf{T}$ is a d-torus if and only if it splits completely over $\mathbb{F}_{q^d}$ and no non-trivial subtorus of it splits over any proper subfield of $\mathbb{F}_{q^d}$. If $\mathbf{T}$ is an F-stable torus in $\mathbf{G}$ we denote by $\mathbf{T}_d$ its maximal d-subtorus. This is well defined as the following description shows: In terms of complete root data, if $\mathbf{T}$ has complete root datum $((X, \varnothing, Y, \varnothing), \varphi)$, then the root datum of $\mathbf{T}_d$ is given by $\big((X', \varnothing, Y', \varnothing), \varphi'\big)$, where X' is the largest quotient of X on which φ has characteristic polynomial a power of Φ_d, Y' is the kernel of $\Phi_d(\varphi)$ on Y, and φ' the map on X' induced by φ.

It is well known (see e.g. [MaTe11, Prop. 12.10]) that the centraliser of a torus in a connected reductive group is a Levi subgroup. The centralisers in $\mathbf{G}$ of d-tori are called *d-split Levi subgroups* of $\mathbf{G}$. So in particular $\mathbf{G}$ itself is d-split as the centraliser of its trivial subtorus. Observe that d-split Levi subgroups are F-stable, being the centralisers of F-stable d-tori. It is clear from the definition that d-tori of $(\mathbf{G}, F)$ become 1-tori in $(\mathbf{G}, F^d)$. Thus, d-split Levi subgroups of $(\mathbf{G}, F)$ are 1-split if considered with respect to F^d.

Example 3.5.2 In the case $d = 1$ Lemma 3.1.3 shows that 1-split Levi subgroups are exactly what we had called (1-)split Levi subgroups in Example 3.1.2.

3.5.3 The above definitions have to be modified slightly if the Steinberg map is very twisted (see 1.6.2). For simplicity assume here that $\mathbf{G}$ is an F-stable Levi subgroup of some simple algebraic group with a very twisted Steinberg map. Let p denote the underlying characteristic of $\mathbf{G}$. We consider the set Ψ_p consisting of $\mathbf{q}^2 - 1$ together with the irreducible factors over $\mathbb{Z}[\sqrt{p}]$ of dth cyclotomic polynomials Φ_d over $\mathbb{Q}$ with $d \geqslant 3$. Then for $\Phi \in \Psi_p$, a *Φ-torus* of $\mathbf{G}$ is an F-stable torus with complete root datum $((X, \emptyset, Y, \emptyset), \varphi)$ such that the characteristic polynomial of φ on $Y \otimes_{\mathbb{Z}} \mathbb{R}$ is a power of Φ, and *Φ-split Levi subgroups* are the centralisers in $\mathbf{G}$ of Φ-tori (see [BrMa92, 3.F] and [EM17]). By a slight abuse of notation we will also call d-tori and d-split Levi subgroups these Φ-tori and Φ-split Levi subgroups respectively, where d is the order of the zeros of Φ, respectively $d = 1$ when $\Phi = \mathbf{q}^2 - 1$.

Example 3.5.4 In the very twisted case, when $p = 2$, the set Ψ_2 contains the polynomials of degree 2

$$\mathbf{q}^2 - 1,\ \mathbf{q}^2 + 1,\ \Phi_8' := \mathbf{q}^2 + \sqrt{2}\mathbf{q} + 1,\ \Phi_8'' := \mathbf{q}^2 - \sqrt{2}\mathbf{q} + 1,$$

with $d = 1, 4, 8, 8$ respectively, as well as the polynomials of degree 4

$$\Phi_{24}' := \mathbf{q}^4 + \sqrt{2}\mathbf{q}^3 + \mathbf{q}^2 + \sqrt{2}\mathbf{q} + 1,\ \Phi_{24}'' := \mathbf{q}^4 - \sqrt{2}\mathbf{q}^3 + \mathbf{q}^2 - \sqrt{2}\mathbf{q} + 1,$$

with $d = 24$. These occur as factors of the order polynomial of groups of type ${}^2F_4$. When $p = 3$, Ψ_3 contains for example the polynomials

$$\Phi_{12}' := \mathbf{q}^2 + \sqrt{3}\mathbf{q} + 1,\ \Phi_{12}'' := \mathbf{q}^2 - \sqrt{3}\mathbf{q} + 1,$$

with $d = 12$. These occur as factors of the order polynomial of groups of type ${}^2G_2$. Observe that evaluating any of these polynomials at the (positive) square root of an odd power p^{2f+1} of p yields positive integral values.

In order to make the exposition simpler we will from now on assume that F is not very twisted. Nevertheless, all statements, suitably modified, continue to hold in the very twisted case.

Any Levi subgroup $\mathbf{L}$ of $\mathbf{G}$ is the centraliser of its connected centre: $\mathbf{L} = C_{\mathbf{G}}(\mathbf{Z}^\circ(\mathbf{L}))$ (see [DiMi20, Prop. 3.4.6]). The d-split Levi subgroups are characterised as follows:

Proposition 3.5.5 *An F-stable Levi subgroup $\mathbf{L}$ of $\mathbf{G}$ is d-split if and only if it satisfies $\mathbf{L} = C_{\mathbf{G}}(\mathbf{Z}^\circ(\mathbf{L})_d)$, where $\mathbf{Z}^\circ(\mathbf{L})_d$ denotes the maximal d-torus of $\mathbf{Z}^\circ(\mathbf{L})$.*

Proof Note that $\mathbf{Z}^\circ(\mathbf{L})$ is an F-stable torus of $\mathbf{G}$, so its d-part is defined and its centraliser is a Levi subgroup. Clearly any Levi subgroup with the stated property is d-split by definition. Conversely, if $\mathbf{L}$ is the centraliser of a d-torus $\mathbf{S}$, then $\mathbf{S} \leqslant \mathbf{Z}^\circ(\mathbf{L})$, so

$$\mathbf{L} = C_{\mathbf{G}}(\mathbf{S}) \geqslant C_{\mathbf{G}}(\mathbf{Z}^\circ(\mathbf{L})_d) \geqslant C_{\mathbf{G}}(\mathbf{Z}^\circ(\mathbf{L})) = \mathbf{L},$$

proving the claim. □

3.5.6 We now describe how to construct d-tori of $\mathbf{G}$ in terms of complete root data. For this let $\big((X, R, Y, R'), \varphi\mathbf{W}\big)$ be the complete root datum of $(\mathbf{G}, F)$. If $w \in \mathbf{W}$ then $Y' = \ker_Y(\Phi_d(\varphi w))$ is a pure sublattice of Y and $\big((X', \varnothing, Y', \varnothing), \varphi w\big)$, with X' the largest quotient of X on which φ has minimal polynomial Φ_d, is the complete root datum of a d-torus of $\mathbf{G}$ since, by construction, $\varphi w|_{Y'}$ has characteristic polynomial a power of Φ_d. The d-tori constructed from elements $w \in \mathbf{W}$ with $\ker_Y(\Phi_d(\varphi w))$ of maximal possible rank are called *Sylow d-tori* of $\mathbf{G}$. It is clear from the order formula in 1.6.8(b) that the order polynomial of a Sylow d-torus of $\mathbf{G}$ is precisely the Φ_d-part of the order polynomial of $\mathbf{G}$.

The justification for this name comes from the fact that, as first observed by Broué–Malle [BrMa92, Thm. 3.4], the d-tori of $\mathbf{G}$ satisfy a Sylow theory. This is a consequence of Springer's theory of eigenspaces of elements in finite reflection groups [Spr74]. For this, let $\zeta_d = \exp(2\pi\sqrt{-1}/d)$ be a primitive dth root of unity. For $w \in \mathbf{W}$ denote by $V(\varphi w, \zeta_d)$ the ζ_d-eigenspace of φw on $V := Y \otimes_{\mathbb{Z}} \mathbb{C}$. Then the eigenspaces $V(\varphi w, \zeta_d)$ of maximal possible dimension are all conjugate under $\mathbf{W}$, and any such eigenspace is contained in one of maximal dimension.

Example 3.5.7 (Regular numbers) A minimal d-split Levi subgroup of $\mathbf{G}$ is, by our definitions, just the centraliser of a Sylow d-torus. Now observe that this is itself a (maximal) torus if and only if the maximal ζ_d-eigenspaces $V(\varphi w, \zeta_d)$ of elements φw (with $w \in \mathbf{W}$) in the natural reflection representation have trivial centraliser in $\mathbf{W}$. This happens if and only if $V(\varphi w, \zeta_d)$ contains a *regular vector*, that is, a vector not fixed by any element of $\mathbf{W}$. By a result of Steinberg [St64, Thm. 1.5], these are exactly the vectors not lying in any reflecting hyperplane of $\mathbf{W}$. Elements $\varphi w \in \varphi\mathbf{W}$ with a ζ_d-eigenspace containing regular vectors are called *d-regular* and d is then called a *regular number* for $\varphi\mathbf{W}$ (see [Spr74]).

As with ordinary Harish-Chandra theory an important ingredient for d-Harish-Chandra theory is given by relative Weyl groups:

Definition 3.5.8 Let $\mathbf{L}$ be an F-stable Levi subgroup of $\mathbf{G}$. The *relative Weyl group* of $\mathbf{L}$ in $\mathbf{G}$ is $W_{\mathbf{G}}(\mathbf{L}) := N_{\mathbf{G}}(\mathbf{L})^F/\mathbf{L}^F$.

The relative Weyl group of an F-stable Levi subgroup can be described purely in terms of the complete root datum. Assume that $\mathbf{L}$ has complete root datum $\big((X, R_{\mathbf{L}}, Y, R_{\mathbf{L}}^\vee), \varphi\mathbf{W}_{\mathbf{L}}\big)$. Then $W_{\mathbf{G}}(\mathbf{L}) = N_{\mathbf{W}}(\varphi\mathbf{W}_{\mathbf{L}})/\mathbf{W}_{\mathbf{L}}$. Thus for 1-split Levi subgroups $L = \mathbf{L}^F$ of $G = \mathbf{G}^F$ this agrees with the relative Weyl group $W_G(L)$ as defined in 3.1.25, see Example 3.1.26.

Now let $d \geqslant 1$ and assume that $\mathbf{L}$ is d-split. Let $\big((X', \varnothing, Y', \varnothing), \varphi w\big)$ be the complete root datum of the Sylow d-subtorus $\mathbf{Z}^\circ(\mathbf{L})_d$ of $\mathbf{Z}^\circ(\mathbf{L})$, so in particular $Y' = \ker_Y(\Phi_d(\varphi w))$ (see 3.5.1). Then $W_{\mathbf{G}}(\mathbf{L})$ acts on $V = Y' \otimes_{\mathbb{Z}} \mathbb{C}$, and thus on the ζ_d-eigenspace $V(\varphi w, \zeta_d)$ of φw in V. A key feature of the theory is the fact that the relative Weyl group $W_{\mathbf{G}}(\mathbf{L})$ of a minimal d-split Levi subgroup $\mathbf{L}$ is again a reflection group on $V(\varphi w, \zeta_d)$, but not necessarily any longer a real one:

Proposition 3.5.9 *Let $\mathbf{S}_d$ be a Sylow d-torus of $\mathbf{G}$ with complete root datum $\big((X, \varnothing, Y, \varnothing), \varphi w\big)$ and $\mathbf{L} = C_{\mathbf{G}}(\mathbf{S}_d)$ be the corresponding minimal d-split Levi subgroup. Then $W_{\mathbf{G}}(\mathbf{L})$ acts faithfully as a complex reflection group on $V(\varphi w, \zeta_d)$. Moreover, if the root system of $\mathbf{G}$ is indecomposable, then $W_{\mathbf{G}}(\mathbf{L})$ is irreducible on $V(\varphi w, \zeta_d)$.*

This was shown first by Springer [Spr74, Thms. 4.2(iii) and 6.4(iii)] for the case that the centraliser of a Sylow d-torus is a maximal torus and then by Lehrer–Springer [LeSp99, Thms. 3.4 and 5.1] for the general case using the invariant theory of reflection groups. Their results are actually more precise; for example they show how the degrees of the complex reflection group $W_{\mathbf{G}}(\mathbf{L})$ can be read off from those of $\mathbf{W}$. The irreducibility statement is due to the same authors [LeSp99b, Thm. A]. See also [LeTa09, Thm. 11.15 and 11.38] for more elementary proofs.

Remark 3.5.10 Some of these proofs rely on case-by-case considerations using the *Shephard–Todd classification* of the finite irreducible complex reflection groups [ST54]. These groups comprise several infinite series and a further 34 exceptional types usually denoted $G_4, \ldots, G_{37}$. Here, $G_4, \ldots, G_{22}$ are in dimension 2, $G_{23}, \ldots, G_{27}$ in dimension 3, $G_{28}, \ldots, G_{32}$ in dimension 4, and $G_{33}, \ldots, G_{37}$ are in dimensions 5,6,6,7,8 respectively. The finite Coxeter groups of exceptional types are contained in this list as

$$W(H_3) = G_{23},\ W(F_4) = G_{28},\ W(H_4) = G_{30},$$

$$W(E_6) = G_{35},\ W(E_7) = G_{36},\ W(E_8) = G_{37}.$$

The three infinite series are the cyclic groups (in dimension 1), the symmetric groups $\mathfrak{S}_n, n \geqslant 3$, in dimension $n-1$, and the imprimitive groups $G(de, e, n)$ with $n, de \geqslant 2$ and $(de, e, n) \neq (2, 2, 2)$. Here $G(de, e, n)$ is a normal subgroup of index e of the wreath product $G(d, 1, n) \cong C_d \wr \mathfrak{S}_n$. The dihedral Weyl groups $W(I_{2m})$ of order $2m$

occur in this series as the groups $G(m, m, 2)$. In particular the group $G(e, e, 2)$ is isomorphic to the Weyl group of type G_2.

The relative Weyl groups of the minimal d-split Levi subgroups of quasi-simple finite reductive groups are described in [BrMa93, Tab. 3.6] and in [BMM93, Tab. 1 and 3].

Example 3.5.11 We illustrate Proposition 3.5.9 by giving the non-cyclic relative Weyl groups (from [BMM93, Tab. 1 and 3]) of minimal d-split Levi subgroups in the simple exceptional groups in Table 3.2 in terms of their Shephard–Todd labels; see Remark 3.5.10.

Table 3.2 *Non-cyclic relative Weyl groups of minimal d-split Levi subgroups in exceptional types*

	$d=1$	2	3	4	5	6	8	10	12
G_2	$G(6,6,2)$	$G(6,6,2)$							
${}^3D_4$	$G(6,6,2)$	$G(6,6,2)$	G_4			G_4			
${}^2F_4$	$G(8,8,2)$			G_{12}			G_8		
F_4	G_{28}	G_{28}	G_5	G_8		G_5			
E_6	G_{35}	G_{28}	G_{25}	G_8		G_5			
${}^2E_6$	G_{28}	G_{35}	G_5	G_8		G_{25}			
E_7	G_{36}	G_{36}	G_{26}	G_8		G_{26}			
E_8	G_{37}	G_{37}	G_{32}	G_{31}	G_{16}	G_{32}	G_9	G_{16}	G_{10}

The relative Weyl groups in simple groups of classical type will be described in Examples 3.5.14 and 3.5.15.

As noted above, minimal d-split Levi subgroups are centralisers of Sylow d-tori. The previous result enables us to describe the complete lattice of d-split Levi subgroups, in analogy to the case $d = 1$ considered in Lemma 3.1.3:

Proposition 3.5.12 *Let $\mathbf{L}_d$ be a minimal d-split Levi subgroup of $\mathbf{G}$. The d-split Levi subgroups of $\mathbf{G}$ containing $\mathbf{L}_d$ are in bijection with the parabolic subgroups of the reflection group $W_{\mathbf{G}}(\mathbf{L}_d)$.*

Here, by definition *parabolic subgroups* of a complex reflection group are the centralisers of subspaces in its reflection representation. By a theorem of Steinberg [St64, Thm. 1.5], any such centraliser is generated by the reflections it contains, hence in particular it is itself a reflection group.

Proof Let $\mathbf{S}_d$ be a Sylow d-torus of $\mathbf{G}$ such that $\mathbf{L}_d = C_{\mathbf{G}}(\mathbf{S}_d)$, and let $\mathbf{L} \geqslant \mathbf{L}_d$ be d-split. Then $\mathbf{S} = \mathbf{Z}^{\circ}(\mathbf{L})_d$ is a d-torus contained in $\mathbf{S}_d$. Thus, if $\big((X_d, \varnothing, Y_d, \varnothing), \varphi w\big)$ is the complete root datum of $\mathbf{S}_d$, then $\mathbf{S}$ has complete root datum $\big((X', \varnothing, Y', \varnothing), \varphi w\big)$ for some pure sublattice $Y' \leqslant Y_d$. Hence,

$$V'(\varphi w, \zeta_d) := \ker_{Y'\otimes\mathbb{C}}(\varphi w - \zeta_d) \leqslant V(\varphi w, \zeta_d) := \ker_{Y_d\otimes\mathbb{C}}(\varphi w - \zeta_d).$$

As $\mathbf{L} = C_{\mathbf{G}}(\mathbf{S})$ by Proposition 3.5.5, the complete root datum of $\mathbf{L}$ has the form $\big((X, R_{\mathbf{L}}, Y, R_{\mathbf{L}}^\vee), \varphi w \mathbf{W}_{\mathbf{L}}\big)$, where $R_{\mathbf{L}}$ consists of the roots of $\mathbf{G}$ that are trivial on Y'. Thus, in its action on $V(\varphi w, \zeta_d)$, $\mathbf{W}_{\mathbf{L}}$ is the centraliser in $W_{\mathbf{G}}(\mathbf{L}_d)$ of the subspace $V'(\varphi w, \zeta_d)$, hence a parabolic subgroup of $W_{\mathbf{G}}(\mathbf{L}_d)$.

Conversely, a parabolic subgroup P of $W_{\mathbf{G}}(\mathbf{L}_d)$ is the centraliser of a subspace V' of $V(\varphi w, \zeta_d)$. This defines a d-torus $\mathbf{S}$ with cocharacters Y' such that $V' = \ker_{Y'\otimes\mathbb{C}}(\varphi w - \zeta_d)$, and thus its centraliser $\mathbf{L} = C_{\mathbf{G}}(\mathbf{S})$, a d-split Levi subgroup of $\mathbf{G}$. □

It is no longer true for general complex reflection groups that the parabolic subgroups can be described in terms of standard parabolic subgroups corresponding to subsets of a suitable fixed minimal generating system. Still, it turns out that this description continues to hold for the complex reflection groups occurring in our setting as $W_{\mathbf{G}}(\mathbf{L}_d)$. The parabolic subgroups of all finite complex reflection groups have been determined; for the exceptional types see for example the lists given in [OrTe92, App. C].

Example 3.5.13 Let $\mathbf{G}$ be simple of type E_7 with standard Frobenius map, and consider $d = 4$. Here, the centraliser $\mathbf{L}_4 = C_{\mathbf{G}}(\mathbf{S}_4)$ of a Sylow 4-torus $\mathbf{S}_4$ of $\mathbf{G}$ has rational form $\Phi_4^2.A_1(q)^3$ and its relative Weyl group is the complex reflection group G_8 in the Shephard–Todd notation, of order 96 (see Table 3.2). This reflection group has just one conjugacy class of proper non-trivial parabolic subgroups, all isomorphic to the cyclic group C_4, and the 4-split Levi subgroups containing $\mathbf{L}_4$ corresponding to these by Proposition 3.5.12 are of rational type $\Phi_4.{}^2D_4(q).A_1(q)$.

Example 3.5.14 We determine the d-split Levi subgroups in $\mathbf{G} = \mathrm{GL}_n$.

(a) First let $\big((X', \varnothing, Y', \varnothing), w\big)$ be the complete root datum of an n-torus (in the sense of 3.5.1) of GL_n with split Frobenius map. Then at least one eigenvalue of w on $V = Y' \otimes_{\mathbb{Z}} \mathbb{C}$ has to be a primitive nth root of unity. The only elements with this property of $\mathbf{W} = \mathfrak{S}_n$ in its natural permutation representation are the n-cycles (which comprise in particular the Coxeter elements of $\mathbf{W}$). Thus any non-trivial n-torus $\mathbf{S}$ of GL_n must have complete root datum such that w is an n-cycle and Y' is the maximal sublattice of $Y = \mathbb{Z}^n$ such that all eigenvalues of w on $Y' \otimes_{\mathbb{Z}} \mathbb{C}$ are primitive nth roots of unity (so Y' has rank $\varphi(n)$). The centraliser $C_{\mathbf{G}}(\mathbf{S})^F$ is then a Coxeter torus of $\mathrm{GL}_n(q)$ of order $q^n - 1$, isomorphic to $\mathrm{GL}_1(q^n)$.

(b) Still taking for F the standard Frobenius map, more generally it is easily seen by considering cycle shapes of elements in the Weyl group $\mathfrak{S}_n$ that for arbitrary

d, the elements with a maximal ζ_d-eigenspace are precisely those with cycle type consisting of a disjoint d-cycles and a disjoint partition of r, where $n = ad + r$ with $0 \leqslant r < d$. Then a Sylow d-torus of GL_n has centraliser $\mathbf{L}_d$ of rational form $\mathrm{GL}_1(q^d)^a.\mathrm{GL}_r(q)$ and its relative Weyl group is $W_{\mathbf{G}}(\mathbf{L}_d) \cong C_d \wr \mathfrak{S}_a$, the imprimitive complex reflection group usually denoted $G(d,1,a)$. The minimal parabolic subgroups of $W_{\mathbf{G}}(\mathbf{L}_d)$ up to conjugacy are C_d and C_2, generated by reflections of order d and 2 respectively, with corresponding d-split Levi subgroups

$$\mathrm{GL}_d(q) \times \mathrm{GL}_1(q^d)^{a-1} \times \mathrm{GL}_r(q), \qquad \mathrm{GL}_1(q^d)^{a-2} \times \mathrm{GL}_2(q^d) \times \mathrm{GL}_r(q),$$

respectively, where the latter only occurs when $a \geqslant 2$. More generally, according to the description given in Proposition 3.5.12 the d-split Levi subgroups of GL_n have rational form

$$\mathrm{GL}_{n_1}(q^d) \times \cdots \times \mathrm{GL}_{n_s}(q^d) \times \mathrm{GL}_t(q)$$

for some $n_1, \ldots, n_s, t \geqslant 0$ with $d(n_1 + \cdots + n_s) + t = n$.

The d-tori and d-split Levi subgroups of SL_n are now obtained by intersecting those of GL_n with SL_n, and the ones of PGL_n are the images under the canonical map of those in GL_n.

(c) From the above it is easy to read off the corresponding results for $\mathrm{GU}_n(q)$ since its complete root datum is obtained from the one of $\mathrm{GL}_n(q)$ by just replacing the automorphism $\varphi = \mathrm{id}$ by $-\mathrm{id}$ (see Example 1.6.18(b)). Now from the transformation property of cyclotomic polynomials $\Phi_d(\mathbf{q})$ upon replacing $\mathbf{q}$ by $-\mathbf{q}$, this immediately implies that the d-tori of $\mathrm{GU}_n(q)$ have rational forms obtained from those of the d'-tori of $\mathrm{GL}_n(q)$ by replacing q by $-q$ (Ennola duality), where

$$d' = \begin{cases} 2d & \text{if } d \text{ is odd,} \\ d/2 & \text{if } d \equiv 2 \pmod 4, \\ d & \text{if } d \equiv 0 \pmod 4, \end{cases}$$

and accordingly for the d-split Levi subgroups. In particular, the (1-)split Levi subgroups of $\mathrm{GU}_n(q)$ are obtained from the 2-split Levi subgroups of $\mathrm{GL}_n(q)$ by replacing q by $-q$, and the Weyl group of $\mathrm{GU}_n(q)$, that is, the relative Weyl group of a Sylow 1-torus of $\mathrm{GU}_n(q)$ is isomorphic to the relative Weyl group of a Sylow 2-torus of $\mathrm{GL}_n(q)$, of type B_m with $m = \lfloor n/2 \rfloor$.

Example 3.5.15 We next describe the d-split Levi subgroups for classical groups. This is most easily done for the matrix group versions with 1-dimensional centre and simple derived subgroup. So let $\mathbf{G}$ be of classical type and $F : \mathbf{G} \to \mathbf{G}$ be a Steinberg map such that $[\mathbf{G}, \mathbf{G}]^F \in \{\mathrm{SO}_{2n+1}(q), \mathrm{Sp}_{2n}(q), \mathrm{SO}_{2n}^{\pm}(q)\}$ is of classical type. First let $d \geqslant 1$ be odd. Then the d-split Levi subgroups in $\mathbf{G}$ have rational forms

$$\mathrm{GL}_{n_1}(q^d) \times \cdots \times \mathrm{GL}_{n_s}(q^d) \times \mathbf{H}^F$$

for some $n_1, \ldots, n_s \geqslant 0$, where $\mathbf{H}$ is of the same classical type as $\mathbf{G}$ and of rank t such that $t + d \sum n_i = n$. This follows by a straightforward calculation in the respective Weyl groups, similar to the case of GL_n discussed in Example 3.5.14, see also [BMM93, p. 49]. In particular the minimal d-split Levi subgroups have rational form $\mathrm{GL}_1(q^d)^a.\mathbf{H}^F$ where $n = ad + r$ and $0 \leqslant r < d$, with relative Weyl group the complex reflection group $G(2d, 1, a) \cong C_{2d} \wr \mathfrak{S}_a$.

If d is even, then set $e := d/2$. In this case the d-split Levi subgroups in $\mathbf{G}$ have rational forms

$$\mathrm{GU}_{n_1}(q^e) \times \cdots \times \mathrm{GU}_{n_s}(q^e) \times \mathbf{H}^F$$

for some $n_1, \ldots, n_s \geqslant 0$, where $\mathbf{H}$ is of the same classical type as $\mathbf{G}$ and of rank t such that $t + e \sum n_i = n$, except that here the twisting induced by F on $\mathbf{H}$ is opposite to the one on $\mathbf{G}$ if $\mathbf{G}$ is of type D_n and $\sum n_i$ is odd (see [BMM93, p. 52]). The minimal d-split Levi subgroups of $\mathbf{G}$ now have rational form $\mathrm{GU}_1(q^e)^a.\mathbf{H}^F$ where $n = ae + r$ and $0 \leqslant r < e$, with relative Weyl group $G(2e, 1, a) \cong C_{2e} \wr \mathfrak{S}_a$, unless $[\mathbf{G}, \mathbf{G}]^F = \mathrm{SO}_{2n}^+(q)$ and $r = 0$, in which case it has type $G(2e, 2, a)$, a normal subgroup of index 2 in $G(2e, 1, a)$.

Example 3.5.16 Finally we give the rational structure of d-split Levi subgroups in simple groups of exceptional type. The 1-split Levi subgroups are just the Levi subgroups of F-stable parabolic subgroups and hence easily described via the Dynkin diagram of $\mathbf{G}$. The 2-split Levi subgroups are obtained from these by Ennola duality. If the relative Weyl group of a Sylow d-torus $\mathbf{T}_d$ is cyclic then according to Proposition 3.5.12 the only d-split Levi subgroups, up to conjugacy, are $C_{\mathbf{G}}(\mathbf{T}_d)$ and $\mathbf{G}$ itself. All the remaining cases are listed in Table 3.3. Here a cyclotomic polynomial Φ_d stands for the group of F-fixed points of a d-torus of generic order Φ_d. The notation for factors of cyclotomic polynomials over $\mathbb{Q}(\sqrt{2})$ and $\mathbb{Q}(\sqrt{3})$ is as introduced in Example 3.5.4. (In Chevie [MiChv] they can be obtained by the command `SplitLevis`.)

We now turn to representation theoretic properties in the d-split setting. First, generalising the 1-split case from Theorem 3.1.11 and the subsequent remarks, the Mackey formula takes a particularly simple form in this setting, see [BMM93, Thm. 1.35]:

Theorem 3.5.17 *Let $\mathbf{M}_1, \mathbf{M}_2$ be d-split Levi subgroups of $\mathbf{G}$ containing the minimal d-split Levi subgroup $\mathbf{L}$. Let $\psi \in \mathrm{Irr}(\mathbf{M}_1^F)$ and assume that one of the assump-*

Table 3.3 *Rational type of proper d-split Levi subgroups, $d \geqslant 3$, in simple exceptional groups*

$\mathbf{G}^F$	d	$\mathbf{L}^F$
${}^3D_4(q)$	3	Φ_3^2, $\Phi_3.A_2(q)$
	6	Φ_6^2, $\Phi_6.{}^2A_2(q)$
$F_4(q)$	3	Φ_3^2, $\Phi_3.A_2(q)$
	4	Φ_4^2, $\Phi_4.B_2(q)$
	6	Φ_6^2, $\Phi_6.{}^2A_2(q)$
${}^2F_4(q^2)$	4	Φ_4^2, $\Phi_4.A_1(q^2)$
	$8'$	$\Phi_8'^2$, $\Phi_8'.{}^2B_2(q^2)$
	$8''$	$\Phi_8''^2$, $\Phi_8''.{}^2B_2(q^2)$
$E_6(q)$	3	Φ_3^3, $\Phi_3^2.A_2(q)$, $\Phi_3.A_2(q)^2$, $\Phi_3.{}^3D_4(q)$
	4	$\Phi_4^2\Phi_1^2$, $\Phi_4\Phi_1.{}^2A_3(q)$
	6	$\Phi_6^2\Phi_3$, $\Phi_6\Phi_3.{}^2A_2(q)$, $\Phi_6.A_2(q^2)$
${}^2E_6(q)$	3	$\Phi_3^2\Phi_6$, $\Phi_3\Phi_6.A_2(q)$, $\Phi_3.A_2(q^2)$
	4	$\Phi_4^2\Phi_2^2$, $\Phi_4\Phi_2.A_3(q)$
	6	Φ_6^3, $\Phi_6^2.{}^2A_2(q)$, $\Phi_6.{}^2A_2(q)^2$, $\Phi_6.{}^3D_4(q)$
$E_7(q)$	3	$\Phi_3^3\Phi_1$, $\Phi_3^2\Phi_1.A_2(q)$, $\Phi_3^2.A_1(q^3)$, $\Phi_3\Phi_1.{}^3D_4(q)$, $\Phi_3.A_5(q)$, $\Phi_3.A_2(q)A_1(q^3)$
	4	$\Phi_4^2.A_1(q)^3$, $\Phi_4.{}^2D_4(q)A_1(q)$
	6	$\Phi_6^3\Phi_2$, $\Phi_6^2\Phi_2.{}^2A_2(q)$, $\Phi_6^2.A_1(q^3)$, $\Phi_6\Phi_2.{}^3D_4(q)$, $\Phi_6.{}^2A_5(q)$, $\Phi_6.{}^2A_2(q)A_1(q^3)$
$E_8(q)$	3	Φ_3^4, $\Phi_3^3.A_2(q)$, $\Phi_3^2.A_2(q)^2$, $\Phi_3^2.{}^3D_4(q)$, $\Phi_3.{}^3D_4(q)A_2(q)$, $\Phi_3.E_6(q)$
	4	Φ_4^4, $\Phi_4^3.A_1(q^2)$, $\Phi_4^2.{}^2A_2(q^2)$, $\Phi_4^2.A_1(q^2)^2$, $\Phi_4^2.D_4(q)$, $\Phi_4.{}^2A_3(q^2)$, $\Phi_4.{}^2D_6(q)$, $\Phi_4.A_1(q^2).{}^2A_2(q^2)$
	5	Φ_5^2, $\Phi_5.A_4(q)$
	6	Φ_6^4, $\Phi_3^6.{}^2A_2(q)$, $\Phi_6^2.{}^2A_2(q)^2$, $\Phi_6^2.{}^3D_4(q)$, $\Phi_6.{}^3D_4(q){}^2A_2(q)$, $\Phi_6.{}^2E_6(q)$
	8	Φ_8^2, $\Phi_8.{}^2D_4(q)$
	10	Φ_{10}^2, $\Phi_{10}.{}^2A_4(q)$
	12	Φ_{12}^2, $\Phi_{12}.{}^3D_4(q)$

tions of Theorem 3.3.7 is satisfied. Then

$$
{}^*R^{\mathbf{G}}_{\mathbf{M}_2} \circ R^{\mathbf{G}}_{\mathbf{M}_1}(\psi) = \sum_{w} R^{\mathbf{M}_2}_{\mathbf{M}_2 \cap {}^w\mathbf{M}_1} \circ \mathrm{ad}(w) \circ {}^*R^{\mathbf{M}_1}_{\mathbf{M}_2^w \cap \mathbf{M}_1}(\psi),
$$

where the sum runs over a system of $W_{\mathbf{M}_2}(\mathbf{L})^F$–$W_{\mathbf{M}_1}(\mathbf{L})^F$ double coset representatives in $W_{\mathbf{G}}(\mathbf{L})^F$.

The proof of that result hinges on the following observation (see [BMM93, Prop. 1.24]), which we will use again later:

Lemma 3.5.18 *Let $\mathbf{M}_1, \mathbf{M}_2$ be d-split Levi subgroups of $\mathbf{G}$. If $\mathbf{M} := \mathbf{M}_1 \cap \mathbf{M}_2$ contains an F-stable maximal torus $\mathbf{T}$ of $\mathbf{G}$ then $\mathbf{M}$ is also d-split and in particular contains a minimal d-split Levi subgroup.*

Proof Let $\mathbf{S}_i$ be the Sylow d-torus of $Z^\circ(\mathbf{M}_i)$, for $i = 1, 2$. This is contained in every maximal torus of $\mathbf{M}_i$, hence in $\mathbf{T}$. So $\mathbf{S} := \mathbf{S}_1\mathbf{S}_2$ is a d-torus in $\mathbf{T}$. It is now easy to see that $\mathbf{M} = C_{\mathbf{G}}(\mathbf{S})$ and hence is a d-split Levi subgroup. □

The following natural generalisation of Definition 3.1.14 of cuspidal characters is fundamental in the setting of modular representation theory of finite reductive groups:

Definition 3.5.19 Let $d \geqslant 1$. An irreducible character $\rho \in \mathrm{Irr}(\mathbf{G}^F)$ is *d-cuspidal* if ${}^*R_{\mathbf{L}}^{\mathbf{G}}(\rho) = 0$ for all proper d-split Levi subgroups $\mathbf{L}$ of $\mathbf{G}$. A pair $(\mathbf{L}, \lambda)$, where $\mathbf{L}$ is a d-split Levi subgroup of $\mathbf{G}$ and $\lambda \in \mathrm{Irr}(\mathbf{L}^F)$ is d-cuspidal is called a *d-cuspidal pair* of $\mathbf{G}$.

Example 3.5.20 (a) With this notation, 1-cuspidal characters of $G = \mathbf{G}^F$ are just characters of cuspidal $\mathbb{K}G$-modules as introduced in Definition 3.1.14.

(b) Assume that F defines an $\mathbb{F}_q$-structure on $\mathbf{G}$. Then the group $\mathbf{G}^{F^d}$ inherits an $\mathbb{F}_q$-structure, but it can also be considered as a group over $\mathbb{F}_{q^d}$. By 3.5.1 the d-split Levi subgroups of $\mathbf{G}$ with respect to the Frobenius map F are the 1-split Levi subgroups of $\mathbf{G}$ with respect to F^d. It follows that the d-cuspidal characters of $\mathbf{G}^{F^d}$ considered as a group defined over $\mathbb{F}_q$ are the (1-)cuspidal characters of $\mathbf{G}^{F^d}$ considered over $\mathbb{F}_{q^d}$.

In analogy to Proposition 3.2.2 d-cuspidal characters enjoy the following uniform property:

Proposition 3.5.21 *Let $\rho \in \mathrm{Irr}(\mathbf{G}^F)$ be d-cuspidal. Then ${}^*R_{\mathbf{T}}^{\mathbf{G}}(\rho) = 0$ for all F-stable maximal tori $\mathbf{T}$ of $\mathbf{G}$ contained in some proper d-split Levi subgroup of $\mathbf{G}$.*

The proof of this statement is identical to that of Proposition 3.2.2, simply replacing (1-)split and (1-)cuspidal by d-split and d-cuspidal throughout. But note that the proof of the reverse direction in Proposition 3.2.2 does not go through here since it might be that ${}^*R_{\mathbf{L}}^{\mathbf{G}}(\rho)(1) = 0$ even though ${}^*R_{\mathbf{L}}^{\mathbf{G}}(\rho) \neq 0$. We will say something about the converse in 3.5.24 below. Again, by Proposition 3.5.5 the condition on $\mathbf{T}$ in Proposition 3.5.21 can be reformulated as follows: An F-stable maximal torus $\mathbf{T}$ of $\mathbf{G}$ does not lie in a proper d-split Levi subgroup of $\mathbf{G}$ if and only if its Sylow d-torus $\mathbf{T}_d$ is central in $\mathbf{G}$.

There is also a weak analogue of the characterisation in Theorem 3.2.22 of cuspidal characters in terms of Jordan decomposition; for this let us say that a character $\rho \in \mathrm{Irr}(\mathbf{G}^F)$ has *property* (U_d) if it satisfies the conclusion of the uniform criterion for d-cuspidality in Proposition 3.5.21. Thus, in particular, any d-cuspidal character has property (U_d).

Lemma 3.5.22 *Let $\mathbf{G} \hookrightarrow \tilde{\mathbf{G}}$ be a regular embedding. Let $\tilde{\mathbf{L}}$ be an F-stable Levi subgroup of $\tilde{\mathbf{G}}$ and $\tilde{\lambda} \in \mathrm{Irr}(\tilde{\mathbf{L}}^F)$. Let $\mathbf{L} = \tilde{\mathbf{L}} \cap \mathbf{G}$ and let λ be an irreducible constituent of $\mathrm{Res}^{\tilde{\mathbf{L}}^F}_{\mathbf{L}^F}(\tilde{\lambda})$. Then $(\mathbf{L}, \lambda)$ has property (U_d) for $\mathbf{G}$ if and only if $(\tilde{\mathbf{L}}, \tilde{\lambda})$ has property (U_d) for $\tilde{\mathbf{G}}$.*

This is shown in [CE99, Prop. 1.10] using a certain amount of case-by-case analysis. As a direct consequence they obtain:

Proposition 3.5.23 *Let $s \in G^*$ be semisimple. Let $\rho \in \mathscr{E}(G, s)$ and let ψ be in a $C_{\mathbf{G}^*}(s)^F$-orbit of unipotent characters of $C^\circ_{\mathbf{G}^*}(s)^F$ corresponding to ρ under Jordan decomposition (see Theorem 2.6.22). Then ρ has property (U_d) if and only if*

(1) ψ is d-cuspidal; and
(2) $\mathbf{Z}^\circ(\mathbf{G}^*)_d = \mathbf{Z}^\circ(C^\circ_{\mathbf{G}^*}(s))_d$, that is, the Sylow d-torus of $\mathbf{Z}^\circ(C^\circ_{\mathbf{G}^*}(s))$ lies in $\mathbf{Z}(\mathbf{G}^*)$.

If the centre of $\mathbf{G}$ is connected, this follows exactly as for the case $d = 1$ shown in the proof of Theorem 3.2.22. The reduction to this situation for property (U_d) is given by Lemma 3.5.22 and for properties (1) and (2) it follows by elementary arguments from the properties of Jordan decomposition, see [CE99, Prop. 1.10].

Characters with the property in Proposition 3.5.23 have been baptised *d-Jordan cuspidal* in [KeMa15, Def. 2.1].

3.5.24 We define two relations on the set of d-cuspidal pairs of $\mathbf{G}$ in terms of Lusztig induction, generalising the relation $\leqslant_1$ from Definition 3.1.14: We say that $(\mathbf{M}, \mu) \leqslant_d (\mathbf{L}, \lambda)$ for d-cuspidal pairs $(\mathbf{M}, \mu), (\mathbf{L}, \lambda)$ of $\mathbf{G}$ if $\mathbf{M} \leqslant \mathbf{L}$ and there is a parabolic subgroup $\mathbf{P}$ of $\mathbf{L}$ with Levi complement $\mathbf{M}$ such that $\langle \lambda, R^{\mathbf{L}}_{\mathbf{M} \leqslant \mathbf{P}}(\mu) \rangle \neq 0$ (see [BMM93, Def. 3.1]). The transitive closure of $\leqslant_d$ is denoted $\ll_d$. So, by definition $\rho \in \mathrm{Irr}(\mathbf{G}^F)$ is d-cuspidal if $(\mathbf{G}, \rho)$ is minimal with respect to $\ll_d$. As with ordinary Harish-Chandra theory, we can now define the *d-Harish-Chandra series* $\mathscr{E}(\mathbf{G}^F, (\mathbf{L}, \lambda))$ above a d-cuspidal pair as consisting of those $\rho \in \mathrm{Irr}(\mathbf{G}^F)$ such that $(\mathbf{L}, \lambda) \ll_d (\mathbf{G}, \rho)$. Note that it is not clear from our definition whether the d-Harish-Chandra series partition $\mathrm{Irr}(\mathbf{G}^F)$ (but see Theorem 4.6.21 for the case of unipotent characters). Still, let us point out the following immediate consequence of Proposition 3.3.20:

Corollary 3.5.25 *Let $(\mathbf{L}, \lambda)$ be a d-cuspidal pair of $\mathbf{G}$ for some $d \geqslant 1$, and assume that $\lambda \in \mathscr{E}(\mathbf{G}^F, s)$ for some $s \in \mathbf{L}^{*F}$. Then the d-Harish-Chandra series $\mathscr{E}(\mathbf{G}^F, (\mathbf{L}, \lambda))$ is contained in $\mathscr{E}(\mathbf{G}^F, s)$.*

It is conjectured in [CE99, 1.11] that the relations $\leqslant_d$ and $\ll_d$ coincide, that the minimal pairs with respect to $\ll_d$ below a fixed $\rho \in \mathrm{Irr}(\mathbf{G}^F)$ are all $\mathbf{G}^F$-conjugate, and that the criterion for d-cuspidality in Proposition 3.5.21 is also sufficient, that is, d-cuspidality is determined by a uniform condition. In contrast to the situation for Harish-Chandra induction and ordinary cuspidality, none of these statements is known to hold in general, though. Apart from the case $d = 1$ (see Corollary 3.1.17), and from the case of unipotent characters (see Theorem 4.6.20) the validity of the stated conjectures has been shown by Cabanes–Enguehard [CE99, Thm. 4.2] for characters lying in Lusztig series $\mathscr{E}(G, s)$ with $s \in \mathbf{G}^{*F}$ a semisimple ℓ'-element, for primes $\ell \geqslant 5$ different from p such that q has order d modulo ℓ, by using sophisticated methods from block theory. Some quasi-isolated cases in exceptional type groups have been settled by Kessar–Malle [KeMa13]. See Corollary 4.7.8 for the case of simple groups with connected centre.

Let us mention another compatibility of d-cuspidality with Jordan decomposition that was shown in [KeMa15, Prop. 4.1]. Let $\mathbf{G}$ be connected reductive and assume that F is a Frobenius map. Let $s \in \mathbf{G}^{*F}$, and $\mathbf{G}_1 \leqslant \mathbf{G}$ an F-stable Levi subgroup with dual $\mathbf{G}_1^*$ containing $C_{\mathbf{G}^*}(s)$. For $(\mathbf{L}_1, \lambda_1)$ a d-cuspidal pair of $\mathbf{G}_1$ below $\mathscr{E}(\mathbf{G}_1^F, s)$ set

$$\Psi_{\mathbf{G}_1}^{\mathbf{G}}(\mathbf{L}_1, \lambda_1) := (\mathbf{L}, \lambda) \quad \text{with } \mathbf{L} := C_{\mathbf{G}}(\mathbf{Z}^\circ(\mathbf{L}_1)_d) \text{ and } \lambda := \varepsilon_{\mathbf{L}}\varepsilon_{\mathbf{L}_1} R_{\mathbf{L}_1}^{\mathbf{L}}(\lambda_1).$$

Observe that $\lambda \in \mathrm{Irr}(\mathbf{L}^F)$ by Theorem 3.3.22.

Proposition 3.5.26 *Assume that the Mackey formula holds for $\mathbf{G}^F$. Then $\Psi_{\mathbf{G}_1}^{\mathbf{G}}$ defines a bijection between the set of d-cuspidal pairs of $\mathbf{G}_1$ below $\mathscr{E}(\mathbf{G}_1^F, s)$ and the set of d-cuspidal pairs of $\mathbf{G}$ below $\mathscr{E}(\mathbf{G}^F, s)$.*

The inverse of $\Psi_{\mathbf{G}_1}^{\mathbf{G}}$ is given as follows: let $(\mathbf{L}, \lambda)$ be d-cuspidal in $\mathbf{G}$ below $\mathscr{E}(\mathbf{G}^F, s)$. Set $\mathbf{L}_1^* := \mathbf{L}^* \cap \mathbf{G}_1^*$ and observe that $\mathbf{L}_1^* \geqslant \mathbf{L}^* \cap C_{\mathbf{G}^*}(s) = C_{\mathbf{L}^*}(s)$, so ${}^*R_{\mathbf{L}_1}^{\mathbf{L}}(\lambda)$ has a unique constituent λ_1 by Theorem 3.3.22. Then $(\mathbf{L}_1, \lambda_1)$ is d-cuspidal.

As in the case $d = 1$ we still have the following consequence of Proposition 3.5.21:

Corollary 3.5.27 *Let $\lambda \in \mathrm{Irr}(G)$ be d-cuspidal. Then the degree polynomial of λ has the form*

$$\mathbb{D}_\lambda = \Phi_d(\mathbf{q})^{a(d)} f(\mathbf{q})$$

where $a(d)$ is the precise power of Φ_d dividing the order polynomial of $[\mathbf{G}, \mathbf{G}]$ and $f \in \mathbb{Q}[\mathbf{q}]$ is not divisible by $\Phi_d(\mathbf{q})$.

Proof The proof is identical to the one for Corollary 3.2.4, only replacing (1-)split and (1-)cuspidal by d-split and d-cuspidal throughout, $\mathbf{q}-1$ by Φ_d, and the reference to Proposition 3.2.2 to one to Proposition 3.5.21. □

Definition 3.5.28 Let $(\mathbf{L}, \lambda)$ be a d-cuspidal pair in $\mathbf{G}$. The *relative Weyl group* of $(\mathbf{L}, \lambda)$ is

$$W_{\mathbf{G}}(\mathbf{L}, \lambda) := N_{\mathbf{G}^F}(\mathbf{L}, \lambda)/\mathbf{L}^F$$

(a subgroup of the relative Weyl group $W_{\mathbf{G}}(\mathbf{L})$ introduced in 3.5.8).

We saw in Theorem 3.2.5 that relative Weyl groups of unipotent (1-)cuspidal pairs are always Coxeter groups, and in Proposition 3.5.9 that relative Weyl groups of d-split Levi subgroups are complex reflection groups. It is a remarkable fact, for which no conceptual proof is known, that the relative Weyl groups of unipotent d-cuspidal pairs $(\mathbf{L}, \lambda)$ are also complex reflection groups. They are described in all cases in [BrMa93, §3] and [BMM93, Tab. 3].

Example 3.5.29 We describe the relative Weyl groups of unipotent d-cuspidal pairs in classical types.

(a) First let $\mathbf{G} = \mathrm{GL}_n$ with $\mathbf{G}^F = \mathrm{GL}_n(q)$. The d-split Levi subgroups $\mathbf{L}$ of $\mathbf{G}$ were described in Example 3.5.14(b): they have rational form

$$\mathbf{L}^F \cong \mathrm{GL}_{n_1}(q^d) \times \cdots \times \mathrm{GL}_{n_s}(q^d) \times \mathrm{GL}_t(q)$$

with $t+d\sum n_i = n$. In order for $\mathbf{L}$ to have a d-cuspidal character, all factors must have such a character. According to the observation in 3.5.1 and Example 3.5.20(b) the d-cuspidal characters of $\mathrm{GL}_{n_i}(q^d)$ are just its cuspidal characters if we consider it as a group over $\mathbb{F}_q^d$. Now it follows from Example 2.4.20 that the group $\mathrm{GL}_n(q)$ only has cuspidal unipotent characters when $n = 0$, so $\mathbf{L}^F$ possesses d-cuspidal unipotent characters only when it has rational form $\mathrm{GL}_1(q^d)^a \times \mathrm{GL}_t(q)$ with $n = ad + t$ for some $a, t \geqslant 0$. We will actually see in Corollary 4.6.5 that the d-cuspidal unipotent characters of $\mathrm{GL}_t(q)$ are exactly the characters parametrised by a partition of t that is a d-core. Since any automorphism of $\mathrm{GL}_t(q)$ fixes all unipotent characters, see Theorem 4.5.11, the relative Weyl group of any such d-cuspidal character $\lambda \in \mathrm{Uch}(\mathbf{L}^F)$ satisfies $W_{\mathbf{G}}(\mathbf{L}, \lambda) = W_{\mathbf{G}}(\mathbf{L})$, hence equals the complex reflection group $G(d, 1, a)$. The same line of argument can be used when $\mathbf{G}^F = \mathrm{GU}_n(q)$ to conclude that again we have $W_{\mathbf{G}}(\mathbf{L}, \lambda) = W_{\mathbf{G}}(\mathbf{L})$ for all d-cuspidal unipotent characters of d-split Levi subgroups $\mathbf{L}$ of $\mathbf{G}$.

(b) Now assume that $\mathbf{G}$ is of classical type as in Example 3.5.15. For $d \geqslant 1$ we set $e := d$ when d is odd, and $e := d/2$ when d is even. Using Example 3.5.15 we conclude as above that the only d-split Levi subgroups possessing a d-cuspidal unipotent character are of type $(q^e + (-1)^d)^a \times \mathbf{H}^F$ with $\mathbf{H}$ of rank t where $n = ae+t$. The d-cuspidal characters of $\mathbf{H}^F$ will be described explicitly in Corollary 4.6.16. By Theorem 4.5.11, unless $\mathbf{H}^F$ is of untwisted type D_n, all unipotent characters are invariant under outer automorphisms, so the relative Weyl groups again satisfy $W_{\mathbf{G}}(\mathbf{L}, \lambda) = W_{\mathbf{G}}(\mathbf{L})$ for all d-cuspidal unipotent characters of d-split Levi subgroups

$\mathbf{L} \leqslant \mathbf{G}$. On the other hand, when $\mathbf{H}^F$ is of untwisted type D_n with n even, there exist d-cuspidal unipotent characters $\lambda \in \mathrm{Uch}(\mathbf{L}^F)$ labelled by so-called degenerate symbols for which $W_{\mathbf{G}}(\mathbf{L}, \lambda)$ is a subgroup of index 2 in $W_{\mathbf{G}}(\mathbf{L})$, see [BMM93, p. 51] and also Section 4.4.

Example 3.5.30 We continue Example 3.5.13 with the principal 4-series in $\mathbf{G}$ of type E_7. We saw that the relative Weyl group $W_{\mathbf{G}}(\mathbf{L}_4)$ of a Sylow 4-centraliser $\mathbf{L}_4$ in $\mathbf{G}$ is isomorphic to the complex reflection group G_8. This group has two orbits of length 3 and two fixed points on the set $\mathrm{Uch}(\mathbf{L}_4^F)$ of unipotent characters of $\mathbf{L}_4^F = \Phi_4^2.A_1(q)^3$. Thus, for any (necessarily 4-cuspidal) unipotent character λ of $\mathbf{L}_4$ in an orbit of length 3, the relative Weyl group $W_{\mathbf{G}}(\mathbf{L}_4, \lambda)$ is a proper subgroup of $W_{\mathbf{G}}(\mathbf{L}_4)$. It turns out to be the imprimitive complex reflection group $C_4 \wr \mathfrak{S}_2 = G(4, 1, 2)$ (in the Shephard–Todd notation) in each case, see [BMM93, Tab. 1].

We will have more to say about d-Harish-Chandra theory for unipotent characters in Section 4.6.

4

Unipotent Characters

Lusztig's Jordan decomposition of characters discussed in Section 2.6 underlines the importance of understanding the unipotent characters of finite reductive groups. It shows that many questions on arbitrary irreducible characters can be reduced to problems on unipotent characters. Thus, unipotent characters carry basic information on the ordinary representation theory of finite reductive groups. In fact, much is known about unipotent characters and they are easier to understand in purely combinatorial terms than other characters.

In this chapter we collect various important properties of unipotent characters of (nearly) simple groups of Lie type. As unipotent characters are well behaved with respect to almost direct products of finite reductive groups, this is sufficient for many purposes.

The parametrisation of unipotent characters and the description of their properties are closely related to the irreducible characters of Weyl groups. We hence first recall in Section 4.1 some of the properties of and concepts around Weyl group characters. In Section 4.2 we introduce the notion of families of unipotent characters and associated Fourier matrices which are then used to state Lusztig's fundamental decomposition theorem (Theorem 4.2.16). We continue by defining the Frobenius eigenvalues attached to unipotent characters and explaining their relation with the Fourier matrices.

In Sections 4.3–4.5 we give more detailed information for the various simple types. For groups of type A, which are considered in Section 4.3, the associated combinatorics is entirely in terms of partitions and highly reminiscent of the representation theory of symmetric groups. As one might expect, the properties of unipotent characters of groups of classical type also have a quite combinatorial flavour. This can be encoded in suitable combinatorial parameters, the so-called Lusztig symbols, in terms of which the properties of unipotent characters of classical groups are then expounded in Section 4.4. On the other hand, the unipotent characters of groups of exceptional type have mainly to be studied case-by-case, and this is done in Section 4.5. We conclude that section by collecting some general properties that can be read off from the explicit classification of unipotent characters, for example concerning rationality and on the action of automorphisms.

In Section 4.6 we discuss the decomposition of Lusztig induction of unipotent characters, giving in particular a new proof of Asai's formula, and from this derive the d-Harish-Chandra theories for unipotent characters. With this knowledge we then return to Lusztig's Jordan decomposition of characters in Section 4.7 and present the known results on its commutation with Lusztig induction. In the final Section 4.8 we give a brief introduction to the as yet quite incomplete representation theory of finite reductive disconnected groups.

Throughout this chapter, p denotes the underlying characteristic of all considered algebraic groups.

4.1 Characters of Weyl Groups

In this section we collect some basic properties of irreducible characters of finite Weyl groups that will be relevant for the description of unipotent characters. We do not aim to develop the theory from scratch, an introduction with proofs can be found for example in [GePf00]. Let us point out that several crucial facts in the representation theory of finite Weyl groups rely on an explicit case-by-case analysis of the various irreducible types as no general proofs have been found to date.

Throughout this section let W be a finite Weyl group, with generating set S of simple reflections and corresponding length function $l\colon W \to \mathbb{Z}_{\geqslant 0}$. Let

$$T = \{wsw^{-1} \mid w \in W,\ s \in S\}$$

denote the set of reflections in W, and let $\mathrm{Irr}(W)$ be the set of (complex) irreducible characters of W. Following Lusztig [Lu79b], we introduce two functions $\phi \mapsto a_\phi$ and $\phi \mapsto b_\phi$ on $\mathrm{Irr}(W)$.

4.1.1 The *b-invariant* b_ϕ of $\phi \in \mathrm{Irr}(W)$ is defined as the smallest integer $i \geqslant 0$ such that ϕ occurs in the character of the ith symmetric power of the natural reflection representation of W. For example, if 1_W denotes the trivial character and ε denotes the sign character of W, then

$$b_{1_W} = 0 \qquad \text{and} \qquad b_\varepsilon = |T| = l(w_0)$$

where $w_0 \in W$ is the longest element (see [GePf00, 5.3.1(a)]). Let $W' \leqslant W$ be a subgroup generated by reflections and let $T' = W' \cap T$. Let ε' be the sign character of W'. By a result due to Macdonald (see [GePf00, 5.2.11]), there is a unique

$\phi \in \operatorname{Irr}(W)$ such that $b_\phi = |T'|$ and

$$\operatorname{Ind}_{W'}^{W}(\varepsilon') = \phi + (\text{sum of various } \psi \in \operatorname{Irr}(W) \text{ such that } b_\psi > b_\phi).$$

We shall denote this character by

$$\phi := \mathbf{j}_{W'}^{W}(\varepsilon').$$

This **j**-*induction* can be used to systematically construct all the irreducible characters of Weyl groups W of type A_{n-1}, B_n and D_n; it also provides a convenient way of labelling the characters of W of exceptional type.

Example 4.1.2 Let $n \geqslant 1$ and $W = \mathfrak{S}_n$ be the symmetric group, where the generators are the basic transpositions $s_i = (i, i+1)$ for $1 \leqslant i \leqslant n-1$. (We also set $\mathfrak{S}_0 = \{1\}$.) It is well known that the irreducible characters of $\mathfrak{S}_n$ are parametrised by the partitions of n; we write this as

$$\operatorname{Irr}(\mathfrak{S}_n) = \{\phi^\alpha \mid \alpha \vdash n\}.$$

This labelling is determined as follows; see, for example, [GePf00, 5.4.7]. Given a partition $\alpha \vdash n$, let α^* denote the transpose partition. Let $\mathfrak{S}_{\alpha^*} \leqslant \mathfrak{S}_n$ be the corresponding Young subgroup; we have $\mathfrak{S}_{\alpha^*} \cong \mathfrak{S}_{\alpha_1^*} \times \cdots \times \mathfrak{S}_{\alpha_k^*}$, where $\alpha_1^*, \ldots, \alpha_k^*$ are the parts of α^*. Let ε_{α^*} be the sign character of $\mathfrak{S}_{\alpha^*}$. Then

$$\phi^\alpha = \mathbf{j}_{\mathfrak{S}_{\alpha^*}}^{\mathfrak{S}_n}(\varepsilon_{\alpha^*}) \qquad \text{and} \qquad b_{\phi^\alpha} = n(\alpha) := \sum_{i=1}^{l} (i-1)\alpha_i$$

(as in Example 2.8.7) where $\alpha = (\alpha_1 \geqslant \cdots \geqslant \alpha_l)$. For example, $\phi^{(n)}$ is the trivial character and $\phi^{(1^n)}$ is the sign character ε. We have $\varepsilon\phi^\alpha = \phi^{\alpha^*}$ for all $\alpha \vdash n$ (see [GePf00, 5.4.9]).

Example 4.1.3 Let $n \geqslant 1$ and W_n be a Coxeter group of type B_n, with generators $\{t, s_1, s_2, \ldots, s_{n-1}\}$ and Coxeter diagram given as follows:

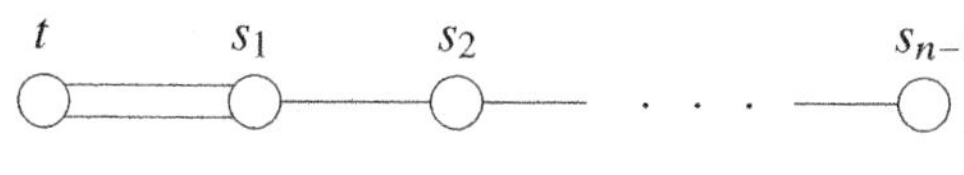

(We also set $W_0 = \{1\}$.) The irreducible characters of W_n are parametrised by pairs of partitions (α, β) such that $|\alpha| + |\beta| = n$. We write this as

$$\operatorname{Irr}(W_n) = \{\phi^{(\alpha,\beta)} \mid (\alpha, \beta) \vdash n\}.$$

For $(\alpha, \beta) \vdash n$, there is a reflection subgroup $W_{\alpha,\beta} \leqslant W_n$ of type

$$D_{\alpha_1} \times D_{\alpha_2} \times \cdots \times D_{\alpha_l} \times B_{\beta_1} \times B_{\beta_2} \times \cdots \times B_{\beta_k}$$

where $\alpha = (\alpha_1 \geqslant \cdots \geqslant \alpha_l)$ and $\beta = (\beta_1 \geqslant \cdots \geqslant \beta_k)$, and D_1 is understood as the empty Dynkin diagram. Let $\varepsilon_{\alpha,\beta}$ be the sign character of $W_{\alpha,\beta}$. Then, by [GePf00, 5.5.1, 5.5.3], we have

$$\phi^{(\alpha,\beta)} = \mathbf{j}_{W_{\alpha,\beta}}^{W_n}(\varepsilon_{\alpha,\beta}) \qquad \text{and} \qquad b_{\phi^{(\alpha,\beta)}} = 2n(\alpha) + 2n(\beta) + |\beta|$$

with $n(\alpha)$ as defined in Example 4.1.2. Note also that $W_n \cong (\mathbb{Z}/2\mathbb{Z})^n \rtimes \mathfrak{S}_n$, which leads to an alternative description of $\mathrm{Irr}(W_n)$ in terms of Clifford theory. Furthermore, we have $\varepsilon\phi^{(\alpha,\beta)} = \phi^{(\beta^*,\alpha^*)}$ for all $(\alpha,\beta) \vdash n$ (see [GePf00, 5.5.6]). To fix the notation, the character table for the case $n = 2$ is printed in Table 4.1; here C denotes the Cartan matrix of $W(B_2)$.

Table 4.1 *The character table of* $W(B_2)$

$S = \{t, s_1\}$

$$C = \begin{pmatrix} 2 & -2 \\ -1 & 2 \end{pmatrix}$$

B_2	b_ϕ	1	t	s_1	ts_1	$(ts_1)^2$
$1_W = \phi^{(2,-)}$	0	1	1	1	1	1
$\epsilon = \phi^{(-,11)}$	4	1	−1	−1	1	1
$\phi^{(11,-)}$	2	1	−1	1	−1	1
$\phi^{(-,2)}$	2	1	1	−1	−1	1
$\phi^{(1,1)}$	1	2	0	0	0	−2

Example 4.1.4 Let $n \geqslant 2$ and W_n' be a Coxeter group of type D_n, with generators $u, s_1, \ldots, s_{n-1}$ and Coxeter diagram given as follows:

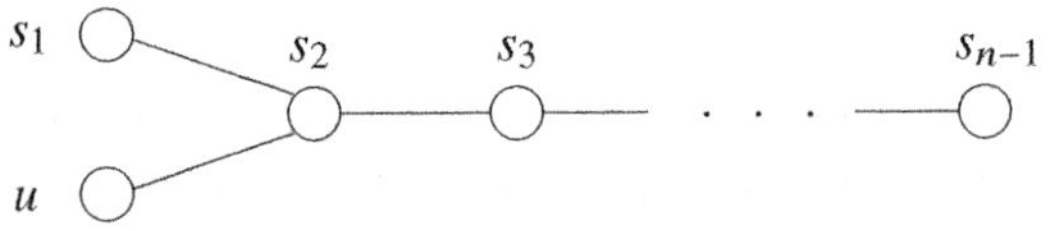

(By convention, we also set $W_0' = W_1' = \{1\}$.) We can identify W_n' with a reflection subgroup of index 2 in a Weyl group of type B_n. Indeed, let W_n be as in Example 4.1.3, with generators $t, s_1, \ldots, s_{n-1}$. Then the assignment

$$u \mapsto ts_1t, \quad s_i \mapsto s_i \quad \text{for } 1 \leqslant i \leqslant n-1,$$

defines an embedding $W_n' \hookrightarrow W_n$. Observe that t acts on W_n' as the non-trivial graph automorphism σ of order 2 interchanging u and s_1, thus identifying W_n with $W_n' \rtimes \langle\sigma\rangle$. This provides a convenient setting for classifying the irreducible characters of W_n'. Given $(\alpha,\beta) \vdash n$, we denote by $\phi^{[\alpha,\beta]}$ the restriction of $\phi^{(\alpha,\beta)} \in \mathrm{Irr}(W_n)$ to W_n'. Then we have (see [GePf00, 5.6.1, 5.6.2]):

(a) If $\alpha \neq \beta$, then $\phi^{[\alpha,\beta]} = \phi^{[\beta,\alpha]} \in \mathrm{Irr}(W_n')$, and

$$b_{\phi^{[\alpha,\beta]}} = 2n(\alpha) + 2n(\beta) + \min\{|\alpha|, |\beta|\} = \min\{b_{\phi^{(\alpha,\beta)}}, b_{\phi^{(\beta,\alpha)}}\}.$$

(b) If $\alpha = \beta$, then $\phi^{[\alpha,\beta]} = \phi^{[\alpha,+]} + \phi^{[\alpha,-]}$ where $\phi^{[\alpha,+]}$, $\phi^{[\alpha,-]}$ are distinct irreducible characters of W_n'. Here,

$$b_{\phi^{[\alpha,+]}} = b_{\phi^{[\alpha,-]}} = 4n(\alpha) + n/2 = b_{\phi^{(\alpha,\alpha)}}.$$

Obviously, by Clifford theory all irreducible characters of W_n' arise in this way. Of course, case (b) can only occur if n is even. In this case, the two characters $\phi^{[\alpha,\pm]}$ can explicitly be specified as follows; see [Lu84a, 4.6.2]. Let

$$H_n^+ = \langle s_1, s_2, \ldots, s_{n-1} \rangle \qquad \text{and} \qquad H_n^- = \langle u, s_2, \ldots, s_{n-1} \rangle$$

be two maximal parabolic subgroups of W_n' isomorphic to $\mathfrak{S}_n$. Let $\alpha \vdash n/2$ and $\mathfrak{S}_{2\alpha^*}$ be the corresponding Young subgroup in $\mathfrak{S}_n$ where $2\alpha^*$ denotes the partition of n obtained by multiplying all parts of α^* by 2. We have corresponding parabolic subgroups $H_{2\alpha^*}^+ \leqslant H_n^+$ and $H_{2\alpha^*}^- \leqslant H_n^-$. Then

$$\phi^{[\alpha,+]} = \mathbf{j}_{H_{2\alpha^*}^+}^{W_n'}(\varepsilon_{2\alpha^*}^+) \qquad \text{and} \qquad \phi^{[\alpha,-]} = \mathbf{j}_{H_{2\alpha^*}^-}^{W_n'}(\varepsilon_{2\alpha^*}^-)$$

where $\varepsilon_{2\alpha^*}^\pm$ denotes the sign character of $H_{2\alpha^*}^\pm$, respectively. (This is also discussed in [GePf00, §5.6] but [GePf00, 5.6.3] has to be reformulated as above.) The effect of tensoring with the sign character ε of W_n' immediately follows from the formulae in Example 4.1.3 for type B_n, except for the cases in (b). So let n be even and $\alpha \vdash n/2$. Then we have

$$\varepsilon\phi^{[\alpha,+]} = \begin{cases} \phi^{[\alpha^*,+]} & \text{if } n/2 \text{ is even,} \\ \phi^{[\alpha^*,-]} & \text{if } n/2 \text{ is odd.} \end{cases}$$

(See [Ge13, 3.5]; this was stated incorrectly in [GePf00, 5.6.5].)

Example 4.1.5 Let W be of exceptional type G_2, F_4, E_6, E_7 or E_8. Then, from the explicit knowledge of the irreducible characters of W one sees that:

- If W is of type E_6, E_7 or E_8, then each $\phi \in \mathrm{Irr}(W)$ is uniquely determined by the pair $(\phi(1), b_\phi)$.
- If W is of type G_2, then each $\phi \in \mathrm{Irr}(W)$ is uniquely determined by the pair $(\phi(1), b_\phi)$ together with the tuple of values $\{\phi(s) \mid s \in S\}$, see Table 4.2.
- For W of type F_4, there are eight pairs of characters with equal invariants $(\phi(1), b_\phi)$; seven of these can be distinguished by their values on reflections, see Table 4.3 where we give one character of each pair. These are precisely those pairs of characters swapped by the non-trivial graph automorphism of the Coxeter diagram of F_4 (see also Theorem 4.5.11). Here, notation is chosen such that the first character in each pair takes smaller value on s_1 than the second (and thus larger value on s_3). Finally, $\phi_{6,6}''$ is defined to be the exterior square of the character $\phi_{4,1}$ of the reflection representation, and $\phi_{6,6}'$ is its tensor product with $\phi_{1,12}'$ (or with $\phi_{1,12}''$), see [Lu84a, 4.10].

Explicit tables defining the notation can also be found in [GePf00, App. C].

Table 4.2 *The character table of* $W(G_2)$

$S = \{s_1, s_2\}$

$$C = \begin{pmatrix} 2 & -1 \\ -3 & 2 \end{pmatrix}$$

G_2	b_ϕ	1	s_1	s_2	s_1s_2	$(s_1s_2)^2$	$(s_1s_2)^3$
1_W	0	1	1	1	1	1	1
ε	6	1	−1	−1	1	1	1
$\phi'_{1,3}$	3	1	−1	1	−1	1	−1
$\phi''_{1,3}$	3	1	1	−1	−1	1	−1
$\phi_{2,1}$	1	2	0	0	1	−1	−2
$\phi_{2,2}$	2	2	0	0	−1	−1	2

Table 4.3 *Part of the character table of* $W(F_4)$

$S = \{s_1, s_2, s_3, s_4\}$

$$C = \begin{pmatrix} 2 & -1 & 0 & 0 \\ -1 & 2 & -1 & 0 \\ 0 & -2 & 2 & -1 \\ 0 & 0 & -1 & 2 \end{pmatrix}$$

F_4	b_ϕ	1	s_1	s_3
$\phi'_{1,12}$	12	1	−1	1
$\phi'_{2,4}$	4	2	.	2
$\phi'_{2,16}$	16	2	−2	.
$\phi'_{4,7}$	7	4	−2	2
$\phi'_{8,3}$	3	8	.	4
$\phi'_{8,9}$	9	8	−4	.
$\phi'_{9,6}$	6	9	−3	3

4.1.6 The b-invariant is also encoded in the so-called fake degrees. To define these let V be the natural reflection representation of W over $\mathbb{R}$. This induces a representation of W on the symmetric algebra $S(V)$ of V. Observe that $S(V)$ carries a natural grading by placing V in degree 1. Denote by $S(V)^W_+$ the ideal generated by the W-invariants in $S(V)$ of positive degree and by $R_W := S(V)/S(V)^W_+$ the *coinvariant algebra*. The grading on $S(V)$ induces a grading on $S(V)^W_+$ and thus also on R_W. Write R^i_W for the ith graded component of R_W, so $R_W = \bigoplus_i R^i_W$. It is known that R_W affords a graded version of the regular representation of W (see [Bou68, V.5.2, Thm. 2(ii)]). The *fake degree* of $\phi \in \mathrm{Irr}(W)$ is the graded multiplicity

$$P_\phi := \sum_{i \geqslant 0} \langle \phi, R^i_W \rangle\, \mathbf{q}^i \in \mathbb{Z}[\mathbf{q}].$$

It is clear from our definitions that the precise power of $\mathbf{q}$ dividing P_ϕ is $\mathbf{q}^{b_\phi}$. We define B_ϕ to be the degree in $\mathbf{q}$ of P_ϕ, so that

$$\begin{aligned} P_\phi(\mathbf{q}) &= c_\phi \mathbf{q}^{b_\phi} + \text{linear combination of higher powers of } \mathbf{q} \\ &= c'_\phi \mathbf{q}^{B_\phi} + \text{linear combination of lower powers of } \mathbf{q} \end{aligned}$$

for some $c_\phi, c'_\phi \in \mathbb{Z}_{>0}$.

The relevance of fake degrees of Weyl groups lies in the fact that, as we will see in Proposition 4.2.5, they are the degree polynomials of the unipotent uniform almost characters of a corresponding (untwisted) finite reductive group.

Example 4.1.7 (Fake degrees for type A_{n-1}) Let us describe the fake degrees of the irreducible characters of the Weyl group $\mathfrak{S}_n$ of type A_{n-1}. The fake degree of $\phi^\alpha \in \mathrm{Irr}(\mathfrak{S}_n)$ labelled by the partition $\alpha = (\alpha_1 \leqslant \cdots \leqslant \alpha_m)$ is given by

$$P_{\phi^\alpha} = \frac{\prod_{i=1}^{n}(\mathbf{q}^i - 1)\prod_{i<j}(\mathbf{q}^{u_j} - \mathbf{q}^{u_i})}{\mathbf{q}^{\binom{r-1}{2}+\binom{r-2}{2}+\ldots}\prod_i\prod_{k=1}^{u_i}(\mathbf{q}^k - 1)}$$

where $(u_1, \ldots, u_r)$, with $u_i := \alpha_i + i - 1$, is the *β-set* associated to α (see e.g. [Ma95, Bem. 2.10]).

Example 4.1.8 (Fake degrees for classical types) Let $\underline{u} = (u_1 < \cdots < u_{m+d})$, $\underline{v} = (v_1 < \cdots < v_m)$ be two strictly increasing sequences of non-negative integers, with $d \in \{0, 1\}$. Define

$$P_d(\underline{u}, \underline{v}) := \frac{\mathbf{q}^{v_1+\ldots+v_m}\prod_{i<j}(\mathbf{q}^{2u_j} - \mathbf{q}^{2u_i})\prod_{i<j}(\mathbf{q}^{2v_j} - \mathbf{q}^{2v_i})}{\mathbf{q}^{\binom{2(m-1)+d}{2}+\binom{2(m-2)+d}{2}+\ldots}\prod_i\prod_{k=1}^{u_i}(\mathbf{q}^{2k} - 1)\prod_i\prod_{k=1}^{v_i}(\mathbf{q}^{2k} - 1)}.$$

Now let W_n be a Weyl group of type B_n. As explained in Example 4.1.3 the irreducible characters of W_n are labelled by pairs of partitions of n. For $(\alpha, \beta) \vdash n$ such a pair, add zeros to the parts of α or β such that α has one more part than β, and denote by $\underline{u} = (u_1 < \cdots < u_{m+1})$, $\underline{v} = (v_1 < \cdots < v_m)$ the corresponding β-sets. Then the fake degree of $\phi^{(\alpha,\beta)}$ is given by

$$P_{\phi^{(\alpha,\beta)}} = \prod_{i=1}^{n}(\mathbf{q}^{2i} - 1)\, P_1(\underline{u}, \underline{v})$$

(see e.g. [Lu77a, Lemma 2.4] or [Ma95, Bem. 2.10]).

Now let W'_n be of type D_n, with $n \geqslant 2$. The irreducible characters of W'_n were described in Example 4.1.4 as $\phi^{[\alpha,\beta]}$ for $(\alpha, \beta) \vdash n$, respectively $\phi^{[\alpha,\pm]}$ for $\alpha \vdash n/2$. Let (α, β) be a pair of partitions of n with corresponding β-sets $\underline{u} = (u_1 < \cdots < u_m)$, $\underline{v} = (v_1 < \cdots < v_m)$. Then the fake degree of $\phi^{[\alpha,\beta]}$ is given by

$$P_{\phi^{[\alpha,\beta]}} = (\mathbf{q}^n - 1)\prod_{i=1}^{n-1}(\mathbf{q}^{2i} - 1)\,(P_0(\underline{u}, \underline{v}) + P_0(\underline{v}, \underline{u}))$$

if $\alpha \neq \beta$, and by

$$P_{\phi[\alpha,\pm]} = (\mathbf{q}^n - 1)\prod_{i=1}^{n-1}(\mathbf{q}^{2i} - 1)\, P_0(\underline{u}, \underline{v})$$

when $\alpha = \beta$ (see e.g. [Lu77a, Lemma 2.7(iii)] or [Ma95, Bem. 5.6]).

The following semi-palindromicity property of fake degrees of Weyl groups with respect to tensoring with the sign character can easily be proved from their definition, see [Ca85, Prop. 11.1.2] for example; here $N = |T|$ denotes the number of reflections in W:

Lemma 4.1.9 *For $\phi \in \mathrm{Irr}(W)$ we have*

$$P_{\varepsilon\phi}(\mathbf{q}) = \mathbf{q}^N\, P_\phi(\mathbf{q}^{-1}).$$

In fact, most fake degrees turn out to be palindromic, see Remark 4.1.16 below.

We now define the second function on $\mathrm{Irr}(W)$. For this, observe that the set of reflections T is a union of rational conjugacy classes of W, so by an elementary result of Burnside from character theory we have

$$\omega_\phi := \sum_{t\in T}\frac{\phi(t)}{\phi(1)} \in \mathbb{Z} \qquad \text{for all } \phi \in \mathrm{Irr}(W).$$

For $J \subseteq S$, let $W_J := \langle J\rangle$ be the corresponding parabolic subgroup. If $I \subseteq J \subseteq S$ then we denote by Ind_I^J the induction of characters from W_I to W_J.

Definition 4.1.10 Define the *a-function* $\mathrm{Irr}(W) \longrightarrow \mathbb{Z}_{\geqslant 0}$, $\phi \mapsto a_\phi$, inductively as follows. If $W = \{1\}$, then $\mathrm{Irr}(W) = \{1_W\}$ and we set $a_{1_W} := 0$. Now assume that $W \neq \{1\}$ and that the function $\psi \mapsto a_\psi$ has already been defined for the irreducible characters of all proper parabolic subgroups of W. Then, for any $\phi \in \mathrm{Irr}(W)$, define

$$a'_\phi := \max\{a_\psi \mid \psi \in \mathrm{Irr}(W_J) \text{ where } J \subsetneqq S \text{ and } \langle \mathrm{Ind}_J^S(\psi), \phi\rangle \neq 0\}.$$

Finally, set

$$a_\phi := \begin{cases} a'_\phi & \text{if } a'_{\varepsilon\phi} - a'_\phi \leqslant \omega_\phi, \\ a'_{\varepsilon\phi} - \omega_\phi & \text{otherwise,}\end{cases}$$

where, as before, ε denotes the sign character of W.

Remark 4.1.11 Note that $\omega_{\varepsilon\phi} = -\omega_\phi$ for all $\phi \in \mathrm{Irr}(W)$. With this one immediately checks that the a-function has the following properties:

$$a_\phi \geqslant a'_\phi \geqslant 0 \quad \text{and} \quad a_{\varepsilon\phi} - a_\phi = \omega_\phi \quad \text{for all } \phi \in \mathrm{Irr}(W).$$

This also shows that $a_\phi \geqslant a_\psi$ if $\psi \in \mathrm{Irr}(W_J)$ with $J \subseteq S$ and $\langle \mathrm{Ind}_J^S(\psi), \phi\rangle \neq 0$.

It is also immediate from the definition that if $W = W_1 \times W_2$ is a direct product of Weyl groups then $a_{\phi_1\boxtimes\phi_2} = a_{\phi_1} + a_{\phi_2}$ for all $\phi_i \in \mathrm{Irr}(W_i)$, $i = 1, 2$ (see [GePf00, 6.5.4]).

Example 4.1.12 The trivial character has a-invariant $a_{1_W} = 0$ and a straightforward computation shows that for the sign character ε it equals $a_\varepsilon = l(w_0) = |T|$ where $w_0 \in W$ is the longest element; we have $0 \leqslant a_\phi \leqslant l(w_0)$ for all $\phi \in \mathrm{Irr}(W)$.

Example 4.1.13 Let $W = \mathfrak{S}_n$. Then, with the notation in Example 4.1.2, we have $a_{\phi^\alpha} = b_{\phi^\alpha} = n(\alpha)$ for all $\alpha \vdash n$, see [GePf00, 6.5.8].

The a-invariants in the other Weyl groups of classical type will be given in 4.4.10.

Lusztig [Lu79b, Lu03b] originally defined 'a-invariants' a_ϕ using the generic degrees of the 1-parameter Iwahori–Hecke algebra associated with W (see 4.1.14). It was shown in [Ge11, Rem. 4.3] that this is equivalent to Definition 4.1.10.

4.1.14 For $\phi \in \mathrm{Irr}(W)$, we define a further invariant by

$$A_\phi := l(w_0) - \omega_\phi - a_\phi = l(w_0) - a_{\varepsilon\phi},$$

with w_0 the longest element in W. The a- and A-invariants are related to the generic degrees $D_\phi \in \mathbb{Q}[\mathbf{q}]$ (where $\mathbf{q}$ is an indeterminate) introduced in Definition 3.2.13 in terms of the associated 1-parameter generic Iwahori–Hecke algebra $\mathscr{H}(W, \mathbf{q})$ (that is, the specialisation of the generic Iwahori–Hecke algebra with parameters $\mathbf{x}_s = \mathbf{q}$ at each reflection $s \in S$, see 3.1.20) in exactly the same way as the b-invariant is related to the fake degrees. Namely, we have

$$\begin{aligned} D_\phi(\mathbf{q}) &= f_\phi^{-1}\mathbf{q}^{a_\phi} + \text{ linear combination of higher powers of } \mathbf{q} \\ &= f_\phi^{-1}\mathbf{q}^{A_\phi} + \text{ linear combination of lower powers of } \mathbf{q}, \end{aligned}$$

where f_ϕ is a positive integer; see Lusztig [Lu84a, 4.1]. In view of Example 3.2.15 this shows that if $W = W_1 \times W_2$ is a direct product of Weyl groups, then $A_{\phi_1\boxtimes\phi_2} = A_{\phi_1} + A_{\phi_2}$ for all $\phi_i \in \mathrm{Irr}(W_i)$, $i = 1, 2$.

The generic degrees satisfy the following semi-palindromicity property, which is completely analogous to that for fake degrees stated in Lemma 4.1.9. It can easily be proved from their definition, see [Ca85, Prop. 11.3.2] for example; here $N = |T|$ again denotes the number of reflections in W:

Lemma 4.1.15 *For $\phi \in \mathrm{Irr}(W)$ we have*

$$D_{\varepsilon\phi}(\mathbf{q}) = \mathbf{q}^N D_\phi(\mathbf{q}^{-1}).$$

Remark 4.1.16 Recall our observation in 3.2.16 that the generic degrees are products of cyclotomic polynomials Φ_d. As they map to the character degrees of W under the specialisation $\mathbf{q} \mapsto 1$, they are not divisible by $\Phi_1 = \mathbf{q} - 1$. Since all cyclotomic polynomials Φ_d with $d \geqslant 2$ are palindromic this shows that we actually also have the palindromicity

$$D_\phi(\mathbf{q}) = \mathbf{q}^{a_\phi + A_\phi} D_\phi(\mathbf{q}^{-1}) \qquad \text{for all } \phi \in \mathrm{Irr}(W).$$

As first observed by Beynon and Lusztig [BeLu78, Prop. A] the corresponding result does not quite hold for the fake degrees. Almost all fake degrees turn out to actually be palindromic, that is, they satisfy $P_\phi(\mathbf{q}) = \mathbf{q}^{N-\omega_\phi} P_\phi(\mathbf{q}^{-1})$. The only characters of irreducible Weyl groups for which this fails to hold are precisely those for which the corresponding irreducible character of the 1-parameter Iwahori–Hecke algebra $\mathscr{H}(W, \mathbf{q})$ does not take values in $\mathbb{Q}(\mathbf{q})$, but in $\mathbb{Q}(\sqrt{\mathbf{q}})$, see Remark 3.1.21. A conceptual explanation for this palindromicity property was subsequently found by Opdam [Op95]. Recall that the non-rational characters in question are the two irreducible characters of degree 512 for W of type E_7 and the four irreducible characters of degree 4096 for W of type E_8, see [BeCu72]. Moreover, unless we are in these exceptional cases, we then have

$$b_\phi + B_\phi = a_\phi + A_\phi \qquad (\phi \in \mathrm{Irr}(W) \text{ non-exceptional})$$

by [Lu79b, (2.2)]. This formula fails to hold in the excluded cases.

The a- and b-invariants can also be defined in an analogous way for finite complex reflection groups. A generalisation of the above observations to these groups is discussed in [Ma99, Thm. 6.5].

Definition 4.1.17 Following [Lu84a, 4.2], [Ge12b, 2.10], define a relation $\preceq$ on $\mathrm{Irr}(W)$ inductively as follows. For $W = \{1\}$, the trivial character is related to itself. Now assume that $W \neq \{1\}$ and that $\preceq$ has already been defined for all proper parabolic subgroups of W. Let $\phi, \phi' \in \mathrm{Irr}(W)$. Then we write $\phi \preceq \phi'$ if there is a sequence $\phi = \phi_0, \phi_1, \ldots, \phi_m = \phi'$ in $\mathrm{Irr}(W)$ such that, for each $i \in \{1, \ldots, m\}$, the following condition is satisfied. There exists a subset $I_i \subsetneq S$ and $\psi_i, \psi_i' \in \mathrm{Irr}(W_{I_i})$, where $\psi_i \preceq \psi_i'$ within $\mathrm{Irr}(W_{I_i})$, such that either

$$\langle \mathrm{Ind}_{I_i}^S(\psi_i), \phi_{i-1}\rangle \neq 0, \quad \langle \mathrm{Ind}_{I_i}^S(\psi_i'), \phi_i\rangle \neq 0 \quad \text{and} \quad a_{\psi_i'} = a_{\phi_i},$$

or

$$\langle \mathrm{Ind}_{I_i}^S(\psi_i), \varepsilon\phi_i\rangle \neq 0, \quad \langle \mathrm{Ind}_{I_i}^S(\psi_i'), \varepsilon\phi_{i-1}\rangle \neq 0 \quad \text{and} \quad a_{\psi_i'} = a_{\varepsilon\phi_{i-1}}.$$

We say that ϕ, ϕ' belong to the same *family in* $\mathrm{Irr}(W)$, if $\phi \preceq \phi'$ and $\phi' \preceq \phi$; this defines an equivalence relation on $\mathrm{Irr}(W)$ and thus a partition of $\mathrm{Irr}(W)$ into the various families. (By [Ge12b, 4.4], this definition of families via the relation $\preceq$ does indeed coincide with Lusztig's original definition in [Lu84a, 4.2].)

Remark 4.1.18 The definition immediately implies that, if $\phi, \phi' \in \mathrm{Irr}(W)$ are such that $\phi \preceq \phi'$, then $\varepsilon\phi' \preceq \varepsilon\phi$. In particular, if $\mathscr{F} \subseteq \mathrm{Irr}(W)$ is a family, then $\varepsilon\mathscr{F} := \{\varepsilon\phi \mid \phi \in \mathscr{F}\}$ also is a family.

It also follows from the definition that if $W = W_1 \times W_2$ is a product of two Weyl groups, then $\phi_1 \boxtimes \phi_2, \phi_1' \boxtimes \phi_2' \in \mathrm{Irr}(W)$ lie in the same family of $\mathrm{Irr}(W)$ if and only if $\phi_i, \phi_i' \in \mathrm{Irr}(W_i)$ are in the same family in $\mathrm{Irr}(W_i)$ for $i = 1, 2$ (see [GePf00, 6.5.4]).

Families are closely related to the a-function:

Proposition 4.1.19 (See [Lu84a, 4.14.1], [Ge12b, 4.4]) *If $\phi, \phi' \in \mathrm{Irr}(W)$ are such that $\phi \preceq \phi'$, then $a_{\phi'} \leqslant a_\phi$. In particular, the function $\phi \mapsto a_\phi$ is constant on the families of* $\mathrm{Irr}(W)$.

The original proof of this result relies on a case-by-case verification. One can give a general proof, based on deep results about Kazhdan–Lusztig cells; see [Lu87b]. A more general statement, involving weight functions on W, is established in [GeIa12, §9].

By 4.1.14 we have $A_\phi = l(w_0) - a_{\varepsilon\phi}$. So Remark 4.1.18 and Proposition 4.1.19 imply that we also have $A_{\phi'} \leqslant A_\phi$ whenever $\phi \preceq \phi'$, with equality if ϕ, ϕ' belong to the same family. Note however that we may well have $a_{\phi'} = a_\phi$ and $A_{\phi'} = A_\phi$ for two characters $\phi, \phi' \in \mathrm{Irr}(W)$ not lying in the same family. The smallest example already occurs in type $A_1 \times A_1$.

The subdivision of $\mathrm{Irr}(W)$ into families has been determined explicitly by Lusztig in each case, see the results given later in this chapter. From this, it is possible to verify the following fact, for which no a priori proof seems to be known:

Proposition 4.1.20 *We have $a_\phi \leqslant b_\phi \leqslant B_\phi \leqslant A_\phi$ for all $\phi \in \mathrm{Irr}(W)$. Moreover, in any family $\mathscr{F} \subseteq \mathrm{Irr}(W)$ there is a unique character $\phi \in \mathscr{F}$ such that $a_\phi = b_\phi$. This character is called the special character in that family.*

See [Lu79b, (2.1)] for the inequalities. In fact, it suffices to check the first inequality; the second is clear by definition, and then the last one follows from the first as

$$A_\phi = N - a_{\varepsilon\phi} \geqslant N - b_{\varepsilon\phi} = B_\phi$$

by Lemmas 4.1.15 and 4.1.9. The notion of special characters was first introduced by Lusztig [Lu79b, p. 324], and in [Lu82a, Sec. 12] he observes that each family contains a unique special character. A new characterisation was given in [Lu18b]. Note that if ϕ is special and non-exceptional (in the sense of Remark 3.1.21) then we also have $A_\phi = B_\phi$ by Remark 4.1.16. The family in type E_7 and the two families in type E_8 containing exceptional characters are called *exceptional*; for those the previous conclusion fails to hold.

Remark 4.1.21 By 4.1.14 and Proposition 4.1.20, $0 \leqslant a_\phi \leqslant A_\phi \leqslant l(w_0)$ for all

$\phi \in \mathrm{Irr}(W)$. The explicit determination of a_ϕ shows that the first inequality is strict unless $\phi = 1_W$ is the trivial character, and thus the last inequality is strict unless $\phi = \varepsilon$ is the sign character, while the middle inequality turns out to be strict unless ϕ is a product of trivial or sign characters of the various irreducible factors of W. Thus, only in the latter cases the generic degree D_ϕ is a monomial.

Example 4.1.22 (Families and special characters in type A_{n-1}) Let $W = \mathfrak{S}_n$ be the Weyl group of type A_{n-1}. Then all families in $\mathrm{Irr}(W)$ are singletons, that is, all $\phi \in \mathrm{Irr}(W)$ lie in a family of their own and (hence) all of them are special [Lu84a, (4.4)]. In particular $a_{\phi^\alpha} = b_{\psi^\alpha} = n(\alpha)$, see Example 4.1.2.

Example 4.1.23 (Families and special characters in type B_n) Let W_n be a Coxeter group of type B_n. As explained in Example 4.1.3 the irreducible characters of W_n are labelled by pairs of partitions (α, β) of n. Given such (α, β), add zero parts to α and β such that $\alpha = (\alpha_1 \leqslant \cdots \leqslant \alpha_{n+1})$ has $n + 1$ parts and $\beta = (\beta_1 \leqslant \cdots \leqslant \beta_n)$ has n parts, and consider the corresponding β-sets

$$x = (x_1, \ldots, x_{n+1}) \qquad \text{with } x_i = \alpha_i + i - 1,$$
$$y = (y_1, \ldots, y_n) \qquad \text{with } y_i = \beta_i + i - 1.$$

Then $\phi^{(\alpha,\beta)}$ is special if and only if $x_1 \leqslant y_1 \leqslant x_2 \leqslant \cdots \leqslant y_n \leqslant x_{n+1}$. Furthermore, two characters $\phi^{(\alpha,\beta)}, \phi^{(\alpha',\beta')} \in \mathrm{Irr}(W_n)$ lie in the same family if and only if the corresponding β-sets (x, y) and (x', y') have the same multiset of entries [Lu84a, (4.5.3) and (4.5.6)].

Example 4.1.24 (Families and special characters in type D_n) Let $n \geqslant 2$ and W_n' a Coxeter group of type D_n. The irreducible characters of W_n' have been described in Example 4.1.4. For (α, β) a pair of partitions of n add parts of size 0 so that both α and β have exactly n parts, and denote by (x, y) the corresponding pair of β-sets. If $\alpha = \beta$ then both $\phi^{[\alpha,+]}$ and $\phi^{[\alpha,-]}$ are special, and each lies in a singleton family. If $\alpha \neq \beta$ then $\phi^{[\alpha,\beta]}$ is special if and only if $x_1 \leqslant y_1 \leqslant \cdots \leqslant x_n \leqslant y_n$ (after possibly interchanging x and y). Furthermore, $\phi^{[\alpha,\beta]}, \phi^{[\alpha',\beta']} \in \mathrm{Irr}(W_n')$ lie in the same family if and only if the corresponding β-sets (x, y) and (x', y') have the same multiset of entries [Lu84a, (4.6.4) and (4.6.10)].

See [Lu84a, (4.8)–(4.13)] for families and special characters for the irreducible Weyl groups of exceptional type.

Remark 4.1.25 It turns out that the relation $a_\phi \leqslant b_\phi$ is no longer satisfied for all characters of complex reflection groups; in fact, validity of this relation is one of several equivalent ways to characterise the so-called *spetsial reflection groups*, see [Ma00, Prop. 8.1]. The notion of families in $\mathrm{Irr}(W)$ and of special characters has been extended to these spetsial reflection groups, see [BrKi02, MaRo03, Chl09, Chl10].

4.1.26 We next extend some of the previous notions to Weyl groups with an automorphism. Recall that V denotes the natural real reflection representation of W so that we can view $W \leqslant \mathrm{GL}(V)$, and let $\sigma \in N_{\mathrm{GL}(V)}(W)$ be such that it induces a *Coxeter group automorphism* of W, also denoted by σ, that is, σ satisfies $\sigma(S) = S$. Then in particular σ respects the length function on W and it induces a graph automorphism of the Coxeter diagram of W. Conversely it can easily be seen that every graph automorphism of the Coxeter diagram of W is induced by a suitable $\sigma \in N_{\mathrm{GL}(V)}(W)$. Following Lusztig [Lu84a, 3.1] we say that σ is *ordinary* if, whenever $s \neq s'$ in S lie in the same σ-orbit, then ss' has order 2 or 3. Equivalently, this happens if and only if σ is not only an automorphism of the Coxeter diagram, but even of any Dynkin diagram associated to (W, S). In particular for irreducible Weyl groups, the non-ordinary graph automorphisms are precisely those for types B_2, G_2 and F_4.

As σ stabilises the reflection representation V of W, in its induced action on $\mathrm{Irr}(W)$ it preserves the b-invariants, and it also acts in a natural way on the symmetric algebra $S(V)$ and on the coinvariant algebra R_W. This allows us to define σ-twisted fake degrees. For this let $\tilde{W}$ be the semidirect product of W with $\langle\sigma\rangle$. Thus, $\tilde{W} = \langle W, \sigma\rangle$ and, in $\tilde{W}$, we have $\sigma(w) = \sigma w \sigma^{-1}$ for all $w \in W$. Let $\phi \in \mathrm{Irr}(W)$ be σ-stable, and denote by $\tilde{\phi}$ a σ-extension of ϕ as in 2.1.7. The *fake degree* of $\tilde{\phi}$ (or *σ-twisted fake degree* of ϕ) is then the graded multiplicity

$$P_{\tilde{\phi}} := \sum_i \langle \tilde{\phi}, R_W^i \rangle_\sigma \, \mathbf{q}^i \quad \in \mathbb{Z}[\mathbf{q}],$$

where $\langle\, , \rangle_\sigma$ is the scalar product on the space of σ-class functions (see 2.1.7). For $\sigma = \mathrm{id}_V$ this specialises to our original definition of the fake degree P_ϕ in 4.1.6. Note that if we choose a different extension $\tilde{\phi}_1$ of ϕ to $\tilde{W}$, then its values on the coset $W\sigma$ differ from those of $\tilde{\phi}$ by a fixed root of unity, and thus $P_{\tilde{\phi}}$ and $P_{\tilde{\phi}_1}$ differ by that same root of unity. In particular the σ-twisted fake degree is determined by ϕ and σ up to a root of unity of order dividing the order of σ. See Proposition 4.2.5 for the interpretation of twisted fake degrees in the representation theory of finite reductive groups.

Example 4.1.27 Consider the situation when $\sigma = -w_0 \in N_{\mathrm{GL}(V)}(W)$. Let $\tilde{\phi}$ be the σ-extension of $\phi \in \mathrm{Irr}(W)$ such that $\tilde{\phi}(ww_0) = \tilde{\phi}(-w\sigma) = \phi(w)$ for all $w \in W$. As $w_0\sigma = -\mathrm{id}_V$, it acts by $(-1)^i$ on the homogeneous component R_W^i of the coinvariant algebra. Then we have

$$P_{\tilde\phi}(\mathbf{q}) = \sum_i \langle \tilde\phi, R^i_W\rangle_\sigma \, \mathbf{q}^i = \frac{1}{|W|}\sum_i \sum_{w\in W} \tilde\phi(ww_0\sigma)\mathrm{Trace}(ww_0\sigma, R^i_W)\,\mathbf{q}^i$$
$$= \frac{1}{|W|}\sum_i \sum_{w\in W} \phi(w)\mathrm{Trace}(w, R^i_W)\,(-\mathbf{q})^i$$
$$= \sum_i \langle \phi, R^i_W\rangle\,(-\mathbf{q})^i = P_\phi(-\mathbf{q}),$$

so the twisted fake degree $P_{\tilde\phi}(\mathbf{q})$ is obtained from $P_\phi(\mathbf{q})$ by replacing $\mathbf{q}$ by $-\mathbf{q}$. This case occurs for the non-trivial graph automorphism of W of type A_n with $n \geqslant 2$, of type D_{2n+1} with $n \geqslant 1$ and of type E_6.

Example 4.1.28 Let W'_n denote the Weyl group of type D_n, $n \geqslant 2$, and let $\sigma : W'_n \to W'_n$ be the automorphism of W'_n induced by the symmetry of order 2 of the Dynkin diagram, interchanging the Coxeter generators u, s_1 in the labelling introduced in Example 4.1.4. According to our description in that example, the σ-invariant characters of W'_n are the ones labelled $\phi^{[\alpha,\beta]}$ with $\alpha \neq \beta$. Let $\tilde\phi^{[\alpha,\beta]}(w) := \phi^{(\alpha,\beta)}(w\sigma)$ for $w \in W'_n$, where $\phi^{(\alpha,\beta)}$ is one of the two extensions of $\phi^{[\alpha,\beta]}$ to $W'_n.\langle\sigma\rangle = W_n$, the Weyl group of type B_n. Let $\underline{u} = (u_1 < \cdots < u_m)$, $\underline{v} = (v_1 < \cdots < v_m)$ be corresponding β-sets of equal length. Then the twisted fake degree of $\tilde\phi^{(\alpha,\beta)}$ is given by

$$P_{\tilde\phi^{[\alpha,\beta]}} = (\mathbf{q}^n + 1)\prod_{i=1}^{n-1}(\mathbf{q}^{2i} - 1)\big(P_0(\underline{u}, \underline{v}) - P_0(\underline{v}, \underline{u})\big)$$

with $P_0(\underline{u}, \underline{v})$ as defined in Example 4.1.8 (see [Lu77a, Lemma 2.7(ii)] or [Ma95, Bem. 5.6]). Observe that this expression does indeed depend on the choice of σ-extension of $\phi^{[\alpha,\beta]}$ to W_n: if we define $\tilde\phi^{[\alpha,\beta]}$ using $\phi^{(\beta,\alpha)}$ instead, the twisted fake degree is multiplied by -1. Also observe that if n is odd, $P_{\tilde\phi^{[\alpha,\beta]}}(\mathbf{q}) = \pm P_{\phi^{[\alpha,\beta]}}(-\mathbf{q})$ as already seen in Example 4.1.27.

As the Coxeter group automorphism σ also sends parabolic subgroups of W to parabolic subgroups, it moreover preserves a-invariants and the relation $\preceq$ on $\mathrm{Irr}(W)$. Hence σ sends families in $\mathrm{Irr}(W)$ to families. Here we have the following observation (see [Lu84a, 4.17]) for which again no a priori proof seems to be known:

Proposition 4.1.29 *Assume that the automorphism σ of W as in 4.1.26 is ordinary. If $\mathscr{F} \subseteq \mathrm{Irr}(W)$ is a σ-stable family then all elements of $\mathscr{F}$ are σ-stable.*

This no longer holds for the non-ordinary graph automorphisms of the Weyl groups W of types B_2, G_2 and F_4: Let $\mathscr{F} \subseteq \mathrm{Irr}(W)$ be a σ-stable family with $|\mathscr{F}| > 1$. If $|\mathscr{F}| = 3$, which occurs in types B_2 and F_4, then $|\mathscr{F}^\sigma| = 1$; if $|\mathscr{F}| = 4$, which only occurs in type G_2, then $|\mathscr{F}^\sigma| = 2$, and if $|\mathscr{F}| = 11$, which happens in F_4, then $|\mathscr{F}^\sigma| = 5$. But in all of these cases, it is still true that the special character in the family is σ-stable.

Remark 4.1.30 The constructions in 4.1.26 depend on the choice of σ-extensions of the characters $\phi \in \mathrm{Irr}(W)^\sigma$. It was already shown in Proposition 2.1.14 that any $\phi \in \mathrm{Irr}(W)^\sigma$ has a σ-extension $\tilde{\phi}$ with values lying in $\mathbb{R}$. Note that $\tilde{\phi}$ may not be uniquely determined but, clearly, there are at most two possibilities for $\tilde{\phi}$ with this property.

It is sometimes convenient to fix a particular choice of σ-extension $\tilde{\phi}$ of ϕ. Lusztig [LuCS, IV, 17.2] has introduced the notion of a *preferred σ-extension* $\tilde{\phi}$. In the cases where W is an irreducible Weyl group and σ is ordinary, these are defined as follows.

- If $\sigma = \mathrm{id}_V$, then $\tilde{\phi} = \phi$.
- If σ acts by conjugation with w_0 and W is of type A_n ($n \geqslant 2$) or E_6, then $\tilde{\phi}$ is the unique σ-extension such that $\tilde{\phi}(w_0) = (-1)^{a_\phi}\phi(1)$ where a_ϕ is the a-invariant of ϕ from Definition 4.1.10.
- If σ has order 3 and W is of type D_4, then $\tilde{\phi}$ is the unique σ-extension with values in $\mathbb{Z}$.
- If σ has order 2 and W is of type D_n ($n \geqslant 4$) then $\tilde{W}$ can be identified with a Weyl group of type B_n (see Example 4.1.4) and the irreducible characters of $\tilde{W}$ that remain irreducible upon restriction to W are parametrised by pairs of partitions (α, β) of n with $\alpha \neq \beta$. The preferred σ-extension of the character $\phi = \phi^{[\alpha,\beta]}$ is then defined as follows. Interchange α, β if necessary such that the following holds: we add parts of size 0 to α or β so that both have the same number of parts. Order the parts of each increasingly and let k denote the first index at which $\alpha_k \neq \beta_k$. Then $\beta_k < \alpha_k$. Then $\tilde{\phi}$ is defined by $\tilde{\phi}(w) := \phi^{(\alpha,\beta)}(w\sigma)$ for $w \in W$.

For W of type B_2, G_2 or F_4 and σ non-ordinary, we take as preferred extensions the ones described in Examples 2.1.12, 2.1.13, 2.1.15 and Remark 2.8.19

4.2 Families of Unipotent Characters and Fourier Matrices

We now explain how the combinatorial notions introduced before connect with the representation theory of finite reductive groups. Let $\mathbf{G}$ be connected reductive and $F : \mathbf{G} \to \mathbf{G}$ a Steinberg endomorphism. Recall from Definition 2.3.8 that an irreducible character $\rho \in \mathrm{Irr}(\mathbf{G}^F)$ is called unipotent if $\langle R_{\mathbf{T}}^{\mathbf{G}}(1), \rho \rangle \neq 0$ for some F-stable maximal torus $\mathbf{T}$ of $\mathbf{G}$. The set of unipotent characters of $\mathbf{G}^F$ is denoted $\mathrm{Uch}(\mathbf{G}^F)$. It was explained in Chapter 2 that these play a distinguished role in the whole theory via Lusztig's Jordan decomposition. In this section we gather several important constructions related to unipotent characters.

Remark 4.2.1 Let us begin this section by explaining how the classification of $\mathrm{Uch}(\mathbf{G}^F)$ is reduced to the case where $\mathbf{G}$ is a simple algebraic group of adjoint type. (See [Lu76c, 1.18] and [Lu77b, 3.15]). First, since $\mathbf{G}/\mathbf{Z}(\mathbf{G})$ is semisimple, there exists a surjective homomorphism of algebraic groups $\pi\colon \mathbf{G} \to \mathbf{G}_{\mathrm{ad}}$ with central kernel and where $\mathbf{G}_{\mathrm{ad}}$ is a semisimple group of adjoint type (see Proposition 1.5.8 and [St67, p. 45/64]). Furthermore, there exists a Steinberg map $F\colon \mathbf{G}_{\mathrm{ad}} \to \mathbf{G}_{\mathrm{ad}}$ such that $F \circ \pi = \pi \circ F$. (See Proposition 1.5.9(b) and [St68, 9.16].) Hence, we obtain a group homomorphism $\pi\colon \mathbf{G}^F \to \mathbf{G}_{\mathrm{ad}}^F$ (but note that this is not necessarily surjective any more). By [DeLu76, 7.10], this induces a bijection

$$\mathrm{Uch}(\mathbf{G}_{\mathrm{ad}}^F) \xrightarrow{\sim} \mathrm{Uch}(\mathbf{G}^F), \qquad \rho \mapsto \rho \circ \pi.$$

Now by Proposition 1.5.10 we can write $\mathbf{G}_{\mathrm{ad}} = \mathbf{G}_1 \times \cdots \times \mathbf{G}_r$ where each $\mathbf{G}_i$ is semisimple of adjoint type, F-stable and F-simple, that is, $\mathbf{G}_i$ is a direct product of simple algebraic groups that are cyclically permuted by F. Let $h_i \geqslant 1$ be the number of simple factors in $\mathbf{G}_i$, and let $\mathbf{H}_i$ be one of these. Then $F^{h_i}(\mathbf{H}_i) = \mathbf{H}_i$ and

$$\iota_i\colon \mathbf{H}_i \to \mathbf{G}_i, \qquad g \mapsto gF(g)\cdots F^{h_i-1}(g),$$

is an injective homomorphism of algebraic groups that restricts to an isomorphism $\iota_i\colon \mathbf{H}_i^{F_i} \xrightarrow{\sim} \mathbf{G}_i^F$ where we denote $F_i := F^{h_i}|_{\mathbf{H}_i}\colon \mathbf{H}_i \to \mathbf{H}_i$ (see Lemma 1.5.15) so that $\mathbf{G}_{\mathrm{ad}}^F \cong \mathbf{H}_1^{F_1} \times \cdots \times \mathbf{H}_r^{F_r}$. Let $f\colon \mathbf{G}_1 \times \cdots \times \mathbf{G}_r \to \mathbf{G}_{\mathrm{ad}}$ be the product map. Then, finally, it is shown in [Lu76c, 1.18] that f and the homomorphisms $\iota_1, \ldots, \iota_r$ induce bijections

$$\mathrm{Uch}(\mathbf{G}_{\mathrm{ad}}^F) \xrightarrow{\sim} \mathrm{Uch}(\mathbf{G}_1^F) \times \cdots \times \mathrm{Uch}(\mathbf{G}_r^F) \xrightarrow{\sim} \mathrm{Uch}(\mathbf{H}_1^{F_1}) \times \cdots \times \mathrm{Uch}(\mathbf{H}_r^{F_r}),$$

such that if χ maps to $\chi_1 \boxtimes \cdots \boxtimes \chi_r$ then $\chi(1) = \chi_1(1)\cdots\chi_r(1)$. By definition the $\mathbf{H}_i$ are simple algebraic groups and $F_i : \mathbf{H}_i \to \mathbf{H}_i$ are Steinberg maps. Thus, the classification of $\mathrm{Uch}(\mathbf{G}^F)$ is reduced to the case where $\mathbf{G}$ is simple. A similar reduction applies to various invariants attached to unipotent characters. For this reason we will, when convenient, just explain the situation for the case when $\mathbf{G}$ is simple. The preceding argument also shows that the classification of unipotent characters is independent of the isogeny type.

4.2.2 In order to describe the multiplicities of unipotent characters in Deligne–Lusztig characters we now introduce an important partition, due to Lusztig, of the set of unipotent characters of a finite reductive group into so-called families. For this, recall that F induces a Coxeter group automorphism σ of the Weyl group $\mathbf{W}$ of $\mathbf{G}$ (see 1.6.1). Now define a graph on the set of vertices $\mathrm{Uch}(\mathbf{G}^F)$ as follows:

two unipotent characters $\rho_1, \rho_2 \in \mathrm{Uch}(\mathbf{G}^F)$ are joined if and only if there is a σ-stable irreducible character $\phi \in \mathrm{Irr}(\mathbf{W})^\sigma$ such that $\langle R_{\tilde{\phi}}, \rho_i \rangle \neq 0$ for $i = 1, 2$ for the almost character $R_{\tilde{\phi}}$ associated to some σ-extension $\tilde{\phi}$ of ϕ (see Definition 2.4.17). (Observe that this does not depend on the chosen σ-extensions.) Then the sets of vertices of the various connected components of this graph are called the *families* in $\mathrm{Uch}(\mathbf{G}^F)$.

Clearly the families play an important role in understanding the relation between almost characters and unipotent characters, so in the decomposition of the Deligne–Lusztig characters $R_{\mathbf{T}}^{\mathbf{G}}(1)$.

The above procedure induces a corresponding subdivision of $\mathrm{Irr}(\mathbf{W})^\sigma$ by declaring that two characters $\phi, \phi' \in \mathrm{Irr}(\mathbf{W})^\sigma$ are equivalent if and only if $R_{\tilde{\phi}}, R_{\tilde{\phi}'}$ have unipotent constituents lying in the same family of $\mathrm{Uch}(\mathbf{G}^F)$, for some (all) σ-extensions $\tilde{\phi}, \tilde{\phi}'$ of ϕ, ϕ' respectively. The choice of notation is justified by the following:

Proposition 4.2.3 *Assume that σ is ordinary. The equivalence classes in* $\mathrm{Irr}(\mathbf{W})^\sigma$ *as defined above are precisely the σ-stable families in* $\mathrm{Irr}(\mathbf{W})$ *defined in 4.1.17.*

This is proved in [Lu84a]. Indeed, in the case where F is a Frobenius map [Lu84a, 6.17] shows that the equivalence relation on $\mathrm{Irr}(\mathbf{W})^\sigma$ defined above is at least as coarse as the subdivision into families. On the other hand, by Corollary 4.2.19 below all unipotent characters in a given family occur as constituents of the almost character of the special character (see 4.1.20) in a σ-stable family of $\mathrm{Irr}(\mathbf{W})$. Thus, both relation must agree.

The fact that a similar statement is also true for the Suzuki and Ree groups is implicitly contained in the remarks in [Lu84a, §14.2]. Here, the equivalence classes in $\mathrm{Irr}(\mathbf{W})^\sigma$ consist exactly of the σ-fixed points in σ-stable families in $\mathrm{Irr}(\mathbf{W})$. This shows that the families in $\mathrm{Uch}(\mathbf{G}^F)$ are closely related to the purely combinatorial notion of families in the Weyl group introduced in Definition 4.1.17.

There is a way to read off the families of unipotent characters from the character table of $\mathbf{G}^F$, in terms of the unipotent support introduced in Theorem 2.7.15:

Theorem 4.2.4 (Geck–Malle) *Two unipotent characters of* $\mathbf{G}^F$ *lie in the same family if and only if they have the same unipotent support.*

This was shown in [GeMa00, Prop. 4.2 and Cor. 5.2] after having been conjectured by Lusztig [Lu80b, Problem II].

The following is an elementary computation in the invariant theory of Weyl groups (see [Lu77b, 3.16 and 3.19] or [BrMa92, Prop. 1.6']):

Proposition 4.2.5 *Let* $\mathbf{G}$ *be connected reductive with a Steinberg map* F. *Let* $\mathbf{W}$

be the Weyl group of $\mathbf{G}$ *and* σ *the automorphism of* $\mathbf{W}$ *induced by* F. *Then for* $\phi \in \mathrm{Irr}(\mathbf{W})^\sigma$ *with* σ*-extension* $\tilde{\phi}$ *we have*

$$P_{\tilde{\phi}} = \mathbb{D}_{\tilde{\phi}},$$

that is, the fake degree coincides with the degree polynomial of the corresponding unipotent almost character (see Remark 2.4.31).

4.2.6 Recall from Definition 2.3.25 that the degrees of unipotent characters can be written as polynomials in q. As in Remark 2.3.26 for $\rho \in \mathrm{Uch}(\mathbf{G}^F)$ we denote by a_ρ the precise power of $\mathbf{q}$ dividing this polynomial, and by A_ρ its degree in $\mathbf{q}$, so that

$$\mathbb{D}_\rho = \sum_{i=a_\rho}^{A_\rho} c_i \mathbf{q}^i \qquad \text{for suitable } c_i \in \mathbb{Q}.$$

These invariants coincide with the a- and A-invariants of the associated family in $\mathrm{Irr}(\mathbf{W})$:

Proposition 4.2.7 *The* a*- and* A*-values on a family of* $\mathrm{Uch}(\mathbf{G}^F)$ *agree with the* a*- and* A*-values of the corresponding family of* $\mathrm{Irr}(\mathbf{W})$. *In particular both are constant on families.*

Proof Let $\mathscr{U} \subseteq \mathrm{Uch}(\mathbf{G}^F)$ be a family of unipotent characters and $\mathscr{F} \subseteq \mathrm{Irr}(\mathbf{W})^\sigma$ the corresponding family in $\mathrm{Irr}(\mathbf{W})$. By Definition 2.3.25 the degree polynomial of ρ is given by

$$\mathbb{D}_\rho = \frac{1}{|\mathbf{W}|} \sum_{(w,\theta)\in\mathfrak{X}(\mathbf{W},\sigma)} (-1)^{l(w)} \langle R_w^\theta, \rho\rangle\, \mathbf{q}^{-N} \frac{|\mathbb{G}|}{|\mathbb{T}_w|}.$$

Now ρ is unipotent so by Definition 2.3.8 only terms with $\theta = 1_\mathbf{T}$ will contribute to the sum. Hence using the definition of the almost characters $R_{\tilde{\phi}}$ in Remark 2.4.17 we obtain

$$\begin{aligned}
\mathbb{D}_\rho &= \frac{1}{|\mathbf{W}|} \sum_{w\in\mathbf{W}} (-1)^{l(w)} \langle R_w, \rho\rangle\, \mathbf{q}^{-N} \frac{|\mathbb{G}|}{|\mathbb{T}_w|} \\
&= \frac{1}{|\mathbf{W}|} \sum_{w\in\mathbf{W}} (-1)^{l(w)} \sum_{\phi\in\mathrm{Irr}(\mathbf{W})^\sigma} \tilde{\phi}(w)\, \langle R_{\tilde{\phi}}, \rho\rangle\, \mathbf{q}^{-N} \frac{|\mathbb{G}|}{|\mathbb{T}_w|} \\
&= \sum_{\phi\in\mathrm{Irr}(\mathbf{W})^\sigma} \langle R_{\tilde{\phi}}, \rho\rangle \frac{1}{|\mathbf{W}|} \sum_{w\in\mathbf{W}} (-1)^{l(w)} \tilde{\phi}(w)\, \mathbf{q}^{-N} \frac{|\mathbb{G}|}{|\mathbb{T}_w|} \\
&= \sum_{\phi\in\mathrm{Irr}(\mathbf{W})^\sigma} \langle R_{\tilde{\phi}}, \rho\rangle\, \mathbb{D}_{R_{\tilde{\phi}}}.
\end{aligned}$$

By Proposition 4.2.5 the degree of $\mathbb{D}_{R_{\tilde{\phi}}}$ is given by the fake degree $P_{\tilde{\phi}}$, and by

Proposition 4.1.20 the special character $\phi \in \mathscr{F}$ is the unique one with $a_\phi = b_\phi$, while all other b-values in $\mathscr{F}$ are strictly bigger, and by Corollary 4.2.19 below we have $\langle R_{\tilde{\phi}}, \rho \rangle \neq 0$ for all $\rho \in \mathscr{U}$. Thus $a_\rho = a_\phi$ is constant on $\mathscr{U}$. The argument for A is entirely similar, using again Proposition 4.1.20 and the subsequent remark, or alternatively Proposition 4.2.8 together with Proposition 3.4.21. □

Proposition 4.2.8 *The Alvis–Curtis–Kawanaka–Lusztig duality* $D_{\mathbf{G}}$ *(see Definition 3.4.1) sends families in* $\mathrm{Uch}(\mathbf{G}^F)$ *to families. Moreover, special unipotent characters are sent to special unipotent characters, except for the exceptional families in types* E_7 *and* E_8.

Here we call *special unipotent character* in a family $\mathscr{U} \subseteq \mathrm{Uch}(\mathbf{G}^F)$ the unipotent principal series character in $\mathscr{U}$ labelled by the special character in the corresponding family of $\mathrm{Irr}(\mathbf{W}^F)$.

Proof The first assertion is immediate from the definition of families and the commutation of $D_{\mathbf{G}}$ with $R_{\mathbf{T}}^{\mathbf{G}}$ from Theorem 3.4.4. For the second claim, if $\mathscr{U} \subseteq \mathrm{Uch}(\mathbf{G}^F)$ is a family and $\rho \in \mathscr{U}$ is labelled by the special character ϕ in the corresponding family $\mathscr{F} \subseteq \mathrm{Irr}(\mathbf{W}^F)$, then $D_{\mathbf{G}}(\rho)$ is labelled by $\epsilon \otimes \phi \in \mathrm{Irr}(\mathbf{W}^F)$ by Theorem 3.4.8, where $\epsilon \in \mathrm{Irr}(\mathbf{W}^F)$ is the sign character. Now $\epsilon \otimes \phi$ is again special by the remarks after Proposition 4.1.20 unless $\mathscr{F}$ is an exceptional family in types E_7 or E_8. □

We now turn to the problem of describing the multiplicities of the unipotent characters in the unipotent almost characters. Observe that by the very definition of families, a unipotent character can occur as a constituent of some almost character $R_{\tilde{\phi}}$ only if the restriction ϕ of $\tilde{\phi}$ lies in the corresponding family of $\mathrm{Irr}(\mathbf{W})$. So the base change matrix from almost characters to unipotent characters is block-diagonal according to families, and it suffices to consider the decomposition problem family by family. Here Lusztig has discovered an intriguing combinatorics which we describe next; see also [Lu87b, 2.5] for a different interpretation in terms of equivariant vector bundles.

4.2.9 (Lusztig's non-abelian Fourier transform) Let $\mathscr{G}$ be a finite group. Write $\mathscr{M}(\mathscr{G})$ for the set of pairs (x, σ), where $x \in \mathscr{G}$ and $\sigma \in \mathrm{Irr}(C_{\mathscr{G}}(x))$, modulo the equivalence relation $(x, \sigma) \sim ({}^g x, \sigma^g)$ for $g \in \mathscr{G}$, where $\sigma^g({}^g y) := \sigma(y)$ for $y \in C_{\mathscr{G}}(x)$. Observe that if $\mathscr{G} = \mathscr{G}_1 \times \mathscr{G}_2$ is a direct product, then $\mathscr{M}(\mathscr{G})$ can naturally be identified with $\mathscr{M}(\mathscr{G}_1) \times \mathscr{M}(\mathscr{G}_2)$. Following Lusztig [Lu79a, §4] we define a pairing $\{\, ,\, \} : \mathscr{M}(\mathscr{G}) \times \mathscr{M}(\mathscr{G}) \to \mathbb{C}$ by

$$\{(x, \sigma), (y, \tau)\} := \frac{1}{|C_{\mathscr{G}}(x)|\,|C_{\mathscr{G}}(y)|} \sum_{\substack{g \in \mathscr{G} \\ xgyg^{-1} = gyg^{-1}x}} \sigma(gyg^{-1})\tau(g^{-1}x^{-1}g).$$

Let $\mathbf{M}_{\mathscr{G}}$ denote the operator on the space of (complex-valued) functions on $\mathscr{M}(\mathscr{G})$ defined by

$$(\mathbf{M}_{\mathscr{G}} f)(x,\sigma) := \sum_{(y,\tau)\in\mathscr{M}(\mathscr{G})} \{(x,\sigma),(y,\tau)\}\, f(y,\tau)$$

for $(x,\sigma) \in \mathscr{M}(\mathscr{G})$. This is called the *non-abelian Fourier transform* associated to $\mathscr{G}$. Alternatively we can also consider the endomorphism of the vector space $\mathbb{C}\mathscr{M}(\mathscr{G})$ defined on the basis by

$$(y,\tau) \mapsto \sum_{(x,\sigma)\in\mathscr{M}(\mathscr{G})} \{(x,\sigma),(y,\tau)\}\,(x,\sigma),$$

but the first point of view will be more convenient later.

Lemma 4.2.10 *The operator $\mathbf{M}_{\mathscr{G}}$ is hermitian, that is,*

$$\{(x,\sigma),(y,\tau)\} = \overline{\{(y,\tau),(x,\sigma)\}} \qquad \textit{for all} \quad (x,\sigma),(y,\tau) \in \mathscr{M}(\mathscr{G}),$$

and $\mathbf{M}_{\mathscr{G}}^2 = 1$.

Proof The first claim follows directly from the definition of $\mathbf{M}(\mathscr{G})$, the second is an easy consequence of the orthogonality relations for irreducible characters. □

4.2.11 Another interpretation of $\mathscr{M}(\mathscr{G})$ can be given in terms of the category $\mathscr{C}(\mathscr{G})$ of finite-dimensional $\mathscr{G}$-graded $\mathbb{C}\mathscr{G}$-modules, that is, the objects of $\mathscr{C}(\mathscr{G})$ are $\mathbb{C}\mathscr{G}$-modules V with a decomposition

$$V = \bigoplus_{x\in\mathscr{G}} V_x$$

such that $g.V_x = V_{{}^g x}$ for all $x, g \in \mathscr{G}$, and the morphisms in $\mathscr{C}(\mathscr{G})$ are $\mathbb{C}\mathscr{G}$-linear maps $f : V \to V'$ such that $f(V_x) \leqslant V'_x$ for all $x \in \mathscr{G}$. Equivalently, this can be described as the category of finite-dimensional modules for the Drinfeld double of $\mathscr{G}$, see [Dr87]. Observe that any graded component V_x, $x \in \mathscr{G}$, of a $\mathscr{G}$-graded $\mathbb{C}\mathscr{G}$-module V carries a representation of the centraliser $C_{\mathscr{G}}(x)$. It is easily seen that $\mathscr{C}(\mathscr{G})$ is semisimple, and the simple objects are obtained as follows (see e.g., [Br17, §8.1]): For $x \in \mathscr{G}$ let U be an irreducible $\mathbb{C}C_{\mathscr{G}}(x)$-module. Then $V(x,U) := \bigoplus_{t\in\mathscr{G}} V(x,U)_t$, with

$$V(x,U)_t := \begin{cases} g\otimes U & \text{if } t = gxg^{-1},\\ 0 & \text{else,}\end{cases}$$

is an irreducible $\mathscr{G}$-graded $\mathbb{C}\mathscr{G}$-module. The $V(x,U)$ provide a complete set of non-isomorphic simple objects in $\mathscr{C}(\mathscr{G})$ when x runs over a system of representatives for the conjugacy classes of $\mathscr{G}$, and U over a complete system of representatives for the isomorphism classes of irreducible $\mathbb{C}C_{\mathscr{G}}(x)$-modules.

Thus, $\mathscr{M}(\mathscr{G})$ as introduced above is in bijection with the set of isomorphism classes of simple objects in $\mathscr{C}(\mathscr{G})$, by sending (x, σ) to $V(x, U)$ where U affords the character σ.

Remark 4.2.12 (Mellin transform) Digne and Michel [DiMi85, VII.3] introduced another basis for the space $\mathbb{C}\mathscr{M}(\mathscr{G})$. Let $\mathscr{G}$ be a finite group and denote by $\mathscr{N}(\mathscr{G})$ the set of pairs $(x, z) \in \mathscr{G} \times \mathscr{G}$ with $xz = zx$, taken modulo simultaneous $\mathscr{G}$-conjugation. Clearly, $|\mathscr{N}(\mathscr{G})| = |\mathscr{M}(\mathscr{G})|$. To $(x, z) \in \mathscr{N}(\mathscr{G})$ we associate

$$e_{(x,z)} := \sum_{\sigma \in \mathrm{Irr}(C_{\mathscr{G}}(x))} \sigma(z)(z, \sigma) \in \mathbb{C}\mathscr{M}(\mathscr{G}).$$

From the orthogonality relations for $\mathrm{Irr}(C_{\mathscr{G}}(x))$ one gets that $\{e_{(x,z)} \mid (x, z) \in \mathscr{N}(\mathscr{G})\}$ is also a basis of $\mathbb{C}\mathscr{M}(\mathscr{G})$. (The base change is called *Mellin transform* by Digne–Michel.) A straightforward computation shows that the matrix of the non-abelian Fourier transform on $\mathbb{C}\mathscr{M}(\mathscr{G})$ induced by $\mathbf{M}_{\mathscr{G}}$ takes a particularly simple form when expressed with respect to this basis (see [DiMi85, Prop. 3.1]):

Proposition 4.2.13 *We have* $\mathbf{M}_{\mathscr{G}} e_{(x,z)} = e_{(z^{-1},x^{-1})}$ *for all* $(x, z) \in \mathscr{N}(\mathscr{G})$.

4.2.14 For twisted groups we need to introduce an extension of the set-up described in 4.2.9, see [Lu84a, 4.16]. Assume that $\tilde{\mathscr{G}}$ is a finite group with a normal subgroup $\mathscr{G} \trianglelefteq \tilde{\mathscr{G}}$ of index c such that $\tilde{\mathscr{G}}$ is a split cyclic extension of $\mathscr{G}$. For a fixed generator of $\tilde{\mathscr{G}}/\mathscr{G}$ let $\mathscr{G}'$ denote its full preimage in $\tilde{\mathscr{G}}$, a coset of $\mathscr{G}$. To this we associate two sets $\mathscr{M} = \mathscr{M}(\mathscr{G} \trianglelefteq \tilde{\mathscr{G}})$ and $\overline{\mathscr{M}} = \overline{\mathscr{M}}(\mathscr{G} \trianglelefteq \tilde{\mathscr{G}})$ as follows. The set $\mathscr{M}$ consists of all pairs (x, σ) where $x \in \mathscr{G}$ is such that $C_{\tilde{\mathscr{G}}}(x) \cap \mathscr{G}' \neq \emptyset$ and $\sigma \in \mathrm{Irr}(C_{\tilde{\mathscr{G}}}(x))$ has irreducible restriction to $C_{\mathscr{G}}(x)$, taken again modulo the equivalence relation $(x, \sigma) \sim ({}^g x, \sigma^g)$ for $g \in \tilde{\mathscr{G}}$. The cyclic group $\mathrm{Irr}(\tilde{\mathscr{G}}/\mathscr{G})$ of linear characters of $\tilde{\mathscr{G}}/\mathscr{G}$ acts on $\mathscr{M}$ by $\varepsilon : (y, \tau) \mapsto (y, \tau \otimes \varepsilon)$ for $\varepsilon \in \mathrm{Irr}(\tilde{\mathscr{G}}/\mathscr{G})$.

The set $\overline{\mathscr{M}}$ consists of all pairs $(x, \bar{\sigma})$ with $x \in \mathscr{G}'$ and $\bar{\sigma} \in \mathrm{Irr}(C_{\mathscr{G}}(x))$ an irreducible character, taken modulo the same equivalence relation as for $\mathscr{M}$. Observe that $C_{\tilde{\mathscr{G}}}(x)$ is a central cyclic extension of $C_{\mathscr{G}}(x)$ by x, so all irreducible characters of $C_{\mathscr{G}}(x)$ extend to $C_{\tilde{\mathscr{G}}}(x)$.

In the case that $\mathscr{G} = \tilde{\mathscr{G}}$ we have $\mathscr{M} = \overline{\mathscr{M}} = \mathscr{M}(\tilde{\mathscr{G}})$ and we recover the situation in 4.2.9.

The pairing $\{,\}$ on $\mathscr{M}(\tilde{\mathscr{G}})$ introduced in 4.2.9 now induces a pairing

$$\{,\} : \overline{\mathscr{M}} \times \mathscr{M} \to \mathbb{C}, \qquad \{(x, \bar{\sigma}), (y, \tau)\} := c\{(x, \sigma), (y, \tau)\},$$

for any fixed extension σ of $\bar{\sigma}$ to $C_{\tilde{\mathscr{G}}}(x)$. Consider the operator $\mathbf{M}_{\mathscr{G} \lessdot \tilde{\mathscr{G}}}$ on the space of functions $\mathscr{M}(\tilde{\mathscr{G}}) \to \mathbb{C}$ with support contained in $\mathscr{M}$, defined by

$$(\mathbf{M}_{\mathscr{G} \lessdot \tilde{\mathscr{G}}} f)(x, \bar{\sigma}) := \sum_{(y,\tau)/\sim} \{(x, \bar{\sigma}), (y, \tau)\}\, f(y, \tau),$$

where the sum runs over a system of representatives of the $\mathrm{Irr}(\tilde{\mathcal{G}}/\mathcal{G})$-orbits on $\mathcal{M}$. It defines an isomorphism onto the space of functions on $\overline{\mathcal{M}}$, with inverse given by

$$(\mathbf{M}^{-1}_{\mathcal{G}\leqslant\tilde{\mathcal{G}}}f)(y,\tau) = \sum_{(x,\bar{\sigma})\in\overline{\mathcal{M}}} \{(x,\bar{\sigma}),(y,\tau)\}\, f(x,\bar{\sigma}).$$

4.2.15 Now return to our setting with $\mathbf{G}$ connected reductive with a Steinberg map $F : \mathbf{G} \to \mathbf{G}$ and assume that the automorphism σ induced by F on the Weyl group $\mathbf{W}$ is ordinary. In [Lu84a, §4] Lusztig defines the following data: to each family $\mathcal{U} \subseteq \mathrm{Uch}(\mathbf{G}^F)$ of unipotent characters with corresponding σ-stable family $\mathcal{F} \subseteq \mathrm{Irr}(\mathbf{W})^\sigma$ he associates finite groups $\mathcal{G}_\mathcal{U} \trianglelefteq \tilde{\mathcal{G}}_\mathcal{U}$, where $|\tilde{\mathcal{G}}_\mathcal{U} : \mathcal{G}_\mathcal{U}| = c$ is the order of the automorphism σ, and defines a bijection $\mathcal{U} \to \overline{\mathcal{M}}(\mathcal{G}_\mathcal{U} \trianglelefteq \tilde{\mathcal{G}}_\mathcal{U}), \rho \mapsto \bar{x}_\rho$, and an injection $\mathcal{F} \to \mathcal{M}(\mathcal{G}_\mathcal{U} \trianglelefteq \tilde{\mathcal{G}}_\mathcal{U}), \phi \mapsto x_\phi$. An interpretation of the group $\mathcal{G}_\mathcal{U}$ is given without proof in [Lu84a, 13.1.3] and then substantiated in [Lu13a].

Moreover, if ρ is a unipotent character lying in the (ordinary) Harish-Chandra series of a cuspidal unipotent character of a split Levi subgroup $\mathbf{L} \leqslant \mathbf{G}$, then let

$$\Delta(\bar{x}_\rho) := \epsilon_{[\mathbf{L},\mathbf{L}]} = (-1)^{r(\mathbf{L}^F)}$$

(see [Lu84a, 6.7 and Prop. 6.20]). By Proposition 3.4.7 this is exactly the sign by which $\mathrm{D}_\mathbf{L}(\rho)$ differs from being a true character. It turns out that for simple groups $\mathbf{G}$ with a split Frobenius F, the sign $\Delta(\bar{x}_\rho)$ is always equal to 1, except for the non-principal series characters in the exceptional families in types E_7 and E_8 already encountered in Remark 3.1.21 and also discussed in 4.1.16: these are exactly the families in which the principal series characters correspond to representations of the equal-parameter Iwahori–Hecke algebra $\mathcal{H}(\mathbf{W}, \mathbf{q})$ that cannot be realised over $\mathbb{Q}(\mathbf{q})$.

With these data the following fundamental decomposition theorem holds:

Theorem 4.2.16 (Lusztig [Lu84a, Thm. 4.23]) *Assume that σ is ordinary. Let $\mathcal{U} \subseteq \mathrm{Uch}(\mathbf{G}^F)$ be a family of unipotent characters with corresponding σ-stable family $\mathcal{F} \subseteq \mathrm{Irr}(\mathbf{W})^\sigma$. Then we have*

$$\langle \rho, R_{\tilde{\phi}} \rangle = \Delta(\bar{x}_\rho)\{\bar{x}_\rho, x_\phi\}$$

for all $\rho \in \mathcal{U}$ and $\phi \in \mathrm{Irr}(\mathbf{W})^\sigma$ with preferred σ-extensions $\tilde{\phi}$.

Observe that the set of unipotent characters $\mathrm{Uch}(\mathbf{G}^F)$ as well as the multiplicities $\langle R_{\tilde{\phi}}, \rho \rangle$ are thus generic, that is, they depend only on the complete root datum underlying $(\mathbf{G}, F)$, so the same is true for the subdivision of $\mathrm{Uch}(\mathbf{G}^F)$ into families. By the definition of almost characters the multiplicities $\Delta(\bar{x}_\rho)\{\bar{x}_\rho, x_\phi\}$ are related to the coefficients $m(w, \bar{x}_\rho)$ mentioned in Theorem 2.4.1 through the σ-character table of $\mathbf{W}$.

Remark 4.2.17 The matrix $\mathbf{M}(\mathscr{U}) := (\{\bar{x}_\rho, x\})$, with $\rho \in \mathscr{U}$, $x \in \mathscr{M}(\mathscr{G}_\mathscr{U})$, is called the *non-abelian Fourier transform matrix* associated to the family $\mathscr{U}$. Theorem 4.2.16 immediately implies the following formula for almost characters (see [Lu84a, Cor. 4.24]):

$$R_{\tilde{\phi}} = \sum_{\rho \in \mathscr{U}} \Delta(\bar{x}_\rho)\{\bar{x}_\rho, x_\phi\}\, \rho,$$

involving the rectangular submatrix $(\{\bar{x}_\rho, x_\phi\})$ of $\mathbf{M}(\mathscr{U})$ with $\rho \in \mathscr{U}$ and $\phi \in \mathscr{F}$. The orthonormal set of class functions

$$R_x = \sum_{\rho \in \mathscr{U}} \Delta(\bar{x}_\rho)\{\bar{x}_\rho, x\}\, \rho \qquad (x \in \mathscr{M}(\mathscr{G}_\mathscr{U}))$$

is called the *unipotent almost characters* of $\mathbf{G}^F$. As already mentioned in Remark 2.7.26, Lusztig conjectured that each R_x should have a geometric interpretation, as the characteristic function of an F-stable character sheaf on $\mathbf{G}$ (suitably normalised); that is, the Fourier matrix $\mathbf{M}(\mathscr{U})$ should, up to scalars, describe the base change from characteristic functions to unipotent characters. This is known to be true by Shoji's work [Sho95]. (The restrictions on the characteristic in [Sho95] can now be removed thanks to [Lu12a].) See [Ge18, §7] for a more in-depth discussion.

The proof of Theorem 4.2.16 (for arbitrary characters in place of unipotent ones) occupies most of the book [Lu84a]. Lusztig had previously already shown this for unipotent characters in the case when the underlying field $\mathbb{F}_q$ is sufficiently large in [Lu80a, Lu81c, Lu82b]. Let us comment somewhat more on this result. The finite group $\mathscr{G}_\mathscr{U}$ turns out in all cases to be a direct product of symmetric groups $\mathfrak{S}_n$, with $1 \leqslant n \leqslant 5$. Moreover, $\mathfrak{S}_5$ only occurs if $\mathbf{G}$ has a factor of type E_8, and then only for one family, and $\mathfrak{S}_4$ only occurs if $\mathbf{G}$ has a factor of type F_4, again for a single family. If $\mathbf{G}$ is of classical type, all groups $\mathscr{G}_\mathscr{U}$ are elementary abelian 2-groups. The sizes of the occurring Fourier matrices are given by

$$|\mathscr{M}(C_2^e)| = 2^{2e}, \; |\mathscr{M}(\mathfrak{S}_3)| = 8, \; |\mathscr{M}(\mathfrak{S}_4)| = 21, \; |\mathscr{M}(\mathfrak{S}_5)| = 39.$$

Let us point out that although all symmetric groups have rational character tables, the corresponding Fourier transform matrices $\mathbf{M}(\mathscr{U})$ may contain non-rational entries (and in fact do so for $\mathscr{G} = \mathfrak{S}_5$ which occurs in type E_8).

Remark 4.2.18 A trivial consequence of this explicit decomposition is the existence of a Jordan decomposition $\mathscr{E}(\mathbf{G}^F, 1) \to \mathscr{E}(\mathbf{G}^{*F}, 1)$ as claimed in Theorem 2.6.4, at least for unipotent characters. Indeed, as pointed out in Remark 2.6.5 there is an immediate reduction to the case of simple groups using the arguments in Remark 4.2.1. Then one uses that the adjoint quotients of $\mathbf{G}$ and $\mathbf{G}^*$ are isomorphic except when $\mathbf{G}$ is of type B or C. In the latter case, though, the combinatorial data describing the multiplicities of unipotent characters in almost characters in Theorem 4.2.16 are the same in both types, thus there exists a bijection as claimed.

As a consequence of the explicit Decomposition Theorem 4.2.16 Lusztig [Lu84a, l. 1 on p. 133] observed:

Corollary 4.2.19 *Let* $\mathscr{U} \subseteq \mathrm{Uch}(\mathbf{G}^F)$ *be a family of unipotent characters. Then there is a* σ*-extension of the special character in the corresponding* σ*-stable family in* $\mathrm{Irr}(\mathbf{W})$ *such that all* $\chi \in \mathscr{U}$ *occur with positive multiplicity in the corresponding almost character.*

Example 4.2.20 (a) It easily follows from the definitions given in 4.2.9 and in 4.2.14 that in the case when $\mathscr{G}_{\mathscr{U}} = C_2^e$ all entries of the corresponding $2^{2e} \times 2^{2e}$-Fourier transform matrix are of the form $\pm 2^{-e}$.

(b) We give the Fourier transform matrices for the two smallest non-trivial cases, the symmetric groups $\mathfrak{S}_2 \cong C_2$ and $\mathfrak{S}_3$ in Tables 4.4 and 4.5. As in previous tables, an entry "." stands for "0". Here we write g_2 for the transposition $(1, 2)$ and g_3 for the 3-cycle $(1, 2, 3)$. Also, ε denotes the sign character of $\mathfrak{S}_2$ or $\mathfrak{S}_3$, r is the reflection character of $\mathfrak{S}_3$ and θ, θ^2 denote the two non-trivial characters of the cyclic group generated by g_3. We see that by Corollary 4.2.19 for families with group $\mathfrak{S}_2$ or $\mathfrak{S}_3$ the special character must always correspond to the pair $(1, 1)$.

Table 4.4 *Fourier transform matrix for* $\mathfrak{S}_2$

$\frac{1}{2}$ ×

	$(1,1)$	$(1,\varepsilon)$	$(g_2,1)$	(g_2,ε)
$(1,1)$	1	1	1	1
$(1,\varepsilon)$	1	1	−1	−1
$(g_2,1)$	1	−1	1	−1
(g_2,ε)	1	−1	−1	1

Table 4.5 *Fourier transform matrix for* $\mathfrak{S}_3$

$\frac{1}{6}$ ×

	$(1,1)$	$(g_2,1)$	$(1,r)$	$(g_3,1)$	$(1,\varepsilon)$	(g_2,ε)	(g_3,θ)	(g_3,θ^2)
$(1,1)$	1	3	2	2	1	3	2	2
$(g_2,1)$	3	3	.	.	−3	−3	.	.
$(1,r)$	2	.	4	−2	2	.	−2	−2
$(g_3,1)$	2	.	−2	4	2	.	−2	−2
$(1,\varepsilon)$	1	−3	2	2	1	−3	2	2
(g_2,ε)	3	−3	.	.	−3	3	.	.
(g_3,θ)	2	.	−2	−2	2	.	4	−2
(g_3,θ^2)	2	.	−2	−2	2	.	−2	4

In the exceptional 4-element families of E_7 and E_8, the sign $\Delta(\bar{x}_\rho)$ equals -1 on the two elements $(g_2, 1), (g_2, \varepsilon)$ (see [Lu84a, 4.14]).

See [Lu84a, 4.15] or [Ca85, 13.6] for those parts of the matrices for $\mathfrak{S}_4$ and $\mathfrak{S}_5$ that are relevant in Theorem 4.2.16. The complete matrices can also easily be obtained in Chevie [MiChv].

4.2.21 A further important invariant of unipotent characters are their so-called Frobenius eigenvalues, which we define now. Let $\delta > 0$ be the smallest integer such that F^δ acts trivially on the Weyl group $\mathbf{W}$ of $\mathbf{G}$. Let $\rho \in \mathrm{Uch}(\mathbf{G}^F)$. First, since ρ occurs with non-zero multiplicity in some Deligne–Lusztig generalised character, it is clear that there exist some $w \in \mathbf{W}$, $i \geqslant 0$ and $\mu \in \overline{\mathbb{Q}}_\ell^\times$ such that ρ occurs in the character of the generalised μ-eigenspace of F^δ on the ℓ-adic cohomology group $H_c^i(\mathbf{X}_w, \overline{\mathbb{Q}}_\ell)$. Now Lusztig [Lu77b, 3.9] showed that μ is uniquely determined by ρ, independently of w and i, up to a factor which is an integral power of q^δ. Furthermore, by [DiMi85, III.2.3], there is a well-defined root of unity $\omega_\rho \in \overline{\mathbb{Q}}_\ell$ and a well-defined element $\lambda_\rho \in \{1, q^{\delta/2}\}$ such that

$$\mu = \omega_\rho \, \lambda_\rho \, q^{s\delta} \quad \text{for some integer } s \geqslant 0.$$

Moreover, λ_ρ is an integral power of q unless ρ lies in an exceptional family in type E_7 or E_8 (see [DiMi85, III.2.3 in conjunction with II.3.4] and also [GeMa03, Lemma 4.4 and Ex. 4.6–4.8]). The root of unity ω_ρ is called the *Frobenius eigenvalue* of ρ. This is closely related to the parametrisation of unipotent characters occurring in Lusztig's Decomposition Theorem 4.2.16. For $(x, \sigma) \in \mathcal{M}(\mathcal{G})$ let $\omega_{(x,\sigma)} := \sigma(x)/\sigma(1)$ be the scalar by which the central element x acts in any representation with character σ.

Proposition 4.2.22 *Assume that F is split. Let $\mathcal{U} \subseteq \mathrm{Uch}(\mathbf{G}^F)$ be a non-exceptional family of unipotent characters with corresponding group $\mathcal{G}_\mathcal{U}$. Then the bijection $\mathcal{U} \xrightarrow{\sim} \mathcal{M}(\mathcal{G}_\mathcal{U})$ in Theorem 4.2.16 can be chosen such that if $\rho \leftrightarrow (x, \sigma) \in \mathcal{M}(\mathcal{G}_\mathcal{U})$ then $\omega_\rho = \omega_{(x,\sigma)}$.*

This shows that Frobenius eigenvalues, at least in the split case, are also generic data, only depending on the underlying complete root datum. The explicit results presented later show that this statement remains true for general Steinberg maps.

See [Lu84a, Thm. 11.2] (and also [DiMi85, Prop. 1.5]) for the proof of the above result. It relies, among other things, on the fact that Frobenius eigenvalues are preserved by Harish-Chandra induction:

Proposition 4.2.23 *Let $\rho \in \mathrm{Uch}(\mathbf{G}^F)$ lie in the Harish-Chandra series of the cuspidal pair $(\mathbf{L}, \lambda)$. Then the Frobenius eigenvalues ω_λ and ω_ρ coincide.*

See [Lu77b, 3.33] and [Lu84a, 11.3]. If instead we consider the d-Harish-Chandra series of a d-cuspidal character of some d-split Levi subgroup $\mathbf{L} \leqslant \mathbf{G}$ (see 3.5.24), then there still is a relation between Frobenius eigenvalues, but in general they are no longer preserved, see [Ma95, Satz 4.21] and also [BMM14, Axiom 4.31(d)].

4.2.24 We now present an axiomatic framework for the Fourier matrices and Frobenius eigenvalues attached to unipotent characters of a series of finite reductive groups which was first spelled out in this form by Geck–Malle [GeMa03], and which was mainly inspired by the empirical data obtained for spetsial imprimitive complex reflection groups in [Ma95] and those for primitive groups which were constructed in 1994 but published only much later in [BMM14].

Let $\mathbf{G}$ be a connected reductive group with a Steinberg endomorphism F and denote by σ the automorphism of the Weyl group $\mathbf{W}$ of $\mathbf{G}$ induced by $\mathbf{G}$. Let $\mathscr{F} \subseteq \mathrm{Irr}(\mathbf{W})$ be a σ-invariant family and $\mathrm{Uch}(\mathscr{F}) \subset \mathrm{Uch}(\mathbf{G}^F)$ the corresponding family of unipotent characters. If σ is ordinary in the sense of 4.1.26, then Lusztig's Decomposition Theorem 4.2.16 defines a Fourier matrix $\mathbf{M} = \mathbf{M}(\mathrm{Uch}(\mathscr{F}))$ for this family, depending only on the family $\mathscr{F}$ and σ. In the remaining cases (i.e., types B_2, G_2 and F_4 with a non-ordinary σ) such a matrix was obtained in [GeMa03], see Theorem 4.5.2. In all cases the columns of $\mathbf{M}$ are indexed by $\mathrm{Uch}(\mathscr{F})$, the rows are indexed by the almost characters $\mathrm{Alm}(\mathscr{F})$ corresponding to this family (that is, characteristic functions of F-stable unipotent character sheaves on $\mathbf{G}$, see Remarks 2.7.26 and 4.2.17). Let us write $\mathbf{M} = \mathbf{M}(\mathscr{F}, \sigma) = (a_{\lambda\mu})$, where $(\lambda, \mu) \in \mathrm{Alm}(\mathscr{F}) \times \mathrm{Uch}(\mathscr{F})$, for this Fourier matrix.

Following Asai [As84c], define the *twisting operator* t_1^* on class functions f on $\mathbf{G}^F$ by

$$t_1^*(f)(x) := f(yxy^{-1}) \qquad \text{for } x \in \mathbf{G}^F,$$

where $y \in \mathbf{G}$ is such that $x = y^{-1}F(y)$. Any uniform almost character is an eigenvector for t_1^* with eigenvalue 1. More generally, for any F-stable character sheaf K on $\mathbf{G}$ (see 2.7.24) the corresponding characteristic function $\chi_{K,\phi}$ is also an eigenvector of t_1^*, where the corresponding eigenvalue does not depend on σ (see Eftekhari [Ef94] and also Shoji [Sho95, I.§3]).

Let $\mathrm{Fr}_1 = \mathrm{Fr}_1(\mathscr{F}, \sigma)$ be the diagonal matrix of eigenvalues of t_1^* on the almost characters in $\mathrm{Alm}(\mathscr{F})$, and $\mathrm{Fr}_2 := \mathrm{Fr}_2(\mathscr{F}, \sigma)$ the diagonal matrix of Frobenius eigenvalues of the unipotent characters in $\mathrm{Uch}(\mathscr{F})$. By the results of Lusztig [Lu84a, 11.2] and Shoji [Sho95, I.§3] we have $\mathrm{Fr}_1 = \mathrm{Fr}_2$ when $\sigma = 1$, in which case we just write Fr for this matrix. These eigenvalues are known explicitly in all cases, see [GeMa03, Rem. 4.9 and Thm. 4.11] and the references given there. Finally, let Δ be the permutation matrix describing complex conjugation on $\mathrm{Alm}(\mathscr{F})$ (hence also on the eigenvalues of t_1^* on $\mathrm{Alm}(\mathscr{F})$). Then we may observe the following general properties satisfied by all these matrices (see [GeMa03, Thm. 6.9]):

Theorem 4.2.25 *In the notation introduced above we have:*

(F1) **M** *transforms the vector of unipotent degrees in* $\mathrm{Uch}(\mathscr{F})$ *to the vector of fake degrees (extended by zeros) associated to* $\mathscr{F}$.
(F2) *All entries* $a_{\lambda\mu}$ *of* **M** *are real.*
(F3) *We have* $\mathbf{M}\cdot\mathbf{M}^{tr}=\mathbf{M}^{tr}\cdot\mathbf{M}=1$.
(F4) *Let* λ_0 *be the row index of the uniform almost character corresponding to the special character in* $\mathscr{F}$. *Then all entries in that row are non-zero.*
(F5) *The structure constants*

$$a_{\lambda\mu}^{\nu} := \sum_{\kappa\in\mathrm{Uch}(\mathscr{F})} \frac{a_{\lambda\kappa}\, a_{\mu\kappa}\, a_{\nu\kappa}}{a_{\lambda_0\kappa}}$$

are rational integers for all $\lambda,\mu,\nu\in\mathrm{Alm}(\mathscr{F})$.
(F6) *If* $\sigma=1$ *then* $\mathbf{M}=\mathbf{M}^{tr}$, $\Delta\cdot\mathbf{M}=\mathbf{M}\cdot\Delta$ *and* $(\mathrm{Fr}\cdot\Delta\cdot\mathbf{M})^3=1$.
(F6') *If* $\sigma\neq 1$, $\sigma^2=1$ *then* $\left(\mathrm{Fr}_2\cdot\mathbf{M}^{tr}\cdot\mathrm{Fr}_1^{-1}\cdot\mathbf{M}\right)^2=1$.
(F6'') *If* $\sigma\neq 1$, $\sigma^3=1$ *then* $\left(\mathrm{Fr}_2\cdot\mathbf{M}^{tr}\cdot\mathrm{Fr}_1^{-1}\cdot\mathbf{M}\right)^3=1$.

This assertion may be derived from the explicitly known data. In fact, the result stated in [GeMa03] is more general as it also applies to the unipotent degrees, Fourier matrices and Frobenius eigenvalues associated to the non-rational finite Coxeter groups of types $I_2(m), H_3$ and H_4. See also [Ma95] and [BMM14] for analogous results pertaining to the unipotent degrees associated to spetsial complex reflection groups.

4.3 Unipotent Characters in Type A

Here we describe the properties of the unipotent characters of groups of type A. That is, we let $\mathbf{G}=\mathrm{SL}_n$ and $F:\mathbf{G}\to\mathbf{G}$ a Frobenius endomorphism, so that the finite group $\mathbf{G}^F$ of fixed points is either a finite special linear group $\mathrm{SL}_n(q)$ or a finite special unitary group $\mathrm{SU}_n(q)$, with q a power of the prime p. According to Remark 4.2.1 all the results stated in this section also hold for all groups isogenous to $\mathbf{G}$, so for example for $\mathrm{PGL}_n(q)$, respectively $\mathrm{PGU}_n(q)$.

We start off with the case that F is an untwisted Frobenius map, so $\mathbf{G}^F=\mathrm{SL}_n(q)$. The unipotent characters of $\mathrm{GL}_n(q)$ were first computed by Steinberg [St51a] (notwithstanding the fact that the notion of unipotent character was only introduced later). According to Remark 4.2.1 this also yields the parametrisation and degrees of unipotent characters of $\mathrm{SL}_n(q)$. By Example 2.4.20 all unipotent characters lie in the principal Harish-Chandra series, so by Theorem 3.2.5 they are in bijection with the irreducible characters of the Weyl group $\mathbf{W}=\mathfrak{S}_n$ of $\mathbf{G}$, the symmetric group of degree n. We will denote the unipotent character labelled by the partition $\alpha \vdash n$ by ρ_α. By the degree formula in Theorem 3.2.18 its degree can be expressed in terms of the corresponding Schur element of the 1-parameter Hecke algebra $\mathscr{H}(\mathbf{W}, q)$, which was given in Example 3.2.17; the a-value was given in Example 4.1.2. We obtain:

Proposition 4.3.1 (Hook formula for type A) *The degree of the unipotent character ρ_α of $\mathrm{SL}_n(q)$ labelled by $\alpha = (\alpha_1 \leqslant \cdots \leqslant \alpha_m) \vdash n$ is given by*

$$\rho_\alpha(1) = q^{a(\alpha)} \prod_{i=1}^{n} \frac{q^i - 1}{q^{l_i} - 1},$$

where l_i is the length of the hook at the ith box of the Young diagram of α, and $a(\alpha) = n(\alpha) = \sum_{i=1}^{m}(m-i)\alpha_i$.

It is obvious from this that under the substitution $q \mapsto 1$ the degree of the unipotent character ρ_α of $\mathrm{SL}_n(q)$ labelled by α specialises to the degree of the irreducible character of the symmetric group $\mathfrak{S}_n$ labelled by the same partition α.

There is another formula in terms of β-sets (see Example 4.1.7: the β-set of the partition $\alpha = (\alpha_1 \leqslant \alpha_2 \leqslant \cdots \leqslant \alpha_m)$ is given by $(\alpha_1, \alpha_2 + 1, \ldots, \alpha_m + m - 1)$).

Proposition 4.3.2 *The degree of the unipotent character ρ_α of $\mathrm{SL}_n(q)$ labelled by $\alpha \vdash n$ with β-set $(x_1, \ldots, x_m)$ is given by*

$$\rho_\alpha(1) = \frac{\prod_{i=1}^{n}(q^i - 1)\prod_{i<j}(q^{x_j} - q^{x_i})}{q^{\binom{m-1}{2}+\binom{m-2}{2}+\cdots}\prod_{i}\prod_{k=1}^{x_i}(q^k - 1)}.$$

This follows by elementary calculations from Proposition 4.3.1, see [Ca85, 13.8].

Example 4.3.3 The degrees of the unipotent characters of $\mathrm{SL}_n(q)$, $2 \leqslant n \leqslant 5$, are as given in Table 4.6, where Φ_d denotes the dth cyclotomic polynomial evaluated at q.

One standard application of the hook formula is the determination of unipotent characters of ℓ-defect 0 for a prime ℓ different from the defining characteristic p. Let e be the order of q modulo ℓ. Then ℓ divides $(q^f - 1)/(q^e - 1)$ if and only if f is divisible by $e\ell$. Now by definition an irreducible character of a finite group is *of ℓ-defect zero* if its degree is divisible by the full ℓ-part of the group order. Proposition 4.3.1 then leads to the following combinatorial criterion:

Corollary 4.3.4 *Let ρ be the unipotent character of $\mathrm{SL}_n(q)$ labelled by the partition $\alpha \vdash n$. Let ℓ be a prime different from p, and e the order of q modulo ℓ. Then ρ is of*

Table 4.6 *Unipotent characters of* $\mathrm{SL}_2(q)$, $\mathrm{SL}_3(q)$, $\mathrm{SL}_4(q)$ *and* $\mathrm{SL}_5(q)$

α	$\rho_\alpha(1)$
2	1
1^2	q

α	$\rho_\alpha(1)$
3	1
21	$q\Phi_2$
1^3	q^3

α	$\rho_\alpha(1)$
4	1
31	$q\Phi_3$
2^2	$q^2\Phi_4$
21^2	$q^3\Phi_3$
1^4	q^6

α	$\rho_\alpha(1)$
5	1
41	$q\Phi_2\Phi_4$
32	$q^2\Phi_5$
31^2	$q^3\Phi_3\Phi_4$
2^21	$q^4\Phi_5$
21^3	$q^6\Phi_2\Phi_4$
1^5	q^{10}

ℓ-defect zero if and only if α is an e-core, that is, it has no hook of length (divisible by) e.

This is in complete analogy to the result that an irreducible character of the symmetric group $\mathfrak{S}_n$ is of ℓ-defect zero if and only if its parametrising partition does not have a hook of length (divisible by) ℓ. The situation for the prime $\ell = 2$ is slightly different, because both $q - 1$ and $q + 1$ are even when q is odd. Yet it is immediate from the hook formula that the degree polynomial of any unipotent character of $\mathrm{SL}_n(q)$ is not divisible by $q - 1$, so that indeed for $\ell = 2$ the group $\mathrm{SL}_n(q)$, $n \geqslant 2$ and q odd, has no unipotent characters of 2-defect zero.

As already stated above, all unipotent characters of $\mathrm{SL}_n(q)$ lie in the principal series, so the only case with a cuspidal unipotent character occurs for $n = 0$, that is, when $\mathbf{G}$ is the trivial group. It then follows from Proposition 4.2.23 that all unipotent characters have their associated Frobenius eigenvalue equal to 1.

We saw in Corollary 2.4.19 that the unipotent characters for groups of type A all lie in singleton families. Thus, all Fourier matrices are the 1-by-1 identity matrix. This also explains why the unipotent degrees (as given in Proposition 4.3.2) agree with the fake degrees given in Example 4.1.7.

All unipotent characters of groups of type A are uniform; so by transitivity the decomposition of the Lusztig functor $R_\mathbf{L}^\mathbf{G}$ on the space spanned by unipotent characters can in principle be computed from the known decomposition of Deligne–Lusztig characters. We will give a description of this decomposition in the situation of d-Harish-Chandra theory, that is, for d-split Levi subgroups $\mathbf{L}$, in Section 4.6.

Now consider the case when F is twisted, so that $\mathbf{G}^F = \mathrm{SU}_n(q)$. Here it was shown by Lusztig–Srinivasan [LuSr77, Thm. 2.2] that the unipotent characters are again naturally labelled by partitions α of n; from their result it follows (see for example [Lu77a, Rem. 9.5]) that the degree of the unipotent character ρ_α of $\mathrm{SU}_n(q)$ labelled by α is obtained from the degree of the corresponding unipotent character of $\mathrm{SL}_n(q)$ by formally replacing q by $-q$, and adjusting the sign if necessary. This result is called *Ennola duality* since it is a particular instance of a conjecture made by Ennola [Enn63], see Theorem 2.8.17. So in particular we have:

Proposition 4.3.5 (Hook formula for type ^2A) *The degree of the unipotent character ρ_α of $\mathrm{SU}_n(q)$ labelled by $\alpha \vdash n$ is given by*

$$\rho_\alpha(1) = q^{a(\alpha)} \prod_{i=1}^{n} \frac{q^i - (-1)^i}{q^{l_i} - (-1)^{l_i}},$$

where l_i is the length of the hook at the ith box of the Young diagram of α, and $a(\alpha) = n(\alpha) = \sum_{i=1}^{m}(m-i)\alpha_i$ as for $\mathrm{SL}_n(q)$.

Clearly, the analogue of the formula in Proposition 4.3.2 obtained by replacing q by $-q$ then also holds for $\mathrm{SU}_n(q)$.

On the other hand, the subdivision of unipotent characters into (ordinary) Harish-Chandra series is more interesting here. The group $\mathrm{SU}_n(q)$ has a cuspidal unipotent character if and only if n is a triangular number $n = \binom{t}{2}$ for some $t \geqslant 1$, and then it is labelled by the triangular partition $\delta_t = (1, 2, \ldots, t-1)$ of n. Now if $n = 2m + \binom{t}{2}$ for some $m \geqslant 0$ and $t \geqslant 1$ there exists an F-stable 1-split Levi subgroup $\mathbf{L}_{n-2m}$ of SL_n of type A_{n-2m-1} corresponding to the F-stable subset of simple reflections $s_{m+1}, \ldots, s_{n-m-1}$ in the set-up of Example 3.1.4. We write ρ_t for the cuspidal unipotent character of $\mathbf{L}_{n-2m}^F$ labelled by δ_t. Then we have:

Proposition 4.3.6 *Let $n = 2m + \binom{t}{2}$ for some $m \geqslant 0, t \geqslant 1$. The Harish-Chandra series $\mathscr{E}(\mathrm{SU}_n(q), (\mathbf{L}_{n-2m}, \rho_t))$ consists of the unipotent characters ρ_α such that $\alpha \vdash n$ has 2-core δ_t. The relative Weyl group of $(\mathbf{L}_{n-2m}, \rho_t)$ has type B_m, and the parameters of the associated Iwahori–Hecke algebra are given by q^2, q^{2t-1}.*

See [FoSr90, App.]. This can also be seen as a consequence of Ennola duality applied to the 2-Harish-Chandra theory for $\mathrm{SL}_n(q)$, see Section 4.6.

Again by Corollary 2.4.19 the unipotent characters for groups $\mathrm{SU}_n(q)$ all lie in singleton families. The Frobenius eigenvalues (see 4.2.21) of unipotent characters of unitary groups are given as follows:

Proposition 4.3.7 *Let ρ be a cuspidal unipotent character of $G = \mathrm{SU}_n(q)$ labelled by the 2-core $\lambda \vdash n = \binom{t}{2}$. Then the Frobenius eigenvalue ω_ρ attached to ρ satisfies $D_{\mathbf{G}}(\rho) = \omega_\rho\, \rho$, that is*

$$\omega_\rho = (-1)^{r(G)} = \begin{cases} 1 & \textit{if } t \equiv -1, 0, 1, 2 \pmod 8, \\ -1 & \textit{if } t \equiv 3, 4, 5, 6 \pmod 8. \end{cases}$$

See Rem. (a) after Thm. 3.34 in [Lu77b] for q sufficiently large. The claim for arbitrary q also follows with the proof of [GeMa03, Thm. 4.11] applied to the case of ^2A_n.

4.4 Unipotent Characters in Classical Types

4.4.1 The unipotent characters of the other classical groups are best parametrised by combinatorial objects introduced by Lusztig, the so-called symbols. These symbols are natural generalisations of partitions, or rather their β-sets, and provide a very intuitive way to label the unipotent characters of groups of classical types, to understand their distribution into families, to encode arithmetical properties of their degrees, but also to describe the decomposition of Lusztig induction, the distribution into ℓ-blocks and further properties. Moreover, symbols also serve as a tool to describe unipotent conjugacy classes of classical groups and to describe the generalised Springer correspondence, in bad as well as in good characteristic, and even for disconnected groups. Thus, symbols provide the right generalisation of the concept of partitions, which govern the representation theory of the symmetric groups as well as of the general linear and unitary groups.

Symbols were first introduced by Lusztig in [Lu77a] to parametrise unipotent characters of groups of classical type. Later, Lusztig [Lu84c, §11] showed that they are also very convenient to describe the generalised Springer correspondence, see also [LuSp85]. The formalism of hooks and cohooks of symbols was introduced by Olsson [Ol84].

Symbols should be thought of as encoding pairs of partitions which are shifted with respect to each other in a well-defined way. Recall that the β-set associated to a partition $(\alpha_1 \leqslant \cdots \leqslant \alpha_r)$ is the strictly increasing sequence $(x_1 < \cdots < x_r)$ with entries $x_i := \alpha_i + i - 1$. Conversely, a finite sequence $(x_1 < \cdots < x_r)$ of non-negative integers encodes the partition $(\alpha_1 \leqslant \cdots \leqslant \alpha_r)$ with $\alpha_i := x_i - i + 1$ (where we first remove leading zeros).

For integers $a, b \geqslant 0$ and for $d \in \mathbb{Z}$, we consider the following set $\tilde{\mathscr{X}}_d^{a,b}$ of *symbols*: The elements of $\tilde{\mathscr{X}}_d^{a,b}$ are pairs $S = (X, Y)$ of finite sequences $X = (x_1, \ldots, x_r)$ and $Y = (y_1, \ldots, y_s)$ of non-negative integers, subject to the conditions

$$\begin{aligned} &x_j \geqslant x_{j-1} + a + b && \text{for } 2 \leqslant j \leqslant r, \\ &y_j \geqslant y_{j-1} + a + b && \text{for } 2 \leqslant j \leqslant s, \\ &y_1 \geqslant b, \\ &r - s = d. \end{aligned}$$

Note that for $b = 0$ the third condition is empty. Symbols are often displayed in the form

$$S = \begin{pmatrix} X \\ Y \end{pmatrix} = \begin{pmatrix} x_1 & x_2 & \ldots \\ y_1 & y_2 & \ldots \end{pmatrix},$$

and we then say that X, Y are the *rows* of S.

For $S = (X, Y)$ a symbol, we call $d = d(S) = |X| - |Y|$ its *defect*. The *content* of S is defined by

$$|S| := |X| + |Y|$$

and the *rank* by

$$\mathrm{rnk}(S) := \sum_{x \in S} x - (a+b) \left\lfloor \frac{(|S|-1)^2}{4} \right\rfloor - b \left\lfloor \frac{|S|}{2} \right\rfloor,$$

where $\lfloor z \rfloor := \max\{n \in \mathbb{Z} \mid n \leqslant z\}$ denotes the Gauß brackets. It is easy to check that $\mathrm{rnk}(S) \in \mathbb{N}_0$ for any $S \in \tilde{\mathscr{X}}_d^{a,b}$.

On $\tilde{\mathscr{X}}_d^{a,b}$ we have the *shift operation* which to $S = (X, Y)$ associates the symbol

$$S' = \big(\{0\} \cup (X + a + b), \{b\} \cup (Y + a + b)\big) \in \tilde{\mathscr{X}}_d^{a,b}.$$

It is clear that the defect of symbols is invariant under shift. We write $\mathscr{X}_d^{a,b}$ for the classes in $\tilde{\mathscr{X}}_d^{a,b}$ under the equivalence relation defined by the symmetric and transitive hull of the shift operation. Similarly, an easy calculation shows that the rank is invariant under shift, and we write $\mathscr{X}_{d,n}^{a,b}$ for the set of classes of symbols in $\mathscr{X}_d^{a,b}$ of rank n. Note that also the contents is invariant modulo 2 under shift. It is easy to see that $\mathscr{X}_{d,n}^{a,b}$ is a finite set for any choice of parameters $a, b \geqslant 0$, $d \in \mathbb{Z}$ and $n \geqslant 0$.

Example 4.4.2 Consider the case that $a = b = 0$. Then for $S \in \mathscr{X}_{d,n}^{0,0}$ we have $\mathrm{rnk}(S) = \sum_{x \in S} x$. So for any $d \in \mathbb{Z}$ and $n \geqslant 0$, the set $\mathscr{X}_{d,n}^{0,0}$ is in natural bijection with the set of bipartitions of n as follows: Let $(\alpha, \beta) \vdash n$. By adding zero parts to α or β we may assume that α has exactly d more entries than β. Then $(\alpha, \beta) \in \mathscr{X}_{d,n}^{0,0}$. It is easy to see that this assignment defines a bijection. Thus, according to Example 4.1.3 this provides a natural parametrisation $\mathscr{X}_{d,n}^{0,0} \to \mathrm{Irr}(W_n)$ of the irreducible characters of the Weyl group of type B_n in terms of symbols of rank n and fixed defect d.

4.4.3 An element $S = (X, Y) \in \mathscr{X}_d^{a,b}$ is called *distinguished* if $d \in \{0, 1\}$ and $x_1 \leqslant y_1 \leqslant x_2 \leqslant \cdots \leqslant y_s$ (and $y_s \leqslant x_{s+1}$ if $d = 1$). Note that this definition is independent of the chosen representative in its class. Two (classes of) symbols are called *similar* if for suitable representatives $S_i = (X_i, Y_i)$, $i = 1, 2$, the multisets of entries $X_1 \cup Y_1$ and $X_2 \cup Y_2$ of entries coincide. Thus for any symbol $S \in \mathscr{X}_{d,n}^{a,b}$, $d \in \{0, 1\}$, there exists a unique distinguished symbol in $\mathscr{X}_{d,n}^{a,b}$ up to shift which is similar to S.

A further important invariant of symbol classes is the *genus of a symbol* $S \in \tilde{\mathscr{X}}_d^{a,b}$

defined by

$$g(S) := \sum_{\{x,y\}} \min(x,y) - \begin{cases} \left(a\binom{t}{2}\frac{4t+1}{3} + b\binom{t+1}{2}\frac{4t-1}{3}\right) & \text{if } |S| = 2t+1, \\ \left(a\binom{t}{2}\frac{4t-5}{3} + b\binom{t}{2}\frac{4t+1}{3}\right) & \text{if } |S| = 2t \end{cases}$$

(see [GeMa00, (2.22)]). Here the sum extends over all 2-element subsets of the multiset $X \cup Y$ of entries of $S = (X,Y)$. Thus we have $g(S) \in \mathbb{N}_0$ for any $S \in \tilde{\mathscr{X}}_d^{a,b}$. Furthermore, the following is easily checked:

Lemma 4.4.4 *The genus is invariant under shift, hence well defined on* $\mathscr{X}_d^{a,b}$.

4.4.5 (Cuspidal symbols, addition) The *core* or *1-core* of a symbol $S = (X,Y)$ in $\tilde{\mathscr{X}}_d^{a,b}$ is the symbol $\mathrm{core}(S) = (X',Y') \in \tilde{\mathscr{X}}_d^{a,b}$ with

$$X' = (0, a+b, \ldots, (r-1)(a+b)), \qquad Y' = (b, a+2b, \ldots, (s-1)a+sb),$$

where $r = |X|$, $s = |Y|$. Thus the defect $d(\mathrm{core}(S)) = d(S)$ remains unchanged while $\mathrm{rnk}(\mathrm{core}(S)) \leqslant \mathrm{rnk}(S)$. If equality holds, then $\mathrm{core}(S) = S$. Such symbols S will be called *cuspidal*. Note that the equivalence class of any cuspidal symbol contains a representative with either $X = \varnothing$ or $Y = \varnothing$.

For any fixed defect d there exists an *addition*

$$+ : \mathscr{X}_{d,n}^{a,b} \times \mathscr{X}_{d,n'}^{a',b'} \longrightarrow \mathscr{X}_{d,n+n'}^{a+a',b+b'}$$

defined by component-wise addition of representatives of the same length. For fixed a, b, d, n and a cuspidal symbol $S \in \mathscr{X}_{d,k}^{a,b}$ of rank $\mathrm{rnk}(S) = k$ let $X(S)$ be the subset of (classes of) symbols in $\mathscr{X}_{d,n}^{a,b}$ with core S. The following observation is then immediate:

Lemma 4.4.6 *Let* $S = (X,Y) \in \mathscr{X}_{d,k}^{a,b}$ *be cuspidal. Then addition of* S *defines a natural bijection*

$$\mathscr{X}_{d,n-k}^{0,0} \xrightarrow{1-1} X(S) \subseteq \mathscr{X}_{d,n}^{a,b}, \qquad S' \mapsto S' + S.$$

Example 4.4.7 By Example 4.4.2 the set $\mathscr{X}_{d,n-k}^{0,0}$ can be naturally identified with the set of pairs of partitions of $n-k$, which in turn by Example 4.1.3 is in natural bijection with the set $\mathrm{Irr}(W_{n-k})$ of irreducible characters of the Weyl group W_{n-k} of type B_{n-k}. Thus Lemma 4.4.6 gives a natural parametrisation of $X(S)$ by irreducible characters of W_{n-k}.

4.4.8 If the parameter b equals 0 then the cyclic group of order 2 acts on $\mathscr{X}_d^{a,0} \cup \mathscr{X}_{-d}^{a,0}$ by interchanging the two rows, that is, $S = (X,Y) \mapsto S^{\mathrm{tr}} := (Y,X)$. The symbol S is called *degenerate* if it is a fixed point under this action, that is, if $X = Y$. Clearly, degenerate symbols necessarily have defect $d = 0$. Let $\mathscr{Y}_d^a$ denote the set of equivalence classes in $\mathscr{X}_d^{a,0} \cup \mathscr{X}_{-d}^{a,0}$ modulo this action. Again, the rank, the genus and the contents modulo 2 are well defined on elements of $\mathscr{Y}_d^a$, the cores of equivalent symbols are equivalent, and if a symbol is cuspidal, then so are all equivalent symbols. The defect is well defined up to sign. We say that $S \in \mathscr{Y}_d^a$ is distinguished if one of $S, S^{\rm tr}$ is distinguished. As before, we denote by $\mathscr{Y}_{d,n}^a$ the subset of symbol classes in $\mathscr{Y}_d^a$ of rank n.

Example 4.4.9 Let $S = (X, Y) \in \mathscr{Y}_{d,k}^a$ be cuspidal of rank k and denote by $Y(S)$ the subset of (classes of) symbols in $\mathscr{Y}_{d,n}^a$ with core S. Note that the only degenerate core in this case is equivalent to $(-;-)$ and this occurs if and only if $d = 0$. First assume that S is not degenerate, that is, $d \neq 0$. Then $\mathscr{Y}_{d,n}^a$ is naturally in bijection with $\mathscr{X}_{d,n}^{a,0}$ and addition again defines a bijection

$$\mathscr{X}_{d,n-k}^{0,0} \xrightarrow{1-1} Y(S) \subseteq \mathscr{Y}_{d,n}^a, \quad S' \mapsto S' + S,$$

which by Example 4.4.7 induces a bijection between $\mathrm{Irr}(W_{n-k})$ and $Y(S)$. On the other hand, when S is degenerate, hence equivalent to the symbol $(-;-)$ of defect $d = 0$ and rank 0, then we obtain a bijection

$$\mathscr{Y}_{0,n}^0 \xrightarrow{1-1} Y(S) \subseteq \mathscr{Y}_{0,n}^a, \quad S' \mapsto S' + S.$$

According to Example 4.1.4 this sets up a map $\mathrm{Irr}(W_n') \to Y(S)$ from the irreducible characters of the Weyl group W_n' of type D_n to the symbol classes in $\mathscr{Y}_{0,n}^a$ with core S which is one-to-one, except that both characters in a pair $\{\phi^{[\alpha,\pm]}\}$ are sent to the same (degenerate) symbol class.

4.4.10 As for partitions, there is a notion of hooks of symbols, but now they come in two different flavours. Let $d \geqslant 1$. A *d-hook h of a symbol* $S = (X, Y)$ is an entry $x \geqslant d$ of S with either $x \in X$, $x - d \notin X$, or $x \in Y$, $x - d \notin Y$. Thus, a hook of S is nothing else but a hook of the partition with associated β-set either X or Y. The *length* of the hook h is $l(h) := d$. *Removing the d-hook h* at x leads to the symbol $S \setminus h := (X', Y)$ with $X' = X \setminus \{x\} \cup \{x - d\}$ (respectively $S \setminus h := (X, Y')$ with $Y' = Y \setminus \{x\} \cup \{x - d\}$). We then also say that S is obtained from $S \setminus h$ by *adding the d-hook h*. Attached to the hook h is the sign

$$\epsilon_h := (-1)^m \quad \text{where } m := \begin{cases} |\{y \in X \mid x - d < y < x\}| & \text{if } x \in X, \\ |\{y \in Y \mid x - d < y < x\}| & \text{if } x \in Y. \end{cases}$$

A symbol having no d-hooks is called a *d-core*. A moment's thought shows that a d-core can not possess any hooks of length divisible by d.

A *d-cohook c of S* is an entry $x \geqslant d$ of S with either $x \in X$, $x - d \notin Y$, or $x \in Y$, $x - d \notin X$. Again its length is defined to be $l(c) = d$. *Removing the d-cohook c* at x from S leads to the symbol $S \setminus c := (X', Y')$ with $X' = X \setminus \{x\}, Y' = Y \cup \{x - d\}$

(respectively $Y' = Y \setminus \{x\}, X' = X \cup \{x - d\}$). Again, S is called the symbol obtained by *adding the d-cohook c* to $S \setminus c$. Attached to the cohook c is the sign $\epsilon_c := (-1)^m$, where

$$m := \begin{cases} |\{y \in X \mid y < x\}| + |\{y \in Y \mid y < x - d\}| & \text{if } x \in X, \\ |\{y \in Y \mid y < x\}| + |\{y \in X \mid y < x - d\}| & \text{if } x \in Y. \end{cases}$$

A symbol without d-cohooks is a *d-cocore*. A d-cocore does not possess any cohooks of length an odd multiple of d, nor any hook of length a multiple of $2d$.

Note that the multiset of hook lengths and of cohook lengths of a symbol is invariant under shift and hence constant on equivalence classes. Clearly, removing a d-hook or a d-cohook from a symbol of rank n yields a symbol of rank $n - d$. The following is obvious from our definitions:

Proposition 4.4.11 *Let S be a symbol and $d \geqslant 1$. Then successively removing d-hooks (respectively d-cohooks) from S as often as possible leads to a uniquely determined symbol, called the d-*core *(respectively d-*cocore*) of S.*

These notions will play a fundamental role in the combinatorial description of unipotent character degrees but also in the decomposition of Lusztig induction.

Remark 4.4.12 The combinatorics of symbols introduced above has a natural generalisation to so-called *e-symbols*, that is, symbols with $e \geqslant 2$ rows. These play a central role in the parametrisation of unipotent characters attached to the imprimitive complex reflection groups $G(d, 1, n)$ and $G(d, d, n)$, extending the combinatorics of unipotent characters of classical groups, see [Ma95] for details.

The unipotent characters of the classical groups of types B, C and D, that is, the symplectic and various orthogonal groups, were determined by Lusztig [Lu77a, Thm. 8.2]. He showed that they can naturally be parametrised by suitable equivalence classes of symbols:

Theorem 4.4.13 (Lusztig) *The unipotent characters of a finite classical group of rank n are labelled by equivalence classes of symbols of rank n in $\mathscr{Y}^1_{d,n}$ with*

(1) *odd defect d for types B_n and C_n,*
(2) *defect $d \equiv 0 \pmod 4$ for type D_n,*
(3) *defect $d \equiv 2 \pmod 4$ for type ${}^2D_n$,*

where in type D_n any class of degenerate symbols labels two unipotent characters.

Example 4.4.14 (a) The symbols in $\mathscr{Y}^1_d$ of rank 2 and odd defect have representatives up to equivalence given by

$$\binom{2}{\ }, \binom{0\ 2}{1}, \binom{1\ 2}{0}, \binom{0\ 1}{2}, \binom{\ }{0\ 1\ 2}, \binom{0\ 1\ 2}{1\ 2}.$$

In particular, $\mathrm{Sp}_4(q)$ and $\mathrm{SO}_5(q)$ both have six unipotent characters.

(b) The symbols in $\mathscr{Y}_d^1$ of rank 4 and defect $d \equiv 2 \pmod 4$ are given up to equivalence by

$$\binom{0\,4}{}, \binom{3\,1}{}, \binom{0\,1\,4}{1}, \binom{0\,1\,3}{2}, \binom{0\,1\,2}{3}, \binom{0\,2\,3}{1},$$
$$\binom{1\,2\,3}{0}, \binom{0\,1\,2\,4}{1\,2}, \binom{0\,1\,2\,3}{1\,3}, \binom{0\,1\,2\,3\,4}{1\,2\,3}.$$

In particular, $\mathrm{SO}_8^-(q)$ has ten unipotent characters.

(c) If $n = 2m$ is even, then there exist degenerate symbols in $\mathscr{Y}_0^1$ of rank n. They are in natural bijection with partitions of m, by sending a partition $\lambda \vdash m$ to the symbol with two equal rows each consisting of the β-set of λ. For example, for $\mathrm{SO}_8^+(q)$ there are two pairs of unipotent characters, labelled by each of the degenerate symbols

$$\binom{2}{2} \quad \text{and} \quad \binom{1\,3}{1\,3}.$$

The degrees of the unipotent characters are given by a formula analogous to the one for type A in Proposition 4.3.2. Again, we state the result for one particular isogeny type, but according to Remark 4.2.1 it is the same for any other group isogenous to this.

Proposition 4.4.15 *The degree of the unipotent character ρ_S of a classical group $G = \mathrm{Sp}_{2n}(q)$, $\mathrm{SO}_{2n+1}(q)$, or $\mathrm{SO}_{2n}^{\pm}(q)$ labelled by a symbol $S = (X, Y) = (x_1 < \cdots < x_r; y_1 < \cdots < y_s)$ is given by*

$$\rho_S(1) = \frac{|G|_{q'} \prod_{i<j}(q^{x_j} - q^{x_i}) \prod_{i<j}(q^{y_j} - q^{y_i}) \prod_{i,j}(q^{x_i} + q^{y_j})}{2^{b(S)} q^{\binom{r+s-2}{2}+\binom{r+s-4}{2}+\cdots} \prod_i \prod_{k=1}^{x_i}(q^{2k}-1) \prod_i \prod_{k=1}^{y_i}(q^{2k}-1)}$$

where

$$b(S) = \begin{cases} \lfloor (r+s-1)/2 \rfloor & \text{if } X \neq Y, \\ r & \text{else.} \end{cases}$$

It is an easy exercise to check that this expression is indeed constant on equivalence classes. It is also clear from this that the unipotent characters in types B_n and C_n are not only labelled by the same set, but also have the same degrees.

Example 4.4.16 The degrees of the unipotent characters of $\mathrm{Sp}_4(q)$ and of $\mathrm{SO}_8^-(q)$ for which the symbols were shown in Example 4.4.14 are as given in Table 4.7.

Table 4.7 *Unipotent characters of* $\mathrm{SO}_8^-(q)$. . .

S	$\rho_S(1)$	S	$\rho_S(1)$
$\binom{0\ 4}{\ }$	1	$\binom{0\ 2\ 3}{1}$	$\frac{1}{2}q^3\Phi_3\Phi_8$
$\binom{3\ 1}{\ }$	$q\Phi_8$	$\binom{1\ 2\ 3}{0}$	$\frac{1}{2}q^3\Phi_6\Phi_8$
$\binom{0\ 1\ 4}{1}$	$q^2\Phi_3\Phi_6$	$\binom{0\ 1\ 2\ 4}{1\ 2}$	$q^6\Phi_3\Phi_6$
$\binom{0\ 1\ 3}{2}$	$\frac{1}{2}q^3\Phi_3\Phi_8$	$\binom{0\ 1\ 2\ 3}{1\ 3}$	$q^7\Phi_8$
$\binom{0\ 1\ 2}{3}$	$\frac{1}{2}q^3\Phi_6\Phi_8$	$\binom{0\ 1\ 2\ 3\ 4}{1\ 2\ 3}$	q^{12}

. . . *and of* $\mathrm{Sp}_4(q)$

S	$\rho_S(1)$
$\binom{2}{\ }$	1
$\binom{0\ 2}{1}$	$\frac{1}{2}q\Phi_2^2$
$\binom{1\ 2}{0}$	$\frac{1}{2}q\Phi_4$
$\binom{0\ 1}{2}$	$\frac{1}{2}q\Phi_4$
$\binom{0\ 1\ 2}{\ }$	$\frac{1}{2}q\Phi_1^2$
$\binom{0\ 1\ 2}{1\ 2}$	q^4

For many applications concerning character degrees an analogue of the hook formula for $\mathrm{SL}_n(q)$ in Proposition 4.3.1 is more useful than the degree formula given in Proposition 4.4.15, since it reflects the arithmetic properties of the degrees much more clearly. Here, for a symbol $S = (X, Y)$ we set

$$a(S) := \sum_{\{x,y\}\subseteq S} \min\{x, y\} - \sum_{i\geqslant 1} \binom{|X| + |Y| - 2i}{2},$$

where the first sum runs over all 2-element subsets of the multiset $X \cup Y$ of entries of S (this is just the genus of S as introduced in 4.4.3, so in particular is independent of the chosen representative in an equivalence class by Lemma 4.4.4).

Proposition 4.4.17 (Hook formula for classical types) *The degree of the unipotent character* ρ_S *of a classical group* $G = \mathrm{Sp}_{2n}(q)$, $\mathrm{SO}_{2n+1}(q)$, *or* $\mathrm{SO}_{2n}^{\pm}(q)$ *labelled by a symbol* $S = (X, Y)$ *is given by*

$$\rho_S(1) = q^{a(S)} \frac{|G|_{q'}}{2^{b'(S)} \prod_h (q^{l(h)} - 1) \prod_c (q^{l(c)} + 1)},$$

where the products in the denominator run over all hooks h, *and cohooks* c *of* S, *and*

$$b'(S) = \begin{cases} \lfloor (|X| + |Y| - 1)/2 \rfloor - |X \cap Y| & \text{if } X \neq Y, \\ 0 & \text{else.} \end{cases}$$

This formula was first given by Olsson [Ol84, Prop. 5]; see also Malle [Ma95, Bem. 3.12 and 6.8] for a proof in the more general setting of imprimitive complex reflection groups. As the multisets of hook lengths and cohook lengths are invariant under equivalence, this expression is indeed constant on equivalence classes.

As in type A, the hook formula yields an easy criterion for a unipotent character of a classical group to be of ℓ-defect 0. Observe for this that for a prime $\ell > 2$ different from the defining characteristic p, if e denotes the order of q modulo ℓ, then $q^f + 1$ is divisible by ℓ if and only if e is even and $2f$ is an odd multiple of e. The following combinatorial criterion is then immediate from Proposition 4.4.17:

Corollary 4.4.18 *Let ρ be a unipotent character of a classical group $\mathrm{Sp}_{2n}(q)$, $\mathrm{SO}_{2n+1}(q)$, or $\mathrm{SO}_{2n}^{\pm}(q)$, labelled by a symbol S. Let $2 < \ell$ be a prime different from p, and e the order of q modulo ℓ. Then ρ is of ℓ-defect zero if and only if*

(1) *S is an e-core, that is, it has no hook of length (divisible by) e, if e is odd;*
(2) *S is an $e/2$-cocore, that is, it has no cohook of length $e/2$, if e is even.*

Again, the situation is slightly more complicated for the prime $\ell = 2$. Here again it turns out that no unipotent character is of 2-defect zero when q is odd.

Example 4.4.19 Let $G = \mathrm{Sp}_{2n}(q)$ or $G = \mathrm{SO}_{2n+1}(q)$ and consider primes ℓ such that q has order $2n$ modulo ℓ. Then from Corollary 4.4.18(1) an easy argument shows that the unipotent characters of G not of ℓ-defect zero are the ones labelled by the symbols

$$S = \begin{pmatrix} 1 & \dots & & a & \\ 0 & 1 & \dots & a-1 & n \end{pmatrix} \qquad (0 \leqslant a \leqslant n),$$

and

$$S = \begin{pmatrix} 0 & 1 & \dots & a & n \\ 1 & \dots & a-1 & & \end{pmatrix} \qquad (1 \leqslant a \leqslant n-1).$$

Thus, there are $2n$ unipotent characters of G not of ℓ-defect 0 (this was first used in [MSW94, Thm. 2.3]; see also [Ma88] for an application of such considerations in constructive Galois theory, and the survey [Ma14] for more recent applications).

Now assume that $G = \mathrm{SO}_{2n}^{-}(q)$, and let ℓ be as above. Then the unipotent characters not of ℓ-defect zero are the n different unipotent characters labelled by

$$S = \begin{pmatrix} 0 & 1 & \dots & a & n \\ 1 & \dots & a & & \end{pmatrix} \qquad (0 \leqslant a \leqslant n-1)$$

(see [MSW94, Thm. 2.5]).

The combinatorics of symbols makes it easy to describe the partition of the unipotent characters into families (introduced in 4.2.2); for this let us say that a symbol S is *reduced* if at most one of its rows contains an entry 0. By applying suitable shifts it is easily seen that any equivalence class of symbols contains (at least) one reduced one.

Proposition 4.4.20 (Families in classical groups) *Let G be a classical group $\mathrm{Sp}_{2n}(q)$, $\mathrm{SO}_{2n+1}(q)$, or $\mathrm{SO}_{2n}^{\pm}(q)$.*

(a) *Two unipotent characters $\rho, \rho' \in \mathrm{Uch}(G)$ lie in the same family if and only if their reduced labelling symbols are similar, that is, if their multisets of entries agree. The two unipotent characters labelled by a degenerate symbol each lie in a singleton family.*

(b) *For $\mathscr{U} \subseteq \mathrm{Uch}(G)$ a non-degenerate family, denote by k the number of entries of any symbol S belonging to $\mathscr{U}$ that occur in exactly one row of S. Then $|\mathscr{U}| = 2^{2m}$ and $\mathscr{G}_{\mathscr{U}} \cong C_2^m$, with $m = \lfloor \frac{k-1}{2} \rfloor$.*

See [Lu84a, (4.5.6) and (4.6.10)] for (a) and [Lu84a, p. 88 and 94] for (b).

4.4.21 (Fourier matrices in classical types) We next describe the Fourier matrices for families of unipotent characters in classical groups. Let $\mathbf{G}$ be isogenous to a simple group with F-fixed points one of $\mathrm{Sp}_{2n}(q)$, $\mathrm{SO}_{2n+1}(q)$, $\mathrm{SO}_{2n}^+(q)$ or $\mathrm{SO}_{2n}^-(q)$. Fix a family $\mathscr{U} \subseteq \mathrm{Uch}(G)$ of unipotent characters of G. If $\mathscr{U}$ contains a unipotent character labelled by a degenerate symbol, then by the preceding proposition we have $|\mathscr{U}| = 1$, and the corresponding Fourier matrix is the 1-by-1 identity matrix. So assume that this is not the case.

Choose, as we may by Proposition 4.4.20, a set $\mathscr{S} = \mathscr{S}(\mathscr{U})$ of representative symbols $S = (X, Y) \in \mathscr{Y}_{d,n}^1$ for the characters $\rho \in \mathscr{U}$ all having the same multiset of entries, and such that $|X| \equiv |Y| + d \pmod 4$, where for types $\mathrm{Sp}_{2n}(q)$ and $\mathrm{SO}_{2n+1}(q)$ we have that $d \equiv 1 \pmod 4$, while for type $\mathrm{SO}_{2n}^-(q)$ we have $d \equiv 2 \pmod 8$. We then have $|\mathscr{S}| = 2^{2m}$ where $2m + 1$ and $2m + 2$, respectively, denote the number of entries that appear only once in any symbol $S \in \mathscr{S}$ (see Proposition 4.4.20(b)). We set $k := \lfloor (|X| + |Y|)/2 \rfloor$.

Let $\mathscr{F}$ be the corresponding family of $\mathrm{Irr}(\mathbf{W})$ (which is σ-stable in the case of $\mathrm{SO}_{2n}^-(q)$). For $\mathrm{Sp}_{2n}(q)$ and $\mathrm{SO}_{2n+1}(q)$ according to Lemma 4.4.6 addition of the cuspidal symbol $(0; -) \in \mathscr{Y}_{1,0}^1$ induces an embedding of $\mathscr{F}$ into $\mathscr{S}$. For $\mathrm{SO}_{2n}^+(q)$ note that no member of $\mathscr{F}$ will be degenerate. Thus again addition of the cuspidal symbol $(-; -) \in \mathscr{Y}_{0,0}^1$ defines an embedding of $\mathscr{F}$ into $\mathscr{S}$. In both cases we set $\mathscr{S}' := \mathscr{S}$. For $G = \mathrm{SO}_{2n}^-(q)$ we let $\mathscr{S}'$ be a set of representative symbols $S = (X, Y) \in \mathscr{Y}_{0,n}^1$ having the same multiset of entries as the symbols in $\mathscr{S}$. Recall from 4.4.8 that the σ-stable irreducible characters of $\mathbf{W}$ are naturally identified (via addition of the cuspidal symbol $(-; -) \in \mathscr{Y}_{0,0}^1$) with elements of $\mathscr{Y}_{0,n}^1$. This defines an injection of $\mathscr{F}$ into $\mathscr{S}'$ in this case.

For $S = (X, Y) \in \mathscr{S} \cup \mathscr{S}'$ define

$$c(S) := |\{(x, y) \in X \times Y \mid x < y\}| + \binom{|Y|}{2}.$$

The Fourier transform matrix of $\mathscr{U}$ is then given by $(\langle S, S' \rangle)_{S \in \mathscr{S}, S' \in \mathscr{S}'}$ (see [Lu77a, §1] and [Ma95, Satz 4.17 and 6.26]), where for $(S = (X, Y), S' = (X', Y')) \in \mathscr{S} \times \mathscr{S}'$

we set

$$\langle S, S' \rangle := \frac{1}{2^m}(-1)^{c(S)+c(S')+k+|Y\cap Y'|}.$$

4.4.22 Let us reformulate the above expression for the Fourier matrices. First of all, assume that the common multiset of entries of the symbols in $\mathcal{S} \cup \mathcal{S}'$ contains an entry z appearing twice. Then denoting by $S_1 = (X_1, Y_1)$ the symbol obtained by removing that entry z from both rows of $S \in \mathcal{S} \cup \mathcal{S}'$, we see that

$$\begin{aligned} c(S)-c(S_1) &= |\{x \in X_1 \mid x < z\}| + |\{y \in Y_1 \mid z < y\}| + \binom{|Y_1|+1}{2} - \binom{|Y_1|}{2} \\ &= |\{x \in X_1 \mid x < z\}| - |\{y \in Y_1 \mid y < z\}| + 2|Y_1| \\ &\equiv |\{x \in X_1 \sqcup Y_1 \mid x < z\}| \pmod 2 \end{aligned}$$

is independent of S. Since $k+|Y\cap Y'|$ also remains unchanged modulo 2, the Fourier matrix does not change if we remove all double entries in all symbols in $\mathcal{S} \cup \mathcal{S}'$. Moreover, the precise value of the entries (which are now all different) does not matter. Hence we may assume that $X \cup Y = \{0, \ldots, |S|-1\}$ for all $(X, Y) \in \mathcal{S} \cup \mathcal{S}'$. Then we get

$$\begin{aligned} c(S) &= |\{(x, y) \in X \times Y \mid x < y\}| + \binom{|Y|}{2} \\ &= |\{(x, y) \in (X \cup Y) \times Y \mid x < y\}| = \sum_{y \in Y} y \end{aligned}$$

which is congruent to the number of odd entries in Y modulo 2. Write $o(Z)$ for the number of odd entries in a finite set $Z \subset \mathbb{Z}$, and $e(Z)$ for the number of even entries. Then for $(X, Y) \in \mathcal{S}$, $(X', Y') \in \mathcal{S}'$ we have $o(Y) + o(Y') = o(Y \cup Y') + o(Y \cap Y')$, $|Y \cap Y'| = o(Y \cap Y') + e(Y \cap Y')$ and $o(X \cup Y) = k$, so we have the following congruences

$$\begin{aligned} c(S) + c(S') + k + |Y \cap Y'| &\equiv o(Y) + o(Y') + k + o(Y \cap Y') + e(Y \cap Y') \\ &\equiv o(Y \cup Y') + o(X \cup Y) + e(Y \cap Y') \\ &\equiv o(X \cap X') + e(Y \cap Y') = |Y^\# \cap Y'^\#| \end{aligned}$$

modulo 2, where we write $Y^\# := Y \triangle \{1, 3, \ldots, 2k-1\}$ for the symmetric difference, so $Y^\#$ contains all even entries in Y and all odd integers in X. This proves Lusztig's original formula [Lu84a, 4.15 and 4.18]

$$\langle S, S' \rangle = \frac{1}{2^m}(-1)^{|Y^\# \cap Y'^\#|}$$

for the coefficients of the Fourier matrix.

An immediate consequence of this explicit description is the following important observation, which was first shown in [DiMi90, Prop. 6.3], and stated without proof in [Lu88, Proof of Prop. 8.1] and in [Lu02, Lemma 1.7]:

Theorem 4.4.23 *Let* $\mathbf{G}$ *be connected reductive such that all simple components are of classical type and* $F : \mathbf{G} \to \mathbf{G}$ *a Frobenius endomorphism. Then the unipotent characters of* $\mathbf{G}^F$ *are uniquely determined by their uniform projections, that is, by their multiplicities in the uniform almost characters.*

Proof First assume $\mathbf{G}$ is simple. There is nothing to prove in type A, since by Corollary 2.4.19 there the unipotent characters coincide with uniform almost characters, and hence in particular are uniform.

For the types B_n, C_n, D_n and ${}^2D_n$ we argue as follows. Since the value on the identity element is a uniform function it is sufficient to see that unipotent characters are uniquely determined in their family by their degrees. Let ρ be labelled by a symbol $S = (X, Y)$. The degree formula in Proposition 4.4.15 shows that it would be enough to see that

$$\prod_{\{i<j\}\subseteq X} (q^j - q^i) \prod_{\{i<j\}\subseteq Y} (q^j - q^i) \prod_{(i,j)\in X\times Y} (q^i + q^j)$$

distinguishes characters in a family $\mathscr{U}$. Now by 4.4.22 the Fourier matrix of $\mathscr{U}$ as well as the set of columns indexed by almost characters in $\mathscr{U}$ do not depend on the exact values of the multi-set $X \cup Y$ of entries of the symbols in $\mathscr{U}$, only on their number. Thus, it suffices to show the above assertion in just one family with a given Fourier, for which we may choose these entries arbitrarily. Since the powers of q are Zariski dense in the integers, it thus suffices to show that the polynomial

$$\prod_{\{i<j\}\subseteq X} (Z_j - Z_i) \prod_{\{i<j\}\subseteq Y} (Z_j - Z_i) \prod_{(i,j)\in X\times Y} (Z_i + Z_j) \qquad (*)$$

with indeterminates Z_i, determines $\{X, Y\}$. But this is clear since any polynomial ring is a UFD.

Finally, for $\mathbf{G}^F$ of type ${}^3D_4$ all unipotent characters have distinct degree (see [Ca85, §13.9]), hence necessarily distinct uniform projection.

It now follows by Corollary 4.2.19 and Example 4.2.20 that for $\mathbf{G}$ simple any two unipotent characters with proportional uniform projection must already be equal. Then using the considerations in Remark 4.2.1 there is a straightforward reduction of the general case to the one treated above. □

Corollary 4.4.24 *Let* $\mathbf{G}$ *be connected reductive such that all simple components are of classical type and* $F : \mathbf{G} \to \mathbf{G}$ *a Frobenius endomorphism. Then all unipotent characters of* $\mathbf{G}$ *are rational valued.*

Proof This is immediate from Theorem 4.4.23 and the fact that the unipotent almost characters, being rational linear combinations of the $R_{\mathbf{T}}^{\mathbf{G}}(1_{\mathbf{T}})$, are rational. □

The conclusion of the previous theorem as well as of its corollary both fail to hold for any group $\mathbf{G}^F$ of exceptional type apart from ${}^3D_4(q)$; see Theorem 4.5.3.

Remark 4.4.25 (a) Digne–Michel [DiMi90, Prop. 6.3] propose a more explicit proof of Theorem 4.4.23: let ρ be a unipotent character, labelled by a symbol S say. Let $R_{\tilde{\phi}}$ be an almost character in the same family corresponding to the σ-extension of a character $\phi \in \mathrm{Irr}(\mathbf{W})^\sigma$ labelled by the symbol S'. Then in the notation introduced above, $\langle S, S' \rangle = 2^{-m}(-1)^{|Y^\# \cap Y'^\#|}$ where 2^{2m} is the size of the corresponding family, and Y, Y' are the second rows of the symbols S, S' respectively. First assume that $\mathbf{G}$ is of type B_n or C_n. For $x \in Z := X \cup Y$ choose a partition $Y_1 \cup Y_2 = Z \setminus \{x\}$ into two subsets of equal cardinality m and let S_1, S_2 be the two symbols in the family with second row Y_1, Y_2 respectively. Then by construction these parametrise uniform almost characters. It is then easy to see that

$$|Y^\# \cap Y_1^\#| \equiv |Y^\# \cap Y_2^\#| \mod 2 \quad \text{if and only if} \quad x \notin Y^\#.$$

Hence Y and so S is uniquely determined by the multiplicities in almost characters.

For types D_n and ${}^2D_n$, for any 2-element subset $\{x, y\} \subseteq Z = X \cup Y$ choose subsets $Y_1, Y_2 \subset Z$ of cardinality $m + 1$ with $Y_1 \triangle Y_2 = \{x, y\}$. Then

$$|Y^\# \cap Y_1^\#| + |Y^\# \cap Y_2^\#| \equiv |Y^\# \cap \{x, y\}| \mod 2.$$

Since the two symbols $(X, Y), (Y, X)$ parametrise the same character, we may always arrange things so that $x \notin Y$, and then again Y is determined by the multiplicities in the almost characters.

(b) For families $\mathscr{U}$ with $|\mathscr{U}| = 2^{2e} \leqslant 2^8 = 64$ elements another easy argument is as follows: one can check that in these cases more than half of the symbols in the family lie in the principal series, that is, the space of uniform class functions in the family has dimension more than half of the whole space of class functions. If two unipotent characters have the same uniform projection, then their difference f will involve less than 2^{2e-1} almost characters in that family, all with multiplicity absolutely bounded by 2^{1-e}, so f has norm less than 2, a contradiction.

We have an even stronger statement that we will need later:

Proposition 4.4.26 *Let $\mathbf{G}$ be connected reductive such that all simple components are of classical type and $F : \mathbf{G} \to \mathbf{G}$ be a Frobenius endomorphism. Let $\mathscr{U} \subset \mathrm{Uch}(\mathbf{G}^F)$ be a family with $|\mathscr{U}| > 4$. Then the various sums of two distinct unipotent characters from $\mathscr{U}$ are uniquely determined by their uniform projections.*

Proof As in the proof of Theorem 4.4.23 we may reduce to the case that $\mathbf{G}$ is simple. For types A and ${}^3D_4$ all families have at most 4 elements. For the remaining types of classical groups we will again argue via the degrees of characters. For this set $I = X \cup Y$ (where again we may assume that X, Y are disjoint) and rewrite the polynomial in (*) above as

$$f(X,Y) := \prod_{\{i<j\}\subseteq I} (Z_j - \epsilon_{ij} Z_i)$$

with $\epsilon_{ij} := -1$ if $(i,j) \in X \times Y \cup Y \times X$ and $\epsilon_{ij} := 1$ otherwise. For $S' = (X', Y')$ the symbol of a second unipotent character in the family we have

$$f(X',Y') = \prod_{\{i<j\}\subseteq I} (Z_j - \epsilon'_{ij} Z_i)$$

with ϵ'_{ij} defined as the ϵ_{ij} but now in terms of X', Y' in place of X, Y. Then we claim that the polynomial $f(X,Y) + f(X',Y')$ uniquely determines the set $\{\{X,Y\},\{X',Y'\}\}$. If $k, l \in I$ with $\epsilon_{kl} = \epsilon'_{kl}$ then the specialisation of both $f(X,Y)$ and $f(X',Y')$ and hence of their sum at $Z_k = \epsilon_{kl} Z_l$ vanishes. On the other hand, if $\epsilon_{kl} \neq \epsilon'_{kl}$ then the specialisation of $f(X,Y) + f(X',Y')$ at $Z_k = \pm\epsilon_{kl} Z_l$ gives

$$f(X',Y')|_{Z_k=\epsilon_{kl}Z_l} \neq 0, \quad \text{respectively} \quad f(X,Y)|_{Z_k=-\epsilon_{kl}Z_l} \neq 0,$$

and from these we can determine the sets X, Y and X', Y' up to the positions of k, l. Observe that $|I| \geqslant 5$ since $|\mathscr{U}| > 4$; but then if $S \neq S'$ there are at least three distinct $k < l$ with $\epsilon_{kl} \neq \epsilon'_{kl}$, so the above allows us to retrieve $\{\{X,Y\},\{X',Y'\}\}$. □

It can easily be seen that the conclusion fails for 4-element families.

Example 4.4.27 (a) Application of Proposition 4.4.20 to the symbols in Example 4.4.14(a) shows that the unipotent characters of $\mathrm{Sp}_4(q)$ fall into three families, two of them singletons containing the trivial and the Steinberg character respectively, and one family $\mathscr{U}$ containing the four characters labelled by

$$\binom{0\ 2}{1}, \binom{1\ 2}{0}, \binom{0\ 1}{2}, \binom{}{0\ 1\ 2}.$$

The corresponding family $\mathscr{F} \subset \mathrm{Irr}(\mathbf{W})$ contains the characters $\phi^{(\alpha,\beta)}$ with $(\alpha,\beta) = (1,1), (1^2,-), (-,2)$ respectively, which by the procedure described in 4.4.21 correspond to the first three symbols listed above in this order. The distinguished symbol, labelling the special character in that family, is the first one. Here, the parameter m attached to this family $\mathscr{U}$ as in Proposition 4.4.20 equals 1 and so $\mathscr{G}_{\mathscr{U}} \cong C_2$ and indeed $|\mathscr{U}| = 4$. By 4.4.22 the Fourier matrix is given by

$$\frac{1}{2}\begin{pmatrix} 1 & 1 & 1 & 1 \\ 1 & 1 & -1 & -1 \\ 1 & -1 & 1 & -1 \\ 1 & -1 & -1 & 1 \end{pmatrix}.$$

(b) The fourteen unipotent characters of $\mathrm{SO}_8^+(q)$ fall into ten singleton families, and one 4-element family $\mathscr{U}$ labelled by the symbols

$$\begin{pmatrix}1\ 3\\0\ 2\end{pmatrix}, \begin{pmatrix}1\ 2\\0\ 3\end{pmatrix}, \begin{pmatrix}2\ 3\\0\ 1\end{pmatrix}, \begin{pmatrix}\\0\ 1\ 2\ 3\end{pmatrix},$$

and $\mathscr{G}_{\mathscr{U}} \cong C_2$. The corresponding family $\mathscr{F} \subset \mathrm{Irr}(\mathbf{W})$ contains the characters $\phi^{[\alpha,\beta]}$ with $(\alpha, \beta) = (21, 1), (1^2, 2), (2^2, -)$, which are sent to the first three symbols in this order. The special character is the first one, and the Fourier matrix is as given in (a).

(c) The ten unipotent characters of $\mathrm{SO}_8^-(q)$ as described in Example 4.4.14(b) fall into six singleton families, and one 4-element family $\mathscr{U}$ labelled by the symbols

$$\begin{pmatrix}0\ 1\ 3\\2\end{pmatrix}, \begin{pmatrix}0\ 1\ 2\\3\end{pmatrix}, \begin{pmatrix}0\ 2\ 3\\1\end{pmatrix}, \begin{pmatrix}1\ 2\ 3\\0\end{pmatrix},$$

where again we have $\mathscr{G}_{\mathscr{U}} \cong C_2$. The corresponding σ-stable family $\mathscr{F} \subset \mathrm{Irr}(\mathbf{W})$ contains the characters $\phi^{[\alpha,\beta]}$ with $(\alpha, \beta) = (21, 1), (1^2, 2), (2^2, -)$, which by the procedure described in 4.4.21 are identified to the symbols

$$\begin{pmatrix}1\ 3\\0\ 2\end{pmatrix}, \begin{pmatrix}1\ 2\\0\ 3\end{pmatrix}, \begin{pmatrix}2\ 3\\0\ 1\end{pmatrix}$$

in this order. The special character is the first one. The Fourier matrix with respect to these orderings and these choices of representatives is again as given in (a).

The description of Harish-Chandra series is also very intuitive in terms of cores of symbols as introduced in 4.4.5. For this, we first describe the cuspidal unipotent characters; they turn out to be very rare (see Lusztig [Lu77a, Thm. 8.2]):

Theorem 4.4.28 *A classical group of rank n has a cuspidal unipotent character if and only if*

(1) $n = s(s+1)$ *for some* $s \geqslant 1$ *for types* B_n *and* C_n $(n \geqslant 2)$,
(2) $n = (2s)^2$ *for some* $s \geqslant 1$ *for type* D_n $(n \geqslant 4)$,
(3) $n = (2s+1)^2$ *for some* $s \geqslant 1$ *for type* ${}^2D_n$ $(n \geqslant 4)$.

In each of these cases, there is a unique such character, and it is labelled by a cuspidal symbol (in the sense of 4.4.5).

It is easy to check from the definition in 4.4.5 that there is a unique cuspidal symbol (up to equivalence) under the conditions stated in the theorem, and none otherwise. Now recall from Example 3.5.15 that 1-split Levi subgroups of simple groups G of classical type have at most one factor not of untwisted type A, which is then of the same type as G. As a group of untwisted type A_n only possesses cuspidal unipotent characters when $n = 0$ (see Section 4.3), that is, when it is a torus, a 1-split Levi subgroup of a simple classical group can have a cuspidal unipotent character only if its derived subgroup is again a simple classical group, whose rank m has the form $m(s) = s(s+1)$, respectively $(2s)^2, (2s+1)^2$, according to the type of group, as given in Theorem 4.4.28. So the Harish-Chandra series of unipotent characters of a simple classical group of rank n are in bijection with the set of integers s such that $m(s) \leqslant n$. The following is in [Lu77a, 8.10]:

Proposition 4.4.29 (Harish-Chandra series in classical groups) *Let G be a classical group of rank n, and S a cuspidal symbol of a Levi subgroup of rank $n-m$ of G. Then the Harish-Chandra series above the corresponding cuspidal character ρ_S consists of the unipotent characters of G labelled by symbols with core S.*

If S is non-degenerate, then

$$\mathscr{Y}^0_{d,m} \longrightarrow \mathscr{Y}^1_{d,n}, \qquad S' \mapsto S + S',$$

defines a natural bijection of $\mathrm{Irr}(W(B_m))$ *with the Harish-Chandra series above ρ_S.*

If $S = (-;-)$ is degenerate, and so G is of type D_n and $m = n$, then

$$\mathscr{Y}^0_{0,n} \longrightarrow \mathscr{Y}^1_{0,n}, \qquad S' \mapsto S + S',$$

defines a natural bijection of $\mathrm{Irr}(W(D_n))$ *with the Harish-Chandra series above ρ_S.*

Example 4.4.30 We continue Examples 4.4.14 and 4.4.27.

(a) An application of Proposition 4.4.29 shows that five out of the six unipotent characters of $\mathrm{Sp}_4(q)$ listed in Example 4.4.14(a) lie in the principal series, while the one labelled by

$$\binom{\ }{0\ 1\ 2}$$

lying in the four-element family is cuspidal.

By Proposition 4.4.29 the Harish-Chandra series in $\mathrm{Sp}_8(q)$ above this cuspidal unipotent character of $\mathrm{Sp}_4(q)$ contains the unipotent characters with labels

$$\binom{0\ 1\ 4}{\ }, \binom{0\ 2\ 3}{\ }, \binom{0\ 1\ 2\ 4}{1}, \binom{0\ 1\ 2\ 3}{2}, \binom{0\ 1\ 2\ 3\ 4}{1\ 2}.$$

(b) On the other hand, all ten unipotent characters of $\mathrm{SO}_8^-(q)$ lie in the Harish-Chandra series of the cuspidal character labelled by

$$\binom{0\ 1}{\ }$$

of a Levi subgroup of type $\mathrm{SO}_2^-(q)$. But the latter is a torus, so all ten unipotent characters lie in the principal series. This shows that even non-singleton families may consist solely of principal series characters. It is easily seen that this latter phenomenon can happen only in groups of twisted type. In untwisted groups, each family does contain at least one.

The parameters of the associated Hecke algebras (for all types) have been determined by Lusztig, see [Lu77b, Tab. II]:

Theorem 4.4.31 (Lusztig) *Let* **G** *be simple with Steinberg endomorphism* F. *Then the parameters of the Iwahori–Hecke algebras associated to cuspidal unipotent characters of 1-split Levi subgroups of* **G** *are as given in Table 4.8.*

Table 4.8 *Parameters for Hecke algebras for cuspidal unipotent characters*

$\mathbf{G}^F$	$[\mathbf{L},\mathbf{L}]^F$	$W_{\mathbf{G}^F}(\mathbf{L},\lambda)$	parameters
untwisted	1	$\mathbf{W}$	all q
${}^2A_{2n+\binom{s+1}{2}-1}(q)$ $(n \geqslant 1,\ s \geqslant 0)$	${}^2A_{\binom{s+1}{2}-1}(q)$	B_n	$q^{2s+1}; q^2, \ldots, q^2$
$B_{n+s^2+s}(q)$ $(n \geqslant 1,\ s \geqslant 1)$	$B_{s^2+s}(q)$	B_n	$q^{2s+1}; q, \ldots, q$
$C_{n+s^2+s}(q)$ $(n \geqslant 1,\ s \geqslant 1)$	$C_{s^2+s}(q)$	B_n	$q^{2s+1}; q, \ldots, q$
$D_{n+s^2}(q)$ $(n \geqslant 1,\ s \geqslant 2$ even$)$	$D_{s^2}(q)$	B_n	$q^{2s}; q, \ldots, q$
${}^2D_{n+s^2}(q)$ $(n \geqslant 1,\ s \geqslant 3$ odd$)$	${}^2D_{s^2}(q)$	B_n	$q^{2s}; q, \ldots, q$
${}^2D_n(q)$ $(n \geqslant 4)$	1	B_{n-1}	$q^2; q, \ldots, q$
${}^3D_4(q)$	1	G_2	$q^3; q$
$F_4(q)$	$B_2(q)$	B_2	$q^3; q^3$
$E_6(q)$	$D_4(q)$	A_2	q^4, q^4
${}^2E_6(q)$	1	F_4	$q^2, q^2; q, q$
	${}^2A_5(q)$	A_1	q^9
$E_7(q)$	$D_4(q)$	B_3	$q; q^4, q^4$
	$E_6(q)$	A_1	q^9
$E_8(q)$	$D_4(q)$	F_4	$q, q; q^4, q^4$
	$E_6(q)$	G_2	$q^9; q$
	$E_7(q)$	A_1	q^{15}
${}^2B_2(q^2)$	1	A_1	q^4
${}^2G_2(q^2)$	1	A_1	q^6
${}^2F_4(q^2)$	1	$I_2(8)$	$q^4; q^2$
	${}^2B_2(q^2)$	A_1	q^{12}

This extends Table 3.1 which gave the parameters for the principal series.

Recall from Proposition 4.2.23 that Frobenius eigenvalues are determined by the Harish-Chandra source of a character. Thus, to describe the Frobenius eigenvalues of unipotent characters it is sufficient to do this for the cuspidal characters.

Proposition 4.4.32 *Let ρ be a cuspidal unipotent character of a classical group labelled by the cuspidal symbol S. Then the Frobenius eigenvalue of ρ equals*

$$\omega_\rho = \begin{cases} (-1)^{\lfloor (s+1)/2 \rfloor} & \text{if } S = \binom{0\ \dots\ 2s}{}, \\ (-1)^{s/2} & \text{if } S = \binom{0\ \dots\ 2s-1}{}, \ s \text{ even}, \\ 1 & \text{if } S = \binom{0\ \dots\ 2s-1}{}, \ s \text{ odd}. \end{cases}$$

This was proved for sufficiently large q in [Lu81c, Prop. 6.6] for symplectic and odd-dimensional orthogonal groups, in [Lu82b, 3.18] for even-dimensional orthogonal groups of split type, and then without restriction on q in [Lu84a, 11.3]. For even-dimensional orthogonal groups of twisted type it was shown in [GeMa03, Thm. 4.11]. In Table 4.9 we give the Frobenius eigenvalues of cuspidal characters for all simple groups of classical type and rank at most 15.

Table 4.9 *Frobenius eigenvalues of some cuspidal unipotent characters*

	B_2	D_4	B_6, C_6	${}^2D_9$	B_{12}, C_{12}
S	$\binom{012}{}$	$\binom{0123}{}$	$\binom{01234}{}$	$\binom{012345}{}$	$\binom{0123456}{}$
ω_ρ	-1	-1	-1	1	1

4.5 Unipotent Characters in Exceptional Types

In view of the fact that the data for unipotent characters of exceptional groups of Lie type are not only given in printed form in [Lu84a, App.] and [Ca85, §13.8], but are also readily available electronically through the CHEVIE system [MiChv] we refrain from giving complete tables of the degrees of the unipotent characters in this case.

On the other hand, neither [Lu84a] nor [Ca85] give explicitly the Frobenius eigenvalues of the unipotent characters, and these are somewhat difficult to find in the literature, so we include them here. According to Proposition 4.2.23 they are constant on Harish-Chandra series, so it is sufficient for this to list the various Harish-Chandra series together with their Frobenius eigenvalues.

The Frobenius eigenvalues of unipotent characters were determined by Lusztig [Lu76c] for the constituents of Deligne–Lusztig characters for Coxeter tori. In [Lu77b, Thm. 3.34] he restricts the possible values in the general case and in [Lu84a, Chap. 11] he settles completely the case of split groups. The remaining eigenvalues were determined in [GeMa03].

Theorem 4.5.1 *The Frobenius eigenvalues of the Harish-Chandra series of unipotent characters in simple groups of exceptional type are as given in Tables 4.10–4.14.*

Here, for $d \geqslant 1$, we denote by $\zeta_d := \exp(2\pi\sqrt{-1}/d)$ a primitive dth root of unity. The Frobenius eigenvalues of the cuspidal characters in groups of classical type occurring as Levi subgroups can be found in Table 4.9.

Table 4.10 *Unipotent Harish-Chandra series*

in $G_2(q)$ and...

$([\mathbf{L},\mathbf{L}],\lambda)$	$W_G(\mathbf{L},\lambda)$	$\lvert\mathscr{E}\rvert$	ω_ρ
$(1,1)$	$W(G_2)$	6	1

+4 cuspidals with $\omega_\rho = 1, -1, \zeta_3, \zeta_3^2$

... in $^3D_4(q)$

$([\mathbf{L},\mathbf{L}],\lambda)$	$W_G(\mathbf{L},\lambda)$	$\lvert\mathscr{E}\rvert$	ω_ρ
$(1,1)$	$W(G_2)$	6	1

+2 cuspidals with $\omega_\rho = 1, -1$

Table 4.11 *Unipotent Harish-Chandra series in $F_4(q)$*

$([\mathbf{L},\mathbf{L}],\lambda)$	$W_G(\mathbf{L},\lambda)$	$\lvert\mathscr{E}(G,(\mathbf{L},\lambda))\rvert$	ω_ρ
$(1,1)$	$W(F_4)$	25	1
$(B_2, B_2[-1])$	$W(B_2)$	5	-1

+7 cuspidals with $\omega_\rho = 1, 1, -1, \zeta_3, \zeta_3^2, \zeta_4, -\zeta_4$.

Table 4.12 *Unipotent Harish-Chandra series*

in $E_6(q)$ and...

$([\mathbf{L},\mathbf{L}],\lambda)$	$W_G(\mathbf{L},\lambda)$	$\lvert\mathscr{E}\rvert$	ω_ρ
$(1,1)$	$W(E_6)$	25	1
$(D_4, D_4[-1])$	$W(A_2)$	3	-1

+2 cuspidals with $\omega_\rho = \zeta_3, \zeta_3^2$

... in $^2E_6(q)$

$([\mathbf{L},\mathbf{L}],\lambda)$	$W_G(\mathbf{L},\lambda)$	$\lvert\mathscr{E}\rvert$	ω_ρ
$(1,1)$	$W(F_4)$	25	1
$(^2A_5, {}^2A_5[-1])$	$W(A_1)$	2	-1

+3 cuspidals with $\omega_\rho = 1, \zeta_3, \zeta_3^2$

We include the complete data for unipotent characters of groups of very twisted type, as the table for $^2F_4(q^2)$ in [Ca85, §13.8] contains misprints (and neither the Frobenius eigenvalues nor the Fourier matrices or labels for cuspidal characters are given there). The Harish-Chandra labels are given via the characters of the relative Hecke algebra as in [MiChv]. Recall our notation for cyclotomic polynomials over $\mathbb{Q}(\sqrt{2})$ and $\mathbb{Q}(\sqrt{3})$ from Example 3.5.4 (which differs from the one used in [Ca85, §13]).

Table 4.13 *Unipotent Harish-Chandra series in* $E_7(q)$

$([\mathbf{L},\mathbf{L}],\lambda)$	$W_G(\mathbf{L},\lambda)$	$\lvert\mathscr{E}(G,(\mathbf{L},\lambda))\rvert$	ω_ρ
$(1,1)$	$W(E_7)$	60	1
$(D_4, D_4[-1])$	$W(C_3)$	10	-1
$(E_6, E_6[\zeta_3])$	$W(A_1)$	2	ζ_3
$(E_6, E_6[\zeta_3^2])$	$W(A_1)$	2	ζ_3^2

+2 cuspidals with $\omega_\rho = \zeta_4, -\zeta_4$.

Table 4.14 *Unipotent Harish-Chandra series in* $E_8(q)$

$([\mathbf{L},\mathbf{L}],\lambda)$	$W_G(\mathbf{L},\lambda)$	$\lvert\mathscr{E}(G,(\mathbf{L},\lambda))\rvert$	ω_ρ
$(1,1)$	$W(E_8)$	112	1
$(D_4, D_4[-1])$	$W(F_4)$	25	-1
$(E_6, E_6[\zeta_3])$	$W(G_2)$	6	ζ_3
$(E_6, E_6[\zeta_3^2])$	$W(G_2)$	6	ζ_3^2
$(E_7, E_7[\zeta_4])$	$W(A_1)$	2	ζ_4
$(E_7, E_7[-\zeta_4])$	$W(A_1)$	2	$-\zeta_4$

+13 cuspidals with $\omega_\rho = 1, 1, -1, \zeta_3, \zeta_3^2, \zeta_4, -\zeta_4, \zeta_5, \zeta_5^2, \zeta_5^3, \zeta_5^4, -\zeta_3, -\zeta_3^2$.

Theorem 4.5.2 *Let* $(\mathbf{G}, F)$ *be such that* $\mathbf{G}^F$ *is a Suzuki group* ${}^2B_2(q^2)$*, a Ree group* ${}^2G_2(q^2)$ *or a Ree group* ${}^2F_4(q^2)$*. Let* $\mathscr{F} \subseteq \mathrm{Irr}(\mathbf{W})$ *be an* F*-stable family. Then the number of unipotent characters in* $\mathscr{E}(\mathbf{G}^F \mid \mathscr{F})$ *is* 1, 2, 6 *or* 13.

(a) *If* $\mathrm{Uch}(\mathbf{G}^F \mid \mathscr{F})$ *contains* 2 *characters, the associated Fourier matrix is given by*

$$\mathbf{M}(\mathscr{F}) = \frac{1}{\sqrt{2}}\begin{pmatrix} 1 & 1 \\ 1 & -1 \end{pmatrix}.$$

(b) *In type* ${}^2G_2(q^2)$ *there is a six-element family of unipotent characters. The associated Fourier matrix is given by*

$$\mathbf{M}(\mathscr{F}) = \frac{1}{2\sqrt{3}}\begin{pmatrix} 1 & \sqrt{3} & . & 2 & 1 & \sqrt{3} \\ 1 & -\sqrt{3} & . & 2 & 1 & -\sqrt{3} \\ 1 & \sqrt{3} & 2 & . & -1 & -\sqrt{3} \\ 1 & -\sqrt{3} & 2 & . & -1 & \sqrt{3} \\ 2 & . & . & -2 & 2 & . \\ 2 & . & -2 & . & -2 & . \end{pmatrix}.$$

Here, the rows are labelled by $\rho_2, \ldots, \rho_7$ *from Table 4.15, and the columns are*

Table 4.15 *Unipotent characters of* $^2G_2(q^2)$...

HC-label	$\rho(1)$	ω_ρ
$\phi_{1,0}$	1	1
$^2G_2^I[\zeta_{12}^5]$	$\frac{1}{2\sqrt{3}}q\Phi_1\Phi_2\Phi'_{12}$	ζ_{12}^5
$^2G_2^I[\zeta_{12}^7]$	$\frac{1}{2\sqrt{3}}q\Phi_1\Phi_2\Phi'_{12}$	ζ_{12}^7
$^2G_2^{II}[\zeta_{12}^5]$	$\frac{1}{2\sqrt{3}}q\Phi_1\Phi_2\Phi''_{12}$	ζ_{12}^5
$^2G_2^{II}[\zeta_{12}^7]$	$\frac{1}{2\sqrt{3}}q\Phi_1\Phi_2\Phi''_{12}$	ζ_{12}^7
$^2G_2[\zeta_4]$	$\frac{1}{\sqrt{3}}q\Phi_1\Phi_2\Phi_4$	ζ_4
$^2G_2[-\zeta_4]$	$\frac{1}{\sqrt{3}}q\Phi_1\Phi_2\Phi_4$	$-\zeta_4$
$\phi_{1,6}$	q^6	1

... *and of* $^2B_2(q^2)$

HC-label	$\rho(1)$	ω_ρ
$\phi_{1,0}$	1	1
$^2B_2[\zeta_8^3]$	$\frac{1}{\sqrt{2}}q\Phi_1\Phi_2$	ζ_8^3
$^2B_2[\zeta_8^5]$	$\frac{1}{\sqrt{2}}q\Phi_1\Phi_2$	ζ_8^5
$\phi_{1,4}$	q^4	1

Table 4.16 *Unipotent characters of* $^2F_4(q^2)$

HC-label	$\rho(1)$	ω_ρ	HC-label	$\rho(1)$	ω_ρ
$\phi_{1,0}$	1	1	$^2F_4^I[\zeta_4]$	$\frac14 q^4\Phi_1^2\Phi_2^2\Phi_4^2\Phi_{12}\Phi''_{24}$	ζ_4
$^2B_2[\zeta_8^3], 1$	$\frac{1}{\sqrt{2}}q\Phi_1\Phi_2\Phi_4^2\Phi_{12}$	ζ_8^3	$^2F_4^I[-\zeta_4]$	$\frac14 q^4\Phi_1^2\Phi_2^2\Phi_4^2\Phi_{12}\Phi''_{24}$	$-\zeta_4$
$^2B_2[\zeta_8^5], 1$	$\frac{1}{\sqrt{2}}q\Phi_1\Phi_2\Phi_4^2\Phi_{12}$	ζ_8^5	$^2F_4^{II}[\zeta_4]$	$\frac14 q^4\Phi_1^2\Phi_2^2\Phi_4^2\Phi_{12}\Phi'_{24}$	ζ_4
$\phi'_{1,4}$	$q^2\Phi_{12}\Phi_{24}$	1	$^2F_4^{II}[-\zeta_4]$	$\frac14 q^4\Phi_1^2\Phi_2^2\Phi_4^2\Phi_{12}\Phi'_{24}$	$-\zeta_4$
$\phi_{2,1}$	$\frac14 q^4\Phi_4^2\Phi_{12}\Phi_8''^2\Phi'_{24}$	1	$^2F_4[-\zeta_3]$	$\frac13 q^4\Phi_1^2\Phi_2^2\Phi_4^2\Phi_8^2$	$-\zeta_3$
$\phi_{2,3}$	$\frac14 q^4\Phi_4^2\Phi_{12}\Phi_8'^2\Phi''_{24}$	1	$^2F_4[-\zeta_3^2]$	$\frac13 q^4\Phi_1^2\Phi_2^2\Phi_4^2\Phi_8^2$	$-\zeta_3^2$
$\phi_{2,2}$	$\frac12 q^4\Phi_8^2\Phi_{24}$	1	$^2F_4^{IV}[-1]$	$\frac13 q^4\Phi_1^2\Phi_2^2\Phi_{12}\Phi_{24}$	-1
$^2F_4^I[-1]$	$\frac16 q^4\Phi_1^2\Phi_2^2\Phi_4^2\Phi_{24}$	-1	$\phi''_{1,4}$	$q^{10}\Phi_{12}\Phi_{24}$	1
$^2F_4^{II}[-1]$	$\frac{1}{12}q^4\Phi_1^2\Phi_2^2\Phi_{12}\Phi_8'^2\Phi'_{24}$	-1	$^2B_2[\zeta_8^3], \epsilon$	$\frac{1}{\sqrt{2}}q^{13}\Phi_1\Phi_2\Phi_4^2\Phi_{12}$	ζ_8^3
$^2F_4^{III}[-1]$	$\frac{1}{12}q^4\Phi_1^2\Phi_2^2\Phi_{12}\Phi_8''^2\Phi''_{24}$	-1	$^2B_2[\zeta_8^5], \epsilon$	$\frac{1}{\sqrt{2}}q^{13}\Phi_1\Phi_2\Phi_4^2\Phi_{12}$	ζ_8^5
			$\phi_{1,8}$	q^{24}	1

labelled by suitable almost characters; with the first two lying in the principal series.

(c) *In type* $^2F_4(q^2)$ *there is a* 13*-element family of unipotent characters. The associated Fourier matrix* $\mathbf{M}(\mathscr{F})$ *is given in Table 4.17.*

This was proved in [GeMa03, Thm. 5.4] by explicitly decomposing the unipotent almost characters into unipotent characters, using the known character tables.

By inspection of the Fourier matrices one obtains the following partial analogue of Theorem 4.4.23 (the statement in [DiMi90, Prop. 6.3] is not quite complete):

Theorem 4.5.3 *Let* $\mathbf{G}$ *be simple such that* $\mathbf{G}^F$ *is of exceptional type. Then the*

Table 4.17 *The Fourier transform matrix in type* ${}^2F_4$

$$\frac{1}{12}\begin{pmatrix}
3 & 3 & . & -6 & 3\sqrt{2} & 3\sqrt{2} & 3\sqrt{2} & 3\sqrt{2} & 3 & -3 & . & . & . \\
3 & 3 & . & -6 & -3\sqrt{2} & -3\sqrt{2} & -3\sqrt{2} & -3\sqrt{2} & 3 & -3 & . & . & . \\
6 & 6 & . & . & . & . & . & . & -6 & 6 & . & . & . \\
2 & -6 & -4 & -4 & . & . & . & . & -6 & -2 & -4 & -4 & . \\
1 & -3 & 4 & -2 & -3\sqrt{2} & 3\sqrt{2} & 3\sqrt{2} & -3\sqrt{2} & -3 & -1 & 4 & 4 & . \\
1 & -3 & 4 & -2 & 3\sqrt{2} & -3\sqrt{2} & -3\sqrt{2} & 3\sqrt{2} & -3 & -1 & 4 & 4 & . \\
3 & -3 & . & . & 3\sqrt{2} & -3\sqrt{2} & 3\sqrt{2} & -3\sqrt{2} & 3 & 3 & . & . & 6 \\
3 & -3 & . & . & 3\sqrt{2} & 3\sqrt{2} & -3\sqrt{2} & -3\sqrt{2} & 3 & 3 & . & . & -6 \\
3 & -3 & . & . & -3\sqrt{2} & 3\sqrt{2} & -3\sqrt{2} & 3\sqrt{2} & 3 & 3 & . & . & 6 \\
3 & -3 & . & . & -3\sqrt{2} & -3\sqrt{2} & 3\sqrt{2} & 3\sqrt{2} & 3 & 3 & . & . & -6 \\
4 & . & 4 & 4 & . & . & . & . & . & -4 & 4 & -8 & . \\
4 & . & 4 & 4 & . & . & . & . & . & -4 & -8 & 4 & . \\
4 & . & -8 & 4 & . & . & . & . & . & -4 & 4 & 4 & .
\end{pmatrix}$$

The rows are labelled by the unipotent characters $\rho_5, \ldots, \rho_{17}$ from Table 4.16. The columns are labelled by the almost characters in the principal series $\widetilde{12}_1$, $\tilde{4}_1, \tilde{6}_1, \tilde{6}_2, \widetilde{16}_1$, and eight further almost characters not in the principal series.

unipotent characters ρ of $\mathbf{G}^F$ are uniquely determined by their uniform projections, that is, by their multiplicities in the uniform almost characters except in the following cases:

(1) *ρ is labelled by a character (x, σ) of the Drinfeld double corresponding to the family of ρ (see 4.2.9) such that σ is non-rational linear of odd order;*
(2) *ρ belongs to an exceptional family in types E_7 or E_8; or*
(3) *$\mathbf{G}^F$ is very twisted.*

For a family corresponding to the symmetric group $\mathfrak{S}_3$ there is one such pair of non-rational characters of the Drinfeld double; for a family corresponding to the symmetric group $\mathfrak{S}_4$ there are two pairs; for a family corresponding to the symmetric group $\mathfrak{S}_5$ there are five pairs and one quadruple. The four unipotent characters in an exceptional family fall into two pairs, in each of which both characters have the same restriction to the space of uniform functions.

Corollary 4.5.4 (Digne–Michel [DiMi90, Prop. 6.4]) *Let $\mathbf{G}$ be simple with a Steinberg map F such that $\mathbf{G}^F$ is of exceptional type. Then the unipotent characters of $\mathbf{G}^F$ are uniquely determined by their uniform projections, together with their eigenvalue of Frobenius and, for characters in the principal series belonging to an exceptional family, by the associated character of the Iwahori–Hecke algebra.*

Proof We only need to discuss the exceptions (1)–(3) in Theorem 4.5.3. For these the statement can be checked from the explicit knowledge of the Frobenius eigenvalues; for example, for the Suzuki and Ree groups the lists given in Tables 4.15 and 4.16 show that their unipotent characters are distinguished by their degrees (hence by their uniform projection) together with the associated Frobenius eigenvalues. □

This result plays a crucial role in Digne and Michel's construction of a unique Jordan decomposition, see Theorem 4.7.1.

Next we consider the rationality properties of unipotent characters. For χ an irreducible character of a finite group we denote by $\mathbb{Q}(\chi)$ its character field, that is, the field generated by the values of χ. The following result first stated by Geck [Ge03b, Thm. 1.4] gives a nice connection to Frobenius eigenvalues:

Proposition 4.5.5 *Let* $\mathbf{G}$ *be simple with a Steinberg map* F *and* $\rho \in \mathrm{Uch}(\mathbf{G}^F)$. *Then* $\mathbb{Q}(\rho) = \mathbb{Q}(\omega_\rho)$, *unless* ρ *lies in the principal series in an exceptional family for* E_7 *or* E_8. *In the latter case we have* $\mathbb{Q}(\rho) = \mathbb{Q}(\sqrt{q})$.

Proof Let $\rho \in \mathrm{Uch}(\mathbf{G}^F)$ be a unipotent character with Frobenius eigenvalue ω_ρ. By definition there exists an F-stable maximal torus $\mathbf{T} \leqslant \mathbf{G}$ such that the Deligne–Lusztig character $R_\mathbf{T}^\mathbf{G}(1_\mathbf{T})$ has ρ as a constituent. Then there is an ℓ-adic cohomology group M of the associated Deligne–Lusztig variety whose ρ-isotypic component M_ρ is non-zero, and F^δ acts on it by ω_ρ times a half-integral power of q. In fact this is an integral power of q unless ρ lies in an exceptional family of a group of type E_7 or E_8 (see 4.2.21). The actions of $\mathbf{G}^F$ and F^δ on M commute, and any Galois automorphism σ of $\overline{\mathbb{Q}}_\ell/\mathbb{Q}_\ell$ sends the ω_ρ-eigenspace of F^δ to the $\sigma(\omega_\rho)$ eigenspace, and the ρ-isotypic component to the $\sigma(\rho)$-isotypic component. Thus $\sigma(\rho) \neq \rho$ if $\sigma(\omega_\rho) \neq \omega_\rho$. Since for any $m > 2$ there is some prime $\ell \neq p$ such that $\mathbb{Q}_\ell$ does not contain a primitive mth root of unity, this argument shows that $\mathbb{Q}(\omega_\rho) \leqslant \mathbb{Q}(\rho)$ unless ρ lies in an exceptional family.

For the converse, note that the Deligne–Lusztig characters $R_\mathbf{T}^\mathbf{G}(1_\mathbf{T})$ are rational valued (see 2.2.1). So the same holds for all unipotent characters that are determined by their uniform projections. This was already used in Corollary 4.4.24 to see that all unipotent characters for $\mathbf{G}$ of classical type and F a Frobenius map are rational. By the same argument this holds for those unipotent characters of groups of exceptional type which are not in one of the exceptions (1)–(3) of Theorem 4.5.3. In case (1) of that result, the number of unipotent characters with non-rational Frobenius eigenvalue is exactly equal to the length of the Galois orbit of that root of unity, so our claim follows. This also applies to case (3) when F is not a Frobenius map. Finally, the characters lying in an exceptional family but not in the principal series also satisfy this property. The last assertion requires some further considerations and is shown in [Ge03b, Prop. 5.6]. □

With this we obtain the following complement to Corollary 4.4.24:

Corollary 4.5.6 *Let* **G** *be simple and* $F : \mathbf{G} \to \mathbf{G}$ *a Steinberg map. Then a unipotent character* ρ *of* **G** *is rational valued if and only if the following hold:*

(1) *its associated Frobenius eigenvalue is* ± 1*; and*
(2) *if* ρ *lies in the principal series in an exceptional family in type* E_7 *or* E_8 *then the associated character of the Iwahori–Hecke algebra is rational.*

Proof This is an immediate consequence of Proposition 4.5.5. □

For **G** an arbitrary connected reductive group, Remark 4.2.1 gives an immediate reduction of this question to the case when **G** is simple.

Remark 4.5.7 The much more difficult problem of determining the Schur index has also been solved completely for unipotent characters, see [Ohm96, Lu02, Ge03b, Ge03c, Ge05a]. The restrictions in the last cited paper can now be removed using [Ta16]. It turns out that the Schur index is always at most 2, and there do exist unipotent characters for which this value is attained.

Rationality properties of general irreducible characters are studied in the papers [Ge03b] and [TiZa04], using rather different methods; see also Theorem 4.7.9.

We now describe some general consequences that can be derived from the explicit knowledge of the properties of unipotent characters. The results of Lusztig described above show in particular that unipotent characters are *generic*, that is, their classification, their degrees and many of their further properties only depend on the underlying complete root datum. This can be formalised as follows:

Theorem 4.5.8 *Let* $\mathbb{G}$ *be a complete root datum. There exists a set* $\mathrm{Uch}(\mathbb{G})$ *and a function*

$$\mathbb{D} : \mathrm{Uch}(\mathbb{G}) \longrightarrow \mathbb{Q}[\mathbf{q}], \qquad \gamma \mapsto \mathbb{D}_\gamma,$$

such that for any admissible choice of prime power q *(and hence of connected reductive group* **G** *and Steinberg map* F *corresponding to* $\mathbb{G}$*) there is a bijection* $\psi_q^{\mathbb{G}} : \mathrm{Uch}(\mathbb{G}) \to \mathrm{Uch}(\mathbf{G}^F)$ *with* $\mathbb{D}_\gamma = \mathbb{D}_{\psi_q^{\mathbb{G}}(\gamma)}$*, so* $\psi_q^{\mathbb{G}}(\gamma)(1) = \mathbb{D}_\gamma(q)$*.*

Here, $\mathrm{Uch}(\mathbb{G})$ is called the set of *generic unipotent characters* of $\mathbb{G}$. As we saw above, the partition of unipotent characters into Harish-Chandra series and into families is also described in combinatorial terms, only depending on the complete root datum, hence these are generic as well; that is, they already exist at the level of $\mathrm{Uch}(\mathbb{G})$. This observation has led Broué, Malle and Michel to introduce similar combinatorial sets of unipotent characters attached to certain types of complex reflection groups (baptised 'spetsial' in [Ma00]), see e.g. [Ma95, BMM99, BMM14], which should play the role of unipotent characters of yet to be discovered new objects called 'spetses' [Ma98]. Independently, Lusztig [Lu93] had defined unipotent degrees for those finite irreducible Coxeter groups that are not Weyl groups, of types H_3, H_4 and $I_2(m)$ with $m = 5$ or $m \geqslant 7$.

We take this opportunity to recall some properties of the degree polynomials introduced in 2.3.25:

Proposition 4.5.9 *The degree polynomial of $\rho \in \mathrm{Uch}(\mathbf{G}^F)$ is of the form*

$$\mathbb{D}_\rho = \frac{1}{n_\rho}(\mathbf{q}^{A_\rho} \pm \cdots \pm \mathbf{q}^{a_\rho})$$

for a real number $n_\rho > 0$, and $n_\rho \mathbb{D}_\rho$ is a product of cyclotomic polynomials.

Here, $a_\rho = 0$ if and only if $\rho = 1_{\mathbf{G}}$, $A_\rho = l(w_0)$ is maximal if and only if $\rho = \mathrm{St}_{\mathbf{G}}$, and $a_\rho < A_\rho$ unless ρ is a tensor product of trivial and Steinberg characters of the various simple factors of $\mathbf{G}^F$. Furthermore, if F is a Frobenius map then n_ρ is an integer dividing the order $|\mathcal{G}_{\mathcal{U}}|$ of the group $\mathcal{G}_{\mathcal{U}}$ attached to the family $\mathcal{U} \subseteq \mathrm{Uch}(\mathbf{G}^F)$ of ρ.

Proof The form of $\mathbb{D}_\rho$ follows from 4.1.14 in conjunction with the fact that a- and A-values are constant on families of unipotent characters by Proposition 4.2.7. By Proposition 4.1.20 only the (trailing term of the) fake degree of the special character in $\mathcal{U}$ contributes to the trailing term of $\mathbb{D}_\rho$. But the coefficient of the trailing term of the fake degree is 1 for special characters and so the coefficient of $\mathbf{q}^{a_\rho}$ equals the entry in the Fourier matrix at the special character. The real number n_ρ was introduced in Remark 2.3.26. For F a Frobenius map its properties then follow from the explicit description of Fourier matrices by inspection.

The second assertion is a consequence of Remark 4.1.21 together with the fact that the trivial and the sign character always lie in singleton families. □

It is relatively straightforward from this to see that for $\mathbf{G}$ simple the Steinberg character is the only unipotent character of p-power degree larger than 1. In fact, all irreducible characters of prime power degree of quasi-simple groups of Lie type have been determined, see [MZ01].

It can be shown that also Lusztig induction and restriction behave generically on unipotent characters:

Theorem 4.5.10 *Let $\mathbb{G}$ be a complete root datum. For each Levi sub-datum $\mathbb{L}$ of $\mathbb{G}$ there exist linear maps*

$$R_{\mathbb{L}}^{\mathbb{G}} : \mathbb{Z}\mathrm{Uch}(\mathbb{L}) \longrightarrow \mathbb{Z}\mathrm{Uch}(\mathbb{G}),$$
$${}^*R_{\mathbb{L}}^{\mathbb{G}} : \mathbb{Z}\mathrm{Uch}(\mathbb{G}) \longrightarrow \mathbb{Z}\mathrm{Uch}(\mathbb{L}),$$

such that for any admissible choice of connected reductive groups $\mathbf{L} \leqslant \mathbf{G}$ and Steinberg map F (and hence prime power q) corresponding to $\mathbb{L}$ and $\mathbb{G}$ we have

$$R_{\mathbf{L}}^{\mathbf{G}} \circ \psi_q^{\mathbb{L}} = \psi_q^{\mathbb{G}} \circ R_{\mathbb{L}}^{\mathbb{G}} \quad \textit{and} \quad {}^*R_{\mathbf{L}}^{\mathbf{G}} \circ \psi_q^{\mathbb{G}} = \psi_q^{\mathbb{L}} \circ {}^*R_{\mathbb{L}}^{\mathbb{G}}.$$

This follows from results of Shoji [Sho87], see [BMM93, Thm. 1.33]. For classical groups it can also be derived from the explicit decomposition formulas of Asai, see Section 4.6.

A further consequence of the explicit descriptions concerns the action of automorphisms on unipotent characters. It turns out to be generic as well.

Theorem 4.5.11 *Let* $\mathbf{G}$ *be simple with Steinberg endomorphism* F. *Let* ρ *be a unipotent character of* $\mathbf{G}^F$. *Then* ρ *is* $\mathrm{Aut}(\mathbf{G}^F)$*-invariant, except in the following cases:*

(a) *In* $\mathbf{G}^F = D_n(q)$ *with even* $n \geqslant 4$, *the graph automorphism of order 2 interchanges the two unipotent characters in all pairs labelled by the same degenerate symbol of defect 0 and rank n.*

(b) *In* $\mathbf{G}^F = D_4(q)$ *the graph automorphism of order 3 has two non-trivial orbits consisting of the unipotent characters labelled by the symbols*

$$\left\{\binom{2}{2}, \binom{2}{2}', \binom{1\,4}{0\,1}\right\}, \quad \left\{\binom{1\,2}{1\,2}, \binom{1\,2}{1\,2}', \binom{1\,2\,4}{0\,1\,4}\right\}.$$

(c) *In* $\mathbf{G}^F = \mathrm{Sp}_4(2^{2f+1})$, $f \geqslant 0$, *the graph automorphism of order 2 interchanges the two unipotent principal series characters labelled by the symbols*

$$\left\{\binom{1\,2}{0}, \binom{0\,1}{2}\right\}.$$

(d) *In* $\mathbf{G}^F = G_2(3^{2f+1})$, $f \geqslant 0$, *the graph automorphism of order 2 interchanges the two unipotent principal series characters labelled by the characters* $\{\phi'_{1,3}, \phi''_{1,3}\}$ *of the Weyl group* $W(G_2)$.

(e) *In* $\mathbf{G}^F = F_4(2^{2f+1})$, $f \geqslant 0$, *the graph automorphism of order 2 has eight orbits of length 2, consisting of the unipotent characters labelled by*

$$\{\phi'_{8,3}, \phi''_{8,3}\}, \ \{\phi'_{8,9}, \phi''_{8,9}\}, \ \{\phi'_{2,4}, \phi''_{2,4}\}, \ \{\phi'_{2,16}, \phi''_{2,16}\},$$

$$\{\phi'_{9,6}, \phi''_{9,6}\}, \ \{\phi'_{1,12}, \phi''_{1,12}\}, \ \{\phi'_{4,7}, \phi''_{4,7}\}, \{(B_2, \epsilon'), (B_2, \epsilon'')\}.$$

Here the notation for the unipotent characters of groups of type F_4 is taken from [Ca85, 13.8].

This result follows rather easily from Lusztig's description of the unipotent characters. For example for $\mathbf{G}$ of classical type, the unipotent characters are uniquely determined by their multiplicities in the Deligne–Lusztig characters $R_{\mathbf{T}}^{\mathbf{G}}(1)$ (see Theorem 4.4.23), and the action of automorphisms on the latter is readily described. For exceptional types, this statement is not always true, but then ad hoc arguments can be used, for example based on Corollary 4.5.4. See e.g. [Ma07, Prop. 3.7 and 3.9].

Based on the previous theorem the following extension property was shown in [Ma08, Thm. 2.4]:

Theorem 4.5.12 (Malle) *Let* $\mathbf{G}$ *be simple with Steinberg endomorphism* F. *Then any unipotent character of* $\mathbf{G}^F$ *extends to its inertia group in* $\mathrm{Aut}(\mathbf{G}^F)$.

For the proof note that by our standard reductions we may choose $\mathbf{G}$ to be of adjoint type. Now the outer automorphism groups of simple groups of Lie type are known (see e.g. [MaTe11, Thm. 24.24]); in particular one finds that the group $\mathrm{Out}(\mathbf{G}^F)$ is cyclic (in which case the claim follows by elementary character theory) unless possibly when F is untwisted and the Dynkin diagram of $\mathbf{G}$ has a non-trivial graph automorphism, that is, $\mathbf{G}$ is of type A_n, D_n or E_6. The latter groups can be dealt with via a case-by-case analysis.

In the general case the action of automorphisms on the irreducible characters of a finite reductive group is not yet known. Problematic here are the characters of groups with disconnected centre that fuse under a diagonal automorphism. Partial results for cuspidal characters can be found for example in [Ma17b].

4.6 Decomposition of $R_{\mathbf{L}}^{\mathbf{G}}$ and d-Harish-Chandra Series

While the decomposition of Deligne–Lusztig characters was determined completely by Lusztig in [Lu84a] — see his Decomposition Theorem 4.2.16 in conjunction with Jordan decomposition in Theorem 2.6.4 — the problem of decomposing Lusztig induction (and restriction) is not yet solved in full generality. In this section we present some partial results, which include in particular the case of unipotent characters. From this, we will deduce the decomposition for characters in arbitrary Lusztig series, at least for classical groups with connected centre, in the next section.

4.6.1 Let us start off by making some easy reductions. Let $\mathbf{G}$ be connected reductive and $F : \mathbf{G} \to \mathbf{G}$ be a Steinberg map. First, by transitivity of Lusztig induction (see Theorem 3.3.6) it is clearly sufficient to know the decomposition of $R_{\mathbf{L}}^{\mathbf{G}}$ for the *maximal* proper F-stable Levi subgroups $\mathbf{L} < \mathbf{G}$. We claim that in this case $\mathbf{L}$ is necessarily d-split for some $d \geqslant 1$. Indeed, since $\mathbf{L} = C_{\mathbf{G}}(\mathbf{Z}^\circ(\mathbf{L}))$ (see [DiMi20, Prop. 3.4.6]) and $\mathbf{L} < \mathbf{G}$ is proper, there must be some d such that the Sylow d-torus $\mathbf{S}_d \leqslant \mathbf{Z}^\circ(\mathbf{L})$ of $\mathbf{Z}^\circ(\mathbf{L})$ (see 3.5.6) is not contained in $\mathbf{Z}(\mathbf{G})$. Then by maximality of $\mathbf{L}$ we have $\mathbf{L} = C_{\mathbf{G}}(\mathbf{S}_d)$, whence $\mathbf{L}$ is d-split. Thus, for the question of decomposing $R_{\mathbf{L}}^{\mathbf{G}}$ we may assume that $\mathbf{L} < \mathbf{G}$ is a maximal d-split Levi subgroup.

Here we first consider the case of unipotent characters. Then we may further reduce to the situation when $\mathbf{G}$ is simple: Write $[\mathbf{G},\mathbf{G}] = \mathbf{G}_1 \cdots \mathbf{G}_r$ with each $\mathbf{G}_i$ an F-simple group. Then $\mathbf{L} = \mathbf{Z}^\circ(\mathbf{G})(\mathbf{L} \cap \mathbf{G}_1) \cdots (\mathbf{L} \cap \mathbf{G}_r)$ where $\mathbf{L}_i := \mathbf{L} \cap \mathbf{G}_i$ is an F-stable Levi subgroup of $\mathbf{G}_i$. Since by assumption $\mathbf{L}$ is maximal in $\mathbf{G}$ there is a unique i such that $\mathbf{L}_i \neq \mathbf{G}_i$, say $i = 1$. By Remark 4.2.1 any $\lambda \in \mathrm{Uch}(\mathbf{L}^F)$ is of the form $\lambda = \lambda_1 \boxtimes \cdots \boxtimes \lambda_r$ with $\lambda_1 \in \mathrm{Uch}(\mathbf{L}_1^F)$ and $\lambda_i \in \mathrm{Uch}(\mathbf{G}_i^F)$ for $i > 1$. Then we have $R_{\mathbf{L}}^{\mathbf{G}}(\lambda) = R_{\mathbf{L}_1}^{\mathbf{G}_1}(\lambda_1) \boxtimes \lambda_2 \boxtimes \cdots \boxtimes \lambda_r$ and thus it is sufficient to understand $R_{\mathbf{L}_1}^{\mathbf{G}_1}(\lambda_1)$. Note that here $||R_{\mathbf{L}}^{\mathbf{G}}(\lambda)|| = ||R_{\mathbf{L}_1}^{\mathbf{G}_1}(\lambda_1)||$ and

$$\pi_{\mathrm{un}}^{\mathbf{G}}(R_{\mathbf{L}}^{\mathbf{G}}(\lambda)) = \pi_{\mathrm{un}}^{\mathbf{G}_1}(R_{\mathbf{L}_1}^{\mathbf{G}_1}(\lambda_1)) \boxtimes \pi_{\mathrm{un}}^{\mathbf{G}_2}(\lambda_2) \boxtimes \cdots \boxtimes \pi_{\mathrm{un}}^{\mathbf{G}_r}(\lambda_r).$$

Now if $\mathbf{G}_1 = \prod_{j=1}^m \mathbf{G}_{1j}$ is a product of m simple groups $\mathbf{G}_{1j}$ permuted transitively by F, then we may write $\mathbf{L}_1 = \prod_{j=1}^m \mathbf{L}_{1j}$ with $\mathbf{L}_{1j} := \mathbf{L}_1 \cap \mathbf{G}_{1j}$ maximal in $\mathbf{G}_{1j}$, and by Proposition 3.3.26 Lusztig induction commutes with restriction of scalars from $\mathbf{L}_1$ to $\mathbf{L}_{11}$, and from $\mathbf{G}_1$ to $\mathbf{G}_{11}$, as does uniform projection. Thus, we may assume that $\mathbf{G}$ is simple.

As already remarked in Section 3.5 the d-Harish-Chandra theories for unipotent characters are rather well understood. In fact, the decomposition of Lusztig induction of unipotent characters from certain maximal d-split Levi subgroups has been determined explicitly. For groups of classical type this is a result of Asai (for which we present a new proof, see Theorem 4.6.9), for groups of exceptional type the decomposition was computed by Broué–Malle–Michel [BMM93], up to some small indeterminacies.

4.6.2 We first discuss the decomposition of Lusztig induction for unipotent characters in groups of type A. Here the d-split Levi subgroups were described in Example 3.5.14. Now by Proposition 3.5.12 the maximal d-split Levi subgroups of $\mathbf{G} = \mathrm{GL}_n$ above a fixed Sylow d-torus $\mathbf{S}_d$ are in bijection with the maximal parabolic subgroups of the relative Weyl group of $\mathbf{S}_d$. Thus for the split Frobenius map they are of rational type $\mathrm{GL}_m(q^d).\mathrm{GL}_{n-dm}(q)$ for some $1 \leqslant m \leqslant n/d$. We give the explicit decomposition for the case when $m = 1$, the others can then in principle be derived from that one inductively.

Let us recall a Murnaghan–Nakayama type rule for restriction of irreducible characters in certain wreath products. The complex reflection group denoted $G(b,1,m)$ is isomorphic to the wreath product of the cyclic group of order b with the symmetric group $\mathfrak{S}_m$. The irreducible characters of $G(b,1,m)$ are parametrised by b-multi-partitions $\underline{\alpha}$ of m (see e.g. [Os54, §2] or [Ma95, 2A]). Observe that for any $1 \leqslant d \leqslant m$, $G(b,1,m)$ has a parabolic subgroup of type $\mathfrak{S}_d \times G(b,1,m-d)$. By a hook of a multi-partition $\alpha = (\alpha_1, \ldots, \alpha_b)$ we mean a hook h of one of the α_i. The *leg length* $f(h)$ of h is defined to be the difference between the row indices of the Young diagram of α_i where the hook ends and where it starts.

The following generalisation of the classical Murnaghan–Nakayama rule (see e.g. [JaKe81, 2.4.7]) relating characters of symmetric groups to those of suitable Young subgroups was shown by Osima [Os54, Thm. 7].

Proposition 4.6.3 *Let $b, d \geqslant 1$, $\phi^{\underline{\alpha}} \in \mathrm{Irr}(G(b,1,m))$ be parametrised by the* *b-multipartition* $\underline{\alpha} \vdash m$. *Then for any d-cycle* $x \in \mathfrak{S}_d$ *and any* $y \in G(b,1,m-d)$

$$\phi^{\underline{\alpha}}(xy) = \sum_h (-1)^{f(h)} \phi^{\underline{\alpha}\setminus h}(y)$$

where the sum runs over all d-hooks h of $\underline{\alpha}$ *and* $\underline{\alpha} \setminus h$ *denotes the multi-partition obtained by removing h from* $\underline{\alpha}$.

From the case $b = 1$ we obtain the following result of Fong and Srinivasan [FoSr86, Thm. (2A)]:

Proposition 4.6.4 *Let* $(\mathbf{G}, F)$ *be simple of rational type* $A_{n-1}(q)$ *and let* $1 \leqslant d \leqslant n$. *Let* $\mathbf{L} = C_{\mathbf{G}}(\mathbf{T}_d)$ *where* $\mathbf{T}_d$ *is an F-stable torus of* $\mathbf{G}$ *with* $\mathbf{T}_d^F \cong \mathrm{GL}_1(q^d)$, *so* $|\mathbf{T}_d^F| = q^d - 1$. *Let* $\rho_\alpha \in \mathrm{Uch}(\mathbf{G}^F)$ *be the unipotent character labelled by the partition* $\alpha \vdash n$. *Then*

$$^*R_{\mathbf{L}}^{\mathbf{G}}(\rho_\alpha) = \sum_{h\ d\text{-hook}} (-1)^{f(h)} \rho_{\alpha\setminus h}$$

where the sum runs over all d-hooks h of α.

Proof The derived subgroup $[\mathbf{L}, \mathbf{L}]$ is simple of rational type $A_{n-d-1}(q)$. Let ρ_β be the unipotent character of $\mathbf{L}^F$ parametrised by $\beta \vdash n - d$. By Example 2.4.20 the unipotent characters of $\mathbf{G}^F$ are almost characters, so

$$\rho_\alpha = \frac{1}{|\mathbf{W}|} \sum_{w \in \mathbf{W}} \phi^\alpha(w) R_w^{\mathbf{G}} \qquad \text{and} \qquad \rho_\beta = \frac{1}{|\mathbf{W}_{\mathbf{L}}|} \sum_{v \in \mathbf{W}_{\mathbf{L}}} \phi^\beta(v) R_v^{\mathbf{L}},$$

where $\mathbf{W} = \mathfrak{S}_n$ is the Weyl group of $\mathbf{G}$, $\mathbf{W}_{\mathbf{L}} = \mathfrak{S}_{n-d}$ is the Weyl group of $\mathbf{L}$ and where $R_v^{\mathbf{L}}$ denotes the Deligne–Lusztig character of $\mathbf{L}^F$ for a maximal torus of the form $\mathbf{T}_d^F \mathbf{T}_v^F$ with $\mathbf{T}_v \leqslant [\mathbf{L}, \mathbf{L}]$ a maximal torus of type v. Thus

$$|\mathbf{W}_{\mathbf{L}}| \langle \rho_\alpha, R_{\mathbf{L}}^{\mathbf{G}}(\rho_\beta) \rangle = \frac{1}{|\mathbf{W}|} \sum_{v \in \mathbf{W}_{\mathbf{L}}} \sum_{w \in \mathbf{W}} \phi^\alpha(w) \phi^\beta(v) \langle R_w^{\mathbf{G}}, R_{\mathbf{L}}^{\mathbf{G}}(R_v^{\mathbf{L}}) \rangle. \qquad (*)$$

By transitivity, $R_{\mathbf{L}}^{\mathbf{G}}(R_v^{\mathbf{L}}) = R_{uv}^{\mathbf{G}}$ where $uv \in \mathfrak{S}_n$ is the product of $v \in \mathfrak{S}_{n-d}$ with a disjoint d-cycle $u \in \mathfrak{S}_d$. By Example 2.3.22 this has non-zero scalar product with $R_w^{\mathbf{G}}$ if and only if w and uv are conjugate in $\mathbf{W}$, so have the same cycle type. Thus $(*)$ becomes

$$\frac{1}{|\mathbf{W}|} \sum_{v \in \mathbf{W}_{\mathbf{L}}} |\mathbf{W} : C_{\mathbf{W}}(uv)| \phi^\alpha(uv) \phi^\beta(v) \langle R_w^{\mathbf{G}}, R_{uv}^{\mathbf{G}} \rangle = \sum_{v \in \mathbf{W}_{\mathbf{L}}} \phi^\alpha(uv) \phi^\beta(v).$$

By Proposition 4.6.3 applied with $b = 1$, this yields

$$\langle {}^*R_{\mathbf{L}}^{\mathbf{G}}(\rho_\alpha), \rho_\beta \rangle = \langle \rho_\alpha, R_{\mathbf{L}}^{\mathbf{G}}(\rho_\beta) \rangle = \sum_{h\ d\text{-hook}} (-1)^{f(h)} \langle \phi^{\alpha\setminus h}, \phi^\beta \rangle$$

and the only non-zero term in the sum occurs for $\alpha \setminus h = \beta$. □

The following description of d-cuspidal unipotent characters and of d-Harish-Chandra series is now immediate:

Corollary 4.6.5 *Let $(\mathbf{G}, F)$ be simple of rational type $A_{n-1}(q)$ and let $d \geqslant 1$.*

(a) *A unipotent character of $\mathbf{G}^F$ is d-cuspidal if and only if it is labelled by a d-core.*
(b) *Let $\mathbf{L} \leqslant \mathbf{G}$ be d-split with $[\mathbf{L}, \mathbf{L}]$ of rational type $A_{n-wd-1}(q)$ for some $w \geqslant 0$, and let λ be a d-cuspidal unipotent character of $\mathbf{L}^F$ labelled by the d-core $\beta \vdash n - wd$. Then the d-Harish-Chandra series $\mathscr{E}(\mathbf{G}^F, (\mathbf{L}, \lambda))$ consists of the unipotent characters of $\mathbf{G}^F$ labelled by partitions of n with d-core β.*

For the unitary groups the same formula as in Proposition 4.6.4 holds but with signs adjusted suitably: using that all unipotent characters are uniform and that the complete root datum for ${}^2A_{n-1}(q)$ is obtained by replacing F by $-F$ in the one for $A_{n-1}(q)$, we obtain the formula from that for $A_{n-1}(q)$ by formally replacing q by $-q$. Here, for a unipotent character ρ, A_ρ denotes the A-value introduced in 4.2.6.

Proposition 4.6.6 *Let $(\mathbf{G}, F)$ be of rational type ${}^2A_{n-1}(q)$ and let $d \geqslant 1$. Let $\mathbf{L} = C_{\mathbf{G}}(\mathbf{T}_d)$ where $\mathbf{T}_d$ is an F-stable torus of $\mathbf{G}$ with $|\mathbf{T}_d^F| = q^d - (-1)^d$. Let $\rho_\alpha \in \mathrm{Uch}(\mathbf{G}^F)$ be labelled by the partition $\alpha \vdash n$. Then*

$$(-1)^{A_{\rho_\alpha}} {}^*R_{\mathbf{L}}^{\mathbf{G}}(\rho_\alpha) = \sum_{h\ d\text{-hook}} (-1)^{f(h)+A_{\rho_{\alpha\setminus h}}} \rho_{\alpha\setminus h}$$

where the sum runs over the d-hooks h of α.

Note that here $\mathbf{L}$ is d'-split in $\mathbf{G}$, where

$$d' = \begin{cases} 2d & \text{if } d \text{ is odd,} \\ d/2 & \text{if } d \equiv 2 \pmod 4, \\ d & \text{if } d \equiv 0 \pmod 4) \end{cases}$$

(compare also Example 3.5.14(c)). Thus to obtain the analogue of Corollary 4.6.5 for the unitary groups one just needs to replace d-hooks by d'-hooks:

Corollary 4.6.7 *Let $(\mathbf{G}, F)$ be of rational type ${}^2A_{n-1}(q)$ and let $d \geqslant 1$.*

(a) *A unipotent character of $\mathbf{G}^F$ is d-cuspidal if and only if it is labelled by a d'-core.*
(b) *Let $\mathbf{L} \leqslant \mathbf{G}$ be d-split with $[\mathbf{L}, \mathbf{L}]$ of rational type ${}^2A_{n-wd'-1}(q)$ for some $w \geqslant 0$, and let λ be a d-cuspidal unipotent character of $\mathbf{L}^F$ labelled by the d'-core $\beta \vdash n - wd'$. Then the d-Harish-Chandra series $\mathscr{E}(\mathbf{G}^F, (\mathbf{L}, \lambda))$ consists of the unipotent characters of $\mathbf{G}^F$ labelled by partitions of n with d'-core β.*

4.6.8 We now turn to the symplectic and orthogonal groups. Let $\mathbf{G}$ be simple of classical type B, C or D and $F : \mathbf{G} \to \mathbf{G}$ a Frobenius endomorphism such that $\mathbf{G}^F$ is not of rational type ${}^3D_4$. Let d be a positive integer. We worked out the structure of d-split Levi subgroups in Example 3.5.15. Again, the maximal ones above a fixed Sylow d-torus $\mathbf{S}_d$ are in bijection with maximal parabolic subgroups in the relative Weyl group of $\mathbf{S}_d$. First, if d is odd then it is easily seen that any maximal d-split Levi subgroup has the rational form

$$\mathrm{GL}_m(q^d).\mathbf{H}^F \qquad \text{for some } m \geqslant 1,$$

where $\mathbf{H}^F$ is a classical group of the same type as $\mathbf{G}^F$ with $\mathrm{rnk}(\mathbf{H}) + dm = \mathrm{rnk}(\mathbf{G})$. If $d = 2e$ is even, the maximal d-split Levi subgroups have rational form

$$\mathrm{GU}_m(q^e).\mathbf{H}^F \qquad \text{for some } m \geqslant 1,$$

where again $\mathbf{H}$ is a classical group of the same type as $\mathbf{G}$ with $\mathrm{rnk}(\mathbf{H}) + em = \mathrm{rnk}(\mathbf{G})$. Here, however, the rational types of $\mathbf{H}^F$ and $\mathbf{G}^F$ differ if $\mathbf{G}$ is of type D_n: $\mathbf{H}^F$ is twisted if and only if $\mathbf{G}^F$ is untwisted. Now observe with Example 3.5.20 that the d-cuspidal unipotent characters of $\mathrm{GL}_m(q^d)$ are in bijection with the 1-cuspidal unipotent characters of $\mathrm{GL}_m(q)$, of which there are none except for $m = 1$ by Example 2.4.20, so the only maximal d-split Levi subgroups of $\mathbf{G}$ possibly possessing a d-cuspidal unipotent character are of the rational form

$$\mathrm{GL}_1(q^d).\mathbf{H}^F, \qquad \text{with } \mathrm{rnk}(\mathbf{H}) = \mathrm{rnk}(\mathbf{G}) - d$$

if d is odd, and

$$\mathrm{GU}_1(q^e).\mathbf{H}^F, \qquad \text{with } \mathrm{rnk}(\mathbf{H}) = \mathrm{rnk}(\mathbf{G}) - e$$

if $d = 2e$ is even, respectively. We will describe Lusztig induction from these d-split Levi subgroups.

We present the fundamental result of Asai on the decomposition of Lusztig induction as formulated in [FoSr86, (3.1), (3.2)]. This is a summary of results from [As84a, 2.8], [As84b, 1.5] and [As85, 2.2.3], see also [LM16, Thm. 3.2]. As pointed out there, a sign is missing in the formula [FoSr86, (3.2)] in the case of D_n, as can be seen for example by evaluating it at the trivial character.

Theorem 4.6.9 (Asai) *Let $\mathbf{G}$ be simple of type B_n, C_n or D_n with a Frobenius map F such that $\mathbf{G}^F$ is not of type ${}^3D_4$. For S a symbol we write ρ_S for the corresponding unipotent character if S is not degenerate, and for the sum of the two corresponding characters if S is degenerate.*

(a) *Let d be odd and $\mathbf{L} = C_{\mathbf{G}}(\mathbf{T}_d)$ where $\mathbf{T}_d \leqslant \mathbf{G}$ is an F-stable torus with*

$|\mathbf{T}_d^F| = q^d - 1$. *Then for* $\rho_S \in \mathrm{Uch}(\mathbf{L}^F)$ *we have*

$$R_{\mathbf{L}}^{\mathbf{G}}(\rho_S) = \sum_{(S',h)} \epsilon_h \, \rho_{S'},$$

where the sum runs over all symbols S' *with a* d*-hook* h *such that* $S = S' \setminus h$.

(b) *Let* $d \geqslant 1$ *and* $\mathbf{L} = C_{\mathbf{G}}(\mathbf{T}_d)$ *where* $\mathbf{T}_d \leqslant \mathbf{G}$ *is an* F*-stable torus with* $|\mathbf{T}_d^F| = q^d + 1$. *Then for* $\rho_S \in \mathrm{Uch}(\mathbf{L}^F)$ *we have*

$$R_{\mathbf{L}}^{\mathbf{G}}(\rho_S) = (-1)^{\delta} \sum_{(S',c)} \epsilon_c \, \rho_{S'},$$

where the sum runs over all symbols S' *with a* d*-cohook* c *such that* $S = S' \setminus c$, *and* $\delta = 0$ *for types* B_n, C_n, $\delta = 1$ *for type* D_n.

Note that in the theorem $[\mathbf{L}, \mathbf{L}]^F$ is a classical group of rank $n - d$ of the same rational type as $\mathbf{G}^F$, except that removing cohooks changes the defect of symbols modulo 4, so in type D_n interchanges the twisted and untwisted types.

We give a purely combinatorial proof of Asai's formula, only using the Mackey formula for unipotent characters for part (b).

First, let us observe that part (a) follows just from ordinary Harish-Chandra theory. For this let $\mathbf{G}$ be as in Theorem 4.6.9 and $d \geqslant 1$ odd. Consider the d-split Levi subgroup $\mathbf{L} = C_{\mathbf{G}}(\mathbf{T}) = \mathbf{T}\mathbf{G}'$ of $\mathbf{G}$ where $\mathbf{T}$ is an F-stable torus of $\mathbf{G}$ with $\mathbf{T}^F \cong \mathrm{GL}_1(q^d)$ and $\mathbf{G}'$ is of the same type as $\mathbf{G}$ but of rank $n - d$. Let $\mathbf{L}_1 = \mathrm{GL}_d\mathbf{G}'$ be the intermediate 1-split Levi subgroup of $\mathbf{G}$ with $\mathbf{T} \leqslant \mathrm{GL}_d$. By transitivity of Lusztig restriction, ${}^*R_{\mathbf{L}}^{\mathbf{G}}(\rho) = {}^*R_{\mathbf{L}}^{\mathbf{L}_1} \, {}^*R_{\mathbf{L}_1}^{\mathbf{G}}(\rho)$ for any class function ρ on $\mathbf{G}^F$.

Proof of Theorem 4.6.9(a) (See [As84a, Lemma 2.8.4].) Let $\rho \in \mathrm{Uch}(\mathbf{G}^F)$ be labelled by the symbol S. Let $(\mathbf{L}_0, \lambda_0)$ be a Harish-Chandra source of ρ, that is, $\mathbf{L}_0$ is a 1-split Levi subgroup of $\mathbf{G}$ and $\lambda_0 \in \mathrm{Uch}(\mathbf{L}_0^F)$ is cuspidal with $\langle \rho, R_{\mathbf{L}_0}^{\mathbf{G}}(\lambda_0) \rangle \neq 0$. By Theorem 4.4.28 then $[\mathbf{L}_0, \mathbf{L}_0]$ is simple of the same classical type as $\mathbf{G}$. If $\mathrm{rnk}([\mathbf{L}_0, \mathbf{L}_0]) > n - d$ then fewer than d 1-hooks can be removed successively from S, and hence S cannot have a d-hook. Moreover, $\mathbf{L}_0$ cannot be contained in $\mathbf{L}$, so ${}^*R_{\mathbf{L}}^{\mathbf{G}}(\rho) = 0$, in accordance with the claim.

Else, we may assume after conjugation that $\mathbf{L}_0$ is contained in the maximal 1-split Levi subgroup $\mathbf{L}_1$. Let S_0 denote the symbol labelling λ_0 and first assume that S_0 is non-degenerate. Let $W_0 := W_{\mathbf{G}}(\mathbf{L}_0, \lambda_0) \cong W(B_{n-k})$ be the relative Weyl group of $(\mathbf{L}_0, \lambda_0)$ in $\mathbf{G}$, and $W_1 := W_{\mathbf{L}_1}(\mathbf{L}_0, \lambda_0) \cong \mathfrak{S}_d \times W(B_{n-k-d})$ the one in $\mathbf{L}_1$ (see Example 3.5.29). The irreducible characters of W_0 are parametrised by bipartitions $\underline{\alpha} \vdash n - k$. By Proposition 4.4.29 addition of S_0 defines the natural labelling

$$\mathrm{Irr}(W_0) \longrightarrow \mathscr{E}(\mathbf{G}^F, (\mathbf{L}_0, \lambda_0)), \quad \phi^{\underline{\alpha}} \mapsto \rho_{\underline{\alpha}},$$

of the Harish-Chandra series above λ_0, with $\rho_{\underline{\alpha}}$ denoting the unipotent character labelled by the symbol $S_0 + \underline{\alpha}$. Any $\psi \in \mathrm{Irr}(W_1)$ is of the form $\psi_1 \boxtimes \psi_2$ with $\psi_1 \in \mathrm{Irr}(\mathfrak{S}_d)$ and $\psi_2 \in \mathrm{Irr}(G(2,1,n-k-d))$, and correspondingly we obtain a parametrisation $\mathrm{Irr}(W_1) \longrightarrow \mathscr{E}(\mathbf{L}_1^F, (\mathbf{L}_0, \lambda_0))$, $\psi_1 \boxtimes \psi_2 \mapsto \rho_{\psi_1} \boxtimes \rho_{\underline{\alpha}}$, if ψ_2 is labelled by $\underline{\alpha} \vdash n-k-d$. Observe that addition of S_0 commutes with adding or removing d-hooks.

Let ρ correspond to the character $\phi \in \mathrm{Irr}(W_0)$. By the Howlett–Lehrer Comparison Theorem 3.2.7 the decomposition of ${}^*R_{\mathbf{L}_1}^{\mathbf{G}}(\rho)$ is given by

$$ {}^*R_{\mathbf{L}_1}^{\mathbf{G}}(\rho) = \sum_{\psi} a_\psi \rho_\psi \quad \text{where} \quad \phi|_{W_1} = \sum_{\psi \in \mathrm{Irr}(W_1)} a_\psi \psi. $$

Now ${}^*R_{\mathbf{L}}^{\mathbf{L}_1}(\rho_{\psi_1} \boxtimes \rho_{\psi_2}) = {}^*R_{\mathbf{T}}^{\mathrm{GL}_d}(\rho_{\psi_1}) \boxtimes \rho_{\psi_2}$, and ${}^*R_{\mathbf{T}}^{\mathrm{GL}_d}(\rho_{\psi_1})$ is non-zero if and only if ψ_1 does not vanish on the d-cycles, that is, if and only if it is parametrised by a hook partition. Thus we may set $a_\psi = 0$ unless $\psi_1(x) \neq 0$ for $x \in \mathfrak{S}_d$ a d-cycle. But then Proposition 4.6.3 applied to $W_0 \cong G(2,1,n-k)$ shows that $a_\psi = (-1)^{f(h)}$ if ψ_2 is obtained from ϕ by removing the d-hook h, and $a_\psi = 0$ else. Since removing d-hooks commutes with the Harish-Chandra parametrisation, our claim follows.

If S_0 is degenerate, then $\mathbf{L}_0$ is of type D_k and the relative Weyl groups $W_0 \cong W(D_{n-k})$, $W_1 \cong \mathfrak{S}_d \times W(D_{n-k-d})$ are normal subgroups of index 2 in the relative Weyl groups in the non-degenerate case. As described in Example 4.1.4 their irreducible characters are now labelled by unordered bipartitions, and an easy Clifford theory argument shows that again the constituents of ${}^*R_{\mathbf{L}}^{\mathbf{G}}(\rho)$ are as claimed. □

For the second part of Theorem 4.6.9 we establish two combinatorial statements. The first requires the Mackey formula. For simplicity we will assume that we are in the situation of part (b) of the theorem, but the arguments similarly apply to part (a).

Let $d \geqslant 1$. We write $a_d(S)$ for the number of different d-cohooks that can be added to a symbol S, and $r_d(S)$ for the number of d-cohooks that can be removed from S.

We can reinterpret these numbers in terms of abacus diagrams. As in [Ma95, Bem. 3.4] the two rows of $S = (X_0, X_1)$ can naturally be encoded in a $2d$-runner abacus diagram A as follows: the ith runner of A has a bead at position j if $X_{i+j \pmod 2}$ has an entry $\lfloor i/2 \rfloor + dj$. Then adding a d-cohook to S corresponds to moving one bead in A one position down. Thus the number $a_d(S)$ of addable d-cohooks of S equals the number of beads in A such that the next position downwards is empty. Clearly, the lowest bead on any runner can always be moved down, which already gives $2d$ possibilities. All other addable d-cohooks correspond to an empty position on some runner such that some later position is occupied. Thus, they are in natural bijection with the removable d-cohooks of S. This shows that

$$ a_d(S) = 2d + r_d(S). $$

The above considerations hold as long as the symbol S is not degenerate. If it is, then we only obtain half that many distinct symbols by adding a hook (or cohook), and also only half that many by removing one. (Note that neither operation can again result in a degenerate symbol.)

Proposition 4.6.10 *Let* $\mathbf{L} \leqslant \mathbf{G}$ *be a* $2d$*-split Levi subgroup as in Theorem 4.6.9(b) and* $\rho \in \mathrm{Uch}(\mathbf{L}^F)$ *parametrised by the symbol* S*. Then* $\|R_{\mathbf{L}}^{\mathbf{G}}(\rho)\|^2 = a_d(S)$ *is as claimed by Asai's formula.*

Proof First assume $\mathbf{G}$ is of type B_n or C_n. We apply the Mackey formula for d-split Levi subgroups in Theorem 3.5.17. Let $\mathbf{M}$ be a minimal d-split Levi subgroup contained in $\mathbf{L}$. Then we have

$$\langle R_{\mathbf{L}}^{\mathbf{G}}(\rho), R_{\mathbf{L}}^{\mathbf{G}}(\rho)\rangle = \langle \rho, {}^*R_{\mathbf{L}}^{\mathbf{G}}(R_{\mathbf{L}}^{\mathbf{G}}(\rho))\rangle = \langle \rho, \sum_{w} R_{\mathbf{L}\cap {}^w\mathbf{L}}^{\mathbf{L}}({}^*R_{\mathbf{L}\cap {}^w\mathbf{L}}^{{}^w\mathbf{L}}({}^w\rho))\rangle$$

where the sum is over $W_{\mathbf{L}}(\mathbf{M})$–$W_{\mathbf{L}}(\mathbf{M})$ double coset representatives w in $W_{\mathbf{G}}(\mathbf{M})$. Here, $W_{\mathbf{G}}(\mathbf{M})$ is the complex reflection group $G(2d, 1, m)$, where $dm = \mathrm{rnk}(\mathbf{G}) - \mathrm{rnk}(\mathbf{M})$, and $W_{\mathbf{L}}(\mathbf{M})$ is its maximal parabolic subgroup $G(2d, 1, m-1)$ (see Example 3.5.15). Since $\mathbf{L}$ is proper in $\mathbf{G}$ we have $m \geqslant 1$. If $m = 1$ then $G(2d, 1, 1)$ is cyclic of order $2d$ and there are $2d$ double coset representatives for the trivial group in $G(2d, 1, 1)$, each with ${}^w\mathbf{L} = \mathbf{L}$, leading to $\langle R_{\mathbf{L}}^{\mathbf{G}}(\rho), R_{\mathbf{L}}^{\mathbf{G}}(\rho)\rangle = 2d$.

Now assume that $m \geqslant 2$. It is easy to see that there are exactly $2d+1$ double coset representatives $w_0, \ldots, w_{2d}$, with $W_{\mathbf{L}}(\mathbf{M}) \cap {}^{w_i}W_{\mathbf{L}}(\mathbf{M}) = W_{\mathbf{L}}(\mathbf{M})$ for $i = 1, \ldots, 2d$ and $W_{\mathbf{L}}(\mathbf{M}) \cap {}^{w_0}W_{\mathbf{L}}(\mathbf{M}) = G(2d, 1, m-2)$, so $\mathbf{L}_0 := \mathbf{L} \cap {}^{w_0}\mathbf{L}$ is a maximal $2d$-split Levi subgroup of $\mathbf{L}$. Thus we find

$$\langle R_{\mathbf{L}}^{\mathbf{G}}(\rho), R_{\mathbf{L}}^{\mathbf{G}}(\rho)\rangle = 2d\langle \rho, \rho\rangle + \langle \rho, R_{\mathbf{L}_0}^{\mathbf{L}}({}^*R_{\mathbf{L}_0}^{\mathbf{L}}(\rho))\rangle.$$

By induction, $\|{}^*R_{\mathbf{L}_0}^{\mathbf{L}}(\rho)\|^2 = r_d(S)$ and so $\|R_{\mathbf{L}}^{\mathbf{G}}(\rho)\|^2 = 2d + r_d(S)$, which is our claim by the considerations prior to this proposition.

The above arguments remain valid for $\mathbf{G}$ of type D_n, except when ρ lies in the $2d$-Harish-Chandra series above a $2d$-cuspidal unipotent character parametrised by a degenerate symbol. (For this observe that if S is degenerate, then so is its d-cocore.) In this case $W_{\mathbf{G}}(\mathbf{M})$ is the complex reflection group $G(2d, 2, m)$, and $W_{\mathbf{L}}(\mathbf{M})$ is its maximal parabolic subgroup $G(2d, 2, m-1)$. If $m = 1$, then $G(2d, 2, 1)$ is cyclic of order d and there are d double coset representatives for the trivial group, leading to $\langle R_{\mathbf{L}}^{\mathbf{G}}(\rho), R_{\mathbf{L}}^{\mathbf{G}}(\rho)\rangle = d$. If $m \geqslant 2$ then it is easy to see that again there are exactly $2d + 1$ double coset representatives $w_0, \ldots, w_{2d}$, all but w_0 normalising $\mathbf{L}$, while $W_{\mathbf{L}}(\mathbf{M}) \cap {}^{w_0}W_{\mathbf{L}}(\mathbf{M}) = G(2d, 2, m-2)$. Arguing as before we again find $\|R_{\mathbf{L}}^{\mathbf{G}}(\rho)\|^2 = 2d + r_d(S)$. □

We consider the $2d$-split Levi subgroup $\mathbf{L} = \mathbf{T}\mathbf{G}' = C_{\mathbf{G}}(\mathbf{T})$ of $\mathbf{G}$ where $\mathbf{T}$ is an F-stable torus with $\mathbf{T}^F \cong \mathrm{GU}_1(q^d)$ and $\mathbf{G}'$ is of the same type as $\mathbf{G}$ but of rank

$n-d$. Let $\mathbf{L}_1 = \mathrm{GL}_d\mathbf{G}'$ be the intermediate F-stable Levi subgroup of $\mathbf{G}$ with $\mathbf{T} \leqslant \mathrm{GL}_d$ and $\mathrm{GL}_d^F = \mathrm{GU}_d(q)$. We first determine Lusztig restriction on uniform almost characters; for this let $\mathbf{W}$ denote the Weyl group of $\mathbf{G}$.

Lemma 4.6.11 *In the above setting, let $\phi \in \mathrm{Irr}(\mathbf{W})^F$ be parametrised by $\underline{\alpha}$ and R_ϕ be the associated unipotent almost character of $\mathbf{G}^F$. Then*

$$^*R_\mathbf{L}^\mathbf{G}(R_\phi) = \sum_h (-1)^{f(h)} R_{\phi^{\underline{\alpha}\setminus h}},$$

where the sum runs over all d-hooks h of $\underline{\alpha}$.

Proof The Levi subgroup $\mathbf{L}_1$ has Weyl group $\mathbf{W}_1 \cong \mathfrak{S}_d \times G(2,1,n-d)$ and $\mathbf{L}$ has Weyl group $\mathbf{W}_\mathbf{L} \cong G(2,1,n-d)$. Then $^*R_{\mathbf{L}_1}^\mathbf{G}(R_\phi) = R_{\phi_1}$, with $\phi_1 = \phi|_{\mathbf{W}_1}$, and by transitivity of Lusztig induction

$$^*R_\mathbf{L}^\mathbf{G}(R_\phi) = {}^*R_\mathbf{L}^{\mathbf{L}_1}\,{}^*R_{\mathbf{L}_1}^\mathbf{G}(R_\phi) = {}^*R_\mathbf{L}^{\mathbf{L}_1}(R_{\phi_1}) = R_{\phi'},$$

with ϕ' the restriction of ϕ_1 to the product $C \times \mathbf{W}_1$, with C the class of d-cycles in $\mathfrak{S}_d$. Applying Proposition 4.6.3 we thus find

$$\phi' = \sum_h (-1)^{f(h)} \phi^{\underline{\alpha}\setminus h},$$

where the sum runs over all d-hooks h of $\underline{\alpha}$, from which the claim follows. □

Proposition 4.6.12 *The uniform projection of Theorem 4.6.9(b) holds.*

Proof Let $\rho \in \mathrm{Uch}(\mathbf{G}^F)$ be parametrised by the symbol S. We write $T \sim S$ if T is the symbol of some $\phi \in \mathrm{Irr}(\mathbf{W})$ lying in the family of ρ. By Proposition 3.3.10 uniform projection commutes with Lusztig restriction, so

$$\pi_{\mathrm{un}}({}^*R_\mathbf{L}^\mathbf{G}(\rho)) = {}^*R_\mathbf{L}^\mathbf{G}(\pi_{\mathrm{un}}(\rho)) = \sum_{T\sim S} \langle S, T\rangle\, {}^*R_\mathbf{L}^\mathbf{G}(R_T),$$

where R_T is the uniform almost character labelled by T and $\langle S, T\rangle$ is the Fourier coefficient. By Lemma 4.6.11 this equals

$$\sum_{T\sim S} \langle S, T\rangle \sum_h (-1)^{f(h)} R_{T\setminus h} \tag{1}$$

with the inner sum ranging over all d-hooks h of T. This should equal $(-1)^\delta$ times

$$\pi_{\mathrm{un}}\Big(\sum_c \epsilon_c \rho_{S\setminus c}\Big) = \sum_c \epsilon_c \pi_{\mathrm{un}}(\rho_{S\setminus c}) = \sum_c \epsilon_c \sum_{T\sim S\setminus c} \langle S\setminus c, T\rangle R_T, \tag{2}$$

where the outer sum runs over d-cohooks c of S. In order to prove this, we will compare coefficients according to the d-cohooks of S at an entry x of S.

Let us first assume that $x - d$ is not an entry of S. Then any symbol $T \sim S$ also has a d-hook h at x, and we need to show that

$$(-1)^{f(h)}\langle S, T\rangle = (-1)^{\delta}\epsilon_c\langle S \setminus c, T \setminus h\rangle$$

for all $T \sim S$. This is a combinatorial exercise using 4.4.22.

Now assume that $x - d$ is an entry of S in the opposite row to x. Then there is no d-cohook in S at x and (2) gives no contribution. We get a contribution to (1) for all $T \sim S$ for which $x, x - d$ lie in different rows. But these come in pairs, according to whether x is in the first row or the second row, and the corresponding terms in the sum, having opposite sign, cancel.

Finally, assume that $x - d$ lies in the same row of S as x. Then the contribution to (2) is the sum over all $T \sim S \setminus c$. Note that here the size of the family of $S \setminus c$ is smaller than that of S, and the coefficients in the Fourier matrix have twice the absolute value. We get a contribution to (1) for all $T \sim S$ that have $x, x - d$ in distinct rows. Again, these come in pairs, and this time the two symbols give contributions with the same sign, thus matching with the doubled Fourier coefficients in (2). The proof is complete. □

Based on Propositions 4.6.10 and 4.6.12 we now prove the second part of Asai's formula. This is inspired by Enguehard [En13, Lemma 5.3.9], but note that the proof given there contains a serious gap, which is here amended by Proposition 4.4.26.

Proof of Theorem 4.6.9(b) It suffices to compare the projection of $R_{\mathbf{L}}^{\mathbf{G}}(\rho)$ and of Asai's formula to any family $\mathscr{U} \subset \mathrm{Uch}(\mathbf{G}^F)$. By Proposition 4.6.12 the uniform projection of $R_{\mathbf{L}}^{\mathbf{G}}(\rho)$ and of Asai's formula agree. In Asai's formula, the symbols parametrising the constituents of $R_{\mathbf{L}}^{\mathbf{G}}(\rho)$ are obtained from the symbol of ρ by adding a d-cohook, that is, by increasing one of the entries by d and moving it to the other row. It is obvious that at most two of the resulting symbols can share the same multi-set of entries, and in the latter case, this multi-set has more distinct elements than the one for ρ, so its associated family in $\mathrm{Uch}(\mathbf{G}^F)$ is bigger.

On the other hand, since all unipotent characters have the same positive multiplicity in the special uniform almost character of a family, we can read off from the uniform projection whether the number of constituents of $R_{\mathbf{L}}^{\mathbf{G}}(\rho)$ is one character or the sum or the difference of two of them, and the result will be the same as for Asai's formula. Now by Proposition 4.6.10 the norm of $R_{\mathbf{L}}^{\mathbf{G}}(\rho)$ agrees with the one in the formula of Asai, so we conclude that the projection to any family of either have the same number of constituents, either zero, one or two.

Thus, the projection f of $R_{\mathbf{L}}^{\mathbf{G}}(\rho)$ onto $\mathscr{U}$ must in fact be a linear combination of at most two unipotent characters. We are thus left to show that f is already determined by its uniform projection $\pi_{\mathrm{un}}^{\mathbf{G}}(f)$ and its norm being $\leqslant 2$. If f has norm one then by Theorem 4.4.23 there is exactly one unipotent character having this uniform projection, and we are done. If f has norm 2 and $\mathscr{U}$ has more than four elements, our claim follows by Proposition 4.4.26. Finally assume that $|\mathscr{U}| = 4$. Then by what we said above, ρ lies in a 1-element family of $\mathrm{Uch}(\mathbf{L}^F)$ and hence is uniform, whence so is $R_{\mathbf{L}}^{\mathbf{G}}(\rho)$. In particular it is orthogonal to the space of non-uniform functions and we conclude again. □

Let us call *(d-)Asai Levi subgroup* any F-stable Levi subgroup of a simple group of classical type with rational form as occurring in the statement of Theorem 4.6.9. As a consequence of the above proof we see (this can be shown by the same methods to also hold for case (a) of Theorem 4.6.9):

Corollary 4.6.13 *Let $\mathbf{L}$ be an Asai Levi subgroup of a simple group $\mathbf{G}$ of classical type with a Frobenius map F and $\rho \in \mathrm{Uch}(\mathbf{L}^F)$. Assume that*

(1) *the uniform projection of $R_{\mathbf{L}}^{\mathbf{G}}(\rho)$, and*
(2) *the norm of $R_{\mathbf{L}}^{\mathbf{G}}(\rho)$*

are as in Asai's formula. Then $R_{\mathbf{L}}^{\mathbf{G}}(\rho)$ is as given by Theorem 4.6.9.

Remark 4.6.14 The result of Corollary 4.6.13 can be reformulated to say: if $\mathbf{L} < \mathbf{G}$ is an Asai Levi subgroup and $\rho \in \mathrm{Uch}(\mathbf{L}^F)$, then $R_{\mathbf{L}}^{\mathbf{G}}(\rho)$ is uniquely determined as the virtual character of minimal norm with the same uniform projection as $R_{\mathbf{L}}^{\mathbf{G}}(\rho)$. This conclusion continues to hold for any Levi subgroup of $\mathbf{G}$ of type A, as there all unipotent characters are uniform, but also for $\mathbf{G}^F = {}^3D_4(q)$. Indeed, in this case there are five maximal Levi subgroups up to conjugation, of rational types $A_1(q^3)(q \pm 1)$, $A_2(q)(q^2 + q + 1)$, ${}^2A_2(q)(q^2 - q + 1)$, and the Coxeter torus. All of them are of type A, so all of their unipotent characters are uniform and thus their Lusztig induction can be computed explicitly as linear combinations of Deligne–Lusztig characters. It turns out that in all cases $R_{\mathbf{L}}^{\mathbf{G}}(\rho)$ is multiplicity-free, has at most two constituents in the 4-element family of $\mathbf{G}^F$, and the 1-dimensional space of functions in that family with zero uniform projection is orthogonal to all of them. Thus indeed $R_{\mathbf{L}}^{\mathbf{G}}(\rho)$ has minimal norm among all virtual characters with the same uniform projection.

Theorem 4.6.9 provides a recursive algorithm to compute the decomposition of $R_{\mathbf{L}}^{\mathbf{G}}$ for any F-stable Levi subgroup $\mathbf{L}$ of a group $\mathbf{G}$ of classical type:

Proposition 4.6.15 *Let $\mathbf{G}$ be simple of classical type with a Frobenius map F and $\mathbf{L} \leqslant \mathbf{G}$ be an F-stable Levi subgroup. Then for any $\rho \in \mathrm{Uch}(\mathbf{L}^F)$ the decomposition of $R_{\mathbf{L}}^{\mathbf{G}}(\rho)$ can be computed from Theorem 4.6.9.*

Proof For the case when $\mathbf{G}^F = {}^3D_4(q)$ see Remark 4.6.14. Else, by the reduction laid out in 4.6.1 it is sufficient to consider maximal d-split Levi subgroups $\mathbf{L}$ of $\mathbf{G}$ for the various relevant d. By 4.6.8 these have rational type $\mathbf{L}^F = \mathrm{GL}_m(\pm q^d).\mathbf{H}^F$

for some $1 \leqslant m \leqslant n/d$, where n denotes the rank of $\mathbf{G}$, $\mathbf{H}$ is simple of the same classical type as $\mathbf{G}$ and where we write $\mathrm{GL}_m(-q^d) := \mathrm{GU}_m(q^d)$.

Let $\rho \in \mathrm{Uch}(\mathbf{L}^F)$. Then $\rho = \rho_1 \boxtimes \rho_2$ with $\rho_1 \in \mathrm{Uch}(\mathrm{GL}_m(\pm q^d))$ and $\rho_2 \in \mathrm{Uch}(\mathbf{H}^F)$. Now any unipotent character of $\mathrm{GL}_m(\pm q^d)$ is uniform, so can be written as an (explicitly known) linear combination of Deligne–Lusztig characters for the various maximal tori of $\mathrm{GL}_m(\pm q^d)$. By transitivity of Lusztig induction, to determine $R_{\mathbf{L}}^{\mathbf{G}}(\rho)$ it is hence sufficient to decompose Lusztig induction from F-stable Levi subgroups of $\mathbf{G}$ of the form $\mathbf{T}.\mathbf{H}$, with $\mathbf{T}^F$ a maximal torus of $\mathrm{GL}_m(\pm q^d)$. By transitivity this can be obtained as a sequence of Lusztig inductions from d-Asai Levi subgroups for various d. The decomposition of the latter is known by Theorem 4.6.9. □

As remarked in 4.6.8, d-Asai Levi subgroups are the only maximal d-split Levi subgroups of simple groups of classical type possibly possessing d-cuspidal unipotent characters, so we obtain from Theorem 4.6.9:

Corollary 4.6.16 *Let $\mathbf{G}$ be simple of classical type with a Frobenius map F such that $(\mathbf{G}, F)$ is not of type ${}^3D_4$, and let $d \geqslant 1$.*

(a) *A unipotent character of $\mathbf{G}^F$ parametrised by a symbol S is d-cuspidal if and only if S is a d-core if d is odd, or a $d/2$-cocore if d is even, respectively.*
(b) *Let $\mathbf{L} \leqslant \mathbf{G}$ be a d-split Levi subgroup, and let $\lambda \in \mathrm{Uch}(\mathbf{L}^F)$ be a d-cuspidal unipotent character labelled by the d-core (respectively $d/2$-cocore) S. Then the d-Harish-Chandra series $\mathscr{E}(\mathbf{G}^F, (\mathbf{L}, \lambda))$ consists of the unipotent characters of $\mathbf{G}^F$ labelled by symbols with d-core (respectively $d/2$-cocore) S.*

In particular, in contrast to the situation when $d = 1$ (considered in Theorem 4.4.28) a classical group will in general have many d-cuspidal unipotent characters. Note the similarity of this result with Corollary 4.4.18: thus, d-cuspidal unipotent characters are exactly those unipotent characters which are of ℓ-defect zero for any prime $\ell \geqslant 3$ such that q has order d modulo ℓ. This fact is a crucial ingredient in the classification of the unipotent ℓ-blocks of the finite classical groups, see Fong–Srinivasan [FoSr86, FoSr89], Broué–Malle–Michel [BMM93], and Cabanes–Enguehard [CE94].

4.6.17 Considerations as in the proof of Theorem 4.6.9 have also been used in [BMM93] to determine the decomposition of $R_{\mathbf{L}}^{\mathbf{G}}$ for d-split Levi subgroups $\mathbf{L}$ of simple groups $\mathbf{G}$ of exceptional type, in conjunction with the Mackey formula Theorem 3.3.7 for unipotent characters. We will not give this decomposition explicitly, it can be found in [BMM93, Tab. 2] (up to some small indeterminacies which we will comment on in Remark 4.6.19). A conceptual description will be given in Theorem 4.6.21. The proof is based upon the following analogue of Corollary 4.6.13:

Proposition 4.6.18 *Let* $\mathbf{G}$ *be simple of exceptional type with a Steinberg morphism* F, *let* $\mathbf{L}$ *be a maximal* F*-stable Levi subgroup of* $\mathbf{G}$ *and* $\lambda \in \mathrm{Uch}(\mathbf{L}^F)$. *Then* $R_{\mathbf{L}}^{\mathbf{G}}(\lambda)$ *is uniquely determined by its uniform projection and its norm, up to ambiguities in* $\mathbf{G}$ *of type* E_7 *or* E_8 *arising from pairs of unipotent characters of* $\mathbf{L}$ *having the same uniform projection.*

The latter only occur when $\mathbf{L} < \mathbf{G}$ *has a component of type* E_6 *or* E_7 *and* λ *involves a unipotent character with non-rational Frobenius eigenvalue, or a principal series character in the exceptional family of* E_7.

Sketch of proof We start off by making some reductions. As pointed out in 4.6.1 a maximal F-stable Levi subgroup is necessarily d-split for some $d \geqslant 1$. If λ is uniform, then so is $R_{\mathbf{L}}^{\mathbf{G}}(\lambda)$, so certainly it is determined by its uniform projection and norm. In particular, this is the case if all simple factors of $\mathbf{L}$ are of type A. This already deals with the groups $\mathbf{G}^F$ of types ${}^2B_2$, ${}^2G_2$, G_2 and ${}^3D_4$.

Assume that $\mathbf{G}$ is of type F_4 and F is a Frobenius map. By Table 3.3 the only maximal d-split Levi subgroups with not all factors of type A are of type B_3 and C_3 (with $d = 1, 2$) and of type B_2 (with $d = 4$). Using that the families define an orthogonal decomposition of the space spanned by unipotent characters, one concludes that for any of these the uniform projection of $R_{\mathbf{L}}^{\mathbf{G}}(\lambda)$, with $\lambda \in \mathrm{Uch}(\mathbf{L}^F)$ not uniform, to a 4-element family in $\mathrm{Uch}(\mathbf{G}^F)$ agrees with that of a single unipotent character and thus is determined by its norm being 1. The uniform projection restricted to the 21-element family $\mathscr{U} \subseteq \mathrm{Uch}(\mathbf{G}^F)$ associated to $\mathscr{G}_{\mathscr{U}} \cong \mathfrak{S}_4$ agrees with a linear combination of at most 5 unipotent characters. From the explicit Fourier matrix it can then be checked that this, together with the norm, determines $R_{\mathbf{L}}^{\mathbf{G}}(\lambda)$: The pair of cuspidal unipotent characters of $F_4(q)$ that have the primitive fourth roots of unity as associated Frobenius eigenvalues (so are algebraically conjugate by Proposition 4.5.5) possess the same uniform projection. From the norm and the uniform projection of $R_{\mathbf{L}}^{\mathbf{G}}(\lambda)$ it follows that $R_{\mathbf{L}}^{\mathbf{G}}(\lambda)$ must contain both characters with the same multiplicity.

In type E_6 the only relevant d-split Levi subgroups are of type ${}^dD_4$ with $d = 1, 2, 3$ and ${}^dD_5$ with $d = 1, 2$. Again, it turns out that the projection of $R_{\mathbf{L}}^{\mathbf{G}}(\lambda)$ to 4-element families with λ not uniform agrees with that of a single unipotent character, while on the 8-element family $\mathscr{U} \subseteq \mathrm{Uch}(\mathbf{G}^F)$ associated to $\mathscr{G}_{\mathscr{U}} \cong \mathfrak{S}_3$ there are at most 4 constituents. Again this determines $R_{\mathbf{L}}^{\mathbf{G}}(\lambda)$ uniquely, up to the individual multiplicities of the two cuspidal unipotent characters of $E_6(q)$ which have the primitive third roots of unity as associated Frobenius eigenvalues. The norm condition shows that they must have the same multiplicity in all $R_{\mathbf{L}}^{\mathbf{G}}(\lambda)$.

The arguments for the other types are analogous. The only cases that can not be settled completely are those configurations mentioned in the statement. □

Remark 4.6.19 (a) There is another way to resolve certain ambiguities about multiplicities of characters with the same uniform projection. If these characters have non-rational Frobenius eigenvalues, hence are Galois conjugate by Proposition 4.5.5, one can argue as follows: assume that $\lambda \in \mathrm{Uch}(\mathbf{L}^F)$ is rational valued; then so is $R_{\mathbf{L}}^{\mathbf{G}}(\lambda)$ by Corollary 3.3.14. Hence the multiplicities in $R_{\mathbf{L}}^{\mathbf{G}}(\lambda)$ of all unipotent characters of $\mathbf{G}^F$ from a fixed Galois orbit must agree. This applies for example in groups $\mathbf{G}$ of type F_4 or E_6, showing that the multiplicities in all $R_{\mathbf{L}}^{\mathbf{G}}(\lambda)$ of the cuspidal unipotent characters with non-rational Frobenius eigenvalues are the same in each pair.

(b) This approach fails, though, when we consider Lusztig induction of the two cuspidal unipotent characters $\lambda_j = {}^2E_6[\zeta_3^j]$, $j = 1, 2$, of ${}^2E_6(q)$, where $\zeta_3 = \exp(2\pi\sqrt{-1}/3)$. These are non-rational, with character field $\mathbb{Q}(\lambda_j) = \mathbb{Q}(\zeta_3)$ (see Proposition 4.5.5). Since λ_1, λ_2 are not conjugate under any group automorphism of ${}^2E_6(q)$ by Theorem 4.5.11, $R_{\mathbf{L}}^{\mathbf{G}}(\lambda_1)$ and $R_{\mathbf{L}}^{\mathbf{G}}(\lambda_1)^\sigma = R_{\mathbf{L}}^{\mathbf{G}}(\lambda_1^\sigma) = R_{\mathbf{L}}^{\mathbf{G}}(\lambda_2)$ differ for any group $\mathbf{G}$ containing a Levi subgroup $\mathbf{L}$ of rational type ${}^2E_6$, where σ is the non-trivial Galois automorphism of $\mathbb{Q}(\zeta_3)/\mathbb{Q}$, that is, they are also non-rational. For $\mathbf{G}^F = E_7(q)$ and $\mathbf{L}$ of rational type ${}^2E_6(q).\Phi_2$ the uniform projections show that $R_{\mathbf{L}}^{\mathbf{G}}(\lambda_1) + R_{\mathbf{L}}^{\mathbf{G}}(\lambda_2)$ contains both pairs of non-rational unipotent characters of $E_7(q)$, but this does not allow us to deduce their subdivision among $R_{\mathbf{L}}^{\mathbf{G}}(\lambda_1)$ and $R_{\mathbf{L}}^{\mathbf{G}}(\lambda_2)$.

(c) In the latter case, the decomposition of $R_{\mathbf{L}}^{\mathbf{G}}(\lambda_i)$ can still be determined by a block-theoretic argument, namely we have

$$R_{\mathbf{L}}^{\mathbf{G}}({}^2E_6[\zeta_3^j]) = E_6[\zeta_3^j], 1 - E_6[\zeta_3^j], \epsilon \quad \text{for } j = 1, 2$$

(see [KeMa19]). Similarly one can argue that for $\mathbf{L}$ of rational type $E_7(q).\Phi_2$ in $\mathbf{G} = E_8$ we have

$$R_{\mathbf{L}}^{\mathbf{G}}(\phi_{512,11}) = \phi_{4096,11} - \phi_{4096,26} \quad \text{and} \quad R_{\mathbf{L}}^{\mathbf{G}}(\phi_{512,12}) = \phi_{4096,12} - \phi_{4096,27}$$

for the two principal series unipotent characters $\phi_{512,11}, \phi_{512,12}$ lying in the exceptional family of $E_7(q)$.

Among d-cuspidal unipotent pairs $(\mathbf{L}, \lambda)$ of simple groups this leaves only one open case, viz. the decomposition of $R_{\mathbf{L}}^{\mathbf{G}}({}^2E_6[\zeta_3^j])$, $j = 1, 2$, for $\mathbf{L}$ of rational type ${}^2E_6(q).\Phi_2^2$ in $\mathbf{G} = E_8$, see case "40+41" in [BMM93, Tab. 2]. If we move away from d-cuspidal pairs, there are further open cases in E_8, such as the decomposition of Lusztig induction to $E_8(q)$ of $\lambda_j = E_6[\zeta_3^j]$, $j = 1, 2$, from a 3-split Levi subgroup of rational type $E_6(q).\Phi_3$, and of $\lambda_j = {}^2E_6[\zeta_3^j]$, $j = 1, 2$, from a 6-split Levi subgroup of rational type ${}^2E_6(q).\Phi_6$.

The explicit results on the decomposition of $R_{\mathbf{L}}^{\mathbf{G}}$ for d-split Levi subgroups of simple groups allow one to verify the following important structural observation on unipotent d-Harish-Chandra series and the relations $\leqslant_d$ and $\ll_d$ (introduced in 3.5.24), notwithstanding the yet unknown decompositions:

Theorem 4.6.20 (Broué–Malle–Michel) *Let* $\mathbf{G}$ *be connected reductive with a Steinberg map* F *and* $d \geqslant 1$.

(a) *The unipotent* d*-Harish-Chandra series partition the set of unipotent characters of* $\mathbf{G}^F$, *that is,*

$$\mathrm{Uch}(\mathbf{G}^F) = \coprod_{(\mathbf{L},\lambda)} \mathscr{E}(\mathbf{G}^F, (\mathbf{L}, \lambda)),$$

where the union runs over a system of representatives of d*-cuspidal unipotent pairs* $(\mathbf{L}, \lambda)$ *in* $\mathbf{G}$ *modulo* $\mathbf{G}^F$*-conjugation.*

(b) *The relation* $\leqslant_d$ *is transitive for unipotent characters, so* $\leqslant_d$ *and* $\ll_d$ *agree.*

See [BMM93, Thms. 3.2 and 3.11]. That is, d-Harish-Chandra series of unipotent characters enjoy the same properties as usual Harish-Chandra series do by Corollary 3.1.17. This turns attention to the structure of the individual d-Harish-Chandra series. Here, we have the following natural generalisation of the Howlett–Lehrer Comparison Theorem 3.2.7. But as is the case for Theorem 4.6.20, at present there is no conceptual proof known for this fact; rather it is obtained by comparing the explicit decomposition of $R_{\mathbf{L}}^{\mathbf{G}}$ with the induction in the corresponding relative Weyl groups $W_{\mathbf{G}}(\mathbf{L}, \lambda) = N_{\mathbf{G}^F}(\mathbf{L}, \lambda)/\mathbf{L}^F$ (see 3.5.8) in each individual case, see [BMM93, Thm. 3.2]:

Theorem 4.6.21 (Comparison Theorem) *Let* $\mathbf{G}$ *be connected reductive with Steinberg map* F *and* $d \geqslant 1$. *For any* d*-cuspidal unipotent pair* $(\mathbf{L}, \lambda)$ *in* $\mathbf{G}$ *there exists a collection of isometries*

$$I_{\mathbf{L},\lambda}^{\mathbf{M}} : \mathbb{Z}\mathrm{Irr}(W_{\mathbf{M}}(\mathbf{L}, \lambda)) \to \mathbb{Z}\mathscr{E}(\mathbf{M}^F, (\mathbf{L}, \lambda)),$$

where $\mathbf{M}$ *runs over* d*-split Levi subgroups* $\mathbf{L} \leqslant \mathbf{M} \leqslant \mathbf{G}$, *such that the diagram*

$$\begin{array}{ccc} \mathbb{Z}\mathrm{Irr}(W_{\mathbf{G}}(\mathbf{L}, \lambda)) & \xrightarrow{I_{\mathbf{L},\lambda}^{\mathbf{G}}} & \mathbb{Z}\mathscr{E}(\mathbf{G}^F, (\mathbf{L}, \lambda)) \\ \uparrow \mathrm{Ind} & & \uparrow R_{\mathbf{M}}^{\mathbf{G}} \\ \mathbb{Z}\mathrm{Irr}(W_{\mathbf{M}}(\mathbf{L}, \lambda)) & \xrightarrow{I_{\mathbf{L},\lambda}^{\mathbf{M}}} & \mathbb{Z}\mathscr{E}(\mathbf{M}^F, (\mathbf{L}, \lambda)) \end{array}$$

commutes for all $\mathbf{M}$, *where* Ind *denotes ordinary induction.*

In fact, the latter result can be thought of as a specialisation of a generic statement. The explicit results on the decomposition of $R_{\mathbf{L}}^{\mathbf{G}}$ show that the classification of d-cuspidal unipotent pairs $(\mathbf{L}, \lambda)$, their relative Weyl groups $W_{\mathbf{G}}(\mathbf{L}, \lambda)$ and the partition of $\mathrm{Uch}(\mathbf{G}^F)$ into its d-Harish-Chandra series $\mathscr{E}(\mathbf{G}^F, (\mathbf{L}, \lambda))$ are all generic, as are the maps $I_{\mathbf{L},\lambda}^{\mathbf{M}}$. So for any complete root datum $\mathbb{G}$ and any d-cuspidal unipotent pair $(\mathbb{L}, \lambda)$

of $\mathbb{G}$ there exists a collection of isometries $I_{\mathbb{L},\lambda}^{\mathbb{M}} : \mathbb{Z}\mathrm{Irr}(W_{\mathbb{M}}(\mathbb{L},\lambda)) \to \mathbb{Z}\mathscr{E}(\mathbb{M},(\mathbb{L},\lambda))$, where $\mathbb{L} \leqslant \mathbb{M} \leqslant \mathbb{G}$ runs over d-split Levi sub-data, such that the diagram

$$\begin{array}{ccc} \mathbb{Z}\mathrm{Irr}(W_{\mathbb{G}}(\mathbb{L},\lambda)) & \xrightarrow{I_{\mathbb{L},\lambda}^{\mathbb{G}}} & \mathbb{Z}\mathscr{E}(\mathbb{G},(\mathbb{L},\lambda)) \\ \uparrow \scriptstyle{\mathrm{Ind}} & & \uparrow \scriptstyle{R_{\mathbb{M}}^{\mathbb{G}}} \\ \mathbb{Z}\mathrm{Irr}(W_{\mathbb{M}}(\mathbb{L},\lambda)) & \xrightarrow{I_{\mathbb{L},\lambda}^{\mathbb{M}}} & \mathbb{Z}\mathscr{E}(\mathbb{M},(\mathbb{L},\lambda)) \end{array}$$

commutes and specialises to the diagram in Theorem 4.6.21 for any choice of connected reductive group $\mathbf{G}$ and Steinberg map F corresponding to $\mathbb{G}$.

Example 4.6.22 Let $\mathbf{L}_d$ denote the centraliser of a Sylow d-torus of $\mathbf{G}$. Since this is a minimal d-split Levi subgroup, all of its characters are d-cuspidal by definition. The d-Harish-Chandra series $\mathscr{E}(\mathbf{G}^F,(\mathbf{L}_d,1_{\mathbf{L}_d}))$ above the trivial character of $\mathbf{L}_d^F$ is called the *d-principal series* of $\mathbf{G}^F$. According to Theorem 4.6.21 it is in bijection with $\mathrm{Irr}(W_{\mathbf{G}}(\mathbf{L}_d))$.

A formal consequence of Theorem 4.6.21 in conjunction with the Mackey formula is as follows (see [BMM93, Prop. 3.15]):

Corollary 4.6.23 *Let $\mathbf{G}$ be connected reductive, $(\mathbf{L},\lambda)$ be a d-cuspidal unipotent pair in $\mathbf{G}$ and $\rho \in \mathscr{E}(\mathbf{G}^F,(\mathbf{L},\lambda))$. Then*

$$^*R_{\mathbf{L}}^{\mathbf{G}}(\rho) = \langle \rho, R_{\mathbf{L}}^{\mathbf{G}}(\lambda)\rangle \sum_{w \in W_{\mathbf{G}}(\mathbf{L})/W_{\mathbf{G}}(\mathbf{L},\lambda)} {}^{w}\lambda.$$

In particular

$$^*R_{\mathbf{L}}^{\mathbf{G}}(\rho)(1) = \langle \rho, R_{\mathbf{L}}^{\mathbf{G}}(\lambda)\rangle\, |W_{\mathbf{G}}(\mathbf{L}) : W_{\mathbf{G}}(\mathbf{L},\lambda)|\, \lambda(1) \neq 0.$$

The analogy of d-Harish-Chandra theory for unipotent characters with ordinary Harish-Chandra theory goes even further, as we have the complete analogue of Theorem 3.2.18 describing the degrees of the constituents of $R_{\mathbf{L}}^{\mathbf{G}}$ in terms of suitable Schur elements. We first present the generic formulation in the case of Frobenius maps:

Theorem 4.6.24 *Let $\mathbb{G}$ be a complete root datum and $(\mathbb{L},\lambda)$ a d-cuspidal pair of $\mathbb{G}$. Then for any $\phi \in \mathrm{Irr}(W_{\mathbb{G}}(\mathbb{L},\lambda))$ there exists a Laurent polynomial $c_\phi \in \mathbb{Q}[\mathbf{q}^{\pm 1}]$ with zeros only at roots of unity or zero, such that for $\rho = I_{\mathbb{L},\lambda}^{\mathbb{G}}(\phi)$ we have*

$$\mathbb{D}_\rho = \pm \mathbb{D}_\lambda \frac{|\mathbb{G}|_{\mathbf{q}'}}{|\mathbb{L}|_{\mathbf{q}'}} c_\phi^{-1}$$

and moreover $c_\phi(\zeta_d) = \phi(1)/|W_{\mathbb{G}}(\mathbb{L},\lambda)|$, where $\zeta_d := \exp(2\pi\sqrt{-1}/d)$.

Thus, applying the maps $\psi_q^{\mathbb{G}}$ from Theorem 4.5.8 we obtain: if $\mathbf{G}$ is a connected reductive group with Frobenius map F corresponding to $\mathbb{G}$, and $(\mathbf{L}, \lambda)$ is a d-cuspidal unipotent pair in $\mathbf{G}$, then for any $\phi \in \mathrm{Irr}(W_{\mathbf{G}}(\mathbf{L}, \lambda))$ the degree of the associated unipotent character $\rho = I_{\mathbf{L},\lambda}^{\mathbf{G}}(\phi)$ is given by

$$\rho(1) = \pm R_{\mathbf{L}}^{\mathbf{G}}(\lambda)(1)\, c_\phi(q)^{-1} = \pm\lambda(1)\frac{|\mathbf{G}^F|_{p'}}{|\mathbf{L}^F|_{p'}}\, c_\phi(q)^{-1}.$$

In the case $d = 1$ (which is the content of Theorem 3.2.18) this was a consequence of the Howlett–Lehrer–Lusztig theory of Hecke algebras of induced cuspidal representations, and the Laurent polynomial c_ϕ is then the Schur element of ϕ of the associated Iwahori–Hecke algebra. In the general case, Broué and Malle [BrMa93] introduced a so-called cyclotomic Hecke algebra attached to the complex reflection group $W_{\mathbf{G}}(\mathbf{L}, \lambda)$. The Laurent polynomials c_ϕ should then be suitable specialisations of the Schur elements of this cyclotomic Hecke algebra with respect to a certain canonical trace form (specified in [BMM99, Thm.-Ass. 2]). The latter statement is conjectured to be true in general (see [BrMa93, (d-HV6)]). It has been proved for all but finitely many types, but a general proof is not known at present. Nevertheless, the existence of the polynomials c_ϕ with the stated properties has been verified in [Ma95, Folg. 3.16 and 6.11] for the groups of classical type, and in [Ma97, Prop. 5.2], [Ma00, Prop. 7.1] for those of exceptional type.

We also have the analogue of Corollary 3.2.21:

Corollary 4.6.25 *Let $(\mathbf{L}, \lambda)$ be a d-cuspidal unipotent pair in $\mathbf{G}$ and let $\rho \in \mathscr{E}(\mathbf{G}^F, (\mathbf{L}, \lambda))$. Then the degree polynomial of ρ satisfies*

$$\mathbb{D}_\rho = \Phi_d^{a_d([\mathbf{L},\mathbf{L}])} f,$$

where $f \in \mathbb{Q}[\mathbf{q}]$ is not divisible by Φ_d, and $a_d([\mathbf{L}, \mathbf{L}])$ denotes the precise power of Φ_d dividing the order polynomial of $[\mathbf{L}, \mathbf{L}]$.

Proof This is entirely analogous to the proof of Corollary 3.2.21. According to Theorem 4.6.24 the Laurent polynomials c_ϕ have no zero (or pole) at ζ_d. Since $\mathbf{L}$ is a d-split Levi subgroup of $\mathbf{G}$, it contains a Sylow d-torus of $\mathbf{G}$ by Proposition 3.5.5 and so its order polynomial is divisible by the same power of Φ_d as the one of $\mathbf{G}$. The claim then follows by Corollary 3.5.27 in conjunction with Theorem 4.6.24. □

The assertions of Theorem 4.6.24 and Corollary 4.6.25 continue to hold for very twisted Steinberg maps F (see [BrMa93, Folg. 5.11]) except that then the Schur elements c_ϕ lie in $\mathbb{Q}(\sqrt{p})[\mathbf{q}^{\pm 1}]$ and the Φ_d have to be replaced by cyclotomic polynomials over $\mathbb{Q}(\sqrt{p})$ as in 3.5.3.

4.7 On Lusztig's Jordan Decomposition

The explicit results on unipotent characters can be used to make Lusztig's Jordan decomposition in Theorem 2.6.22 unique at least in certain cases. For example, in classical groups with connected centre, Theorem 4.4.23 shows that any irreducible character is uniquely determined by its uniform projection, so in this case there can exist at most one Jordan decomposition. More generally we have the following uniqueness result from [DiMi90, Thm. 7.1]:

Theorem 4.7.1 (Digne–Michel) *There exists a unique collection of bijections*

$$J_s^{\mathbf{G}} : \mathscr{E}(\mathbf{G}^F, s) \longrightarrow \mathrm{Uch}(C_{\mathbf{G}^*}(s)^F),$$

where $\mathbf{G}$ *runs over connected reductive groups with connected centre and Frobenius map* F, *and* $s \in \mathbf{G}^{*F}$ *is semisimple, satisfying the following, where we write* $\mathbf{H} := C_{\mathbf{G}^*}(s)$:

(1) *For any* F*-stable maximal torus* $\mathbf{T}^* \leqslant \mathbf{H}$,

$$\langle R_{\mathbf{T}^*}^{\mathbf{G}}(s), \rho \rangle = \varepsilon_{\mathbf{G}} \varepsilon_{\mathbf{H}} \langle R_{\mathbf{T}^*}^{\mathbf{H}}(1_{\mathbf{T}^*}), J_s^{\mathbf{G}}(\rho) \rangle \quad \textit{for all } \rho \in \mathscr{E}(\mathbf{G}^F, s).$$

(2) *If* $s = 1$ *and* $\rho \in \mathscr{E}(\mathbf{G}^F, 1)$ *is unipotent then*
   (a) *the Frobenius eigenvalues* ω_ρ *and* $\omega_{J_1^{\mathbf{G}}(\rho)}$ *are equal, and*
   (b) *if* ρ *lies in the principal series then* ρ *and* $J_1^{\mathbf{G}}(\rho)$ *correspond to the same character of the Iwahori–Hecke algebra.*

(3) *If* $z \in \mathbf{Z}(\mathbf{G}^{*F})$ *then* $J_{sz}^{\mathbf{G}}(\rho \otimes \hat{z}) = J_s^{\mathbf{G}}(\rho)$ *for* $\rho \in \mathscr{E}(\mathbf{G}^F, s)$, *where* $\hat{z}$ *is the linear character of* $\mathbf{G}^F$ *corresponding to* z *(see Proposition 2.5.20).*

(4) *For any* F*-stable Levi subgroup* $\mathbf{L}^*$ *of* $\mathbf{G}^*$ *such that* $\mathbf{H} \leqslant \mathbf{L}^*$, *with dual* $\mathbf{L} \leqslant \mathbf{G}$, *the following diagram commutes:*

$$\begin{array}{ccc} \mathscr{E}(\mathbf{G}^F, s) & \xrightarrow{J_s^{\mathbf{G}}} & \mathrm{Uch}(\mathbf{H}^F) \\ \uparrow R_{\mathbf{L}}^{\mathbf{G}} & & \uparrow \mathrm{id} \\ \mathscr{E}(\mathbf{L}^F, s) & \xrightarrow{J_s^{\mathbf{L}}} & \mathrm{Uch}(\mathbf{H}^F) \end{array}$$

(5) *If* $\mathbf{G}$ *is of type* E_8 *and* $\mathbf{H}$ *is of type* E_7A_1 *(resp.* E_6A_2*) and* $\mathbf{L} \leqslant \mathbf{G}$ *is a Levi subgroup of type* E_7 *(resp.* E_6*) with dual* $\mathbf{L}^* \leqslant \mathbf{H}$ *then the following diagram commutes:*

$$\begin{array}{ccc} \mathbb{Z}\mathscr{E}(\mathbf{G}^F, s) & \xrightarrow{J_s^{\mathbf{G}}} & \mathbb{Z}\mathrm{Uch}(\mathbf{H}^F) \\ \uparrow R_{\mathbf{L}}^{\mathbf{G}} & & \uparrow R_{\mathbf{L}^*}^{\mathbf{H}} \\ \mathbb{Z}\mathscr{E}(\mathbf{L}^F, s)_c & \xrightarrow{J_s^{\mathbf{L}}} & \mathbb{Z}\mathrm{Uch}(\mathbf{L}^{*F})_c \end{array}$$

where the index c *denotes the subspace spanned by the cuspidal part of the corresponding Lusztig series.*

(6) *For any F-stable central torus $\mathbf{T}_1 \leqslant \mathbf{Z}(\mathbf{G})$ with corresponding natural epimorphism $\varphi : \mathbf{G} \to \mathbf{G}_1 := \mathbf{G}/\mathbf{T}_1$ and for $s_1 \in \mathbf{G}_1^*$ with $s = \varphi^*(s_1)$ the following diagram commutes:*

$$\begin{array}{ccc} \mathscr{E}(\mathbf{G}^F, s) & \xrightarrow{J_s^{\mathbf{G}}} & \mathrm{Uch}(\mathbf{H}^F) \\ \uparrow & & \downarrow \\ \mathscr{E}(\mathbf{G}_1^F, s_1) & \xrightarrow{J_{s_1}^{\mathbf{G}_1}} & \mathrm{Uch}(\mathbf{H}_1^F) \end{array}$$

with $\mathbf{H}_1 = C_{\mathbf{G}_1^*}(s)$ and where the vertical maps are just the inflation map along $\mathbf{G}^F \to \mathbf{G}_1^F$ and the restriction along the embedding $\mathbf{H}_1^F \hookrightarrow \mathbf{H}^F$ respectively.

(7) *If $\mathbf{G}$ is a direct product $\prod_i \mathbf{G}_i$ of F-stable subgroups $\mathbf{G}_i$ then $J_{\prod s_i}^{\mathbf{G}} = \prod J_{s_i}^{\mathbf{G}_i}$.*

Sketch of proof Let us explain the main points in the proof by Digne and Michel. Their statement and proof are given in terms of the alternative description of Lusztig induction discussed in 3.3.5. In particular, they do not (need to) assume the Mackey formula since in that setting there's always a well-defined standard parabolic subgroup containing a given standard Levi subgroup. In our case, the independence of Lusztig induction and restriction from the parabolic subgroup is guaranteed by Theorem 3.3.8.

First, observe that $C_{\mathbf{L}^*}(s)$ is connected for all Levi subgroups $\mathbf{L}^*$ of $\mathbf{G}^*$, since $\mathbf{G}$ has connected centre, see e.g. [Bo06, §8.B]. Now one needs to show that conditions (1)–(7) are compatible with each other and do specify J_s uniquely. One starts with the case when $\mathbf{G}_{\mathrm{der}} := [\mathbf{G}, \mathbf{G}]$ is F-simple, that is, $\mathbf{G}_{\mathrm{der}}$ is a central product of simple algebraic groups permuted transitively by F.

First, if $s = 1$ then conditions (1) and (2) do determine a unique bijection J_1 by Theorem 4.4.23 when $\mathbf{G}_{\mathrm{der}}$ has classical type, and by Corollary 4.5.4 when $\mathbf{G}_{\mathrm{der}}$ is exceptional. If $s \neq 1$ is central, then J_s is determined by (3); note that (2) is not relevant here and that (1) continues to hold as $R_{\mathbf{T}}^{\mathbf{G}}(\hat{s} \otimes \hat{z}) = R_{\mathbf{T}}^{\mathbf{G}}(\hat{s}) \otimes \hat{z}$ (see Proposition 2.5.21).

Next suppose that $\mathbf{H}$ is contained in a proper F-stable Levi subgroup $\mathbf{L}^* < \mathbf{G}^*$ with dual $\mathbf{L} < \mathbf{G}$. In this case $R_{\mathbf{L}}^{\mathbf{G}}$ is an isometry from $\mathscr{E}(\mathbf{L}^F, s)$ to $\mathscr{E}(\mathbf{G}^F, s)$ by Theorem 3.3.22 which respects (1). Thus, as we may assume by induction on the rank of $\mathbf{G}$ that $J_s^{\mathbf{L}}$ exists and is unique, (4) specifies $J_s^{\mathbf{G}}$ uniquely. Again (3) holds as $R_{\mathbf{L}}^{\mathbf{G}}(\hat{s} \otimes \hat{z}) = R_{\mathbf{L}}^{\mathbf{G}}(\hat{s}) \otimes \hat{z}$ by Proposition 3.3.16.

In the case when $\mathbf{G}_{\mathrm{der}}$ is F-simple it only remains to consider non-central elements s that are quasi-isolated in $\mathbf{G}^*$ and in fact isolated, as we assume that $\mathbf{Z}(\mathbf{G})$ and hence $C_{\mathbf{G}^*}(s)$ is connected. In this case Digne–Michel show by explicit case-by-case computations [DiMi90, Lemma 7.2] that there is a unique bijection $J_s^{\mathbf{G}}$ satisfying (1)–(5). Note that (5) is only relevant for $\mathbf{G}$ of type E_8.

Observe that the bijections J_s for $\mathbf{G}$ obtained above do satisfy (6). This is clear when J_s has been determined by (1); it holds for J_1 found with (2) since in this case φ induces an isomorphism between the relevant Deligne–Lusztig varieties, as well as between the Iwahori–Hecke algebras; in case we constructed J_s through (3) it is again clear as $\hat{z}$ factors through $\mathbf{G}_1^F$; and finally when we use (4) or (5) it follows as φ commutes with $R_{\mathbf{L}}^{\mathbf{G}}$ by Proposition 3.3.24.

Now consider the case of a general $\mathbf{G}$. Then there are connected reductive groups $\mathbf{G}_i$ with Frobenius maps again denoted F such that $(\mathbf{G}_i)_{\mathrm{der}}$ is F-simple, and an F-equivariant epimorphism $\prod_i \mathbf{G}_i \to \mathbf{G}$ with kernel a central torus. In this case, J_s is uniquely determined by (6) and (7). Here one needs to show that the map obtained in this way does not depend on the choice of central isotypy $\prod_i \mathbf{G}_i \to \mathbf{G}$, which uses the fact that (6) holds for the $\mathbf{G}_i$. Finally, one proves that this J_s also satisfies the requirements (1)–(5), which is easy. □

It is not known in general whether Lusztig induction and restriction commute with Jordan decomposition of characters, or more precisely, whether a Jordan decomposition can always be chosen such that it has this property. For classical groups with connected centre, this commutation was first proved by Fong and Srinivasan [FoSr89, App. A] based on results of Shoji. Later, Enguehard [En13, Prop. 5.3] sketched a proof which only uses the Mackey formula and Asai's decomposition formula for Lusztig induction of unipotent characters. We present a slightly stronger version (whose proof is essentially the same):

Theorem 4.7.2 *Let $\mathbf{G}$ be connected reductive with a Frobenius endomorphism $F : \mathbf{G} \to \mathbf{G}$ and assume that the Mackey formula holds for $\mathbf{G}^F$. Let $s \in \mathbf{G}^{*F}$ be a semisimple element such that $C_{\mathbf{G}^*}(s)$ is connected and only has components of classical type A, B, C and D. Then for all F-stable Levi subgroups $\mathbf{L}^* \leqslant \mathbf{M}^* \leqslant \mathbf{G}^*$ satisfying $s \in \mathbf{L}^*$, with duals $\mathbf{L} \leqslant \mathbf{M} \leqslant \mathbf{G}$, the diagram*

$$\begin{array}{ccc} \mathbb{Z}\mathcal{E}(\mathbf{M}^F, s) & \xrightarrow{J_s^{\mathbf{M}}} & \mathbb{Z}\mathrm{Uch}(C_{\mathbf{M}^*}(s)^F) \\ \uparrow R_{\mathbf{L}}^{\mathbf{M}} & & \uparrow R_{C_{\mathbf{L}^*}(s)}^{C_{\mathbf{M}^*}(s)} \\ \mathbb{Z}\mathcal{E}(\mathbf{L}^F, s) & \xrightarrow{J_s^{\mathbf{L}}} & \mathbb{Z}\mathrm{Uch}(C_{\mathbf{L}^*}(s)^F) \end{array}$$

commutes, where $J_s^\bullet$ denotes Jordan decompositions as in Theorem 2.6.22.

Proof Since $C_{\mathbf{G}^*}(s)$ is connected by assumption, the centraliser of s in any Levi subgroup of $\mathbf{G}^*$ is also connected (see [Bo06, §8.B]). Thus we have corresponding Jordan decompositions $J_s^\bullet$ as in Theorem 2.6.22. We let $\Psi^\bullet$ denote its inverse.

If $C_{\mathbf{M}^*}(s) = C_{\mathbf{L}^*}(s)$ then we have $\mathbf{Z}^\circ(\mathbf{L}^*) \leqslant \mathbf{Z}^\circ(C_{\mathbf{M}^*}(s))$ since $\mathbf{L}^* = C_{\mathbf{G}^*}(\mathbf{Z}^\circ(\mathbf{L}^*))$ and so $C_{\mathbf{L}^*}(s) = C_{C_{\mathbf{G}^*}(s)}(\mathbf{Z}^\circ(\mathbf{L}^*))$. Then $R_{\mathbf{L}}^{\mathbf{M}}$ restricts to a bijection $\mathcal{E}(\mathbf{L}^F, s) \to \mathcal{E}(\mathbf{M}^F, s)$ (up to a global sign) which commutes with Jordan decomposition by Theorem 3.3.22, that is, we have $R_{\mathbf{L}}^{\mathbf{M}} \circ \Psi^L = \Psi^{\mathbf{M}}$.

So we may assume $\mathbf{L} < \mathbf{M}$ is proper. We next claim that we may assume $\mathbf{L}$ is maximal in $\mathbf{M}$. Indeed, let $\mathbf{K}$ be an F-stable Levi subgroup of $\mathbf{G}$ with $\mathbf{L} < \mathbf{K} < \mathbf{M}$, with dual $\mathbf{L}^* < \mathbf{K}^* < \mathbf{M}^*$. Then $\Psi^{\mathbf{M}} \circ R_{C_{\mathbf{K}^*}(s)}^{C_{\mathbf{M}^*}(s)} = R_{\mathbf{K}}^{\mathbf{M}} \circ \Psi^{\mathbf{K}}$ and $\Psi^{\mathbf{K}} \circ R_{C_{\mathbf{L}^*}(s)}^{C_{\mathbf{K}^*}(s)} = R_{\mathbf{L}}^{\mathbf{K}} \circ \Psi^{\mathbf{L}}$, which implies the claim.

By definition Jordan decomposition commutes with uniform projection, so we have

$$\pi_{\text{un}}^{\mathbf{M}} \circ \Psi^{\mathbf{M}} \circ R_{C_{\mathbf{L}^*}(s)}^{C_{\mathbf{M}^*}(s)} = \Psi^{\mathbf{M}} \circ \left(\pi_{\text{un}}^{C_{\mathbf{M}^*}(s)} \circ R_{C_{\mathbf{L}^*}(s)}^{C_{\mathbf{M}^*}(s)}\right) = \left(\pi_{\text{un}}^{\mathbf{M}} \circ R_{\mathbf{L}}^{\mathbf{M}}\right) \circ \Psi^{\mathbf{L}}.$$

That is, $\Psi^{\mathbf{M}}(R_{C_{\mathbf{L}^*}(s)}^{C_{\mathbf{M}^*}(s)}(\lambda))$ has the same uniform projection as $R_{\mathbf{L}}^{\mathbf{M}}(\Psi^{\mathbf{L}}(\lambda))$ for any $\lambda \in \mathrm{Uch}(C_{\mathbf{L}^*}(s)^F)$, and by Proposition 4.7.3, $\psi := \Psi^{\mathbf{M}}(R_{C_{\mathbf{L}^*}(s)}^{C_{\mathbf{M}^*}(s)}(\lambda))$ has the same norm as $R_{C_{\mathbf{L}^*}(s)}^{C_{\mathbf{M}^*}(s)}(\lambda)$. Now by Corollary 4.6.13 and Remark 4.6.14, ψ is uniquely determined as the element of minimal norm in $\mathbb{Z}\mathscr{E}(\mathbf{M}^F, s)$ with the same uniform projection as ψ, so our claim follows. □

The following was used in the previous proof:

Proposition 4.7.3 *Let G be connected reductive with a Steinberg map F and assume that the Mackey formula holds for $\mathbf{G}^F$. Let $\mathbf{L} \leqslant \mathbf{G}$ be a d-split Levi subgroup for some $d \geqslant 1$ and $s \in \mathbf{L}^{*F}$ be semisimple with $C_{\mathbf{G}^*}(s)$ connected. Assume that Lusztig induction commutes with Jordan decomposition for any proper d-split Levi subgroup of $\mathbf{L}$. Then for any Jordan decomposition $J_s : \mathscr{E}(\mathbf{L}^F, s) \to \mathrm{Uch}(C_{\mathbf{L}^*}(s)^F)$ as in Theorem 2.6.22 we have*

$$\|R_{\mathbf{L}}^{\mathbf{G}}(\chi)\| = \|R_{C_{\mathbf{L}^*}(s)}^{C_{\mathbf{G}^*}(s)}(J_s(\chi))\| \qquad \textit{for all } \chi \in \mathscr{E}(\mathbf{L}^F, s).$$

Proof Let $\chi \in \mathscr{E}(\mathbf{L}^F, s)$. By the Mackey formula 3.3.7

$$\|R_{\mathbf{L}}^{\mathbf{G}}(\chi)\|^2 = \langle {}^*R_{\mathbf{L}}^{\mathbf{G}}(R_{\mathbf{L}}^{\mathbf{G}}(\chi)), \chi \rangle = \sum_{g \in S} n(g) \qquad (*)$$

where g runs over a system S of $\mathbf{L}^F$–$\mathbf{L}^F$ double coset representatives in $\mathbf{G}^F$ such that $\mathbf{L}_g := \mathbf{L} \cap {}^g\mathbf{L}$ contains a maximal torus of $\mathbf{G}$, and we have put $n(g) := \langle R_{\mathbf{L}_g}^{\mathbf{L}}({}^*R_{\mathbf{L}_g}^{{}^g\mathbf{L}}({}^g\chi)), \chi \rangle$. Let $\mathbf{T}_0$ be a maximally split torus of a minimal d-split Levi subgroup of $\mathbf{L}$ (and hence of $\mathbf{G}$). By Lemma 3.5.18 we may assume that the elements of S are chosen in $N_{\mathbf{G}}(\mathbf{T}_0)^F$. Set $\mathbf{W} := N_{\mathbf{G}}(\mathbf{T}_0)/\mathbf{T}_0$, the Weyl group of $\mathbf{G}$.

Let $g \in S$. If

$$0 \neq n(g) = \langle R_{\mathbf{L}_g}^{\mathbf{L}}({}^*R_{\mathbf{L}_g}^{{}^g\mathbf{L}}({}^g\chi)), \chi \rangle = \langle {}^*R_{\mathbf{L}_g}^{{}^g\mathbf{L}}({}^g\chi), {}^*R_{\mathbf{L}_g}^{\mathbf{L}}(\chi) \rangle$$

then there exists some common constituent $\psi \in \mathrm{Irr}(\mathbf{L}_g^F)$, that is,

$$\langle {}^*R_{\mathbf{L}_g}^{\mathbf{L}}(\chi), \psi \rangle \neq 0 \quad \text{and} \quad \langle {}^*R_{\mathbf{L}_g}^{{}^g\mathbf{L}}({}^g\chi), \psi \rangle = \langle {}^*R_{\mathbf{L}_g^g}^{\mathbf{L}}(\chi), \psi^g \rangle \neq 0.$$

Choose $(\mathbf{T}, \theta) \in \mathfrak{X}(\mathbf{L}_g, F)$ with $\langle R_{\mathbf{T}}^{\mathbf{L}_g}(\theta), \psi \rangle \neq 0$, then also $\langle R_{\mathbf{T}^g}^{\mathbf{L}_g^g}(\theta^g), \psi^g \rangle \neq 0$, and hence $(\mathbf{T}, \theta)$ and $(\mathbf{T}^g, \theta^g)$ are geometrically conjugate in $\mathfrak{X}(\mathbf{L}, F)$. Let $g^*\mathbf{T}_0^{*F} \in \mathbf{W}^{*F}$ be the image of $g\mathbf{T}_0^F$ under the isomorphism $\mathbf{W} \to \mathbf{W}^*$ (see Remark 1.5.19). Then by Proposition 2.5.5 the corresponding pairs $(\mathbf{T}^*, s_1)$ and $((\mathbf{T}^*)^{g^*}, s_1^{g^*})$ in $\mathfrak{Y}(\mathbf{L}^*, F)$ are geometrically conjugate. That is, by Theorem 2.6.2 there are $y^*, z^* \in N_{\mathbf{L}^*}(\mathbf{T}_0^*)^F$ such that $y^*g^*z^* \in C_{\mathbf{G}^*}(s)^F$. Replacing the representative $g \in S$ by ygz we have thus shown: only those elements of S contribute to $(*)$ for which $g^* \in C_{\mathbf{G}^*}(s)^F$.

Now note that $C_{\mathbf{L}^*}(s) = C_{C_{\mathbf{G}^*}(s)}(\mathbf{Z}(\mathbf{L}^*)_d^\circ)$ is d-split in $C_{\mathbf{G}^*}(s)$, so arguing as before we see that we can choose a system S_s of $C_{\mathbf{L}^*}(s)^F$–$C_{\mathbf{L}^*}(s)^F$ double coset representatives in $C_{\mathbf{G}^*}(s)^F$ lying in the normaliser of a maximally split torus $\mathbf{T}_s^*$ of a minimal d-split Levi subgroup of $C_{\mathbf{L}^*}(s)$. The preceding construction then shows that we may replace our reference torus $\mathbf{T}_0$ for S above such that $\mathbf{T}_0^* = \mathbf{T}_s^*$. The above construction thus defines a surjective map $g \mapsto g^*$ from $\{g \in S \mid n(g) \neq 0\}$ to S_s. It is now easy to check that it is also injective. Thus we obtain our claim once we have shown that $n(g) = n(g^*)$ for any pair (g, g^*) as above.

Let (g, g^*) be such a pair. As g induces an F-equivariant isomorphism $\mathbf{L} \to {}^g\mathbf{L}$, g^* induces an F-equivariant isomorphism $\mathbf{L}^* \to {}^{g^*}\mathbf{L}^*$ fixing s so that we may assume $J_s({}^g\chi) = {}^{g^*}\lambda$, where $\lambda = J_s(\chi)$. Now if $g \in N_{\mathbf{G}}(\mathbf{L})^F$ then $\mathbf{L}_g = \mathbf{L}$ and $\mathbf{L}_g^* = \mathbf{L}^*$, and $n(g) \neq 0$ if and only if ${}^g\chi = \chi$, if and only if ${}^{g^*}\lambda = \lambda$, and in that case

$$n(g) = \langle {}^g\chi, \chi \rangle = 1 = \langle {}^{g^*}\lambda, \lambda \rangle = n(g^*).$$

Otherwise $\mathbf{L}_g = \mathbf{L} \cap {}^g\mathbf{L} < \mathbf{L}$ is proper and d-split, so the inductive hypothesis applies to give

$$J_s^{\mathbf{L}_g}({}^*R_{\mathbf{L}_g}^{{}^g\mathbf{L}}({}^g\chi)) = {}^*R_{C_{\mathbf{L}_g^*}(s)}^{C_{{}^{g^*}\mathbf{L}^*}(s)}({}^{g^*}\lambda) \quad \text{and} \quad J_s^{\mathbf{L}_g}({}^*R_{\mathbf{L}_g}^{\mathbf{L}}(\chi)) = {}^*R_{C_{\mathbf{L}_g^*}(s)}^{C_{\mathbf{L}^*}(s)}(\lambda),$$

whence

$$n(g) = \left\langle {}^*R_{C_{\mathbf{L}_g^*}(s)}^{C_{{}^{g^*}\mathbf{L}^*}(s)}({}^{g^*}\lambda), {}^*R_{C_{\mathbf{L}_g^*}(s)}^{C_{\mathbf{L}^*}(s)}(\lambda) \right\rangle = n(g^*). \qquad \square$$

Remark 4.7.4 Theorem 4.7.2 shows, using Proposition 4.7.3, that the equality of norms stated in the latter actually holds for all Levi subgroups, not just for d-split ones.

Notice that in the proof of Theorem 4.7.2 the assumption on the components of $C_{\mathbf{G}^*}(s)$ being of classical type is only used in the very last paragraph when invoking Corollary 4.6.13. All of the preceding steps work in complete generality. In view of this, the above proof can be adapted to the case of exceptional groups; since here there may exist several Jordan decompositions, we need to fix one.

Theorem 4.7.5 *Let $\mathbf{G}$ be simple with connected centre, $F : \mathbf{G} \to \mathbf{G}$ a Steinberg map, and assume that the Mackey formula holds for $\mathbf{G}^F$. Let $s \in \mathbf{G}^{*F}$ be a semisimple* *element. Then for all F-stable Levi subgroups $\mathbf{L}^* \leqslant \mathbf{M}^* \leqslant \mathbf{G}^*$ satisfying $s \in \mathbf{L}^*$, with duals $\mathbf{L} \leqslant \mathbf{M} \leqslant \mathbf{G}$, the diagram*

$$\begin{array}{ccc} \mathbb{Z}\mathscr{E}(\mathbf{M}^F, s) & \xrightarrow{J_s^{\mathbf{M}}} & \mathbb{Z}\mathrm{Uch}(C_{\mathbf{M}^*}(s)^F) \\ \uparrow R_{\mathbf{L}}^{\mathbf{M}} & & \uparrow R_{C_{\mathbf{L}^*}(s)}^{C_{\mathbf{M}^*}(s)} \\ \mathbb{Z}\mathscr{E}(\mathbf{L}^F, s) & \xrightarrow{J_s^{\mathbf{L}}} & \mathbb{Z}\mathrm{Uch}(C_{\mathbf{L}^*}(s)^F) \end{array}$$

commutes, where $J_s^\bullet$ denotes the Jordan decomposition specified in Theorem 4.7.1, except possibly when $\mathbf{G} = \mathbf{M}$ is of type E_8, $C_{\mathbf{G}^*}(s)$ is of type E_6A_2 or E_7A_1 and $C_{\mathbf{L}^*}(s)$ has a factor of type E_6 or E_7.

Proof First, if F is not a Frobenius map, and hence $\mathbf{G}$ is of type B_2, G_2 or F_4 in characteristic 2, 3 or 2 respectively, then the claim follows with Theorem 3.3.22 unless $1 \neq s \in \mathbf{G}^{*F}$ is isolated. But the only such semisimple elements are involutions in type G_2, with centraliser A_1^2, and elements of order 3 in F_4 with centraliser of type A_2^2. In either case, $C_{\mathbf{G}^*}(s)$ only has factors of type A, and thus all of its unipotent characters are uniform. The commutation follows from Theorem 4.7.1(1).

So now assume that F is a Frobenius endomorphism. Since $\mathbf{G}$ has connected centre, $C_{\mathbf{G}^*}(s)$ is connected and thus Theorem 4.7.2 applies. Hence we may assume that $C_{\mathbf{M}^*}(s)$ has some factor of exceptional type and thus in particular $\mathbf{G}$ itself is of exceptional type. If $s = 1$ then we have $C_{\mathbf{M}^*}(s) = \mathbf{M}^*$ and the commutation follows by (2) of Theorem 4.7.1 and the fact that Lusztig induction of unipotent characters in $\mathbf{M}^F$ and $\mathbf{M}^{*F}$ are given by the same formulas (by Proposition 4.6.18 and Remark 4.6.19) unless we are in the excluded case that $\mathbf{M} = \mathbf{G} = E_8$ and $\mathbf{L}$ has rational type ${}^2E_6(q).\Phi_2^2$. But in the latter case, since $\mathbf{G}^* = \mathbf{G}$, whatever the precise decomposition of $R_{\mathbf{L}}^{\mathbf{G}}$ is, it will be the same on both sides. If $1 \neq s \in \mathbf{Z}(\mathbf{M}^*)$ (and thus also $s \in \mathbf{Z}(\mathbf{L}^*)$) we have $C_{\mathbf{M}^*}(s) = \mathbf{M}^*$ and the commutation follows by property (3) in Theorem 4.7.1 together with the previous case.

Thus s is non-central in $\mathbf{M}^*$ and since $C_{\mathbf{M}^*}(s)$ has an exceptional factor, $\mathbf{M}^*$ can not be of type G_2, F_4 or E_6. If $C_{\mathbf{M}^*}(s)$ lies in a proper Levi subgroup of $\mathbf{M}^*$ we are again done by property (4) in Theorem 4.7.1. Thus we may assume that s is isolated but not central in $\mathbf{M}^*$. Such elements are classified in [Bo05]. If $\mathbf{M}$ is of type E_7 then there are no such centralisers with an exceptional factor. The only cases in $\mathbf{M}$ of type E_8 (which then also forces $\mathbf{G} = E_8$) are those for elements of order 2 with centraliser structure E_7A_1, respectively of order 3 with centraliser structure E_6A_2.

Set $\mathbf{C} := C_{\mathbf{M}^*}(s)$, $\mathbf{C}_1 := C_{\mathbf{L}^*}(s)$. Proposition 4.6.18 shows that the decomposition of $R_{\mathbf{C}_1}^{\mathbf{C}}(\lambda)$ for any $\lambda \in \mathrm{Uch}(\mathbf{C}_1^F)$ is uniquely determined by its norm and its uniform projection, except for indeterminacies coming from pairs of unipotent characters with the same uniform projection, and by Proposition 4.7.3 the same holds for $R_{\mathbf{L}}^{\mathbf{G}}(\Psi_s^{\mathbf{L}}(\lambda))$. If $\mathbf{C}_1$ is of classical type, then the indeterminacies can be resolved again by using the arguments given in the proof of Proposition 4.6.18. Thus we may further assume that $C_{\mathbf{L}^*}(s)$ also has a factor of exceptional type and we reach the excluded configurations. □

Something more can be said even in the excluded cases of Theorem 4.7.5. In fact, the proof of Proposition 4.6.18 shows that the decomposition of $R_{\mathbf{L}}^{\mathbf{G}}(\lambda)$ is the same on the two sides of our diagram unless λ belongs to a pair of characters with the same uniform projection: those lying above a cuspidal unipotent character of $E_6(q)$, ${}^2E_6(q)$ or $E_7(q)$ or involving the two principal series characters $\phi_{511,11}, \phi_{512,12}$ of $E_7(q)$.

Corollary 4.7.6 *Let $\mathbf{G}$ be simple with connected centre and $F : \mathbf{G} \to \mathbf{G}$ be a Steinberg map. Then Harish-Chandra induction commutes with some Jordan decomposition $J_s^\bullet$ as in Theorem 2.6.22.*

Proof We need to show that for all semisimple elements $s \in \mathbf{G}^{*F}$ the diagram in Theorem 4.7.5 commutes for all F-stable Levi subgroups $\mathbf{L}^* \leqslant \mathbf{M}^* \leqslant \mathbf{G}^*$, with $s \in \mathbf{L}^*$ and $\mathbf{L}^*$ 1-split in $\mathbf{M}^*$, for some choice of $J_s^{\mathbf{L}}$. Now the Mackey formula holds for 1-split Levi subgroups by Theorem 3.1.11. So by Theorem 4.7.5 this claim holds unless $\mathbf{M} = \mathbf{G}$ is of type E_8, s has centraliser $C_{\mathbf{G}^*}(s)$ of type E_6A_2 or E_7A_1 and $C_{\mathbf{L}^*}(s)$ has a factor of type E_6 or E_7. The only Harish-Chandra series not covered by our arguments are the principal series for $C_{\mathbf{G}^*}(s) = E_7(q).A_1(q)$ and those series listed in Table 4.18. Let's discuss these in turn.

Table 4.18 *Critical 1-Harish-Chandra series*

$C_{\mathbf{G}^*}(s)$	$C_{\mathbf{L}^*}(s)$	λ	$W_{C_{\mathbf{G}^*}(s)}(C_{\mathbf{L}^*}(s), \lambda)^F$
$E_6(q).A_2(q)$	$E_6(q)\Phi_1^2$	$E_6[\zeta_3^j]$, $j = 1, 2$	$W(A_2)$
${}^2E_6(q).{}^2A_2(q)$	${}^2E_6(q)\Phi_1\Phi_2$	${}^2E_6[\zeta_3^j]$, $j = 1, 2$	$W(A_1)$
$E_7(q).A_1(q)$	$E_6(q)\Phi_1^2$	$E_6[\zeta_3^j]$, $j = 1, 2$	$W(A_1) \times W(A_1)$
$E_7(q).A_1(q)$	$E_7(q)\Phi_1$	$E_7[\pm\zeta_4]$	$W(A_1)$

For λ as in the table let $\chi \in \mathscr{E}(\mathbf{L}^F, s)$ be such that $J_s^{\mathbf{L}}(\chi) = \lambda$ (for the Jordan decomposition for $\mathbf{L}$ from Theorem 4.7.1). Then χ is also cuspidal by Theorem 3.2.22. Furthermore the relative Weyl groups of χ and of its Jordan correspondent λ agree. By the Comparison Theorem 3.2.7, the multiplicities in the decompositions of $R_{\mathbf{L}}^{\mathbf{G}}(\chi)$ and $R_{C_{\mathbf{L}^*}(s)}^{C_{\mathbf{G}^*}(s)}(\lambda)$ are the same for characters on both sides whose uniform projections correspond under Jordan decomposition. It follows that there is some choice of Jordan decomposition between $\mathscr{E}(\mathbf{G}^F, s)$ and $\mathrm{Uch}(C_{\mathbf{G}^*}(s)^F)$ that makes the diagram commute.

Finally, for the principal series in $E_7(q).A_1(q)$ we need to choose a bijection for the two pairs of characters in $\mathrm{Uch}(C_{\mathbf{G}^*}(s)^F)$ lying in an exceptional family and having the same uniform projection. Again the relative Weyl groups on both sides agree and thus also the multiplicities. So there exists a compatible choice of bijection. □

Let us define an equivalence relation on the sets of unipotent characters of reductive subgroups **H** of a simple group of type E_8, as follows: if **H** is of classical type, all unipotent characters form an equivalence class on their own. If **H** has (one) simple factor of type E_6, E_7 or E_8 then any two unipotent characters with same uniform projection and with Frobenius eigenvalue either a third or sixth root of unity are equivalent, the two cuspidal unipotent characters of $E_7(q)$ with Frobenius eigenvalue $\pm\sqrt{-1}$ are equivalent, as are the pairs of principal series characters of $E_7(q)$ and $E_8(q)$ lying in an exceptional family. We then let $\tilde{\mathscr{E}}(\mathbf{H}^F, 1)$ be the set of sums over equivalence classes of unipotent characters of $\mathbf{H}^F$. Thus, in particular, all elements of $\tilde{\mathscr{E}}(\mathbf{H}^F, 1)$ have norm either 1 or 2. Via the unique Jordan decomposition from Theorem 4.7.1 this also defines corresponding sets $\tilde{\mathscr{E}}(\mathbf{L}^F, s)$, for **L** an F-stable Levi subgroup of a simple group of exceptional type with connected centre. We then have:

Corollary 4.7.7 *Let* **G** *be simple with connected centre,* $F : \mathbf{G} \to \mathbf{G}$ *a Steinberg map, and assume that the Mackey formula holds for* $\mathbf{G}^F$. *Then for all* F*-stable Levi subgroups* $\mathbf{L}^* \leqslant \mathbf{M}^* \leqslant \mathbf{G}^*$ *with duals* $\mathbf{L} \leqslant \mathbf{M} \leqslant \mathbf{G}$ *and semisimple elements* $s \in \mathbf{L}^{*F}$ *the diagram*

$$\begin{array}{ccc} \mathbb{Z}\tilde{\mathscr{E}}(\mathbf{M}^F, s) & \xrightarrow{J_s^{\mathbf{M}}} & \mathbb{Z}\tilde{\mathscr{E}}(C_{\mathbf{M}^*}(s)^F, 1) \\ \uparrow R_{\mathbf{L}}^{\mathbf{M}} & & \uparrow R_{C_{\mathbf{L}^*}(s)}^{C_{\mathbf{M}^*}(s)} \\ \mathbb{Z}\tilde{\mathscr{E}}(\mathbf{L}^F, s) & \xrightarrow{J_s^{\mathbf{L}}} & \mathbb{Z}\tilde{\mathscr{E}}(C_{\mathbf{L}^*}(s)^F, 1) \end{array}$$

commutes, where $J_s^\bullet$ *denotes the Jordan decomposition specified in Theorem 4.7.1.*

Proof This follows from the previous Theorem 4.7.5 except for the pairs of equivalent characters. But for those, Lusztig induction of the sum over an equivalence class is again uniquely determined by its norm and its uniform projection. □

Corollary 4.7.8 *Let* **G** *and* $s \in \mathbf{G}^{*F}$ *be as in Theorem 4.7.5. Then for any* $d \geqslant 1$ *the relations* $\leqslant_d$ *and* $\ll_d$ *agree on* $\mathscr{E}(\mathbf{G}^F, s)$. *In particular,* $\rho \in \mathscr{E}(\mathbf{G}^F, s)$ *is* d*-cuspidal if and only if* ${}^*R_{\mathbf{T}}^{\mathbf{G}}(\rho) = 0$ *for all* F*-stable maximal tori* $\mathbf{T} \leqslant \mathbf{G}$ *contained in some proper* d*-split Levi subgroup of* **G**.

The first claim follows from Corollary 4.7.7, using that it holds for unipotent characters by Theorem 4.6.20, and the second is a direct consequence of this by the discussion after Corollary 3.5.25. As already pointed out there, this corollary had been obtained by Cabanes–Enguehard [CE99, Thm. 4.2] in a wide range of cases using block theoretic methods.

Arguments as in the proof of Theorem 4.7.5 had been used by Kessar and Malle [KeMa13] and Hollenbach [Ho19] to obtain partial results for isolated series in exceptional groups.

We end this section by stating another application of Digne and Michel's unique Jordan decomposition for groups with connected centre, namely a result of Srinivasan and Vinroot [SrVi18, Thm. 5.1] on Galois action (see also the earlier [SrVi15] for the case of complex conjugation):

Theorem 4.7.9 (Srinivasan–Vinroot) *Let* $\mathbf{G}$ *be connected reductive with connected centre with a Frobenius endomorphism* F *such that the Mackey formula holds for* $\mathbf{G}^F$. *Then for any semisimple element* $s \in \mathbf{G}^{*F}$ *and field automorphism* σ *of* $\mathbb{K}$ *the diagram*

$$\begin{array}{ccc} \mathscr{E}(\mathbf{G}^F, s) & \xrightarrow{J_s^{\mathbf{G}}} & \mathrm{Uch}(C_{\mathbf{G}^*}(s)^F) \\ \Big\downarrow \sigma & & \Big\downarrow \sigma \\ \mathscr{E}(\mathbf{G}^F, s^r) & \xrightarrow{J_{s^r}^{\mathbf{G}}} & \mathrm{Uch}(C_{\mathbf{G}^*}(s)^F) \end{array}$$

commutes, where $r \in \mathbb{N}$ *is such that* $\sigma(\zeta) = \zeta^r$ *for all* $|\mathbf{G}^F|$*th roots of unity* ζ, $J_s^{\mathbf{G}}$ *denotes the Jordan decomposition from Theorem 4.7.1 and the downward arrows mean conjugation by* σ.

Sketch of proof Let σ be as in the statement. By Proposition 3.3.15 the Lusztig series $\mathscr{E}(\mathbf{G}^F, s)$ is mapped to $\mathscr{E}(\mathbf{G}^F, s^r)$ by σ. As r is prime to the order of s by assumption, s and s^r have the same centraliser in $\mathbf{G}^*$, so the diagram makes sense. Now consider the map

$$\tilde{J}_s^{\mathbf{G}} : \mathscr{E}(\mathbf{G}^F, s) \to \mathrm{Uch}(C_{\mathbf{G}^*}(s)^F), \quad \rho \mapsto \sigma^{-1}(J_s^{\mathbf{G}}(\sigma(\rho))).$$

If we can show that $\tilde{J}_s^{\mathbf{G}}$ satisfies the same conditions as $J_s^{\mathbf{G}}$ in Theorem 4.7.1, then by the uniqueness statement we must have $\tilde{J}_s^{\mathbf{G}} = J_s^{\mathbf{G}}$, that is, $\sigma \circ J_s^{\mathbf{G}} = J_s^{\mathbf{G}} \circ \sigma$, as claimed. For properties (1), (4) and (5) in Theorem 4.7.1 this follows from Corollary 3.3.14. For property (2) one can use Proposition 4.5.5, and for property (7) the claim is straightforward. We omit further details. □

4.8 Disconnected Groups, Groups with Disconnected Centre

The notion of connected reductive groups is a very useful one in the character theory of finite groups of Lie type, but it has two important shortcomings. First, Lusztig's Jordan decomposition of characters of groups with disconnected centre naturally leads to having to consider unipotent characters of groups that are no longer connected. And secondly, for many applications of character theory to problems that have a reduction to some sort of decorated finite simple groups one would need to know about representations of automorphism groups of the finite quasi-simple groups, and the latter do include extensions of simple groups of Lie type by their graph automorphisms.

For both reasons it seems necessary to develop a character theory of finite disconnected reductive groups. Some important steps in this direction have been made, and Lusztig has started an investigation of character sheaves on disconnected groups (see [Lu09e] and the references there), but the theory is still far away from as complete a picture as, for example, given by Lusztig's Jordan Decomposition Theorem 2.6.4 in the connected case. Here, we explain some of the available results, and indicate in particular their relevance to the first problem mentioned above: how to obtain the commutation of (some) Jordan decomposition map with Lusztig induction for groups with disconnected centre, at least in certain situations. In particular for type A, or more precisely, for $\mathbf{G} = \mathrm{SL}_n$, which in some sense constitutes the worst case as its disconnected centre may become arbitrarily large when n increases, a definite result has been obtained by Bonnafé [Bo06] and Cabanes [Ca13].

4.8.1 Let us start by establishing some notions from the structure theory of disconnected groups. Let $\mathbf{G}$ be a (not necessarily connected) reductive group. A *parabolic subgroup* of $\mathbf{G}$ is any closed subgroup $\mathbf{P} \leqslant \mathbf{G}$ containing a Borel subgroup of $\mathbf{G}^\circ$. Its connected component of the identity $\mathbf{P}^\circ$ is then a parabolic subgroup of $\mathbf{G}^\circ$ and thus has a semidirect product decomposition $\mathbf{P}^\circ = R_u(\mathbf{P}) \rtimes \mathbf{L}^\circ$ with $\mathbf{L}^\circ$ a Levi complement of $\mathbf{P}^\circ$. Then $\mathbf{P} = R_u(\mathbf{P}) \rtimes \mathbf{L}$ with $\mathbf{L} := N_{\mathbf{G}}(\mathbf{L}^\circ)$ (see [Bo99, §6.1]). The subgroup $\mathbf{L}$ is called a *quasi-Levi subgroup* of $\mathbf{G}$. A special class of parabolic subgroups are the *quasi-Borel subgroups*, the normalisers in $\mathbf{G}$ of Borel subgroups of $\mathbf{G}^\circ$, and their quasi-Levi subgroups are called *maximal quasi-tori*. Since all Borel subgroups of $\mathbf{G}^\circ$ are $\mathbf{G}^\circ$-conjugate, all quasi-Borel subgroups of $\mathbf{G}$ are $\mathbf{G}$-conjugate, and similarly all maximal quasi-tori of $\mathbf{G}$ are $\mathbf{G}$-conjugate.

Example 4.8.2 In the disconnected case quasi-tori may contain non-trivial unipotent elements. Indeed, let $\mathbf{G} = \mathrm{SL}_3\langle a\rangle$ in characteristic 2 with a inducing the transpose-inverse automorphism of order 2. Then the maximal torus $\mathbf{T}_0$ of $\mathbf{G}^\circ$ of diagonal matrices is a-invariant, so the quasi-torus $N_{\mathbf{G}}(\mathbf{T}_0) = \mathbf{T}_0\langle a\rangle$ contains the unipotent element $a \neq 1$. This can only happen, though, for so-called quasi-central elements.

Following Steinberg [St68, §9], an automorphism a of a connected reductive group $\mathbf{G}$ is called *quasi-semisimple* if it stabilises a pair $\mathbf{T} \leqslant \mathbf{B}$ consisting of a maximal torus contained in a Borel subgroup of $\mathbf{G}$, and *quasi-central* if it is quasi-semisimple and its centraliser $C_{\mathbf{G}}(a)$ in $\mathbf{G}$ has maximal dimension among the quasi-semisimple elements in its Inn($\mathbf{G}$)-coset. Such a quasi-central element a naturally acts on the Weyl group $\mathbf{W} = N_{\mathbf{G}}(\mathbf{T})/\mathbf{T}$ of $\mathbf{G}$. Quasi-central elements behave somewhat as a replacement for the identity element in the non-trivial components. For example, a unipotent element can be quasi-semisimple only if it is quasi-central [Spa82a, Cor. II.2.21].

Example 4.8.3 An important class of examples of disconnected groups are the almost simple groups $\mathbf{G}$ obtained from a simple algebraic group $\mathbf{G}^\circ$ by extension with a non-trivial graph automorphism a. The various cases are (up to isogeny):

(1) for $\mathbf{G}^\circ = \mathrm{SL}_n$, $n \geqslant 3$, with a inducing the transpose-inverse automorphism, any outer element with centraliser either Sp_n (when n is even) or one of GO_n or $\mathrm{Sp}_{n-1} \times \{\pm 1\}$ (if n is odd) is quasi-central;
(2) for $\mathbf{G}^\circ = \mathrm{SO}_{2n}$ with $n \geqslant 3$ and $\mathbf{G} = \mathrm{GO}_{2n}$ the quasi-central elements in $\mathrm{GO}_{2n} \setminus \mathrm{SO}_{2n}$ are those with centraliser GO_{2n-1};
(3) for $\mathbf{G}^\circ$ of simply connected or adjoint type D_4 with a inducing triality, the quasi-central elements in the outer cosets have centraliser of type G_2;
(4) for $\mathbf{G}^\circ$ of type E_6 and a inducing the non-trivial graph automorphism of order 2, the outer quasi-central elements have centraliser F_4.

See [DiMi94, pp. 356–357] for a further discussion of the quasi-central elements.

Whenever convenient we will restrict ourselves to a situation when $\mathbf{G}/\mathbf{G}^\circ$ is cyclic, as in the above examples. This is certainly sufficient if one is just interested in character values, since any element $a \in \mathbf{G}$ lies in the cyclic extension $\mathbf{G}^\circ\langle a\rangle$ of $\mathbf{G}^\circ$. In this case parabolic subgroups and quasi-Levi subgroups are parametrised in terms of the centraliser of a quasi-central element (see [DiMi94, Cor. 1.25]):

Proposition 4.8.4 *Let $\mathbf{G} = \mathbf{G}^\circ\langle a\rangle$ be reductive with $a \in \mathbf{G}$ quasi-central.*

(a) *The map $\mathbf{P} \mapsto C^\circ_{\mathbf{P}}(a)$ defines a bijection between parabolic subgroups of $\mathbf{G}$ containing a and parabolic subgroups of $C^\circ_{\mathbf{G}}(a)$.*
(b) *The map $\mathbf{L} \mapsto C^\circ_{\mathbf{L}}(a)$ defines a bijection between a-stable quasi-Levi subgroups of parabolic subgroups of $\mathbf{G}$ containing a and Levi subgroups of $C^\circ_{\mathbf{G}}(a)$. Its inverse is given by $\mathbf{L} \mapsto C_{\mathbf{G}}(\mathbf{Z}^\circ(\mathbf{L}))\langle a\rangle$.*

We now consider reductive groups with a Frobenius map. Here the first basic result is as follows (see [Bo99, Lemme 6.2.1]):

Lemma 4.8.5 *Let $\mathbf{G}$ be reductive with a Frobenius map $F : \mathbf{G} \to \mathbf{G}$. Then there exists an F-stable supplement $A \leqslant \mathbf{G}$ to $\mathbf{G}^\circ$ (that is, we have $\mathbf{G} = \mathbf{G}^\circ A$) such that $A \cap \mathbf{G}^\circ$ is central and all elements of A induce quasi-central automorphisms on $\mathbf{G}^\circ$.*

In particular, if $\mathbf{G} = \mathbf{G}^\circ\langle a\rangle$ with a quasi-central then we may always assume that $\langle a\rangle$ is F-stable. The classification of F-stable maximal quasi-tori can be derived from Proposition 4.8.4 and is again in terms of the centraliser of a quasi-central element:

Proposition 4.8.6 *Let $\mathbf{G} = \mathbf{G}^\circ\langle a\rangle$ be a reductive group with a Frobenius map $F : \mathbf{G} \to \mathbf{G}$ such that $a \in \mathbf{G}^F$ is quasi-central. Then:*

(a) *There exists an $\langle F, a\rangle$-stable maximal torus $\mathbf{T}_0$ of $\mathbf{G}^\circ$.*
(b) *Any F-stable maximal quasi-torus of $\mathbf{G}$ has a $\mathbf{G}^F$-conjugate containing a.*
(c) *If $\mathbf{T}$ is an F-stable maximal quasi-torus of $\mathbf{G}$ containing a then $C^\circ_{\mathbf{T}}(a)$ is a maximal torus of $C_{\mathbf{G}^\circ}(a)$, and $\mathbf{T} \mapsto C^\circ_{\mathbf{T}}(a)$ induces a bijection between $\mathbf{G}^F$-classes of F-stable maximal quasi-tori of $\mathbf{G}$ and $C^\circ_{\mathbf{G}}(a)^F$-classes of F-stable maximal tori of $C^\circ_{\mathbf{G}}(a)$.*

Proof See [DiMi94, 1.36(ii)] for (a), [DiMi94, 1.40] for (b) and (c). □

The $C^\circ_{\mathbf{G}}(a)^F$-classes of F-stable maximal tori of $C^\circ_{\mathbf{G}}(a)$ are parametrised by the F-conjugacy classes of $C_{\mathbf{W}^\circ}(a)$, where $\mathbf{W}^\circ := W_{\mathbf{G}^\circ}(\mathbf{T}_0)$ (see, e.g., Remark 2.3.21). Then an F-stable maximal quasi-torus of $\mathbf{G}$ is said to be *of type* $w \in C_{\mathbf{W}^\circ}(a)$ with respect to $\mathbf{T}_0$ if it corresponds to a maximal torus of type w under the map in Proposition 4.8.6(c).

Example 4.8.7 Let $\mathbf{G}$ be the extension of SL_n, $n \geqslant 3$, with the transpose-inverse automorphism. By Example 4.8.3 the Weyl group of the centraliser of a quasi-central outer element is of type B_m, where $m = \lfloor n/2\rfloor$, and thus by Proposition 4.8.6 the classes of F-stable maximal quasi-tori of $\mathbf{G}$ are in bijection with the conjugacy classes of this Weyl group, hence with pairs of partitions of m (see [GePf00, Prop. 3.4.7])). Note that there are fewer of these than there are partitions of n, unless $n = 4$.

Definition 4.8.8 Let $\mathbf{L}$ be an F-stable quasi-Levi subgroup of the parabolic subgroup $\mathbf{P}$ of $\mathbf{G}$. The unipotent radical $\mathbf{Y} := R_u(\mathbf{P})$ satisfies condition (∗) in 2.2.4 with respect to $\mathbf{L}^F$, and as in 3.3.2 we define (generalised) *Lusztig induction* (or *twisted induction*)

$$R^{\mathbf{G}}_{\mathbf{L}\leqslant\mathbf{P}} : \mathbb{Z}\mathrm{Irr}(\mathbf{L}^F) \to \mathbb{Z}\mathrm{Irr}(\mathbf{G}^F)$$

by

$$R^{\mathbf{G}}_{\mathbf{L}\leqslant\mathbf{P}}(\lambda)(g) := \sum_{i\geqslant 0}(-1)^i\,\mathrm{Trace}\big(g^*, \mathrm{H}^i_c(\mathscr{L}^{-1}(\mathbf{Y}), \overline{\mathbb{Q}}_\ell)_\lambda\big) \quad \text{for } g \in \mathbf{G}^F$$

for any $\lambda \in \mathrm{Irr}(\mathbf{L}^F)$ (see [DiMi94, Def. 2.2], and [Bo99, 6.3]). We again denote by ${}^*R^{\mathbf{G}}_{\mathbf{L}\leqslant\mathbf{P}}$ the adjoint map, *Lusztig restriction*, sending characters of $\mathbf{G}^F$ to virtual characters of $\mathbf{L}^F$.

The definition shows that

$$\mathrm{Res}^{\mathbf{G}^F}_{\mathbf{G}^{\circ F}} \circ R^{\mathbf{G}}_{\mathbf{L}\leqslant\mathbf{P}} = R^{\mathbf{G}^\circ}_{\mathbf{L}^\circ\leqslant\mathbf{P}^\circ} \circ \mathrm{Res}^{\mathbf{L}^F}_{\mathbf{L}^{\circ F}},$$

so in the case of connected groups the above specialises to the ordinary Lusztig induction introduced in 3.3.2, and in fact many of the basic properties from that case can be shown to carry over to the more general setting:

Proposition 4.8.9 *Let $\mathbf{Q} \leqslant \mathbf{P}$ be parabolic subgroups of $\mathbf{G}$ with F-stable quasi-Levi subgroups $\mathbf{M}, \mathbf{L}$ respectively, such that $\mathbf{M} \leqslant \mathbf{L}$. Then*

$$R^{\mathbf{G}}_{\mathbf{L}\leqslant\mathbf{P}} \circ R^{\mathbf{L}}_{\mathbf{M}\leqslant\mathbf{L}\cap\mathbf{Q}} = R^{\mathbf{G}}_{\mathbf{M}\leqslant\mathbf{Q}} \quad \text{and} \quad {}^*R^{\mathbf{L}}_{\mathbf{M}\leqslant\mathbf{L}\cap\mathbf{Q}} \circ {}^*R^{\mathbf{G}}_{\mathbf{L}\leqslant\mathbf{P}} = {}^*R^{\mathbf{G}}_{\mathbf{M}\leqslant\mathbf{Q}}.$$

This is shown in [Bo99, Prop. 6.3.3]. As a particular case this includes the situation when the quasi-Levi subgroup $\mathbf{L}$ contains $\mathbf{G}^\circ$ and hence $\mathbf{P} = \mathbf{L}$, where we get

$$R^{\mathbf{G}}_{\mathbf{L}\leqslant\mathbf{L}} = \mathrm{Ind}^{\mathbf{G}^F}_{\mathbf{L}^F} \quad \text{and} \quad {}^*R^{\mathbf{G}}_{\mathbf{L}\leqslant\mathbf{L}} = \mathrm{Res}^{\mathbf{G}^F}_{\mathbf{L}^F}$$

(see [Bo99, Prop. 6.3.2]). As in Proposition 3.3.3 in the connected case, we have the following:

Proposition 4.8.10 *Let $\mathbf{L}$ be an F-stable quasi-Levi subgroup of an F-stable parabolic subgroup $\mathbf{P}$ of $\mathbf{G}$. Then*

$$R^{\mathbf{G}}_{\mathbf{L}\leqslant\mathbf{P}} = \mathrm{Ind}^{\mathbf{G}^F}_{\mathbf{P}^F} \circ \mathrm{Infl}^{\mathbf{P}^F}_{\mathbf{L}^F}.$$

That is, Lusztig induction specialises to (generalised) Harish-Chandra induction, see [Bo99, Prop. 6.3.4]. In fact, there is a generalisation of Harish-Chandra theory to disconnected groups with suitable attached Hecke algebras of extended Weyl groups similar to what we described in Section 3.2 but we will not go into this here (see [DiMi85] and [Ma91, §1] for details).

There is also a character formula for $R^{\mathbf{G}}_{\mathbf{L}}$ in the spirit of Theorem 3.3.12, see [DiMi94, Prop. 2.6], which in the case when p divides $|\mathbf{G}^F : \mathbf{G}^{\circ F}|$ involves a new 2-parameter Green function. On the identity element we obtain (see [DiMi94, Cor. 2.5]):

Proposition 4.8.11 *Let $\mathbf{L}$ be an F-stable quasi-Levi subgroup of $\mathbf{G}$. Then*

$$R^{\mathbf{G}}_{\mathbf{L}}(\lambda)(1) = \epsilon_{\mathbf{G}^\circ}\epsilon_{\mathbf{L}^\circ} \frac{|\mathbf{G}^F : \mathbf{L}^F|}{|\mathbf{G}^{\circ F} : \mathbf{L}^{\circ F}|_p} \lambda(1) \qquad \text{for } \lambda \in \mathrm{Irr}(\mathbf{L}^F).$$

The character formula also immediately yields the following extension of Proposition 3.3.16:

Proposition 4.8.12 *Let $\mathbf{G} = \mathbf{G}^\circ\langle a\rangle$ and $f \in \mathrm{CF}(\mathbf{G}^F)_{p'}$ be p-constant. Then for every F-stable quasi-Levi subgroup $\mathbf{L}$ containing a of some parabolic subgroup $\mathbf{P}$ of $\mathbf{G}$ we have:*

(a) $R^{\mathbf{G}}_{\mathbf{L}\leqslant\mathbf{P}}(\pi\otimes\mathrm{Res}^{\mathbf{G}^F}_{\mathbf{L}^F}(f)) = R^{\mathbf{G}}_{\mathbf{L}\leqslant\mathbf{P}}(\pi)\otimes f$ *for all* $\pi\in\mathrm{CF}(\mathbf{L}^F)$,
(b) ${}^*R^{\mathbf{G}}_{\mathbf{L}\leqslant\mathbf{P}}(\eta)\otimes\mathrm{Res}^{\mathbf{G}^F}_{\mathbf{L}^F}(f) = {}^*R^{\mathbf{G}}_{\mathbf{L}\leqslant\mathbf{P}}(\eta\otimes f)$ *for all* $\eta\in\mathrm{CF}(\mathbf{G}^F)$.

In particular, ${}^*R^{\mathbf{G}}_{\mathbf{L}\leqslant\mathbf{P}}(f) = \mathrm{Res}^{\mathbf{G}^F}_{\mathbf{L}^F}(f)$.

See [DiMi94, Prop. 2.11].

The Mackey formula is at present only proved in a few cases, see [DiMi94, Thms. 3.2 and 4.5]:

Theorem 4.8.13 *Let* $\mathbf{G}$ *be reductive with* $\mathbf{G}/\mathbf{G}^\circ$ *cyclic. Then the Mackey formula for Lusztig induction holds if*

(1) *one of the quasi-Levi subgroups is a quasi-torus; or*
(2) *both quasi-Levi subgroups are contained in* F*-stable parabolic subgroups.*

In particular, in these cases $R^{\mathbf{G}}_{\mathbf{L}\leqslant\mathbf{P}}$ *does not depend on the choice of* $\mathbf{P}$.

Here, the last assertion follows from the preceding one by the same formal argument as in the connected case. We will write $R^{\mathbf{G}}_{\mathbf{L}}$ if this operator does not depend on the choice of parabolic subgroup. The Mackey formula can be used to derive the following extension of Proposition 4.8.11 to arbitrary quasi-central elements, see [DiMi94, Prop. 4.15]:

Proposition 4.8.14 *Let* $\mathbf{G} = \mathbf{G}^\circ\langle a\rangle$ *be reductive with a quasi-central element* a *and* $\mathbf{L}$ *an* F*-stable quasi-Levi subgroup of* $\mathbf{G}$ *containing* a. *Then*

$$R^{\mathbf{G}}_{\mathbf{L}}(\lambda)(a) = R^{C_{\mathbf{G}^\circ}(a)}_{C^\circ_{\mathbf{L}}(a)}(1_{C_{\mathbf{L}^\circ}(a)})\,\lambda(a) \qquad \text{for all } \lambda\in\mathrm{Irr}(\mathbf{L}^F).$$

Digne and Michel [DiMi94, Def. 3.10] also introduce a duality operator:

Definition 4.8.15 Assume that $\mathbf{G} = \mathbf{G}^\circ\langle a\rangle$. The *duality operator* on class functions on $\mathbf{G}^F$ is defined as

$$\mathrm{D}_{\mathbf{G}} := \sum_{\mathbf{P}\supseteq\mathbf{B}} \epsilon_{C^\circ_{\mathbf{L}}(a)} R^{\mathbf{G}}_{\mathbf{L}}\circ {}^*R^{\mathbf{G}}_{\mathbf{L}}$$

where $\mathbf{B}$ is a fixed F-stable quasi-Borel subgroup of $\mathbf{G}$ containing a and the sum runs over all F-stable parabolic subgroups $\mathbf{P}$ of $\mathbf{G}$ containing $\mathbf{B}$ with an F-stable Levi subgroup $\mathbf{L}$ containing a. (This will not depend on the choices for $\mathbf{L}$.)

This operator is a self-adjoint involution and satisfies properties analogous to those of the Alvis–Curtis–Kawanaka–Lusztig duality for connected groups (see [DiMi94, §3.1]). For example by [DiMi94, Prop. 3.13] we have the following analogue of Proposition 3.4.7:

Proposition 4.8.16 *Assume that* $\mathbf{G} = \mathbf{G}^\circ\langle a\rangle$. *Then for any* $\rho \in \mathrm{Irr}(\mathbf{G}^F)$ *restricting irreducibly to* $\mathbf{G}^{\circ F}$, *the class function* ψ *on* $\mathbf{G}^F$ *defined by*

$$\psi(g) := \mathrm{D}_{\mathbf{G}^\circ\langle a^i\rangle}(\mathrm{Res}^{\mathbf{G}^F}_{\mathbf{G}^{\circ F}\langle a^i\rangle}(\rho))(g) \qquad \text{for } g \in \mathbf{G}^{\circ F}\langle a^i\rangle,\ i \in \mathbb{N},$$

is up to sign an irreducible character of $\mathbf{G}^F$.

It is then natural to define (see [Ma93a, §2] and [DiMi94, Def. 3.16]):

Definition 4.8.17 The irreducible character $\mathrm{St}_{\mathbf{G}} := \mathrm{D}_{\mathbf{G}}(1_{\mathbf{G}})$, where $1_{\mathbf{G}}$ is the trivial character of $\mathbf{G}^F$, is called the *Steinberg character* of $\mathbf{G}^F$.

The Steinberg character is an actual character; its values are given by a formula completely similar to the connected case:

Proposition 4.8.18 *Let* $\mathbf{G} = \mathbf{G}^\circ\langle a\rangle$ *and* $g \in a\mathbf{G}^F$. *Then*

$$\mathrm{St}_{\mathbf{G}}(g) = \begin{cases} \varepsilon_{C^\circ_{\mathbf{G}}(a)}\varepsilon_{C^\circ_{\mathbf{G}}(g)}|C^\circ_{\mathbf{G}}(g)^F|_p & \text{if } g \text{ is quasi-semisimple,} \\ 0 & \text{else.} \end{cases}$$

See [DiMi94, Prop. 3.18]. The Steinberg character can be used to prove an analogue for disconnected groups of Steinberg's formula for the number of unipotent elements (see the remarks after Corollary 3.4.19). This is worked out in [LLS14, Thm. 1.1].

The character $\mathrm{St}_{\mathbf{G}}$ defined above is an extension of the Steinberg character of $\mathbf{G}^{\circ F}$ to the disconnected group $\mathbf{G}^F$. In Remark 2.6.26 the *unipotent characters* $\mathrm{Uch}(\mathbf{G}^F)$ of $\mathbf{G}^F$ were defined as the constituents of $\mathrm{Ind}^{\mathbf{G}^F}_{\mathbf{G}^{\circ F}}(\rho)$, for $\rho \in \mathrm{Uch}(\mathbf{G}^{\circ F})$. We have the following alternative characterisation, which is far from obvious given that the $R^{\mathbf{G}}_{\mathbf{T}}(1_{\mathbf{T}})$ are virtual characters:

Proposition 4.8.19 *Let* $\mathbf{G}$ *be reductive with a Frobenius map* F. *The unipotent characters of* $\mathbf{G}^F$ *are exactly the irreducible constituents of the various* $R^{\mathbf{G}}_{\mathbf{T}}(1_{\mathbf{T}})$ *where* $\mathbf{T}$ *ranges over the* F*-stable maximal quasi-tori of* $\mathbf{G}$.

Proof This can be deduced from the formula

$$\sum_{\rho \in \mathrm{Uch}(\mathbf{G}^{\circ F})} \rho(1)\,\mathrm{Ind}^{\mathbf{G}^F}_{\mathbf{G}^{\circ F}}(\rho) = \frac{1}{|\mathbf{W}^\circ|}\sum_{w \in \mathbf{W}^\circ} R_w(1)\, R^{\mathbf{G}}_{\mathbf{T}_w},$$

which in turn follows from Proposition 2.3.24 together with Proposition 4.8.9, see [Bo99, Lemme 6.4.2]. □

It follows from this that as in the connected case, unipotent characters of $\mathbf{G}^F$ are trivial on $\mathbf{Z}^\circ(\mathbf{G})^F$ [Bo99, Prop. 6.4.3].

Example 4.8.20 The parametrisation of unipotent characters of disconnected groups is rather straightforward, given the result for the connected case, by using the knowledge of the action of automorphisms from Theorem 4.5.11 in conjunction with the extendability result in Theorem 4.5.12. For example, consider $\mathbf{G} = \mathrm{GO}_{2n}$ where $\mathbf{G}^\circ = \mathrm{SO}_{2n}$. If F is an untwisted Frobenius map, then the unipotent characters of $\mathbf{G}^F = \mathrm{GO}_{2n}^+(q)$ are parametrised by equivalence classes of symbols in $\mathscr{X}_{d,n}^{1,0}$ of rank n and defect $d \equiv 0 \pmod 4$. Here the symbols with two equal rows label those unipotent characters of $\mathbf{G}^F$ whose restriction to $\mathrm{SO}_{2n}^+(q)$ is reducible.

The decomposition of $R_{\mathbf{T}}^{\mathbf{G}}(1_{\mathbf{T}})$ into unipotent characters is known. For $\mathbf{G}^\circ$ simple either of exceptional type, or of classical type of small rank, it was determined in [Ma93a], which also gave a conjecture for the general case ([Ma93a, §7]). This was shown in [DiMi94, Thm. 5.11], based on results of Asai [As84c] on Shintani descent to deal with the case of classical groups. Lusztig [Lu12b, Thm. 2.4(ii)] gives a rather different proof in the general case in the spirit of his approach in the connected case. The first step is again a base change via the character table of a suitable Weyl group; here again σ denotes the automorphism of $\mathbf{W}^\circ$ induced by F:

Definition 4.8.21 Assume that $\mathbf{G} = \mathbf{G}^\circ\langle a\rangle$ with a quasi-central element a. Let $\phi \in \mathrm{Irr}(C_{\mathbf{W}^\circ}(a))^\sigma$ and $\tilde\phi$ a σ-extension of ϕ. Then

$$R_{\tilde\phi}^{\mathbf{G}} := \frac{1}{|C_{\mathbf{W}^\circ}(a)|} \sum_{w\in C_{\mathbf{W}^\circ}(a)} \tilde\phi(w)\, R_{\mathbf{T}_w}^{\mathbf{G}}(1_{\mathbf{T}_w}),$$

with $\mathbf{T}_w \leqslant \mathbf{G}$ a quasi-torus of type w, is called the corresponding *almost character*.

Let us point out the following subtlety: even if the restriction of such an almost character $R_{\tilde\phi}^{\mathbf{G}}$ to any coset of $\mathbf{G}^{\circ F}$ in $\mathbf{G}^F$ coincides with the restriction of an irreducible character of $\mathbf{G}^F$, $R_{\tilde\phi}^{\mathbf{G}}$ itself need not be an irreducible character. We refer to this as the gluing problem.

As in the connected case, a class function on a $\mathbf{G}^{\circ F}$-coset $x\mathbf{G}^{\circ F}$ in $\mathbf{G}^F$ is called *uniform* if it is a linear combination of the restriction to $x\mathbf{G}^{\circ F}$ of characters $R_{\mathbf{T}}^{\mathbf{G}}(\theta)$, for F-stable maximal quasi-tori $\mathbf{T}$ and $\theta \in \mathrm{Irr}(\mathbf{T}^F)$. Again, the trivial character on any coset turns out to be uniform by [DiMi94, Prop. 4.12].

Example 4.8.22 A noteworthy difference to the connected case is that there can be unipotent characters of $\mathbf{G}^F$ whose restriction to some coset of $\mathbf{G}^{\circ F}$ is orthogonal to all $R_{\mathbf{T}}^{\mathbf{G}}(1_{\mathbf{T}})$, that is, its uniform projection is zero:

(a) For $\mathbf{G}$ the extension of SL_n, $n \geqslant 3$, by the transpose-inverse automorphism, by [DiMi94, Thm. 5.2] the almost characters yield extensions of those unipotent characters of $\mathrm{SL}_n(q)$ or $\mathrm{SU}_n(q)$ labelled by partitions with a 2-core of size 0 or 1 (depending on the parity of n), that is, for those which lie in the principal 2-Harish-Chandra series of $\mathrm{SL}_n(q)$ (respectively in the principal series for $\mathrm{SU}_n(q)$). All other extensions of unipotent characters are orthogonal to the space of uniform functions, and no construction of these seems to be known.

(b) More concretely, consider the case $n = 3$, when $\mathbf{G}$ is the extension of SL_3 by the transpose-inverse graph automorphism a and F a Frobenius map commuting with a. The three unipotent characters of $\mathbf{G}^{\circ F}$ extend to $\mathbf{G}^F$ by Theorem 4.5.11, but the restrictions of the extensions of the unipotent character ρ_{21} of $\mathbf{G}^{\circ F}$ to the coset $\mathbf{G}^{\circ F}a$ are orthogonal to the restrictions of all $R_{\mathbf{T}}^{\mathbf{G}}(1_{\mathbf{T}})$ to this coset [Ma93a, Thm. 4] (or, to put it differently, the two extensions of ρ_{21} to $\mathbf{G}^F$ have the same multiplicity in all $R_{\mathbf{T}}^{\mathbf{G}}(1_{\mathbf{T}})$). For F such that $\mathbf{G}^{\circ F} = \mathrm{SU}_3(q)$ this can also be deduced from the fact that the character field of these extensions contains $\sqrt{-q}$ (see [Ma90b, Tab. 2]) while the $R_{\mathbf{T}}^{\mathbf{G}}(1_{\mathbf{T}})$ are rational valued.

4.8.23 When the index $|\mathbf{G} : \mathbf{G}^\circ|$ is divisible by the characteristic p of k, outer cosets of $\mathbf{G}^\circ$ may contain non-trivial unipotent conjugacy classes. One can then define 1-parameter *Green functions* as in the connected case as the restriction of the $R_{\mathbf{T}}^{\mathbf{G}}(\theta)$ to the unipotent elements in a coset, see [Ma93a, Sect. 8–10]: Let $\mathbf{G}_{\mathrm{uni}}$ denote the set of unipotent elements of $\mathbf{G}$ and let $a \in \mathbf{G}$ be quasi-central. For $\mathbf{T} \leqslant \mathbf{G}$ an F-stable maximal quasi-torus with $a \in \mathbf{T}^F$ set

$$Q_{\mathbf{T}}^{\mathbf{G}}(u) := R_{\mathbf{T}}^{\mathbf{G}}(1_{\mathbf{T}})(u) \qquad \text{for } u \in (a\mathbf{G})_{\mathrm{uni}}^F.$$

These Green functions have been computed for groups of exceptional type [Ma93b] and for groups of classical type and small rank [Ma93a]. Sorlin [So04] has defined a Springer correspondence which was then determined in [MaSo04] for groups of classical type. A generalised Springer correspondence in this setting was introduced by Lusztig [Lu04a] and determined completely for groups of classical type. In [Ma05] it is calculated for the groups of exceptional type and it is shown that the obvious generalisation of the Lusztig–Shoji algorithm (see 2.8.9) applied with this Springer correspondence yields the Green functions computed in [Ma93a, Ma93b]. It is to be expected that this holds in general.

After this short excursion into the representation theory of disconnected groups, we now prove a generalisation of Theorem 4.7.2 to groups with cyclic centre.

Theorem 4.8.24 *Let $\mathbf{G}$ be a Levi subgroup of a simple group with cyclic centre, $F : \mathbf{G} \to \mathbf{G}$ a Frobenius endomorphism, and assume that the Mackey formula holds for $\mathbf{G}^F$. Then for all F-stable Levi subgroups $\mathbf{L}^* \leqslant \mathbf{G}^*$ with dual $\mathbf{L} \leqslant \mathbf{G}$ and all semisimple $s \in \mathbf{L}^{*F}$ such that $C_{\mathbf{L}^*}(s)$ is connected and only has components of classical types A, B, C and D, there is a Jordan decomposition $J_s^\bullet$ as in Theorem 2.6.22* *for which the diagram*

$$\begin{array}{ccc}
\mathbb{Z}\mathscr{E}(\mathbf{G}^F, s) & \xrightarrow{J_s^{\mathbf{G}}} & \mathbb{Z}\mathrm{Uch}(C_{\mathbf{G}^*}(s)^F) \\
\uparrow R_{\mathbf{L}}^{\mathbf{G}} & & \uparrow R_{C_{\mathbf{L}^*}(s)}^{C_{\mathbf{G}^*}(s)} \\
\mathbb{Z}\mathscr{E}(\mathbf{L}^F, s) & \xrightarrow{J_s^{\mathbf{L}}} & \mathbb{Z}\mathrm{Uch}(C_{\mathbf{L}^*}(s)^F)
\end{array}$$

commutes.

Observe that here Lusztig induction on the right-hand upwards arrow is to a possibly disconnected group as introduced in Definition 4.8.8.

Proof Let $i : \mathbf{G} \hookrightarrow \tilde{\mathbf{G}}$ be a regular embedding with dual epimorphism $i^* : \tilde{\mathbf{G}}^* \to \mathbf{G}^*$. Let $\tilde{\mathbf{L}} = \mathbf{L}\mathbf{Z}(\tilde{\mathbf{G}})$ and $\tilde{\mathbf{L}}^*$ the full preimage of $\mathbf{L}^*$ in $\tilde{\mathbf{G}}^*$, dual to $\tilde{\mathbf{L}}$. Let $\tilde{s} \in \tilde{\mathbf{L}}^{*F}$ be a preimage of s. By Theorem 4.7.5 the middle square of the following diagram commutes:

$$\begin{array}{ccccccc}
\mathbb{Z}\mathscr{E}(\mathbf{G}^F, s) & \xrightarrow{\mathrm{Ind}} & \mathbb{Z}\mathscr{E}(\tilde{\mathbf{G}}^F, \tilde{s}) & \xrightarrow{J_{\tilde{s}}^{\tilde{\mathbf{G}}}} & \mathbb{Z}\mathrm{Uch}(C_{\tilde{\mathbf{G}}^*}(\tilde{s})^F) & \xrightarrow{\mathrm{Ind}} & \mathbb{Z}\mathrm{Uch}(C_{\mathbf{G}^*}(s)^F) \\
\uparrow R_{\mathbf{L}}^{\mathbf{G}} & & \uparrow R_{\tilde{\mathbf{L}}}^{\tilde{\mathbf{G}}} & & \uparrow R_{C_{\tilde{\mathbf{L}}^*}(\tilde{s})}^{C_{\tilde{\mathbf{G}}^*}(\tilde{s})} & & \uparrow R_{C_{\mathbf{L}^*}(s)}^{C_{\mathbf{G}^*}(s)} \\
\mathbb{Z}\mathscr{E}(\mathbf{L}^F, s) & \xrightarrow{\mathrm{Ind}} & \mathbb{Z}\mathscr{E}(\tilde{\mathbf{L}}^F, \tilde{s}) & \xrightarrow{J_{\tilde{s}}^{\tilde{\mathbf{L}}}} & \mathbb{Z}\mathrm{Uch}(C_{\tilde{\mathbf{L}}^*}(\tilde{s})^F) & = & \mathbb{Z}\mathrm{Uch}(C_{\mathbf{L}^*}(s)^F)
\end{array}$$

Let us explain the other parts of the diagram. The horizontal arrows on the left mean induction and then intersecting with the Lusztig series on the right. On the right-hand side bottom row, since $C_{\mathbf{L}^*}(s)$ is connected, the unipotent characters of $C_{\mathbf{L}^*}(s)^F$ are just the (deflations of the) unipotent characters of $C_{\tilde{\mathbf{L}}^*}(\tilde{s})^F$. Similarly, on the top row, the unipotent characters of $C^\circ_{\mathbf{G}^*}(s)^F$ are just the (deflations of the) unipotent characters of $C_{\tilde{\mathbf{G}}^*}(\tilde{s})^F$, and the arrow is ordinary induction from $C^\circ_{\mathbf{G}^*}(s)^F$ to $C_{\mathbf{G}^*}(s)^F$. Now the left-hand square commutes by Corollary 3.3.25, and the right-hand one by Proposition 4.8.10.

Let $\lambda \in \mathscr{E}(\mathbf{L}^F, s)$. Since $C_{\mathbf{L}^*}(s)$ is connected, λ is $\tilde{\mathbf{L}}^F$-invariant, so extends to $\tilde{\mathbf{L}}^F$ (since by assumption $\mathbf{Z}(\mathbf{L})/\mathbf{Z}^\circ(\mathbf{L})$ is cyclic), and all constituents of $\mathrm{Ind}_{\mathbf{L}^F}^{\tilde{\mathbf{L}}^F}(\lambda)$ lie in distinct Lusztig series corresponding to the $|\tilde{\mathbf{G}}^F : \mathbf{G}^F|$ distinct preimages of s. So there is a unique constituent $\tilde{\lambda}$ in $\mathscr{E}(\tilde{\mathbf{L}}^F, \tilde{s})$, occurring with multiplicity 1 by the Multiplicity-Freeness Theorem 1.7.15. Let ρ be a constituent of $R_{\mathbf{L}}^{\mathbf{G}}(\lambda)$, and $g \in \tilde{\mathbf{L}}^F$. Then $R_{\mathbf{L}}^{\mathbf{G}}(\lambda) = R_{\mathbf{L}}^{\mathbf{G}}(\lambda^g) = R_{\mathbf{L}}^{\mathbf{G}}(\lambda)^g$, so all constituents of $R_{\mathbf{L}}^{\mathbf{G}}(\lambda)$ in one $\tilde{\mathbf{G}}^F$-orbit appear with the same multiplicity.

Since $\mathbf{Z}(\mathbf{G})/\mathbf{Z}^\circ(\mathbf{G})$ is cyclic, so is $C_{\mathbf{G}^*}(s)/C^\circ_{\mathbf{G}^*}(s)$ and thus the induction from $C^\circ_{\mathbf{G}^*}(s)^F$ to $C_{\mathbf{G}^*}(s)^F$ is multiplicity-free. By Jordan decomposition 2.6.22, the number of characters in $\mathbb{Z}\mathrm{Uch}(C_{\mathbf{G}^*}(s)^F)$ above $J_{\tilde{s}}^{\tilde{\mathbf{G}}}(\tilde{\rho})$ is the same as the number of constituents of $\mathrm{Res}_{\mathbf{G}^F}^{\tilde{\mathbf{G}}^F}(\tilde{\rho})$, for any $\tilde{\rho} \in \mathscr{E}(\tilde{\mathbf{G}}^F, \tilde{s})$. Choosing any bijection between these two sets, the commutation of the above diagram then shows our claim. □

In the last step of this proof, one could make a more specific choice of bijection by fixing the image of one character, and then using the action of the cyclic group $\tilde{\mathbf{G}}^F/\mathbf{G}^F\mathbf{Z}(\tilde{\mathbf{G}})^F$ on the left-hand side, and of the isomorphic group $\mathrm{Irr}(C_{\mathbf{G}^*}(s)^F/C^\circ_{\mathbf{G}^*}(s)^F)$ on the right to assign the other images.

Example 4.8.25 Let us consider the simplest example: assume that n is prime and $\mathbf{G}^F = \mathrm{SL}_n(q)$ with n dividing $q-1$. Then it is easily seen that there is a unique conjugacy class of semisimple elements in $\mathbf{G}^* = \mathrm{PGL}_n$ with disconnected centraliser, namely the class of the image s in PGL_n of $\tilde{s} = \mathrm{diag}(1, \zeta, \ldots, \zeta^{n-1}) \in \mathrm{GL}_n$, where $\zeta \in \mathbb{F}_q^\times$ has order n (see, e.g., [Bo05]). Here $C_{\mathbf{G}^*}(s)/C^\circ_{\mathbf{G}^*}(s)$ is cyclic of order n and F acts trivially on it, so the $\mathbf{G}^*$-class of s splits into n distinct $\mathbf{G}^{*F}$-classes, with representatives $s_0, \ldots, s_{n-1}$, with $C^\circ_{\mathbf{G}^*}(s_i)^F$ a torus of order $(q-1)^{n-1}$ for $i = 0$ and of order $(q^n-1)/(q-1)$ when $i > 0$. Hence $|\mathrm{Uch}(C_{\mathbf{G}^*}(s_i)^F)| = n = |\mathcal{E}(\mathbf{G}, s_i)|$ by Jordan decomposition (Theorem 2.6.22).

For any F-stable proper Levi subgroup $\mathbf{L}^* < \mathbf{G}^*$ containing s_i we have that $C_{\mathbf{L}^*}(s_i)$ is connected, hence we are in the situation of Theorem 4.8.24. Then any choice of bijection between the n elements of $\mathcal{E}(\mathbf{G}^F, s_i)$ and of $\mathrm{Uch}(C_{\mathbf{G}^*}(s_i)^F)$ will make Lusztig induction $R_\mathbf{L}^\mathbf{G}$ commute with this Jordan decomposition.

4.8.26 The previous example has been vastly generalised by Bonnafé [Bo99] and subsequently Cabanes [Ca13]. For this let $\mathbf{G} = \mathrm{SL}_n$ with a Frobenius map $F : \mathbf{G} \to \mathbf{G}$. Here the dual group is $\mathbf{G}^* = \mathrm{PGL}_n$. Since $\mathbf{Z}(\mathrm{SL}_n)$ is cyclic, the centraliser $\mathbf{H} := C_{\mathbf{G}^*}(s)$ of any semisimple element $s \in \mathbf{G}^* = \mathrm{PGL}_n$ has the property that $\mathbf{H}/\mathbf{H}^\circ$ is cyclic of order dividing $|\mathbf{Z}(\mathrm{SL}_n)| = n_{p'}$ and thus $\mathbf{H}$ is a possibly disconnected group as considered above. Bonnafé works in a rather more general setting than ours here, in that he allows groups $\mathbf{H}$ such that $\mathbf{H}/\mathbf{H}^\circ$ acts on the simple factors of $\mathbf{H}^\circ$ in a wreath fashion. We restrict ourselves to the case above where $\mathbf{H} = \mathbf{H}^\circ \rtimes A$, with A cyclic, permuting the simple factors of $\mathbf{H}^\circ$ and no element of A inducing a non-trivial graph automorphism on any simple factor.

For every $a \in A$ and a-invariant unipotent character ρ of $\mathbf{H}^{\circ F}$ he defines an almost character as in Definition 4.8.21 on the stabiliser $\mathbf{H}^F_\rho$, which turns out to be (up to sign) an irreducible character of $\mathbf{H}^F_\rho$ extending ρ, see [Bo99, Thm. 7.3.2] for $\mathrm{SL}_n(q)$ and [Ca13, Thm. 3.6] for $\mathrm{SU}_n(q)$. This implies in particular:

Corollary 4.8.27 *In the above situation, all unipotent characters of $\mathbf{H}^F$ are uniform and thus the Mackey formula holds for $\mathbf{H}^F$ on the subspace of unipotent class functions.*

This is as in the connected case, but in contrast to the situation when non-trivial graph automorphisms are involved, see Example 4.8.22. The explicit description of the unipotent characters in terms of characters $R_\mathbf{T}^\mathbf{H}(1)$, together with transitivity, also allows one to work out the decomposition of all $R_{\mathbf{L}}^{\mathbf{H}}(1)$ in this case, see [Bo99, Thm. 7.6.1] for $\mathrm{SL}_n(q)$, and [Ca13, Thm. 4.4] for $\mathrm{SU}_n(q)$.

Theorem 4.8.28 (Bonnafé [Bo06, §27], Cabanes [Ca13, Thm. 4.9]) *Let F be a Frobenius map on $\mathbf{G} = \mathrm{SL}_n$ and assume q is large enough. Then for every semisimple element $s \in \mathbf{G}^{*F}$ and any F-stable Levi subgroup $\mathbf{L}^* \leqslant \mathbf{G}^*$ containing s, with dual $\mathbf{L}$, there exists a bijection $J_s^{\mathbf{L}} : \mathscr{E}(\mathbf{L}^F, s) \to \mathrm{Uch}(C_{\mathbf{L}^*}(s)^F)$ such that for all F-stable Levi subgroups $\mathbf{L}^* \leqslant \mathbf{M}^* \leqslant \mathbf{G}^*$ the following diagram commutes:*

$$\begin{array}{ccc} \mathbb{Z}\mathscr{E}(\mathbf{M}^F, s) & \xrightarrow{J_s^{\mathbf{M}}} & \mathbb{Z}\mathrm{Uch}(C_{\mathbf{M}^*}(s)^F) \\ \uparrow R_{\mathbf{L}}^{\mathbf{M}} & & \uparrow R_{C_{\mathbf{L}^*}(s)}^{C_{\mathbf{M}^*}(s)} \\ \mathbb{Z}\mathscr{E}(\mathbf{L}^F, s) & \xrightarrow{J_s^{\mathbf{L}}} & \mathbb{Z}\mathrm{Uch}(C_{\mathbf{L}^*}(s)^F) \end{array}$$

The assumption on q is introduced by Lusztig's result [Lu88, Thm. 1.14] which at present is only known under this condition.

About the proof For the diagram to make sense one first needs to observe that $C_{\mathbf{L}^*}(s)$ is a quasi-Levi subgroup of $C_{\mathbf{M}^*}(s)$ in the sense introduced above, so that the Lusztig induction on the right-hand side is defined.

Then, it should be noted that neither Lusztig induction in the statement mentions a parabolic subgroup containing the (quasi-)Levi subgroup. This is justified by the validity of the Mackey formula for the groups in question, see Theorem 3.3.7 and Corollary 4.8.27.

The main task is then to define the maps $J_s^{\mathbf{L}}$ such that the commutation holds. They are obtained by identifying both sides with the set of irreducible characters of a suitable relative Weyl group. □

The commutation remains open in general for groups with disconnected centre.

Appendix

Further Reading and Open Questions

In this appendix, we loosely present, somewhat informally and in no particular order, topics and problems that were touched upon in the main text but, for one reason or another, could not be thoroughly dealt with. Thus, we mainly collect open ends and problems for which the theory is not complete or not even existent (yet), as well as questions it would be important to solve for various applications. We also mention related projects and topics, with indications for further reading.

A.1 Representatives of Conjugacy Classes

As already mentioned in the preface, we do not discuss in this book any particular aspects of the problem of determining the conjugacy classes of $\mathbf{G}^F$. Nevertheless, it may be useful to indicate some critical issues related to this problem. On a general level, the Jordan decomposition of elements provides the framework in which the conjugacy classes are described: first one determines the classes of semisimple elements and then, for each semisimple class representative $s \in \mathbf{G}^F$, one determines the unipotent classes of $C_{\mathbf{G}}(s)^F$. (Recall from 2.2.13 that $C_{\mathbf{G}}^\circ(s)$ is connected reductive, and that all unipotent elements of $C_{\mathbf{G}}(s)$ are already contained in $C_{\mathbf{G}}^\circ(s)$.) See also [Spa82a], [Der84], [Ca85], [Hum95] for further information and detailed references. Thus, the problem of obtaining a classification of the conjugacy classes of $\mathbf{G}^F$ is essentially under control. As a model case see [Miz77], where the conjugacy classes of $\mathbf{G}^F = E_6(q)$ are determined.

As briefly discussed in Remark 2.7.4, there is an issue of choosing class representatives $g \in \mathbf{G}^F$ such that the action of F on $A(g) = C_{\mathbf{G}}(g)/C_{\mathbf{G}}^\circ(g)$ is as simple as possible. Specific choices of class representatives are needed, for example, in order to fix characteristic functions of F-stable character sheaves, or for the computation of Green functions (see Example 2.7.27 and Remark 2.8.8).

More concretely, let us assume that $\mathbf{G}$ is a simple algebraic group. Then $\mathbf{G}$ =

$\langle x_\alpha(t) \mid \alpha \in \Phi, t \in k\rangle$, where Φ is the root system of $\mathbf{G}$ with respect to a maximally split torus $\mathbf{T}_0$ of $\mathbf{G}$. Representatives for the unipotent classes are typically described by explicit expressions as products of elements $x_\alpha(t)$, for various α and t. On the other hand, it is known that every semisimple element of $\mathbf{G}$ is conjugate to an element of $\mathbf{T}_0$, but the conjugation takes place in the algebraic group $\mathbf{G}$. In order to obtain representatives in $\mathbf{G}^F$, one could use Steinberg's cross section; see [Hum95, §4.10 and §4.15]. This produces a set of representatives for the semisimple classes, but it is totally unrelated to $\mathbf{T}_0$. Consequently, we will also have a problem for all mixed classes, that is, classes that are neither unipotent nor semisimple. For such classes, it appears to be very difficult – both conceptually and in concrete terms – to specify representatives with 'good' properties.

A.2 Green Functions for Type E_8 and $p = 2, 3, 5$

There is a long tradition of work on Green functions; the principal ideas and methods are summarised in Shoji's survey [Sho86]. At that point, the Green functions were known in all cases where q is a power of a good prime for $\mathbf{G}$, or where q is arbitrary, $\mathbf{G}$ is of small rank and the whole table of unipotent character values is available (like for G_2, ${}^3D_4$, ${}^2B_2$, ${}^2G_2$). A precursor of the full Lusztig–Shoji algorithm was employed in [Sho82], [BeSp84] to determine the Green functions for $\mathbf{G}$ of type F_4, E_6, E_7 and E_8 in good characteristic. Explicit results for the larger groups of exceptional type in characteristic 2 are obtained in [Ma90], [Ma93b]. Classical groups in characteristic 2 are dealt with in [Sho07]. In view of [Ge19c], the only open cases (at the time of this writing) are groups of type E_8 in bad characteristic. There is some hope that extensions of the methods employed in [Ge19c] will lead to a solution of these cases as well.

The methods in [Ge19c] rely on a theoretical result which also led to a solution of an old problem concerning the Glauberman correspondence in the general character theory of finite groups; see [Ge19b].

A.3 Character Tables of $F_4(3)$ and $E_7(2)$

One of the ultimate computational challenges is the determination of the 'generic' character tables of the large groups of exceptional type F_4, E_6, ${}^2E_6$, E_7 and E_8. (Complete tables for the smaller groups of exceptional type are already known and contained in the library of CHEVIE; see Table 2.4, p. 103.) Note that for each type of group, there will be several such tables according to the congruence classes of q modulo a certain integer that is determined by the underlying complete root datum.

(For example, in Section 2.1 we have seen that two such tables are required for $\mathbf{G}^F = \mathrm{SL}_2(q)$, one for q odd and one for q a power of 2; for type E_8 one needs to distinguish congruence classes mod 60.) A considerable amount of information (on conjugacy classes, Green functions, ...) is already available electronically, via Michel's extensions [MiChv] of CHEVIE and Lübeck's data webpage [Lue07]. However, a number of technically difficult problems related to the decomposition of the Lusztig induction functor $R_{\mathbf{L}}^{\mathbf{G}}$ and the evaluation of characteristic functions of character sheaves remain to be addressed. For various reasons (see, e.g., [Ge19a], [Ge19c]) it is extremely useful to have explicit character tables for concrete, small values of q available; in particular, this includes the cases where $q = p$ is a bad prime for $\mathbf{G}$. The Cambridge ATLAS [CCNPW] contains the character tables of $F_4(2)$ and of the non-abelian composition factor of ${}^2E_6(2)_{\mathrm{sc}}$. In addition, the character table library of GAP (see [Bre18]) contains the character tables of $E_6(2)$ and the whole group ${}^2E_6(2)_{\mathrm{sc}}$. As far as the remaining exceptional groups are concerned, we find the following numbers using Lübeck's online data [Lue07]:

$$|\mathrm{Irr}(F_4(3))| = 273, \quad |\mathrm{Irr}(E_6(3))| = 1269, \quad |\mathrm{Irr}({}^2E_6(3))| = 1389,$$
$$|\mathrm{Irr}(E_7(2))| = 531, \quad |\mathrm{Irr}(E_7(3)_{\mathrm{sc}})| = 5052,$$
$$|\mathrm{Irr}(E_8(2))| = 1156, \quad |\mathrm{Irr}(E_8(3))| = 12825, \quad |\mathrm{Irr}(E_8(5))| = 519071.$$

Thus, it should be within reach to determine the individual character tables of $F_4(3)$, $E_7(2)$ and, perhaps, even of $E_8(2)$. The knowledge of these tables would be very useful for solving the open problems on characteristic functions of cuspidal character sheaves mentioned in Example 2.7.27.

A.4 Mackey Formula

The Mackey formula for Lusztig induction is known in most cases (see Theorem 3.3.7) but its proof is far from satisfactory and requires difficult and tricky case-by-case arguments. Knowledge of the Mackey formula implies in particular independence of Lusztig induction from the choice of parabolic subgroup, a result which would be very important to have even independently.

A.5 Jordan Decomposition

For connected reductive groups with connected centre, Digne and Michel defined a unique Jordan decomposition (see Theorem 4.7.1). Yet, this Jordan decomposition is not constructed (or proved) by an intrinsic method, but by some ad hoc assumptions and conventions. It would be highly desirable to find a general approach to a unique Jordan decomposition. This would most likely also lead to a general result on commutation of Jordan decomposition with Lusztig induction, which we could here only arrive at by some detailed case-by-case discussion, and even then not in complete generality (see Theorem 4.7.5).

In the general case of groups with not necessarily connected centre, there has so far not been a method to define a unique Jordan decomposition except for groups of type A (see Theorem 4.8.28), and consequently the very important commutation problem is widely open. It might be hoped that Lusztig's approach via categorical centres [Lu16] could give a conceptual solution. A first important open question here is already whether Jordan decomposition can be chosen such that it commutes with Harish-Chandra induction (see Corollary 4.7.6 for the case of connected centre).

This is also related to the much deeper question on existence of Morita equivalences for Brauer ℓ-blocks of characters in Lusztig series corresponding under Jordan decomposition as discussed in [BoRo93, BDR15].

A.6 d-Harish-Chandra Theory and Cyclotomic Hecke Algebras

The d-Harish-Chandra theory presented in Section 3.5 should be explained by the so-called cyclotomic Hecke algebras, certain deformations of group algebras of complex reflection groups, that conjecturally occur as endomorphism algebras of Lusztig induced d-cuspidal modules, see the conjectures in Broué–Malle [BrMa93]. That had first been shown (not under this point of view) by Lusztig [Lu76c] in certain cyclic cases, and in some further particular situations by Digne, Michel and Rouquier [DMR07], and Digne–Michel [DiMi14]. It should lead to a conceptual proof of the Comparison Theorem 4.6.21 and the analogue of Howlett–Lehrer–Lusztig theory stated in Theorem 4.6.24. Moreover this would give a unified approach to not only proving but also 'understanding' Broué's abelian defect group conjecture in modular representation theory for groups of Lie type (see [BrMa93]).

This is closely connected to the so-called spetses programme which strives to construct analogues of finite reductive groups whose Weyl group is a 'spetsial' complex reflection group (see [Ma98]), begun in [Ma95, BMM99, BMM14].

A.7 Disconnected Groups

The representation theory of disconnected groups is still at its beginnings; we sketched some results of Digne–Michel [DiMi94] and Bonnafé [Bo99] in Section 4.8. Probably one would need a continuation of Lusztig's approach via character sheaves on disconnected groups [Lu09e], leading, for example, to an algorithm for the computation of Green functions extending the Lusztig–Shoji algorithm discussed in Section 2.8, to a construction of those unipotent characters with zero uniform projection (see Example 4.8.22), and more generally to an analogue of Jordan decomposition for characters on a coset.

A.8 Shintani Descent

Another interesting direction, which we did not touch upon in this book, is the character theory of finite groups obtained by extending a finite reductive group (connected or not) by a group of field automorphisms. This is one of the themes of Shintani descent, originating in [Shi76] (see also the survey [Di87]). Shintani descent is an important tool for studying various powers of a Frobenius map $F\colon \mathbf{G} \to \mathbf{G}$ at the same time; see [Lu84a, Chap. 2], [DiMi85, Chap. III], [Kaw87, §1]. The 'Shintani descent identity', due independently to Asai, Digne–Michel and Lusztig, connects zeta functions of Deligne–Lusztig varieties to Hecke algebras; see Curtis' survey [Cu88, §2] and further references there. Shintani descent also plays an important role in determining the Lusztig induction functor $R_{\mathbf{L}}^{\mathbf{G}}$; see Asai's papers (as discussed in Section 4.6) as well as [Sho85], [Sho87].

A.9 Automorphisms and Galois Action

Both of the latter questions, on characters of disconnected groups as well as on extensions by field automorphisms, are closely related to the more general question of the action of outer automorphisms on the irreducible characters of finite reductive groups. While this is solved for groups with connected centre (see e.g. Theorems 4.5.11 and 4.5.12), the question is open in the general case. See Cabanes–Späth [CaSp17] and Taylor [Ta18b] for the symplectic groups and [Ma17b] for partial results for cuspidal characters in quasi-isolated series. For example it would be desirable to have a Jordan decomposition equivariant with respect to outer automorphisms, and, in fact, with respect to field automorphisms (see Theorem 4.7.9 for connected centre groups).

A.10 Some Applications

Many applications of the character theory of finite groups are related to the following two purely group-theoretical problems. Let Γ be a finite group.

- If C_1, C_2, C_3 are conjugacy classes of Γ (not necessarily distinct), then compute the number of pairs $(x, y) \in C_1 \times C_2$ with $xy \in C_3$.
- If $\Gamma' \leqslant \Gamma$ is a subgroup and C is a conjugacy class of Γ, then compute the cardinality $|C \cap \Gamma'|$ or, more generally, $|C \cap \Gamma' g \Gamma'|$ for any $g \in \Gamma$.

It is well known that both problems can be solved efficiently, once sufficient information about the character table of Γ is known. (See the exercises in [Is76, Chap. 3] for the first problem, and [CuRe81, §11.D] for the second problem.) In the case where $\Gamma = \mathbf{G}^F$ is a finite group of Lie type, this character-theoretic approach to group-theoretical problems has been applied successfully in a variety of situations: to mention but a few, see Malle–Matzat [MaMa18] (realisation of finite groups as Galois groups); Malle–Saxl–Weigel [MSW94] (generation of finite classical groups); Lusztig [Lu03a] (subgroups isomorphic to the alternating group $\mathfrak{A}_5$ inside groups of Lie type of type E_8); Lusztig [Lu11] (construction of a surjective map from conjugacy classes in the Weyl group $\mathbf{W}$ to unipotent classes of the underlying algebraic group $\mathbf{G}$); the solution of a conjecture of P. Neumann on the existence of elements with small fixed spaces in linear groups by Guralnick–Malle [GM12a]; their construction of Beauville surfaces for all but one finite non-abelian simple group [GM12b]; the resolution of cases of the Arad–Herzog conjecture on products of conjugacy classes by Guralnick–Malle–Tiep [GMT13]; and the solution of the Ore conjecture on commutators in simple groups by Liebeck–O'Brien–Shalev–Tiep (see [Ma14], which also describes further applications to images of word maps).

For further applications of a somewhat different nature, see Liebeck–Shalev [LiSh05] (character degrees and random walks on finite groups), Lusztig [Lu00] (existence of $\mathbf{G}^F$-invariant vectors in representations of $\mathbf{G}^{F^2}$, a problem which has previously been studied in the context of Lie groups and p-adic groups), Reeder [Re07] (study of the restriction of Deligne–Lusztig characters, motivated by a problem for finite orthogonal groups).

A.11 Related Topics and Projects

In this book, we are exclusively dealing with algebraic groups defined over finite fields, and representations of these groups over fields of characteristic 0, following the approach initiated by Deligne and Lusztig in the 1970s. One of the features of this theory is that it can be, and has been, distilled into finite combinatorial terms (root data, Hecke algebras, Lusztig's Fourier matrices and so on), that lead to efficient algorithms and implementations on a computer: see the CHEVIE project. These programs have turned out to be highly useful, both in a number of applications, and in the explicit verification of certain general properties for groups of exceptional type. (An example is mentioned in the remarks concerning the proof of Theorem 2.7.25.)

There is an analogous, multi-author project for real reductive groups and unitary representations, under the title 'Atlas of Lie groups and representations'; see `www.liegroups.org/`. The article by Adams et al. [ALTV] provides a good overview and also gives a detailed description of the atlas algorithms for computing unitary representations. Again, the general theory is translated into finite combinatorial terms and efficient computer implementations. One of the major successes of this project attracted a huge interest in the mathematical community and beyond; see [Vo07].

In another direction, the Harish-Chandra philosophy discussed in Chapter 3 has been appearing in a number of different contexts which, although sometimes based on entirely different theoretical foundations, eventually lead to similar combinatorial patterns. A recent example is the work of Achar et al. [AHJR], where the Springer correspondence (which we encountered in this book in Section 2.8, in connection with the problem of computing Green functions) has been generalised to a modular setting where one considers sheaves with coefficients in a field of characteristic $\ell > 0$ (and not just in $\overline{\mathbb{Q}}_\ell$). There are results about the classification of cuspidal objects and the parametrisation of Harish-Chandra series which are highly reminiscent of results in [DD97], [GHM94], [GHM96], [GeHi97] (where ℓ-modular representations of finite groups of Lie type are considered). Compare, for example, the descriptions of modular Harish-Chandra series for GL_n in [AHJR, Example 6.1] and in [DD97, §4]; or the characterisation of a modular Steinberg object in [AHJR, Theorem 7.1] and in [GHM94, Theorem 4.2].

References

[AHJR] P. N. Achar, A. Henderson, D. Juteau and S. Riche, Modular generalized Springer correspondence: an overview. See arXiv:1510.08962.

[Ac02] B. Ackermann, A short note on Howlett-Lehrer theory. *Arch. Math. (Basel)* **79** (2002), 161–166.

[ALTV] J. D. Adams, M. van Leeuwen, P. E. Trapa and D. A. Vogan, Unitary representations of real reductive groups. See arXiv:1212.2192.

[Al09] D. Allcock, A new approach to rank one linear algebraic groups. *J. Algebra* **321** (2009), 2540–2544.

[Al79] D. Alvis, The duality operation in the character ring of a finite Chevalley group. *Bull. Amer. Math. Soc. (N.S.)* **1** (1979), 907–911.

[Al82] D. Alvis, Duality and character values of finite groups of Lie type. *J. Algebra* **74** (1982), 211–222.

[AlLu82] D. Alvis and G. Lusztig, The representations and generic degrees of the Hecke algebras of type H_4. *J. Reine Angew. Math.* **336** (1982), 201–212; Erratum, **449** (1994), 217–218.

[As] T. Asai, Endomorphism algebras of the reductive groups over $\mathbb{F}_q$ of classical type. Unpublished manuscript (around 1980).

[As84a] T. Asai, Unipotent class functions of split special orthogonal groups SO_{2n}^+ over finite fields. *Comm. Algebra* **12** (1984), 517–615.

[As84b] T. Asai, The unipotent class functions on the symplectic groups and odd orthogonal groups over finite fields. *Comm. Algebra* **12** (1984), 617–645.

[As84c] T. Asai, The unipotent class functions of exceptional groups over finite fields. *Comm. Algebra* **12** (1984), 2729–2857.

[As85] T. Asai, The unipotent class functions of non-split finite special orthogonal groups. *Comm. Algebra* **13** (1985), 845–924.

[Asch00] M. Aschbacher, The classification of the finite simple groups. *Jber. d. Dt. Math.-Verein* **102** (2000), 95–101.

[Asch04] M. Aschbacher, The status of the classification of the finite simple groups. *Notices of the AMS* **51** (2004), 736–740.

[BBD82] A. A. Beilinson, J. Bernstein and P. Deligne, Faisceaux pervers. *Astérisque* No. 100, Soc. Math. France, 1982.

[Be79] C. Benson, The generic degrees of the irreducible characters of E_8. *Comm. Algebra* **7** (1979), 1199–1209.

[BeCu72] C. Benson and C. W. Curtis, On the degrees and rationality of certain characters of finite Chevalley groups. *Trans. Amer. Math. Soc.* **165** (1972), 251–273; Corrections and additions: ibid. **202** (1975), 405–406.

[BeLu78] W. M. Beynon and G. Lusztig, Some numerical results on the characters of exceptional Weyl groups. *Math. Proc. Cambridge Philos. Soc.* **84** (1978), 417–426.

[BeSp84] W. M. Beynon and N. Spaltenstein, Green functions of finite Chevalley groups of type E_n $(n = 6, 7, 8)$. *J. Algebra* **88** (1984), 584–614.

[Bo98] C. Bonnafé, Formule de Mackey pour q grand. *J. Algebra* **201** (1998), 207–232.

[Bo99] C. Bonnafé, Produits en couronne de groupes linéaires. *J. Algebra* **211** (1999), 57–98.

[Bo00] C. Bonnafé, Mackey formula in type A. *Proc. London Math. Soc.* **80** (2000), 545–574; Erratum, *ibid.* **86** (2003), 435–442.

[Bo05] C. Bonnafé, Quasi-isolated elements in reductive groups. *Comm. Algebra* **33** (2005), 2315–2337.

[Bo06] C. Bonnafé, Sur les caractères des groupes réductifs finis à centre non connexe: applications aux groupes spéciaux linéaires et unitaires. *Astérisque* No. 306 (2006).

[Bo11] C. Bonnafé, *Representations of* $\mathrm{SL}_2(F_q)$. Algebra and Applications, 13. Springer-Verlag London, 2011.

[BDR15] C. Bonnafé, J.-F. Dat and R. Rouquier, Derived categories and Deligne–Lusztig varieties II. *Ann. of Math. (2)* **185** (2017), 609–670.

[BoMi11] C. Bonnafé and J. Michel, Computational proof of the Mackey formula for $q > 2$. *J. Algebra* **327** (2011), 506–526.

[BoRo93] C. Bonnafé and R. Rouquier, Catégories dérivées et variétés de Deligne-Lusztig. *Publ. Math. Inst. Hautes Études Sci.* **97** (2003), 1–59.

[Bor91] A. Borel, *Linear Algebraic Groups*. Second enlarged edition. Graduate Texts in Mathematics vol. 126, Springer Verlag, Berlin–Heidelberg–New York, 1991.

[Bor70] A. Borel, R. Carter, C. W. Curtis, N. Iwahori, T. A. Springer and R. Steinberg, *Seminar on Algebraic Groups and Related Algebraic Groups*. Lecture Notes in Mathematics, vol. 131. Springer Verlag, Berlin–New York, 1970.

[BoTi65] A. Borel and J. Tits, Groupes réductifs. *Publ. Math. IHES* **27** (1965), 55–150.

[Bou68] N. Bourbaki, *Groupes et algèbres de Lie, Chap. 4, 5 et 6*. Hermann, Paris, 1968.

[Bou07] N. Bourbaki, *Algèbre, Chap. 9*. Springer Verlag, Berlin Heidelberg, 2007.

[Bre18] T. Breuer, *The GAP character table library*. Online available at `www.math.rwth-aachen.de/~Thomas.Breuer/ctbllib/index.html`.

[Br17] M. Broué, *On Characters of Finite Groups*. Springer Verlag, Heidelberg, 2017.

[BrKi02] M. Broué and S. Kim, Familles de caractères des algèbres de Hecke cyclotomiques. *Adv. Math.* **172** (2002), 53–136.

[BrMa92] M. Broué and G. Malle, Théorèmes de Sylow génériques pour les groupes réductifs sur les corps finis. *Math. Ann.* **292** (1992), 241–262.

[BrMa93] M. Broué and G. Malle, Zyklotomische Hecke-Algebren. *Astérisque* No. 212 (1993), 119–189.

[BMM93] M. Broué, G. Malle and J. Michel, Generic blocks of finite reductive groups. *Astérisque* No. 212 (1993), 7–92.

[BMM99] M. Broué, G. Malle and J. Michel, Towards spetses, I. *Transform. Groups* **4** (1999), 157–218.

[BMM14] M. Broué, G. Malle and J. Michel, Split spetses for primitive reflection groups. *Astérisque* No. 359 (2014).

[BrMi89] M. Broué and J. Michel, Blocs et séries de Lusztig dans un groupe réductif fini. *J. Reine Angew. Math.* **395** (1989), 56–67.

[BrLu12] O. Brunat and F. Lübeck, On defining characteristic representations of finite reductive groups. *J. Algebra* **395** (2013), 121–141.

[Ca88] M. Cabanes, Brauer morphism between modular Hecke algebras. *J. Algebra* **115** (1988), 1–31.

[Ca13] M. Cabanes, On Jordan decomposition of characters for $\mathrm{SU}(n, q)$. *J. Algebra* **374** (2013), 216–230.

[CE94] M. Cabanes and M. Enguehard, On unipotent blocks and their ordinary characters. *Invent. Math.* **117** (1994), 149–164.

[CE99] M. Cabanes and M. Enguehard, On blocks of finite reductive groups and twisted induction. *Adv. Math.* **145** (1999), 189–229.

[CE04] M. Cabanes and M. Enguehard, *Representation Theory of Finite Reductive Groups*. New Mathematical Monographs, 1. Cambridge University Press, 2004.

[CaSp17] M. Cabanes, B. Späth, Inductive McKay condition for finite simple groups of type C. *Represent. Theory* **21** (2017), 61–81.

[Car55-56] H. Cartan, *Variétés Algébriques Affines*. Séminaire Henri Cartan, vol. 8 (1955-1956), Exp. No. 3, 12 pp. Available at `www.numdam.org/item?id=SHC_1955-1956__8__A3_0`

[Ca72] R. W. Carter, *Simple Groups of Lie Type*. Wiley, New York, 1972; reprinted 1989 as Wiley Classics Library Edition.

[Ca85] R. W. Carter, *Finite Groups of Lie Type: Conjugacy Classes and Complex Characters*. Wiley, New York, 1985; reprinted 1993 as Wiley Classics Library Edition.

[Ca95] R. W. Carter, On the representation theory of the finite groups of Lie type over an algebraically closed field of characteristic 0. *Algebra, IX*, pp. 1–120, 235–239, Encyclopaedia Math. Sci., 77, Springer, Berlin, 1995.

[Cha68] B. Chang, The conjugate classes of Chevalley groups of type G_2. *J. Algebra* **9** (1968), 190–211.

[ChRe74] B. Chang and R. Ree, The characters of $G_2(q)$. *Symposia Mathematica, Vol. XIII (Convegno di Gruppi e loro Rappresentazioni, INDAM, Rome, 1972)*, pp. 395–413. Academic Press, London, 1974.

[Ch55] C. Chevalley, Sur certains groups simples. *Tôhoku Math. J.* **7** (1955), 14–66.

[Ch05] C. Chevalley, *Classification des Groupes Algébriques Semi-simples*. Collected works. Vol. 3. Edited and with a preface by P. Cartier. With the collaboration of Cartier, A. Grothendieck and M. Lazard. Springer-Verlag, Berlin, 2005.

[Chl09] M. Chlouveraki, *Blocks and Families for Cyclotomic Hecke Algebras*. Lecture Notes in Mathematics, 1981. Springer-Verlag, Berlin, 2009.

[Chl10] M. Chlouveraki, Rouquier blocks of the cyclotomic Hecke algebras of $G(de, e, r)$. *Nagoya Math. J.* **197** (2010), 175–212.

[ClPr13] M. C. Clarke and A. Premet, The Hesselink stratification of nullcones and base change. *Invent. Math.* **191** (2013), 631–669.

[CMT04] A. M. Cohen, S. H. Murray and D. E. Taylor, Computing in groups of Lie type. *Math. Comp.* **73** (2004), 1477–1498.

[Con14] B. Conrad, Reductive group schemes. *Autour des Schémas en Groupes*. Vol. I, pp. 93–444, Panor. Synthèses, 42/43, Soc. Math. France, Paris, 2014.

[CCNPW] J. H. Conway, R. T. Curtis, S. P. Norton, R. A. Parker and R. A. Wilson, *Atlas of Finite Groups*. Oxford University Press, London/New York, 1985.

[Cu66] C. W. Curtis, The Steinberg character of a finite group with a (B, N)-pair. *J. Algebra* **4** (1966), 433–441.

[Cu80] C. W. Curtis, Truncation and duality in the character ring of a finite group of Lie type. *J. Algebra* **62** (1980), 320–332.

[Cu88] C. W. Curtis, Representations of Hecke algebras. *Astérisque* **168** (1988), 13–60.

[CuRe81] C. W. Curtis and I. Reiner, *Methods of Representation Theory Vol. I*. Wiley, New York, 1981.

[CuRe87] C. W. Curtis and I. Reiner, *Methods of Representation Theory Vol. II*. Wiley, New York, 1987.

[DeLu76] P. Deligne and G. Lusztig, Representations of reductive groups over finite fields. *Ann. of Math. (2)* **103** (1976), 103–161.

[DeLu82] P. Deligne and G. Lusztig, Duality for representations of a reductive group over a finite field. *J. Algebra* **74** (1982), 284–291.

[DeLu83] P. Deligne and G. Lusztig, Duality for representations of a reductive group over a finite field. II. *J. Algebra* **81** (1983), 540–545.

[DG70/11] M. Demazure and A. Grothendieck (eds.), *Séminaire de Géométrie Algébrique du Bois Marie - 1962-64 - Schémas en groupes - (SGA 3) - Tome III (Structure des schémas en groupes réductifs)*. Édition recomposée et annotée du volume 153 des Lecture Notes in Mathematics publié en 1970 par Springer-Verlag. Documents Mathématiques, vol. 8, Soc. Math. France, 2011.

[DN19] T. De Medts and K. Naert, Suzuki–Ree groups as algebraic groups over $\mathbb{F}_{\sqrt{p}}$. See `arXiv:1904.10741`.

[Der84] D. I. Deriziotis, *Conjugacy Classes and Centralizers of Semisimple Elements in Finite Groups of Lie Type*. Vorlesungen aus dem Fachbereich Mathematik der Universität GH Essen, 11. Universität Essen, Fachbereich Mathematik, Essen, 1984.

[DeMi87] D. I. Deriziotis and G. O. Michler, Character table and blocks of finite simple triality groups ${}^3D_4(q)$. *Trans. Amer. Math. Soc.* **303** (1987), 39–70.

[Dieu74] J. Dieudonné, *La Géométrie des Groupes Classiques* (troisième édition). Springer Verlag, Berlin–Heidelberg–New York, 1974.

[Di87] F. Digne, Shintani descent and $\mathscr{L}$ functions on Deligne-Lusztig varieties. *The Arcata Conference on Representations of Finite Groups (Arcata, Calif., 1986)*, pp. 61–68, Proc. Sympos. Pure Math., 47, Part 1, Amer. Math. Soc., Providence, RI, 1987.

[DiMi83] F. Digne and J. Michel, Foncteur de Lusztig et fonctions de Green généralisées. *C. R. Acad. Sci. Paris Sér. I Math.* **297** (1983), 89–92.

[DiMi85] F. Digne and J. Michel, *Fonctions $\mathscr{L}$ des variétés de Deligne–Lusztig et descente de Shintani*. Mém. Soc. Math. France, no. 20, suppl. au Bull. S. M. F. **113**, 1985.

[DiMi90] F. Digne and J. Michel, On Lusztig's parametrization of characters of finite groups of Lie type. *Astérisque* No. 181–182 (1990), 113–156.

[DiMi94] F. Digne and J. Michel, Groupes réductifs non connexes. *Ann. scient. Éc. Norm. Sup.* **27** (1994), 345–406.

[DiMi14] F. Digne and J. Michel, Parabolic Deligne-Lusztig varieties. *Adv. Math.* **257** (2014), 136–218.

[DiMi15] F. Digne and J. Michel, Complements on disconnected reductive groups. *Pacific J. Math.* **279** (2015), 203–228.

[DiMi20] F. Digne and J. Michel, *Representations of Finite Groups of Lie Type*. London Mathematical Society Student Texts, 21. 2nd Edition, Cambridge University Press, 2020, to appear.

[DMR07] F. Digne, J. Michel and R. Rouquier, Cohomologie des variétés de Deligne–Lusztig. *Adv. Math.* **209** (2007), 749–822.

[DD93] R. Dipper and J. Du, Harish-Chandra vertices. *J. Reine Angew. Math.* **437** (1993), 101–130.

[DD97] R. Dipper and J. Du, Harish-Chandra vertices and Steinberg's tensor product theorems for finite greneral linear groups. *Proc. London Math. Soc.* **75** (1997), 559–599.

[DF92] R. Dipper and P. Fleischmann, Modular Harish-Chandra theory. I. *Math. Z.* **211** (1992), 49–71.

[Dr87] V. G. Drinfeld, Quantum groups. *Proceedings of the International Congress of Mathematicians, Vol. 1, 2 (Berkeley, Calif., 1986)*, pp. 798–820, Amer. Math. Soc., Providence, RI, 1987.

[DuMa18] O. Dudas and G. Malle, Modular irreducibility of cuspidal unipotent characters. *Invent. Math.* **211** (2018), 579–589.

[Ef94] M. Eftekhari, Descente de Shintani des faisceaux caractères. *C. R. Acad. Sci. Paris Sér. I Math.* **318** (1994), 305–308.

[En13] M. Enguehard, Towards a Jordan decomposition of blocks of finite reductive groups. Available at `arxiv:1312.0106`.

[EM17] M. Enguehard and J. Michel, The Sylow subgroups of a finite reductive group. *Bull. Inst. Math. Acad. Sinica* **13** (2018), 227–247.

[Enn63] V. Ennola, On the characters of the finite unitary groups. *Ann. Acad. Sci. Fenn. Ser. A I* **323** (1963), 120–155.

[Eno72] H. Enomoto, The characters of the finite symplectic group $\mathrm{Sp}(4, q)$, $q = 2^f$. *Osaka J. Math.* **9** (1972), 75–94.

[Eno76] H. Enomoto, The characters of the finite Chevalley group $G_2(q)$, $q = 3^f$. *Japan. J. Math.* **2** (1976), 191–248.

[EnYa86] H. Enomoto and H. Yamada, The characters of $G_2(2^n)$. *Japan. J. Math.* **12** (1986), 325–377.

[Et11] P. Etingof, O. Golberg, S. Hensel, T. Liu, A. Schwendner, D. Vaintrob and E. Yudovina, *Introduction to Representation Theory. With Historical Interludes by Slava Gerovitch*. Student Math. Library, vol. 59, Amer. Math. Soc., Providence, RI, 2011.

[Fei82] W. Feit, *The Representation Theory of Finite Groups*. North–Holland, Amsterdam, 1982.

[Fo69] J. Fogarty, *Invariant Theory*. W. A. Benjamin, New York, 1969.

[FoSr82] P. Fong and B. Srinivasan, The blocks of finite general linear and unitary groups. *Invent. Math.* **69** (1982), 109–153.

[FoSr86] P. Fong and B. Srinivasan, Generalized Harish-Chandra theory for unipotent characters of finite classical groups. *J. Algebra* **104** (1986), 301–309.

[FoSr89] P. Fong and B. Srinivasan, The blocks of finite classical groups. *J. Reine Angew. Math.* **396** (1989), 122–191.

[FoSr90] P. Fong and B. Srinivasan, Brauer trees in classical groups. *J. Algebra* **131** (1990), 179–225.

[Fro96] F. G. Frobenius, Über Gruppencharaktere. *Sitzungsberichte der Königlich Preußischen Akademie der Wissenschaften zu Berlin* (1896), 985–1021.

[FuHa91] W. Fulton and J. Harris, *Representation Theory. A First Course*. Graduate Texts in Mathematics, 129. Readings in Mathematics. Springer Verlag, New York, 1991.

[GAP4] The GAP Group, *GAP – Groups, Algorithms, and Programming*. Version 4.8.10, 2018. (www.gap-system.org)

[Ge92] M. Geck, On the classification of l-blocks of finite groups of Lie type. *J. Algebra* **151** (1992), 180–191.

[Ge93a] M. Geck, Basic sets of Brauer characters of finite groups of Lie type, II. *J. London Math. Soc. (2)* **47** (1993), 255–268.

[Ge93b] M. Geck, A note on Harish-Chandra induction. *Manuscripta Math.* **80** (1993), 393–401.

[Ge95] M. Geck, *Beiträge zur Darstellungstheorie von Iwahori-Hecke-Algebren*. Habilitationsschrift, Aachener Beiträge zur Mathematik **11**, Verlag der Augustinus Buchhandlung, Aachen, 1995.

[Ge96] M. Geck, On the average values of the irreducible characters of finite groups of Lie type on geometric unipotent classes. *Doc. Math. J. DMV* **1** (1996), 293–317.

[Ge00] M. Geck, On the representation theory of Iwahori–Hecke algebras of extended finite Weyl groups. *Represent. Theory* **4** (2000), 370–397.

[Ge01] M. Geck, Modular Harish-Chandra series, Hecke algebras and (generalized) q-Schur algebras. *Modular Representation Theory of Finite Groups (Charlottesville, VA, 1998)*, pp. 1–66, Walter de Gruyter, Berlin, 2001.

[Ge03a] M. Geck, *An Introduction to Algebraic Geometry and Algebraic Groups*. Oxford Graduate Texts in Mathematics, vol. 10, Oxford University Press, New York, 2003.

[Ge03b] M. Geck, Character values, Schur indices and character sheaves. *Represent. Theory* **7** (2003), 19–55.

[Ge03c] M. Geck, On the Schur indices of cuspidal unipotent characters. *Finite Groups 2003 (Gainesville, FL, 2003)*, pp. 87–104, Walter de Gruyter, Berlin, 2004.

[Ge03d] M. Geck, On the p-defects of character degrees of finite groups of Lie type. *Carpathian J. Math.* **19** (2003), 97–100.

[Ge05a] M. Geck, The Schur indices of the cuspidal unipotent characters of the finite Chevalley groups $E_7(q)$. *Osaka J. Math.* **42** (2005), 201–215.

[Ge11] M. Geck, Some applications of CHEVIE to the theory of algebraic groups. *Carpath. J. Math.* **27** (2011), 64–94.

[Ge12a] M. Geck, Remarks on modular representations of finite groups of Lie type in non-defining characteristic. *Algebraic Groups and Quantum Groups*, pp. 71–80, Contemp. Math., 565, Amer. Math. Soc., Providence, RI, 2012.

[Ge12b] M. Geck, On the Kazhdan–Lusztig order on cells and families. *Comment. Math. Helv.* **87** (2012), 905–927.

[Ge13] M. Geck, On Kottwitz' conjecture for twisted involutions. *J. Lie Theory* **25** (2015), 395–429.

[Ge17] M. Geck, On the construction of semisimple Lie algebras and Chevalley groups. *Proc. Amer. Math. Soc.* **145** (2017), 3233–3247.

[Ge18] M. Geck, A first guide to the character theory of finite groups of Lie type. *Local Representation Theory and Simple Groups (eds. R. Kessar, G. Malle, D. Testerman)*, pp. 63–106, EMS Lecture Notes Series, EMS Publ. House, 2018.

[Ge19a] M. Geck, On the values of unipotent characters in bad characteristic. *Rend. Cont. Sem. Mat. Univ. Padova* **141** (2019), 37–63.

[Ge19b] M. Geck, Green functions and Glauberman degree-divisibility. See `arXiv: 1904.04586`.

[Ge19c] M. Geck, Computing Green functions in small characteristic. *J. Algebra* (2020), to appear.

[GeHe08] M. Geck and D. Hézard, On the unipotent support of character sheaves. *Osaka J. Math.* **45** (2008), 819–831.

[GeHi91] M. Geck and G. Hiss, Basic sets of Brauer characters of finite groups of Lie type. *J. Reine Angew. Math.* **418** (1991), 173–188.

[GeHi97] M. Geck and G. Hiss, Modular representations of finite groups of Lie type in non-defining characteristic. *Finite Reductive Groups: Related Structures and Representations (Luminy, 1994)*, pp. 195–249, Birkhäuser Boston, Ma, 1997.

[GHLMP] M. Geck, G. Hiß, F. Lübeck, G. Malle and G. Pfeiffer, CHEVIE–A system for computing and processing generic character tables for finite groups of Lie type, Weyl groups and Hecke algebras. *Appl. Algebra Engrg. Comm. Comput.* **7** (1996), 175–210; see also `www.math.rwth-aachen.de/~CHEVIE`.

[GHM94] M. Geck, G. Hiss and G. Malle, Cuspidal unipotent Brauer characters. *J. Algebra* **168** (1994), 182–220.

[GHM96] M. Geck, G. Hiss and G. Malle, Towards a classification of the irreducible representations in non-describing characteristic of a finite group of Lie type. *Math. Z.* **221** (1996), 353–386.

[GeIa12] M. Geck and L. Iancu, Ordering Lusztig's families in type B_n. *J. Algebraic Comb.* **38** (2013), 457–489.

[GIM00] M. Geck, L. Iancu and G. Malle, Weights of Markov traces and generic degrees. *Indag. Math. (N.S.)* **11** (2000), 379–397.

[GeJa11] M. Geck and N. Jacon, *Representations of Hecke Algebras at Roots of Unity*. Algebra and Applications, vol. 15, Springer-Verlag, London, 2011.

[GKP00] M. Geck, S. Kim and G. Pfeiffer, Minimal length elements in twisted conjugacy classes of finite Coxeter groups. *J. Algebra* **229** (2000), 570–600.

[GeMa99] M. Geck and G. Malle, On special pieces in the unipotent variety. *Experiment. Math.* **8** (1999), 281–290.

[GeMa00] M. Geck and G. Malle, On the existence of a unipotent support for the irreducible characters of finite groups of Lie type. *Trans. Amer. Math. Soc.* **352** (2000), 429–456.

[GeMa03] M. Geck and G. Malle, Fourier transforms and Frobenius eigenvalues for finite Coxeter groups. *J. Algebra* **260** (2003), 162–193.

[GePf92] M. Geck and G. Pfeiffer, The unipotent characters of the Chevalley groups $D_4(q)$, q odd. *Manuscripta Math.* **76** (1992), 281–304.

[GePf00] M. Geck and G. Pfeiffer, *Characters of Finite Coxeter Groups and Iwahori–Hecke Algebras*. London Mathematical Society Monographs. New Series, 21. The Clarendon Press, Oxford University Press, New York, 2000.

[GoWa98] R. Goodman and N. R. Wallach, *Symmetry, Representations and Invariants*. Graduate Texts in Mathematics vol. 255, Springer Verlag, Berlin–Heidelberg–New York, 2009.

[GLS94] D. Gorenstein, R. Lyons and R. Solomon, *The Classification of the Finite Simple Groups*. Math. Surveys and Monographs vol. 40, no. 1, Amer. Math. Soc., 1994.

[GLS96] D. Gorenstein, R. Lyons and R. Solomon, *The Classification of the Finite Simple Groups*. Math. Surveys and Monographs vol. 40, no. 2, Amer. Math. Soc., 1996.

[GLS98] D. Gorenstein, R. Lyons and R. Solomon, *The Classification of the Finite Simple Groups*. Math. Surveys and Monographs vol. 40, no. 3, Amer. Math. Soc., 1998.

[Gre55] J. A. Green, The characters of the finite general linear groups. *Trans. Amer. Math. Soc.* **80** (1955), 402–447.

[Gre99] J. A. Green, Discrete series characters for $\mathrm{GL}(n, q)$. *Algebr. Represent. Theory* **2** (1999), 61–82.

[Gro02] L. C. Grove, *Classical Groups and Geometric Algebra*. Graduate Studies in Math., vol. 39, Amer. Math. Soc., Providence, RI, 2002.

[GM12a] R. M. Guralnick and G. Malle, Products of conjugacy classes and fixed point spaces. *J. Amer. Math. Soc.* **25** (2012), 77–121.

[GM12b] R. M. Guralnick and G. Malle, Simple groups admit Beauville structures. *J. London Math. Soc.* **85** (2012), 694–721.

[GMT13] R. M. Guralnick, G. Malle and P. Tiep, Products of conjugacy classes in finite and algebraic simple groups. *Adv. in Math.* **234** (2013), 618–652.

[HaCh] Harish-Chandra, Eisenstein series over finite fields. *Functional Analysis and Related Fields (Proc. Conf. M. Stone, Chicago, 1968)*, pp. 76–88, Springer, New York, 1970.

[He19] J. Hetz, On the values of unipotent characters of finite Chevalley groups of type E_6 in characteristic 3. *J. Algebra* **536** (2019), 242–255.

[Hi90a] G. Hiss, Regular and semisimple blocks of finite reductive groups. *J. London Math. Soc.* **41** (1990), 63–68.

[Hi90b] G. Hiss, *Zerlegungszahlen endlicher Gruppen vom Lie-Typ in nicht-definierender Charakteristik*. Habilitationsschrift, RWTH Aachen, 1990.

[Hi93] G. Hiss, Harish-Chandra series of Brauer characters in a finite group with a split BN-pair. *J. London Math. Soc.* **48** (1993), 219–228.

[Ho74] P. Hoefsmit, *Representations of Hecke Algebras of Finite Groups with BN-Pairs of Classical Type*. Ph. D. Thesis, The University of British Columbia (Canada), 1974.

[Ho19] R. Hollenbach, *Quasi-Isolated Blocks and the Malle–Robinson Conjecture*. Dissertation, TU Kaiserslautern, 2019.

[HoSh79] R. Hotta and H. Shimomura, The fixed point subvarieties of unipotent transformations on generalized flag varieties and the Green functions. *Math. Ann.* **241** (1979), 193–208.

[HoSp77] R. Hotta and T. A. Springer, A specialization theorem for certain Weyl group representations and an application to the Green polynomials of unitary groups. *Invent. Math.* **41** (1977), 113–127.

[Ho80] R. B. Howlett, Normalizers of parabolic subgroups of reflection groups. *J. London Math. Soc. (2)* **21** (1980), 62–80.

[HoLe80] R. B. Howlett and G. I. Lehrer, Induced cuspidal representations and generalised Hecke rings. *Invent. Math.* **58** (1980), 37–64.

[HoLe83] R. B. Howlett and G. I. Lehrer, Representations of generic algebras and finite groups of Lie type. *Trans. Amer. Math. Soc.* **280** (1983), 753–779.

[HoLe94] R. B. Howlett and G. I. Lehrer, On Harish-Chandra induction for modules of Levi subgroups. *J. Algebra* **165** (1994), 172–183.

[Hum91] J. E. Humphreys, *Linear Algebraic Groups*. Graduate Texts in Mathematics vol. 21, Springer Verlag, Berlin–Heidelberg–New York, 1975.

[Hum95] J. E. Humphreys, *Conjugacy Classes in Semisimple Algebraic Groups*. Math. Surveys and Monographs, vol. 43, Amer. Math. Soc., RI, 1995.

[Hup67] B. Huppert, *Endliche Gruppen I*. Grundlehren der math. Wissenschaften vol. 134, Springer, Heidelberg, 1967.

[HuBl82] B. Huppert and N. Blackburn, *Finite Groups II*. Grundlehren der math. Wissenschaften vol. 242, Springer, Heidelberg, 1982.

[Ia01] L. Iancu, Markov traces and generic degrees in type B_n. *J. Algebra* **236** (2001), 731–744.

[Is76] I. M. Isaacs, *Character Theory of Finite Groups*. Academic Press, New York, 1976; corrected reprint by Dover Publ., New York, 1994.

[Iw64] N. Iwahori, On the structure of a Hecke ring of a Chevalley group over a finite field. *J. Fac. Sci. Univ. Tokyo* **10** (1964), 215–236.

[Ja86] G. D. James, The irreducible representations of the finite general linear groups. *Proc. London Math. Soc.* **52** (1986), 236–268.

[JaKe81] G. D. James and A. Kerber, *The Representation Theory of the Symmetric Group*. Encyclopedia of Mathematics and its Applications, 16. Addison-Wesley Publishing Co., Reading, MA, 1981.

[Jan03] J. C. Jantzen, *Representations of Algebraic Groups*. Second edition. Mathematical Surveys and Monographs, 107. Amer. Math. Soc., Providence, RI, 2003.

[Jor07] H. Jordan, Group-characters of various types of linear groups. *Amer. J. Math.* **29** (1907), 387–405.

[Kac85] V. Kac, *Infinite Dimensional Lie Algebras*. Cambridge University Press, 1985.

[Kaw82] N. Kawanaka, Fourier transforms of nilpotently supported invariant functions on a simple Lie algebra over a finite field. *Invent. Math.* **69** (1982), 411–435.

[Kaw85] N. Kawanaka, Generalized Gel'fand-Graev representations and Ennola duality. *Algebraic Groups and Related Topics (Kyoto/Nagoya, 1983)*, pp. 175–206, Adv. Stud. Pure Math., 6, North-Holland, Amsterdam, 1985.

[Kaw86] N. Kawanaka, Generalized Gelfand-Graev representations of exceptional simple algebraic groups over a finite field I. *Invent. Math.* **84** (1986), 575–616.

[Kaw87] N. Kawanaka, Shintani lifting and Gel'fand-Graev representations. *The Arcata Conference on Representations of Finite Groups (Arcata, Calif., 1986)*, pp. 147–163, Proc. Sympos. Pure Math., vol. 47, Part 1, Amer. Math. Soc., Providence, RI, 1987.

[Kaz77] D. A. Kazhdan, Proof of Springer's hypothesis. *Israel J. Math.* **28** (1977), 272–286.

[KaLu79] D. A. Kazhdan and G. Lusztig, Representations of Coxeter groups and Hecke algebras. *Invent. Math.* **53** (1979), 165–184.

[KeMa13] R. Kessar and G. Malle, Quasi-isolated blocks and Brauer's height zero conjecture. *Ann. of Math. (2)* **178** (2013), 321–384.

[KeMa15] R. Kessar and G. Malle, Lusztig induction and ℓ-blocks of finite reductive groups. *Pacific J. Math.* **279** (2015), 267–296.

[KeMa19] R. Kessar and G. Malle, Characters in quasi-isolated blocks. In preparation.

[KlTi09] A. Kleshchev and P. H. Tiep, Representations of finite special linear groups in non-defining characteristic. *Adv. Math.* **220** (2009), 478–504.

[Klu16] M. Klupsch, *On Cuspidal and Supercuspidal Representations of Finite Reductive Groups*. Dissertation, RWTH Aachen University, 2016.

[Ko66] B. Kostant, Groups over $\mathbb{Z}$. *Algebraic Groups and Their Discontinuous Subgroups (Boulder, Colorado, 1965)*, pp. 90–98, Proc. Sympos. Pure Math., vol. 8, Amer. Math. Soc., Providence, RI, 1966.

[LaSr90] L. Lambe and B. Srinivasan, A computation of Green functions for some classical groups. *Comm. Algebra* **18** (1990), 3507–3545.

[La56] S. Lang, Algebraic groups over finite fields. *Amer. J. Math.* **78** (1956), 555–563.

[Lau89] G. Laumon, Faisceaux caractères (d'après Lusztig). *Séminaire Bourbaki*, Vol. 1988/89. Astérisque No. 177–178 (1989), Exp. No. 709, 231–260.

[LLS14] R. Lawther, M. Liebeck and G. Seitz, Outer unipotent classes in automorphism groups of simple algebraic groups. *Proc. Lond. Math. Soc. (3)* **109** (2014), 553–595.

[Leh73] G. I. Lehrer, The characters of the finite special linear groups. *J. Algebra* **26** (1973), 564–583.

[Leh78] G. I. Lehrer, On the values of characters of semi-simple groups over finite fields. *Osaka J. Math.* **15** (1978), 77–99.

[LeSp99] G. I. Lehrer and T. Springer, Intersection multiplicities and reflection subquotients of unitary reflection groups. I. *Geometric Group Theory Down Under (Canberra, 1996)*, 181–193, de Gruyter, Berlin, 1999.

[LeSp99b] G. I. Lehrer and T. Springer, Reflection subquotients of unitary reflection groups. *Canad. J. Math.* **51** (1999), 1175–1193.

[LeTa09] G. I. Lehrer and D. E. Taylor, *Unitary Reflection Groups*. Australian Mathematical Society Lecture Series, 20. Cambridge University Press, 2009.

[LiSh05] M. Liebeck and A. Shalev, Character degrees and random walks in finite groups of Lie type. *Proc. London Math. Soc.* **90** (2005), 61–86.

[Loc77] J. Locker, *The complex irreducible characters of $Sp(6, q)$, q even*. Ph. D. Thesis, University of Sydney, 1977.

[Lue93] F. Lübeck, *Charaktertafeln für die Gruppen $CSp_6(q)$ mit ungeradem q und* $\mathrm{Sp}_6(q)$ *mit geradem q*. Dissertation, Universität Heidelberg, 1993.

[Lue07] F. Lübeck, Data for finite groups of Lie type and related algebraic groups. See `www.math.rwth-aachen.de/~Frank.Luebeck/chev/index.html`.

[LM16] F. Lübeck and G. Malle, A Murnaghan-Nakayama rule for values of unipotent characters in classical groups. *Represent. Theory* **20** (2016), 139–161; Corrigendum: ibid. **21** (2017), 1–3.

[Lu75] G. Lusztig, Sur la conjecture de Macdonald. *C. R. Acad. Sci. Paris, Ser. A* **280** (1975), 371–320.

[Lu76b] G. Lusztig, On the finiteness of the number of unipotent classes. *Invent. Math.* **34** (1976), 201–213.

[Lu76c] G. Lusztig, Coxeter orbits and eigenspaces of Frobenius. *Invent. Math.* **38** (1976), 101–159.

[Lu77a] G. Lusztig, Irreducible representations of finite classical groups. *Invent. Math.* **43** (1977), 125–175.

[Lu77b] G. Lusztig, *Representations of Finite Chevalley Groups*. C.B.M.S. Regional Conference Series in Mathematics, vol. 39, Amer. Math. Soc., Providence, RI, 1977.

[Lu79a] G. Lusztig, Unipotent representations of a finite Chevalley group of type E_8. *Quart. J. Math. Oxford* **30** (1979), 315–338.

[Lu79b] G. Lusztig, A class of irreducible representations of a finite Weyl group. *Indag. Math.* **41** (1979), 323–335.

[Lu80a] G. Lusztig, On the unipotent characters of the exceptional groups over finite fields. *Invent. Math.* **60** (1980), 173–192.

[Lu80b] G. Lusztig, Some problems in the representation theory of finite Chevalley groups. *The Santa Cruz Conference on Finite Groups (Univ. California, Santa Cruz, Calif., 1979)*, pp. 313–317, Proc. Sympos. Pure Math. Vol. 37, Amer. Math. Soc., Providence, RI, 1980.

[Lu81a] G. Lusztig, On a theorem of Benson and Curtis. *J. Algebra* **71** (1981), 490–498.

[Lu81c] G. Lusztig, Unipotent characters of the symplectic and odd orthogonal groups over a finite field. *Invent. Math.* **64** (1981), 263–296.

[Lu82a] G. Lusztig, A class of irreducible representations of a finite Weyl group II. *Indag. Math.* **44** (1982), 219–226.

[Lu82b] G. Lusztig, Unipotent characters of the even orthogonal groups over a finite field. *Trans. Amer. Math. Soc.* **272** (1982), 733–751.

[Lu84a] G. Lusztig, *Characters of Reductive Groups Over a Finite Field*. Annals of Mathematics Studies, 107. Princeton University Press, Princeton, NJ, 1984.

[Lu84b] G. Lusztig, Characters of reductive groups over finite fields. *Proceedings of the International Congress of Mathematicians, Vol. 1, 2 (Warsaw, 1983)*, pp. 877–880, PWN, Warsaw, 1984.

[Lu84c] G. Lusztig, Intersection cohomology complexes on a reductive group. *Invent. Math.* **75** (1984), 205–272.

[LuCS] G. Lusztig, Character sheaves. *Adv. Math.* **56** (1985), 193–237; II, **57** (1985), 226–265; III, **57** (1985), 266–315; IV, **59** (1986), 1–63; V, **61** (1986), 103–155.

[Lu86a] G. Lusztig, On the character values of finite Chevalley groups at unipotent elements. *J. Algebra* **104** (1986), 146–194.

[Lu87a] G. Lusztig, Introduction to character sheaves. *The Arcata Conference on Representations of Finite Groups (Arcata, Calif., 1986)*, pp. 165–179, Proc. Symp. Pure Math., vol. 47, Amer. Math. Soc., Providence, RI, 1987.

[Lu87b] G. Lusztig, Leading coefficients of character values of Hecke algebras. *The Arcata Conference on Representations of Finite Groups (Arcata, Calif., 1986)*, pp. 235–262, Proc. Symp. Pure Math., vol. 47, Amer. Math. Soc., Providence, RI, 1987.

[Lu88] G. Lusztig, On the representations of reductive groups with disconnected centre. *Astérisque* No. 168 (1988), 157–166.

[Lu89] G. Lusztig, Affine Hecke algebras and their graded versions. *J. Amer. Math. Soc.* **2** (1989), 599–635.

[Lu90] G. Lusztig, Green functions and character sheaves. *Ann. of Math. (2)* **131** (1990), 355–408.

[Lu91] G. Lusztig, Intersection cohomology methods in representation theory. *Proceedings of the International Congress of Mathematics (Kyoto, 1990)*, pp. 155–174, Springer-Verlag, 1991.

[Lu92a] G. Lusztig, A unipotent support for irreducible representations. *Adv. Math.* **94** (1992), 139–179.

[Lu92b] G. Lusztig, Remarks on computing irreducible characters. *J. Amer. Math. Soc.* **5** (1992), 971–986.

[Lu93] G. Lusztig, Appendix: Coxeter groups and unipotent representations. *Astérisque* No. 212 (1993), 191–203.

[Lu94] G. Lusztig, Exotic Fourier transform. With an appendix by Gunter Malle. *Duke Math. J.* **73** (1994), 227–241, 243–248.

[Lu00] G. Lusztig, $G(F_q)$-invariants in irreducible $G(F_{q^2})$-modules. *Represent. Theory* **4** (2000), 446–465.

[Lu02] G. Lusztig, Rationality properties of unipotent representations. *J. Algebra* **258** (2002), 1–22.

[Lu03a] G. Lusztig, Homomorphisms of the alternating group A_5 into reductive groups. *J. Algebra* **260** (2003), 298–322.

[Lu03b] G. Lusztig, *Hecke Algebras with Unequal Parameters*. CRM Monographs Ser., vol. 18, Amer. Math. Soc., Providence, RI, 2003. Enlarged and updated version at `arXiv:0208154v2`.

[Lu04a] G. Lusztig, Character sheaves on disconnected groups. II. *Represent. Theory* **8** (2004), 72–124.

[Lu04b] G. Lusztig, Character sheaves on disconnected groups. IV. *Represent. Theory* **8** (2004), 145–178.

[Lu05] G. Lusztig, Unipotent elements in small characteristic. *Transform. Groups* **10** (2005), 449–487.

[Lu06] G. Lusztig, Character sheaves and generalizations. *The Unity of Mathematics*, pp. 443–455, Progr. Math. 244, Birkhäuser, Boston, 2006.

[Lu08a] G. Lusztig, Irreducible representations of finite spin groups. *Represent. Theory* **12** (2008), 1–36.

[Lu09a] G. Lusztig, Unipotent classes and special Weyl group representations. *J. Algebra* **321** (2009), 3418–3449.

[Lu09c] G. Lusztig, Study of a **Z**-form of the coordinate ring of a reductive group. *J. Amer. Math. Soc.* **22** (2009), 739–769.

[Lu09d] G. Lusztig, Twelve bridges from a reductive group to its Langlands dual. *Representation Theory*, pp. 125–143, Contemp. Math., 478, Amer. Math. Soc., Providence, RI, 2009.

[Lu09e] G. Lusztig, Character sheaves on disconnected groups. X. *Represent. Theory* **13** (2009), 82–140.

[Lu10] G. Lusztig, *Bruhat Decomposition and Applications*. Notes of a lecture at a one day conference in memory of F. Bruhat held at the Institut Poincaré, Paris; available at `arXiv:1006.5004`.

[Lu11] G. Lusztig, From conjugacy classes in the Weyl group to unipotent classes. *Represent. Theory* **15** (2011), 494–530.

[Lu12a] G. Lusztig, On the cleanness of cuspidal character sheaves. *Mosc. Math. J.* **12** (2012), 621–631.

[Lu12b] G. Lusztig, On the representations of disconnected reductive groups over F_q. *Recent Developments in Lie Algebras, Groups and Representation Theory, ed. K. Misra*, pp. 227–246, Proc. Symp. Pure Math. 86, Amer. Math. Soc., 2012.

[Lu13a] G. Lusztig, Families and Springer's correspondence. *Pacific J. Math.* **267** (2014), 431–450.

[Lu14a] G. Lusztig, Restriction of a character sheaf to conjugacy classes. *Bull. Math. Soc. Sci. Math. Roumanie (N.S.)* **58 (106)** (2015), 297–309.

[Lu14b] G. Lusztig, Unipotent representations as a categorical centre. *Represent. Theory* **19** (2015), 211–235.

[Lu14c] G. Lusztig, *Algebraic and Geometric Methods in Representation Theory*. Lecture at the Chinese University of Hong Kong, Sep. 25, 2014, `arXiv:1409.8003`.

[Lu16] G. Lusztig, Non-unipotent representations and categorical centres. *Bull. Inst. Math. Acad. Sinica (N.S.)* **12** (2017), 205–296.

[Lu17] G. Lusztig, Comments on my papers. See `arXiv:1707.09368v3`.

[Lu18a] G. Lusztig, On the definition of almost characters. *Lie Groups, Geometry, and Representation Theory, ed. V. Kac et al.*, pp. 367–379, Progr. Math. 326, Birkhäuser, Boston, 2018.

[Lu18b] G. Lusztig, Special representations of Weyl groups: a positivity property. *Adv. Math.* **327** (2018), 161–172.

[Lu19] G. Lusztig, On the generalized Springer correspondence. *Representations of Reductive Groups*, pp. 219–253, Proc. Sympos. Pure Math., 101, Amer. Math. Soc., Providence, RI, 2019.

[LuSp79] G. Lusztig and N. Spaltenstein, Induced unipotent classes. *J. London Math. Soc. (2)* **19** (1979), 41–52.

[LuSp85] G. Lusztig and N. Spaltenstein, On the generalized Springer correspondence for classical groups. *Algebraic Groups and Related Topics (Kyoto/Nagoya, 1983)*, pp. 289–316, Adv. Stud. Pure Math., 6, North-Holland, Amsterdam, 1985.

[LuSr77] G. Lusztig and B. Srinivasan, The characters of the finite unitary groups. *J. Algebra* **49** (1977), 167–171.

[LuYu19] G. Lusztig and Z. Yun, Endoscopy for Hecke categories and character sheaves. See `arXiv:1904.01176`.

[LuPa10] K. Lux and H. Pahlings, *Representations of Groups. A Computational Approach*. Cambridge Stud. Adv. Math., vol. 124. Cambridge University Press, 2010.

[Mac95] I. G. Macdonald, *Symmetric Functions and Hall Polynomials*. 2nd edition, Oxford University Press, 1995.

[Ma88] G. Malle, Exceptional groups of Lie type as Galois groups. *J. Reine Angew. Math.* **392** (1988), 70–109.

[Ma90] G. Malle, Die unipotenten Charaktere von ${}^2F_4(q^2)$. *Comm. Algebra* **18** (1990), 2361–2381.

[Ma90b] G. Malle, Some unitary groups as Galois groups over $\mathbb{Q}$. *J. Algebra* **131** (1990), 476–482.

[Ma91] G. Malle, Darstellungstheoretische Methoden bei der Realisierung einfacher Gruppen vom Lie Typ als Galoisgruppen. *Representation Theory of Finite Groups and Finite Dimensional Algebras*, pp. 443–459, Progr. Math. 95, Birkhäuser, Basel, 1991.

[Ma93a] G. Malle, Generalized Deligne-Lusztig characters. *J. Algebra* **159** (1993), 64–97.

[Ma93b] G. Malle, Green functions for groups of types E_6 and F_4 in characteristic 2. *Comm. Algebra* **21** (1993), 747–798.

[Ma95] G. Malle, Unipotente Grade imprimitiver komplexer Spiegelungsgruppen. *J. Algebra* **177** (1995), 768–826.

[Ma97] G. Malle, Degrés relatifs des algèbres cyclotomiques associées aux groupes de réflexions complexes de dimension deux. *Finite Reductive Groups: Related Structures and Representations (Luminy, 1994)*, pp. 311–332, Progr. Math. 141, Birkhäuser Boston, MA, 1997.

[Ma98] G. Malle, Spetses. *Doc. Math., Extra Volume ICM II* (1998), 87–96.

[Ma99] G. Malle, On the rationality and fake degrees of characters of cyclotomic algebras. *J. Math. Sci. Univ. Tokyo* **6** (1999), 647–677.

[Ma00] G. Malle, On the generic degrees of cyclotomic algebras. *Represent. Theory* **4** (2000), 342–369.

[Ma05] G. Malle, Springer correspondence for disconnected exceptional groups. *Bull. London Math. Soc.* **37** (2005), 391–398.

[Ma07] G. Malle, Height 0 characters of finite groups of Lie type. *Represent. Theory* **11** (2007), 192–220.

[Ma08] G. Malle, Extensions of unipotent characters and the inductive McKay condition. *J. Algebra* **320** (2008), 2963–2980.

[Ma14] G. Malle, The proof of Ore's conjecture [after Ellers–Gordeev and Liebeck–O'Brien–Shalev–Tiep]. *Sém. Bourbaki, Astérisque* No. 361 (2014), Exp. 1069, 325–348.

[Ma17] G. Malle, Local-global conjectures in the representation theory of finite groups. *Representation Theory — Current Trends and Perspectives*, pp. 519–539, EMS Ser. Congr. Rep., Eur. Math. Soc., Zürich, 2017.

[Ma17b] G. Malle, Cuspidal characters and automorphisms. *Adv. Math.* **320** (2017), 887–903.

[MaMa18] G. Malle and B. H. Matzat, *Inverse Galois Theory*. 2nd edition. Springer, Berlin, 2018.

[MaRo03] G. Malle and R. Rouquier, Familles de caractères de groupes de réflexions complexes. *Represent. Theory* **7** (2003), 610–640.

[MSW94] G. Malle, J. Saxl and T. Weigel, Generation of classical groups. *Geom. Dedicata* **49** (1994), 85–116.

[MaSo04] G. Malle and K. Sorlin, Springer correspondence for disconnected groups. *Math. Z.* **246** (2004), 291–319.

[MaTe11] G. Malle and D. Testerman, *Linear Algebraic Groups and Finite Groups of Lie Type*. Cambridge Studies in Advanced Mathematics, 133. Cambridge University Press, 2011.

[MZ01] G. Malle and A. E. Zalesskii, Prime power degree representations of quasi-simple groups. *Archiv Math.* **77** (2001), 461–468.

[MaSp89] J. G. M. Mars and T. A. Springer, Character sheaves. *Orbites Unipotentes et Représentations, III*. Astérisque No. 173–174 (1989), 111–198.

[Mas10] J. Maslowski, *Equivariant Character Bijections in Groups of Lie Type*. Dissertation, Technische Universität Kaiserslautern, 2010. Available at `https://kluedo.ub.uni-kl.de/frontdoor/index/index/docId/2229`.

[Mat04] A. Mathas, Matrix units and generic degrees for the Ariki-Koike algebras. *J. Algebra* **281** (2004), 695–730.

[MiChv] J. Michel, The development version of the CHEVIE package of GAP3. *J. Algebra* **435** (2015), 308–336. See also `http://people.math.jussieu.fr/~jmichel/chevie/chevie.html`.

[Miz77] K. Mizuno, The conjugate classes of Chevalley groups of type E_6. *J. Fac. Sci. Univ. Tokyo* **24** (1977), 525–563.

[Miz80] K. Mizuno, The conjugate classes of unipotent elements of the Chevalley groups E_7 and E_8. *Tokyo J. Math.* **3** (1980), 391–461.

[Mor63] A. O. Morris, The characters of the group $\mathrm{GL}(n, q)$. *Math. Z.* **81** (1963), 112–123.

[Mu03] P. Müller, Algebraic groups over finite fields, a quick proof of Lang's theorem. *Proc. Amer. Math. Soc.* **131** (2003), 369–370.

[Na18] G. Navarro, *Character Theory and the McKay Conjecture*. Cambridge Studies in Advanced Mathematics, 175. Cambridge University Press, 2018.

[Na19] G. Navarro, On a question of C. Bonnafé on characters and multiplicity free constituents. *J. Algebra* **520** (2019), 517–519.

[Ohm77] Z. Ohmori, On the Schur indices of $\mathrm{GL}(n, q)$ and $\mathrm{SL}(2n + 1, q)$. *J. Math. Soc. Japan* **29** (1977), 693–707.

[Ohm96] Z. Ohmori, The Schur indices of the cuspidal unipotent characters of the finite unitary groups. *Proc. Japan Acad. Ser. A Math. Sci.* **72** (1996), 111–113.

[Ol84] J. B. Olsson, Remarks on symbols, hooks and degrees of unipotent characters. *J. Combin. Theory Ser. A* **42** (1986), 223–238.

[Op95] E. Opdam, A remark on the irreducible characters and fake degrees of finite real reflection groups. *Invent. Math.* **120** (1995), 447–454.

[OrTe92] P. Orlik and H. Terao, *Arrangements of Hyperplanes*. Grundlehren der Mathematischen Wissenschaften, 300. Springer-Verlag, Berlin, 1992.

[Os54] M. Osima, On the representations of the generalized symmetric group. *Math. J. Okayama Univ.* **4** (1954), 39–56.

[PaVi10] D. I. Panyushev and E. B. Vinberg, The work of Vladimir Morozov on Lie algebras. *Transform. Groups* **15** (2010), 1001–1013.

[Pr82] A. Przygocky, Schur indices of symplectic groups. *Comm. Algebra* **10** (1982), 279–310.

[Re07] M. Reeder, On the restriction of Deligne-Lusztig characters. *J. Amer. Math. Soc.* **20** (2007), 573–602.

[Rou08] R. Rouquier, q-Schur algebras and complex reflection groups. *Mosc. Math. J.* **8** (2008), 119–158.

[Ru01] H. Rui, Weights of Markov traces on cyclotomic Hecke algebras. *J. Algebra* **238** (2001), 762–775.

[Sa71] I. Satake, *Classification Theory of Semi-Simple Algebraic Groups*. With an appendix by M. Sugiura. Lecture Notes in Pure and Applied Mathematics, 3. Marcel Dekker, Inc., New York, 1971.

[Sche85] K. D. Schewe, *Blöcke exzeptioneller Chevalley-Gruppen*. Bonner Mathematische Schriften **165**, Universität Bonn, Math. Institut, 1985.

[Scho97] M. Schönert et. al., GAP – *Groups, algorithms, and programming – version 3 release 4 patchlevel 4*. Lehrstuhl D für Mathematik, RWTH Aachen, Germany, 1997.

[Schu07] I. Schur, Untersuchungen über die Darstellung der endlichen Gruppen durch gebrochene lineare Substitutionen. *J. Reine Angew. Math.* **132** (1907), 85–137.

[ST54] G. C. Shephard and J. A. Todd, Finite unitary reflection groups. *Canadian J. Math.* **6** (1954), 274–304.

[Shi75] K. Shinoda, The conjugacy classes of the finite Ree groups of type (F_4). *J. Fac. Sci. Univ. Tokyo* **22** (1975), 1–15.

[Shi82] K. Shinoda, The characters of the finite conformal symplectic group CSp(4, q). *Comm. Algebra* **10** (1982), 1369–1419.

[Shi76] T. Shintani, Two remarks on irreducible characters of finite general linear groups. *J. Math. Soc. Japan* **28** (1976), 396–414.

[Sho82] T. Shoji, On the Green polynomials of a Chevalley group of type F_4. *Comm. Algebra* **10** (1982), 505–543.

[Sho83] T. Shoji, On the Green polynomials of classical groups. *Invent. Math.* **74** (1983), 239–267.

[Sho85] T. Shoji, Some generalization of Asai's result for classical groups. *Algebraic Groups and Related Topics (Kyoto/Nagoya, 1983)*, pp. 207–229, Adv. Stud. Pure Math., 6, North-Holland, Amsterdam, 1985.

[Sho87] T. Shoji, Shintani descent for exceptional groups over a finite field. *J. Fac. Sci. Univ. Tokyo Sect. IA Math.* **34** (1987), 599–653.

[Sho86] T. Shoji, Green functions of reductive groups over a finite field. *The Arcata Conference on Representations of Finite Groups (Arcata, Calif., 1986)*, pp. 289–302, Proc. Sympos. Pure Math., 47, Part 1, Amer. Math. Soc., Providence, RI, 1987.

[Sho88] T. Shoji, Geometry of orbits and Springer correspondence. *Orbites Unipotentes et Représentations, I. Astérisque* No. 168 (1988), 61–140.

[Sho95] T. Shoji, Character sheaves and almost characters of reductive groups, I, II. *Adv. Math.* **111** (1995), 244–313, 314–354.

[Sho96] T. Shoji, On the computation of unipotent characters of finite classical groups. *Appl. Algebra Engrg. Comm. Comp.* **7** (1996), 165–174.

[Sho97] T. Shoji, Unipotent characters of finite classical groups. *Finite Reductive Groups: Related Structures and Representations (Luminy, 1994)*, pp. 373–413, Progr. Math. 141, Birkhäuser Boston, MA, 1997.

[Sho98] T. Shoji, Irreducible characters of finite Chevalley groups. *Sugaku expositions* **11** (1998), 19–37.

[Sho06a] T. Shoji, Lusztig's conjecture for finite special linear groups. *Represent. Theory* **10** (2006), 164–222.

[Sho06b] T. Shoji, Generalized Green functions and unipotent classes for finite reductive groups. I. *Nagoya Math. J.* **184** (2006), 155–198.

[Sho07] T. Shoji, Generalized Green functions and unipotent classes for finite reductive groups. II. *Nagoya Math. J.* **188** (2007), 133–170.

[Sho09] T. Shoji, Lusztig's conjecture for finite classical groups with even characteristic. *Representation Theory*, pp. 207–236, Contemp. Math., 478, Amer. Math. Soc., Providence, RI, 2009.

[SiFr73] W. A. Simpson and J. S. Frame, The character tables for SL(3, q), SU(3, q^2), PSL(3, q) and PSU(3, q^2). *Canadian J. Math.* **25** (1973), 486–494.

[So04] K. Sorlin, Springer correspondence in non connected reductive groups. *J. Reine Angew. Math.* **568** (2004), 197–234.

[Spa82a] N. Spaltenstein, *Classes Unipotentes et Sous-Groupes de Borel*. Lecture Notes in Mathematics, vol. 946, Springer, 1982.

[Spa82b] N. Spaltenstein, Charactères unipotents de ${}^3D_4(q^3)$. *Comment. Math. Helvetici* **57** (1982), 676–691.

[Spa85] N. Spaltenstein, On the generalized Springer correspondence for exceptional groups. *Algebraic Groups and Related Topics (Kyoto/Nagoya, 1983)*, pp. 317–338, Adv. Stud. Pure Math., 6, North-Holland, Amsterdam, 1985.

[Spr70] T. A. Springer, Cusp forms for finite groups. *Seminar on Algebraic Groups and Related Finite Groups (The Institute for Advanced Study, Princeton, N.J., 1968/69)*, pp. 97–120, Lecture Notes in Mathematics, Vol. 131 Springer, Berlin, 1970.

[Spr74] T. A. Springer, Regular elements of finite reflection groups. *Invent. Math.* **25** (1974), 159–198.

[Spr76] T. A. Springer, Trigonometric sums, Green functions of finite groups and representations of Weyl groups. *Invent. Math.* **36** (1976), 173–207.

[Spr78] T. A. Springer, A construction of representations of Weyl groups. *Invent. Math.* **44** (1978), 279–293.

[Spr82] T. A. Springer, Quelques applications de la cohomologie d'intersection. *Bourbaki Seminar, Vol. 1981/1982*, pp. 249–273, Astérisque No. 92–93, Soc. Math. France, Paris, 1982

[Spr98] T. A. Springer, *Linear Algebraic Groups*. Second Edition, Progr. Math. 9, Birkhäuser, Boston, 1998.

[SpSt70] T. A. Springer and R. Steinberg, Conjugacy classes. *Seminar on Algebraic Groups and Related Finite Groups (The Institute for Advanced Study, Princeton, N.J., 1968/69)*, pp. 167–266, Lecture Notes in Mathematics, Vol. 131, Springer, Berlin, 1970.

[Sr68] B. Srinivasan, The characters of the finite symplectic group $\mathrm{Sp}(4, q)$. *Trans. Amer. Math. Soc.* **131** (1968), 488–525.

[Sr79] B. Srinivasan, *Representations of Finite Chevalley Groups*. Lecture Notes in Mathematics, vol. 764, Springer, 1979.

[Sr91] B. Srinivasan, Character sheaves: applications to finite groups. *Algebraic Groups and Their Generalizations: Classical Methods (University Park, PA, 1991)*, pp. 183–194, Proc. Sympos. Pure Math., 56, Part 1, Amer. Math. Soc., Providence, RI, 1994.

[SrVi15] B. Srinivasan and C. R. Vinroot, Jordan decomposition and real-valued characters of finite reductive groups with connected center. *Bull. Lond. Math. Soc.* **47** (2015), 427–435.

[SrVi18] B. Srinivasan and C. R. Vinroot, Galois group action and Jordan decomposition of characters of finite reductive groups with connected center. *J. Algebra*, to appear.

[St51a] R. Steinberg, A geometric approach to the representations of the full linear group over a Galois field. *Trans. Amer. Math. Soc.* **71** (1951), 274–282. (See also [St97, pp. 1–9].)

[St51b] R. Steinberg, The representations of $\mathrm{GL}(3, q)$, $\mathrm{GL}(4, q)$, $\mathrm{PGL}(3, q)$, and $\mathrm{PGL}(4, q)$. Canadian J. Math. **3** (1951), 225–235. (See also [St97, pp. 11–21].)

[St57] R. Steinberg, Prime power representations of finite linear groups II. *Canadian J. Math.* **9** (1957), 347–351. (See also [St97, pp. 35–39].)

[St64] R. Steinberg, Differential equations invariant under finite reflection groups. *Trans. Amer. Math. Soc.* **112** (1964), 392–400.

[St67] R. Steinberg, *Lectures on Chevalley groups*. Mimeographed notes, Department of Math., Yale University, 1967. Now available as vol. 66 of the University Lecture Series, Amer. Math. Soc., Providence, RI, 2016.

[St68] R. Steinberg, *Endomorphisms of Linear Algebraic Groups*. Memoirs of the American Mathematical Society, No. 80 American Mathematical Society, Providence, RI, 1968. (See also [St97, pp. 229–285].)

[St74] R. Steinberg, *Conjugacy Classes in Algebraic Groups*. Notes by Vinay V. Deodhar. Lecture Notes in Mathematics, vol. 366. Springer-Verlag, Berlin–New York, 1974.

[St77] R. Steinberg, On theorems of Lie–Kolchin, Borel and Lang. *Contributions to Algebra: A Collection of Papers Dedicated to Ellis Kolchin*, 349–354, Academic Press, New York, 1977. (See also [St97, pp. 467–472].)

[St97] R. Steinberg, *Robert Steinberg*. Collected Works, 7. American Mathematical Society, Providence, RI, 1997.

[St99] R. Steinberg, The isomorphism and isogeny theorems for reductive algebraic groups. *Algebraic Groups and Their Representations (Cambridge, 1997)*, pp. 233–240, Kluwer Academic, Dordrecht, 1998.

[St99] R. Steinberg, The isomorphism and isogeny theorems for reductive algebraic groups. *J. Algebra* **216** (1999), 366–383.

[Su77] D. Surowski, Representations of a subalgebra of the generic algebra corresponding to the Weyl group of type F_4. *Comm. Algebra* **5** (1977), 873–888.

[Su78] D. Surowski, Degrees of irreducible characters of (B, N)-pairs of types E_6 and E_7. *Trans. Amer. Math. Soc.* **243** (1978), 235–249.

[Suz62] M. Suzuki, On a class of doubly transitive groups. *Ann. of Math.* **75** (1962), 105–145.

[Tay92] D. E. Taylor, *The Geometry of Classical Groups*. Sigma Series in Pure Math. vol. 9, Heldermann-Verlag, Berlin, 1992.

[Ta13] J. Taylor, On unipotent supports of reductive groups with a disconnected centre. *J. Algebra* **391** (2013), 41–61.

[Ta14b] J. Taylor, Evaluating characteristic functions of character sheaves at unipotent elements. *Represent. Theory* **18** (2014), 310–340.

[Ta16] J. Taylor, Generalised Gelfand–Graev representations in small characteristics. *Nagoya Math. J.* **224** (2016), 93–167.

[Ta18a] J. Taylor, On the Mackey formula for connected centre groups. *J. Group Theory* **21** (2018), 439–448.

[Ta18b] J. Taylor, Action of automorphisms on irreducible characters of symplectic groups. *J. Algebra* **505** (2018), 211–246.

[Ta19] J. Taylor, The structure of root data and smooth regular embeddings of reductive groups. *Proc. Edinb. Math. Soc.* **62** (2019), 523–552.

[TiZa04] P. H. Tiep and A. E. Zalesskii, Unipotent elements of finite groups of Lie type and realization fields of their complex representations. *J. Algebra* **271** (2004), 327–390.

[Ti62] J. Tits, Théorème de Bruhat et sous-groupes paraboliques. *C. R. Acad. Sci. Paris* **254** (1962), 2910–2912.

[Vo07] D. Vogan, The character table for E_8. *Notices Amer. Math. Soc.* **54** (2007), 1122–1134.

[Wal04] J.-L. Waldspurger, *Une Conjecture de Lusztig Pour les Groupes Classiques*. Mém. Soc. Math. France (N.S.), no. 96, 2004.

[War66] H. N. Ward, On Ree's series of simple groups. *Trans. Amer. Math. Soc.* **121** (1966), 62–89.

Index